MONOGRAPHS ON NUMERICAL ANALYSIS

General Editors
L. FOX, J. WALSH

THE ALGEBRAIC EIGENVALUE PROBLEM

BY

J. H. WILKINSON

M.A. (Cantab.), Sc.D., D.Tech., F.R.S.

CLARENDON PRESS · OXFORD

Oxford University Press, Walton Street, Oxford OX2 6DP

OXFORD LONDON GLASGOW
NEW YORK TORONTO MELBOURNE WELLINGTON
IBADAN NAIROBI DAR ES SALAAM LUSAKA CAPE TOWN
KUALA LUMPUR SINGAPORE JAKARTA HONG KONG TOKYO
DELHI BOMBAY CALCUTTA MADRAS KARACHI

ISBN 0 19 853403 5

© *Oxford University Press*, 1965

FIRST PUBLISHED 1965
REPRINTED LITHOGRAPHICALLY IN GREAT BRITAIN
FROM CORRECTED SHEETS OF THE FIRST EDITION
BY WILLIAM CLOWES & SONS LIMITED
LONDON, BECCLES AND COLCHESTER
1967, 1969, 1972, 1977, 1978

Preface

THE solution of the algebraic eigenvalue problem has for long had a particular fascination for me because it illustrates so well the difference between what might be termed classical mathematics and practical numerical analysis. The eigenvalue problem has a deceptively simple formulation and the background theory has been known for many years; yet the determination of accurate solutions presents a wide variety of challenging problems.

The suggestion that I might write a book on this topic in the series of Monographs on Numerical Analysis was made by Professor L. Fox and Dr. E. T. Goodwin after my early experience with automatic digital computers. It is possible that its preparation would have remained a pious aspiration had it not been for an invitation from Professor J. W. Givens to attend a Matrix Symposium at Detroit in 1957 which led to successive invitations to lecture on 'Practical techniques for solving linear equations and computing eigenvalues and eigenvectors' at Summer Schools held in the University of Michigan. The discipline of providing a set of notes each year for these lectures has proved particularly valuable and much of the material in this book has been presented during its formative stage in this way.

It was originally my intention to present a description of most of the known techniques for solving the problem, together with a critical assessment of their merits, supported wherever possible by the relevant error analysis, and by 1961 a manuscript based on these lines was almost complete. However, substantial progress had been made with the eigenvalue problem and error analysis during the period in which this manuscript was prepared and I became increasingly dissatisfied with the earlier chapters. In 1962 I took the decision to redraft the whole book with a modified objective. I felt that it was no longer practical to cover almost all known methods and to give error analyses of them and decided to include mainly those methods of which I had extensive practical experience. I also inserted an additional chapter giving fairly general error analyses which apply to almost all the methods given subsequently. Experience over the years had convinced me that it is by no means easy to give a reliable assessment of a method without using it, and that often comparatively minor changes in the details of a practical process have a disproportionate influence on its effectiveness.

The writer of a book on numerical analysis faces one particularly difficult problem, that of deciding to whom it should be addressed. A practical treatment of the eigenvalue problem is potentially of interest to a wide variety of people, including among others, design engineers, theoretical physicists, classical applied mathematicians and numerical analysts who wish to do research in the matrix field. A book which was primarily addressed to the last class of readers might prove rather inaccessible to those in the first. I have not omitted anything purely on the grounds that some readers might find it too difficult, but have tried to present everything in as elementary terms as the subject matter warrants. The dilemma was at its most acute in the first chapter. I hope the elementary presentation used there will not offend the serious mathematician and that he will treat this merely as a rough indication of the classical material with which he must be familiar if he is to benefit fully from the rest of the book. I have assumed from the start that the reader is familiar with the basic concepts of vector spaces, linear dependence and rank. An admirable introduction to the material in this book is provided by L. Fox's 'An introduction to numerical linear algebra' (Oxford, this series). Research workers in the field will find A. S. Householder's 'The theory of matrices in numerical analysis' (Blaisdell, 1964) an invaluable source of information.

My decision mentioned earlier to treat only those methods of which I have extensive experience has inevitably resulted in the omission of a number of important algorithms. Their omission is perhaps less serious than it might have been, since the treatises by Durand, 'Solutions numériques des équations algébriques' (Masson, 1960, 1961) and by D. K. Faddeev and V. N. Faddeeva, 'Vycislitel'nye metody lineinoi algebry' (Moscow, 1963) give fairly extensive coverage. However, two omissions call for special mention. The first is the Jacobi-type method developed by P. J. Eberlein and various modifications of this method developed independently by H. Rutishauser. These seem to me to show exceptional promise and may well provide the most satisfactory solution to the general eigenvalue problem. I was reluctant to include them without first giving them the detailed study which they deserve. My decision was reinforced by the realization that they are not directly covered by the general error analyses I have given. The second omission is a general treatment of Rutishauser's QD algorithm. This is of very wide application and I did not feel I could do full justice to this work within the restricted field of the eigenvalue problem. The reader is referred to papers by Rutishauser and Henrici. The literature in the eigenvalue field is very extensive and I have restricted the bibliography mainly to those papers to which explicit reference is made in the text. Very detailed bibliographies are available in the books by Faddeev and Faddeeva and Householder mentioned earlier.

. When redrafting the book I was tempted to present the algorithms in ALGOL, but decided that the difficulties of providing procedures which were correct in every detail were prohibitive at this stage. Accordingly I have used the language of classical mathematics but have adopted a form of words which facilitates translation into ALGOL and related computer languages.

The material for this book is drawn from many sources but I have been particularly influenced by the work of colleagues at the University of Michigan Summer Schools, especially that of F. L. Bauer, G. E. Forsythe, J. W. Givens, P. Henrici, A. S. Householder, O. Taussky, J. Todd and R. S. Varga. Apart from this it is the algorithmic genius of H. Rutishauser which has been my main source of inspiration.

It is a pleasure to acknowledge the assistance I have received from numerous friends in this enterprise. I would like to thank Professors J. W. Carr and R. C. F. Bartels for successive invitations to lecture at the Summer Schools in Ann Arbor, thereby providing me with an annual stimulus for completing the manuscript. I am particularly indebted to Professors G. E. Forsythe and J. W. Givens who generously provided duplicating facilities for the two main drafts of the manuscript. This has enabled me to benefit from the opinions of a much wider circle of readers than would otherwise have been possible. Many valuable comments were received from Professor G. E. Forsythe, Dr. J. M. Varah and Mr. E. A. Albasiny. The whole of the final manuscript was read by Dr. D. W. Martin and Professor L. Fox and the first proofs by Dr. E. T. Goodwin. Their comments and suggestions rescued me from numerous blunders and infelicities of style.

I am a notoriously poor proof reader and I was fortunate indeed to have the expert guidance of Mrs. I. Goode in seeing the book through the press. Finally sincere thanks are due to my wife who typed successive versions of the manuscript. Without her patience and pertinacity the project would almost certainly have foundered.

J. H. Wilkinson
Teddington, March, 1965

Contents

3. ERROR ANALYSIS

5. HERMITIAN MATRICES

7. EIGENVALUES OF MATRICES OF CONDENSED FORMS

1

Theoretical Background

Introduction

1. IN THIS introductory chapter we shall give a brief account of the classical theory of canonical forms and of several other theoretical topics which will be required in the remainder of the book.

Since a complete treatment is beyond the scope of this chapter we have concentrated on showing how the eigensystem of a matrix is related to its various canonical forms. The emphasis is therefore somewhat different from that which would be found in a standard treatise. No attempt has been made to give rigorous proofs of all the fundamental theorems; instead we have contented ourselves, in general, with the statement of the results and have given proofs only when they are particularly closely related to practical techniques which are used subsequently.

In order to avoid repeated explanation we refer here to some conventions of notation to which we have adhered throughout the book.

We denote matrices by capital letters and, unless there is an indication to the contrary, matrices may be assumed to be square and of order n. The (i, j) element of a matrix A will be denoted by a_{ij}. Vectors will be denoted by lower case letters and will generally be assumed to be of order n. The ith component of a vector x will be denoted by x_i.

The identity matrix will be denoted by I and its (i, j) element by δ_{ij}, the Kronecker symbol, rather than i_{ij}. We have then

$$\delta_{ij} = 0 \quad (i \neq j), \qquad \delta_{ii} = 1. \tag{1.1}$$

The ith column of I will be denoted by e_i and e will be used to denote the vector $e_1 + e_2 + \ldots + e_n$, having all its components equal to unity.

We shall frequently be concerned with systems of n vectors which will be denoted by $x_1, x_2, \ldots, x_n$. When this is done we shall refer to the jth element of x_i as x_{ji} and to the matrix having x_i as its ith column as X, without always making specific reference to this fact.

The notation $|A|$ will be reserved for the matrix which has $|a_{ij}|$ for its (i, j) element and the notation

will imply that
$$|A| < |B| \tag{1.2}$$
$$|a_{ij}| < |b_{ij}| \quad \text{for all } i, j. \tag{1.3}$$

As a result of this decision and the use of $\|A\|$ for the norm of A (cf. § 52) the determinant of A will be denoted by det (A). In line with this, an $(n \times n)$ diagonal matrix with (i, i) element equal to λ_i will be denoted by diag (λ_i).

The transpose of a matrix will be denoted by A^T and is such that its (i, j) element is equal to a_{ji}. Similarly x^T will denote a row vector having its ith component equal to x_i. Since it is more convenient to print a row vector than a column vector, we shall frequently introduce a column vector x and then define it by giving the explicit expression for x^T.

We assume throughout that we are working in the field of complex numbers, though in computational procedures there is much to be gained from remaining in the real field whenever possible.

Definitions

2. The fundamental algebraic eigenproblem is the determination of those values of λ for which the set of n homogeneous linear equations in n unknowns

$$Ax = \lambda x \tag{2.1}$$

has a non-trivial solution. Equation (2.1) may be written in the form

$$(A - \lambda I)x = 0, \tag{2.2}$$

and for arbitrary λ this set of equations has only the solution $x = 0$. The general theory of simultaneous linear algebraic equations shows that there is a non-trivial solution if, and only if, the matrix $(A - \lambda I)$ is singular, that is

$$\det(A - \lambda I) = 0. \tag{2.3}$$

The determinant on the left hand side of equation (2.3) may be expanded to give the explicit polynomial equation

$$\alpha_0 + \alpha_1 \lambda + \cdots + \alpha_{n-1} \lambda^{n-1} + (-1)^n \lambda^n = 0. \tag{2.4}$$

Equation (2.4) is called the *characteristic equation* of the matrix A and the polynomial on the left-hand side of the equation is called the *characteristic polynomial*. Since the coefficient of λ^n is not zero and we are assuming throughout that we are working in the complex number field, the equation always has n roots. In general the roots may be complex, even if the matrix A is real, and there may be roots of any multiplicities up to n. The n roots are called the *eigenvalues*, *latent roots*, *characteristic values* or *proper values* of the matrix A.

Corresponding to any eigenvalue λ, the set of equations (2.2) has at least one non-trivial solution x. Such a solution is called an *eigenvector, latent vector, characteristic vector* or *proper vector* corresponding to that eigenvalue. If the matrix $(A - \lambda I)$ is of rank less than $(n-1)$ then there will be more than one independent vector satisfying equation (2.2). We discuss this point in detail in §§ 6–9. It is evident that if x is a solution of (2.2), then kx is also a solution for any value of k, so that even when $(A - \lambda I)$ is of rank $(n-1)$, the eigenvector corresponding to λ is arbitrary to the extent of a constant multiplier. It is convenient to choose this multiplier so that the eigenvector has some desirable numerical property, and such vectors are called *normalized* vectors. The most convenient forms of normalization are those for which

(i) the sum of the squares of the moduli of the components is equal to unity,

(ii) a component of the vector which has the largest modulus is equal to unity, and

(iii) the sum of the moduli of the components is equal to unity.

The normalizations (i) and (iii) have the disadvantage of being arbitrary to the extent of a complex multiplier of modulus unity.

Many of the processes we describe in this book lead, in the first instance, to eigenvectors which are not in normalized form. It is often convenient to multiply such vectors by a power of 10 (or 2) so that the largest element lies between 0·1 and 1·0 (or $\frac{1}{2}$ and 1) in modulus. We shall refer to vectors which have been treated in this way as *standardized* vectors.

Eigenvalues and eigenvectors of the transposed matrix

3. The eigenvalues and eigenvectors of the transpose of a matrix are of considerable importance in the general theory. The eigenvalues of A^T are, by definition, those values of λ for which the set of equations

$$A^T y = \lambda y \tag{3.1}$$

has a non-trivial solution. These are the values for which

$$\det(A^T - \lambda I) = 0, \tag{3.2}$$

and since the determinant of a matrix is equal to that of its transpose, the eigenvalues of A^T are the same as those of A. The eigenvectors are, in general, different. We shall usually denote the eigenvector of A^T corresponding to λ_i by y_i, so that we have

$$A^T y_i = \lambda_i y_i. \tag{3.3}$$

Equation (3.3) may be written in the form

$$y_i^T A = \lambda_i y_i^T, \tag{3.4}$$

so that y_i^T is sometimes called a *left-hand* eigenvector of A corresponding to λ_i, while the vector x_i such that

$$A x_i = \lambda_i x_i \tag{3.5}$$

is called a *right-hand* eigenvector of A corresponding to λ_i. The importance of the eigenvectors of the transposed matrix springs mainly from the following result.

If x_i is an eigenvector of A corresponding to the eigenvalue λ_i, and y_j is an eigenvector of A^T corresponding to λ_j, then

$$x_i^T y_j = 0 \quad (\text{if } \lambda_i \neq \lambda_j). \tag{3.6}$$

For we may write (3.5) in the form

$$x_i^T A^T = \lambda_i x_i^T \tag{3.7}$$

and we have

$$A^T y_j = \lambda_j y_j. \tag{3.8}$$

Post-multiplying (3.7) by y_j and pre-multiplying (3.8) by x_i^T and subtracting we have

$$0 = \lambda_i x_i^T y_j - \lambda_j x_i^T y_j, \tag{3.9}$$

leading to equation (3.6).

Note that since x_i and/or y_j may be complex vectors, $x_i^T y_j$ is not an *inner-product* as usually understood. We have in fact

$$x_i^T y_j = y_j^T x_i, \tag{3.10}$$

and not

$$x_i^T y_j = \overline{(y_j^T x_i)}. \tag{3.11}$$

When x is complex we can have

$$x^T x = 0 \tag{3.12}$$

whereas a true inner-product is positive for all non-zero x.

Distinct eigenvalues

4. The general theory of the eigensystem is simplest when the n eigenvalues are distinct, and we consider this case first. We denote the eigenvalues by $\lambda_1, \lambda_2, ..., \lambda_n$. Equation (2.2) certainly has at least one solution for each value of λ_i and we can therefore assume the existence of a set of eigenvectors which we denote by $x_1, x_2, ..., x_n$. We shall show that each of these vectors is unique, apart from an arbitrary multiplier.

We first show that the vectors must be linearly independent. For, if not, let r be the smallest number of linearly dependent vectors. We may number the vectors so that these are $x_1, x_2,..., x_r$ and, from our assumption, there is a relation between them of the form

$$\sum_1^r \alpha_i x_i = 0 \qquad (4.1)$$

in which none of the α_i is zero. Clearly $r \geqslant 2$ because, by definition, none of the x_i is the null vector. Pre-multiplying equation (4.1) by A, we have

$$\sum_1^r \alpha_i \lambda_i x_i = 0. \qquad (4.2)$$

Multiplying equation (4.1) by λ_r and subtracting (4.2) gives

$$\sum_1^{r-1} \alpha_i (\lambda_r - \lambda_i) x_i = 0 \qquad (4.3)$$

and none of the coefficients in this relation is zero. Now equation (4.3) implies that $x_1, x_2,..., x_{r-1}$ are linearly dependent, which is contrary to hypothesis. The n eigenvectors are therefore independent and span the whole n space. Consequently they may be used as a base in which to express an arbitrary vector.

From the last result the uniqueness of the eigenvectors follows immediately. For if there is a second eigenvector z_1 corresponding to λ_1 we may write

$$z_1 = \sum_1^n \alpha_i x_i, \qquad (4.4)$$

where at least one of the α_i is non-zero. Multiplying (4.4) by A gives

$$\lambda_1 z_1 = \sum_1^n \alpha_i \lambda_i x_i. \qquad (4.5)$$

Multiplying (4.4) by λ_1 and subtracting from (4.5) we have

$$0 = \sum_2^n \alpha_i (\lambda_i - \lambda_1) x_i, \qquad (4.6)$$

and hence from the independence of the x_i

$$\alpha_i (\lambda_i - \lambda_1) = 0 \quad (i = 2, 3,..., n),$$

giving

$$\alpha_i = 0 \quad (i = 2, 3,..., n). \qquad (4.7)$$

The non-zero α_i must therefore be α_1, showing that z_1 is a multiple of x_1.

Similarly we may show that the eigenvectors of A^T are unique and linearly independent. We have proved earlier that

$$x_i^T y_j = 0 \quad (\lambda_i \neq \lambda_j) \qquad (4.8)$$

so that the two sets of vectors form a *bi-orthogonal system*. We have further

$$x_i^T y_i \neq 0 \quad (i = 1, 2, \ldots, n). \tag{4.9}$$

For if x_i were orthogonal to y_i it would be orthogonal to $y_1, y_2, \ldots, y_n$ and hence to the whole n-space. This is impossible, since x_i is not the null vector.

Equations (4.8) and (4.9) enable us to obtain an explicit expression for an arbitrary vector v in terms of the x_i or the y_i. For if we write

$$v = \sum_{i=1}^{n} \alpha_i x_i, \tag{4.10}$$

then

$$y_j^T v = \sum_{i=1}^{n} \alpha_i y_j^T x_i = \alpha_j y_j^T x_j. \tag{4.11}$$

Hence

$$\alpha_j = y_j^T v / y_j^T x_j, \tag{4.12}$$

the division being permissible because the denominator is non-zero.

Similarity transformations

5. If we choose the arbitrary multipliers in the x_i and y_j so that

$$y_i^T x_i = 1 \quad (i = 1, 2, \ldots, n), \tag{5.1}$$

the relations (4.8) and (5.1) imply that the matrix Y^T, which has y_i^T as its ith row, is the inverse of the matrix, X, which has x_i as its ith column. The n sets of equations

$$A x_i = \lambda_i x_i \tag{5.2}$$

may be written in matrix form

$$AX = X \operatorname{diag}(\lambda_i). \tag{5.3}$$

Now we have seen that the inverse of the matrix X exists and is equal to Y^T. Equation (5.3) therefore gives

$$X^{-1}AX = Y^T AX = \operatorname{diag}(\lambda_i). \tag{5.4}$$

Transforms, $H^{-1}AH$, of the matrix A, where H is non-singular, are of fundamental importance from both the theoretical and the practical standpoints and are known as *similarity transforms* while A and $H^{-1}AH$ are said to be *similar*. Obviously HAH^{-1} is also a similarity transform of A. We have shown that when the eigenvalues of A are distinct then there is a *similarity transformation* which reduces A to diagonal form and that the *matrix of the transformation* has its columns equal to the eigenvectors of A. Conversely if we have

$$H^{-1}AH = \mathrm{diag}(\mu_i) \tag{5.5}$$

then

$$AH = H\,\mathrm{diag}(\mu_i). \tag{5.6}$$

This last equation implies that the μ_i are the eigenvalues of A in some order, and that the ith column of H is an eigenvector corresponding to μ_i.

The eigenvalues of a matrix are invariant under a similarity transformation. For if

$$Ax = \lambda x \tag{5.7}$$

then

$$H^{-1}Ax = \lambda H^{-1}x,$$

giving

$$H^{-1}A(HH^{-1})x = \lambda H^{-1}x, \qquad (H^{-1}AH)H^{-1}x = \lambda H^{-1}x. \tag{5.8}$$

The eigenvalues are therefore preserved and the eigenvectors are multiplied by H^{-1}.

Many of the numerical methods for finding the eigensystem of a matrix consist essentially of the determination of a similarity transformation which reduces a matrix A of general form to a matrix B of special form, for which the eigenproblem may be more simply solved.

Similarity is a transitive property, for if

$$B = H_1^{-1}AH_1 \quad \text{and} \quad C = H_2^{-1}BH_2 \tag{5.9}$$

then

$$C = H_2^{-1}H_1^{-1}AH_1H_2 = (H_1H_2)^{-1}A(H_1H_2). \tag{5.10}$$

In general the reduction of a general matrix to one of special form will be effected by a sequence of simple similarity transformations.

Multiple eigenvalues and canonical forms for general matrices

6. The structure of the system of eigenvectors for a matrix which has one or more multiple eigenvalues may be far less simple than that described above. It may still be true that there is a similarity transformation which reduces A to diagonal form. When this is true, we have, for some non-singular H,

$$H^{-1}AH = \mathrm{diag}(\lambda_i). \tag{6.1}$$

The λ_i must then be the eigenvalues of A and each λ_i must occur with the appropriate multiplicity. For

$$H^{-1}(A-\lambda I)H = H^{-1}AH-\lambda H^{-1}IH = \mathrm{diag}(\lambda_i-\lambda), \qquad (6.2)$$

and taking determinants of both sides we have

$$\det(H^{-1})\det(A-\lambda I)\det(H) = \prod(\lambda_i-\lambda), \qquad (6.3)$$

giving

$$\det(A-\lambda I) = \prod(\lambda_i-\lambda). \qquad (6.4)$$

The λ_i are therefore the roots of the characteristic equation of A. Writing (6.1) in the form

$$AH = H\,\mathrm{diag}(\lambda_i), \qquad (6.5)$$

we see that the columns of H are eigenvectors of A. Since H is non-singular, its columns are independent. If, for example, λ_1 is a double root, then, denoting the first two columns of H by x_1 and x_2 we have

$$Ax_1 = \lambda_1 x_1, \quad Ax_2 = \lambda_1 x_2, \qquad (6.6)$$

where x_1 and x_2 are independent. From equations (6.6), any vector in the subspace spanned by x_1 and x_2 is also an eigenvector. For we have

$$A(\alpha_1 x_1 + \alpha_2 x_2) = \alpha_1 \lambda_1 x_1 + \alpha_2 \lambda_1 x_2 = \lambda_1(\alpha_1 x_1 + \alpha_2 x_2). \qquad (6.7)$$

Coincident roots for a matrix which can be reduced to diagonal form by a similarity transformation are associated with an indeterminacy in the eigenvectors corresponding to the multiple root. We can still select a set of eigenvectors which span the whole n-space and can therefore be used as a base in which to express an arbitrary vector.

Thus the diagonal matrix Λ defined by

$$\Lambda = \begin{bmatrix} 1 & & & & \\ & 1 & & & \\ & & 2 & & \\ & & & 2 & \\ & & & & 3 \end{bmatrix} \qquad (6.8)$$

has the 5 independent eigenvectors e_1, e_2, e_3, e_4 and e_5. Any combination of e_1 and e_2 is an eigenvector corresponding to $\lambda = 1$ and any combination of e_3 and e_4 is an eigenvector corresponding to $\lambda = 2$. For any matrix similar to Λ there is a non-singular H such that

$$H^{-1}AH = \Lambda \qquad (6.9)$$

and the eigenvectors of A are He_1, He_2, He_3, He_4 and He_5, any combination of the first two being an eigenvector corresponding to $\lambda = 1$ and any combination of the third and fourth an eigenvector corresponding to $\lambda = 2$.

Defective system of eigenvectors

7. The properties of a matrix which is similar to a diagonal matrix but has multiple eigenvalues are in most respects like those of a matrix with distinct eigenvalues. However, not all matrices with multiple eigenvalues are of this type.

The matrix $C_2(a)$ defined by

$$C_2(a) = \begin{bmatrix} a & 1 \\ 0 & a \end{bmatrix} \tag{7.1}$$

has the double eigenvalue $\lambda = a$. If x is an eigenvector of $C_2(a)$ with components x_1 and x_2, then we must have

$$\left. \begin{matrix} 0x_1 + x_2 = 0 \\ 0x_1 + 0x_2 = 0 \end{matrix} \right\}. \tag{7.2}$$

These equations imply that $x_2 = 0$. There is thus only one eigenvector corresponding to $\lambda = a$, the vector e_1. We may regard $C_2(a)$ as the limit as $b \to a$ of a matrix B_2 defined by

$$B_2 = \begin{bmatrix} a & 1 \\ 0 & b \end{bmatrix}. \tag{7.3}$$

For $b \neq a$ this matrix has two eigenvalues a and b, and the corresponding vectors are

$$\begin{bmatrix} 1 \\ 0 \end{bmatrix} \quad \text{and} \quad \begin{bmatrix} 1 \\ b-a \end{bmatrix}.$$

As $b \to a$ the two eigenvalues and eigenvectors become coincident.

More generally if we define matrices $C_r(a)$ by the relations

$$C_1(a) = [a], \quad C_r(a) = \begin{bmatrix} a & 1 & & & & \\ & a & 1 & & & \\ & & a & 1 & & \\ & & \cdots\cdots\cdots\cdots & & \\ & & & & a & 1 \\ & & & & & a \end{bmatrix}, \tag{7.4}$$

then $C_r(a)$ has an eigenvalue, a, of multiplicity r, but it has only one eigenvector, $x = e_1$. We may prove this as we proved the corresponding result for $C_2(a)$. Alternatively it follows from the fact that the rank of $[C_r(a)-aI]$ is $(r-1)$. (The matrix of order $(r-1)$ in the top right-hand corner has determinant 1.) The transposed matrix $[C_r(a)]^T$ also has only one eigenvector, $y = e_r$. We have immediately

$$x^T y = e_1^T e_r = 0 \qquad (7.5)$$

which contrasts with the result obtained for the eigenvectors of matrices with distinct eigenvalues.

The matrix $C_r(a)$, $(r > 1)$, cannot be reduced to diagonal form by a similarity transformation. For if there exists an H for which

$$H^{-1}C_r H = \mathrm{diag}(\lambda_i), \qquad (7.6)$$

that is,

$$C_r H = H\,\mathrm{diag}(\lambda_i), \qquad (7.7)$$

then, as shown in § 6, the λ_i must be equal to the eigenvalues of $C_r(a)$ and must therefore all have the value a. Equation (7.7) then shows that the columns of H are all eigenvectors of $C_r(a)$. These columns are independent, since H is non-singular. The assumption that such an H exists must therefore be false.

The matrix $C_r(a)$ is called a *simple Jordan submatrix* of order r, or sometimes a *simple classical submatrix* of order r. Matrices of this form are of special importance in the theory, as will be seen in the next section.

The Jordan (classical) canonical form

8. We see then that a matrix with multiple eigenvalues is not necessarily similar to a diagonal matrix. It is natural to ask what is the most compact form to which a matrix A may be reduced by means of a similarity transformation. The answer is contained in the following theorem.

Let A be a matrix of order n with r distinct eigenvalues $\lambda_1, \lambda_2, ..., \lambda_r$ of multiplicities $m_1, m_2, ..., m_r$ so that

$$m_1 + m_2 + ... + m_r = n. \qquad (8.1)$$

Then there is a similarity transformation with matrix H such that the matrix $H^{-1}AH$ consists of simple Jordan submatrices isolated along the diagonal with all other elements equal to zero. The sum of the orders of the submatrices associated with λ_i is equal to m_i. Apart from the ordering of the submatrices along the diagonal, the transformed matrix is unique. We say that $H^{-1}AH$ is the *direct sum* of the simple Jordan submatrices.

Thus a matrix of order six with five eigenvalues equal to λ_1 and one equal to λ_2, which is similar to C defined by

$$C = \begin{bmatrix} C_2(\lambda_1) & & & \\ & C_2(\lambda_1) & & \\ & & C_1(\lambda_1) & \\ & & & C_1(\lambda_2) \end{bmatrix}, \tag{8.2}$$

cannot also be similar to a matrix of the form

$$\begin{bmatrix} C_3(\lambda_1) & & \\ & C_2(\lambda_1) & \\ & & C_1(\lambda_2) \end{bmatrix}, \tag{8.3}$$

although this too has five roots equal to λ_1 and one equal to λ_2. The matrix of simple Jordan submatrices is called the *Jordan canonical form* or the *classical canonical form* of the matrix A.

In the special case when all the Jordan submatrices in the Jordan canonical form are of order unity, this form becomes diagonal. We have already seen that if the eigenvalues of a matrix are distinct then the Jordan canonical form is necessarily the matrix $\mathrm{diag}(\lambda_i)$.

The total number of independent eigenvectors is equal to the number of submatrices in the Jordan canonical form. Thus a matrix A which can be reduced to C, defined by equation (8.2), has four independent eigenvectors. The eigenvectors of C are e_1, e_3, e_5 and e_6 and those of A are He_1, He_3, He_5 and He_6. There are only three eigenvectors corresponding to the eigenvalue λ_1, which is of multiplicity five.

We shall not prove the existence and the uniqueness of the Jordan canonical form since the method of proof has little relevance to the methods we describe in later chapters. In the simple Jordan submatrix of order r we have taken the elements equal to unity to lie on the super-diagonal line. There is a corresponding form in which these elements are taken on the sub-diagonal. We may refer to the form of equation (7.4) as the *upper* Jordan form and the alternative form as the *lower* Jordan form. The choice of the value unity for these elements is of no special significance. *Clearly by a suitable similarity transformation with a diagonal matrix these elements could be given arbitrary non-zero values.*

The elementary divisors

9. If C is the Jordan canonical form of A, then consider the matrix $(C-\lambda I)$. Corresponding to the matrix C of equation (8.2), for example, we have

$$(C-\lambda I) = \begin{bmatrix} C_2(\lambda_1-\lambda) & & & \\ & C_2(\lambda_1-\lambda) & & \\ & & C_1(\lambda_1-\lambda) & \\ & & & C_1(\lambda_2-\lambda) \end{bmatrix}. \quad (9.1)$$

Clearly, in general, $(C-\lambda I)$ is the direct sum of matrices of the form $C_r(\lambda_i-\lambda)$. The determinants of the submatrices of the canonical form $(C-\lambda I)$ are called the *elementary divisors* of the matrix A. Thus the elementary divisors of any matrix similar to C of equation (8.2) are

$$(\lambda_1-\lambda)^2, \ (\lambda_1-\lambda)^2, \ (\lambda_1-\lambda) \quad \text{and} \quad (\lambda_2-\lambda). \quad (9.2)$$

It is immediately obvious that the product of the elementary divisors of a matrix is equal to its characteristic polynomial. When the Jordan canonical form is diagonal, then the elementary divisors are

$$(\lambda_1-\lambda), \ (\lambda_2-\lambda), \ldots, \ (\lambda_n-\lambda), \quad (9.3)$$

and are therefore all linear. A matrix with distinct eigenvalues always has *linear elementary divisors* but we have seen that if a matrix has one or more multiple eigenvalues, then it may or may not have linear elementary divisors.

If a matrix has one or more *non-linear* elementary divisors, this means that at least one of the submatrices in its Jordan canonical form is of order greater than unity, and hence it has fewer than n independent eigenvectors. Such a matrix is called *defective*.

Companion matrix of the characteristic polynomial of A

10. Let us denote the characteristic equation of A by

$$(-1)^n[\lambda^n - p_{n-1}\lambda^{n-1} - p_{n-2}\lambda^{n-2} - \ldots - p_0] = 0. \quad (10.1)$$

Then it will be readily verified that the characteristic equation of the matrix

$$C = \begin{bmatrix} p_{n-1} & p_{n-2} & \cdots & p_1 & p_0 \\ 1 & 0 & \cdots & 0 & 0 \\ 0 & 1 & \cdots & 0 & 0 \\ \cdot & \cdot & \cdots & \cdot & \cdot \\ 0 & 0 & \cdots & 1 & 0 \end{bmatrix} \quad (10.2)$$

is identical to the characteristic equation of A. The matrix C is called the *companion matrix* of the characteristic polynomial of A. Since C and A have the same characteristic polynomial, it is natural to ask whether they are, in general, similar. We give the answer to this question in the next paragraph. Notice that we could have given the companion matrix in three other forms. For example, we could have

$$
\begin{bmatrix}
0 & 0 & 0 & \cdots & 0 & p_0 \\
1 & 0 & 0 & \cdots & 0 & p_1 \\
0 & 1 & 0 & \cdots & 0 & p_2 \\
\cdot & \cdot & \cdot & \cdots & \cdot & \cdot \\
\cdot & \cdot & \cdot & \cdots & \cdot & \cdot \\
0 & 0 & 0 & \cdots & 1 & p_{n-1}
\end{bmatrix},
\tag{10.3}
$$

or the elements p_i could have been taken in the first column or the last row with the unity elements on the superdiagonal. We shall refer to any of these forms as the companion matrix, using whichever may be the most convenient.

Non-derogatory matrices

11. Suppose the matrix A has distinct eigenvalues $\lambda_1, \lambda_2, ..., \lambda_n$; then we know that A is similar to $\mathrm{diag}(\lambda_i)$. We shall show that in this case $\mathrm{diag}(\lambda_i)$ is similar to the companion matrix. We prove this by exhibiting the matrix that transforms C into $\mathrm{diag}(\lambda_i)$. The general case will be sufficiently illustrated by considering the case $n = 4$.

Consider the matrix H defined by

$$
H = \begin{bmatrix}
\lambda_1^3 & \lambda_2^3 & \lambda_3^3 & \lambda_4^3 \\
\lambda_1^2 & \lambda_2^2 & \lambda_3^2 & \lambda_4^2 \\
\lambda_1 & \lambda_2 & \lambda_3 & \lambda_4 \\
1 & 1 & 1 & 1
\end{bmatrix}.
\tag{11.1}
$$

This matrix is non-singular if the λ_i are distinct and we have

$$
H\,\mathrm{diag}(\lambda_i) = \begin{bmatrix}
\lambda_1^4 & \lambda_2^4 & \lambda_3^4 & \lambda_4^4 \\
\lambda_1^3 & \lambda_2^3 & \lambda_3^3 & \lambda_4^3 \\
\lambda_1^2 & \lambda_2^2 & \lambda_3^2 & \lambda_4^2 \\
\lambda_1 & \lambda_2 & \lambda_3 & \lambda_4
\end{bmatrix}.
\tag{11.2}
$$

Further, the relation

$$\begin{bmatrix} p_3 & p_2 & p_1 & p_0 \\ 1 & 0 & 0 & 0 \\ 0 & 1 & 0 & 0 \\ 0 & 0 & 1 & 0 \end{bmatrix} H = \begin{bmatrix} \lambda_1^4 & \lambda_2^4 & \lambda_3^4 & \lambda_4^4 \\ \lambda_1^3 & \lambda_2^3 & \lambda_3^3 & \lambda_4^3 \\ \lambda_1^2 & \lambda_2^2 & \lambda_3^2 & \lambda_4^2 \\ \lambda_1 & \lambda_2 & \lambda_3 & \lambda_4 \end{bmatrix} \tag{11.3}$$

follows immediately from the fact that

$$\lambda_i^4 = p_3\lambda_i^3 + p_2\lambda_i^2 + p_1\lambda_i + p_0. \tag{11.4}$$

Hence

$$CH = H \, \mathrm{diag}(\lambda_i), \tag{11.5}$$

so that C is similar to $\mathrm{diag}(\lambda_i)$ and hence to any matrix with eigenvalues λ_i.

Turning now to the case when there are some multiple eigenvalues, we see immediately that if A has more than one independent eigenvector corresponding to λ_i then it cannot be similar to C. For $(C-\lambda I)$ is of rank $(n-1)$ for any value of λ, since the determinant of the matrix of order $(n-1)$ in the bottom left-hand corner is obviously 1. The eigenvector x_i corresponding to the eigenvalue λ_i is given by

$$x_i^T = [\lambda_i^{n-1}, \lambda_i^{n-2}, \ldots, \lambda_i, 1], \tag{11.6}$$

as can be seen by solving

$$(C-\lambda_i I)x_i = 0 . \tag{11.7}$$

For a matrix to be similar to the companion matrix of its characteristic polynomial then, it is necessary that its Jordan canonical form should have only one Jordan submatrix associated with each distinct eigenvalue. We now prove that this condition is also sufficient and again it will suffice to demonstrate it for a simple case.

12. Consider a matrix of order 4 having $(\lambda_1-\lambda)^2$, $(\lambda_3-\lambda)$, $(\lambda_4-\lambda)$ for its elementary divisors. Such a matrix is similar to A_4 given by

$$A_4 = \begin{bmatrix} \lambda_1 & 1 & & \\ 0 & \lambda_1 & & \\ & & \lambda_3 & \\ & & & \lambda_4 \end{bmatrix} . \tag{12.1}$$

Now the matrix H defined by

$$H = \begin{bmatrix} \lambda_1^3 & 3\lambda_1^2 & \lambda_3^3 & \lambda_4^3 \\ \lambda_1^2 & 2\lambda_1 & \lambda_3^2 & \lambda_4^2 \\ \lambda_1 & 1 & \lambda_3 & \lambda_4 \\ 1 & 0 & 1 & 1 \end{bmatrix} \qquad (12.2)$$

is clearly non-singular if λ_1, λ_3, λ_4 are distinct, and we have

$$HA_4 = \begin{bmatrix} \lambda_1^4 & 4\lambda_1^3 & \lambda_3^4 & \lambda_4^4 \\ \lambda_1^3 & 3\lambda_1^2 & \lambda_3^3 & \lambda_4^3 \\ \lambda_1^2 & 2\lambda_1 & \lambda_3^2 & \lambda_4^2 \\ \lambda_1 & 1 & \lambda_3 & \lambda_4 \end{bmatrix}. \qquad (12.3)$$

Since λ_1 is a double root of the characteristic equation $f(\lambda) = 0$, we have $f'(\lambda_1) = 0$, giving

$$4\lambda_1^3 = 3p_3\lambda_1^2 + 2p_2\lambda_1 + p_1, \qquad (12.4)$$

and hence

$$CH = HA_4. \qquad (12.5)$$

In an exactly similar way we can deal with a Jordan submatrix of order r associated with a matrix of order n. If λ_1 is the relevant eigenvalue then the corresponding matrix H, has the r columns

$$\begin{bmatrix} \lambda_1^{n-1} & \binom{n-1}{1}\lambda_1^{n-2} & \cdots & \binom{n-1}{r-1}\lambda_1^{n-r} \\ \lambda_1^{n-2} & \binom{n-2}{1}\lambda_1^{n-3} & \cdots & \binom{n-2}{r-1}\lambda_1^{n-r-1} \\ \cdot & \cdot & \cdots & \cdot \\ \cdot & \cdot & \cdots & \cdot \\ \cdot & \cdot & \cdots & \cdot \\ \lambda_1 & 1 & \cdots & 0 \\ 1 & 0 & \cdots & 0 \end{bmatrix}. \qquad (12.6)$$

Hence, provided there is only one Jordan matrix associated with each distinct λ_i, and therefore only one eigenvector associated with each distinct λ_i, the matrix is similar to the companion matrix of its characteristic polynomial. Such a matrix is called *non-derogatory*.

The Frobenius (rational) canonical form

13. A matrix is called *derogatory* if there is more than one Jordan submatrix (and therefore more than one eigenvector) associated with

λ_i for some i. We have seen that a derogatory matrix cannot be similar to the companion matrix of its characteristic polynomial, and we now investigate the canonical form analogous to the companion form for a derogatory matrix.

We define a *Frobenius matrix* B_r of order r to be a matrix of the form

$$B_r = \begin{bmatrix} b_{r-1} & b_{r-2} & \cdots & b_1 & b_0 \\ 1 & 0 & \cdots & 0 & 0 \\ 0 & 1 & \cdots & 0 & 0 \\ \cdot & \cdot & \cdot & \cdot & \cdot \\ 0 & 0 & \cdots & 1 & 0 \end{bmatrix}. \tag{13.1}$$

The fundamental theorem for general matrices may then be expressed in the following form.

Every matrix A may be reduced by a similarity transformation to the direct sum of a number s of Frobenius matrices which may be denoted by $B_{r_1}, B_{r_2}, \ldots, B_{r_s}$. The characteristic polynomial of each B_{r_i} divides the characteristic polynomial of all the preceding B_{r_j}. In the case of a non-derogatory matrix, $s = 1$ and $r_1 = n$. The direct sum of the Frobenius matrices is called the *Frobenius* (or *rational*) canonical form. The name rational canonical form springs from the fact that it can be derived by transformations which are rational in the field of the elements of A.

We shall not give a direct proof of the existence of the Frobenius canonical form but instead we show how it is related to the Jordan canonical form. This does not illustrate the rational nature of the transformation since the operations giving the Jordan form are, in general, irrational in the field of A.

Relationship between the Jordan and Frobenius canonical forms

14. The relationship between the two canonical forms is probably best illustrated by an example. Suppose a matrix A, of order 10, has the following elementary divisors:

$$\left. \begin{array}{lll} (\lambda_1 - \lambda)^3 & (\lambda_2 - \lambda)^2 & (\lambda_3 - \lambda) \\ (\lambda_1 - \lambda) & (\lambda_2 - \lambda)^2 & \\ (\lambda_1 - \lambda) & & \end{array} \right\}. \tag{14.1}$$

Its Jordan canonical form may be written in the form

$$
\begin{bmatrix}
C_3(\lambda_1) & & & & & \\
& C_2(\lambda_2) & & & & \\
& & C_1(\lambda_3) & & & \\
\hline
& & & C_1(\lambda_1) & & \\
& & & & C_2(\lambda_2) & \\
\hline
& & & & & C_1(\lambda_1)
\end{bmatrix}
=
\begin{bmatrix}
G_1 & & \\
\hline
& G_2 & \\
\hline
& & G_3
\end{bmatrix},
\tag{14.2}
$$

where we have grouped together the Jordan submatrices of highest order in each eigenvalue to give the matrix G_1, then those of next highest order to give G_2 and so on.

Each of the matrices G_i therefore has only one Jordan submatrix associated with each λ_i and can therefore be reduced by a similarity transformation H_i to a single Frobenius matrix of the corresponding order. In our example

$$
\left.
\begin{aligned}
H_1^{-1}G_1H_1 &= B_6 \quad \text{(a Frobenius matrix of order 6)} \\
H_2^{-1}G_2H_2 &= B_3 \quad \text{(a Frobenius matrix of order 3)} \\
H_3^{-1}G_3H_3 &= B_1 \quad \text{(a Frobenius matrix of order 1)}
\end{aligned}
\right\}.
\tag{14.3}
$$

Hence

$$
\begin{bmatrix}
H_1^{-1} & & \\
& H_2^{-1} & \\
& & H_3^{-1}
\end{bmatrix}
\begin{bmatrix}
G_1 & & \\
& G_2 & \\
& & G_3
\end{bmatrix}
\begin{bmatrix}
H_1 & & \\
& H_2 & \\
& & H_3
\end{bmatrix}
=
\begin{bmatrix}
B_6 & & \\
& B_3 & \\
& & B_1
\end{bmatrix}.
\tag{14.4}
$$

The characteristic polynomial of B_6 is $(\lambda_1-\lambda)^3(\lambda_2-\lambda)^2(\lambda_3-\lambda)$, of B_3 is $(\lambda_1-\lambda)(\lambda_2-\lambda)^2$, and of B_1 is $(\lambda_1-\lambda)$. Each of these polynomials is a factor of the one which precedes it, and it follows from the manner in which the G_i were defined that this will be true in the general case. It is evident that the elementary divisors of a matrix are the factors of the characteristic polynomials of the submatrices in the Frobenius canonical form. Notice that if the matrix A is real, then any complex eigenvalues occur as conjugate pairs. Hence each G_i will obviously be real. This is to be expected since we have stated that the Frobenius canonical form may be derived by transformations which are rational in the field of A.

Equivalence transformations

15. We have seen that the eigenvalues of a matrix are invariant with respect to similarity transformations. We consider now briefly a more general class of transformations. The matrix B is said to be

equivalent to the matrix A if there exist non-singular (necessarily square) matrices P and Q such that

$$PAQ = B. \qquad (15.1)$$

Note that A and B need not be square but they must be of the same dimensions. From the Theorem of Corresponding Matrices (sometimes known as the Binet-Cauchy theorem, see for example Gantmacher, 1959a, Vol. 1), it follows that equivalent matrices have the same rank.

In solving sets of linear equations it is essentially the relationship of equivalence with which we are concerned. For if x satisfies

$$Ax = b, \qquad (15.2)$$

then clearly it satisfies

$$PAx = Pb \qquad (15.3)$$

for any conformable matrix P. If further P is non-singular then any solution of (15.3) is a solution of (15.2) as may be seen by multiplying (15.3) by P^{-1}.

Now consider the change of variables

$$x = Qy, \qquad (15.4)$$

where Q is non-singular. We then have

$$PAQy = Pb, \qquad (15.5)$$

and from any solution y of (15.5) we can derive a solution x of (15.2) and vice versa. The equations (15.2) and (15.5) may therefore be regarded as equivalent.

Lambda matrices

16. We now extend the concept of equivalence to matrices having elements which are polynomials in λ; such matrices are called *λ-matrices*. If $A(\lambda)$ is an $(m \times n)$ matrix having polynomial elements of degrees not greater than k, then we may write

$$A(\lambda) = A_k\lambda^k + A_{k-1}\lambda^{k-1} + \ldots + A_0, \qquad (16.1)$$

where $A_k, A_{k-1}, \ldots, A_0$ are $(m \times n)$ matrices with elements which are independent of λ. The rank of a λ-matrix is the maximum order of those minors which are non-null polynomials in λ.

Two λ-matrices $A(\lambda)$ and $B(\lambda)$ are called equivalent if there exist square λ-matrices $P(\lambda)$ and $Q(\lambda)$ *with non-zero determinants independent of λ* such that

$$P(\lambda)\, A(\lambda)\, Q(\lambda) = B(\lambda). \qquad (16.2)$$

Note that since $P(\lambda)$ has a non-zero determinant independent of λ, $[P(\lambda)]^{-1}$ is also a λ-matrix having the same property.

Elementary operations

17. We shall be interested in simple non-singular λ-matrices $P(\lambda)$ of the following types:

(i) $p_{ii}(\lambda) = 1 \ (i \neq p, q); \ p_{pq}(\lambda) = p_{qp}(\lambda) = 1; \ p_{ij}(\lambda) = 0$
$\qquad\qquad\qquad\qquad\qquad\qquad\qquad\qquad\qquad$ otherwise,

(ii) $p_{ii}(\lambda) = 1 \ (i = 1, ..., n); \ p_{pq}(\lambda) = f(\lambda); \ p_{ij}(\lambda) = 0$ otherwise,

(iii) $p_{ii}(\lambda) = 1 \ (i \neq p); \ p_{pp}(\lambda) = k \neq 0; \ p_{ij}(\lambda) = 0$ otherwise.

It is evident that the determinant of a matrix of type (i) is -1 and of type (ii) is $+1$ while that of a matrix of type (iii) is k.

Pre-multiplication of $A(\lambda)$ by a matrix of type (i), (ii) or (iii) respectively, has the following effect.

(i) Rows p and q are interchanged,

(ii) $f(\lambda)$ times row q is added to row p,

(iii) Row p is multiplied by the non-zero constant k.

Post-multiplication has a corresponding effect upon columns. We refer to these transformations of $A(\lambda)$ as *elementary operations*.

Smith's canonical form

18. The fundamental theorem on equivalent λ-matrices is the following.

Every λ-matrix $A(\lambda)$ of rank r can be reduced by elementary transformations (rational in the field of the elements of $A(\lambda)$) to the equivalent diagonal form D given by

$$
D = \begin{bmatrix}
E_1(\lambda) & & & & & & \\
& E_2(\lambda) & & & & & \\
& & \ddots & & & & \\
& & & E_r(\lambda) & & & \\
& & & & 0 & & \\
& & & & & \ddots & \\
& & & & & & 0
\end{bmatrix}, \qquad (18.1)
$$

where each $E_i(\lambda)$ is a *monic* polynomial in λ such that $E_i(\lambda)$ divides $E_{i+1}(\lambda)$. (In a monic polynomial the coefficient of the term of highest degree is unity.) All elements of D other than the $E_i(\lambda)$ are zero.

The proof is as follows. We shall produce the diagonal form by means of a series of elementary operations, and shall refer to the (i, j) element of the *current* transformed matrix at any stage as $a_{ij}(\lambda)$.

Let $a_{ij}(\lambda)$ be one of the elements of lowest degree in $A(\lambda)$. By interchanging rows 1 and i and columns 1 and j this element may be made the element $a_{11}(\lambda)$. We now write

$$a_{i1}(\lambda) = a_{11}(\lambda)q_{i1}(\lambda) + r_{i1}(\lambda) \quad (i = 2, 3, \ldots, m) \atop a_{1j}(\lambda) = a_{11}(\lambda)q_{1j}(\lambda) + r_{1j}(\lambda) \quad (j = 2, 3, \ldots, n) \right\}$$ (18.2)

where the $r_{i1}(\lambda)$ and $r_{1j}(\lambda)$ are remainders and therefore of degrees lower than that of $a_{11}(\lambda)$. If not all of the $r_{i1}(\lambda)$ and $r_{1j}(\lambda)$ are zero, suppose $r_{k1}(\lambda)$ is non-zero. Subtract $q_{k1}(\lambda)$ times the first row from the kth row and interchange rows 1 and k. The element $a_{11}(\lambda)$ then becomes $r_{k1}(\lambda)$ which is non-zero and of lower degree than the previous element in this position.

Continuing in this way the element $a_{11}(\lambda)$ is always non-zero and its degree steadily diminishes. Ultimately the current $a_{11}(\lambda)$ must divide all the current $a_{i1}(\lambda)$ and $a_{1j}(\lambda)$, either when $a_{11}(\lambda)$ becomes independent of λ or at an earlier stage. When this occurs the matrix may be reduced to the form

$$\begin{bmatrix} a_{11}(\lambda) & O \\ O & A_2(\lambda) \end{bmatrix}$$ (18.3)

by subtracting appropriate multiples of the first row from rows 2 to m and then appropriate multiples of the first column from columns 2 to n.

At this stage $a_{11}(\lambda)$ may divide all elements of $A_2(\lambda)$. If not, suppose for some element $a_{ij}(\lambda)$

$$a_{ij}(\lambda) = a_{11}(\lambda)q_{ij}(\lambda) + r_{ij}(\lambda) \quad (r_{ij}(\lambda) \neq 0).$$ (18.4)

Then, after adding row i to the first row we may apply the previous procedure and obtain the form (18.3) again with a non-zero $a_{11}(\lambda)$ of lower degree. Continuing in this way we must reach a form (18.3) in which $a_{11}(\lambda)$ divides every element of $A_2(\lambda)$. This must happen either when $a_{11}(\lambda)$ becomes independent of λ or at some earlier stage.

If $A_2(\lambda)$ is not identically zero, we may apply to this matrix a process similar to that just applied to $A(\lambda)$. We may think of this as involving operations on rows 2 to m and columns 2 to n only of the matrix (18.3); the first row and column remain unaffected throughout. At all stages in this process clearly all elements of $A_2(\lambda)$ remain multiples of $a_{11}(\lambda)$. Hence we obtain the form

$$\begin{bmatrix} a_{11}(\lambda) & 0 & \vdots & \\ 0 & a_{22}(\lambda) & \vdots & O \\ \cdots & \cdots & & \\ & O & \vdots & A_3(\lambda) \end{bmatrix} \tag{18.5}$$

in which $a_{11}(\lambda)$ divides $a_{22}(\lambda)$ and $a_{22}(\lambda)$ divides all elements of $A_3(\lambda)$.

Continuing in this way we obtain the form

$$\begin{bmatrix} a_{11}(\lambda) & & & & \vdots & \\ & a_{22}(\lambda) & & & \vdots & \\ & & \cdot & & \vdots & O \\ & & & \cdot & \vdots & \\ & & & & a_{ss}(\lambda) & \vdots \\ \cdots & \cdots & \cdots & \cdots & \cdots & \\ & & O & & \vdots & O \end{bmatrix} \tag{18.6}$$

where each $a_{ii}(\lambda)$ divides its successor, the process terminating when s equals m or n, or $A_{s+1}(\lambda)$ is the null matrix. However, we must have $s = r$ since, from the Theorem of Corresponding Matrices, the rank of a λ-matrix is invariant under multiplication by non-singular matrices. Since the $a_{ii}(\lambda)$ are non-zero polynomials, by multiplying the rows by appropriate constants, the coefficients of the highest powers in each of the polynomials may be made unity. The $a_{ii}(\lambda)$ then become the monic polynomials $E_i(\lambda)$ and each $E_i(\lambda)$ divides $E_{i+1}(\lambda)$. This completes the proof and it is clear that only rational operations are used. Since D was obtained by elementary transformations we have

$$P(\lambda)\, A(\lambda)\, Q(\lambda) = D \tag{18.7}$$

for non-singular $P(\lambda)$ and $Q(\lambda)$ with determinants independent of λ. The form D of (18.1) is known as the *Smith's canonical form for equivalent λ-matrices*.

The highest common factor of k-rowed minors of a λ-matrix

19. Consider now the k-rowed minors of the Smith's canonical form. If $k > r$ then all k-rowed minors are zero. Otherwise the only non-zero k-rowed minors are products of k of the $E_i(\lambda)$. Since each $E_i(\lambda)$ is a factor of $E_{i+1}(\lambda)$, the highest common factor (H.C.F.) $G_k(\lambda)$ of k-rowed minors is $E_1(\lambda) E_2(\lambda)...E_k(\lambda)$. Hence if we write $G_0(\lambda) = 1$

$$E_i(\lambda) = \frac{G_i(\lambda)}{G_{i-1}(\lambda)} \qquad (i = 1, 2,..., r). \tag{19.1}$$

We now show that the H.C.F. of k-rowed minors of a matrix (we take the coefficient of the highest power of λ to be unity) is invariant under equivalence transformations. For if

$$P(\lambda) A(\lambda) Q(\lambda) = B(\lambda), \tag{19.2}$$

then any k-rowed minor of $B(\lambda)$ may be expressed as the sum of terms of the form $P_\alpha A_\beta Q_\gamma$ where P_α, A_β, Q_γ denote a k-rowed minor of $P(\lambda)$, $A(\lambda)$, $Q(\lambda)$ respectively. Hence the H.C.F. of the k-rowed minors of $A(\lambda)$ divides any k-rowed minor of $B(\lambda)$. Since, however,

$$A(\lambda) = [P(\lambda)]^{-1} B(\lambda) [Q(\lambda)]^{-1}$$

and $[P(\lambda)]^{-1}$ and $[Q(\lambda)]^{-1}$ are also λ-matrices, the H.C.F. of k-rowed minors of $B(\lambda)$ divides any k-rowed minor of $A(\lambda)$. The two H.C.F.'s must therefore be equal.

All equivalent λ-matrices therefore have the same Smith's canonical form. The polynomials $E_i(\lambda)$ are called the *invariant factors* or *invariant polynomials* of $A(\lambda)$.

Invariant factors of $(A-\lambda I)$

20. If A is a square matrix, then $(A-\lambda I)$ is a λ-matrix of rank n, and hence has n invariant factors, $E_1(\lambda), E_2(\lambda),..., E_n(\lambda)$, each of which divides its successor. We now relate the invariant factors of $(A-\lambda I)$ with the determinants of the Frobenius matrices in the Frobenius canonical forms. Suppose B is the Frobenius canonical form of A, so that there exists an H such that

$$H^{-1}AH = B, \tag{20.1}$$

and therefore

$$H^{-1}(A-\lambda I)H = B-\lambda I. \tag{20.2}$$

Since the matrices H^{-1} and H are non-singular and independent of λ, $(B-\lambda I)$ is equivalent to $(A-\lambda I)$.

We consider now the invariant factors of $(B-\lambda I)$. Their general form will be sufficiently illustrated by considering a simple example. Suppose $(B-\lambda I)$ is of the form

$$
\begin{bmatrix}
p_3-\lambda & p_2 & p_1 & p_0 & & & & & \\
1 & -\lambda & 0 & 0 & & & & & \\
0 & 1 & -\lambda & 0 & & & & & \\
0 & 0 & 1 & -\lambda & & & & & \\
& & & & q_2-\lambda & q_1 & q_0 & & \\
& & & & 1 & -\lambda & 0 & & \\
& & & & 0 & 1 & -\lambda & & \\
& & & & & & & r_1-\lambda & r_0 \\
& & & & & & & 1 & -\lambda
\end{bmatrix}
=
$$

$$
=
\begin{bmatrix}
B_1-\lambda I & & \\
& B_2-\lambda I & \\
& & B_3-\lambda I
\end{bmatrix}. \quad (20.3)
$$

From § 14 we know that in the sequence $\det(B_1-\lambda I)$, $\det(B_2-\lambda I)$, $\det(B_3-\lambda I)$, each of these polynomials is divisible by its successor.

By examination of non-vanishing minors we see that the $G_i(\lambda)$, the H.C.F.'s of the i-rowed minors, are given by

$$
\left.
\begin{aligned}
&G_0(\lambda) = 1 \text{ (by definition)}; \ G_1(\lambda) = G_2(\lambda) = \ldots = G_6(\lambda) = 1; \\
&G_7(\lambda) = \det(B_3-\lambda I); \ G_8(\lambda) = \det(B_3-\lambda I)\det(B_2-\lambda I); \\
&G_9(\lambda) = \det(B_3-\lambda I)\det(B_2-\lambda I)\det(B_1-\lambda I).
\end{aligned}
\right\} \quad (20.4)
$$

Hence the invariant factors are given by

$$
\left.
\begin{aligned}
&E_1(\lambda) = E_2(\lambda) = \ldots = E_6(\lambda) = 1; \\
&E_7(\lambda) = \det(B_3-\lambda I); \ E_8(\lambda) = \det(B_2-\lambda I); \ E_9(\lambda) = \det(B_1-\lambda I).
\end{aligned}
\right\} \quad (20.5)
$$

It is evident that our result is quite general so that the invariant factors of $(A-\lambda I)$ which are not equal to unity are equal to the characteristic polynomials of the submatrices in the Frobenius canonical form. The fundamental result relating similarity and λ-equivalence is the following.

A necessary and sufficient condition for A to be similar to B is that $(A - \lambda I)$ *and* $(B - \lambda I)$ *are equivalent.* For this reason the invariant factors of $(A - \lambda I)$ are sometimes called the *similarity invariants of A*.

Again from § 14 we see that if we write

$$E_i(\lambda) = (\lambda_1 - \lambda)^{i_1}(\lambda_2 - \lambda)^{i_2}\ldots(\lambda_s - \lambda)^{i_s}, \tag{20.6}$$

where $\lambda_1, \lambda_2,\ldots, \lambda_s$ are the distinct eigenvalues of A, then the factors on the right of (20.6) are the elementary divisors of A.

The triangular canonical form

21. Reduction to Jordan canonical form provides complete information on both the eigenvalues and the eigenvectors. This form is a special case of the *triangular canonical form*. The latter is very much simpler from the theoretical standpoint and also for practical computation. Its importance is a consequence of the following theorem and that of § 25.

Every matrix is similar to a matrix of *triangular* form, that is, a matrix in which all the elements above the diagonal are zero. We refer to such a matrix as a *lower triangular matrix*. As with the Jordan form there is an alternative transformation leading to a matrix with zeros below the diagonal (*upper triangular*). The proof is comparatively elementary but is deferred until § 42 when the appropriate transformation matrices have been introduced.

Clearly the diagonal elements are the eigenvalues with the appropriate multiplicities. For if these elements are μ_i then the characteristic polynomial of the triangular matrix is $\prod (\mu_i - \lambda)$ and we have seen that the characteristic polynomial is invariant under similarity transformations.

Hermitian and symmetric matrices

22. Matrices which are real and symmetric are of special importance in practice as we shall see in § 31. We could extend this class of matrices by including all matrices which are symmetric and have complex elements. However, such an extension proves to be of little value because matrices of this extended class do not have many of the more important properties of real symmetric matrices. The most useful extension is to the class of matrices of the form $(P + iQ)$ where P is real and symmetric and Q is real and *skew* symmetric, that is

$$Q^T = -Q. \tag{22.1}$$

Matrices of this class are called *Hermitian* matrices. If A is Hermitian we have from the definition

$$\bar{A}^T = A, \tag{22.2}$$

where $\bar{A}$ denotes the matrix whose elements are the complex conjugates of those of A. The matrix $\bar{A}^T$ is frequently denoted by A^H and is referred to as the *Hermitian transposed matrix*. Similarly x^H is the row vector with components equal to the complex conjugates of those of the column vector x. The notation is a convenient one and has the advantage that results proved for Hermitian matrices may be converted to the corresponding results for real symmetric matrices by replacing A^H by A^T and x^H by x^T. If a is a scalar, a^H is defined to be the complex conjugate of a. It follows immediately from the definitions that

$$\left.\begin{aligned} (A^H)^H &= A \\ (x^H)^H &= x \\ (ABC)^H &= C^H B^H A^H \end{aligned}\right\}, \tag{22.3}$$

while

$$A^H = A \tag{22.4}$$

if A is Hermitian.

Note that $y^H x$ is the inner-product of y and x as normally understood and

$$\overline{y^H x} = (y^H x)^H = x^H y. \tag{22.5}$$

The inner-product $x^H x$ is real and positive for all x other than the null vector since it is equal to the sum of the squares of the moduli of the elements of x.

Elementary properties of Hermitian matrices

23. The eigenvalues of a Hermitian matrix are all real. For if

$$Ax = \lambda x \tag{23.1}$$

then

$$x^H A x = \lambda x^H x. \tag{23.2}$$

Now we have remarked that $x^H x$ is real and positive for a non-null x. Further we have

$$(x^H A x)^H = x^H A^H x^{HH} = x^H A x, \tag{23.3}$$

and since $x^H A x$ is a scalar this means that it is real. Hence λ is real. For real symmetric matrices, real eigenvalues imply real eigenvectors, but the eigenvectors of a complex Hermitian matrix are in general complex.

If $Ax = \lambda x$ we have for a Hermitian matrix

$$A^H x = \lambda x \tag{23.4}$$

and taking the complex conjugate of both sides

$$A^T \bar{x} = \lambda \bar{x}. \tag{23.5}$$

The eigenvectors of the transpose of a Hermitian matrix are therefore the complex conjugates of those of the matrix itself. In the notation of § 3

$$y_i = \bar{x}_i. \tag{23.6}$$

The quantities $y_i^T x_i$ which are of great importance in the theory therefore become $x_i^H x_i$ for Hermitian matrices. It follows immediately from the general theory of § 3 that if a Hermitian matrix has distinct eigenvalues then its eigenvectors satisfy

$$x_i^H x_j = 0 \quad (i \neq j). \tag{23.7}$$

If we normalize the x_i so that

$$x_i^H x_i = 1, \tag{23.8}$$

equations (23.7) and (23.8) imply that the matrix X formed by the eigenvectors satisfies $X^H X = I$ i.e. $X^H = X^{-1}.$ (23.9)

A matrix which satisfies equations (23.9) is called a *unitary* matrix. A real unitary matrix is called an *orthogonal* matrix.

We have therefore proved the following for Hermitian and real symmetric matrices with distinct eigenvalues.

(I) If A is Hermitian there exists a unitary matrix U such that $U^H A U = \text{diag}(\lambda_i)$ (real λ_i).

(II) If A is real and symmetric, there exists an orthogonal matrix U such that $U^T A U = \text{diag}(\lambda_i)$ (real λ_i).

Perhaps the most important property of Hermitian (real symmetric) matrices is that I (II), which we have so far proved only in the case of distinct eigenvalues, is true whatever multiplicities there may be. The proof is deferred until § 47.

Immediate consequences of this fundamental result are.

(i) The elementary divisors of a Hermitian matrix are all linear.

(ii) If a Hermitian matrix has any multiple eigenvalues then it is derogatory.

(iii) A Hermitian matrix cannot be defective.

Complex symmetric matrices

24. That none of the important properties of real symmetric matrices is shared by complex symmetric matrices may be seen by considering the matrix A defined by

$$A = \begin{bmatrix} 2i & 1 \\ 1 & 0 \end{bmatrix}. \tag{24.1}$$

The characteristic equation is $\lambda^2 - 2i\lambda - 1 = 0$, so that $\lambda = i$ is an eigenvalue of multiplicity 2. The components x_1, x_2 of a corresponding eigenvector satisfy

$$ix_1 + x_2 = 0, \qquad x_1 - ix_2 = 0, \tag{24.2}$$

so that there is only one eigenvector, and that has components $(1, -i)$.

We conclude that there is a quadratic divisor associated with $\lambda = i$. In fact the Jordan canonical form is

$$\begin{bmatrix} i & 1 \\ 0 & i \end{bmatrix}, \tag{24.3}$$

and we have

$$\begin{bmatrix} 2i & 1 \\ 1 & 0 \end{bmatrix} \begin{bmatrix} 1 & 0 \\ -i & 1 \end{bmatrix} = \begin{bmatrix} 1 & 0 \\ -i & 1 \end{bmatrix} \begin{bmatrix} i & 1 \\ 0 & i \end{bmatrix}. \tag{24.4}$$

Reduction to triangular form by unitary transformations

25. Since a Hermitian matrix may always be reduced to diagonal form by a unitary similarity transformation, it is interesting to consider what reduction of a general matrix may be achieved by such a transformation. It can be shown that any matrix may be reduced by a unitary similarity transformation to triangular form. Since the triangular form provides us immediately with the eigenvalues, and the eigenvectors may be found comparatively simply, this result is of practical importance because unitary transformations have desirable numerical properties. We defer the proof until § 47 when the appropriate transformation matrices have been introduced.

Quadratic forms

26. We may associate with a real symmetric matrix A, a *quadratic form* $x^T A x$ in the n variables $x_1, x_2, \ldots, x_n$ defined by

$$x^T A x = \sum_{i,j=1}^{n} a_{ij} x_i x_j, \tag{26.1}$$

where x is the vector with components x_i. Corresponding to each quadratic form we may define a *bilinear form* $x^T A y$ defined by

$$x^T A y = \sum_{i,j=1}^{n} a_{ij} x_i y_j = y^T A x. \tag{26.2}$$

Certain types of quadratic form are of particular interest. A quadratic form with matrix A is defined to be

$$
\left.\begin{array}{l}
\text{positive definite} \\[4pt]
\text{negative definite} \\[4pt]
\text{non-negative definite} \\[4pt]
\text{non-positive definite}
\end{array}\right\}
\text{according as } x^T A x
\left\{\begin{array}{l}
> 0 \\[4pt]
< 0 \\[4pt]
\geqslant 0 \\[4pt]
\leqslant 0
\end{array}\right.
\text{for all real } x \neq 0.
$$

The matrix of a positive definite quadratic form is called a positive definite matrix.

A necessary and sufficient condition for $x^T A x$ to be positive definite is that all the eigenvalues of A should be positive. For we know that there exists an orthogonal matrix R such that

$$R^T A R = \operatorname{diag}(\lambda_i). \tag{26.3}$$

Hence writing

$$x = Rz \quad \text{i.e.} \quad z = R^T x, \tag{26.4}$$

we have

$$x^T A x = z^T R^T A R z = \sum_{i=1}^{n} \lambda_i z_i^2. \tag{26.5}$$

A necessary and sufficient condition for $\sum_{i=1}^{n} \lambda_i z_i^2$ to be positive for all non-null z is that all λ_i should be positive. From (26.4) we see that non-null z correspond to non-null x and conversely; hence the result is established. Since the determinant of a matrix is equal to the product of its eigenvalues the determinant of a positive definite matrix must be positive.

From equations (26.4) and (26.5) we deduce that the problem of determining the orthogonal matrix R, and hence the eigenvectors of A, is identical with that of determining the principal axes of the quadratic

$$\sum_{i,j} a_{ij} x_i x_j = 1 \tag{26.6}$$

and that the λ_i are the reciprocals of the squares of the principal axes. A multiple eigenvalue is associated with an indeterminacy of the principal axes. As a typical example we may take an ellipsoid in three dimensions with two equal principal axes. One of its principal sections is then a circle and we may take any two perpendicular diameters of the circle as principal axes.

Necessary and sufficient conditions for positive definiteness

27. We denote by $x^T A_r x$ the quadratic form derived from $x^T A x$ by setting $x_{r+1}, x_{r+2}, \ldots, x_n$ equal to zero. $x^T A_r x$ is therefore a quadratic form in the r variables $x_1, x_2,..., x_r$, and the matrix A_r of this form is the leading principal submatrix of order r of A. Clearly if $x^T A x$ is positive definite, $x^T A_r x$ must be positive definite. For if there exists a non-null set of values $x_1, x_2,..., x_r$ for which $x^T A_r x$ is non-positive, then $x^T A x$ is non-positive for the vector x given by

$$x^T = (x_1, x_2,..., x_r, 0,..., 0). \tag{27.1}$$

Hence all leading principal minors of a positive definite matrix must be positive.

Note that an exactly similar method of proof shows that *any* principal submatrix of a positive definite matrix is positive definite and hence has positive determinant. However, the possession of n positive leading principal minors is also a *sufficient* condition for positive definiteness. The proof is by induction.

Suppose it is true for matrices of order $(n-1)$. We then show that it is true for matrices of order n. For suppose on the contrary that A is a matrix of order n for which the n leading principal minors are positive and that A is not positive definite. If R is the orthogonal matrix for which $R^T A R = \mathrm{diag}(\lambda_i)$ then writing $x = Rz$ we have

$$x^T A x = z^T R^T A R z = \sum \lambda_i z_i^2. \tag{27.2}$$

Now by hypothesis A is not positive definite and therefore, since $\prod_{i=1}^{n} (\lambda_i) = \det(A) > 0$, there must be an even number r of negative λ_i. Suppose the first r are negative. Then consider the set of equations

$$z_{r+1} = z_{r+2} = ... = z_n = x_n = 0. \tag{27.3}$$

Since the z_i are linear functions of the x_i we have $(n-r+1)$ linear homogeneous equations in the n unknowns $x_1, x_2,..., x_n$. There is therefore at least one non-null solution x which we may denote by $(x_1, x_2,..., x_{n-1}, 0)$. The corresponding z is also non-null and we may denote this by $(z_1, z_2,..., z_r, 0, 0,..., 0)$. For this x and z we have

$$x^T A x = \sum_1^n \lambda_i z_i^2 = \sum_1^r \lambda_i z_i^2 < 0. \tag{27.4}$$

Now since $x_n = 0$, this implies that $x^T A_{n-1} x$ is negative for the vector $(x_1, x_2,..., x_{n-1})$. This contradicts our inductive hypothesis. The result is obviously true for matrices of order one and is therefore true for matrices of any order.

In the same way as we have associated a quadratic form with a real symmetric matrix, we may associate with a Hermitian matrix A, a *Hermitian form* defined by

$$x^H A x = \sum_{i,j=1}^{n} a_{ij} x_i \bar{x}_j. \tag{27.5}$$

We have

$$(x^H A x)^H = x^H A^H x^{HH} = x^H A x, \tag{27.6}$$

and hence $x^H A x$ is real. The concepts of positive definiteness etc. carry over immediately to Hermitian forms and there are results analogous to those we have just proved for real quadratic forms.

Differential equations with constant coefficients

28. The solution of the algebraic eigenvalue problem is closely related to the solution of a set of linear simultaneous ordinary differential equations with constant coefficients. The general set of first-order homogeneous equations in n unknowns $y_1, y_2, \ldots, y_n$ may be written in the form

$$B \frac{d}{dt}(y) = Cy, \tag{28.1}$$

where t is the independent variable, y is the vector with components y_i, and B and C are $n \times n$ matrices. If B is singular the n left-hand sides of the system of equations satisfy a linear relation. The right-hand side must satisfy the same linear relation and hence the y_i are not independent. The set of equations may therefore be reduced to a system of lower order. We shall not pursue this further and shall assume that B is non-singular.

Equation (28.1) may then be written in the form

$$\frac{d}{dt}(y) = Ay, \tag{28.2}$$

where

$$A = B^{-1}C. \tag{28.3}$$

Let us assume that (28.2) has a solution of the form

$$y = xe^{\lambda t}, \tag{28.4}$$

where x is a vector independent of t. Then we must have

$$\lambda x e^{\lambda t} = A x e^{\lambda t},$$

and therefore

$$\lambda x = Ax. \tag{28.5}$$

Equation (28.4) will give a solution of (28.2) if λ is an eigenvalue of A and x is the corresponding eigenvector. If there are r independent eigenvectors of A then we obtain r independent solutions of A, this being true regardless of whether all the corresponding eigenvalues are distinct. However, there must be n independent solutions of

(28.2), so that if $r < n$, that is if some of the elementary divisors of A are non-linear, equation (28.2) must have some solutions which are not of pure exponential type.

Solutions corresponding to non-linear elementary divisors

29. The nature of the solution when there are non-linear elementary divisors can be examined in terms of the Jordan canonical form. Let X be the matrix of the similarity transformation which reduces A to the Jordan form. If we now define new variables z by the relation

$$y = Xz, \tag{29.1}$$

then (28.2) becomes

$$X \frac{d}{dt}(z) = AXz \tag{29.2}$$

i.e.

$$\frac{d}{dt}(z) = X^{-1}AXz. \tag{29.3}$$

Note that the matrix of the equations undergoes a similarity transformation. The matrix of the system of equations is now the Jordan canonical form of A.

Suppose, for example, A is a matrix of order 6 with the lower Jordan canonical form

$$\begin{bmatrix} C_3(\lambda_1) & & \\ & C_2(\lambda_1) & \\ & & C_1(\lambda_2) \end{bmatrix}.$$

Then equations (29.3) are

$$\begin{bmatrix} \dfrac{dz_1}{dt} = \lambda_1 z_1 & & \\[2mm] \dfrac{dz_2}{dt} = z_1 + \lambda_1 z_2 & & \\[2mm] \dfrac{dz_3}{dt} = \quad z_2 + \lambda_1 z_3 & & \\[2mm] \dfrac{dz_4}{dt} = & \lambda_1 z_4 & \\[2mm] \dfrac{dz_5}{dt} = & z_4 + \lambda_1 z_5 & \\[2mm] \dfrac{dz_6}{dt} = & & \lambda_2 z_6 \end{bmatrix} \tag{29.4}$$

The solutions of the first, fourth, and sixth of these equations are

$$z_1 = a_1 e^{\lambda_1 t}, \quad z_4 = a_4 e^{\lambda_1 t}, \quad z_6 = a_6 e^{\lambda_2 t}, \tag{29.5}$$

where a_1, a_4 and a_6 are arbitrary constants. The general solution of the second equation is

$$z_2 = a_2 e^{\lambda_1 t} + a_1 t e^{\lambda_1 t}, \tag{29.6}$$

of the third is

$$z_3 = a_3 e^{\lambda_1 t} + a_2 t e^{\lambda_1 t} + \tfrac{1}{2} a_1 t^2 e^{\lambda_1 t}, \tag{29.7}$$

and of the fifth,

$$z_5 = a_5 e^{\lambda_1 t} + a_4 t e^{\lambda_1 t}, \tag{29.8}$$

where a_2, a_3 and a_5 are arbitrary constants. The general solution of the system is therefore given by

$$\left.
\begin{aligned}
z_1 &= a_1 e^{\lambda_1 t} \\
z_2 &= a_1 t e^{\lambda_1 t} + a_2 e^{\lambda_1 t} \\
z_3 &= \tfrac{1}{2} a_1 t^2 e^{\lambda_1 t} + a_2 t e^{\lambda_1 t} + a_3 e^{\lambda_1 t} \\
z_4 &= \qquad\qquad\qquad\qquad a_4 e^{\lambda_1 t} \\
z_5 &= \qquad\qquad\qquad\qquad a_4 t e^{\lambda_1 t} + a_5 e^{\lambda_1 t} \\
z_6 &= \qquad\qquad\qquad\qquad\qquad\qquad a_6 e^{\lambda_2 t}
\end{aligned}
\right\}. \tag{29.9}$$

Note that there are only three independent solutions which are pure exponentials. These are

$$z = a_3 e^{\lambda_1 t} e_3, \quad z = a_5 e^{\lambda_1 t} e_5, \quad z = a_6 e^{\lambda_2 t} e_6, \tag{29.10}$$

as we would expect, since e_3, e_5 and e_6 are the only eigenvectors of the Jordan canonical form.

The general solution of the original equations is now obtained by substituting for z in equation (29.1). If the columns of X are denoted by $x_1, x_2, \ldots, x_6$ then the general solution of (28.2) is

$$y = a_1 e^{\lambda_1 t}(x_1 + t x_2 + \tfrac{1}{2} t^2 x_3) + a_2 e^{\lambda_1 t}(x_2 + t x_3) + a_3 e^{\lambda_1 t} x_3 +$$
$$+ a_4 e^{\lambda_1 t}(x_4 + t x_5) + a_5 e^{\lambda_1 t} x_5 + a_6 e^{\lambda_2 t} x_6. \tag{29.11}$$

The result in the general case is immediately obvious from this.

Differential equations of higher order

30. We consider now the system of n simultaneous homogenous differential equations of order r by

$$A_r \frac{d^r}{dt^r}(y_0) + A_{r-1} \frac{d^{r-1}}{dt^{r-1}}(y_0) + \ldots + A_0 y_0 = 0, \tag{30.1}$$

where $A_r, A_{r-1}, \ldots, A_0$ are $(n \times n)$ matrices with constant elements and $\det(A_r) \neq 0$, and we use y_0 to denote the vector of order n formed by the n dependent variables. (We use y_0 instead of y for reasons which will soon become apparent.) The system (30.1) may be reduced to a system of first-order equations by introducing $r(n-1)$ new variables defined by

$$\left.\begin{aligned} y_1 &= \frac{d}{dt}(y_0) \\[2mm] y_2 &= \frac{d}{dt}(y_1) \\ &\cdots\cdots\cdots \\ y_{r-1} &= \frac{d}{dt}(y_{r-2}) \end{aligned}\right\}. \tag{30.2}$$

If A_r is non-singular we may write equations (30.1) in the form

$$B_0 y_0 + B_1 y_1 + \ldots + B_{r-1}y_{r-1} = \frac{d}{dt}(y_{r-1}), \tag{30.3}$$

where

$$B_s = -A_r^{-1}A_s. \tag{30.4}$$

Equations (30.2) and (30.3) may be regarded as a single set of nr equations in the nr variables which are the components of $y_0, y_1, \ldots, y_{r-1}$. We may write them in the form

$$\begin{bmatrix} 0 & I & 0 & 0 & \ldots & 0 \\ 0 & 0 & I & 0 & \ldots & 0 \\ 0 & 0 & 0 & I & \ldots & 0 \\ \cdots & \cdots & \cdots & \cdots & \cdots & \cdots \\ 0 & 0 & 0 & 0 & \ldots & I \\ B_0 & B_1 & B_2 & B_3 & \ldots & B_{r-1} \end{bmatrix} \begin{bmatrix} y_0 \\ y_1 \\ y_2 \\ \cdot \\ y_{r-2} \\ y_{r-1} \end{bmatrix} = \frac{d}{dt} \begin{bmatrix} y_0 \\ y_1 \\ y_2 \\ \cdot \\ y_{r-2} \\ y_{r-1} \end{bmatrix}. \tag{30.5}$$

This system may now be solved by the methods of the previous section and requires the determination of the Jordan canonical form of the matrix B of order nr on the left-hand side of (30.5). Although this matrix is of high order it is of a particularly simple form and has a high percentage of zero elements. If λ is an eigenvalue of B, and x is an eigenvector which is partitioned into the r vectors $x_0, x_1, \ldots, x_{r-1}$, then

$$\left.\begin{aligned} x_i &= \lambda x_{i-1} \qquad (i = 1, \ldots, r-1) \\ B_0 x_0 + B_1 x_1 + \ldots + B_{r-1}x_{r-1} &= \lambda x_{r-1} \end{aligned}\right\}, \tag{30.6}$$

and hence

$$(B_0 + B_1 \lambda + \ldots + B_{r-1}\lambda^{r-1})x_0 = \lambda^r x_0. \tag{30.7}$$

These equations have a non-zero solution if

$$\det (B_0 + B_1\lambda + \ldots + B_{r-1}\lambda^{r-1} - I\lambda^r) = 0. \qquad (30.8)$$

In practical work we have a choice of working with the matrix B of order rn and solving the standard eigenproblem, or alternatively of working with the determinantal equation (30.8) which is no longer standard, but in which the matrix is of order n only.

Second-order equations of special form

31. The equations of motion for small vibrations about a position of stable equilibrium of a mechanical system operated upon by a conservative system of forces are of the form

$$B\ddot{y} = -Ay, \qquad (31.1)$$

where A and B are symmetric and positive definite. If we assume that the solutions are of the form $y = xe^{i\mu t}$ then we must have

$$\mu^2 Bx = Ax. \qquad (31.2)$$

Writing $\mu^2 = \lambda$ this becomes

$$\lambda Bx = Ax. \qquad (31.3)$$

We shall now show that when A and B are symmetric and B is positive definite, then the roots of $\det(A - \lambda B) = 0$ are all real.

Since B is positive definite there exists an orthogonal R such that

$$R^T B R = \mathrm{diag}(\beta_i^2), \qquad (31.4)$$

where the β_i^2 are the eigenvalues, necessarily positive, of B. We may write

$$\mathrm{diag}(\beta_i) = D, \qquad \mathrm{diag}(\beta_i^2) = D^2. \qquad (31.5)$$

Then we have

$$(A - \lambda B) = RD(D^{-1}R^T ARD^{-1} - \lambda I)DR^T, \qquad (31.6)$$

and hence

$$\det(A - \lambda B) = (\det R)^2 (\det D)^2 \det(P - \lambda I), \qquad (31.7)$$

where

$$P = D^{-1}R^T ARD^{-1}. \qquad (31.8)$$

Since $(\det R)^2 = 1$ and $(\det D)^2 = \prod \beta_i^2$, the zeros of $\det(A - \lambda B)$ are those of $\det(P - \lambda I)$. The latter are the eigenvalues $\lambda_1, \lambda_2, \ldots, \lambda_n$ of P, a symmetric matrix, and are therefore real.

The matrix P has a complete set of eigenvectors z_i which may be taken to be orthogonal. We have therefore

giving

$$Pz_i = \lambda_i z_i, \qquad D^{-1}R^T ARD^{-1}z_i = \lambda_i z_i, \qquad (31.9)$$

$$A(RD^{-1}z_i) = \lambda_i RDz_i = \lambda_i RD(DR^T RD^{-1})z_i = \lambda_i B(RD^{-1}z_i). \quad (31.10)$$

Hence $x_i = RD^{-1}z_i$ is an 'eigenvector' corresponding to λ_i of the generalized eigenproblem represented by equation (31.3). Note that since the z_i are orthogonal and RD^{-1} is real and non-singular, the x_i form a complete set of real eigenvectors of equation (31.3). Further we have

$$0 = z_i^T z_j = (DR^T x_i)^T DR^T x_j = x_i^T RDDR^T x_j = x_i^T B x_j, \quad (31.11)$$

showing that the vectors x_i are *orthogonal with respect to B*.

If A is also positive definite, then the λ_i are positive. For we have

$$A x_i = \lambda_i B x_i, \qquad x_i^T A x_i = \lambda_i x_i^T B x_i, \quad (31.12)$$

and both $x_i^T A x_i$ and $x_i^T B x_i$ are positive.

Explicit solution of $B\ddot{y} = -Ay$

32. We can now derive an explicit expression for the solution of equations (31.1) when A and B are positive definite. Let $x_1, x_2, ..., x_n$ be n independent solutions of (31.3) corresponding to the eigenvalues $\lambda_1, \lambda_2, ..., \lambda_n$. The latter must be positive but are not necessarily distinct. Writing
$$\lambda_i = \mu_i^2, \quad (32.1)$$

where the μ_i are taken to be positive, the set of equations (31.1) has the general solution
$$y = \sum_1^n (a_i e^{i\mu_i t} + b_i e^{-i\mu_i t}) x_i, \quad (32.2)$$

where the a_i and b_i are the $2n$ arbitrary constants. Assuming the solution to be real, we may write it in the form

$$y = \sum_1^n c_i \cos(\mu_i t + \epsilon_i) x_i, \quad (32.3)$$

where the c_i and ϵ_i are now the arbitrary constants.

Equations of the form $(AB - \lambda I)x = 0$

33. In theoretical physics the problem we have just considered often arises in the form
$$(AB - \lambda I)x = 0, \quad (33.1)$$

where A and B are symmetric and one (or both) is positive definite. If A is positive definite, then (33.1) is equivalent to

$$(B - \lambda A^{-1})x = 0 \quad (33.2)$$

since a positive definite matrix has a positive determinant and is therefore non-singular. If B is positive definite, (33.1) may be written in the form
$$(A - \lambda B^{-1})(Bx) = 0. \quad (33.3)$$

Now the inverse of a positive definite matrix is itself positive definite. For if

$$A = R^T \operatorname{diag}(\alpha_i^2) R \qquad (33.4)$$

then

$$A^{-1} = R^T \operatorname{diag}(\alpha_i^{-2}) R. \qquad (33.5)$$

Hence equation (33.1) is not essentially different from equation (31.3).

The importance of the positive definiteness of one of the matrices should not be underestimated. If A and B are real symmetric matrices, but neither is positive definite, then the eigenvalues of AB need not be real and the same applies to the roots of $\det(A - \lambda B) = 0$. A simple example will illustrate this. Let

$$A = \begin{bmatrix} a & 0 \\ 0 & b \end{bmatrix}, \quad B = \begin{bmatrix} 0 & 1 \\ 1 & 0 \end{bmatrix}. \qquad (33.6)$$

Then

$$AB = \begin{bmatrix} 0 & a \\ b & 0 \end{bmatrix}. \qquad (33.7)$$

The eigenvalues of AB satisfy the equation

$$\lambda^2 = ab \qquad (33.8)$$

and are complex if a and b have different signs.

If neither A nor B is positive definite and $(\alpha + i\beta)$ is a complex root of $\det(A - \lambda B) = 0$ then we have

$$Ax = (\alpha + i\beta) Bx \qquad (33.9)$$

for some non-null x. Hence

$$x^H A x = (\alpha + i\beta) x^H B x. \qquad (33.10)$$

Since $x^H A x$ and $x^H B x$ are real, equation (33.10) implies that

$$x^H A x = x^H B x = 0. \qquad (33.11)$$

The minimum polynomial of a vector

34. The eigenvectors x of a matrix are characterized by the fact that there exists a linear relation between x and Ax. For an arbitrary vector b no such relation exists, but if we form the sequence b, Ab, $A^2 b, \ldots, A^r b$, then there must be a minimum value m of r for which these vectors are dependent, and clearly $m < n$. We may write the corresponding relation in the form

$$(A^m + c_{m-1} A^{m-1} + \ldots + c_0 I) b = 0. \qquad (34.1)$$

From the definition of m there is no relation of the form

$$(A^r + d_{r-1}A^{r-1} + \ldots + d_0 I)b = 0 \qquad (r < m). \qquad (34.2)$$

The monic polynomial $c(\lambda)$ corresponding to the left-hand side of (34.1) is called the *minimum polynomial of b with respect to A*. If $s(A)$ is any other polynomial in A for which $s(A)b = 0$ then $c(A)$ must divide $s(A)$. For if not, then by the Euclidean algorithm there exist polynomials $p(A)$, $q(A)$, $r(A)$, such that

$$p(A)s(A) - q(A)c(A) = r(A) \qquad (34.3)$$

where $r(A)$ is of lower degree than $c(A)$. But from (34.3) it follows that $r(A)b = 0$ and this contradicts the hypothesis that $c(A)$ is the polynomial of lowest degree for which this is true.

The minimum polynomial is unique, for if $d(A)b = 0$ for a second monic polynomial of degree m, then $[c(A) - d(A)]b = 0$ and $c(A) - d(A)$ is of degree less than m. The degree of $c(A)$ is called the *grade* of b with respect to A.

The minimum polynomial of a matrix

35. In a similar way we may consider the sequence of matrices $I, A, A^2, \ldots, A^r$. Again there is a smallest value s of r for which this set of matrices is linearly dependent. We may write the relation in the form

$$A^s + m_{s-1}A^{s-1} + \ldots + m_0 I = 0. \qquad (35.1)$$

For example if

$$A = \begin{bmatrix} p_1 & p_0 \\ 1 & 0 \end{bmatrix} \qquad (35.2)$$

then

$$A^2 = \begin{bmatrix} p_1^2 + p_0 & p_1 p_0 \\ p_1 & p_0 \end{bmatrix} \qquad (35.3)$$

and hence

$$A^2 - p_1 A - p_0 I = \begin{bmatrix} p_1^2 + p_0 & p_1 p_0 \\ p_1 & p_0 \end{bmatrix} - \begin{bmatrix} p_1^2 & p_1 p_0 \\ p_1 & 0 \end{bmatrix} - \begin{bmatrix} p_0 & 0 \\ 0 & p_0 \end{bmatrix}$$

$$= \begin{bmatrix} 0 & 0 \\ 0 & 0 \end{bmatrix}. \qquad (35.4)$$

There is no relation of lower order because

$$A + c_0 I = \begin{bmatrix} p_1 + c_0 & p_0 \\ 1 & c_0 \end{bmatrix} \qquad (35.5)$$

and this is not null for any c_0. This example is of significance in later sections.

The polynomial $m(\lambda)$ corresponding to the left-hand side of equation (35.1) is called the *minimum polynomial of the matrix A*. Exactly as in § 34 we may show that it is unique and divides any other polynomial $p(\lambda)$ for which $p(A) = 0$.

The minimum polynomial $c(\lambda)$ of any vector b with respect to A divides $m(\lambda)$ since we obviously have $m(A)b = 0$. Let $c_1(\lambda)$, $c_2(\lambda)$,..., $c_n(\lambda)$ be the minimum polynomials of e_1, e_2,..., e_n with respect to A and let $g(\lambda)$ be the least common multiple (L.C.M.) of the $c_i(\lambda)$. Then we have

$$g(A)e_i = 0 \qquad (i = 1,..., n) \tag{35.6}$$

and hence $g(A) = 0$. This implies that $g(A)b = 0$ for any vector and therefore $g(\lambda)$ is the L.C.M. of the minimum polynomials of all vectors.

In fact we must have $g(\lambda) = m(\lambda)$. For by definition $m(A) = 0$ and hence $m(A)e_i = 0$. The L.C.M. of the minimum polynomials of the e_i must therefore divide $m(\lambda)$ and hence $g(\lambda)$ divides $m(\lambda)$. But, by definition, $m(\lambda)$ is the polynomial of lowest degree for which $m(A) = 0$.

We have developed this theory from the system of column vectors b, Ab, A^2b,.... In a similar way, by considering the system of row vectors b^T, b^TA, b^TA^2,... we can determine a minimum polynomial of b^T with respect to A. Clearly the minimum polynomial of b^T with respect to A is the same as the minimum polynomial of b with respect to A^T.

If we denote the L.C.M. of the minimum polynomials of e_1^T, e_2^T,..., e_n^T by $h(\lambda)$, then we may show as before, that $h(A)$ is the null matrix and hence that $h(\lambda)$ is also identical with $m(\lambda)$. We arrive at the same minimum polynomial of A whether we consider row or column vectors. It will be more convenient later (§ 37) to work with row vectors.

Cayley-Hamilton theorem

36. Since there are n^2 elements in each power of A it is evident that the minimum polynomial of A cannot be of degree greater than n^2. We shall now show that if the characteristic equation of A is given by

$$\lambda^n + c_{n-1}\lambda^{n-1} + ... + c_0 = 0 \tag{36.1}$$

then

$$A^n + c_{n-1}A^{n-1} + ... + c_0I = 0. \tag{36.2}$$

This means that the minimum polynomial of a matrix cannot be of degree greater than n.

To prove this, we introduce the *adjoint B* of a matrix A. This is the matrix having for its (i, j) element the co-factor of a_{ji}, that is the coefficient of a_{ji} in the expansion of $\det(A)$ by its jth row. We have therefore

$$AB = \det(A)I. \tag{36.3}$$

Now this result is true for any matrix and is therefore true in particular for the matrix $(\lambda I - A)$. The adjoint $B(\lambda)$ of $(\lambda I - A)$ is a λ-matrix, each element of which is a polynomial of degree $(n-1)$ or less in λ, since it is a minor of order $(n-1)$ of $(\lambda I - A)$. We may write

$$B(\lambda) = B_{n-1}\lambda^{n-1} + B_{n-2}\lambda^{n-2} + \dots + B_0, \tag{36.4}$$

where the B_i are $(n \times n)$ matrices with constant elements. Applying the result (36.3) to $(\lambda I - A)$ we have

$$(\lambda I - A)(B_{n-1}\lambda^{n-1} + \dots + B_0) = \det(\lambda I - A)I$$
$$= (\lambda^n + c_{n-1}\lambda^{n-1} + \dots + c_0)I. \tag{36.5}$$

This relation is an identity in λ and equating coefficients of powers of λ we have

$$\left.\begin{aligned}
B_{n-1} &= I \\
B_{n-2} - AB_{n-1} &= c_{n-1}I \\
B_{n-3} - AB_{n-2} &= c_{n-2}I \\
\dots\dots\dots\dots\dots \\
-AB_0 &= c_0I
\end{aligned}\right\}. \tag{36.6}$$

Multiplying these equations by A^n, $A^{n-1}, \dots, I$ gives

$$0 = A^n + c_{n-1}A^{n-1} + \dots + c_0 I. \tag{36.7}$$

Every square matrix therefore satisfies its own characteristic equation. This result is known as the *Cayley-Hamilton theorem*. We immediately conclude that the minimum polynomial divides the characteristic polynomial and hence the former is of degree not greater than n.

Relation between minimum polynomial and canonical forms

37. The considerations of § 34 are sometimes used in order to derive the Frobenius canonical form. (See Turnbull and Aitken, 1932). We shall not do this, but it is important from the point of view of later work to understand the relationship between the minimum polynomial of a matrix and its Frobenius and Jordan canonical forms.

We observe first that all similar matrices have the same minimum polynomial. For if $H^{-1}AH = B$ we have

$$H^{-1}A^2H = H^{-1}AHH^{-1}AH = B^2 \tag{37.1}$$

and similarly

$$H^{-1}A^rH = B^r,$$

giving

$$H^{-1}f(A)H = f(B), \tag{37.2}$$

where $f(A)$ is any polynomial in A. If $f(A) = 0$ we have $f(B) = 0$ and vice versa. Hence the minimum polynomial of A divides that of B and vice versa. These polynomials must therefore be identical.

Consider now a Frobenius matrix B_r of order r given by equation (13.1). We have

$$\left.\begin{aligned}
e_{s+1}^T B_r &= e_s^T \qquad (s = 1, 2, \ldots, r-1) \\
e_1^T B_r &= (b_{r-1}, b_{r-2}, \ldots, b_0)
\end{aligned}\right\}, \tag{37.3}$$

giving

$$e_r^T(B_r)^k = e_{r-k}^T \qquad (k = 0, 1, \ldots, r-1). \tag{37.4}$$

We have therefore

$$e_r^T(\alpha_k B_r^k + \ldots + \alpha_0 I) = (0, \ldots, 0, \alpha_k, \alpha_{k-1}, \ldots, \alpha_0) \quad \text{for } k \leqslant r-1, \tag{37.5}$$

and the vector on the right-hand side cannot be null unless all α_i are zero. The minimum polynomial of e_r^T with respect to B_r is therefore not of degree less than r. We have in fact

$$e_r^T B_r^r = e_r^T(b_{r-1}B_r^{r-1} + b_{r-2}B_r^{r-2} + \ldots + b_0 I), \tag{37.6}$$

and hence the minimum polynomial of e_r^T is the characteristic polynomial of B_r. The minimum polynomial of B_r is therefore also its characteristic polynomial. The example of § 35 provides an illustration of this for the case $r = 2$.

Turning now to the general Frobenius canonical form B, we may denote this by

$$B = \begin{bmatrix} B_{r_1} & & & & \\ & B_{r_2} & & & \\ & & \cdot & & \\ & & & \cdot & \\ & & & & \cdot \\ & & & & & B_{r_s} \end{bmatrix}, \tag{37.7}$$

where the characteristic polynomial of each B_{r_i} is divisible by that of its successor. For any polynomial $f(B)$ in B we have

$$f(B) = \begin{bmatrix} f(B_{r_1}) & & & & \\ & f(B_{r_2}) & & & \\ & & \cdot & & \\ & & & \cdot & \\ & & & & \cdot \\ & & & & & f(B_{r_s}) \end{bmatrix}. \tag{37.8}$$

If $f(B)$ is the null matrix then each of the matrices $f(B_{r_i})$ must be null. Now $f(B_{r_i})$ is the null matrix if and only if $f(\lambda)$ is a multiple of $f_i(\lambda)$, the characteristic polynomial of B_{r_i}, and since every $f_i(\lambda)$ divides $f_1(\lambda)$ it follows that $f_1(\lambda)$ is the minimum polynomial of B. Hence $f_1(\lambda)$ is the minimum polynomial of any matrix similar to B.

Our proof shows that the minimum polynomial of a matrix is equal to its characteristic polynomial if and only if there is only one Frobenius matrix in its Frobenius canonical form, that is if it is non-derogatory. This is true in particular when the eigenvalues are distinct.

38. Turning now to the Jordan canonical form we consider first a simple Jordan matrix, for example $C_3(a)$ given by

$$C_3(a) = \begin{bmatrix} a & 0 & 0 \\ 1 & a & 0 \\ 0 & 1 & a \end{bmatrix}. \tag{38.1}$$

Denoting the matrix by C for brevity we have

$$C^2 = \begin{bmatrix} a^2 & 0 & 0 \\ 2a & a^2 & 0 \\ 1 & 2a & a^2 \end{bmatrix} \qquad C^3 = \begin{bmatrix} a^3 & 0 & 0 \\ 3a^2 & a^3 & 0 \\ 3a & 3a^2 & a^3 \end{bmatrix}. \tag{38.2}$$

(Here for definiteness we have assumed the lower Jordan canonical form.) It is evident that C^2 is not a linear combination of C and I, and hence the degree of the minimum polynomial of $C_3(a)$ is greater than 2. The minimum polynomial is therefore the characteristic polynomial $(a - \lambda I)^3$. Now if C is a Jordan canonical form it is the direct sum of Jordan submatrices. We denote the general submatrix in this form by $C_{i_j}(\lambda_i)$ where there may, of course, be several submatrices associated with a particular value of λ_i. As with the Frobenius form we see that $f(C)$, a polynomial in C, is the direct sum of the matrices $f[C_{i_j}(\lambda_i)]$. Hence $f(C)$ is null if and only if every $f[C_{i_j}(\lambda_i)]$ is null. If we denote by $C_{i_1}(\lambda_i)$ the Jordan matrix of highest order associated with λ_i then the minimum polynomial of C must therefore be $\prod_i (\lambda - \lambda_i)^{i_1}$. Reference to the relationship established between the Jordan and Frobenius canonical forms in § 14 shows that we have confirmed that similar Jordan and Frobenius canonical forms all have the same minimum polynomials.

It should be verified that the vector having unit elements in each of the leading positions corresponding to the Jordan submatrices of highest order in each λ_i, and zeros elsewhere, has for its minimum polynomial the minimum polynomial of the Jordan form itself.

Principal vectors

39. For a matrix A with linear elementary divisors there exist n eigenvectors spanning the whole n-space. If A has a non-linear divisor, however, this is not true since there are then fewer than n independent eigenvectors. It is convenient nevertheless to have a set of vectors which span the whole n-space and to choose these in such a way that they reduce to the n eigenvectors when A has linear divisors. Now we have seen that in the latter case the eigenvectors may be taken as the columns of a matrix X, such that

$$X^{-1}AX = \mathrm{diag}(\lambda_i). \qquad (39.1)$$

A natural extension when the matrix has non-linear divisors is to take, as base vectors the n columns of a matrix X which reduces A to the Jordan canonical form.

These vectors satisfy important relations. It will suffice to demonstrate them for a simple example of order 8. Suppose the matrix A is such that

$$AX = X\begin{bmatrix} C_3(\lambda_1) & & & \\ & C_2(\lambda_1) & & \\ & & C_2(\lambda_2) & \\ & & & C_1(\lambda_3) \end{bmatrix}, \qquad (39.2)$$

then if $x_1, x_2, \ldots, x_8$ are the columns of X, we have on equating columns

$$Ax_1 = \lambda_1 x_1 + x_2, \quad Ax_4 = \lambda_1 x_4 + x_5, \quad Ax_6 = \lambda_2 x_6 + x_7, \quad Ax_8 = \lambda_3 x_8$$
$$Ax_2 = \lambda_1 x_2 + x_3, \quad Ax_5 = \lambda_1 x_5, \quad\quad Ax_7 = \lambda_2 x_7$$
$$Ax_3 = \lambda_1 x_3$$

$$\left. \right\} ,$$

(39.3)

from which we deduce

$$(A - \lambda_1 I)^3 x_1 = 0, \quad (A - \lambda_1 I)^2 x_4 = 0, \quad (A - \lambda_2 I)^2 x_6 = 0, \quad (A - \lambda_3 I) x_8 = 0$$
$$(A - \lambda_1 I)^2 x_2 = 0, \quad (A - \lambda_1 I) x_5 = 0, \quad (A - \lambda_2 I) x_7 = 0$$
$$(A - \lambda_1 I) x_3 = 0$$

$$\left. \right\} .$$

(39.4)

Each of the vectors therefore satisfies a relation of the form

$$(A - \lambda_i I)^j x_k = 0. \tag{39.5}$$

A vector which satisfies equation (39.5) but does not satisfy a relation of lower degree in $(A - \lambda_i I)$ is called a *principal vector of grade j* corresponding to λ_i. Note that $(\lambda - \lambda_i)^j$ is the minimum polynomial of x_k with respect to A. Any eigenvector is therefore a principal vector of grade one, and columns of the matrix X are all principal vectors of A. Since the columns of X are independent, there exists a set of principal vectors which span the whole n-space. In general the principal vectors are not unique, since if x is a principal vector of grade r corresponding to the eigenvalue λ_i, then the same is true of any vector obtained by adding multiples of any vectors of grades not greater than r corresponding to the same eigenvalue. For a matrix in Jordan canonical form the vectors $e_1, e_2, \ldots, e_n$ form a complete set of principal vectors.

Elementary similarity transformations

40. The reduction of a matrix of general form to one of canonical form, or indeed to any condensed form, will usually be achieved progressively by performing a sequence of simple similarity transformations. We have already discussed some elementary transformation matrices in § 17, but it is convenient to generalize these somewhat, and to introduce a standard notation for each type to which we shall adhere in the remainder of the book. We shall refer to transformations based on any of the following matrices as *elementary transformations* and the matrices as *elementary matrices*. The matrices of these transformations will usually be of the following forms.

(i) The matrices I_{ij}, equal to the identity matrix except in rows i and j and columns i and j, which are of the form

$$\begin{array}{cc} \text{col. } i & \text{col. } j \end{array}$$

$$\begin{bmatrix} 0 & 1 \\ 1 & 0 \end{bmatrix} \begin{array}{l} \text{row } i \\ \text{row } j. \end{array} \tag{40.1}$$

In particular $I_{ii} = I$. It is evident that $I_{ij}I_{ij} = I$ so that I_{ij} is its own inverse; it is also orthogonal. Pre-multiplication by I_{ij} results in the interchange of rows i and j, and post-multiplication in the interchange of columns i and j. A similarity transformation with matrix I_{ij} therefore interchanges rows i and j and columns i and j.

(ii) The matrices $P_{\alpha_1, \alpha_2, \ldots, \alpha_n}$ defined by

$$p_{i\alpha_i} = 1, \qquad p_{ij} = 0 \text{ otherwise,} \tag{40.2}$$

where $(\alpha_1, \alpha_2, \ldots, \alpha_n)$ is some permutation of $(1, 2, \ldots, n)$. These matrices are called *permutation matrices*. Notice that in a permutation matrix every row and every column contains just one unity element. The transpose of a permutation matrix is therefore a permutation matrix, and we have further

$$(P_{\alpha_1, \alpha_2, \ldots, \alpha_n})^{-1} = (P_{\alpha_1, \alpha_2, \ldots, \alpha_n})^T. \tag{40.3}$$

Every permutation matrix may be expressed as a product of matrices of type I_{ij}. Pre-multiplication by P has the effect of transferring the initial row α_i into the new row i; post-multiplication by P^T transfers the initial column α_i into the new column i.

(iii) The matrices R_i, equal to the identity matrix except for the ith row which is

$$(-r_{i1}, -r_{i2}, \ldots, -r_{i,i-1}, 1, -r_{i,i+1}, \ldots, -r_{in}). \tag{40.4}$$

Pre-multiplication of A by R_i results in the *subtraction* of multiples of all the other rows from the ith row, the multiple of the jth row being r_{ij}. Post-multiplication on the other hand results in the subtraction of multiples of column i from each of the other columns. The reciprocal of R_i is a matrix of the same form in which each element $-r_{ij}$ is replaced by $+r_{ij}$. The similarity transform $R^{-1}AR$ of A is obtained by *subtracting* the appropriate multiples of the ith column of A from each of the other columns and then *adding* the same multiples of all other rows to the ith row. Note that whereas in the first step we changed all columns except the ith, in the second we change only the ith row.

(iv) The matrices S_i, of the form obtained by transposing R_i. The ith column of S_i written as a row vector is

$$(-s_{1i}, -s_{2i},\ldots, -s_{i-1,i}, 1, -s_{i+1,i},\ldots, -s_{ni}). \qquad (40.5)$$

Their properties follow from those of the R_i with the appropriate interchanges.

(v) The matrices M_i, equal to the identity matrix except for the ith row, which is

$$(0, 0,\ldots, 0, 1, -m_{i,i+1},\ldots, -m_{in}). \qquad (40.6)$$

M_i is therefore an R_i in which the elements $r_{i1},\ldots, r_{i,i-1}$ are all zero.

(vi) The matrices N_i, of the form obtained by transposing M_i. The ith column written as a row vector is

$$(0, 0,\ldots, 0, 1, -n_{i+1,i},\ldots, -n_{ni}). \qquad (40.7)$$

N_i is therefore an S_i with a number of leading zero elements.

(vii) The matrices M_{ij}, equal to the identity matrix apart from the (i, j) element which is $-m_{ij}$. M_{ij} is thus a special form of the earlier matrices.

Properties of elementary matrices

41. The reader will readily verify the following important properties of the elementary matrices.

(i) The product $M_{n-1}M_{n-2}\ldots M_1$ is an upper triangular matrix having unit diagonal elements. The element in the (i, j) position for $j > i$ is $-m_{ij}$. *No products of the elements m_{pq} occur.*

(ii) The product $M_1M_2\ldots M_{n-1}$ is also an upper triangular matrix with unit diagonal elements, but products of the m_{pq} occur in the other elements and its structure is far more complex than that of the other product.

(iii) Similarly both $N_1N_2\ldots N_{n-1}$ and $N_{n-1}N_{n-2}\ldots N_1$ are lower triangular matrices with unit diagonal elements. In the former, the element in the (i, j) position for $i > j$ is $-n_{ij}$ and no product terms occur. The latter has a complex structure, comparable with that of $M_1M_2\ldots M_{n-1}$.

(iv) The matrix

$$[I_{n-1,(n-1)'}\cdots I_{r+2,(r+2)'}I_{r+1,(r+1)'}] M_r [I_{r+1,(r+1)'}I_{r+2,(r+2)'}\cdots I_{n-1,(n-1)'}]$$

with $(r+1)' > r+1$, $(r+2)' > r+2,..., (n-1)' > n-1$, is a matrix of the same type as M_r in which the elements in row r positions $r+1, r+2,..., n$ are a permutation of those in M_r itself. The permutation is that resulting from the successive substitutions $[(r+1)', r+1]$, $[(r+2)', r+2],..., [(n-1)', n-1]$. A similar result holds for the matrix N_r.

Reduction to triangular canonical form by elementary similarity transformations

42. We may illustrate the use of the elementary similarity transformations by showing that any matrix may be reduced to triangular form by a similarity transformation, the matrix of which is the product of a number of elementary matrices. The proof is by induction.

Suppose it is true for a matrix of order $(n-1)$. Let A be a matrix of order n and let λ be an eigenvalue. There must be at least one eigenvector x corresponding to λ such that

$$Ax = \lambda x. \tag{42.1}$$

Now for any non-singular X the eigenvector of XAX^{-1} is Xx. We show how to choose X so that Xx is e_1. Let us write

$$x^T = (x_1, x_2,..., x_{r-1}, 1, x_{r+1},..., x_n), \tag{42.2}$$

where the rth is the first non-zero element of x and the arbitrary multiplier has been chosen to make it equal to unity. We have

$$(I_{1r}x)^T = (1, y_2, y_3,..., y_n) = y^T(\text{say}), \tag{42.3}$$

where the y_i are the x_i in some order. Note that the case $r = 1$ is covered, since $I_{1.1} = I$. Now if we pre-multiply y by an N_1 matrix in which $n_{i1} = y_i$ then all elements vanish except the first. Hence

$$N_1y = N_1I_{1r}x = e_1 \tag{42.4}$$

and the matrix $N_1I_{1r}AI_{1r}N_1^{-1}$ has an eigenvalue λ with e_1 as a corresponding eigenvector. Consequently it must be of the form

$$\begin{bmatrix} \lambda & b^T \\ O & B \end{bmatrix}, \tag{42.5}$$

where B is a square matrix of order $(n-1)$. Now by hypothesis there is a square matrix H, which is a product of elementary matrices, such that

$$HBH^{-1} = T, \tag{42.6}$$

where T is of triangular form. Hence

$$\left[\begin{array}{c|c} 1 & O \\ \hline O & H \end{array}\right] \left[\begin{array}{c|c} \lambda & b^T \\ \hline O & B \end{array}\right] \left[\begin{array}{c|c} 1 & O \\ \hline O & H^{-1} \end{array}\right] = \left[\begin{array}{c|c} \lambda & b^T H^{-1} \\ \hline O & HBH^{-1} \end{array}\right] = \left[\begin{array}{c|c} \lambda & b^T H^{-1} \\ \hline O & T \end{array}\right], \quad (42.7)$$

and the matrix on the right is of triangular form. The result therefore follows in general since it is clearly true for $n = 1$.

Elementary unitary transformations

43. In the last section we used the elementary matrices to transform a general non-zero vector into ke_1. We now describe the first of two classes of elementary *unitary* matrices which can be used to achieve the same effect.

Consider the matrix R defined by

$$\left.\begin{array}{ll} r_{ii} = e^{i\alpha} \cos\theta, & r_{ij} = e^{i\beta} \sin\theta \\[4pt] r_{ji} = e^{i\gamma} \sin\theta, & r_{jj} = e^{i\delta} \cos\theta \\[4pt] r_{pp} = 1 \quad (p \neq i, j), & r_{pq} = 0 \quad \text{otherwise} \end{array}\right\}, \quad (43.1)$$

where θ, α, β, γ, δ are real. This matrix is unitary if

$$\alpha - \gamma \equiv \beta - \delta + \pi \quad (\text{mod } 2\pi), \quad (43.2)$$

as may be verified by forming $R^H R$.

We shall not need the full generality permitted by relation (43.2) and shall take

$$\gamma = \pi - \beta, \qquad \delta = -\alpha, \quad (43.3)$$

giving

$$\left.\begin{array}{ll} r_{ii} = e^{i\alpha} \cos\theta, & r_{ij} = e^{i\beta} \sin\theta \\[4pt] r_{ji} = -e^{-i\beta} \sin\theta, & r_{jj} = e^{-i\alpha} \cos\theta \end{array}\right\}. \quad (43.4)$$

We shall refer to a matrix of this restricted class as a *plane rotation*. Clearly from (43.4) we may write

$$r_{ii} = \bar{c}, \quad r_{ij} = \bar{s}, \quad r_{ji} = -s, \quad r_{jj} = c, \quad (43.5)$$

where $|c|^2 + |s|^2 = 1$.

R^H is itself a plane rotation corresponding to α' and β' satisfying

$$\alpha' = -\alpha, \qquad \beta' = \beta + \pi. \quad (43.6)$$

The inverse of a plane rotation is therefore also a plane rotation. When $\alpha = \beta = 0$ the matrix is real and therefore orthogonal, and its relationship with a rotation through the angle θ in the (i, j) plane is obvious.

44. If x is any vector, then given i and j we can choose a plane rotation such that $(Rx)_i$ is real and non-negative and $(Rx)_j$ is zero. Clearly $(Rx)_s = x_s$ for all $s \neq i, j$. We have

$$(Rx)_j = -sx_i + cx_j. \qquad (44.1)$$

If we write $r^2 = |x_i|^2 + |x_j|^2$ then, provided r is not zero, we may take

$$s = \frac{x_j}{r}, \qquad c = \frac{x_i}{r} \qquad (44.2)$$

where r is chosen to be positive. Then clearly $(Rx)_j = 0$ and we have

$$(Rx)_i = \bar{c}x_i + \bar{s}x_j = \frac{|x_i|^2}{r} + \frac{|x_j|^2}{r} = r > 0. \qquad (44.3)$$

If $r = 0$, then we may take $c = 1$ and $s = 0$. If we do not require $(Rx)_i$ to be real then we can drop the parameter α in equations (43.1). Notice that if x_i is real, equations (44.2) give a real value for c in any case.

The vector x may be transformed into ke_1 by successive pre-multiplication with $(n-1)$ plane rotations in the planes $(1, 2)$, $(1, 3), \ldots, (1, n)$. The rotation in the $(1, i)$ plane is chosen so as to reduce the ith element to zero. It is evident from what we have just proved that the rotations may be chosen so that k is real and non-negative. Clearly

$$k = (|x_1|^2 + |x_2|^2 + \ldots + |x_n|^2)^{\frac{1}{2}} = (x^H x)^{\frac{1}{2}}. \qquad (44.4)$$

An immediate consequence of this is that if x and y are such that $(x^H x) = (y^H y)$ then there exists a product S of rotation matrices such that

$$Sx = y. \qquad (44.5)$$

This follows because both x and y may be transformed into ke_1 by a product of plane rotation matrices and the inverse of a plane rotation is a plane rotation.

Elementary unitary Hermitian matrices

45. The second class of elementary unitary matrices consists of matrices P defined by

$$P = I - 2ww^H, \qquad w^H w = 1. \qquad (45.1)$$

The matrices are unitary because we have

$$PP^H = (I - 2ww^H)(I - 2ww^H)$$
$$= I - 4ww^H + 4w(w^H w)w^H = I. \qquad (45.2)$$

They are also obviously Hermitian. We shall refer to matrices of this type as *elementary Hermitian matrices*. When w is a real vector, P is orthogonal and symmetric.

If y and x satisfy the relation

$$y = Px = x - 2w(w^H x), \tag{45.3}$$

then we have

$$y^H y = x^H P^H P x = x^H x, \tag{45.4}$$

and

$$x^H y = x^H P x, \tag{45.5}$$

and the right-hand side of (45.5) is real since P is Hermitian. Hence given an x and y we cannot expect to be able to find a P such that $y = Px$ unless $x^H x = y^H y$ *and* $x^H y$ *is real*. When these conditions are satisfied we can indeed find a suitable P as we now show.

If $x = y$, then any w such that $w^H x = 0$ will give an appropriate P. Otherwise we see from (45.3) that any suitable w must lie in the direction of $y - x$ and hence we need consider only the w given by

$$w = e^{i\alpha}(y - x)/[(y - x)^H(y - x)]^{\frac{1}{2}}. \tag{45.6}$$

It is obvious that the factor $e^{i\alpha}$ does not affect the corresponding P. That any such w is satisfactory may be immediately demonstrated. We have in fact

$$(I - 2ww^H)x = x - \frac{2(y - x)(y - x)^H x}{(y - x)^H(y - x)}, \tag{45.7}$$

and

$$2(y - x)^H x = 2(y^H x - x^H x)$$
$$= y^H x + x^H y - x^H x - y^H y$$
$$= -(y - x)^H(y - x), \tag{45.8}$$

since $x^H y$ is real and $x^H x = y^H y$ from our assumptions. Hence

$$(I - 2ww^H)x = x + (y - x) = y. \tag{45.9}$$

46. It is not, in general, possible to convert x into ke_1 *with k real* by pre-multiplying with a suitable P. If we denote the components of x by $r_i e^{i\theta_i}$ we can however convert x into y defined by

$$y = \pm(x^H x)^{\frac{1}{2}} e^{i\theta_1} e_1 \tag{46.1}$$

since $x^H y$ is then obviously real. For the appropriate P we may take

$$w = (x - y)/[(x - y)^H(x - y)]^{\frac{1}{2}}. \tag{46.2}$$

If $r_1 = 0$ then θ_1 may be regarded as arbitrary, and if we take $\theta_1 = 0$ the corresponding P converts x into a real multiple of e_1. We shall often be interested in P for which the corresponding w has zero components in the first r positions for some value of r from 0 to $n-1$. Pre-multiplication of a vector x by such a P leaves the first r components of x unchanged.

If x is real, y and w of (46.1) and (46.2) are real and hence P is real. Note that with unitary matrices of the form $e^{i\theta}P$ we *can* reduce an arbitrary vector to the form ke_1 with k real.

Reduction to triangular form by elementary unitary transformations

47. We now prove the two results mentioned earlier on the reduction to canonical form by orthogonal (unitary) transformations (§§ 23, 25). We show first that any matrix may be reduced to triangular form by a unitary transformation. The proof is by induction.

Suppose it is true for matrices of order $(n-1)$. Let A be a matrix of order n, and λ and x be an eigenvalue and eigenvector of A, so that

$$Ax = \lambda x. \qquad (47.1)$$

We know from § 44 and § 46 that there is a unitary matrix P such that

$$Px = ke_1. \qquad (47.2)$$

P may be taken to be either the product of $(n-1)$ plane rotation matrices or one of the elementary Hermitian matrices. From (47.1)

$$PAx = P\lambda x, \qquad PA(P^H P)x = \lambda Px,$$

$$(PAP^H)ke_1 = \lambda ke_1, \qquad (PAP^H)e_1 = \lambda e_1 \quad \text{since } k \neq 0. \qquad (47.3)$$

The matrix PAP^H must therefore be of the form $A^{(1)}$ given by

$$A^{(1)} = \left[\begin{array}{c|c} \lambda & b^T \\ \hline O & A_{n-1} \end{array}\right], \qquad (47.4)$$

where A_{n-1} is of order $(n-1)$. Now by hypothesis there is a unitary matrix R, such that

$$RA_{n-1}R^H = T_{n-1}, \qquad (47.5)$$

where T_{n-1} is triangular. Hence

$$\left[\begin{array}{c|c} 1 & O \\ \hline O & R \end{array}\right] A^{(1)} \left[\begin{array}{c|c} 1 & O \\ \hline O & R^H \end{array}\right] = \left[\begin{array}{c|c} \lambda & b^T R^H \\ \hline O & T_{n-1} \end{array}\right] \qquad (47.6)$$

and the matrix on the right of (47.6) is triangular. Since $\left[\begin{array}{c|c} 1 & O \\ \hline O & R \end{array}\right] P$ is a unitary matrix, the result is now established. Note that if A is real and its eigenvalues are real, then its eigenvectors are also real, and consequently the matrix P is orthogonal and the whole transformation can be completed using orthogonal matrices.

If A is Hermitian, the triangular matrix is diagonal. For we have seen that there is a unitary matrix R such that

$$RAR^H = T, \tag{47.7}$$

where T is triangular. The matrix on the left is Hermitian and hence T is Hermitian and therefore diagonal.

Normal matrices

48. We have just proved that for any Hermitian matrix there exists a unitary matrix R such that

$$RAR^H = D, \tag{48.1}$$

where D is diagonal and real. We now consider the extension of this class of matrices which is obtained by permitting D to be complex. In other words we consider matrices A which may be expressed in the form $R^H D R$ where R is unitary and D is diagonal. If

$$A = R^H D R, \tag{48.2}$$

then

$$AA^H = R^H D R R^H D^H R = R^H (DD^H) R, \tag{48.3}$$

$$A^H A = R^H D^H R R^H D R = R^H (D^H D) R. \tag{48.4}$$

Hence $AA^H = A^H A$ since DD^H is clearly equal to $D^H D$.

We now show conversely that if

$$AA^H = A^H A, \tag{48.5}$$

then A may be factorized as in (48.2). For any matrix A may be expressed in the form $R^H T R$ where R is unitary and T upper triangular. Hence we have

$$R^H T R R^H T^H R = R^H T^H R R^H T R,$$

giving

$$R^H T T^H R = R^H T^H T R. \tag{48.6}$$

Hence pre-multiplying by R and post-multiplying by R^H we have

$$T T^H = T^H T. \tag{48.7}$$

Equating the diagonal elements, we find that all off-diagonal elements of T are zero, so that T is diagonal. Hence the class of matrices which can be factorized in the form $R^H D R$ is the same as the class of matrices for which $A A^H = A^H A$. Such matrices are called *normal* matrices. From the relation (48.5) it is evident that included among normal matrices are those for which

(i) $\qquad A^H = A \qquad$ (Hermitian matrices),

(ii) $\qquad A^H = -A \qquad$ (Skew-Hermitian matrices),

(iii) $\qquad A^H = A^{-1} \qquad$ (Unitary matrices).

In practice we have found that it is comparatively uncommon for the eigenvalues of skew-Hermitian matrices or of unitary matrices to be required.

Commuting matrices

49. Normal matrices have the property that A and A^H commute. From the relation

$$A = R^H D R, \tag{49.1}$$

we have

$$A R^H = R^H D \tag{49.2}$$

and similarly

$$A^H R^H = R^H D^H. \tag{49.3}$$

The matrices A and A^H therefore have a common complete system of eigenvectors, namely that formed by the columns of R^H.

We now show that if any two matrices A and B have a common complete system of eigenvectors then $AB = BA$. For if this system of eigenvectors forms the columns of the matrix H, then we have

$$A = H D_1 H^{-1}, \qquad B = H D_2 H^{-1}. \tag{49.4}$$

Hence

$$AB = H D_1 H^{-1} H D_2 H^{-1} = H D_1 D_2 H^{-1} \tag{49.5}$$

and

$$BA = H D_2 H^{-1} H D_1 H^{-1} = H D_2 D_1 H^{-1}, \tag{49.6}$$

so that $AB = BA$, since diagonal matrices commute.

50. Conversely if A and B commute *and have linear elementary divisors* then they share a common system of eigenvectors. For suppose the eigenvectors of A are $h_1, h_2, \ldots, h_n$ and that these form the matrix H. Then

$$H^{-1} A H = \mathrm{diag}(\lambda_i) \tag{50.1}$$

where the λ_i may or may not be distinct. The nature of the proof is obvious if we consider a matrix of order eight for which $\lambda_1 = \lambda_2 = \lambda_3$,

$\lambda_4 = \lambda_5$ and λ_6, λ_7 and λ_8 are isolated. Equation (50.1) may then be written

$$H^{-1}AH = \begin{bmatrix} \lambda_1 I_3 & & & & & \\ & \lambda_4 I_2 & & & & \\ & & \lambda_6 & & \\ & & & \lambda_7 & \\ & & & & \lambda_8 \end{bmatrix}. \qquad (50.2)$$

Now since $Ah_i = \lambda_i h_i$, we have

$$BAh_i = \lambda_i Bh_i, \qquad (50.3)$$

giving

$$A(Bh_i) = \lambda_i(Bh_i). \qquad (50.4)$$

This shows that if h_i is an eigenvector of A corresponding to λ_i then Bh_i lies in the subspace of eigenvectors corresponding to λ_i. Hence

$$\left. \begin{aligned} Bh_1 &= p_{11}h_1 + p_{21}h_2 + p_{31}h_3, \\ Bh_2 &= p_{12}h_1 + p_{22}h_2 + p_{32}h_3, & Bh_6 &= \mu_6 h_6 \\ Bh_3 &= p_{13}h_1 + p_{23}h_2 + p_{33}h_3, & Bh_7 &= \mu_7 h_7 \\ Bh_4 &= q_{11}h_4 + q_{21}h_5, & Bh_8 &= \mu_8 h_8 \\ Bh_5 &= q_{12}h_4 + q_{22}h_5, \end{aligned} \right\}. \qquad (50.5)$$

These equations are equivalent to the single matrix equation

$$H^{-1}BH = \begin{bmatrix} P & & & & \\ & Q & & & \\ & & \mu_6 & & \\ & & & \mu_7 & \\ & & & & \mu_8 \end{bmatrix}. \qquad (50.6)$$

Now since B has linear elementary divisors the same must be true of the matrix on the right of equation (50.6) and hence of P and Q. Matrices K and L of orders three and two therefore exist such that

$$K^{-1}PK = \begin{bmatrix} \mu_1 & & \\ & \mu_2 & \\ & & \mu_3 \end{bmatrix} \quad \text{and} \quad L^{-1}QL = \begin{bmatrix} \mu_4 & \\ & \mu_5 \end{bmatrix}. \qquad (50.7)$$

If we write

$$G = \begin{bmatrix} K & & \\ \hline & L & \\ \hline & & I_3 \end{bmatrix}.$$

(50.8)

then

$$G^{-1}H^{-1}BHG = \operatorname{diag}(\mu_i).$$

(50.9)

On the other hand from (50.2) we see that

$$G^{-1}H^{-1}AHG = \operatorname{diag}(\lambda_i).$$

(50.10)

Hence B and A share a common system of eigenvectors, namely the vectors which are the columns of HG.

Eigenvalues of AB

51. Notice, however, that the eigenvalues of AB and of BA are identical for *all* square A and B. The proof is as follows. We have

$$\begin{bmatrix} I & O \\ \hline -B & \mu I \end{bmatrix}\begin{bmatrix} \mu I & A \\ \hline B & \mu I \end{bmatrix} = \begin{bmatrix} \mu I & A \\ \hline O & \mu^2 I - BA \end{bmatrix}$$

(51.1)

and

$$\begin{bmatrix} \mu I & -A \\ \hline O & I \end{bmatrix}\begin{bmatrix} \mu I & A \\ \hline B & \mu I \end{bmatrix} = \begin{bmatrix} \mu^2 I - AB & O \\ \hline B & \mu I \end{bmatrix}.$$

(51.2)

Taking determinants of both sides of (51.1) and (51.2) and writing

$$\begin{bmatrix} \mu I & A \\ \hline B & \mu I \end{bmatrix} = X,$$

(51.3)

we have

$$\mu^n \det(X) = \mu^n \det(\mu^2 I - BA)$$

(51.4)

and

$$\mu^n \det(X) = \mu^n \det(\mu^2 I - AB).$$

(51.5)

Equations (51.4) and (51.5) are identities in μ^2 and writing $\mu^2 = \lambda$ we have

$$\det(\lambda I - BA) = \det(\lambda I - AB),$$

(51.6)

showing that AB and BA have the same eigenvalues.

The same proof shows that if A is an $m \times n$ matrix and B is an $n \times m$ matrix, then AB and BA have the same eigenvalues except that the product which is of higher order has $|m-n|$ extra zero eigenvalues.

Vector and matrix norms

52. It is useful to have a single number which gives an overall assessment of the size of a vector or of a matrix and plays the same role as the modulus in the case of a complex number. For this purpose we shall use certain functions of the elements of a vector and a matrix which are called *norms*.

The *norm of a vector* will be denoted by $\|x\|$ and the norms we use all satisfy the relations

(i) $\qquad \|x\| > 0$ unless $x = 0$

(ii) $\qquad \|kx\| = |k|\,\|x\|$ for any complex scalar k $\left.\vphantom{\begin{matrix}1\\1\\1\end{matrix}}\right\}$. $\qquad$ (52.1)

(iii) $\qquad \|x+y\| < \|x\| + \|y\|$

From (iii) we have
$$\|x-y\| > |\ \|x\| - \|y\|\ |.$$

We use only three simple vector norms. They are defined by

$$\|x\|_p = (|x_1|^p + |x_2|^p + \ldots + |x_n|^p)^{\frac{1}{p}} \quad (p = 1, 2, \infty) \qquad (52.2)$$

where $\|x\|_\infty$ is interpreted as max $|x_i|$. The norm $\|x\|_2$ is the Euclidean *length* of the vector x as normally understood. We have also

$$x^H x = \|x\|_2^2 \qquad (52.3)$$

and
$$|x^H y| < \sum |x_i|\,|y_i|$$
$$< \|x\|_2\,\|y\|_2 \qquad (52.4)$$

by Cauchy's theorem. We may therefore write

$$x^H y = \|x\|_2\,\|y\|_2 e^{i\mu} \cos\theta \quad \left(0 < \theta < \frac{\pi}{2}\right). \qquad (52.5)$$

When x and y are real, θ is the angle between the two vectors; in the complex case we may regard θ, defined by (52.5) as the generalization of the angle between the two vectors.

Similarly the norm of a matrix A will be denoted by $\|A\|$ and the norms we use all satisfy the relations

(i) $\qquad \|A\| > 0$ unless $A = 0$

(ii) $\qquad \|kA\| = |k|\,\|A\|$ for any complex scalar k

(iii) $\qquad \|A+B\| < \|A\| + \|B\|$ $\left.\vphantom{\begin{matrix}1\\1\\1\\1\end{matrix}}\right\}$. $\qquad$ (52.6)

(iv) $\qquad \|AB\| < \|A\|\,\|B\|$

Subordinate matrix norms

53. Corresponding to any vector norm a non-negative quantity, defined by $\sup_{x \neq 0} \|Ax\|/\|x\|$, may be associated with any matrix A. From relation (ii) of (52.1) we see that this is equivalent to $\sup_{\|x\|=1} \|Ax\|$. This quantity is obviously a function of the matrix A and it may readily be verified that it satisfies the conditions required of a matrix norm. It is called the matrix norm *subordinate* to the vector norm. Since

$$\|A\| = \sup_{x \neq 0} \|Ax\|/\|x\| \tag{53.1}$$

we have

$$\|Ax\| < \|A\| \ \|x\| \tag{53.2}$$

for all non-zero x, and the relation is obviously true when $x = 0$. Matrix and vector norms for which (53.2) is true for all A and x are said to be *compatible*. A vector norm and its subordinate matrix norm are therefore always compatible.

For any subordinate norm we have, from the definition, $\|I\| = 1$. Further there always exists a non-zero vector such that

$$\|Ax\| = \|A\| \ \|x\|.$$

This follows because from (52.1) a vector norm is a continuous function of the components of its argument and the domain $\|x\| = 1$ is closed. Hence instead of $\sup \|Ax\|/\|x\|$ we may write $\max \|Ax\|/\|x\|$.

The matrix norm subordinate to $\|x\|_p$ is denoted by $\|A\|_p$. These norms satisfy the relations

$$\|A\|_1 = \max_j \sum_i |a_{ij}|, \tag{53.3}$$

$$\|A\|_\infty = \max_i \sum_j |a_{ij}|, \tag{53.4}$$

$$\|A\|_2 = (\text{maximum eigenvalue of } A^H A)^{\frac{1}{2}}. \tag{53.5}$$

The first two results are trivial, and clearly $\|A\|_1 = \|A^H\|_\infty$. The third may be proved as follows.

The matrix $A^H A$ is Hermitian and its eigenvalues are all non-negative because $x^H A^H A x = (Ax)^H (Ax) \geqslant 0$. We denote its eigenvalues by σ_i^2 where $\sigma_1^2 > \sigma_2^2 > \ldots > \sigma_n^2 > 0$. Now from its definition,

$$\|A\|_2 = \max_{x \neq 0} \frac{\|Ax\|_2}{\|x\|_2}, \tag{53.6}$$

$$\|A\|_2^2 = \max \frac{\|Ax\|_2^2}{\|x\|_2^2} = \max \frac{x^H A^H A x}{x^H x}. \tag{53.7}$$

Now if $u_1, u_2, ..., u_n$ is an orthonormal system of eigenvectors of $A^H A$, we may write

$$x = \sum \alpha_i u_i. \qquad (53.8)$$

Hence

$$\frac{x^H A^H A x}{x^H x} = \frac{\sum |\alpha_i|^2 \sigma_i^2}{\sum |\alpha_i|^2} \leqslant \sigma_1^2. \qquad (53.9)$$

The value σ_1^2 is attained when $x = u_1$. The non-negative quantities σ_i are called the *singular values* of A. The 2-norm is often called the *spectral* norm.

The Euclidean and spectral norms

54. There is a second important norm which is compatible with the vector norm $\|x\|_2$. This is the so-called *Euclidean* or *Schur* norm which we denote by $\|A\|_E$. It is defined by

$$\|A\|_E = \left(\sum_{i,j} |a_{ij}|^2 \right)^{\frac{1}{2}}. \qquad (54.1)$$

It is obvious that it cannot be subordinate to any vector norm because we have

$$\|I\|_E = n^{\frac{1}{2}}. \qquad (54.2)$$

The Euclidean norm is a very useful norm for practical purposes since it may readily be computed. Further, we have from the definition

$$\||A|\|_E = \|A\|_E. \qquad (54.3)$$

The 1 and ∞ matrix norms satisfy similar relations but in general

$$\||A|\|_2 \neq \|A\|_2. \qquad (54.4)$$

In fact since $\|A\|_E^2$ is equal to the trace of $A^H A$ which is itself equal to $\sum \sigma_i^2$ we have

$$\|A\|_2 < \|A\|_E < n^{\frac{1}{2}} \|A\|_2, \qquad (54.5)$$

where both bounds may be attained. Hence we have

$$\||A|\|_2 < \||A|\|_E = \|A\|_E < n^{\frac{1}{2}} \|A\|_2. \qquad (54.6)$$

An inconvenient feature of the Euclidean norm is that

$$\|\text{diag}(\lambda_i)\|_E = (\sum |\lambda_i|^2)^{\frac{1}{2}} \qquad (54.7)$$

whereas

$$\|\text{diag}(\lambda_i)\|_p = \max |\lambda_i| \quad (p = 1, 2, \infty). \qquad (54.8)$$

Both the Euclidean matrix norm and the 2-vector and 2-matrix norms have important invariant properties under unitary transformations. For arbitrary x and A and unitary R we have

$$\|Rx\|_2 = \|x\|_2 \qquad (54.9)$$

since

$$\|Rx\|_2^2 = (Rx)^H Rx = x^H R^H Rx = x^H x = \|x\|_2^2. \qquad (54.10)$$

Similarly, it follows directly from the definition that

$$\|RA\|_2 = \|A\|_2 \quad \text{and} \quad \|RAR^H\|_2 = \|A\|_2. \qquad (54.11)$$

Finally

$$\|RA\|_E = \|A\|_E. \qquad (54.12)$$

This follows because the Euclidean length of each column of RA is equal to that of the corresponding column of A.

Suppose now that $A = R^H \operatorname{diag}(\lambda_i) R$ is a normal matrix, then we have

$$\|AB\|_E = \|R^H \operatorname{diag}(\lambda_i) RB\|_E = \|\operatorname{diag}(\lambda_i) RB\|_E < \max |\lambda_i| \, \|RB\|_E$$
$$= \max |\lambda_i| \, \|B\|_E. \qquad (54.13)$$

In future, whenever we refer to a norm, one of the specific norms we have introduced is implied, and we shall not generalize the concept any further. It is convenient to have a uniform treatment of norms even with this restricted definition and it is surprising how much is gained from a facility in their use.

Norms and limits

55. It will be convenient to establish a number of simple criteria for the convergence of sequences of matrices. Where the results follow immediately from the definitions, the proofs have been omitted.

(A) $\operatorname*{Lim}_{r \to \infty} A^{(r)} = A$ if, and only if, $\|A - A^{(r)}\| \to 0$. If

$$A^{(r)} \to A \text{ then } \|A^{(r)}\| \to \|A\|.$$

(B) $\operatorname*{Lim}_{r \to \infty} A^r = 0$ if $\|A\| < 1$. For

$$\|A^r\| = \|AA^{r-1}\| < \|A\| \, \|A^{r-1}\| < \|A\|^2 \, \|A^{r-2}\| < \dots < \|A\|^r.$$

(C) If λ is an eigenvalue of A, then $\lambda < \|A\|$. For we have $Ax = \lambda x$ for some non-null x. Hence for any of the consistent pairs of matrix and vector norms

$$|\lambda| \, \|x\| = \|\lambda x\| = \|Ax\| < \|A\| \, \|x\|,$$
$$|\lambda| < \|A\|. \qquad (55.1)$$

From this result we have

$$\|A\|_2^2 = \max \text{ eigenvalue of } A^H A$$
$$< \|A^H A\|_1 < \|A^H\|_1 \|A\|_1 = \|A\|_\infty \|A\|_1. \qquad (55.2)$$

(D) $\lim_{r \to \infty} A^r = 0$ if, and only if, $|\lambda_i| < 1$ for all eigenvalues. This follows by considering the Jordan canonical form. For a typical Jordan submatrix we have

$$\begin{bmatrix} a & & \\ 1 & a & \\ & 1 & a \end{bmatrix}^r = a^{r-2} \begin{bmatrix} a^2 & & \\ ra & a^2 & \\ \tfrac{1}{2}r(r-1) & ra & a^2 \end{bmatrix} \qquad (55.3)$$

and clearly this tends to zero if, and only if, $|a| < 1$. It is common to prove result (C) from (D), though since we are interested only in consistent pairs of matrix and vector norms, the proof we have given seems more natural. The quantity $\rho(A)$ defined by

$$\rho(A) = \max |\lambda_i| \qquad (55.4)$$

is called the *spectral radius* of A. We have seen that

$$\rho(A) < \|A\|. \qquad (55.5)$$

(E) The series $I + A + A^2 + \dots$ converges if, and only if, $A^r \to 0$. The sufficiency of the condition follows from (D). For if it is satisfied the eigenvalues of A all lie inside the unit circle and hence $(I - A)$ is non-singular. Now the relation

$$(I - A)(I + A + A^2 + \dots + A^r) = I - A^{r+1} \qquad (55.6)$$

is satisfied identically and hence

$$I + A + A^2 + \dots + A^r = (I - A)^{-1} - (I - A)^{-1} A^{r+1}. \qquad (55.7)$$

The second term on the right tends to zero and therefore

$$I + A + A^2 + \dots \to (I - A)^{-1}.$$

The necessity of the condition is self-evident.

(F) A sufficient criterion for the convergence of the series $I + A + A^2 + \dots$ is that any of the norms of A should be less than unity. For if this is true we have from (C)

$$|\lambda_i| < 1$$

for all eigenvalues. Hence from (D) $A^r \to 0$ and therefore from (E) the series is convergent.

Note that the existence of a norm such that $\|A\| > 1$ does not preclude the convergence of the series. For example if

$$A = \begin{bmatrix} 0 \cdot 9 & 0 \\ 0 \cdot 3 & 0 \cdot 8 \end{bmatrix} \qquad (55.8)$$

then $\|A\|_\infty = 1\cdot 1$, $\|A\|_1 = 1\cdot 2$, and $\|A\|_E = (1\cdot 54)^{\frac{1}{2}}$, yet $A^r \to 0$ since its eigenvalues are $0\cdot 9$ and $0\cdot 8$.

When attempting to establish the convergence of a series in practice we shall be free to choose whichever of the norms is the smallest.

(G) A sufficient criterion for the convergence of the series in (E) is that any norm of any similarity transform of A is less than unity. For if $A = HBH^{-1}$, then

$$I + A + \ldots + A^r = H(I + B + \ldots + B^r)H^{-1}. \tag{55.9}$$

The series on the right converges if $\|B\| < 1$.

We may use a similarity transformation to produce a matrix for which one of our standard norms is smaller than that of the original matrix. For the matrix given in (F) we may take

$$B = \begin{bmatrix} 3 & 0 \\ 0 & 1 \end{bmatrix} A \begin{bmatrix} \tfrac{1}{3} & 0 \\ 0 & 1 \end{bmatrix} = \begin{bmatrix} 0\cdot 9 & 0 \\ 0\cdot 1 & 0\cdot 8 \end{bmatrix}, \tag{55.10}$$

and we have $\|B\|_\infty = 0\cdot 9$.

As a more convincing example consider

$$A = \begin{bmatrix} 0\cdot 1 & 0 & 1\cdot 0 \\ 0\cdot 2 & 0\cdot 3 & 0 \\ 0\cdot 2 & 0\cdot 1 & 0\cdot 4 \end{bmatrix}. \tag{55.11}$$

We have $\|A\|_\infty = 1\cdot 1$, $\|A\|_1 = 1\cdot 4$, $\|A\|_E = (1\cdot 35)^{\frac{1}{2}}$. Defining B by

$$\cdot B = \begin{bmatrix} \tfrac{1}{2} & 0 & 0 \\ 0 & 1 & 0 \\ 0 & 0 & 1 \end{bmatrix} A \begin{bmatrix} 2 & 0 & 0 \\ 0 & 1 & 0 \\ 0 & 0 & 1 \end{bmatrix} = \begin{bmatrix} 0\cdot 1 & 0 & 0\cdot 5 \\ 0\cdot 4 & 0\cdot 3 & 0 \\ 0\cdot 4 & 0\cdot 1 & 0\cdot 4 \end{bmatrix}, \tag{55.12}$$

we see that $\|B\|_\infty = 0\cdot 9$, so that the series in A converges. Similarity transformations with diagonal matrices are frequently of great value in error analysis.

Avoiding use of infinite matrix series

56. We may often avoid the use of infinite series of matrices by making use of the elementary properties of norms. For example, we may obtain a bound for $\|(I+A)^{-1}\|$ if $\|A\| < 1$ as follows.

Since $\|A\| < 1$ all eigenvalues of A are numerically less than unity. $(I+A)$ is therefore non-singular and $(I+A)^{-1}$ exists. We have

$$I = (I+A)^{-1}(I+A)$$

$$= (I+A)^{-1}+(I+A)^{-1}A. \tag{56.1}$$

Hence

$$\|I\| \;>\; \|(I+A)^{-1}\| - \|(I+A)^{-1}A\|$$

$$\geqslant \|(I+A)^{-1}\| - \|(I+A)^{-1}\|\,\|A\| \tag{56.2}$$

$$\|(I+A)^{-1}\| \;<\; \|I\|/(1-\|A\|) = (1-\|A\|)^{-1} \text{ if } \|I\| = 1. \tag{56.3}$$

The alternative is to write

$$(I+A)^{-1} = I - A + A^2 - ..., \tag{56.4}$$

convergence being assured if $\|A\| < 1$. From (56.4) we conclude

$$\|(I+A)^{-1}\| \;<\; \|I\| + \|A\| + \|A\|^2 + ...$$

$$= (1-\|A\|)^{-1} \text{ if } \|I\| = 1. \tag{56.5}$$

Additional notes

In this chapter I have attempted merely to give a sketch of the background material that will be needed in the remainder of the book. The serious research worker in this field will need to supplement this chapter with further reading. A very complete coverage of the classical theory is contained in Volume 1 of Gantmacher's *The theory of matrices* (1959a). In the bibliography I have included a number of standard text books in this field. Householder's *The theory of matrices in numerical analysis* (1964) deserves special mention since it covers virtually the whole of the theoretical material discussed in later chapters, in addition to much of that discussed in this chapter.

2

Perturbation Theory

Introduction

1. IN THE first chapter we were concerned with the mathematical background to the algebraic eigenproblem. The remainder of the book is devoted to a consideration of the practical problems involved in computing eigenvalues and eigenvectors on a digital computer and in determining the accuracy of the computed eigensystems. A major preoccupation will be that of estimating the effect of the various errors which are inherent in the formulation of the problem and its solution. There are three main considerations.

(i) The elements of the given matrix may be determined directly from physical measurements, and therefore be subject to the errors inherent in all observations. In this case the matrix A with which we are presented is an approximation to the matrix which corresponds to exact measurements. If it can be asserted that the error in every element of A is bounded by δ, then we can say that the true matrix is $(A+E)$, where E is some matrix for which

$$|e_{ij}| < \delta. \tag{1.1}$$

A complete solution of the practical problem involves not only the determination of the eigenvalues of A, but also an assessment of the range of variation of the eigenvalues of all matrices of the class $(A+E)$, where E satisfies equation (1.1). We are thus led immediately to the consideration of the perturbations of the eigenvalues of a matrix corresponding to perturbations in its elements.

(ii) The elements of the matrix may be defined exactly by mathematical formulae but we may still be prevented by practical considerations from presenting a digital computer with this exact matrix. If the elements of the matrix are irrational this will almost inevitably be true. Even if they are not irrational their exact representation may require more digits than we are prepared to use. The matrix A, for instance, may be defined as the product of several matrices each of which has elements with the full number of digits normally used on the computer. It is evident that we are faced with much the same problem as in (i). The given matrix will be $(A+E)$ where E is the error matrix.

(iii) Even if we can regard the matrix presented to the digital computer as exact, the same will not be true, in general, of the computed solution. Most commonly the procedure by which the solution is computed will involve the calculation of a sequence of similarity transforms $A_1, A_2, A_3, \ldots$, of the original matrix A, and rounding errors will be made in carrying out each of the transformations. At first sight the errors arising from these transformations are of a different nature from those discussed in (i) and (ii). However, this is not, in general, true. Frequently we shall be able to show that the computed matrices A_i are exactly similar to matrices $(A+E_i)$ where the E_i have small elements which are functions of the rounding errors. Often we will be able to find strict bounds for the E_i. The eigenvalues of the A_i will therefore be exactly those of $(A+E_i)$ and again we are led to consider the perturbations of the eigenvalues and eigenvectors of a matrix corresponding to perturbations in its elements.

Ostrowski's theorem on continuity of the eigenvalues

2. Upper bounds for the perturbations in the eigenvalues have been given by Ostrowski (1957). His results may be expressed as follows.

Let A and B be matrices with elements which satisfy the relations

$$|a_{ij}| < 1, \quad |b_{ij}| < 1. \tag{2.1}$$

Then if λ' is an eigenvalue of $(A+\epsilon B)$, there is an eigenvalue λ of A such that

$$|\lambda'-\lambda| < (n+2)(n^2\epsilon)^{\frac{1}{n}}. \tag{2.2}$$

Further the eigenvalues λ and λ' of A and $(A+\epsilon B)$ can be put in one-to-one correspondence so that

$$|\lambda'-\lambda| < 2(n+1)^2(n^2\epsilon)^{\frac{1}{n}}. \tag{2.3}$$

Although these results are of great theoretical importance they are of little practical value. Suppose we have a matrix A of order 20 with irrational elements satisfying the relation

$$|a_{ij}| < 1. \tag{2.4}$$

Consider now the matrix $(A+E)$ obtained by rounding the elements of A to 10 decimal places, so that

$$|e_{ij}| < \tfrac{1}{2}10^{-10}, \tag{2.5}$$

where the upper bounds are attainable. From Ostrowski's second result we have

$$|\lambda' - \lambda| < 2 \times 21^2 \times (400 \times \tfrac{1}{2} 10^{-10})^{\frac{1}{20}}. \tag{2.6}$$

This bound is far poorer than could be obtained by simple norm theory. For we have

$$\left.\begin{array}{l} |\lambda| < \|A\|_\infty < 20 \\[6pt] |\lambda'| < \|A + E\|_\infty < 20 + 10 \times 10^{-10} \end{array}\right\}, \tag{2.7}$$

and hence

$$|\lambda - \lambda'| < |\lambda| + |\lambda'| < 40 + 10^{-9}. \tag{2.8}$$

Ostrowski's result is an improvement on that obtained by simple norm theory only when

$$2(n+1)^2 (n^2 \epsilon)^{\frac{1}{n}} < 2n,$$

giving

$$\epsilon < \frac{n^{n-2}}{(n+1)^{2n}} \doteq \frac{1}{e^2 n^{n+2}}. \tag{2.9}$$

For $n = 20$ this means that ϵ must be less than 10^{-27}. To obtain a *useful* bound from Ostrowski's result ϵ must be far smaller even than this.

It is evident that the factor $\epsilon^{\frac{1}{n}}$ is the main weakness of Ostrowski's result. Unfortunately, as an example due to Forsythe shows, its presence is inevitable in any general result. Consider for example $C_n(a)$, the Jordan submatrix of order n, which has its n eigenvalues equal to a. If we change the $(1, n)$ element from 0 to ϵ, then the characteristic equation becomes

$$(a - \lambda)^n + (-1)^{n-1}\epsilon = 0, \tag{2.10}$$

giving

$$\lambda = a + \omega^r \epsilon^{\frac{1}{n}} \quad (r = 0, \dots, n-1) \tag{2.11}$$

for the n eigenvalues, where ω is any primitive nth root of unity.

Algebraic functions

3. In the analysis which follows we shall need two results from the theory of algebraic functions (see for example Goursat, 1933). These we now state without proof.

Let $f(x, y)$ be defined by

$$f(x, y) = y^n + p_{n-1}(x)y^{n-1} + p_{n-2}(x)y^{n-2} + \dots + p_1(x)y + p_0(x), \tag{3.1}$$

where the $p_i(x)$ are polynomials in x. Corresponding to any value of x the equation $f(x, y) = 0$ has n roots $y_1(x), y_2(x), \dots, y_n(x)$, each root

This is immediately obvious if we examine the explicit expression for $\det(\lambda I - A - \epsilon B)$. We may write

$$c_r(\epsilon) = c_r + c_{r1}\epsilon + c_{r2}\epsilon^2 + \ldots + c_{r,n-r}\epsilon^{n-r}. \tag{5.4}$$

Now since λ_1 is a simple root of (5.1) we know from Theorem 1 of § 3 that for sufficiently small ϵ there is a simple root $\lambda_1(\epsilon)$ of (5.2), given by a convergent power series

$$\lambda_1(\epsilon) = \lambda_1 + k_1\epsilon + k_2\epsilon^2 + \ldots. \tag{5.5}$$

Clearly $\lambda_1(\epsilon) \to \lambda_1$ as $\epsilon \to 0$. Note that

$$|\lambda_1(\epsilon) - \lambda_1| = O(\epsilon) \tag{5.6}$$

independent of the multiplicities of other eigenvalues.

Perturbation of corresponding eigenvectors

6. Turning now to the perturbation in the corresponding eigenvector, we first develop explicit expressions for the components of an eigenvector x_1 corresponding to a simple eigenvalue λ_1 of A. Since λ_1 is a simple eigenvalue, $(A - \lambda_1 I)$ has at least one non-vanishing minor of order $(n-1)$. Suppose, without loss of generality, that this lies in the first $(n-1)$ rows of $(A - \lambda_1 I)$. Then from the theory of linear equations the components of x may be taken to be

$$(A_{n1}, A_{n2}, \ldots, A_{nn}), \tag{6.1}$$

where A_{ni} denotes the cofactor of the (n, i) element of $(A - \lambda_1 I)$, and hence is a polynomial in λ_1 of degree not greater than $(n-1)$.

We now apply this result to the simple eigenvalue $\lambda_1(\epsilon)$ of $(A + \epsilon B)$. We denote the eigenvector of A by x_1 and the eigenvector of $(A + \epsilon B)$ by $x_1(\epsilon)$. Clearly the elements of $x_1(\epsilon)$ are polynomials in $\lambda_1(\epsilon)$ and ϵ, and since the power series for $\lambda_1(\epsilon)$ is convergent for all sufficiently small ϵ we see that each element of $x_1(\epsilon)$ is represented by a convergent power series in ϵ, the constant term in which is the corresponding element of x_1. We may write

$$x_1(\epsilon) = x_1 + \epsilon z_1 + \epsilon^2 z_2 + \ldots, \tag{6.2}$$

where each component of the vector series on the right is a convergent power series in ϵ. Corresponding to the result of (5.6) for the eigenvalue, we have for the eigenvector the result

$$|x_1(\epsilon) - x_1| = O(\epsilon), \tag{6.3}$$

and again there are no fractional powers of ϵ.

Matrix with linear elementary divisors

7. If the matrix A has linear elementary divisors, then we may assume the existence of complete sets of right-hand and left-hand eigenvectors $x_1, x_2, ..., x_n$ and $y_1, y_2, ..., y_n$ respectively, such that

$$y_i^T x_j = 0 \quad (i \neq j), \qquad (7.1)$$

though these vectors are unique only if all the eigenvalues are simple.

Expressing each of the vectors z_i of (6.2) in terms of the x_j in the form

$$z_i = \sum_{j=1}^n s_{ji} x_j, \qquad (7.2)$$

we have

$$x_1(\epsilon) = x_1 + \epsilon \sum_{j=1}^n s_{j1} x_j + \epsilon^2 \sum_{j=1}^n s_{j2} x_j + ..., \qquad (7.3)$$

and collecting together the terms in the vector x_i

$$x_1(\epsilon) = (1 + \epsilon s_{11} + \epsilon^2 s_{12} + ...)x_1 + (\epsilon s_{21} + \epsilon^2 s_{22} + ...)x_2 +$$
$$+ ... + (\epsilon s_{n1} + \epsilon^2 s_{n2} + ...)x_n. \quad (7.4)$$

The convergence of the n power series in the brackets is a simple consequence of the absolute convergence of the series in (6.2).

We are not interested in a multiplying factor in $x_1(\epsilon)$ and we may divide by $(1 + \epsilon s_{11} + \epsilon^2 s_{12} + ...)$ since this is non-zero for sufficiently small ϵ. We may then redefine the new vector as $x_1(\epsilon)$. This gives

$$x_1(\epsilon) = x_1 + (\epsilon t_{21} + \epsilon^2 t_{22} + ...)x_2 + ... + (\epsilon t_{n1} + \epsilon^2 t_{n2} + ...)x_n, \quad (7.5)$$

where the expressions in brackets are still convergent power series for sufficiently small ϵ.

This relation has been obtained under the assumption that the components of the x_i were obtained directly from their determinantal expressions. However, if we replace these x_i by new x_i which are normalized so that

$$\|x_i\|_2 = 1, \qquad (7.6)$$

this merely introduces a constant factor into each of the expressions in brackets and this may be absorbed into the t_{ij}. Equation (7.5) then holds for normalized x_i though $x_i(\epsilon)$ will not, of course, be normalized for $\epsilon \neq 0$.

First-order perturbations of eigenvalues

8. We now derive explicit expressions for the first-order perturbations in terms of the x_i and y_i. We first introduce quantities s_i which will be used repeatedly in subsequent analysis; they are defined by the relations

$$s_i = y_i^T x_i \quad (i = 1, 2, ..., n), \qquad (8.1)$$

where the y_i and x_i are *normalized* left-hand and right-hand vectors. If y_i and x_i are real, then s_i is the cosine of the angle between these vectors.

If there are any multiple eigenvalues the corresponding x_i and y_i will not be uniquely determined; in this case we assume that the s_i we are using correspond to some particular choice of the x_i and y_i. Even when x_i and y_i correspond to a simple λ_i they are arbitrary to the extent of a complex multiplier of modulus unity, but in this case $|s_i|$ *is* fully determined.

We have in any case

$$|s_i| = |y_i^T x_i| < \|y_i\|_2 \|x_i\|_2 = 1. \tag{8.2}$$

We define also quantities β_{ij} by the relations

$$\beta_{ij} = y_i^T B x_j \tag{8.3}$$

and, since $\|B\|_2 < n$, we have

$$|\beta_{ij}| = |y_i^T(Bx_j)| < \|y_i\|_2 \|Bx_j\|_2 < \|B\|_2 \|y_i\|_2 \|x_j\|_2 < n. \tag{8.4}$$

9. From their definitions we have

$$(A + \epsilon B)x_1(\epsilon) = \lambda_1(\epsilon)x_1(\epsilon), \tag{9.1}$$

and since $\lambda_1(\epsilon)$ and all components of $x_1(\epsilon)$ are represented by convergent power series we may equate terms in the same power of ϵ on each side of this equation. Equating terms in ϵ gives, from (5.5) and (7.5)

$$A\left(\sum_{i=2}^{n} t_{i1}x_i\right) + Bx_1 = \lambda_1\left(\sum_{i=2}^{n} t_{i1}x_i\right) + k_1 x_1, \tag{9.2}$$

or

$$\sum_{i=2}^{n}(\lambda_i - \lambda_1)t_{i1}x_i + Bx_1 = k_1 x_1. \tag{9.3}$$

Pre-multiplying this by y_1^T and remembering that $y_1^T x_i = 0$, $(i \neq 1)$, we obtain

$$k_1 = y_1^T Bx_1 / y_1^T x_1 = \beta_{11}/s_1, \tag{9.4}$$

and hence

$$|k_1| < n/|s_1| \quad \text{from (8.4)}. \tag{9.5}$$

For sufficiently small ϵ the main term in the perturbation of λ_1 is $k_1\epsilon$, so we see that the sensitivity of this eigenvalue is primarily dependent on s_1. Unfortunately an s_i may be arbitrarily small.

First-order perturbations of eigenvectors

10. Pre-multiplying (9.3) by y_i^T we have

$$(\lambda_i - \lambda_1)t_{i1}s_i + \beta_{i1} = 0 \quad (i = 2, 3, ..., n), \tag{10.1}$$

and hence from (7.5) the first order term in the perturbation of x_1 is given by

$$\epsilon \left[\frac{\beta_{21}x_2}{(\lambda_1-\lambda_2)s_2} + \frac{\beta_{31}x_3}{(\lambda_1-\lambda_3)s_3} + \cdots + \frac{\beta_{n1}x_n}{(\lambda_1-\lambda_n)s_n} \right]. \tag{10.2}$$

Notice now that all the s_i other than s_1 are involved and we also have the factors $(\lambda_1-\lambda_i)$ in the denominators. The presence of the s_i is a little misleading as we shall show in § 26. However, we expect the eigenvector corresponding to the simple eigenvalue λ_1 to be very sensitive to perturbations in A if λ_1 is *close* to any of the other eigenvalues, and this is indeed true.

When λ_1 is well separated from the other eigenvalues and none of the s_i $(i = 2, 3,..., n)$ is small we can certainly say that the eigenvector x_1 is comparatively insensitive to perturbations in A.

Higher-order perturbations

11. We may obtain expressions for higher-order perturbations by equating coefficients of the higher powers of ϵ in equation (9.1). Equating the coefficients of ϵ^2, for example, we have

$$A\left(\sum_{i=2}^{n} t_{i2}x_i\right) + B\left(\sum_{i=2}^{n} t_{i1}x_i\right) = k_2x_1 + k_1\left(\sum_{i=2}^{n} t_{i1}x_i\right) + \lambda_1\left(\sum_{i=2}^{n} t_{i2}x_i\right),$$

or

$$\sum_{i=2}^{n} t_{i2}(\lambda_i-\lambda_1)x_i + B\left(\sum_{i=2}^{n} t_{i1}x_i\right) = k_2x_1 + k_1\left(\sum_{i=2}^{n} t_{i1}x_i\right). \tag{11.1}$$

Pre-multiplication by y_1^T gives

$$\sum_{i=2}^{n} t_{i1}\beta_{1i} = k_2s_1, \tag{11.2}$$

and hence from (10.1)

$$k_2 = \frac{1}{s_1}\sum_{i=2}^{n} \frac{\beta_{i1}\beta_{1i}}{s_i(\lambda_1-\lambda_i)}. \tag{11.3}$$

In the same way we may obtain the second-order terms in the perturbation of x_1 and so on, but these higher-order terms are of little practical value.

Multiple eigenvalues

12. Turning to the perturbation of an eigenvalue λ_1 of multiplicity m, an analysis similar to that of § 5 now requires the use of Theorem 2 of § 3. This shows that, corresponding to such an eigenvalue, there will be a set of m eigenvalues which, in general, will fall into groups which can be expanded in terms of fractional power series of ϵ. If there is only one group, then each eigenvalue will be representable by a power series in $\epsilon^{\frac{1}{m}}$, but if there are several groups, other fractional powers will be involved. In particular we could have m groups, in which case no fractional powers would be involved and for each of the m perturbed eigenvalues $\lambda_i(\epsilon)$, a relation of the form

$$|\lambda_i(\epsilon) - \lambda_1| = 0(\epsilon) \tag{12.1}$$

would hold.

Analysis in terms of the characteristic polynomial is very clumsy for multiple zeros because it is too remote from the original matrix form. In the next few sections we reconsider the perturbation theory using a much more powerful tool which is also of great practical value.

Gerschgorin's theorems

13. For this alternative approach we require two theorems which are due to Gerschgorin (1931).

THEOREM 3. *Every eigenvalue of the matrix A lies in at least one of the circular discs with centres a_{ii} and radii $\sum\limits_{j \neq i} |a_{ij}|$.*

The proof is quite simple. Let λ be any eigenvalue of A. Then there is certainly at least one non-null x such that

$$Ax = \lambda x. \tag{13.1}$$

Suppose the rth component of x has the largest modulus; we may normalize x so that

$$x^T = (x_1, x_2, \ldots, x_{r-1}, 1, x_{r+1}, \ldots, x_n), \tag{13.2}$$

where

$$|x_i| < 1 \quad (i \neq r). \tag{13.3}$$

Equating the rth element on each side of (13.1) we have

$$\sum_{j=1}^{n} a_{rj} x_j = \lambda x_r = \lambda. \tag{13.4}$$

Hence

$$|\lambda - a_{rr}| < \sum_{j \neq r} |a_{rj} x_j| < \sum_{j \neq r} |a_{rj}| \, |x_j| < \sum_{j \neq r} |a_{rj}|, \tag{13.5}$$

and λ lies in one of the circular discs.

The second theorem gives more detailed information concerning the distribution of the eigenvalues among the discs.

THEOREM 4. *If s of the circular discs of Theorem 3 form a connected domain which is isolated from the other discs, then there are precisely s eigenvalues of A within this connected domain.*

The proof is dependent on the notion of continuity. We write

$$A = \text{diag}(a_{ii}) + C = D + C \text{ (say)}, \tag{13.6}$$

where C is the matrix of off-diagonal elements, and define quantities r_i by the relations
$$r_i = \sum_{j \neq i} |a_{ij}|. \tag{13.7}$$

Consider now the matrices $(D+\epsilon C)$ for values of ϵ satisfying
$$0 \leqslant \epsilon \leqslant 1. \tag{13.8}$$

For $\epsilon = 0$ this is the matrix D, while for $\epsilon = 1$ it is the matrix A. The coefficients of the characteristic polynomial of $(D+\epsilon C)$ are polynomials in ϵ, and by the theory of algebraic functions the roots of the characteristic equation are continuous functions of ϵ. By Theorem 3, for any value of ϵ the eigenvalues all lie in the circular discs with centres a_{ii} and radii ϵr_i, and if we let ϵ vary steadily from 0 to 1 the eigenvalues all traverse continuous paths.

Without loss of generality we may assume that it is the first s discs which form the connected domain. Then, since the $(n-s)$ discs with radii $r_{s+1}, r_{s+2}, \ldots, r_n$ are isolated from those with radii $r_1, r_2, \ldots, r_s$, the same is true for corresponding discs with radii ϵr_i for all ϵ in the range (13.8). Now when $\epsilon = 0$ the eigenvalues are $a_{11}, a_{22}, \ldots, a_{nn}$, and of these the first s lie in the domain corresponding to the first s discs and the remaining $(n-s)$ lie outside this domain. It follows that this is true for all ϵ up to and including $\epsilon = 1$.

In particular, if any of the Gerschgorin discs is isolated, it contains precisely one eigenvalue. Note that corresponding results may be obtained by working with A^T instead of with A.

Perturbation theory based on Gerschgorin's theorems

14. We now investigate the eigenvalues of $(A+\epsilon B)$ by making use of the Jordan canonical form. We distinguish five main cases.

Case 1. *Perturbation of a simple eigenvalue λ_1 of a matrix having linear elementary divisors.*

The Jordan canonical form is then of diagonal form. There is a matrix H such that
$$H^{-1}AH = \mathrm{diag}(\lambda_i), \tag{14.1}$$

having its columns parallel to a complete set of right-hand eigenvectors x_i, and such that H^{-1} has its rows parallel to a complete set of left-hand eigenvectors y_i^T. If we normalize the vectors so that
$$\|x_i\|_2 = \|y_i\|_2 = 1, \tag{14.2}$$

then we may take the ith column of H to be x_i and the ith row of H^{-1} to be y_i^T/s_i in the notation of § 8. (Note that since λ_1 is simple

x_1 and y_1 are unique within a factor of modulus unity.) With H in this form we see immediately that

$$H^{-1}(A+\epsilon B)H = \text{diag}(\lambda_i)+\epsilon \begin{bmatrix} \beta_{11}/s_1 & \beta_{12}/s_1 & \cdots & \beta_{1n}/s_1 \\ \beta_{21}/s_2 & \beta_{22}/s_2 & \cdots & \beta_{2n}/s_2 \\ \cdot & \cdot & \cdots & \cdot \\ \beta_{n1}/s_n & \beta_{n2}/s_n & \cdots & \beta_{nn}/s_n \end{bmatrix}. \qquad (14.3)$$

An application of Theorem 3 shows that the eigenvalues lie in circular discs with centres $(\lambda_i+\epsilon\beta_{ii}/s_i)$ and radii $\epsilon \sum_{j\neq i} |\beta_{ij}/s_i|$. Using relation (8.4) we see that if $|b_{ij}| < 1$, the ith disc is of radius less than $n(n-1)\epsilon/|s_i|$, and since λ_1 has been assumed to be simple, for sufficiently small ϵ the first disc is isolated and therefore contains precisely one eigenvalue.

15. The result we have just obtained is a little disappointing because we would expect from our previous analysis that the eigenvalue corresponding to λ_1 could be located in a circle centre $(\lambda_1+\epsilon\beta_{11}/s_1)$ but having a radius which is $O(\epsilon^2)$ as $\epsilon \to 0$. We can reduce the radius of the Gerschgorin circular discs by the following simple device.

If we multiply the ith column of any matrix by m and its ith row by $1/m$, then its eigenvalues are unaltered. Let us apply this for $i = 1$ to the matrix on the right-hand side of (14.3), taking $m = k/\epsilon$. This then becomes

$$\text{diag}(\lambda_i)+ \begin{bmatrix} \epsilon\beta_{11}/s_1 & \epsilon^2\beta_{12}/ks_1 & \epsilon^2\beta_{13}/ks_1 & \cdots & \epsilon^2\beta_{1n}/ks_1 \\ k\beta_{21}/s_2 & \epsilon\beta_{22}/s_2 & \epsilon\beta_{23}/s_2 & \cdots & \epsilon\beta_{2n}/s_2 \\ \cdot & \cdot & \cdot & \cdots & \cdot \\ k\beta_{n1}/s_n & \epsilon\beta_{n2}/s_n & \epsilon\beta_{n3}/s_n & \cdots & \epsilon\beta_{nn}/s_n \end{bmatrix}. \qquad (15.1)$$

The elements in the first row other than $(1, 1)$ now contain the factor ϵ^2 while those in the first column other than $(1, 1)$ are independent of ϵ. All other elements are unchanged. We wish to choose k so as to make the first Gerschgorin disc as small as possible while keeping the other discs sufficiently small to avoid overlapping the first.

Clearly this will be true for all sufficiently small ϵ if we choose k to have the largest value consistent with the inequalities

$$|k\beta_{i1}/s_i| < \tfrac{1}{2} |\lambda_1-\lambda_i| \quad (i = 2, 3,..., n). \qquad (15.2)$$

(The factor $\frac{1}{2}$ in relation (15.2) has no special significance; it may be replaced by any other number independent of ϵ and less than unity.) This requirement will be met if

$$k = \min |(\lambda_1 - \lambda_i)s_i/2\beta_{i1}| \quad (i = 2, 3, ..., n). \tag{15.3}$$

(If all the β_{i1} are zero, $\lambda_1 + \epsilon\beta_{11}/s_1$ is an exact eigenvalue and we need not pursue the matter further.) With this definition of k we have for the radius r_1 of the first Gerschgorin disc

$$r_1 = \sum_{j=2}^{n} |\epsilon^2\beta_{1j}/ks_1| \leqslant n(n-1)\epsilon^2/k \, |s_1|. \tag{15.4}$$

We have also

$$k^{-1} = \max |2\beta_{i1}/(\lambda_1 - \lambda_i)s_i| \leqslant \max |2n/(\lambda_1 - \lambda_i)s_i|. \tag{15.5}$$

At this stage we remind the reader that these bounds are based on the assumption that B is normalized so that $|b_{ij}| < 1$.

16. The explicit expressions for the bounds tend, on the whole, to conceal the simplicity of the basic technique. This simplicity becomes immediately apparent if we consider a numerical example. By way of illustration we consider first the matrix X defined by

$$X = \begin{bmatrix} 0 \cdot 9 & & \\ & 0 \cdot 4 & \\ & & 0 \cdot 4 \end{bmatrix} + 10^{-5} \begin{bmatrix} 0 \cdot 1234 & 0 \cdot 4132 & -0 \cdot 2167 \\ -0 \cdot 1342 & 0 \cdot 4631 & 0 \cdot 1276 \\ 0 \cdot 1567 & 0 \cdot 1432 & 0 \cdot 3125 \end{bmatrix}. \tag{16.1}$$

The first matrix on the right has one double eigenvalue but has linear divisors.

On multiplying the first row by 10^{-5} and the first column by 10^5 we see that the eigenvalues of X are those of

$$\begin{bmatrix} 0 \cdot 9 & & \\ & 0 \cdot 4 & \\ & & 0 \cdot 4 \end{bmatrix} + \begin{bmatrix} 10^{-5} \, (0 \cdot 1234) & 10^{-10} \, (0 \cdot 4132) & 10^{-10} \, (-0 \cdot 2167) \\ -0 \cdot 1342 & 10^{-5} \, (0 \cdot 4631) & 10^{-5} \, (0 \cdot 1276) \\ 0 \cdot 1567 & 10^{-5} \, (0 \cdot 1432) & 10^{-5} \, (0 \cdot 3125) \end{bmatrix}. \tag{16.2}$$

The corresponding Gerschgorin discs have

centre $0 \cdot 9 + 10^{-5} \, (0 \cdot 1234)$ radius $10^{-10} \, (0 \cdot 4132 + 0 \cdot 2167) = 10^{-10}$
$(0 \cdot 6299),$

centre $0 \cdot 4 + 10^{-5} \, (0 \cdot 4631)$ radius $0 \cdot 1342 + 10^{-5} \, (0 \cdot 1276),$

centre $0 \cdot 4 + 10^{-5} \, (0 \cdot 3125)$ radius $0 \cdot 1567 + 10^{-5} \, (0 \cdot 1432).$

The first disc is obviously isolated and hence contains just one eigenvalue; the radius of this disc is of order 10^{-10}. The fact that the other two diagonal elements are identical does not affect the result at all.

However if we make one of the other two eigenvalues approach the first, then the location of this eigenvalue is adversely affected. Consider the matrix Y defined by

$$Y = \begin{bmatrix} 0\cdot9 & & \\ & 0\cdot89 & \\ & & 0\cdot4 \end{bmatrix} + 10^{-5} \begin{bmatrix} 0\cdot1234 & 0\cdot4132 & -0\cdot2167 \\ -0\cdot1342 & 0\cdot4631 & 0\cdot1276 \\ 0\cdot1567 & 0\cdot1432 & 0\cdot3125 \end{bmatrix}.$$

(16.3)

We can no longer use the multiplying factor 10^{-5} because the first two Gerschgorin discs would then overlap. Using the multiplier 10^{-3} the matrix Y becomes

$$\begin{bmatrix} 0\cdot9 & & \\ & 0\cdot89 & \\ & & 0\cdot4 \end{bmatrix} + \begin{bmatrix} 10^{-5} & (0\cdot1234) & 10^{-8} & (0\cdot4132) & 10^{-8} & (-0\cdot2167) \\ 10^{-2} & (-0\cdot1342) & 10^{-5} & (0\cdot4631) & 10^{-5} & (0\cdot1276) \\ 10^{-2} & (0\cdot1567) & 10^{-5} & (0\cdot1432) & 10^{-5} & (0\cdot3125) \end{bmatrix}$$

(16.4)

and the first Gerschgorin disc is isolated. There is therefore an eigenvalue in the disc with centre $0\cdot9 + 10^{-5}$ $(0\cdot1234)$ and radius 10^{-8} $(0\cdot6299)$.

17. *Case 2. Perturbation of a multiple eigenvalue λ_1 of a matrix having linear elementary divisors.*

This case is sufficiently illustrated by considering a simple example of order 6 having $\lambda_1 = \lambda_2 = \lambda_3$ and $\lambda_4 = \lambda_5$. Since the matrix A has linear divisors, $(A + \epsilon B)$ is similar to

$$\begin{bmatrix} \lambda_1 & & & & & \\ & \lambda_1 & & & & \\ & & \lambda_1 & & & \\ & & & \lambda_4 & & \\ & & & & \lambda_4 & \\ & & & & & \lambda_6 \end{bmatrix} + \epsilon \begin{bmatrix} \beta_{11}/s_1 & \beta_{12}/s_1 & \cdots & \beta_{16}/s_1 \\ \beta_{21}/s_2 & \beta_{22}/s_2 & \cdots & \beta_{26}/s_2 \\ \cdot & \cdot & \cdots & \cdot \\ \beta_{61}/s_6 & \beta_{62}/s_6 & \cdots & \beta_{66}/s_6 \end{bmatrix}.$$

(17.1)

We have already dealt with λ_6 under case 1. By multiplying column six by k/ϵ and row six by ϵ/k we can locate a simple eigenvalue in a disc with centre $\lambda_6 + \epsilon\beta_{66}/s_6$ and having a radius which is $O(\epsilon^2)$ as $\epsilon \to 0$. Consider now the other five Gerschgorin discs. There are three with centres $\lambda_1 + \epsilon\beta_{ii}/s_i$ ($i = 1, 2, 3$) and two with centres $\lambda_4 + \epsilon\beta_{ii}/s_i$ ($i = 4, 5$), and the corresponding radii are all of order ϵ. Obviously for sufficiently small ϵ the group of three will be isolated from the group of two, but we cannot, in general, claim that the individual discs in a group are isolated. The perturbations in the triple eigenvalue λ_1 are all of order ϵ as $\epsilon \to 0$ but it is not in general true that there is one eigenvalue in each of three discs with centres $\lambda_1 + \epsilon\beta_{ii}/s_i$ ($i = 1, 2, 3$) and radii of order ϵ^2.

18. We can however reduce the radii of these circular discs to some extent. Let us concentrate our attention on the triple eigenvalue and denote (17.1) by

$$\mathrm{diag}(\lambda_i) + \epsilon \begin{bmatrix} P & Q \\ \hline R & S \end{bmatrix}, \tag{18.1}$$

where P, Q, R, S are (3×3) matrices. Multiplying the first three rows by ϵ/k and the first three columns by k/ϵ we obtain

$$\mathrm{diag}(\lambda_i) + \begin{bmatrix} \epsilon P & k^{-1}\epsilon^2 Q \\ \hline kR & \epsilon S \end{bmatrix}. \tag{18.2}$$

We may choose a value of k, independent of ϵ, so that the first three discs are isolated from the others. Hence for some suitable k there are three eigenvalues in the union of the three discs:

centre $\lambda_1 + \epsilon\beta_{11}/s_1$
    radius $\epsilon(|\beta_{12}| + |\beta_{13}|)/|s_1| + \epsilon^2(|\beta_{14}| + |\beta_{15}| + |\beta_{16}|)/k\,|s_1|$,
centre $\lambda_1 + \epsilon\beta_{22}/s_2$
    radius $\epsilon(|\beta_{21}| + |\beta_{23}|)/|s_2| + \epsilon^2(|\beta_{24}| + |\beta_{25}| + |\beta_{26}|)/k\,|s_2|$,
centre $\lambda_1 + \epsilon\beta_{33}/s_3$
    radius $\epsilon(|\beta_{31}| + |\beta_{32}|)/|s_2| + \epsilon^2(|\beta_{34}| + |\beta_{35}| + |\beta_{36}|)/k\,|s_3|$.

For a matrix of order six the reduction in the radius is not impressive but for matrices of higher order it is more substantial. We shall see later (Chapter 9, § 68) that by reducing the matrix P of (18.1) to canonical form we can obtain sharper results, but this does not concern us here.

Note that we have demonstrated that when A has a multiple eigenvalue but has linear divisors, the coefficients in the characteristic polynomial must be specially related in such a way that no proper fractional powers of ϵ occur in the perturbations.

19. *Case* 3. *Perturbation of a simple eigenvalue of a matrix having one or more non-linear elementary divisors.*

Since A has some non-linear elementary divisors there is no longer a complete set of eigenvectors. We first investigate the values that can be attributed to the s_i for such a matrix. Consider the simple matrix A given by

$$A = \begin{bmatrix} a & 1 \\ 0 & b \end{bmatrix}. \tag{19.1}$$

The eigenvalues are $\lambda_1 = a$ and $\lambda_2 = b$ and the corresponding right-hand and left-hand vectors are given by

$$\begin{aligned} x_1^T &= (1, 0), & \alpha x_2^T &= (1, b-a) \\ \alpha y_1^T &= (a-b, 1), & y_2^T &= (0, 1) \end{aligned} \right\} \tag{19.2}$$

$$\text{where } \alpha = [1 + (a-b)^2]^{\frac{1}{2}}.$$

We have therefore

$$s_1 = y_1^T x_1 = (a-b)/\alpha, \qquad s_2 = y_2^T x_2 = (b-a)/\alpha, \tag{19.3}$$

and as $b \to a$ both s_1 and s_2 tend to zero. We have seen that the sensitivity to perturbations of a simple eigenvalue is proportional to s_i^{-1} and hence as b tends to a both eigenvalues become more and more sensitive. Note however that we have $s_1^{-1} + s_2^{-1} = 0$ for any value of b, so that although s_1^{-1} and s_2^{-1} both tend to infinity they are not independent.

20. For matrices with non-linear elementary divisors, an analysis similar to that of the previous sections may be carried out using the Jordan canonical form of A. There is a non-singular matrix H such that

$$H^{-1}AH = C \tag{20.1}$$

where C is the upper Jordan canonical form. We are no longer free to assume that all columns of H are normalized since the scaling factors of those columns associated with non-linear elementary divisors are restricted by the requirement that the relevant super-diagonal elements in C are unity (Chapter 1, § 8). Nevertheless, it is still convenient to assume that the columns associated with linear divisors are normalized. We shall denote by G the matrix which consists of normalized rows of H^{-1} and shall write

$g_i^T h_i = s_i$ if g_i and h_i correspond to a linear divisor,

$g_i^T h_i = t_i$ if g_i and h_i are part of the system corresponding to a non-linear divisor.

The perturbation of a simple eigenvalue will be adequately illustrated by means of a simple example. Consider a matrix of order four having elementary divisors $(\lambda_1 - \lambda)^2$, $(\lambda_3 - \lambda)$, $(\lambda_4 - \lambda)$. We may write

$$H^{-1}(A + \epsilon B)H = \begin{bmatrix} \lambda_1 & 1 & & \\ 0 & \lambda_1 & & \\ & & \lambda_3 & \\ & & & \lambda_4 \end{bmatrix} + \epsilon \left[\begin{array}{cc|cc} \beta_{11}/t_1 & \beta_{12}/t_1 & \beta_{13}/t_1 & \beta_{14}/t_1 \\ \beta_{21}/t_2 & \beta_{22}/t_2 & \beta_{23}/t_2 & \beta_{24}/t_2 \\ \hline \beta_{31}/s_3 & \beta_{32}/s_3 & \beta_{33}/s_3 & \beta_{34}/s_3 \\ \beta_{41}/s_4 & \beta_{42}/s_4 & \beta_{43}/s_4 & \beta_{44}/s_4 \end{array} \right],$$

(20.2)

where
$$\beta_{ij} = g_i^T B h_j \quad \text{(all } i,j\text{)}, \qquad |\beta_{ij}| < n \quad (i, j > 3). \tag{20.3}$$

It is natural to ask whether the perturbations in the simple eigenvalues λ_3 and λ_4 are such that

$$\left. \begin{array}{l} |\lambda_3(\epsilon) - \lambda_3| = 0(\epsilon) \\ |\lambda_4(\epsilon) - \lambda_4| = 0(\epsilon) \end{array} \right\} \text{ as } \epsilon \to 0. \tag{20.4}$$

At first sight the presence of the element unity in the Jordan form appears to be a handicap, since the first Gerschgorin disc remains of radius greater than one for all ϵ. Hence if $|\lambda_1 - \lambda_3|$ is less than unity the third disc will not be isolated from the first for any value of ϵ. However, if we multiply the first row of (20.2) by m and the first column by $1/m$ it becomes

$$\begin{bmatrix} \lambda_1 & m & & \\ 0 & \lambda_1 & & \\ & & \lambda_3 & \\ & & & \lambda_4 \end{bmatrix} + \epsilon \begin{bmatrix} \gamma_{11} & \gamma_{12} & \gamma_{13} & \gamma_{14} \\ \gamma_{21} & \gamma_{22} & \gamma_{23} & \gamma_{24} \\ \gamma_{31} & \gamma_{32} & \gamma_{33} & \gamma_{34} \\ \gamma_{41} & \gamma_{42} & \gamma_{43} & \gamma_{44} \end{bmatrix} \text{(say)}, \tag{20.5}$$

where the γ_{ij} are independent of ϵ. The choice

$$m = \min \{ \tfrac{1}{4} |\lambda_1 - \lambda_3|, \ \tfrac{1}{4} |\lambda_1 - \lambda_4| \} \text{ (say)}, \tag{20.6}$$

removes the difficulty. To isolate the third Gerschgorin disc while making its radius of order ϵ^2, we multiply the third column and third row by k/ϵ and ϵ/k respectively, choosing k so that

$$|k\gamma_{i3}| < \tfrac{1}{2} |\lambda_3 - \lambda_i| \quad (i \neq 3). \tag{20.7}$$

Exactly as in § 15 we see that the disc with centre $\lambda_3 + \epsilon\gamma_{33}$ and radius $\epsilon^2(|\gamma_{31}| + |\gamma_{32}| + |\gamma_{34}|)$ is isolated for sufficiently small ϵ and therefore contains just one eigenvalue. Hence the perturbation in the simple eigenvalue λ_3 is such that

$$|\lambda_3(\epsilon) - \lambda_3 - \epsilon\gamma_{33}| = 0(\epsilon^2) \text{ as } \epsilon \to 0. \tag{20.8}$$

Now γ_{33} is, in fact, β_{33}/s_3, and from (20.2) we have $|\beta_{33}| < n$. The presence of the divisor $(\lambda_1 - \lambda)^2$ makes no essential difference to the behaviour of *a simple eigenvalue different from* λ_1. The reader will readily convince himself that this result is perfectly general.

21. Case 4. *Perturbations of the eigenvalues corresponding to a non-linear elementary divisor of a non-derogatory matrix.*

We may denote the non-linear divisor by $(\lambda_1 - \lambda)^r$ and since the matrix is assumed to be non-derogatory there are no other divisors involving $(\lambda_1 - \lambda)$. The matrix of § 20 will serve to illustrate this case. We consider the perturbations in the double eigenvalue $\lambda = \lambda_1$. In (20.2) the element unity is in one of the rows occupied by a λ_1 element. If we are to obtain a Gerschgorin disc with a radius which tends to zero with ϵ, some multiplying factor must be applied to this element. In fact, multiplying column two by $\epsilon^{\frac{1}{2}}$ and row two by $\epsilon^{-\frac{1}{2}}$ the matrix sum becomes

$$\begin{bmatrix} \lambda_1 & \epsilon^{\frac{1}{2}} & & \\ 0 & \lambda_1 & & \\ & & \lambda_3 & \\ & & & \lambda_4 \end{bmatrix} + \begin{bmatrix} \epsilon & \epsilon & \epsilon^{\frac{3}{2}} & \epsilon \\ \epsilon^{\frac{1}{2}} & \epsilon & \epsilon^{\frac{1}{2}} & \epsilon^{\frac{1}{2}} \\ \epsilon & \epsilon^{\frac{3}{2}} & \epsilon & \epsilon \\ \epsilon & \epsilon^{\frac{3}{2}} & \epsilon & \epsilon \end{bmatrix}, \tag{21.1}$$

where we have omitted the constant multipliers in the elements of the second matrix. It is clear that the terms in $\epsilon^{\frac{1}{2}}$ dominate the radii of the Gerschgorin discs. For sufficiently small ϵ there are two eigenvalues in a circle of centre λ_1 having a radius which is of order $\epsilon^{\frac{1}{2}}$. Similarly for a cubic divisor we may, for example, transform the matrix sum

$$\begin{bmatrix} \lambda_1 & 1 & 0 & \\ 0 & \lambda_1 & 1 & \\ 0 & 0 & \lambda_1 & \\ & & & \lambda_4 \end{bmatrix} + \begin{bmatrix} \epsilon & \epsilon & \epsilon & \epsilon \\ \epsilon & \epsilon & \epsilon & \epsilon \\ \epsilon & \epsilon & \epsilon & \epsilon \\ \epsilon & \epsilon & \epsilon & \epsilon \end{bmatrix} \tag{21.2}$$

to one of the form

$$\begin{bmatrix} \lambda_1 & \epsilon^{\frac{1}{3}} & 0 & \\ 0 & \lambda_1 & \epsilon^{\frac{1}{3}} & \\ 0 & 0 & \lambda_1 & \\ & & & \lambda_4 \end{bmatrix} + \begin{bmatrix} \epsilon & \epsilon^{\frac{2}{3}} & \epsilon^{\frac{1}{3}} & \epsilon \\ \epsilon^{\frac{4}{3}} & \epsilon & \epsilon^{\frac{2}{3}} & \epsilon^{\frac{2}{3}} \\ \epsilon^{\frac{5}{3}} & \epsilon^{\frac{4}{3}} & \epsilon & \epsilon^{\frac{4}{3}} \\ \epsilon & \epsilon^{\frac{2}{3}} & \epsilon^{\frac{1}{3}} & \epsilon \end{bmatrix} \qquad (21.3)$$

by multiplying columns two and three by $\epsilon^{\frac{1}{3}}$ and $\epsilon^{\frac{2}{3}}$ and rows two and three by $\epsilon^{-\frac{1}{3}}$ and $\epsilon^{-\frac{2}{3}}$. The Gerschgorin theorems then show that for sufficiently small ϵ there are three eigenvalues in a disc centre λ_1 and radius proportional to $\epsilon^{\frac{1}{3}}$.

The general result for a divisor of degree n is now obvious and the example we gave in § 2 shows that we cannot avoid the factor $\epsilon^{\frac{1}{n}}$ if B is to be a general matrix. Of course there will be specific perturbations of order ϵ for which the perturbations in the eigenvalues are of order ϵ or even zero.

22. *Case* 5. *Perturbations of eigenvalues λ_i when there is more than one divisor involving $(\lambda_i - \lambda)$ and at least one of them is non-linear.*

Before giving general results for this case we consider the simple example of a matrix having the Jordan canonical form

$$\begin{bmatrix} \lambda_1 & 1 & 0 & & \\ 0 & \lambda_1 & 1 & & \\ 0 & 0 & \lambda_1 & & \\ \hline & & & \lambda_1 & 1 \\ & & & 0 & \lambda_1 \end{bmatrix}. \qquad (22.1)$$

It is natural to ask if there will always be two eigenvalues in a disc with centre λ_1 and radius which is $0(\epsilon^{\frac{1}{2}})$ as $\epsilon \to 0$. The answer to this is clearly 'no', for if we place perturbations ϵ in positions (3, 4) and (5, 1) the characteristic equation is

$$(\lambda_1 - \lambda)^5 + \epsilon^2 = 0, \qquad (22.2)$$

so that all perturbations are proportional to $\epsilon^{\frac{2}{5}}$.

We can prove, however, that there is always at least one perturbation which is not of an order of magnitude *greater* than $\epsilon^{\frac{1}{2}}$. For consider all possible perturbations of order ϵ in the elements of the matrix (22.1). We may expand the characteristic equation in powers of $(\lambda_1 - \lambda)$ and from consideration of the determinants involved, it is evident that the constant term is of the form

$$A_2\epsilon^2 + A_3\epsilon^3 + A_4\epsilon^4 + A_5\epsilon^5. \tag{22.3}$$

There can be no term in ϵ or independent of ϵ. Now if the roots of this equation are given by

$$(\lambda - \lambda_i) = p_i, \tag{22.4}$$

we have

$$\prod_{i=1}^{5} p_i = A_2\epsilon^2 + A_3\epsilon^3 + A_4\epsilon^4 + A_5\epsilon^5. \tag{22.5}$$

Hence it is not possible for all the p_i to be of greater order of magnitude than $\epsilon^{\frac{2}{5}}$. The addition of eigenvalues of any multiplicity but distinct from λ_1 clearly does not affect this result.

Perturbations corresponding to the general distribution of non-linear divisors

23. The presence of elementary divisors of higher degree in $(\lambda_1 - \lambda)$ therefore affects the sensitivity even of the eigenvalues corresponding to linear divisors $(\lambda_1 - \lambda)$, but not that of eigenvalues distinct from λ_1. We leave it as an exercise to prove the following two results.

Let the elementary divisors corresponding to λ_1 be

$$(\lambda_1 - \lambda)^{r_1}, (\lambda_1 - \lambda)^{r_2}, ..., (\lambda_1 - \lambda)^{r_s}, \tag{23.1}$$

where

$$r_1 \geqslant r_2 \geqslant ... \geqslant r_s, \qquad \sum r_i = t. \tag{23.2}$$

Then for sufficiently small ϵ the perturbations $p_1, p_2, ..., p_t$ satisfy

$$\prod_{i=1}^{t} p_i = A_s\epsilon^s + A_{s+1}\epsilon^{s+1} + ... + A_t\epsilon^t, \tag{23.3}$$

and hence at least one λ lies in a disc centre λ_1 and radius $K_1\epsilon^{s/t}$ for some value of K_1. On the other hand all the corresponding perturbed eigenvalues lie in a disc centre λ_1 of radius $K_2\epsilon^{1/r_1}$ for some K_2.

Perturbation theory for the eigenvectors from Jordan canonical form

24. The perturbations in the eigenvectors may also be studied by working with the Jordan canonical form. We shall not consider the problem in its full generality but shall restrict ourselves to the case when A has linear elementary divisors. We have therefore

$$H^{-1}AH = \operatorname{diag}(\lambda_i) \tag{24.1}$$

and the columns of H form a complete set of eigenvectors $x_1, x_2, \ldots, x_n$ of A. We denote the 'corresponding' eigenvectors of $(A + \epsilon B)$ by $x_1(\epsilon)$, $x_2(\epsilon), \ldots, x_n(\epsilon)$, and the eigenvectors of $H^{-1}(A + \epsilon B)H$ by $z_1(\epsilon)$, $z_2(\epsilon), \ldots,$ $z_n(\epsilon)$ so that

$$x_i(\epsilon) = H z_i(\epsilon). \tag{24.2}$$

Since the x_i form a complete set of eigenvectors, we have

$$H z_i(\epsilon) = \alpha_1(\epsilon) x_1 + \alpha_2(\epsilon) x_2 + \ldots + \alpha_n(\epsilon) x_n \tag{24.3}$$

for some set of $\alpha_i(\epsilon)$.

To simplify the notation we consider a matrix of order six used in § 17. We have

$$H^{-1}(A + \epsilon B)H = \begin{bmatrix} \lambda_1 & & & & & \\ & \lambda_1 & & & & \\ & & \lambda_1 & & & \\ & & & \lambda_4 & & \\ & & & & \lambda_4 & \\ & & & & & \lambda_6 \end{bmatrix} +$$

$$+ \epsilon \begin{bmatrix} \beta_{11}/s_1 & \beta_{12}/s_1 & \beta_{13}/s_1 & \cdots & \beta_{16}/s_1 \\ \beta_{21}/s_2 & \beta_{22}/s_2 & \beta_{23}/s_2 & \cdots & \beta_{26}/s_2 \\ \cdot & \cdot & \cdot & \cdots & \cdot \\ \beta_{61}/s_6 & \beta_{62}/s_6 & \beta_{63}/s_6 & \cdots & \beta_{66}/s_6 \end{bmatrix}. \tag{24.4}$$

Concentrating on the simple eigenvalue λ_6 for the moment, we see that $z_6(0) = e_6$. Let the corresponding $z_6(\epsilon)$ be normalized so that its largest component is unity. For sufficiently small ϵ this must be the sixth component. For suppose it were the fourth, say, then putting $z_{46}(\epsilon) = 1$ in the relation

$$\lambda_6(\epsilon)[z_{46}(\epsilon)] = \lambda_4 z_{46}(\epsilon) + \frac{\epsilon}{s_4} \sum_{i=1}^{6} \beta_{4i} z_{i6}(\epsilon) \tag{24.5}$$

we have

$$\lambda_6(\epsilon) - \lambda_4 = \frac{\epsilon}{s_4} \sum_{i=1}^{6} \beta_{4i} z_{i6}(\epsilon). \tag{24.6}$$

Now as $\epsilon \to 0$ the left-hand side tends to $\lambda_6 - \lambda_4$ and the right-hand side tends to zero. We therefore have a contradiction and hence, for sufficiently small ϵ, $z_{66}(\epsilon) = 1$.

We now show that the other components are all less than $K\epsilon$ as $\epsilon \to 0$ for some K. For we have

$$\lambda_6(\epsilon)z_{i6}(\epsilon) = \lambda_i z_{i6}(\epsilon) + \frac{\epsilon}{s_i}\sum_{j=1}^{6}\beta_{ij}z_{j6}(\epsilon), \qquad (24.7)$$

and hence

$$|\lambda_6(\epsilon)-\lambda_i|\,|z_{i6}(\epsilon)| \leqslant \frac{\epsilon}{|s_i|}\sum_{j=1}^{6}|\beta_{ij}|, \qquad (24.8)$$

giving

$$|z_{i6}(\epsilon)| \leqslant \frac{2\epsilon\sum\limits_{j=1}^{6}|\beta_{ij}|}{|s_i|\,|\lambda_6-\lambda_i|} \quad (i = 1,...,5) \quad \text{when } |\lambda_6(\epsilon)-\lambda_i| \geqslant \tfrac{1}{2}|\lambda_6-\lambda_i|. \qquad (24.9)$$

Knowing that $z_{i6}(\epsilon)$ $(i = 1,...,5)$ are of order ϵ we can now obtain an improved bound. In fact (24.7) gives

$$\{\lambda_6(\epsilon)-\lambda_i\}z_{i6}(\epsilon) = \frac{\epsilon\beta_{i6}}{s_i} + \frac{\epsilon}{s_i}\sum_{j=1}^{5}\beta_{ij}z_{j6}(\epsilon), \qquad (24.10)$$

and the second term on the right is of order ϵ^2. We have therefore

$$\left|z_{i6}(\epsilon) - \frac{\epsilon\beta_{i6}}{s_i(\lambda_6-\lambda_i)}\right| = 0(\epsilon^2). \qquad (24.11)$$

The result is essentially that of (10.2), but we can now see how to obtain a strict bound for the ϵ^2 term and this is important in numerical examples.

Perturbations of eigenvectors corresponding to a multiple eigenvalue (linear elementary divisors)

25. We cannot prove so much for eigenvectors corresponding to a multiple eigenvalue. For if $z_5(\epsilon)$ is the normalized eigenvector of $H^{-1}(A+\epsilon B)H$ corresponding to $\lambda_5(\epsilon)$ we can show as in § 24 that the maximum element of $z_5(\epsilon)$ cannot be one of the elements 1, 2, 3 or 6, and indeed that these elements are of order ϵ. The normalized $z_5(\epsilon)$ must therefore take one of the forms

$$[z_{15}(\epsilon), z_{25}(\epsilon), z_{35}(\epsilon), k(\epsilon), 1, z_{65}(\epsilon)] \qquad (25.1)$$

and

$$[z_{15}(\epsilon), z_{25}(\epsilon), z_{35}(\epsilon), 1, k(\epsilon), z_{65}(\epsilon)], \qquad (25.2)$$

where $|k(\epsilon)| \leqslant 1$. The corresponding eigenvector $x_5(\epsilon)$ of $(A+\epsilon B)$ has components of order ϵ in the directions x_1, x_2, x_3, x_6, but the components in the subspace spanned by x_4 and x_5 are either $x_4+k(\epsilon)x_5$ or

$k(\epsilon)x_4 + x_5$. We cannot expect to improve on this since any vector in this subspace is indeed an eigenvector of A.

Limitations of perturbation theory

26. Although the perturbation theory we have developed in the last twelve sections is of very great practical value it is salutary at this stage to draw attention to some shortcomings. We have concentrated on the nature of the perturbations as $\epsilon \to 0$, but consider the simple matrix

$$\begin{bmatrix} a & 10^{-10} \\ 0 & a \end{bmatrix}. \tag{26.1}$$

This matrix certainly has the non-linear elementary divisor $(a-\lambda)^2$, and if we add a perturbation ϵ in the $(2, 1)$ element the eigenvalues become $a \pm (10^{-10}\epsilon)^{\frac{1}{2}}$. The derivatives with respect to ϵ of the eigenvalues therefore tend to infinity as ϵ tends to zero. However, if we are interested in perturbations of the order of magnitude 10^{-10} this is irrelevant. If the perturbation matrix is

$$\begin{bmatrix} \epsilon_1 & \epsilon_2 \\ \epsilon_3 & \epsilon_4 \end{bmatrix}, \tag{26.2}$$

it is much more natural to think of this problem in terms of a perturbation

$$\begin{bmatrix} 0 & 10^{-10} \\ 0 & 0 \end{bmatrix} + \begin{bmatrix} \epsilon_1 & \epsilon_2 \\ \epsilon_3 & \epsilon_4 \end{bmatrix} \tag{26.3}$$

of the matrix

$$\begin{bmatrix} a & 0 \\ 0 & a \end{bmatrix}, \tag{26.4}$$

and this last matrix no longer has a non-linear divisor.

Turning now to the eigenvectors, we have resolved their perturbations into their components in the directions $x_1, x_2, \ldots, x_n$. When the x_i are orthogonal this is very satisfactory, but if some of the x_i are 'almost' linearly dependent then we may have large components in the directions of individual x_i even when the perturbation vector is itself quite small. We have seen in § 10 that the perturbation corresponding to a simple eigenvalue λ_1 is given by

$$\epsilon \sum_{i=2}^{n} \frac{\beta_{i1}}{s_i(\lambda_1 - \lambda_i)} x_i + O(\epsilon^2) \tag{26.5}$$

and hence we might expect, in general, that if s_i is small there would be a large component in the direction of x_i. Just how misleading this can be, will be seen if we consider the matrix

$$\begin{bmatrix} 2 & 0 & 0 \\ 0 & 1 & 1 \\ 0 & 0 & 1+10^{-10} \end{bmatrix}. \tag{26.6}$$

It is very easy to show that

$$s_1 = 1, \quad s_2 = -10^{-10}/(1+10^{-20})^{\frac{1}{2}}, \quad s_3 = 10^{-10}/(1+10^{-20})^{\frac{1}{2}}. \tag{26.7}$$

However the vector x_1 corresponding to $\lambda = 2$ is by no means sensitive to perturbations in the matrix. The two vectors x_2 and x_3 are almost identical (as are also, y_2 and y_3) and the factors $\beta_{21}/s_2(\lambda_1-\lambda_2)$ and $\beta_{31}/s_3(\lambda_1-\lambda_3)$ are almost equal and opposite. Hence the perturbations in the directions x_2 and x_3 almost cancel.

Relationships between the s_i

27. In this example $1/s_2$ and $1/s_3$ were both very large but were equal and opposite. We now show that the s_i are interdependent in a way which precludes the possibility of just one of the $1/s_i$ being large. If we write

$$x_i = \sum_{j=1}^{n} \alpha_{ij} y_j, \quad y_i = \sum_{j=1}^{n} \beta_{ij} x_j, \quad \text{where} \quad \|x_j\|_2 = \|y_j\|_2 = 1, \tag{27.1}$$

then we have

$$\alpha_{ij} = x_j^T x_i / x_j^T y_j = x_j^T x_i / s_j, \quad \beta_{ij} = y_j^T y_i / y_j^T x_j = y_j^T y_i / s_j \tag{27.2}$$

$$y_i^T x_i = \sum_{j=1}^{n} \beta_{ij} x_j^T \sum_{j=1}^{n} \alpha_{ij} y_j = \sum_{j=1}^{n} \beta_{ij} \alpha_{ij} s_j = \sum_{j=1}^{n} (x_j^T x_i)(y_j^T y_i) s_j^{-1}, \tag{27.3}$$

giving

$$s_i = s_i^{-1} + \sum_{j \neq i} (\cos \theta_{ij} \cos \varphi_{ij}) s_j^{-1}, \tag{27.4}$$

where θ_{ij} is the angle between x_i and x_j and φ_{ij} the angle between y_i and y_j. Equation (27.4) holds for each value of i. From it we deduce

$$|s_i^{-1}| \leqslant |s_i| + \sum_{j \neq i} |(\cos \theta_{ij} \cos \varphi_{ij}) s_j^{-1}| \leqslant 1 + \sum_{j \neq i} |s_j^{-1}|. \tag{27.5}$$

This shows, for example, that we cannot have a set of s_i for a matrix of order 3 satisfying

$$|s_1^{-1}| = 10, \quad |s_2^{-1}| = 10^7, \quad |s_3^{-1}| = 10^8,$$

since

$$|s_3^{-1}| \gg 1 + |s_1^{-1}| + |s_2^{-1}|.$$

The condition of a computing problem

28. We shall describe any computing problem as *ill-conditioned* if
the values to be computed are very sensitive to small changes in the
data. The considerations of the previous sections have already given
considerable insight into the factors which determine the sensitivity
of the eigensystem of a matrix. It is evident that a matrix may have
some eigenvalues which are very sensitive to small changes in its
elements while others are comparatively insensitive. Similarly some
of the eigenvectors may be well-conditioned while others are ill-
conditioned, while an eigenvector may be ill-conditioned when the
corresponding eigenvalue is not.

The considerations which determine whether a matrix is ill-
conditioned with respect to the solution of the eigenproblem are
different from those which determine its condition with respect
to the computation of its inverse. As we shall see in Chapter 4, § 11,
a normalized matrix must certainly be regarded as ill-conditioned
in the latter context if it has a very small eigenvalue, but this
is irrelevant to the condition of the eigenproblem. By a suitable choice of k we may make $(A-kI)$ exactly singular, but the
sensitivity of the eigensystem of $(A-kI)$ is the same as that
of the eigensystem of A. However, we shall show that there is
some connexion between the conditions of the two computing
problems.

Condition numbers

29. It is convenient to have some number which defines the condition of a matrix with respect to a computing problem and to call
such a number a 'condition number'. Ideally it should give some
'overall assessment' of the rate of change of the solution with respect
to changes in the coefficients and should therefore be in some way
proportional to this rate of change.

It is evident from what we have said in § 28 that even if we restrict
ourselves to the problem of computing the eigenvalues alone, then
such a single number would have severe limitations. If any one of
the eigenvalues were very sensitive then the condition number
would have to be large, even if some other eigenvalues were very
insensitive. In spite of its obvious limitations a single condition
number has been widely used for matrices with linear elementary
divisors.

Spectral condition number of A with respect to its eigenproblem

30. Let A be a matrix having linear elementary divisors and H be a matrix such that

$$H^{-1}AH = \mathrm{diag}(\lambda_i). \qquad (30.1)$$

If λ is an eigenvalue of $(A+\epsilon B)$ then the matrix $(A+\epsilon B-\lambda I)$ is singular, and hence its determinant is zero. We have then

$$H^{-1}(A+\epsilon B-\lambda I)H = \mathrm{diag}(\lambda_i-\lambda)+\epsilon H^{-1}BH, \qquad (30.2)$$

and taking determinants we see that the matrix on the right of (30.2) must also be singular. We distinguish two cases.

Case 1. $\lambda = \lambda_i$ for some i.

Case 2. $\lambda \neq \lambda_i$ for any i, so that we may write

$$\mathrm{diag}(\lambda_i-\lambda)+\epsilon H^{-1}BH = \mathrm{diag}(\lambda_i-\lambda)[I+\epsilon\, \mathrm{diag}(\lambda_i-\lambda)^{-1}H^{-1}BH],$$
$$\qquad (30.3)$$

and taking determinants again we see that the matrix in brackets must be singular. Now if $(I+X)$ is singular we must have $\|X\| \geqslant 1$ for all norms, for if $\|X\| < 1$ none of the eigenvalues of $(I+X)$ can be zero. Hence we have, in particular,

$$\|\epsilon\, \mathrm{diag}(\lambda_i-\lambda)^{-1}H^{-1}BH\|_2 \geqslant 1, \qquad (30.4)$$

giving

$$\epsilon \max |(\lambda_i-\lambda)^{-1}|\, \|H^{-1}\|_2\, \|B\|_2\, \|H\|_2 \geqslant 1, \qquad (30.5)$$

i.e.

$$\min |\lambda_i-\lambda| \leqslant \epsilon\, \|H^{-1}\|_2\, \|H\|_2\, \|B\|_2. \qquad (30.6)$$

Hence in either case

$$|\lambda_i-\lambda| \leqslant \epsilon\kappa(H)\, \|B\|_2 \quad \text{for at least one value of } i, \qquad (30.7)$$

where

$$\kappa(H) = \|H^{-1}\|_2\, \|H\|_2. \qquad (30.8)$$

Because of its dependence on the spectral norm $\kappa(H)$ is often called *the spectral condition number* (cf. Chapter 4, § 3). We have shown that the overall sensitivity of the eigenvalues of A is dependent on the size of $\kappa(H)$, so that $\kappa(H)$ may be regarded as a condition number of A with respect to its eigenvalue problem. The result we have just proved is due to Bauer and Fike (1960).

Note that (30.6) is true for any norm for which

$$\|\mathrm{diag}(\lambda_i-\lambda)^{-1}\| = \max |\lambda_i-\lambda|^{-1}, \qquad (30.9)$$

and hence it is also true for the 1-norm and ∞-norm. Since it is true of the 2-norm it is true *a fortiori* for the Euclidean norm.

Using the concept of continuity we may localize the roots more accurately by the same method as we used to prove Gerschgorin's second theorem. This leads to the following result. If s of the circular discs

$$|\lambda_i - \lambda| < \epsilon \kappa(H) \, \|B\|_2 \tag{30.10}$$

form a connected domain which is isolated from the remainder, then there are precisely s eigenvalues in this domain. *Note that we do not require ϵ to be small in these results.*

Properties of spectral condition number

31. Since the matrix H is not unique (even if the eigenvalues are distinct, each column may be multiplied by an arbitrary factor), we define the spectral condition number of A with respect to its eigenvalue problem to be the smallest value of $\kappa(H)$ for all permissible H. We have in any case

$$\kappa(H) = \|H^{-1}\|_2 \, \|H\|_2 \geqslant \|H^{-1}H\|_2 = 1. \tag{31.1}$$

If A is normal (Chapter 1 § 48) and therefore if in particular it is Hermitian or unitary, we may take H to be unitary and we then have

$$\kappa(H) = 1. \tag{31.2}$$

The eigenvalue problem is therefore always well-conditioned for a normal matrix, though this is not necessarily true of the eigenvector problem.

We consider now the relationship of $\kappa(H)$ to the quantities $|s_i^{-1}|$ which govern the sensitivity of the individual eigenvalues. Normalized right-hand and left-hand eigenvectors corresponding to λ_i are given by

$$x_i = He_i/\|He_i\|_2, \qquad y_i = (H^{-1})^T e_i/\|(H^{-1})^T e_i\|_2. \tag{31.3}$$

Hence from its definition

$$|s_i| = |y_i^T x_i| = \frac{|e_i^T H^{-1} He_i|}{\|He_i\|_2 \, \|(H^{-1})^T e_i\|_2} = \frac{1}{\|He_i\|_2 \, \|(H^{-1})^T e_i\|_2}, \tag{31.4}$$

and we have

$$\|He_i\|_2 \leqslant \|H\|_2 \, \|e_i\|_2 = \|H\|_2, \tag{31.5}$$

$$\|(H^{-1})^T e_i\| \leqslant \|(H^{-1})^T\|_2 \, \|e_i\|_2 = \|(H^{-1})^T\|_2 = \|H^{-1}\|_2, \tag{31.6}$$

giving

$$|s_i^{-1}| \leqslant \kappa(H). \tag{31.7}$$

On the other hand we may take the columns of H to be $x_i/s_i^{\frac{1}{2}}$ and the rows of H^{-1} to be $y_i^T/s_i^{\frac{1}{2}}$, and with this choice of H

$$\kappa(H) = \|H\|_2 \|H^{-1}\|_2 \leqslant \|H\|_E \|H^{-1}\|_E = \left(\sum_1^n |s_r^{-1}|\right)^{\frac{1}{2}} \left(\sum_1^n |s_r^{-1}|\right)^{\frac{1}{2}} = \sum_1^n |s_r^{-1}|.$$

$$(31.8)$$

For a full knowledge of the sensitivity of the eigenvalues of A with respect to perturbations in its elements we require the n^3 quantities $\partial\lambda_i/\partial a_{kj}$, since the sensitivity of the individual eigenvalues with respect to changes in the individual elements of A may vary very widely. Such a large aggregate is clearly impossible for practical work. A reasonable compromise is provided by the n numbers $|s_i^{-1}|$, and we shall refer to these as the n *condition numbers of A* with respect to the eigenvalue problem. It is perhaps worth mentioning that in order to find an approximate value of $\kappa(H)$ in practice we are usually obliged to compute a set of approximate eigenvectors. Having obtained these it is simpler to obtain approximations to the individual s_i than to obtain an approximation to $\kappa(H)$.

Invariant properties of condition numbers

32. Both κ and the n condition numbers $|s_i^{-1}|$ have important invariant properties with respect to unitary similarity transformations. For if R is unitary and

$$B = RAR^H, \tag{32.1}$$

then corresponding to

$$H^{-1}AH = \text{diag}(\lambda_i) \tag{32.2}$$

we have

$$H^{-1}R^H BRH = \text{diag}(\lambda_i) \quad \text{or} \quad (RH)^{-1}B(RH) = \text{diag}(\lambda_i).$$

$$(32.3)$$

Hence the spectral condition number κ' of B is given by

$$\kappa' = \|(RH)^{-1}\|_2 \|RH\|_2 = \|H^{-1}\|_2 \|H\|_2 = \kappa. \tag{32.4}$$

Similarly if x_i and y_i are right-hand and left-hand eigenvectors of A then those of B are given by

$$x_i' = Rx_i, \qquad y_i' = Ry_i,$$

and hence

$$s_i' = (y_i')^T x_i' = y_i^T \bar{R}^T R x_i = y_i^T R^H R x_i = y_i^T x_i = s_i. \tag{32.5}$$

Hence the sensitivities of the individual eigenvalues are invariant under unitary transformations.

Very ill-conditioned matrices

33. The eigenvalues corresponding to non-linear elementary divisors must, in general, be regarded as ill-conditioned though we must bear in mind the considerations of § 26. However we must not be misled into thinking that this is the main form of ill-conditioning. Even if the eigenvalues are distinct and well separated they may still be very ill-conditioned. This is well illustrated by the following example.

Consider the matrix A of order 20 defined by

$$A = \begin{bmatrix} 20 & 20 & & & & & \\ & 19 & 20 & & & & \\ & & 18 & 20 & & & \\ & & & \cdot & \cdot & \cdot & \cdot \\ & & & & & 2 & 20 \\ & & & & & & 1 \end{bmatrix}. \tag{33.1}$$

This is a triangular matrix and its eigenvalues are therefore its diagonal elements. If the element ϵ is added in position (20, 1) then the characteristic equation becomes

$$(20-\lambda)(19-\lambda)...(1-\lambda) = 20^{19}\epsilon. \tag{33.2}$$

We have seen in § 9 that for sufficiently small ϵ the perturbation of the eigenvalue $\lambda = r$ does not involve fractional powers of ϵ, and if we write

$$\lambda_r(\epsilon) - r \sim K_r\epsilon, \tag{33.3}$$

it is evident from (33.2) that K_r is given by

$$K_r = 20^{19}(-1)^r/(20-r)! \, (r-1)!. \tag{33.4}$$

This constant is large for all values of r. The smallest values are K_1 and K_{20} and the largest are K_{10} and K_{11}. We have in fact

$$-K_1 = K_{20} \doteq 10^7(4\cdot31), \qquad -K_{11} = K_{10} \doteq 10^{12}(3\cdot98). \tag{33.5}$$

The K_i are so large that the linearized theory does not become applicable until ϵ is very small. For $\epsilon = 10^{-10}$ the eigenvalues are displayed in Table 1, in a way that illustrates symmetry about the value $10\cdot5$.

TABLE 1

Eigenvalues of perturbed matrix ($\epsilon = 10^{-10}$)

0·99575439	3·96533070 $\pm i$ 1·08773570
20·00424561	17·03466930 $\pm i$ 1·08773570
2·10924184	5·89397755 $\pm i$ 1·94852927
18·89075816	15·10602245 $\pm i$ 1·94852927
2·57488140	8·11807338 $\pm i$ 2·52918173
18·42511860	12·88192662 $\pm i$ 2·52918173
	10·50000000 $\pm i$ 2·73339736

34. The changes in the eigenvalues corresponding to this single perturbation are as great as the changes in the eigenvalues of $C_{10}(a)$ for a similar perturbation, although the latter matrix has an elementary divisor of degree ten. The fact that the perturbations are ultimately of order ϵ for the former and $\epsilon^{\frac{1}{10}}$ for the latter, as $\epsilon \to 0$, does not become important until we consider perturbations much smaller than 10^{-10}.

The sensitivity of the eigenvalues of the matrix given by (33.1) shows that the s_i must be small. In fact the eigenvector x_r of A corresponding to $\lambda = r$ has components

$$\left[1; \quad \frac{20-r}{-20}; \quad \frac{(20-r)(19-r)}{(-20)^2}; \quad \ldots; \quad \frac{(20-r)!}{(-20)^{20-r}}; \quad 0; \quad \ldots; \quad 0 \right], \quad (34.1)$$

while that of y_r has components

$$\left[0; \quad \ldots; \quad 0; \quad \frac{(r-1)!}{20^{r-1}}; \quad \ldots; \quad \frac{(r-1)(r-2)}{20^2}; \quad \frac{r-1}{20}; \quad 1 \right]. \quad (34.2)$$

These vectors are not quite 2-normalized vectors but $y_r^T x_r$ does give a good estimate of the magnitude of s_r. We have in fact

$$|y_r^T x_r| = (20-r)! \, (r-1)!/20^{19}, \quad (34.3)$$

and the reciprocal of the right-hand of (34.3) is precisely the $|K_r|$ of equation (33.4). We observe that the corresponding eigenvectors of A and A^T are almost orthogonal. Although strict orthogonality is found only with non-linear divisors, we can clearly come very close to it even in a matrix which has well separated eigenvalues.

It might be expected that there is a matrix close to A which has some non-linear elementary divisors and we can readily see that this is true. We can in fact produce such a matrix by means of perturbations in the (20, 1) element. Corresponding to a perturbation ϵ the eigenvalues are the roots of equation (33.2). Now the function on the left-hand side of this equation is symmetric about $\lambda = 10\cdot5$ and has 9 maxima and 10 minima. If we increase ϵ starting from 0 then the roots 10 and 11 move together and ultimately coincide at $10\cdot5$ when

$$(\tfrac{1}{2}\,\tfrac{3}{2}\ldots\tfrac{19}{2})^2 = 20^{19}\epsilon, \qquad (34.4)$$

giving a value of ϵ of approximately $7\cdot8 \times 10^{-14}$.

The corresponding divisor must be quadratic since the perturbed matrix $(A - \lambda I)$ is of rank 19 for any λ, the matrix of order 19 in the upper right-hand corner having determinant 20^{19}. If ϵ is increased further the roots 8 and 9 and 12 and 13 move together and from symmetry they become coincident for the same value of ϵ. The matrix then has two double eigenvalues and both correspond to quadratic divisors. Similarly increasing ϵ further we may bring the roots 6 and 7 and 14 and 15, respectively, together and so on.

35. All the eigenvalues of the matrix we have just considered were ill-conditioned though some were more ill-conditioned than others. We now give an example of a set of matrices each of which has eigenvalues of widely varying condition. The matrices are those of the class B_n defined by

$$
B_n =
\begin{bmatrix}
n & (n-1) & (n-2) & \ldots & 3 & 2 & 1 \\
(n-1) & (n-1) & (n-2) & \ldots & 3 & 2 & 1 \\
 & (n-2) & (n-2) & \ldots & 3 & 2 & 1 \\
 & & & \cdot & \cdot & \cdot & \cdot \\
 & & & & 2 & 2 & 1 \\
 & & & & & 1 & 1
\end{bmatrix}. \qquad (35.1)
$$

The largest eigenvalues of this matrix are very well-conditioned and the smallest very ill-conditioned. For $n = 12$ the first few s_i are of order unity while the last three are of order 10^{-7}. With increasing values of n the smallest eigenvalues become progressively worse conditioned.

That some of the eigenvalues must be very sensitive to small changes in some matrix elements is evident from the following consideration. The determinant of the matrix is unity for all values of n, as can be shown by simple operations on the rows of the matrix. If the $(1, n)$ element is changed to $(1+\epsilon)$ the determinant becomes

$$1 \pm (n-1)!\,\epsilon. \qquad (35.2)$$

If $\epsilon = 10^{-10}$ and $n = 20$ the determinant is changed from 1 to $(1-19!\,10^{-10})$, that is, approximately $-1\cdot 216 \times 10^7$. Now the determinant of a matrix is equal to the product of its eigenvalues. At least one eigenvalue of the perturbed matrix must be very different from that of the original. Again, when $n = 12$ a perturbation of less than 10^{-9} in the $(1, 12)$ element produces a quadratic divisor.

Ill-conditioning of the type we have just discussed is, in our experience, far more important than that associated directly with non-linear elementary divisors. Matrices having exact non-linear divisors are almost non-existent in practical work. Even in theoretical work they are mainly associated with matrices of a very special form usually with small integer coefficients. If the elements of such a matrix are irrational or cannot be represented on the digital computer which is being used, rounding errors will usually lead to a matrix which no longer has non-linear elementary divisors. Matrices with some small s_i, on the other hand, are quite common.

Perturbation theory for real symmetric matrices

36. The next few sections are devoted to an examination of perturbation theory for real symmetric matrices. This theory is simpler than for general matrices because the two sets of vectors x_i and y_i may be taken to be identical (even if there are some multiple eigenvalues) and hence all the s_i are unity. As we have remarked already in § 31, the eigenvalue problem for real symmetric matrices is always well-conditioned. Some of the results we shall prove in the following sections extend immediately in an obvious way to the whole class of normal matrices, but others extend only to Hermitian matrices or not at all. We have indicated the relevant extension in each case without proof.

Unsymmetric perturbations

37. We shall sometimes be interested in unsymmetric perturbations of a symmetric matrix and accordingly we examine the eigensystem of $(A + \epsilon B)$ where A is symmetric but B is not. From equation (30.6) we know that every eigenvalue of $(A + \epsilon B)$ lies in at least one of the circular discs given by

$$|\lambda_i - \lambda| < \epsilon \, \|H^{-1}\|_2 \, \|H\|_2 \, \|B\|_2, \tag{37.1}$$

and since H may be taken to be orthogonal, these reduce to

$$|\lambda_i - \lambda| < \epsilon \, \|B\|_2 < n\epsilon \qquad \text{(if } |b_{ij}| < 1\text{)}. \tag{37.2}$$

We know further that if any s of these discs form a connected domain isolated from the others, then there are precisely s eigenvalues in this domain. These results are obviously true for all normal matrices.

In the case of real symmetric matrices, we shall usually be interested in real B and hence any complex eigenvalues of $(A + \epsilon B)$ must occur as conjugate pairs. If $|\lambda_i - \lambda_j| > 2n\epsilon$ $(j \neq i)$, then the ith circular disc is isolated and contains only one eigenvalue. This eigenvalue must therefore be real. Hence if all the eigenvalues of A are separated by more than $2n\epsilon$, random unsymmetric perturbations of up to ϵ in modulus in individual elements still leave the eigenvalues real and simple and hence the eigenvectors are also real and form a complete system.

Symmetric perturbations

38. Most of the more accurate techniques for computing eigensystems of real symmetric matrices are based on the use of orthogonal similarity transformations. Exact symmetry is preserved in the successive transformed matrices by computing only their upper triangles and completing the matrix by taking elements below the diagonal to be equal to corresponding elements above the diagonal. For this reason we shall most commonly be concerned with symmetric perturbations of symmetric matrices.

When the perturbations are symmetric, the perturbed matrix necessarily has a real eigensystem and the same is true of the matrix of perturbations. It is natural to seek relations between the eigenvalues of the original matrix, the perturbation matrix and the perturbed matrix. Most of the results we have obtained hitherto have been weak unless the perturbations were small. We shall find on the other hand that many of the results we are about to prove are not subject to such limitations. We therefore dispense with the ϵ and write

$$C = A + B, \tag{38.1}$$

where A, B and hence C are real and symmetric.

Classical techniques

39. There are extremely effective techniques, based on the minimax principle, for investigating the eigenvalues of the sum of two symmetric matrices. Partly in order to emphasize their power we start with an analysis based on more classical lines.

We first derive a simple result for a bordered diagonal matrix. Let the symmetric matrix X be defined by

$$X = \left[\begin{array}{c|c} \alpha & a^T \\ \hline a & \mathrm{diag}(\alpha_i) \end{array}\right] \quad (i = 1,\dots, n-1), \qquad (39.1)$$

where a is a vector of order $(n-1)$. We wish to relate the eigenvalues of X with the α_i.

Suppose that only s of the components of a are non-zero; if a_j is zero, then α_j is an eigenvalue of X. By a suitable choice of a permutation matrix P involving the last $(n-1)$ rows only, we may derive a matrix Y such that

$$Y = P^T X P = \left[\begin{array}{c|c|c} \alpha & b^T & O \\ \hline b & \mathrm{diag}(\beta_i) & O \\ \hline O & O & \mathrm{diag}(\gamma_i) \end{array}\right], \qquad (39.2)$$

where no component of b is zero, $\mathrm{diag}(\beta_i)$ is of order s, $\mathrm{diag}(\gamma_i)$ is of order $(n-1-s)$ and the β_i and γ_i together are a rearrangement of the α_i. Note that included in the γ_i there may be some or all of the multiple eigenvalues, if any, of $\mathrm{diag}(\alpha_i)$.

The eigenvalues of X are therefore the γ_i together with those of the matrix Z defined by

$$Z = \left[\begin{array}{c|c} \alpha & b^T \\ \hline b & \mathrm{diag}(\beta_i) \end{array}\right]. \qquad (39.3)$$

If $s = 0$, Z is the single element α and hence the eigenvalues of X are those of $\mathrm{diag}(\alpha_i)$ and the value α. Otherwise let us examine the characteristic polynomial of Z which is

$$(\alpha - \lambda) \prod_{i=1}^{s} (\beta_i - \lambda) - \sum_{j=1}^{s} b_j^2 \prod_{i \neq j} (\beta_i - \lambda) = 0. \qquad (39.4)$$

Suppose there are only t distinct values among the β_i which, without loss of generality, we may take to be $\beta_1, \beta_2, \dots, \beta_t$, and that these are of multiplicities $r_1, r_2, \dots, r_t$ respectively, so that

$$r_1 + r_2 + \dots + r_t = s. \qquad (39.5)$$

(We may, of course, have $t = s$, all the β_i being simple.) Then clearly the left-hand side of (39.4) has the factor

$$\prod_{i=1}^{t} (\beta_i - \lambda)^{r_i - 1}, \qquad (39.6)$$

so that β_i is an eigenvalue of Z of multiplicity $(r_i - 1)$.

Dividing (39.4) by $\prod\limits_{i=1}^{t} (\beta_i - \lambda)^{r_i}$ we see that the remaining eigenvalues of Z are the roots of

$$0 = (\alpha - \lambda) - \sum_1^t c_i^2 (\beta_i - \lambda)^{-1} \equiv \alpha - f(\lambda) \quad \text{say,} \qquad (39.7)$$

where each c_i^2 is the sum of the r_i values b_j^2 associated with β_i and is therefore strictly positive. A graph of $f(\lambda)$ against λ is given in Fig. 1, where we have assumed that distinct β_i are in *decreasing* order. It is

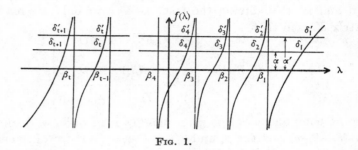

Fig. 1.

immediately evident that the $t+1$ roots of $\alpha = f(\lambda)$ which we denote by $\delta_1, \delta_2, \ldots, \delta_{t+1}$ satisfy the relations

$$\infty > \delta_1 > \beta_1; \quad \beta_{i-1} > \delta_i > \beta_i \quad (i = 2, 3, \ldots, t); \quad \beta_t > \delta_{t+1} > -\infty.$$
$$(39.8)$$

The n eigenvalues of X therefore fall into three sets.

(i) The eigenvalues $\gamma_1, \gamma_2, \ldots, \gamma_{n-1-s}$ corresponding to the zero a_i. These are equal to $n-1-s$ of the α_i.

(ii) $s-t$ eigenvalues consisting of $r_i - 1$ values equal to β_i ($i = 1, 2, \ldots, t$). These are equal to a further $s-t$ of the α_i.

(iii) $t+1$ eigenvalues equal to δ_i satisfying the relations (39.8). If $t = 0$ then $\delta_1 = \alpha$.

Note that either or both of the sets (i) and (ii) may be empty. *In any case the members of these two sets are independent of α.* If the eigenvalues of X be denoted by λ_i ($i = 1, 2, \ldots, n$) where the arrangement is in non-increasing order, then it is an immediate consequence of our enumeration of the λ_i above that if the α_i are also arranged in non-increasing order then

$$\lambda_1 > \alpha_1 > \lambda_2 > \alpha_2 > \ldots > \alpha_{n-1} > \lambda_n. \qquad (39.9)$$

In other words the α_i separate the λ_i at least in the weak sense.

40. Consider now the eigenvalues of X' derived from X by replacing α by α'. The eigenvalues of X' will be equal to those of X as far as sets (i) and (ii) are concerned. Let us denote those in set (iii) by $\delta_1', \delta_2', ..., \delta_{t+1}'$. Now for $\lambda > 0$ we have

$$\frac{df}{d\lambda} = 1 + \sum_1^t c_i^2 / (\beta_i - \lambda)^2 > 1 \quad \text{(all } \lambda\text{)}, \tag{40.1}$$

and hence each $\delta_i' - \delta_i$ lies between 0 and $\alpha' - \alpha$ (see Fig. 1). We may therefore write

$$\delta_i' - \delta_i = m_i(\alpha' - \alpha), \tag{40.2}$$

where

$$0 < m_i < 1 \quad \text{and} \quad \sum m_i = 1. \tag{40.3}$$

If $t = 0$ then $\delta_1' = \alpha'$ and $\delta_1 = \alpha$ and

$$\delta_1' - \delta_1 = \alpha' - \alpha. \tag{40.4}$$

Hence we may write in all cases

$$\delta_i' - \delta_i = m_i(\alpha' - \alpha), \tag{40.5}$$

where

$$0 < m_i < 1, \quad \sum m_i = 1. \tag{40.6}$$

Since the other eigenvalues of X and X' are equal, we may say that we have established a correspondence between the n eigenvalues of X and X', which we rename $\lambda_1, \lambda_2, ..., \lambda_n$ and $\lambda_1', \lambda_2', ..., \lambda_n'$ such that

$$\left.\begin{array}{c} \lambda_i' - \lambda_i = m_i(\alpha' - \alpha) \\ 0 \leqslant m_i \leqslant 1, \quad \sum m_i = 1 \end{array}\right\}, \tag{40.7}$$

where

where we obviously have $m_i = 0$ for the eigenvalues from sets (i) and (ii).

Moreover if the relations (40.7) are satisfied by the λ_i and λ_i' with some ordering, they are satisfied *a fortiori* when the λ_i and λ_i' are each arranged in non-increasing order.

Symmetric matrix of rank unity

41. The result of the previous paragraph may be applied to assess the eigenvalues of C defined by

$$C = A + B, \tag{41.1}$$

where A and B are symmetric and B is of rank unity. In this case there exists an orthogonal matrix R such that

$$R^T B R = \left[\begin{array}{c|c} \rho & O \\ \hline O & O \end{array}\right], \tag{41.2}$$

where ρ is the unique non-zero eigenvalue of B. If we write

$$R^T A R = \left[\begin{array}{c|c} \alpha & a^T \\ \hline a & A_{n-1} \end{array}\right], \tag{41.3}$$

then there is an orthogonal matrix S of order $(n-1)$ such that

$$S^T A_{n-1} S = \text{diag}(\alpha_i), \tag{41.4}$$

and if we define Q by the relation

$$Q = R \left[\begin{array}{c|c} 1 & O \\ \hline O & S \end{array}\right] \tag{41.5}$$

then Q is orthogonal and

$$Q^T(A+B)Q = \left[\begin{array}{c|c} \alpha & b^T \\ \hline b & \text{diag}(\alpha_i) \end{array}\right] + \left[\begin{array}{c|c} \rho & O \\ \hline O & O \end{array}\right], \tag{41.6}$$

where $b = S^T a$. The eigenvalues of A and of $(A + B)$ are therefore those of

$$\left[\begin{array}{c|c} \alpha & b^T \\ \hline b & \text{diag}(\alpha_i) \end{array}\right] \quad \text{and} \quad \left[\begin{array}{c|c} \alpha+\rho & b^T \\ \hline b & \text{diag}(\alpha_i) \end{array}\right], \tag{41.7}$$

and if we denote these eigenvalues by λ_i and λ_i' in decreasing order, then they satisfy the relations

$$\lambda_i' - \lambda_i = m_i \rho$$
$$\left. \right\} . \tag{41.8}$$
where $\qquad 0 < m_i < 1, \quad \sum m_i = 1$

Hence when B is added to A, all eigenvalues of the latter are shifted by an amount which lies between zero and the eigenvalue ρ of B.

Incidentally, from the remark at the end of § 40 and the transform of A given in (41.7) it is evident that the eigenvalues of the principal minor A_{n-1} of A separate those of A at least in the weak sense. If A has an eigenvalue λ_i of multiplicity k, then A_{n-1} must have an eigenvalue λ_i of multiplicity $k-1$, k, or $k+1$.

Extremal properties of eigenvalues

42. Before presenting the minimax characterization of the eigenvalues we consider the problem of determining the maximum value M of the function $x^T A x$ subject to the restriction $x^T x = 1$. Clearly M also satisfies the relation

$$M = \max_{x \neq 0}(x^T A x / x^T x). \tag{42.1}$$

Since A is a symmetric matrix there is an orthogonal matrix R such that

$$R^T A R = \text{diag}(\lambda_i).\tag{42.2}$$

If we introduce new variables y defined by

$$x = Ry \quad \text{i.e.} \quad y = R^T x\tag{42.3}$$

then $\quad x^T A x = y^T R^T A R y = y^T \text{diag}(\lambda_i) y = \sum_1^n \lambda_i y_i^2,\tag{42.4}$

$$x^T x = y^T R^T R y = \sum_1^n y_i^2.\tag{42.5}$$

Now since (42.3) establishes a one-to-one correspondence between x and y, the original problem is equivalent to finding the maximum value of $\sum_1^n \lambda_i y_i^2$ subject to the condition $\sum_1^n y_i^2 = 1$. If we assume that the λ_i are in non-increasing order, then clearly the maximum value is λ_1, this being attained when $y = e_1$. The corresponding vector x is r_1, the first column of R which is, in fact, an eigenvector of A corresponding to λ_1. (Note that if $\lambda_1 = \lambda_2 = \ldots = \lambda_r \neq \lambda_{r+1}$, then any unit vector y in the subspace spanned by $e_1, e_2, \ldots, e_r$ gives the value λ_1.) Similarly λ_n is the minimum value of $x^T A x$ under the same condition.

Let us now consider the same maximization problem subject to the extra restriction that x is orthogonal to r_1. From the relations

$$0 = r_1^T x = r_1^T R y = e_1^T R^T R y = e_1^T y,\tag{42.6}$$

we see that the corresponding restriction on y is that it should have a zero first component. Hence the maximum value of $x^T A x$ subject to this restriction is λ_2, and this is attained for $y = e_2$ and hence for $x = r_2$. Similarly we find that the maximum value of $x^T A x$ subject to the additional restrictions $r_1^T x = r_2^T x = \ldots = r_s^T x$ is λ_{s+1} and this is attained when $x = r_{s+1}$.

The weakness of this characterization of the eigenvalues is that the determination of each λ_s is dependent on a knowledge of $r_1, r_2, \ldots, r_{s-1}$, which are eigenvectors of A corresponding to $\lambda_1, \lambda_2, \ldots, \lambda_{s-1}$.

Minimax characterization of eigenvalues

43. We now turn to a characterization of the eigenvalues (see, for example, Courant and Hilbert, 1953) which does not have the disadvantage we have just mentioned. Let us consider the maximum value of $x^T A x$ subject to the conditions

$$\left.\begin{array}{l} x^T x = 1, \\ p_i^T x = 0, \quad p_i \neq 0 \quad (i = 1, 2, \ldots, s; \; s < n) \end{array}\right\}, \tag{43.1}$$

where the p_i are any s non-null vectors. In other words, we consider only values of x satisfying s linear relations. For all values of x we have

$$\lambda_n \leqslant x^T A x \leqslant \lambda_1, \tag{43.2}$$

so that $x^T A x$ is bounded, and its maximum is clearly a function of the ns components of the p_i. We now ask ourselves 'what is the minimum value assumed by this maximum for all possible choices of the s vectors p_i?'

As before we may work with y defined by (42.3). The relations (43.1) become

$$\left.\begin{array}{l} y^T y = 1, \\ q_i^T y = 0, \quad q_i^T = p_i^T R \neq 0 \end{array}\right\}. \tag{43.3}$$

Now consider any particular choice of the $p_1, p_2, \ldots, p_s$. This gives us a corresponding set of q_i. The n variables y_j therefore satisfy s linear homogeneous equations. If we add the further relations

$$y_{s+2} = y_{s+3} = \ldots = y_n = 0, \tag{43.4}$$

then we have altogether $n-1$ homogeneous equations in the n unknowns $y_1, y_2, \ldots, y_n$ and hence there is at least one non-zero solution

$$(y_1, y_2, \ldots, y_{s+1}, 0, 0, \ldots, 0) \tag{43.5}$$

which may be normalized so that $\sum_{i=1}^{s+1} y_i^2 = 1$. With this choice of y we have

$$y^T \operatorname{diag}(\lambda_i) y = \sum_{i=1}^{s+1} \lambda_i y_i^2 \geqslant \lambda_{s+1}. \tag{43.6}$$

This shows that whatever the choice of the p_i there will always be a y, and therefore an x, for which

$$x^T A x = y^T \operatorname{diag}(\lambda_i) y \geqslant \lambda_{s+1}, \tag{43.7}$$

and hence

$$\max(x^T A x) \geqslant \lambda_{s+1} \tag{43.8}$$

for any choice of the p_i. This means that

$$\min \max x^T A x \geqslant \lambda_{s+1}. \tag{43.9}$$

However, if we take $p_i = Re_i$, then $q_i^T = e_i^T$ and the relations (43.3) become
$$y_i = 0 \quad (i = 1, 2, \dots, s), \tag{43.10}$$
and hence for any y subject to these particular relations we have
$$x^T A x = y^T \operatorname{diag}(\lambda_i) y = \sum_{s+1}^{n} \lambda_i y_i^2 \leqslant \lambda_{s+1}, \tag{43.11}$$
giving
$$\max (x^T A x) \leqslant \lambda_{s+1}. \tag{43.12}$$
Relations (43.8) and (43.12) together imply that with this choice of the p_i
$$\max (x^T A x) = \lambda_{s+1} \tag{43.13}$$
and indeed this value is attained by the permissible y defined by
$$y_{s+1} = 1, \qquad y_i = 0 \quad (i \neq s+1). \tag{43.14}$$
Hence $\min \max (x^T A x)$, when x is subject to s linear relations, is λ_{s+1}. *This result is known as the Courant-Fischer theorem.* We emphasize that we have shown that there is at least one set of p_i and one corresponding x for which the minimax value is attained.

In an exactly similar way we may prove that
$$\left. \begin{array}{l} \lambda_s = \max \min (x^T A x) \\ x^T x = 1, \qquad p_i^T x = 0 \quad (i = 1, 2, \dots, n-s) \end{array} \right\}. \tag{43.15}$$

Notice that in both characterizations the sets of p_i include those in which some of the p_i are equal, although, of course, in general the actual minimization of the maximum will occur for a set of distinct p_i. If then we have *any* s vectors p_i, we may assert that
$$\left. \begin{array}{l} \lambda_{s+1} \leqslant \max (x^T A x) \\ x^T x = 1, \qquad p_i^T x = 0 \quad (i = 1, 2, \dots, s) \end{array} \right\} \tag{43.16}$$
whether or not all the p_i are distinct.

Eigenvalues of the sum of two symmetric matrices

44. The minimax characterization of the eigenvalues may be used to establish a relation between the eigenvalues of symmetric matrices A, B and C where
$$C = A + B. \tag{44.1}$$
We denote the eigenvalues of A, B and C by α_i, β_i and γ_i respectively where all three sets are arranged in non-increasing order. By the minimax theorem we have
$$\left. \begin{array}{l} \gamma_s = \min \max (x^T C x) \\ x^T x = 1, \qquad p_i^T x = 0 \quad (i = 1, 2, \dots, s-1) \end{array} \right\}. \tag{44.2}$$

Hence, if we take any particular set of p_i, we have for all corresponding x

$$\gamma_s \leqslant \max(x^T C x) = \max(x^T A x + x^T B x). \tag{44.3}$$

If R is the orthogonal matrix such that

$$R^T A R = \operatorname{diag}(\alpha_i) \tag{44.4}$$

then if we take $p_i = R e_i$ the relations to be satisfied are

$$0 = p_i^T x = e_i^T y \quad (i = 1, 2, \ldots, s-1).$$

With this choice of the p_i then, the first $(s-1)$ components of y are zero, and from (44.3) we have

$$\gamma_s \leqslant \max\,(x^T A x + x^T B x) = \max\left(\sum_{i=s}^{n} \alpha_i\, y_i^2 + x^T B x\right). \tag{44.6}$$

However,

$$\sum_{i=s}^{n} \alpha_i\, y_i^2 \leqslant \alpha_s, \tag{44.7}$$

while

$$x^T B x \leqslant \beta_1 \tag{44.8}$$

for any x. Hence the expression in brackets is not greater than $\alpha_s + \beta_1$ for any x corresponding to this choice of the p_i. Therefore its maximum is not greater than $\alpha_s + \beta_1$ and we have

$$\gamma_s \leqslant \alpha_s + \beta_1. \tag{44.9}$$

Since $A = C + (-B)$ and the eigenvalues of $(-B)$ in non-increasing order are $-\beta_n, -\beta_{n-1}, \ldots, -\beta_1$, an application of the result we have just proved gives

$$\alpha_s \leqslant \gamma_s + (-\beta_n) \quad \text{or} \quad \gamma_s \geqslant \alpha_s + \beta_n. \tag{44.10}$$

Relations (44.9) and (44.10) imply that when B is added to A all of its eigenvalues are changed by an amount which lies between the smallest and greatest of the eigenvalues of B. This agrees with the result we proved in § 41 by analytic methods for a B of rank unity. Note that we are not concerned here specifically with small perturbations and the results are not affected by multiplicities in the eigenvalues of A, B, or C.

Practical applications

45. The result of the last section is one of the most useful in practice. Frequently B will be small and the only information we shall have is some upper bound for its elements, or perhaps for some norm. If for example we have

$$|b_{ij}| \leqslant \epsilon \tag{45.1}$$

then

$$-n\epsilon \leqslant \beta_n \leqslant \beta_1 \leqslant n\epsilon, \tag{45.2}$$

and hence

$$|\gamma_r - \alpha_r| \leqslant n\epsilon. \tag{45.3}$$

This is stronger than our earlier result of § 37 for unsymmetric perturbations, since there are now no restrictions on the separations of the α_i, β_i and γ_i. Previously we were able to prove the relation (45.3) only when the eigenvalues α_r were all separated by more than $2n\epsilon$.

Further applications of minimax principle

46. The result obtained in § 44 may be generalized to give

$$\gamma_{r+s-1} \leqslant \alpha_r + \beta_s \quad (r+s-1 \leqslant n). \tag{46.1}$$

The proof is as follows.

There is at least one set of p_i for which

$$\max(x^T A x) = \alpha_r \text{ subject to } p_i^T x = 0 \quad (i = 1, 2, \ldots, r-1), \tag{46.2}$$

and a set of q_i for which

$$\max(x^T B x) = \beta_s \text{ subject to } q_i^T x = 0 \quad (i = 1, 2, \ldots, s-1). \tag{46.3}$$

Consider now the set of x satisfying both $p_i^T x = 0$ and $q_i^T x = 0$. Such non-zero x exist because the total number of equations is $r+s-2$ which from (46.1) is not greater than $n-1$. For any such x we have

$$x^T C x = x^T A x + x^T B x \leqslant \alpha_r + \beta_s. \tag{46.4}$$

Hence

$$\max(x^T C x) \leqslant \alpha_r + \beta_s \tag{46.5}$$

for all x satisfying the particular set of $r+s-2$ linear equations of (46.2) and (46.3). Therefore

$$\gamma_{r+s-1} = \min \max(x^T C x) \leqslant \alpha_r + \beta_s, \tag{46.6}$$

since the minimization is taken over all sets of $r+s-2$ relations.

Separation theorem

47. As a further illustration of the minimax theorem we prove that the eigenvalues $\lambda_1', \lambda_2', \ldots, \lambda_{n-1}'$ of the leading principal minor A_{n-1} of the matrix A_n separate the eigenvalues $\lambda_1, \lambda_2, \ldots, \lambda_n$ of A_n. (This result was proved in § 41 by analytical methods for the lowest principal minor of order $(n-1)$; clearly if true at all, it is true for all principal minors of order $(n-1)$ as may be seen immediately by applying similarity transformations with permutation matrices.)

The set of values assumed by $x^T A_{n-1} x$ for all unit vectors x of order $(n-1)$ is the same as the set of values assumed by $x^T A_n x$ for unit vectors of order n subject to the condition $x_n = 0$. Hence

$$\left.\begin{aligned} \lambda'_s &= \min \max(x^T A_n x), \qquad x^T x = 1 \\ x_n &= 0, \qquad p_i^T x = 0 \quad (i = 1, 2,..., s-1) \end{aligned}\right\} \qquad (47.1)$$

Now this maximum value will be attained for some set of p_i. For this set of p_i, λ'_s is the maximum value of $x^T A_n x$ subject to the s linear relations in the second line of (47.1). However, λ_{s+1} is the minimum value taken by the maximum subject to any s relations. Hence

$$\lambda_{s+1} \leqslant \lambda'_s. \qquad (47.2)$$

Now consider any particular set of $s-1$ vectors p_i. Denote the maximum value of $x^T A_n x$, for unit x subject to the corresponding set of linear relations, by $f_n(p_i)$. Let the maximum subject to the extra restriction $x_n = 0$ be denoted by $f_{n-1}(p_i)$, then obviously

$$f_{n-1}(p_i) \leqslant f_n(p_i), \qquad (47.3)$$

and hence

$$\min f_{n-1}(p_i) \leqslant \min f_n(p_i), \qquad (47.4)$$

giving

$$\lambda'_s \leqslant \lambda_s. \qquad (47.5)$$

The whole of the theory of the last six sections extends immediately to Hermitian matrices in general; the results and the proofs are obtained by replacing the superscript T by H wherever it occurs.

The Wielandt-Hoffman theorem

48. There is a theorem of a rather different type from those we have just considered, which relates the perturbations in the eigenvalues to the Euclidean norm of the perturbation matrix. This theorem, which is due to Hoffman and Wielandt (1953), is as follows.

If $C = A + B$, where A, B and C are symmetric matrices having the eigenvalues α_i, β_i, γ_i respectively arranged in non-increasing order, then

$$\sum_{i=1}^{n} (\gamma_i - \alpha_i)^2 \leqslant \|B\|_E^2 = \sum_{i=1}^{n} \beta_i^2. \qquad (48.1)$$

Hoffman and Wielandt gave a very elegant proof which was dependent on the theory of linear programming. The proof that follows is less

elegant but more elementary. It is essentially due to Givens (1954).
We shall require two simple lemmas.

(i) If a symmetric matrix X has eigenvalues λ_i, then

$$\|X\|_E^2 = \sum_1^n \lambda_i^2. \tag{48.2}$$

For there is an orthogonal matrix R such that

$$R^T X R = \text{diag}(\lambda_i) \tag{48.3}$$

and taking the Euclidean norm of both sides, the result follows from
the invariance of the Euclidean norm with respect to orthogonal
transformations.

(ii) If R_1, R_2,... is an infinite sequence of orthogonal matrices, then
there exists a subsequence R_{s_1}, R_{s_2},... such that

$$\lim_{i \to \infty} R_{s_i} = R. \tag{48.4}$$

This follows from the n^2-dimensional version of the Bolzano-Weier-
strass theorem, since all elements of all R_i lie in the range $(-1, 1)$.
The matrix R must be orthogonal since

$$R_{s_i}^T R_{s_i} = I, \tag{48.5}$$

and hence taking limits we have

$$R^T R = I. \tag{48.6}$$

49. Now let R_1 and R_2 be orthogonal matrices such that

$$R_1^T B R_1 = \text{diag}(\beta_i), \qquad R_2^T C R_2 = \text{diag}(\gamma_i). \tag{49.1}$$

Then
$$\begin{aligned}
\text{diag}(\beta_i) &= R_1^T B R_1 = R_1^T [C - A] R_1 \\
&= R_1^T [R_2 \, \text{diag}(\gamma_i) R_2^T - A] R_1 \\
&= R_1^T R_2 \, [\text{diag}(\gamma_i) - R_2^T A R_2] R_2^T R_1, \tag{49.2}
\end{aligned}$$

and taking norms of both sides

$$\sum_{i=1}^n \beta_i^2 = \|\text{diag}(\gamma_i) - R_2^T A R_2\|_E^2. \tag{49.3}$$

Consider now the set of values taken by $\|\text{diag}(\gamma_i) - R^T A R\|_E^2 = f(R)$
say, for all orthogonal matrices R. Equation (49.3) shows that $\sum_{i=1}^n \beta_i^2$
belongs to this set. The set is obviously bounded and therefore has
finite upper and lower bounds, u and l. These bounds must be attained
for some R, since $f(R)$ is a continuous function on the compact set
of orthogonal matrices.

50. We now show that l must be achieved for an R such that $R^T A R$ is diagonal. For suppose that there are r distinct values of γ_i which we denote by δ_i $(i = 1, 2, ..., r)$ ordered so that

$$\delta_1 > \delta_2 > ... > \delta_r. \tag{50.1}$$

We may write

$$\operatorname{diag}(\gamma_i) = \begin{bmatrix} \delta_1 I & & & & \\ & \delta_2 I & & & \\ & & \cdot & & \\ & & & \cdot & \\ & & & & \cdot \\ & & & & & \delta_r I \end{bmatrix}, \tag{50.2}$$

where the identity matrices are of appropriate orders. If $R^T A R$ be partitioned to conform with (50.2) we may write

$$R^T A R = \begin{bmatrix} X_{11} & X_{12} & ... & X_{1r} \\ X_{21} & X_{22} & ... & X_{2r} \\ . & . & ... & . \\ X_{r1} & X_{r2} & ... & X_{rr} \end{bmatrix} \equiv X \text{ (say).} \tag{50.3}$$

We show first that l can only be achieved for an R for which all off-diagonal blocks in (50.3) are null. For suppose there is a non-zero element x in row p and column q of $R^T A R$, and that this is in the block X_{ij} $(i \neq j)$. Then the elements at the intersection of rows and columns p and q of $\operatorname{diag}(\gamma_i)$ and $R^T A R$ are of the forms

$$
\begin{array}{cc}
\operatorname{diag}(\gamma_i) & X = R^T A R \\
\text{col. } p \quad \text{col. } q & \text{col. } p \quad \text{col. } q \\
\end{array}
$$

$$
\begin{array}{c}
\cdots\cdots\delta_i\cdots\cdots 0\cdots\text{row } p \qquad \cdots\cdots a\cdots\cdots x\cdots\text{row } p \\
\\
\cdots 0\cdots\cdots\delta_j\cdots\text{row } q \qquad \cdots\cdots x\cdots\cdots b\cdots\text{row } q
\end{array}
\tag{50.4}
$$

where for simplicity we have avoided using suffices in the relevant elements of X. We show that we can choose an elementary orthogonal matrix S corresponding to a rotation in the (p, q) plane such that

$$g(S) \equiv \|\operatorname{diag}(\gamma_i) - S^T R^T A R S\|_E^2 - \|\operatorname{diag}(\gamma_i) - R^T A R\|_E^2 < 0. \tag{50.5}$$

Let us denote by a', x', b' the corresponding elements of $S^T R^T A RS$ at the intersections of rows and columns p and q. Since

$$\|S^T R^T A RS\|_E^2 = \|R^T A R\|_E^2, \tag{50.6}$$

it is easy to see directly from the definition of the Euclidean norm that

$$g(S) = -2a' \delta_i - 2b' \delta_j + 2a \delta_i + 2b \delta_j$$
$$= 2(a - a') \delta_i + 2(b - b') \delta_j, \tag{50.7}$$

all other terms cancelling. If the angle of the rotation is θ then

$$\left.\begin{aligned} a' &= a \cos^2\theta - 2x \cos \theta \sin \theta + b \sin^2\theta \\ b' &= a \sin^2\theta + 2x \cos \theta \sin \theta + b \cos^2\theta \end{aligned}\right\}. \tag{50.8}$$

This gives

$$h(\theta) \equiv g(S) = 2\delta_i[(a-b)\sin^2\theta + x \sin 2\theta] + 2\delta_j[(b-a)\sin^2\theta - x \sin 2\theta]$$
$$= P \sin^2\theta + Q \sin 2\theta \text{ (say)}, \tag{50.9}$$

where $Q \neq 0$ because x is non-zero by hypothesis, and $\delta_i - \delta_j$ is non-zero because δ_i and δ_j are in different diagonal blocks. Now we have

$$\frac{d}{d\theta} h(\theta) = P \sin 2\theta + 2Q \cos 2\theta, \tag{50.10}$$

and hence the derivative is $2Q$ when $\theta = 0$, showing that we can make $g(S) > 0$ and $g(S) < 0$ for appropriate choices of θ.

51. We have therefore shown that $f(R)$ cannot be a maximum or minimum for any R for which X has any non-zero elements in any X_{ij} $(i \neq j)$. Suppose now that R is an orthogonal matrix for which $f(R)$ achieves its minimum value, then we must have

$$\text{diag}(\gamma_i) - R^T A R = \begin{bmatrix} \delta_1 I & & & \\ & \delta_2 I & & \\ & & \cdot & \\ & & & \cdot \\ & & & & \delta_r I \end{bmatrix} - \begin{bmatrix} X_{11} & & & \\ & X_{22} & & \\ & & \cdot & \\ & & & \cdot \\ & & & & X_{rr} \end{bmatrix}. \tag{51.1}$$

If Q_i are the orthogonal matrices such that

$$Q_i^T X_{ii} Q_i = D_i, \tag{51.2}$$

where the D_i are diagonal, then denoting the direct sum of the Q_i by Q we have

$$Q^T \operatorname{diag}(\gamma_i)Q - Q^T R^T A\, RQ$$

$$= \begin{bmatrix} \delta_1 I & & & & \\ & \delta_2 I & & & \\ & & \cdot & & \\ & & & \cdot & \\ & & & & \delta_r I \end{bmatrix} - \begin{bmatrix} D_1 & & & & \\ & D_2 & & & \\ & & \cdot & & \\ & & & \cdot & \\ & & & & D_r \end{bmatrix}$$

$$= \operatorname{diag}(\gamma_i) - Q^T R^T A\, RQ. \qquad (51.3)$$

Hence

$$f(RQ) = f(R), \qquad (51.4)$$

and the minimum must always be attained for a matrix RQ which reduces A to diagonal form. Clearly the elements of the D_i are the α_j in some order. We have therefore, remembering that the δ_i with the appropriate multiplicities are the γ_i,

$$\min_R f(R) = \sum (\gamma_i - \alpha_{p_i})^2 \qquad (51.5)$$

where $p_1, p_2, \ldots, p_n$ is a permutation of $1, 2, \ldots, n$.

We show now that the minimum occurs when $p_i = i$. Let us write

$$x = \sum (\gamma_i - \alpha_{p_i})^2 \qquad (51.6)$$

for some particular permutation. If $p_1 \neq 1$ then suppose that $p_s = 1$, so that γ_s is paired with α_1. Let us interchange p_1 and p_s in the permutation. The change in x is given by

$$(\gamma_1 - \alpha_1)^2 + (\gamma_s - \alpha_{p_1})^2 - (\gamma_1 - \alpha_{p_1})^2 - (\gamma_s - \alpha_1)^2$$
$$= -2(\gamma_s - \gamma_1)(\alpha_{p_1} - \alpha_1) \leqslant 0, \quad (51.7)$$

and hence the sum is diminished. Similarly, keeping α_1 in the first position and aligning γ_2 and α_2, we see again that the sum is not increased. Hence finally when each α_i is aligned with the corresponding γ_i the sum is not greater than its initial value.

Our final result then, is that the minimum value of $f(R)$ is $\sum (\gamma_i - \alpha_i)^2$. However in (49.3) we showed that $\sum \beta_i^2$ is one of the values attained by $f(R)$ and hence

$$\|B\|_E^2 = \sum_{i=1}^n \beta_i^2 \geqslant \sum_{i=1}^n (\gamma_i - \alpha_i)^2. \qquad (51.8)$$

This completes the proof.

52. The proof extends trivially to Hermitian matrices, but the result is also true for any normal matrix. A similar method of proof can be used in the general case, though now the elements in the (p, q) and (q, p) positions in (50.4) will not be complex conjugates. Using the plane-rotation matrix of equations (43.4) of Chapter 1 we can show that the relevant minimum is attained when $R^H A R$ is diagonal. The details are left as an exercise.

Additional notes

Although the material of §§ 5–12 is essentially classical perturbation theory, a simple but rigorous treatment does not appear to be readily available in the literature.

Gerschgorin's theorem has been widely recommended for the location of eigenvalues, Taussky having been a particularly ardent advocate; an interesting discussion of this theorem and its extensions is given by Taussky (1949). The use of diagonal similarity transforms to improve the location of the eigenvalues was discussed in Gerschgorin's original paper (1931) but the extension of this device to give the results of classical perturbation theory together with rigorous error bounds appears to be new.

A comprehensive account of those inclusion and exclusion domains for eigenvalues which can be derived by the application of norms is given in Chapter 3 of Householder's *The theory of matrices in numerical analysis* (1964). This chapter also gives a discussion of results obtained from extensions of the minimax principle.

The Wielandt-Hoffman theorem does not seem to have attracted as much attention as those arising from the direct application of norms. In my experience it is the most useful result for the error analysis of techniques based on orthogonal transformations in floating-point arithmetic.

3

Error Analysis

Introduction

1. SINCE the main object of this book is to give a comparative assessment of the various techniques for solving the algebraic eigenvalue problem, we have included error analyses of a number of the more important methods. Such analyses naturally require an understanding of the fundamental arithmetic operations as performed on a digital computer. In Wilkinson (1963b) Chapter 1, I have given a fairly detailed account of the rounding errors in these operations and have discussed the main variations which are to be found in current computers.

I shall assume that the reader is familiar with this work and will give only a brief resumé of the results which we shall need here. It will be simpler if we restrict ourselves to a specific set of rounding procedures, since the bounds we obtain are not very sensitive to the variations which exist. We shall be concerned with both fixed-point and floating-point computation and shall assume throughout that binary arithmetic is used. However, it is useful to exhibit simple examples from time to time, and in nearly all cases these have been carried out on a desk computer in the decimal scale. It is rather unsatisfactory to perform such computations on an automatic computer since rounding errors may be involved in the conversion from binary to decimal representation.

Fixed-point operations

2. In fixed-point computation we shall assume that scaling factors are introduced if necessary, so that every number x lies in the standard range

$$-1 \leqslant x \leqslant 1, \tag{2.1}$$

and shall not bother to exclude either end-point since this often complicates error analysis in an inessential manner. We shall assume that there are t digits after the binary point and shall say that the computer has 'a t-digit word' even though $t+1$ digits may be needed for its representation.

An equation of the form
$$z = fi\left(x \overset{\times}{\underset{\div}{\pm}} y\right) \tag{2.2}$$

will imply that x, y and z are standard fixed-point numbers and that z is obtained by performing the appropriate fixed-point operation on x and y. In the case of addition, subtraction and division we assume that the computed z does not lie outside of the permitted range.

No rounding errors are involved in addition or subtraction so that we have
$$z = fi(x \pm y) \equiv x \pm y, \tag{2.3}$$

where the last term in (2.3) represents the exact sum or difference. We use the equivalence sign to emphasize that rounding errors have been taken into account. This helps to distinguish the computational equations from the mathematical equations which may be used to describe an algorithm. Multiplication and division in general involve rounding errors. We assume that the rounding procedure is such that
$$z = fi(x \overset{\times}{\underset{\div}{}} y) \equiv x \overset{\times}{\underset{\div}{}} y + \epsilon, \tag{2.4}$$
where
$$|\epsilon| < 2^{-t-1}. \tag{2.5}$$

We emphasize that in (2.4) the terms $x \overset{\times}{\underset{\div}{}} y$ on the right-hand side represent the exact product or quotient, so that the signs '$\times$' and '$\div$' have their normal arithmetic significance.

Accumulation of inner-products

3. On many computers it is possible without special programming to calculate the inner-product
$$x_1 \times y_1 + x_2 \times y_2 + \ldots + x_n \times y_n \tag{3.1}$$
exactly in fixed-point arithmetic (provided no overspill occurs). In general the exact representation of the sum in (3.1) requires $2t$ digits after the binary place. If z is the number obtained by accumulating the inner-product (3.1) exactly and then rounding the result to give a standard fixed-point number, we write
$$z = fi_2(x_1 \times y_1 + x_2 \times y_2 + \ldots + x_n \times y_n), \tag{3.2}$$
where the suffix '2' denotes that exact accumulation was used. With our rounding procedure we have
$$z = fi_2(x_1 \times y_1 + x_2 \times y_2 + \ldots + x_n \times y_n) \equiv \sum_{i=1}^{n} x_i y_i + \epsilon, \tag{3.3}$$
where
$$|\epsilon| < 2^{-t-1}, \tag{3.4}$$
while
$$fi(x_1 \times y_1 + x_2 \times y_2 + \ldots + x_n \times y_n) \equiv \sum_{i=1}^{n} x_i y_i + \epsilon, \tag{3.5}$$
where
$$|\epsilon| < n2^{-t-1}. \tag{3.6}$$

A computer possessing the facility for accumulating inner-products is usually designed to accept a dividend with $2t$ digits after the binary place and to divide this by a standard fixed-point number to give a rounded fixed-point quotient. We shall write

$$w = fi_2[(x_1 \times y_1 + x_2 \times y_2 + \ldots + x_n \times y_n)/z] \equiv \left[\left(\sum_{i=1}^{n} x_i y_i\right)\bigg/ z\right] + \epsilon, \quad (3.7)$$

where

$$|\epsilon| \leqslant 2^{-t-1}. \quad (3.8)$$

Hence

$$\left| wz - \sum_{i=1}^{n} x_i y_i \right| \equiv |z\epsilon| \leqslant 2^{-t-1} |z|. \quad (3.9)$$

We shall use the notation $fi(\ )$ and $fi_2(\ )$ very freely. For example if we write

$$C = fi_2(A \times B), \quad (3.10)$$

where A, B, C are matrices, we imply that all elements of A, B, C are standard fixed-point numbers and that each c_{ij} is defined by

$$c_{ij} = fi_2(a_{i1} \times b_{1j} + a_{i2} \times b_{2j} + \ldots + a_{in} \times b_{nj}). \quad (3.11)$$

Provided C is representable without overflow we have

$$\left.\begin{array}{l} C = fi_2(A \times B) \equiv AB + F \\ |f_{ij}| \leqslant 2^{-t-1} \end{array}\right\}, \quad (3.12)$$

where AB denotes, of course, the exact matrix product.

It should be noted that if the expression inside the parentheses in $fi(\ )$ is very complex, the result may depend on the order in which the operations are carried out and this will then need to be specified.

Floating-point operations

4. In floating-point computation each standard number x is represented by an ordered pair a and b such that $x = 2^b a$. Here b is an integer, positive or negative, and a is a number satisfying

$$-\tfrac{1}{2} > a \geqslant -1 \quad \text{or} \quad \tfrac{1}{2} \leqslant a \leqslant 1. \quad (4.1)$$

We refer to b as the *exponent* and to a as the *mantissa*. We shall assume that a has t figures after the binary place and say that the computer has 'a t-digit mantissa'. We shall use the same symbol 't' as in fixed-point computation, though in practice, on some computers which can work in both modes, the number of digits in a and b together is the same as in a fixed-point number. It is important to bear this in mind when comparing error bounds obtained for fixed-point and floating-point computation. We shall assume that the number zero has a non-standard representation in which $a = b = 0$.

An equation of the form

$$z = fl(x \overset{+}{\underset{\div}{\times}} y) \qquad (4.2)$$

will imply that x, y and z are standard floating-point numbers and that z is obtained from x and y by performing the appropriate floating-point operation. We assume that the rounding errors in these operations are such that

$$z = fl\left(x \overset{+}{\underset{\div}{\times}} y\right) \equiv \left(x \overset{+}{\underset{\div}{\times}} y\right)(1 + \epsilon), \qquad (4.3)$$

where

$$|\epsilon| \leqslant 2^{-t}, \qquad (4.4)$$

so that each of the individual arithmetic operations gives a result having a low rounding error. As pointed out in Wilkinson (1963b), p. 11, some computers have rounding procedures such that addition and subtraction do not necessarily give results having low relative errors, but this makes very little difference to bounds obtained for extended computations.

Simplified expressions for error bounds

5. A number of results will be required repeatedly throughout this book and we assemble them here for convenience. They are direct consequences of the relations (4.3) and (4.4). However, the application of these relations frequently leads in the first instance to bounds of the form

$$(1 - 2^{-t})^r \leqslant 1 + \epsilon \leqslant (1 + 2^{-t})^r, \qquad (5.1)$$

and these are somewhat inconvenient. In order to simplify such bounds *we make the assumption that in all practical applications in which we shall be interested, r will be subject to the restriction*

$$r2^{-t} < 0 \cdot 1. \qquad (5.2)$$

For any reasonable value of t this is unlikely to involve any restriction on r which would not in any case have been forced on us by considerations of storage capacity. With the restriction (5.2) we have

$$\left.\begin{aligned}(1 + 2^{-t})^r &< 1 + (1 \cdot 06)r2^{-t} \\ (1 - 2^{-t})^r &> 1 - (1 \cdot 06)r2^{-t}\end{aligned}\right\}. \qquad (5.3)$$

In order to simplify these expressions still further we introduce t_1 defined by
$$2^{-t_1} = (1{\cdot}06)2^{-t}, \qquad (5.4)$$
i.e.
$$t_1 = t - \log_2(1{\cdot}06) = t - 0{\cdot}08406, \qquad (5.5)$$
so that t_1 is only marginally different from t. *In all future bounds then, we shall replace the relations* (5.1) *by*
$$|\epsilon| < r2^{-t_1}, \qquad (5.6)$$
whenever this is advantageous, the restriction (5.2) *being tacitly understood.*

Error bounds for some basic floating-point computations

6. We state now the following results without proof.

(i)
$$fl(x_1 \times x_2 \times \dots \times x_n) \equiv (1+E) \prod_{i=1}^{n} x_i, \qquad (6.1)$$
where
$$|E| < (n-1)2^{-t_1}. \qquad (6.2)$$

(ii) $fl(x_1 + x_2 + \dots + x_n) \equiv x_1(1+\epsilon_1) + x_2(1+\epsilon_2) + \dots + x_n(1+\epsilon_n),$ (6.3)
where
$$|\epsilon_1| < (n-1)2^{-t_1}, \qquad |\epsilon_r| < (n+1-r)2^{-t_1} \quad (r = 2,\dots, n). \qquad (6.4)$$
It is assumed that the operations take place in the order
$$s_2 = fl(x_1 + x_2), \qquad s_r = fl(s_{r-1} + x_r) \quad (r = 3,\dots, n).$$

(iii) $fl(x^T y) = fl(x_1 y_1 + x_2 y_2 + \dots + x_n y_n) \equiv x_1 y_1(1+\epsilon_1) +$
$$+ x_2 y_2(1+\epsilon_2) + \dots + x_n y_n(1+\epsilon_n), \qquad (6.5)$$
where
$$|\epsilon_1| < n2^{-t_1}, \qquad |\epsilon_r| < (n+2-r)2^{-t_1} \quad (r = 2,\dots, n). \qquad (6.6)$$
We may write
$$fl(x^T y) - x^T y = x^T z, \qquad (6.7)$$
where
$$z = \text{diag}(\epsilon_i)y. \qquad (6.8)$$
Useful consequences of this basic result are
$$|fl(x^T y) - x^T y| < \sum |x_i|\,|y_i|\,|\epsilon_i| \qquad (6.9)$$
$$\leqslant n2^{-t_1}\,|x|^T\,|y| \qquad (6.10)$$
$$\leqslant n2^{-t_1}\,\|x\|_2\,\|y\|_2. \qquad (6.11)$$

(iv) $C = fl(A+B) \equiv A+B+F$ (say), (6.12)
where
$$c_{ij} = (a_{ij} + b_{ij})(1+\epsilon_{ij}), \qquad |\epsilon_{ij}| < 2^{-t}. \qquad (6.13)$$

Hence

$$|f_{ij}| = \epsilon_{ij} |a_{ij} + b_{ij}|. \qquad (6.14)$$

We have therefore

$$|fl(A+B) - (A+B)| \leqslant 2^{-t} |A+B|. \qquad (6.15)$$

This result is true only for the particular rounding procedure we have chosen and is far from true for other rounding procedures. We shall make little use of it. However (6.15) certainly implies that

$$|fl(A+B) - (A+B)| \leqslant 2^{-t}(|A| + |B|), \qquad (6.16)$$

and this extends immediately to all rounding procedures I have examined if an extra factor of four (say) is included in the bound.

(v) $$\qquad fl(AB) \equiv AB + F, \qquad (6.17)$$

where $\qquad f_{ij} = a_{i1}b_{1j}\epsilon_1^{(ij)} + a_{i2}b_{2j}\epsilon_2^{(ij)} + \ldots + a_{in}b_{nj}\epsilon_n^{(ij)} \qquad (6.18)$

and $\quad |\epsilon_1^{(ij)}| < n2^{-t_1}, \qquad |\epsilon_r^{(ij)}| < (n+2-r)2^{-t_1} \quad (r = 2,\ldots, n). \quad (6.19)$

Hence $$\qquad |F| < |A| |D| |B| < n2^{-t_1} |A| |B|, \qquad (6.20)$$

where D is a diagonal matrix having the elements

$$[n2^{-t_1}, n2^{-t_1}, (n-1)2^{-t_1}, \ldots, 3.2^{-t_1}, 2.2^{-t_1}]. \qquad (6.21)$$

(vi) $$\qquad fl(\alpha A) \equiv \alpha A + F, \qquad (6.22)$$

where $$\qquad f_{ij} = \alpha a_{ij}\epsilon_{ij} \quad \text{and} \quad |\epsilon_{ij}| \leqslant 2^{-t}. \qquad (6.23)$$

Hence $$\qquad |fl(\alpha A) - \alpha A| \leqslant 2^{-t} |\alpha| |A|. \qquad (6.24)$$

(vii) $$\qquad fl(xy^T) \equiv xy^T + F, \qquad (6.25)$$

where $$\qquad f_{ij} = x_i y_j \epsilon_{ij} \quad \text{and} \quad |\epsilon_{ij}| \leqslant 2^{-t}. \qquad (6.26)$$

Hence $$\qquad |fl(xy^T) - xy^T| \leqslant 2^{-t} |x| |y|^T. \qquad (6.27)$$

Bounds for norms of the error matrices

7. Often in applications of the results we have just obtained we shall need bounds for some norm of the error matrix. If we take for example the bound (6.20) for the error in matrix multiplication we have

$$\| F \| \leqslant \| |F| \| \leqslant n2^{-t_1} \| |A| \| \| |B| \|, \qquad (7.1)$$

and hence *for the Euclidean, 1-norm or ∞-norm we have*

$$\| F \| \leqslant n2^{-t_1} \| A \| \| B \|, \qquad (7.2)$$

but *for the 2-norm we have only*

$$\| F \|_2 \leqslant n^2 2^{-t_1} \| A \|_2 \| B \|_2. \qquad (7.3)$$

The bound in (7.3) is in fact weak compared with

$$\|F\|_E \leqslant n2^{-t_1} \|A\|_E \|B\|_E \qquad (7.4)$$

since

$$\|F\|_2 \leqslant \|F\|_E \quad \text{and} \quad \|A\|_E \|B\|_E \leqslant n \|A\|_2 \|B\|_2. \qquad (7.5)$$

We shall often find that the bounds obtained using the 2-norm are weaker than those obtained by using the Euclidean norm. Note, however, that if A and B have non-negative elements then we have

$$\|F\|_2 \leqslant n2^{-t_1} \|A\|_2 \|B\|_2, \qquad (7.6)$$

and this result is stronger than that of (7.4).

Accumulation of inner-products in floating-point arithmetic

8. On some computers which are now under construction special facilities are provided for the addition and subtraction of numbers having a mantissa of $2t$ binary digits. On such computers the extended sum $fl(x_1+x_2+\ldots+x_n)$ may be accumulated with a $2t$-digit mantissa, as may also the scalar product $fl(x_1y_1+x_2y_2+\ldots+x_ny_n)$. In the latter it is assumed that the x_i and y_i are standard numbers with t-digit mantissae. Similarly a dividend with a $2t$-digit mantissa may be accepted and divided by a standard number with a t-digit mantissa. On ACE, one of the computers at the National Physical Laboratory, subroutines exist which perform these operations in much the same times as the corresponding standard floating-point operations. We use the notation $fl_2(\ )$ to denote these operations in analogy with our use of $fi_2(\ )$. Note however that $fl_2(x_1+x_2+\ldots+x_n)$ does not, in general, give the exact sum.

The rounding errors in the basic $fl_2(\ )$ operations were analysed in Wilkinson (1963b), pp. 23–25, and we shall not repeat the analysis here, but again we summarise the most useful results. A direct application of the error bounds for the basic arithmetic operations gives rise to bounds of the form

$$(1-\tfrac{3}{2}2^{-2t})^r \leqslant 1+\epsilon \leqslant (1+\tfrac{3}{2}2^{-2t})^r, \qquad (8.1)$$

and these are rather inconvenient. *We shall always assume that*

$$\tfrac{3}{2}r2^{-2t} < 0{\cdot}1, \qquad (8.2)$$

and with this restriction (8.1) implies that

$$|\epsilon| < \tfrac{3}{2}(1{\cdot}06)r2^{-2t}. \qquad (8.3)$$

Corresponding to the definition of t_1 in § 5 we define t_2 so that

$$2^{-2t_2} = (1\cdot06)2^{-2t}, \qquad 2t_2 = 2t - 0\cdot08406. \tag{8.4}$$

The resulting t_2 is only marginally different from t and (8.1) now implies that

$$|\epsilon| < \tfrac{3}{2}r2^{-2t_2}. \tag{8.5}$$

If both $fl(\ )$ and $fl_2(\ )$ are used in the same computation, the assumption (5.2) renders (8.2) superfluous.

Error bounds for some basic $fl_2(\ )$ computations

9. We now state without proof the following basic results.

(i) $fl_2(x_1 + x_2 + \ldots + x_n) \equiv$
$$\equiv [x_1(1+\epsilon_1) + x_2(1+\epsilon_2) + \ldots + x_n(1+\epsilon_n)](1+\epsilon) \tag{9.1}$$

$$\left.\begin{aligned} |\epsilon| &\leqslant 2^{-t}, \qquad |\epsilon_1| < \tfrac{3}{2}(n-1)2^{-2t_2} \\ |\epsilon_r| &< \tfrac{3}{2}(n+1-r)2^{-2t_2} \quad (r = 2, \ldots, n) \end{aligned}\right\}. \tag{9.2}$$

In (9.1) the sum is accumulated using a $2t$-digit mantissa and is rounded on completion. We have further

$$fl_2(x_1 + x_2 + \ldots + x_n) - (x_1 + x_2 + \ldots + x_n)(1+\epsilon) \equiv$$
$$\equiv x_1\epsilon_1 + x_2\epsilon_2 + \ldots + x_n\epsilon_n, \tag{9.3}$$

where the bounds (9.2) are still adequate.

(ii) $fl_2(x_1 y_1 + x_2 y_2 + \ldots + x_n y_n) \equiv$
$$\equiv [x_1 y_1(1+\epsilon_1) + x_2 y_2(1+\epsilon_2) + \ldots + x_n y_n(1+\epsilon_n)](1+\epsilon), \tag{9.4}$$

$$\left.\begin{aligned} |\epsilon| &\leqslant 2^{-t}, \qquad |\epsilon_1| < \tfrac{3}{2}n2^{-2t_2} \\ |\epsilon_r| &< \tfrac{3}{2}(n+2-r)2^{-2t_2} \quad (r = 2, \ldots, n) \end{aligned}\right\}. \tag{9.5}$$

We have further

$$fl_2(x^T y) - x^T y(1+\epsilon) \equiv x_1 y_1 \epsilon_1 + x_2 y_2 \epsilon_2 + \ldots + x_n y_n \epsilon_n, \tag{9.6}$$

where the bounds (9.5) are still adequate. It is important to realise the full significance of (9.6). It does not enable us to assert that $fl_2(x^T y)$ has a low *relative* error because severe cancellation may take place among the terms of $x^T y$. However (9.6) implies that

$$|fl_2(x^T y) - x^T y| \leqslant |\epsilon|\,|x^T y| + \sum |x_i|\,|y_i|\,|\epsilon_i|$$
$$\leqslant 2^{-t}|x^T y| + \tfrac{3}{2}n2^{-2t_2}|x|^T|y| \tag{9.7}$$
$$\leqslant 2^{-t}|x^T y| + \tfrac{3}{2}n2^{-2t_2}\|x\|_2\,\|y\|_2. \tag{9.8}$$

Unless exceptionally severe cancellation takes place among the terms of $x^T y$ the second term on the right-hand side of (9.7) is negligible compared with the first.

(iii) Taking y equal to x in the results in (ii) we have

$$x^T x = |x|^T |x|, \qquad (9.9)$$

and hence

$$|fl_2(x^T x) - x^T x| < 2^{-t}(x^T x) + \tfrac{3}{2}n2^{-2t_2}(x^T x) =$$
$$= (2^{-t} + \tfrac{3}{2}n2^{-2t_2})(x^T x). \quad (9.10)$$

If $\tfrac{3}{2}n2^{-t} < 0{\cdot}1$ we have

$$2^{-t} + \tfrac{3}{2}n2^{-2t_2} < 2^{-t}[1 + (0{\cdot}1)2^{2t-2t_2}] < 2^{-t}(1{\cdot}11) \text{ (say)}, \quad (9.11)$$

so that $fl_2(x^T x)$ always has a low relative error.

(iv) If x and y are vectors and z is a scalar then

$$fl_2\left(\frac{x^T y}{z}\right) \equiv \frac{x_1 y_1(1 + \epsilon_1) + \ldots + x_n y_n(1 + \epsilon_n)}{z/(1 + \epsilon)}, \qquad (9.12)$$

where the ϵ and ϵ_i satisfy the relations (9.5). We may express this in the form:

The computed value of $x^T y/z$ is the exact value of $\bar{x}^T \bar{y}/\bar{z}$ where the elements of $\bar{x}$ and $\bar{y}$ differ from those of x and y by relative errors of order 2^{-2t_2} while $\bar{z}$ differs from z by a relative error of order 2^{-t}.

(v)
$$fl_2(AB) \equiv AB + F, \qquad (9.13)$$
where
$$|F| < 2^{-t}|AB| + \tfrac{3}{2}n2^{-2t_2}|A|\,|B|. \qquad (9.14)$$

Computation of square roots

10. Square roots are usually unavoidable at some stage or other in techniques involving the use of elementary unitary transformations. The bound for the error made in extracting a square root naturally depends to some extent on the algorithm which is used. We do not wish to enter into a detailed discussion of such algorithms. We shall assume that

$$fi(x^{\frac{1}{2}}) \equiv x^{\frac{1}{2}} + \epsilon, \qquad \text{where } |\epsilon| < (1{\cdot}00001)2^{-t-1}, \qquad (10.1)$$

$$fl(x^{\frac{1}{2}}) \equiv x^{\frac{1}{2}}(1 + \epsilon), \quad \text{where } |\epsilon| < (1{\cdot}00001)2^{-t}, \qquad (10.2)$$

since these are true on ACE. In most matrix algorithms the number of square roots is small compared with the number of other operations, and even appreciably larger errors than those of (10.1) and (10.2) would make little significant difference to the overall error bounds.

Block-floating vectors and matrices

11. In fixed-point arithmetic it is common to use a method of representation of vectors and matrices which to some extent combines the virtues of fixed and floating-point.

In this scheme a single scale factor in the form of a power of 2 is associated with all components of a vector or a matrix. This factor is chosen so that the modulus of the largest component lies between $\frac{1}{2}$ and 1. Such a vector is called a *standardized block-floating vector*. Thus, for example in (11.1), a is a standardized block-floating vector and A is a *standardized block-floating matrix*. (We have naturally used the decimal scale for our illustration.)

$$a = 10^{-4} \begin{bmatrix} 0\cdot0013 \\ -0\cdot2763 \\ 0\cdot0002 \\ 0\cdot0013 \end{bmatrix}, \quad A = 10^5 \begin{bmatrix} 0\cdot0067 & 0\cdot2175 & 0\cdot4132 \\ -0\cdot1352 & 0\cdot3145 & -0\cdot5173 \\ -0\cdot0167 & -0\cdot0004 & 0\cdot5432 \end{bmatrix}.$$

$$(11.1)$$

Sometimes the columns of a matrix differ so much in magnitude that each column is represented as a block-floating vector. Similarly we may represent each row by a block-floating vector. Finally we could have a scale factor associated with each row and each column.

The advantage of this scheme is that only one word is allocated to the exponent and each individual element is allowed a full word for its storage, no digits being wasted on the exponent as they would be in floating-point computation.

Often at intermediate stages in a computation a block-floating vector is computed which does not have any component of modulus as large as $\frac{1}{2}$. Such a vector is called *a non-standardized* block-floating vector and illustrated by a in (11.2).

$$a = 10^3 \begin{bmatrix} -0\cdot0023 \\ 0\cdot0123 \\ 0\cdot0003 \\ -0\cdot0021 \end{bmatrix}.$$

$$(11.2)$$

In some situations the scale factor associated with a vector is of no importance; this is usually true, for example, for an eigenvector. We define a *normalized block-floating vector* to be a standardized block-floating vector having the scale factor 2^0. The scale factor is usually omitted in such a vector.

Fundamental limitations of t-digit computation

12. We have already remarked in Chapter 2 that when a matrix A requires more than t digits for its representation then we cannot present a computer using a t-digit word with the exact matrix. We must *ab initio* restrict ourselves to a 't-digit approximation' to the matrix.

Suppose we agree for the moment to regard the original matrix stored in the computer as exact, then we might well ask ourselves what is the optimum accuracy that we can reasonably expect from any technique when we employ, say, floating-point arithmetic with a t-digit mantissa. (In future we shall refer to this as *t-digit floating-point arithmetic*.) We can shed some light on this question if we consider a very simple operation on A, the calculation of the transform $D^{-1}AD$ where $D = \mathrm{diag}(d_i)$. If we write

$$B = fl(D^{-1}AD), \qquad (12.1)$$

we see that

$$b_{ij} = d_i^{-1}a_{ij}d_j(1+\epsilon_{ij}), \qquad (12.2)$$

where

$$|\epsilon_{ij}| < 2 \cdot 2^{-t_1}. \qquad (12.3)$$

The relations (12.2) show that B is an exact similarity transformation of the matrix with elements $a_{ij}(1+\epsilon_{ij})$, and hence the eigenvalues of B are those of $(A+F)$ (say), where

$$|F| < 2 \cdot 2^{-t_1} |A|. \qquad (12.4)$$

It is not difficult to construct examples for which this bound is almost attained.

The eigenvalues λ_i' of B will therefore have departed to some extent from the eigenvalues λ_i of A. For an isolated eigenvalue we know from Chapter 2, § 9, that as $t \to \infty$

$$\lambda_i' - \lambda_i \sim y_i^T F x_i / s_i, \qquad (12.5)$$

where y_i and x_i are normalized left-hand and right-hand eigenvectors of A. From (12.4) we have

$$|y_i^T F x/s_i| < 2 \cdot 2^{-t_1} \|A\|_E / |s_i| \leqslant 2 \cdot 2^{-t_1} n^{\frac{1}{2}} \|A\|_2 / |s_i|. \qquad (12.6)$$

We deduce that even the simplest of similarity transformations introduces an error in the eigenvalue λ_i which may be expected to be proportional to 2^{-t} and $1/|s_i|$. If $1/|s_i|$ is large it is likely that the eigenvalue λ_i will to a corresponding extent suffer from rounding errors made in a transformation. Note, however, that if the original matrix were not given exactly, errors of this type have already been introduced.

Now most eigenvalue techniques involve the computation of a considerable number of similarity transformations, each of which is rather more complex than the simple transformation we have just considered. Suppose k such transformations are involved. It seems inherently improbable that we would be able to obtain an *a priori* bound for the error which was less than k times the size of that we have just given. Indeed, if the transformation were in any way 'unstable' we might expect that a much larger bound would be relevant.

If then a technique involves k similarity transformations, and we can prove that the final matrix is exactly similar to $(A+F)$ and can obtain a bound of the form

$$|F| < 2k \cdot 2^{-t_1} |A| \tag{12.7}$$

which holds for all A, then we take the view that such a technique must be regarded as extraordinarily numerically stable. It should be emphasized that even such a technique as this might well give inaccurate results for eigenvalues for which the corresponding values of $1/|s_i|$ are large. We regard any inaccuracy which arises from this factor as inevitable and in no way to be regarded as attributable to shortcomings of the technique under consideration.

The reader may well feel that we have little justification at this stage for the comments of the last paragraph. We shall not attempt to defend them here but we hope that the analyses given in the remaining chapters will convince the reader of their validity. However to avoid a fundamental misunderstanding of our view point we add a few comments.

13. (i) We do not imply that if *any* matrix has a large value for any of its n-condition numbers then the corresponding eigenvalue is *certain* to suffer a loss of accuracy of the order of magnitude suggested by the bounds in (12.5) and (12.6) when any technique is used. Consider, for example, the ill-conditioned matrix defined in § 33 of Chapter 2; this belongs to the class of matrices which are known as *tri-diagonal matrices*. A matrix A is tri-diagonal if

$$a_{ij} = 0 \quad \text{unless } j = i, i+1, \text{ or } i-1. \tag{13.1}$$

Now we shall develop techniques for dealing specifically with tri-diagonal matrices and we shall be able to show that the rounding errors made in these techniques are equivalent to perturbatinos *in the non-zero elements of A*. Now although the matrix of § 33 has ill-conditioned eigenvalues, they are not at all sensitive to perturbations in the elements in the three lines centred on the diagonal and hence we may well obtain very accurate results with this matrix, using these techniques, notwithstanding the large values of $1/|s_i|$.

(ii) If we accumulate inner-products or use any of the operations of the $fi_2(\ )$ or $fl_2(\ )$ type, then although we may think of this as t-digit arithmetic, we are effectively obtaining some results which have $2t$ correct figures. If extensive use is made of such operations we may well obtain appreciably better results than we have indicated.

(iii) In our discussion of the previous section we have referred to the possibility that some of the transformations might be 'unstable' and implied that the errors could then be much larger. We may illustrate this concept of instability in a very simple manner. Consider the matrix A_0 defined by

$$A_0 = \begin{bmatrix} 10^{-3}\,(0\cdot3000) & 0\cdot4537 \\ 0\cdot3563 & 0\cdot5499 \end{bmatrix}. \tag{13.2}$$

Using 4-digit floating-point arithmetic, let us multiply A_0 on the right by a matrix of the type M_1 (Chapter 1, § 40), chosen so as to reduce the (1, 2) element to zero. We have

$$M_1 = \begin{bmatrix} 1 & 10^4\,(-0\cdot1512) \\ 0 & 1 \end{bmatrix},$$

$$A_0M_1 = \begin{bmatrix} 10^{-3}\,(0\cdot3000) & 0 \\ 0\cdot3563 & 10^3\,(-0\cdot5382) \end{bmatrix} \tag{13.3}$$

(where we have entered zero for the (1, 2) element of A_0M_1) followed by

$$M_1^{-1} = \begin{bmatrix} 1 & 10^4\,(0\cdot1512) \\ 0 & 1 \end{bmatrix}$$

$$M_1^{-1}A_0M_1 = \begin{bmatrix} 10^3\,(0\cdot5387) & 10^6\,(-0\cdot8138) \\ 0\cdot3563 & 10^3\,(-0\cdot5382) \end{bmatrix} = A_1 \text{ (say).} \tag{13.4}$$

Since the trace of the computed transform is $0\cdot5000$ while that of the original matrix is $0\cdot5502$, A_1 cannot be exactly similar to any (A_0+F) for which

$$|F| < 2\,.\,10^{-4}\,|A_0|. \tag{13.5}$$

The very large element in M_1 has resulted in an A_1 having much larger elements than A_0, and the rounding errors are not equivalent to 'small' perturbations in the elements of A_0. In fact the harmful effects of the transformation are more severe than may be appreciated merely from examining the error in the trace. The eigenvalues of A_0 are $0 \cdot 7621$ and $-0 \cdot 2119$ to four decimal places, while those of A_1 are $0 \cdot 2500 \pm i(5 \cdot 342)$. In general we shall find that when we use a transformation matrix with large elements, the rounding errors made in the transformation are equivalent to large perturbations in the original matrix.

Eigenvalue techniques based on reduction by similarity transformations

14. Bearing in mind the considerations of the previous section, we now give a general analysis of eigenvalue techniques based on similarity transformations. Let us consider an algorithm which, starting from a matrix A_0, defines a sequence of matrices $A_1, A_2,..., A_s$ defined by

$$A_p = H_p^{-1} A_{p-1} H_p \quad (p = 1,..., s). \tag{14.1}$$

Usually, but not always, each A_p is simpler than its predecessor in that H_p is defined in such a way that A_p retains all the zero elements introduced by previous steps and has some additional zero elements. There are some techniques of an iterative character, however, for which the sequence is essentially infinite and the zero elements occur only in the limiting matrix of the sequence. Techniques of this type must be sufficiently rapidly convergent in practice for elements to become 'zero to working accuracy' in a reasonable number of steps. The analysis which follows applies in any case; if the technique is of this iterative type then the sets of zero elements to which we refer from time to time may be empty.

The set of matrices A_p defined by (14.1) corresponds, of course, to exact computation. In practice rounding errors are made, and as a result a set of matrices is obtained which we may denote by A_0, $\bar{A}_1, \bar{A}_2,..., \bar{A}_p$. The $\bar{A}_p$ have one particular feature which deserves comment. The exact algorithm is designed so as to produce a certain set of zero elements in each A_p; *in the practical procedure the matrices $\bar{A}_p$ are also given the same set of zero elements.* In other words no attempt is made to compute these elements; they are automatically set equal to zero at each stage of the process.

Consider now the typical stage in the practical procedure at which we have derived the matrix $\bar{A}_{p-1}$. There exists a matrix H'_p (say) which is obtained by applying the algorithm exactly to $\bar{A}_{p-1}$ and this matrix is not, of course, the same as H_p. Moreover, in practice we obtain a matrix $\bar{H}_p$ which will not, in general, be the same as H'_p because we shall make rounding errors in applying the algorithm to $\bar{A}_{p-1}$. There are thus three sets of matrices.

(i) The matrices H_p and A_p satisfying (14.1) which correspond to exact execution of the algorithm throughout.

(ii) The matrices H'_p and A'_p satisfying

$$A'_p = (H'_p)^{-1}\bar{A}_{p-1}H'_p, \tag{14.2}$$

where H'_p is the matrix corresponding to the exact application of the pth step of the algorithm to $\bar{A}_{p-1}$.

(iii) The matrices $\bar{H}_p$ and $\bar{A}_p$ which are obtained in practice.

We shall find that the first set of matrices plays no part in any of the error analyses which we give. *It might be expected that algorithms which are numerically stable would lead to matrices $\bar{A}_p$ which are close to A_p*, and that it would be fruitful to attempt to find a bound for $\bar{A}_p - A_p$, but this proves to be untrue. At first sight this may seem scarcely credible since one might feel that unless $\bar{A}_p$ were close to A_p there would be no reason to suppose that the eigenvalues of $\bar{A}_p$ were close to those of A_p. However we must remember that we do not really care if $\bar{A}_p$ is close to A_p provided it *is* exactly similar to some matrix which is close to A_0. If this is true, then the error in each eigenvalue will depend primarily on its sensitivity to perturbations in the original matrix and we have already remarked that we cannot expect to avoid the adverse effect of ill-condition of the original matrix.

Error analysis of methods based on elementary non-unitary transformations

15. The considerations involved in the general analysis are different according as the matrices H_p are unitary or non-unitary elementary matrices. We consider the non-unitary case first.

At the pth stage we calculate the computed matrix $\bar{H}_p$ and from it and $\bar{A}_{p-1}$ we derive the matrix $\bar{A}_p$. The elementary non-unitary matrices are all of such a nature that having obtained $\bar{H}_p$, the inverse $\bar{H}_p^{-1}$ *is immediately available without rounding errors* as will be seen from Chapter 1, § 40. In general the elements of $\bar{A}_p$ fall into three main classes.

(i) Elements which have already been eliminated by previous steps and remain zero. Even if the pth step were performed exactly starting with $\bar{A}_{p-1}$ these elements would remain zero since the new elements are usually defined to be linear combinations of zero elements of $\bar{A}_{p-1}$.

(ii) Elements which the algorithm is designed to eliminate in the pth step. These are given the value zero automatically even though the corresponding elements of $\bar{H}_p^{-1}\bar{A}_{p-1}\bar{H}_p$ are not exactly zero, because of the rounding errors made in computing $\bar{H}_p$. Note that the elements of $(H_p')^{-1}\bar{A}_{p-1}H_p'$ in these positions *are* exactly zero because by definition H_p' is the matrix corresponding to the exact application of the pth step of the algorithm to $\bar{A}_{p-1}$.

(iii) Elements which are derived by computation of $\bar{H}_p^{-1}\bar{A}_{p-1}\bar{H}_p$.

We define F_p by the relation

$$\bar{A}_p \equiv \bar{H}_p^{-1}\bar{A}_{p-1}\bar{H}_p + F_p \quad (p = 1,\ldots, s), \tag{15.1}$$

so that it denotes the difference between the matrix which we accept as $\bar{A}_p$ and the exact product $\bar{H}_p^{-1}\bar{A}_{p-1}\bar{H}_p$.

The equations (15.1) can be combined to give

$$\bar{A}_s \equiv F_s + G_s^{-1}F_{s-1}G_s + G_{s-1}^{-1}F_{s-2}G_{s-1} + \ldots + G_2^{-1}F_1G_2 + G_1^{-1}A_0G_1, \tag{15.2}$$

where

$$G_p = \bar{H}_p\bar{H}_{p+1}\ldots\bar{H}_s. \tag{15.3}$$

If we define F by the equation

$$F = F_s + G_s^{-1}F_{s-1}G_s + G_{s-1}^{-1}F_{s-2}G_{s-1} + \ldots + G_2^{-1}F_1G_2, \tag{15.4}$$

then (15.2) becomes

$$\bar{A}_s \equiv F + G_1^{-1}A_0G_1, \quad \text{or} \quad \bar{A}_s - F \equiv G_1^{-1}A_0G_1. \tag{15.5}$$

This implies that $\bar{A}_s$ differs from an exact similarity transformation of A_0 by the matrix F. If on the other hand we define K by the relation

$$K = G_1FG_1^{-1}, \tag{15.6}$$

then (15.2) becomes

$$\bar{A}_s \equiv G_1^{-1}(K + A_0)G_1. \tag{15.7}$$

This shows that $\bar{A}_s$ is exactly similar to $(A_0 + K)$ and hence we have obtained an expression for the perturbation in A_0 which is equivalent to the errors made in the computation. From (15.3), (15.4) and (15.6) we have

$$K = L_sF_sL_s^{-1} + L_{s-1}F_{s-1}L_{s-1}^{-1} + \ldots + L_1F_1L_1^{-1}, \tag{15.8}$$

where

$$L_p = \bar{H}_1\bar{H}_2\ldots\bar{H}_p. \tag{15.9}$$

Now, from what we have said earlier we may regard $\|K\|/\|A_0\|$ as providing a measure of the effectiveness of the algorithm under examination. The smaller the *a priori* bound we can find for this ratio the more effective we must regard the algorithm.

We shall find that it is extremely difficult to find really useful *a priori* bounds for the perturbation matrix K for methods based upon non-unitary elementary transformations. However equation (15.8) gives a strong indication that it is desirable to use elementary matrices which do not have large norms. We shall most commonly use matrices of the type M_p or N_p of Chapter 1, § 40, and in the interests of numerical stability shall usually arrange our algorithms so that the elements of the M_p and N_p are bounded in modulus by unity if this is possible.

This conclusion is reinforced if we observe that a transformation with a matrix $\bar{H}_p$ having large elements is likely to lead to an $\bar{A}_p$ with much larger elements than $\bar{A}_{p-1}$, and the rounding errors in this stage are therefore likely to be comparatively large, leading to an F_p with a large norm (cf. the example of § 13).

We have given expressions for F and K, the equivalent perturbations in the final *computed* matrix $\bar{A}_s$ and the initial matrix A_0 respectively. Bounds for F are likely to be useful *a posteriori* when we have computed the eigensystem of $\bar{A}_s$ (at least approximately) and have estimates for the conditions of its eigenvalues. *A priori* bounds for K, on the other hand, enable us to assess the algorithm.

Turning now to the individual equations (15.1) we see that if we wish to follow the history of an eigenvalue λ_i we need to have estimates of $s_i^{(p)}$ at each stage. In general $s_i^{(p)}$ will be different for each value of p. If we could guarantee that the conditions of the eigenvalues were not deteriorating then we would know that the effect of the various F_p was no more than additive.

Error analysis of methods based on elementary unitary transformations

16. In general we can obtain much better error bounds for methods based on elementary unitary transformations but there is one difficulty which did not arise in the case of non-unitary transformations.

At the pth stage we compute a matrix $\bar{H}_p$ which, if we made no errors at this stage, would be the exact unitary transformation H'_p defined by the algorithm for the matrix $\bar{A}_{p-1}$. Having computed $\bar{H}_p$ we assume that it is indeed unitary and we take $\bar{H}_p^H$ as its inverse.

It is *desirable* that we should do this in all cases, but when the initial A_0 is Hermitian it is almost mandatory, since we will naturally wish to preserve the Hermitian nature of the $\bar{A}_p$. *However in using the formula $\bar{H}_p^H \bar{A}_{p-1} \bar{H}_p$ we are not even attempting to compute a similarity transformation of $\bar{A}_{p-1}$.*

It is important that the method of computing $\bar{H}_p$ should be such that it gives a matrix which is close to an exact unitary matrix. Usually we shall ensure that $\bar{H}_p$ is very close to H'_p, the exact unitary matrix defined by the pth step of the algorithm for $\bar{A}_{p-1}$, but this is not the only course as we show in § 19.

Suppose we have shown that

$$\bar{H}_p \equiv H'_p + X_p, \qquad (16.1)$$

and can find a satisfactorily small bound for X_p. Then we define F_p as in (15.1) by the relation

$$\bar{A}_p \equiv \bar{H}_p^H \bar{A}_{p-1} \bar{H}_p + F_p, \qquad (16.2)$$

so that F_p is again the difference between the accepted $\bar{A}_p$ and the exact product $\bar{H}_p^H \bar{A}_{p-1} \bar{H}_p$. We have then

$$\bar{A}_p \equiv (H'_p + X_p)^H \bar{A}_{p-1}(H'_p + X_p) + F_p = (H'_p)^H \bar{A}_{p-1} H'_p + Y_p, \qquad (16.3)$$

where

$$Y_p = (H'_p)^H \bar{A}_{p-1} X_p + (X_p)^H \bar{A}_{p-1} H'_p + X_p^H \bar{A}_{p-1} X_p + F_p, \qquad (16.4)$$

and $(H'_p)^H \bar{A}_{p-1} H'_p$ *is* an exact unitary similarity transformation of $\bar{A}_{p-1}$.

Combining the equations (16.3) for values of p from 1 to s we have

$$\bar{A}_s \equiv Y_s + G_s^H Y_{s-1} G_s + G_{s-1}^H Y_{s-2} G_{s-1} + \dots + G_2^H Y_1 G_2 + G_1^H A_0 G_1, \qquad (16.5)$$

where

$$G_p = H'_p H'_{p+1} \dots H'_s. \qquad (16.6)$$

Comparing this with (15.2) and (15.3) we see that H'_p has taken the place of $\bar{H}_p$ and Y_p has taken the place of F_p. We may write

$$\bar{A}_s \equiv Y + G_1^H A_0 G_1, \qquad (16.7)$$

where

$$Y = Y_s + G_s^H Y_{s-1} G_s + G_{s-1}^H Y_{s-2} G_{s-1} + \dots + G_2^H Y_1 G_2, \qquad (16.8)$$

or alternatively

$$\bar{A}_s = G_1^H (Z + A_0) G_1, \qquad (16.9)$$

where

$$Z = L_s Y_s L_s^H + L_{s-1} Y_{s-1} L_{s-1}^H + \dots + L_1 Y_1 L_1^H \qquad (16.10)$$

and

$$L_p = H'_1 H'_2 \dots H'_p. \qquad (16.11)$$

Superiority of the unitary transformation

17. Now in spite of the greater complexity of the definition of Y_p compared with that of F_p in § 15, the equations (16.5) to (16.11) are much more informative than the corresponding equations of § 15. All the matrices G_p and L_p are exactly unitary and hence we have

$$\|Y\|_2 \leqslant \|Y_s\|_2 + \|Y_{s-1}\|_2 + \ldots + \|Y_1\|_2 \qquad (17.1)$$

from the invariance of the 2-norm under unitary transformations. Similarly, since $Z = G_1 Y G_1^H$, we have

$$\|Z\|_2 = \|Y\|_2. \qquad (17.2)$$

Equation (16.7) shows that $(\bar{A}_s - Y)$ is an exact unitary similarity transform of A_0, while (16.9) shows that $\bar{A}_s$ is an exact unitary similarity transform of $(A_0 + Z)$.

In practice we shall use either plane rotations or the elementary Hermitian transformations, and we shall find that in all cases we shall be able to obtain a most satisfactory bound for the norm of X_p defined by (16.1). If we can prove, for example, that

$$\|X_p\|_2 \leqslant a2^{-t}, \qquad (17.3)$$

then (16.4) gives

$$\|Y_p\|_2 \leqslant \|\bar{A}_{p-1}\|_2 \, (2a2^{-t} + a^2 2^{-2t}) + \|F_p\|_2, \qquad (17.4)$$

and from (16.3)

$$\|\bar{A}_p\|_2 \leqslant \|\bar{A}_{p-1}\|_2 + \|Y_p\|_2 \leqslant (1 + a2^{-t})^2 \|\bar{A}_{p-1}\|_2 + \|F_p\|_2. \qquad (17.5)$$

Now F_p is the matrix of errors made in attempting to calculate $\bar{H}_p^H \bar{A}_{p-1} \bar{H}_p$, and because $\bar{H}_p$ is almost a unitary matrix the general order of magnitude of the elements of $\bar{A}_p$ will be much the same as that of the elements of $\bar{A}_{p-1}$. Hence it is usually not difficult in practice to find a bound for $\|F_p\|_2$ of the form

$$\|F_p\|_2 \leqslant f(p, n)2^{-t} \|\bar{A}_{p-1}\|_2, \qquad (17.6)$$

where $f(p, n)$ is some simple function of p and n. From (17.4), (17.5), (17.6) we have

$$\|\bar{A}_p\|_2 \leqslant [(1 + a2^{-t})^2 + f(p, n)2^{-t}] \|\bar{A}_{p-1}\|_2, \qquad (17.7)$$

and

$$\|Y_p\|_2 \leqslant [2a + a^2 2^{-t} + f(p, n)]2^{-t} \|\bar{A}_{p-1}\|_2, \qquad (17.8)$$

and thence we can obtain an *a priori* bound for the norm of the equivalent perturbation Z in A_0 of the form

$$\|Z\|_2 \leqslant 2^{-t} \|A_0\|_2 \sum_{p=1}^{s} x_p, \qquad (17.9)$$

where
$$x_p = [2a + a^2 2^{-t} + f(p, n)] y_{p-1} \qquad (17.10)$$

$$y_p = \prod_{i=1}^{p} [(1 + a2^{-t})^2 + f(i, n)2^{-t}]. \qquad (17.11)$$

The analysis of any algorithm based on unitary transformations reduces then to the problem of finding values for a and $f(p, n)$.

Real symmetric matrices

18. When A_0 is real the algorithms we shall use will give real (and therefore orthogonal) matrices H_p. When in addition A_0 is symmetric we shall maintain exact symmetry in all the $\bar{A}_p$ by computing only their upper triangles and completing the matrices by symmetry. Note that in this case the matrices F_p and Y_p in (16.3) are exactly symmetric, since they represent the differences between $\bar{A}_p$ and $(H'_p + X_p)^T \bar{A}_{p-1}(H'_p + X_p)$ and $(H'_p)^T \bar{A}_{p-1} H'_p$ respectively, and each of the last two matrices is exactly symmetric. The relation (17.9) now implies that Z, which is exactly symmetric, satisfies

$$\|Z\|_2 \leqslant 2^{-t} \max |\lambda_i| \sum_{p=1}^{s} x_p. \qquad (18.1)$$

If the computed eigenvalues are $\lambda_1 + \delta\lambda_1, \lambda_2 + \delta\lambda_2, \ldots, \lambda_n + \delta\lambda_n$ then from Chapter 2, § 44,

$$\frac{|\delta\lambda_j|}{\max |\lambda_i|} \leqslant 2^{-t} \sum_{p=1}^{s} x_p, \qquad (18.2)$$

so that our analysis gives us a bound for the relative error in any λ_j in terms of the eigenvalue of maximum modulus.

We have given the analysis throughout in terms of the 2-norm; a similar analysis may be carried out using the Euclidean norm. An application of the Wielandt-Hoffman theorem (Chapter 2, § 48) then gives

$$[\sum(\delta\lambda_i)^2]^{\frac{1}{2}} \leqslant \|Z\|_E. \qquad (18.3)$$

We would like to emphasize that the error analysis for unsymmetric matrices is not, in general, more difficult than that for symmetric matrices. It is the implications of the bound for Z which are so different in the two cases.

Limitations of unitary transformations

19. The discussion of the previous section may well have given the impression that any algorithm based on unitary transformations is necessarily stable, but this is far from true. We have discussed only techniques in which each matrix A_p is obtained by a single elementary unitary transformation of A_{p-1} and have assumed that the extra zero elements (if any) that were introduced in A_p came as a result of that single transformation. We showed that provided we can prove that in practice $\bar{H}_p$ is close to H_p' then stability is guaranteed. There were only two sources of error.

(i) Those made in attempting to compute elements of $\bar{A}_p$ from $\bar{H}_p^H \bar{A}_{p-1} \bar{H}_p$.

(ii) Those corresponding to the insertion of certain new zero elements in $\bar{A}_p$.

In practice it is always easy to show that the first set of errors causes no trouble. As far as the second set is concerned the situation is even more satisfactory. Our analysis was based on finding a small bound for $Y_p = \bar{A}_p - (H_p')\bar{A}_{p-1}H_p'$. For the special zero elements the corresponding elements of Y_p are exactly zero, from the definition of H_p'. However, this particular way of conducting the analysis does not always give the sharpest bound. If we can prove that $\bar{H}_p$ is close to any exact unitary matrix R_p, can obtain a good bound for $(R_p - \bar{H}_p)$ and, moreover, can show that the elements of $(R_p^H \bar{A}_{p-1} R_p)$ in the positions corresponding to the zero elements of $\bar{A}_p$ are small, then we can carry out an analysis similar to that of §§ 16, 17.

However, there are some algorithms based on elementary unitary transformations *in which strategic zero elements are produced only as the end product of several successive transformations*; the emergence of these zeros usually depends in a highly implicit manner on the definition of the H_p matrices contained in the algorithm.

Suppose, starting from A_0, that these zeros ought first to arise in the matrix A_q, that is after q transformations. If we could prove that the product $\bar{H}_1 \bar{H}_2 ... \bar{H}_q$ is close to $H_1 H_2 ... H_q$ then the method would be stable. However, even if we can show that each $\bar{H}_p$ is close to the corresponding H_p' (as we usually can), this is of little value since the H_p' may well be arbitrarily different from the H_p. As implied in § 14 the matrices A_p and $\bar{A}_p$ may diverge completely from each other and it was an essential feature of our analysis that no comparisons are made between A_p and $\bar{A}_p$ or H_p and $\bar{H}_p$.

If the success of such an algorithm depends critically on our being able to accept zeros in those elements of $\bar{A}_q$ corresponding to the zero elements in A_q, then it may be quite useless for practical purposes. We shall give examples of the catastrophic failure of such algorithms.

Error analysis of floating-point computation of plane rotations

20. Although there exists a considerable number of techniques based on unitary transformations, there are a few basic computations which occur repeatedly, and in the next few sections we give error analyses of these computations for future use. We start with plane rotations in standard floating-point arithmetic since these are probably the simplest; we restrict ourselves to the real case, so that the matrices involved are orthogonal.

In many algorithms the transformation at a typical stage is defined in the following manner.

We have a vector $\bar{x}$ (usually a column or row of the current transformed matrix) and we wish to determine a rotation R in the (i, j) plane so that $(R\bar{x})_j = 0$.

If we write

$$\bar{x}^T = (\bar{x}_1, \bar{x}_2, ..., \bar{x}_n), \qquad (20.1)$$

then the components $\bar{x}_p$ are standard floating-point numbers. Referring to Chapter 1, § 43, we see that the exact rotation defined by the algorithm for the vector $\bar{x}$ is such that

$$\left.\begin{array}{ll} r_{ii} = \cos\theta, & r_{ij} = \sin\theta \\ r_{ji} = -\sin\theta, & r_{jj} = \cos\theta \end{array}\right\}, \qquad (20.2)$$

where

$$\bar{x}_i/(\bar{x}_i^2 + \bar{x}_j^2)^{\frac{1}{2}} = \cos\theta = c, \qquad \bar{x}_j/(\bar{x}_i^2 + \bar{x}_j^2)^{\frac{1}{2}} = \sin\theta = s.$$

If $\bar{x}_j = 0$ we take $c = 1$, $s = 0$ whether or not $\bar{x}_i = 0$.

We shall refer to the matrix which is defined exactly by equations (20.2) as R, although strictly in accordance with our previous sections it should be called R', since it is the exact matrix defined by the algorithm for the current values of computed elements. However, since the matrix which would have been obtained by exact computation throughout never enters into our considerations we can now drop the prime in the interests of a simpler notation.

In practice we obtain a matrix $\bar{R}$ with elements $\bar{c}$ and $\bar{s}$ in place of c and s where

$$\bar{c} = fl[\bar{x}_i/(\bar{x}_i^2 + \bar{x}_j^2)^{\frac{1}{2}}], \qquad \bar{s} = fl[\bar{x}_j/(\bar{x}_i^2 + \bar{x}_j^2)^{\frac{1}{2}}]. \qquad (20.3)$$

The computation of $\bar{c}$ and $\bar{s}$ takes place in the steps

$$a = fl(\bar{x}_i^2 + \bar{x}_j^2), \quad b = fl(a^{\frac{1}{2}}), \quad c = fl(\bar{x}_i/b), \quad s = fl(\bar{x}_j/b). \qquad (20.4)$$

If $\bar{x}_j = 0$ we take $\bar{c} = 1$, $\bar{s} = 0$, so that in this case $\bar{R} = R$.

In order to illustrate the application of our method of floating-point analysis we analyse (20.4) in detail. In later applications which are as simple as this we shall give only the final results. In order to obtain as sharp a bound as possible we revert to bounds of the form

$$(1-2^{-t})^r \leqslant 1+\epsilon \leqslant (1+2^{-t})^r, \tag{20.5}$$

rather than using the simpler bound in terms of 2^{-t_1}. We have

(i) $a = fl(\bar{x}_i^2 + \bar{x}_j^2) \equiv \bar{x}_i^2(1+\epsilon_1) + \bar{x}_j^2(1+\epsilon_2) \equiv (\bar{x}_i^2 + \bar{x}_j^2)(1+\epsilon_3),$

where
$$(1-2^{-t})^2 \leqslant 1+\epsilon_i \leqslant (1+2^{-t})^2 \quad (i = 1, 2, 3).$$

(ii) $b = fl(a^{\frac{1}{2}}) \equiv a^{\frac{1}{2}}(1+\epsilon_4)$ where $|\epsilon_4| < (1 \cdot 00001)2^{-t}$ from (10.2).

(iii) $\bar{c} = fl(\bar{x}_i/b) \equiv \bar{x}_i(1+\epsilon_5)/b, \qquad \bar{s} = fl(\bar{x}_j/b) \equiv \bar{x}_j(1+\epsilon_6)/b,$
$$|\epsilon_i| \leqslant 2^{-t} \quad (i = 5, 6).$$

Combining (i), (ii), and (iii) we have

$$\bar{c} \equiv c(1+\epsilon_5)/(1+\epsilon_3)^{\frac{1}{2}}(1+\epsilon_4), \qquad \bar{s} \equiv s(1+\epsilon_6)/(1+\epsilon_3)^{\frac{1}{2}}(1+\epsilon_4), \tag{20.6}$$

and simple computation shows that certainly

$$\bar{c} \equiv c(1+\epsilon_7), \qquad \bar{s} = s(1+\epsilon_8), \tag{20.7}$$

$$|\epsilon_i| < (3 \cdot 003)2^{-t} \quad (i = 7, 8), \tag{20.8}$$

where the factor $3 \cdot 003$ is extremely generous. The computed $\bar{c}$ and $\bar{s}$ therefore have low relative errors. It must be emphasized, however, that the use of floating-point computation does not always guarantee that we obtain results having low relative errors even when the computations are as simple as those we have just analysed (Chapter 5, § 12).

We may write
$$\bar{c} = c+\delta c, \qquad \bar{s} = s+\delta s, \qquad \bar{R} = R+\delta R, \tag{20.9}$$

where δR is null except at the intersections of the ith and jth rows and columns where it has elements

$$\begin{bmatrix} \delta c & \delta s \\ -\delta s & \delta c \end{bmatrix} \tag{20.10}$$

in this pattern. Hence we have

$$\|\delta R\|_2 = (\delta c^2 + \delta s^2)^{\frac{1}{2}} < (3 \cdot 003)2^{-t}, \tag{20.11}$$

$$\|\delta R\|_E = [2(\delta c^2 + \delta s^2)]^{\frac{1}{2}} < (3 \cdot 003)2^{\frac{1}{2}}2^{-t}, \tag{20.12}$$

showing that the computed $\bar{R}$ is close to the exact R corresponding to the given $\bar{x}_i$ and $\bar{x}_j$. If we are given any two floating-point numbers $\bar{x}_i$ and $\bar{x}_j$ we shall call $\bar{R}$ the *corresponding approximate rotation in the (i, j) plane*.

Multiplication by a plane rotation

21. Consider now a composite operation which we define as follows: Given two floating-point numbers $\bar{x}_i$ and $\bar{x}_j$ and a matrix A having standard floating-point components, compute the approximate rotation $\bar{R}$ in the (i, j) plane and then compute $fl(\bar{R}A)$.

For simplicity of notation we first consider the computation of $fl(\bar{R}a)$ where a is a vector. Only two elements of a are modified, those in positions i and j. If we write

$$b = fl(\bar{R}a) \tag{21.1}$$

then

$$\left. \begin{aligned}
b_p &= a_p \quad (p \neq i, j) \\
b_i &= fl(\bar{c}a_i + \bar{s}a_j) \equiv \bar{c}a_i(1+\epsilon_1) + \bar{s}a_j(1+\epsilon_2) \\
b_j &= fl(-\bar{s}a_i + \bar{c}a_j) \equiv -\bar{s}a_i(1+\epsilon_3) + \bar{c}a_j(1+\epsilon_4) \\
(1-2^{-t})^2 &\leqslant 1+\epsilon_i \leqslant (1+2^{-t})^2 \quad (i = 1, 2, 3, 4)
\end{aligned} \right\} . \tag{21.2}$$

We may express this in the form

$$\begin{bmatrix} b_i \\ b_j \end{bmatrix} = \begin{bmatrix} c & s \\ -s & c \end{bmatrix} \begin{bmatrix} a_i \\ a_j \end{bmatrix} + \begin{bmatrix} \delta c & \delta s \\ -\delta s & \delta c \end{bmatrix} \begin{bmatrix} a_i \\ a_j \end{bmatrix} + \begin{bmatrix} \bar{c}a_i\epsilon_1 + \bar{s}a_j\epsilon_2 \\ -\bar{s}a_i\epsilon_3 + \bar{c}a_j\epsilon_4 \end{bmatrix}. \tag{21.3}$$

Using the bounds for δc, δs and the ϵ_i, and writing

$$fl(\bar{R}a) - Ra = f \text{ (say)}, \tag{21.4}$$

we have

$$\begin{aligned}
\|f\|_2 &< (\delta c^2 + \delta s^2)^{\frac{1}{2}}(a_i^2 + a_j^2)^{\frac{1}{2}} + (2 \cdot 0002)2^{-t}[2(\bar{c}^2 + \bar{s}^2)]^{\frac{1}{2}}(a_i^2 + a_j^2)^{\frac{1}{2}} \\
&\leqslant 6 \cdot 2^{-t}(a_i^2 + a_j^2)^{\frac{1}{2}}, \tag{21.5}
\end{aligned}$$

and again the factor 6 is quite generous. Notice that this means that if a_i and a_j are particularly small relative to some of the other components, then $\|f\|_2/\|a\|_2$ is particularly small. We shall find that this is not true for fixed-point computation. From (21.3) and the relation $\|x+y+z\|_2 \leqslant \|x\|_2 + \|y\|_2 + \|z\|_2$ we have

$$(b_i^2 + b_j^2)^{\frac{1}{2}} \leqslant (a_i^2 + a_j^2)^{\frac{1}{2}} + 6 \cdot 2^{-t}(a_i^2 + a_j^2)^{\frac{1}{2}} = (1 + 6 \cdot 2^{-t})(a_i^2 + a_j^2)^{\frac{1}{2}}. \tag{21.6}$$

This relation is important in the subsequent analysis.

An examination of (20.11) *and* (21.5) *shows that half the error comes from the errors made in computing* $\bar{R}$ *and the other half from the errors made in multiplying by the computed* $\bar{R}$.

When the vector a is the vector $\bar{x}$ from which the elements $\bar{x}_i$ and $\bar{x}_j$ were obtained, then in practice one does not usually compute $fl(\bar{R}x)$ but instead takes the ith component to be $fl[(\bar{x}_i^2 + \bar{x}_j^2)^{\frac{1}{2}}]$ and the jth component to be zero. This expression defining the ith component will already have been computed when determining $\bar{R}$. The two elements of f are therefore $fl[(\bar{x}_i^2 + \bar{x}_j^2)^{\frac{1}{2}}] - (\bar{x}_i^2 + \bar{x}_j^2)^{\frac{1}{2}}$ and zero, and it may readily be verified that the bound (21.5) is more than adequate in this case. The special treatment of the vector $\bar{x}$ cannot therefore cause any trouble.

Our result extends immediately to the computation of $fl(\bar{R}A)$ where A is an $(n \times m)$ matrix. The m vectors of $\bar{R}A$ are treated independently and we have

$$fl(\bar{R}A) - RA \equiv F, \qquad (21.7)$$

where F is null except in rows i and j, and

$$\|F\|_E \leqslant 6.2^{-t}(a_{i1}^2 + a_{j1}^2 + a_{i2}^2 + a_{j2}^2 + \dots + a_{im}^2 + a_{jm}^2)^{\frac{1}{2}}. \qquad (21.8)$$

Notice that the bound is for $\|F\|_E$ *and not for* $\|F\|_2$. The squares of the norms of rows i and j replace a_i^2 and a_j^2 in our analysis of the vector case. The bound (21.8) is adequate even if one of the columns of A is our vector $\bar{x}$ and is treated specially.

Multiplication by a sequence of plane rotations

22. Suppose we are given a set of s pairs of floating-point numbers $\bar{x}_{i_p}, \bar{x}_{j_p}$ $(p = 1, \dots, s)$ and a matrix A_0. Corresponding to each pair we may compute an approximate plane rotation $\bar{R}_p$ in the (i_p, j_p) plane and thence we may compute the s matrices defined by

$$\left.\begin{array}{l} \bar{A}_1 = fl(\bar{R}_1 A_0) \equiv R_1 A_0 + F_1 \\[6pt] \bar{A}_p = fl(\bar{R}_p \bar{A}_{p-1}) \equiv R_p \bar{A}_{p-1} + F_p \quad (p = 2, \dots, s) \end{array}\right\}. \qquad (22.1)$$

Combining equations (22.1) for a sequence of values of p

$$\bar{A}_p \equiv Q_1 A_0 + Q_2 F_1 + Q_3 F_2 + \dots + Q_p F_{p-1} + F_p, \qquad (22.2)$$

where
$$Q_k = R_p R_{p-1} \dots R_k.$$

Since, by definition, the R_k are exactly orthogonal, the Q_k are also exactly orthogonal. Hence

$$\|\bar{A}_p - R_p R_{p-1} \dots R_1 A_0\| \leqslant \|F_1\| + \|F_2\| + \dots + \|F_p\| \quad (p = 1, \dots, s), \qquad (22.3)$$

for either the 2-norm or the Euclidean norm.

Now in practice we shall often be concerned with a set of $\frac{1}{2}n(n-1)$ rotations in the successive planes $(1, 2)$ $(1, 3)...(1, n)$; $(2, 3)$ $(2, 4)...$ $(2, n);...; (n-1, n)$. We show that for a special set of rotations of this kind we can obtain a very satisfactory bound for

$$\|\bar{A}_N - R_N R_{N-1}...R_1 A_0\|_E \quad (N = \tfrac{1}{2}n(n-1)). \tag{22.4}$$

23. The main difficulty associated with the analysis is that of notation, so we shall start first with a single vector in place of the matrix A_0 and for further simplicity we shall take $n = 5$ (and hence $N = 10$), the proof in the general case being sufficiently evident from this. We denote the initial vector by v_0 and the successive derived vectors by $\bar{v}_1, \bar{v}_2,..., \bar{v}_N$. In Table 1 we display the elements of the vectors $\bar{v}_i$ and to avoid a confusing multiplicity of suffices we have used a different letter for the elements in each position. We have changed the suffix only when the corresponding element is altered. In general the first element is changed in each of the first $(n-1)$ transformations and then undergoes no further change. Similarly the second element is changed in each of the next $(n-2)$ transformations and then undergoes no further change, and so on. We have

TABLE 1

v_0	$\bar{v}_1$	$\bar{v}_2$	$\bar{v}_3$	$\bar{v}_4$	$\bar{v}_5$	$\bar{v}_6$	$\bar{v}_7$	$\bar{v}_8$	$\bar{v}_9$	$\bar{v}_{10}$
a_0	$\bar{a}_1$	$\bar{a}_2$	$\bar{a}_3$	$\bar{a}_4$	$\bar{a}_4$	$\bar{a}_4$	$\bar{a}_4$	$\bar{a}_4$	$\bar{a}_4$	$\bar{a}_4$
b_0	$\bar{b}_1$	$\bar{b}_1$	$\bar{b}_1$	$\bar{b}_1$	$\bar{b}_2$	$\bar{b}_3$	$\bar{b}_4$	$\bar{b}_4$	$\bar{b}_4$	$\bar{b}_4$
c_0	c_0	$\bar{c}_1$	$\bar{c}_1$	$\bar{c}_1$	$\bar{c}_2$	$\bar{c}_2$	$\bar{c}_2$	$\bar{c}_3$	$\bar{c}_4$	$\bar{c}_4$
d_0	d_0	d_0	d_1	d_1	d_1	d_2	d_2	d_3	d_3	d_4
e_0	e_0	e_0	e_0	$\bar{e}_1$	$\bar{e}_1$	$\bar{e}_1$	$\bar{e}_2$	$\bar{e}_2$	$\bar{e}_3$	$\bar{e}_4$

indicated this grouping in Table 1. Every element is changed $n-1$ times in all, so that the final suffix is $n-1$ for every element.

If we write
$$\bar{v}_p = fl(\bar{R}_p \bar{v}_{p-1}) \equiv R_p \bar{v}_{p-1} + f_p, \tag{23.1}$$

then our previous analysis shows that

$$\|\bar{v}_N - R_N R_{N-1}...R_1 v_0\|_2 \leqslant \|f_1\|_2 + \|f_2\|_2 + ... + \|f_N\|_2. \tag{23.2}$$

Now the analysis of § 21 applies immediately to each transformation, since each $\bar{v}_p$ is a vector with floating-point components, and writing $6.2^{-t} = x$ we have

$$\|f_1\| < x(a_0^2 + b_0^2)^{\frac{1}{2}}; \quad \|f_2\| < x(\bar{a}_1^2 + c_0^2)^{\frac{1}{2}}; \quad \|f_3\| < x(\bar{a}_2^2 + d_0^2)^{\frac{1}{2}}; \quad \|f_4\| < x(\bar{a}_3^2 + e_0^2)^{\frac{1}{2}} \Big\rangle$$

$$\|f_5\| < x(\bar{b}_1^2 + \bar{c}_1^2)^{\frac{1}{2}}; \quad \|f_6\| < x(\bar{b}_2^2 + \bar{d}_1^2)^{\frac{1}{2}}; \quad \|f_7\| < x(\bar{b}_3^2 + \bar{e}_1^2)^{\frac{1}{2}}$$

$$\|f_8\| < x(\bar{c}_2^2 + \bar{d}_2^2)^{\frac{1}{2}}; \quad \|f_9\| < x(\bar{c}_3^2 + \bar{e}_2^2)^{\frac{1}{2}} \Bigg\rangle ,$$

$$\|f_{10}\| < x(\bar{d}_3^2 + \bar{e}_3^2)^{\frac{1}{2}} \Bigg/$$

$$\tag{23.3}$$

where the grouping is designed to emphasize the general form. If we write

$$f = \| f_1 \|_2 + \dots + \| f_N \|_2 = xS, \tag{23.4}$$

then applying the inequality

$$(q_1 + q_2 + \dots + q_N)^2 \leqslant N(q_1^2 + q_2^2 + \dots + q_N^2), \tag{23.5}$$

which holds for any positive q_i, we have from (23.3)

$$S^2 \leqslant 10 \begin{bmatrix} (a_0^2 + b_0^2) + (\bar{a}_1^2 + c_0^2) + (\bar{a}_2^2 + d_0^2) + (\bar{a}_3^2 + e_0^2) \\ + (\bar{b}_1^2 + \bar{c}_1^2) + (\bar{b}_2^2 + \bar{d}_1^2) + (\bar{b}_3^2 + \bar{e}_1^2) \\ + (\bar{c}_2^2 + \bar{d}_2^2) + (\bar{c}_3^2 + \bar{e}_2^2) \\ + (\bar{d}_3^2 + \bar{e}_3^2) \end{bmatrix}, \tag{23.6}$$

where the factor 10 is N in the general case.

Writing $k = (1+x)^2$ for brevity, the relations corresponding to (21.6) for each of the N transformations are

$$\left. \begin{aligned} &\bar{a}_1^2 + \bar{b}_1^2 < k(a_0^2 + b_0^2); \\ &\bar{a}_2^2 + \bar{c}_1^2 < k(\bar{a}_1^2 + c_0^2); \; \bar{b}_2^2 + \bar{c}_2^2 < k(\bar{b}_1^2 + \bar{c}_1^2); \\ &\bar{a}_3^2 + \bar{d}_1^2 < k(\bar{a}_2^2 + d_0^2); \bar{b}_3^2 + \bar{d}_2^2 < k(\bar{b}_2^2 + \bar{d}_1^2); \bar{c}_3^2 + \bar{d}_3^2 < k(\bar{c}_2^2 + \bar{d}_2^2); \\ &\bar{a}_4^2 + \bar{e}_1^2 < k(\bar{a}_3^2 + e_0^2); \; \bar{b}_4^2 + \bar{e}_2^2 < k(\bar{b}_3^2 + \bar{e}_1^2); \; \bar{c}_4^2 + \bar{e}_3^2 < k(\bar{c}_3^2 + \bar{e}_2^2); \; \bar{d}_4^2 + \bar{e}_4^2 < k(\bar{d}_3^2 + \bar{e}_3^2) \end{aligned} \right\} \tag{23.7}$$

We wish to establish a relation between S and $\| v_0 \|_2$. From the relations in the first column of (23.7) we have

$$\left. \begin{aligned} \bar{a}_1^2 &\leqslant k(a_0^2 + b_0^2) && -\bar{b}_1^2 \\ \bar{a}_2^2 &\leqslant k^2(a_0^2 + b_0^2) + kc_0^2 && -k\bar{b}_1^2 - \bar{c}_1^2 \\ \bar{a}_3^2 &\leqslant k^3(a_0^2 + b_0^2) + k^2 c_0^2 + kd_0^2 - k^2 \bar{b}_1^2 - k\bar{c}_1^2 - \bar{d}_1^2 \end{aligned} \right\} \tag{23.8}$$

and hence, defining S_r by the relation

$$S_r = 1 + k + k^2 + \dots + k^{r-1}, \tag{23.9}$$

the sum in the first row of the right-hand side of (23.6) is bounded by

$$S_4 a_0^2 + S_4 b_0^2 + S_3 c_0^2 + S_2 d_0^2 + S_1 e_0^2 - S_3 \bar{b}_1^2 - S_2 \bar{c}_1^2 - S_1 \bar{d}_1^2. \tag{23.10}$$

In exactly the same way the remaining three sums in (23.6) are bounded by

$$\left. \begin{aligned} &S_3 \bar{b}_1^2 + S_3 \bar{c}_1^2 + S_2 \bar{d}_1^2 + S_1 \bar{e}_1^2 && -S_2 \bar{c}_2^2 - S_1 \bar{d}_2^2 \\ &S_2 \bar{c}_2^2 + S_2 \bar{d}_2^2 + S_1 \bar{e}_2^2 && -S_1 \bar{d}_3^2 \\ &S_1 \bar{d}_3^2 + S_1 \bar{e}_3^2 \end{aligned} \right\} \tag{23.11}$$

where our alignment indicates the corresponding result for the general case. Summing the equations in (23.10) and (23.11) we see that (23.6) is equivalent to

$$S^2 \leqslant 10 \begin{bmatrix} S_4 a_0^2 + S_4 b_0^2 + S_3 c_0^2 + S_2 d_0^2 + S_1 e_0^2 \\ + k^2 \bar{c}_1^2 + k \bar{d}_1^2 + \bar{e}_1^2 \\ + k \bar{d}_2^2 + \bar{e}_2^2 \\ + \bar{e}_3^2 \end{bmatrix}, \qquad (23.12)$$

considerable cancellation taking place between the negative terms in each row and the positive terms in the succeeding row. The first row is expressed entirely in terms of the elements of v_0. To reduce the second row to a comparable form we multiply the inequalities in the first column of (23.7) by k^3, k^2, k, and 1 respectively and add. This gives

$$\bar{a}_4^2 + k^3 \bar{b}_1^2 + k^2 \bar{c}_1^2 + k \bar{d}_1 + \bar{e}_1^2 \;<\; k^4(a_0^2 + b_0^2) + k^3 c_0^2 + k^2 d_0^2 + k e_0^2, \qquad (23.13)$$

and hence *a fortiori*

$$k^3 \bar{b}_1^2 + (k^2 \bar{c}_1^2 + k \bar{d}_1^2 + \bar{e}_1^2) \;<\; k^4(a_0^2 + b_0^2) + k^3 c_0^2 + k^2 d_0^2 + k e_0^2. \qquad (23.14)$$

Similarly from the second column of (23.7)

$$k^2 \bar{c}_2^2 + (k \bar{d}_2^2 + \bar{e}_2^2) \;<\; k^3(\bar{b}_1^2 + \bar{c}_1^2) + k^2 \bar{d}_1^2 + k \bar{e}_1^2$$
$$<\; k^5(a_0^2 + b_0^2) + k^4 c_0^2 + k^3 d_0^2 + k^2 e_0^2, \qquad (23.15)$$

where the last line follows *a fortiori* on multiplying (23.14) by k. Similarly from the third column of (23.7) and using (23.15)

$$k \bar{d}_3^2 + (\bar{e}_3^2) \;<\; k^2(\bar{c}_2^2 + \bar{d}_2^2) + k \bar{e}_2^2$$
$$<\; k^6(a_0^2 + b_0^2) + k^5 c_0^2 + k^4 d_0^2 + k^3 e_0^2. \qquad (23.16)$$

The expressions in parentheses on the left-hand sides of (23.14), (23.15) and (23.16) are the second, third and fourth rows of (23.13) and hence we have *a fortiori*

$$S^2 \;<\; 10(S_7 a_0^2 + S_7 b_0^2 + S_6 c_0^2 + S_5 d_0^2 + S_4 e_0^2), \qquad (23.17)$$

and from the method of proof it is clear that in the general case the coefficients are $S_{2n-3}, S_{2n-3}, S_{2n-4}, \ldots, S_{n-1}$. Replacing each of the coefficients in (23.17) by the largest, S_7, we have certainly

$$S^2 \;<\; 10 S_7 (a_0^2 + b_0^2 + c_0^2 + d_0^2 + e_0^2), \qquad (23.18)$$

and hence in the general case

$$S^2 \;<\; \tfrac{1}{2} n(n-1) S_{2n-3} \|v_0\|_2^2$$
$$\leqslant \tfrac{1}{2} n(n-1)(2n-3)(1+x)^{4n-8} \|v_0\|_2^2 \quad (\text{since } S_r < r\,(1+x)^{2r-2})$$
$$\leqslant n^3 (1+x)^{4n-8} \|v_0\|_2^2. \qquad (23.19)$$

Referring to (23.2) and (23.4), we see finally that this means that

$$\|\bar{v}_N - R_N R_{N-1} \dots R_1 v_0\|_2 \leqslant xn^{\frac{3}{2}}(1+x)^{2n-4} \|v_0\|_2. \qquad (23.20)$$

24. The analysis extends immediately to the pre-multiplication of an $(n \times m)$ matrix by the same N approximate rotations. We have only to replace the square of the ith element of the current vector by the sum of the squares of the elements in the ith row of the current transformed matrices at every stage in the argument. The expression corresponding to (23.6) still has only $\frac{1}{2}n(n-1)$ terms in parenthesis within the large bracket. We have then

$$\|\bar{A}_N - R_N R_{N-1} \dots R_1 A_0\|_E \leqslant xn^{\frac{3}{2}}(1+x)^{2n-4} \|A_0\|_E$$
$$= 6 \cdot 2^{-t} n^{\frac{3}{2}}(1+6 \cdot 2^{-t})^{2n-4} \|A_0\|_E. \qquad (24.1)$$

The term $(1+6 \cdot 2^{-t})^{2n-4}$ tends to infinity exponentially with n but fortunately this is unimportant, since the computed $\bar{A}_N$ is likely to be of value only as long as the ratio $\|\text{error}\|_E / \|A_0\|_E$ is small. Suppose we wish, for example, to keep this ratio below 10^{-6}. Then we must have

$$6 \cdot 2^{-t} n^{\frac{3}{2}}(1+6 \cdot 2^{-t})^{2n-4} \leqslant 10^{-6}. \qquad (24.2)$$

If we regard t as fixed then this gives us the largest value of n for which we can guarantee an acceptable result. Clearly we must have $6 \cdot 2^{-t} n^{\frac{3}{2}} < 10^{-6}$ and hence

$$(1+6 \cdot 2^{-t})^{2n-4} < (1+10^{-6} n^{-\frac{3}{2}})^{2n} < \exp(2 \cdot 10^{-6} n^{-\frac{1}{2}}) < (1+2 \cdot 10^{-6}). \qquad (24.3)$$

This factor is therefore quite insignificant in its effect on the range of applicability.

The right-hand side of (24.1) represents an extreme upper bound and there are several points in the analysis at which inequalities were used which are quite weak for some distributions of the matrix elements. The statistical distribution of the rounding errors can also be expected to reduce the error well below the level of our bound, and for this reason alone a factor of the order of $n^{\frac{3}{4}}$ in place of the $n^{\frac{3}{2}}$ might well be more realistic for large n.

Notice that in obtaining this bound we assumed no relationship between the $\bar{x}_i$ and $\bar{x}_j$ determining the rotations and the matrices A_0 and $\bar{A}_p$. Our result holds whatever the origin of these pairs of floating-point numbers. Each pair could be derived from the current $\bar{A}_p$ or it might be derived from some completely independent source. Our comparison, though, is between the exact transform corresponding to the *given* pairs of numbers and the computed transform. If each pair is drawn from the current $\bar{A}_p$, no comparison is made between the final computed $\bar{A}_N$ and the matrix that would have been obtained if exact computation had been performed throughout, *including the selection of the pairs of numbers from the exact transforms.*

Error in product of approximate plane rotations

25. So far we have considered only pre-multiplication and not similarity transformations. Before extending our result, we remark that these one-sided transformations are of interest in themselves. In the next chapter we shall consider the reduction of a square matrix to triangular form by pre-multiplication with elementary orthogonal matrices. The result of § 24 is immediately applicable to this computation.

In the eigenvector problem we shall need to compute the product of sets of approximate plane rotations and we shall be interested in the departure of the computed product from orthogonality. Taking $A_0 = I$ in (24.1) we have

$$\|\bar{P}_N - R_N R_{N-1} \dots R_1\|_E \leqslant x n^{\frac{3}{2}}(1+x)^{2n-4} \|I\|_E = x n^2 (1+x)^{2n-4},$$

$$(25.1)$$

where $\bar{P}_N$ is the computed product; the matrix $R_N R_{N-1} \dots R_1$ is, of course, exactly orthogonal. If we denote the ith columns of $\bar{P}_N$ and $R_N R_{N-1} \dots R_1$ by $\bar{p}_i$ and p_i respectively, then since each column of the matrix I is treated independently, an application of (23.20) shows that

$$\|\bar{p}_i - p_i\|_2 < x n^{\frac{3}{2}}(1+x)^{2n-4} \|e_i\|_2 = x n^{\frac{3}{2}}(1+x)^{2n-4}, \qquad (25.2)$$

and hence

$$1 - x n^{\frac{3}{2}}(1+x)^{2n-4} \leqslant \|\bar{p}_i\|_2 \leqslant 1 + x n^{\frac{3}{2}}(1+x)^{2n-4}. \qquad (25.3)$$

Also if we write

$$\bar{p}_i = p_i + g_i, \qquad (25.4)$$

then

$$\bar{p}_j^T \bar{p}_i = (p_j^T + g_j^T)(p_i + g_i) = g_j^T p_i + p_j^T g_i + g_j^T g_i \qquad (25.5)$$

$$|\bar{p}_j^T \bar{p}_i| \leqslant \|g_j\|_2 + \|g_i\|_2 + \|g_j\|_2 \|g_i\|_2$$
$$< x n^{\frac{3}{2}}(1+x)^{2n-4}[2 + x n^{\frac{3}{2}}(1+x)^{2n-4}]. \qquad (25.6)$$

This inequality gives bounds for the departure of the columns of P_N from orthonormality. As an example, if $t = 40$ and $n = 100$ we have

$$x n^{\frac{3}{2}} = 6 \times 2^{-40} \times 1000 < 6.10^{-9},$$

and hence the computed product of 100 approximate rotations will certainly be an orthogonal matrix to about eight decimal places.

Statistical considerations indicate that it will probably be an orthogonal matrix to about ten decimal places.

Errors in similarity transforms

26. If B_0 is an $(m \times n)$ matrix, then there is clearly an exactly similar error analysis for the post-multiplication of B_0 by the transposes of a set of N approximate rotations. In fact if $\bar{B}_N$ is the final matrix we have, as in (24.1),

$$\| \bar{B}_N - B_0 R_1^T R_2^T \ldots R_N^T \|_E \leqslant x n^{\frac{3}{2}} (1+x)^{2n-4} \| B_0 \|_E. \tag{26.1}$$

Now suppose we start with A_0 and perform the N pre-multiplications to obtain $\bar{A}_N$. This matrix has floating-point elements and hence may be used as B_0 in (26.1) giving

$$\| \bar{B}_N - \bar{A}_N R_1^T R_2^T \ldots R_N^T \|_E \leqslant x n^{\frac{3}{2}} (1+x)^{2n-4} \| \bar{A}_N \|_E = y \| \bar{A}_N \|_E$$
$$\text{(say).} \quad (26.2)$$

Also from (24.1) we have, with the same abbreviated notation,

$$\| \bar{A}_N - R_N R_{N-1} \ldots R_1 A_0 \|_E \leqslant y \| A_0 \|_E, \tag{26.3}$$

and hence
$$\| \bar{A}_N \|_E \leqslant (1+y) \| A_0 \|_E. \tag{26.4}$$

Combining these results we have

$$\begin{aligned}
\| \bar{B}_N - R_N \ldots R_1 A_0 R_1^T \ldots R_N^T \|_E &\leqslant \| \bar{B}_N - \bar{A}_N R_1^T \ldots R_N^T \|_E + \\
&\quad + \| (\bar{A}_N - R_N \ldots R_1 A_0) R_1^T \ldots R_N^T \|_E \\
&\leqslant y \| \bar{A}_N \|_E + \| \bar{A}_N - R_N \ldots R_1 A_0 \|_E \\
&\leqslant y(1+y) \| A_0 \|_E + y \| A_0 \|_E \\
&= y(2+y) \| A_0 \|_E, \tag{26.5}
\end{aligned}$$

showing that the final matrix differs from the exact similarity transform of A_0 by a matrix with its norm bounded by

$$6 \cdot 2^{-t} n^{\frac{3}{2}} (1 + 6 \cdot 2^{-t})^{2n-4} [2 + 6 \cdot 2^{-t} n^{\frac{3}{2}} (1 + 6 \cdot 2^{-t})^{2n-4}] \| A_0 \|_E. \tag{26.6}$$

Since the computed matrix $\bar{B}_N$ will be of interest only if y is small, the presence of the y term in the parentheses in (26.5) is of no importance and the relative error is only twice as big as for the one-sided transformation, as we would expect.

27. There are *some* algorithms in which the pre-multiplications are completed before performing any of the right-hand transformations, but frequently this is not possible because the pth pair of floating-point numbers which determines the pth rotation is taken from the matrix obtained by performing the first $(p-1)$ similarity transformations.

Starting with A_0 the sequence may then be defined by

$$\left.\begin{array}{ll} \bar{X}_1 = fl(\bar{R}_1 A_0), & \bar{A}_1 = fl(\bar{X}_1 \bar{R}_1^T) \\ \bar{X}_p = fl(\bar{R}_p \bar{A}_{p-1}), & \bar{A}_p = fl(\bar{X}_p \bar{R}_p^T) \quad (p = 2,...,N) \end{array}\right\}. \quad (27.1)$$

At first sight it appears possible that the insertion of a post-multiplication between two pre-multiplications might disturb the rather special relationship between the norms of the rows of the $\bar{A}_p$ which were essential in deriving the bound. However, consider the effect of inserting an exact post-multiplication by an exact rotation between two pre-multiplications. The post-multiplication operates on two columns, leaving the norm of each row undisturbed; now the contribution to the error made by any pre-multiplication depends only on the norms of the two rows which it modifies, and hence the insertion of any number of exact post-multiplications has no effect on the bound obtained for the error. This shows immediately that the order in which the multiplications takes place has only a second-order effect. An argument of a similar nature to that of § 23 shows that

$$\|\bar{A}_N - R_N...R_1 A_0 R_1^T...R_N^T\|_E < 12.2^{-t} n^{\frac{3}{2}} (1+6.2^{-t})^{4n-7} \|A_0\|_E. \quad (27.2)$$

As far as the factor $(1+6.2^{-t})^{4n-7}$ is concerned the argument is a little weak and could possibly be improved, but since the computation is only of value if $12.2^{-t} n^{\frac{3}{2}}$ is appreciably less than unity this factor can never be of practical importance.

This final result justifies for plane rotations our remark in § 17 that the rounding errors made in the actual multiplications do not lead to numerical instability. We have given an analysis of considerable generality in order to cover a wide range of possible applications. In most algorithms the successive transforms will have more and more zero elements and hence a somewhat better bound is usually obtainable in any specific application. However, such improvements usually do not affect the main factor $n^{\frac{3}{2}} 2^{-t}$ in the bound, but only the constant multiplier.

Symmetric matrices

28. When the original matrix is symmetric we shall take all $\bar{A}_p$ to be exactly symmetric. Our analysis does not therefore apply immediately since when a rotation is performed in the (i, j) plane the elements below the diagonal in each transform are not actually computed but are automatically assigned the same values as the elements above the diagonal. If we assume that $\bar{A}_{p-1}$ is symmetric, then even if the computation were actually performed using an approximate plane rotation, exact symmetry would be preserved except possibly in the (i, j) and (j, i) elements.

It does not seem to be possible to show in any simple manner that the error bound we have obtained in the unsymmetric case is adequate to cover the symmetric case, but it is comparatively simple to show that an extra factor of $2^{\frac{1}{2}}$ is adequate. The main point is as follows:

If we perform the complete $(1, 2)$ transformation using either the pre-multiplication or the post-multiplication first, the matrix of errors will be null except in rows and columns 1 and 2 and will be symmetric except for the elements at the intersections of these rows and columns. We may denote the errors in these positions by

$$\begin{bmatrix} p & q \\ r & s \end{bmatrix} \quad \text{and} \quad \begin{bmatrix} p & r \\ q & s \end{bmatrix} \tag{28.1}$$

respectively. The two error matrices will be the same elsewhere in these rows and columns. Now if we assume complete symmetry the errors in the positions corresponding to (28.1) will be

$$\begin{bmatrix} p & q \\ q & s \end{bmatrix} \quad \text{or} \quad \begin{bmatrix} p & r \\ r & s \end{bmatrix}. \tag{28.2}$$

The sum of the squares of the errors corresponding to each of the four cases is

(i) $\qquad p^2 + q^2 + r^2 + s^2 + \Sigma^2 = k_1$

(ii) $\qquad p^2 + q^2 + r^2 + s^2 + \Sigma^2 = k_1$

(iii) $\qquad p^2 + q^2 + q^2 + s^2 + \Sigma^2 = k_3$

(iv) $\qquad p^2 + r^2 + r^2 + s^2 + \Sigma^2 = k_4$

$$\left.\begin{matrix}\\\\\\\\\end{matrix}\right\} \tag{28.3}$$

respectively, where Σ^2 is the sum of the squares of the other errors and is the same in all four cases. We have therefore certainly

$$k_3, k_4 < 2k_1, \tag{28.4}$$

and this gives the factor $2^{\frac{1}{2}}$ in the Euclidean norm. It is obvious that this result is quite weak. Indeed a factor of $2^{-\frac{1}{2}}$ rather than $2^{\frac{1}{2}}$ might well be more realistic because the error matrix corresponding to both pre-multiplication and post-multiplication combined is of the form

$$
\begin{bmatrix}
p & q & a_1 & a_2 & \cdots & a_n \\
q & s & \beta_1 & \beta_2 & \cdots & \beta_n \\
\hline
a_1 & \beta_1 & & & & \\
a_2 & \beta_2 & & O & & \\
\cdot & \cdot & & & & \\
a_n & \beta_n & & & &
\end{bmatrix} . \tag{28.5}
$$

In the bound of § 27 we have taken the Euclidean norms of the row and column contributions separately. If it were not for the elements p, q, r, s we could gain a factor of $2^{-\frac{1}{2}}$ by considering the contributions together.

Plane rotations in fixed-point arithmetic

29. We now turn our attention to fixed-point computation; since every element of an orthogonal matrix lies in the range -1 to $+1$ and the norms of the transformed matrices remain constant (except for the small drift resulting from rounding errors), the use of fixed-point arithmetic does not present any severe scaling problems.

We consider first the problem of § 20. We are given a vector $\bar{x}$ with fixed-point components and we wish to determine a rotation in the (i, j) plane such that $(R\bar{x})_j = 0$. We shall assume that

$$
\sum \bar{x}_i^2 < 1. \tag{29.1}
$$

Now in floating-point computation an almost orthogonal matrix is obtained using equations (20.2) without any special computational tricks. This is no longer true in fixed-point arithmetic. Suppose, for example, we have

$$
\bar{x}_i = 0 \cdot 000002, \qquad \bar{x}_j = 0 \cdot 000003, \tag{29.2}
$$

and we use 6-decimal digit arithmetic. We have then
$$
fl(\bar{x}_i^2 + \bar{x}_j^2)^{\frac{1}{2}} = 0 \cdot 000004 \text{ (correct to 6 decimal places),}
$$
giving
$$
\cos \theta = 0 \cdot 500000, \qquad \sin \theta = 0 \cdot 750000. \tag{29.3}
$$

It is immediately apparent that the corresponding approximate plane rotation is far from orthogonal.

We shall describe two methods of overcoming this, the first of which is very convenient on a computer on which inner-products may be accumulated exactly. In either method we take $\cos\theta = 1$ and $\sin\theta = 0$ if $\bar{x}_j = 0$. We write

$$S^2 = \bar{x}_i^2 + \bar{x}_j^2, \tag{29.4}$$

and S^2 is computed exactly; from (29.1) it cannot exceed capacity. We then determine Σ^2 where $\Sigma^2 = 2^{2k}S^2$ and k is the smallest integer for which

$$\tfrac{1}{4} < \Sigma^2 = 2^{2k}S^2 \leqslant 1. \tag{29.5}$$

Finally the required $\bar{s}$ and $\bar{c}$ are computed using the relations

$$\overline{\Sigma} = fl[(\Sigma^2)^{\frac{1}{2}}], \qquad \bar{c} = fl(2^k\bar{x}_i/\overline{\Sigma}), \qquad \bar{s} = fl(2^k\bar{x}_j/\overline{\Sigma}). \tag{29.6}$$

The quantities $2^k\bar{x}_i$ and $2^k\bar{x}_j$ may be computed without rounding error and we have

$$\overline{\Sigma} \equiv \Sigma + \epsilon_1, \qquad |\epsilon_1| \leqslant (1{\cdot}00001)2^{-t-1}, \tag{29.7}$$

$$\bar{c} \equiv \frac{2^k\bar{x}_i}{\overline{\Sigma}} + \epsilon_2, \qquad \bar{s} = \frac{2^k\bar{x}_j}{\overline{\Sigma}} + \epsilon_3, \tag{29.8}$$

$$|\epsilon_i| \leqslant 2^{-t-1} \qquad (i = 2, 3). \tag{29.9}$$

Combining these results

$$\bar{c} = c + \epsilon_4, \qquad \bar{s} = s + \epsilon_5, \tag{29.10}$$

where

$$\epsilon_4 = \epsilon_1 c/\overline{\Sigma} + \epsilon_2, \qquad \epsilon_5 = \epsilon_1 s/\overline{\Sigma} + \epsilon_3. \tag{29.11}$$

The scaling of S, $\bar{x}_i$ and $\bar{x}_j$ ensure that $\bar{c}$ and $\bar{s}$ have small absolute errors, though one or other may have a high relative error. If we write $\bar{R} = R + \delta R$ then δR is null except at the intersections of rows and columns i and j where it has the pattern of values

$$\begin{bmatrix} \epsilon_4 & \epsilon_5 \\ -\epsilon_5 & \epsilon_4 \end{bmatrix}, \tag{29.12}$$

and from this and the additive property of norms we have

$$\|\delta R\|_2 \leqslant |\epsilon_1|/\overline{\Sigma} + (\epsilon_2^2 + \epsilon_3^2)^{\frac{1}{2}} < (1{\cdot}71)2^{-t} \text{ (say)} \tag{29.13}$$

for any reasonable t. Similarly

$$\|\delta R\|_E < (1{\cdot}71)2^{\frac{1}{2}}2^{-t} < (2{\cdot}42)2^{-t}. \tag{29.14}$$

Alternative computation of $\sin\theta$ and $\cos\theta$

30. On some computers the technique we have just described may
not be very convenient. The following scheme is satisfactory on almost
any computer. Since

$$\cos\theta = \bar{x}_i/(\bar{x}_i^2+\bar{x}_j^2)^{\frac{1}{2}}, \qquad \sin\theta = \bar{x}_j/(\bar{x}_i^2+\bar{x}_j^2)^{\frac{1}{2}}, \qquad (30.1)$$

if we write

$$\bar{x}_i/\bar{x}_j = m \quad\text{and}\quad \bar{x}_j/\bar{x}_i = n, \qquad (30.2)$$

we have

$$\cos\theta = m/(m^2+1)^{\frac{1}{2}} = 1/(n^2+1)^{\frac{1}{2}}, \qquad (30.3)$$

$$\sin\theta = 1/(m^2+1)^{\frac{1}{2}} = n/(n^2+1)^{\frac{1}{2}}. \qquad (30.4)$$

We may use the expressions involving m or n according as $|m|$ is
or is not less than unity. In order to keep all numbers within range,
(30.3) and (30.4) may be written in the form

$$\cos\theta = \tfrac{1}{2}m/(\tfrac{1}{4}m^2+\tfrac{1}{4})^{\frac{1}{2}} = \tfrac{1}{2}/(\tfrac{1}{4}n^2+\tfrac{1}{4})^{\frac{1}{2}}, \qquad (30.5)$$

$$\sin\theta = \tfrac{1}{2}/(\tfrac{1}{4}m^2+\tfrac{1}{4})^{\frac{1}{2}} = \tfrac{1}{2}n/(\tfrac{1}{4}n^2+\tfrac{1}{4})^{\frac{1}{2}}. \qquad (30.6)$$

The application of this technique again gives values of $\bar{c}$ and $\bar{s}$ having
small absolute errors.

Pre-multiplication by an approximate fixed-point rotation

31. The operation considered in § 21 is somewhat simpler in fixed-
point computation and we pass immediately to the estimate of
$fi(\bar{R}A)-RA = F$ (say) without considering a vector multiplication
first. We have

$$F = fi(\bar{R}A)-RA = fi(\bar{R}A)-\bar{R}A+\bar{R}A-RA = fi(\bar{R}A)-\bar{R}A+\delta RA. \tag{31.1}$$

We shall need to make some assumption about the size of the elements
of A if fixed-point work is to be practicable. *We shall assume that the*
2-norms of A and all derived matrices are bounded by unity and shall
examine the significance of this assumption only when the analysis is
complete (cf. § 32).

Only rows i and j are modified by the pre-multiplication so that
$fi(\bar{R}A)-\bar{R}A$ is null everywhere except in these rows. If we assume
that we use $fi_2(\ )$ operations then we have

$$[fi_2(\bar{R}A)]_{ik} = fi_2(\bar{c}a_{ik}+\bar{s}a_{jk}) \equiv \bar{c}a_{ik}+\bar{s}a_{jk}+\epsilon_{ik}, \qquad (31.2)$$

$$[fi_2(\bar{R}A)]_{jk} = fi_2(-\bar{s}a_{ik}+\bar{c}a_{jk}) \equiv -\bar{s}a_{ik}+\bar{c}a_{jk}+\epsilon_{jk}, \qquad (31.3)$$

$$|\epsilon_{ik}|, |\epsilon_{jk}| < 2^{-t-1} \quad (i = 1,...,n). \qquad (31.4)$$

For $fi(\ )$ operations the error bounds are twice as big. Relations (31.2) to (31.4) show that we may write

$$fi_2(\bar{R}A) - \bar{R}A = G, \tag{31.5}$$

where G is null except in rows i and j where it has elements which are bounded in modulus by 2^{-t-1}. Hence we have

$$\|G\|_2 \leqslant \| |G| \|_2 \leqslant 2^{-t-1}(2n)^{\frac{1}{2}}, \tag{31.6}$$

$$\|G\|_E \leqslant 2^{-t-1}(2n)^{\frac{1}{2}}, \tag{31.7}$$

and (31.1) and (29.13) give

$$\|F\|_2 = \|fi_2(\bar{R}A) - RA\|_2 \leqslant \|G\|_2 + \|\delta R\|_2 \|A\|_2$$
$$\leqslant 2^{-t-1}(2n)^{\frac{1}{2}} + (1 \cdot 71)2^{-t}. \tag{31.8}$$

Notice that the bound for the contribution to the error coming from the actual multiplication contains the factor $n^{\frac{1}{2}}$ whereas the contribution arising from the departure of $\bar{R}$ from R is independent of n. In floating-point computation the two contributions were almost exactly equal and both were independent of n. Notice also that even if rows i and j consist entirely of elements which are small compared with $\|A\|_2$ we do not obtain a more favourable bound. This contrasts with the result which we obtained for floating-point computation.

If the $\bar{x}_i$ and $\bar{x}_j$ which determine a rotation are elements in some column of rows i and j respectively of $\bar{A}_{p-1}$ then one does not usually compute the new values of these two elements as in (31.2) and (31.3). Instead the values $fi(2^{-k}\bar{\Sigma})$ (cf. § 29) and zero are inserted. It is not difficult to see that this is not serious. The two corresponding elements in RA are $2^{-k}\Sigma$ and zero respectively and hence the errors in these positions are
$$fi(2^{-k}\bar{\Sigma}) - 2^{-k}\Sigma \quad \text{and} \quad \text{zero,}$$
and we have

$$fi(2^{-k}\bar{\Sigma}) - 2^{-k}\Sigma = 2^{-k}\bar{\Sigma} + \epsilon - 2^{-k}\Sigma \quad (|\epsilon| \leqslant 2^{-t-1}),$$
giving

$$|fi(2^{-k}\bar{\Sigma}) - 2^{-k}\Sigma| \leqslant 2^{-k} |\bar{\Sigma} - \Sigma| + |\epsilon| \leqslant 2^{-t-1}(1 + 2^{-k}) \leqslant 2^{-t}. \tag{31.9}$$

It is clear that we need change only the factor in (31.8) from $1 \cdot 71$ to $2 \cdot 21$. If then we take

$$\|F\|_2 \leqslant 2^{-t-1}(2n)^{\frac{1}{2}} + (2 \cdot 21)2^{-t} \tag{31.10}$$

the special case is covered.

Multiplication by a sequence of plane rotations (fixed-point)

32. We now consider the fixed-point version of the problem of § 22. Analogous to equations (22.1) and (22.2) we have

$$\left.\begin{array}{l} \bar{A}_1 = fi_2(\bar{R}_1 A_0) \equiv R_1 A_0 + F_1 \\ \bar{A}_p = fi_2(\bar{R}_p \bar{A}_{p-1}) \equiv R_p \bar{A}_{p-1} + F_p \quad (p = 2,\ldots, s) \end{array}\right\} \tag{32.1}$$

and

$$\bar{A}_p = Q_1 A_0 + Q_2 F_1 + \ldots + Q_p F_{p-1} + F_p, \tag{32.2}$$

where

$$Q_k = R_p R_{p-1} \ldots R_k. \tag{32.3}$$

Hence

$$\|\bar{A}_p - R_p R_{p-1} \ldots R_1 A_0\|_2 \leqslant \|F_1\|_2 + \ldots + \|F_p\|_2. \tag{32.4}$$

If we assume $\|\bar{A}_i\|_2 \leqslant 1$ for all of the transforms, then the analysis of the last section shows that

$$\|F_i\|_2 \leqslant [(2n)^{\frac{1}{2}} + 4 \cdot 42]2^{-t-1} = x \text{ (say)}, \tag{32.5}$$

and hence

$$\|\bar{A}_p - R_p R_{p-1} \ldots R_1 A_0\|_2 \leqslant px, \tag{32.6}$$

giving

$$\|\bar{A}_p\|_2 \leqslant \|A_0\|_2 + px. \tag{32.7}$$

We can now verify that if s transformations are contemplated, then provided

$$\|A_0\|_2 < 1 - sx, \tag{32.8}$$

$$\|\bar{A}_p\|_2 < 1 \quad (p = 0, 1,\ldots, s). \tag{32.9}$$

For a set of N transformations such as were considered in § 22 our final result is

$$\|\bar{A}_N - R_N R_{N-1} \ldots R_1 A_0\|_2 \leqslant Nx \sim \tfrac{1}{4} n^{\frac{3}{2}} 2^{\frac{1}{2}} 2^{-t}, \tag{32.10}$$

provided A_0 is scaled so that

$$\|A_0\|_2 \leqslant 1 - Nx. \tag{32.11}$$

33. Since we are interested in the result only when Nx is small, the initial scaling requirement is scarcely more severe in practice than that which would be necessary even if no rounding errors were involved. However the requirement of (32.11) is impractical since the computation of $\|A_0\|_2$ is about as difficult a problem as the computation of the eigenvalues of A_0. On the other hand we have

$$\|A_0\|_2 \leqslant \|A_0\|_E, \tag{33.1}$$

so that if

$$\|A_0\|_E < 1 - Nx \tag{33.2}$$

the same is true *a fortiori* of $\|A_0\|_2$, and $\|A_0\|_E$ *is* easily computed. If the initial scaling *is* such that (33.2) is satisfied then we can show in much the same way that

$$\|\bar{A}_N - R_N R_{N-1} \dots R_1 A_0\|_E \leqslant Ny, \tag{33.3}$$

where

$$y = [(2n)^{\frac{1}{2}} + 5 \cdot 84] 2^{-t-1}. \tag{33.4}$$

Since x and y are almost equal when n is large, if A_0 has been scaled so as to satisfy (33.2), then (33.3) is the sharper result.

The computed product of an approximate set of plane rotations

34. We may apply these results to the computation of the product of a sequence of approximate plane rotations as in § 25. However, we cannot take $A_0 = I$ in (32.10) because this violates (32.11). If we take $A_0 = \frac{1}{2}I$ then (32.11) will be satisfied for any value of n for which the product is of any value. If $\bar{P}$ is the final computed matrix we have

$$\|\bar{P} - \tfrac{1}{2} R_N R_{N-1} \dots R_1\|_2 \leqslant Nx, \tag{34.1}$$

$$\|2\bar{P} - R_N R_{N-1} \dots R_1\|_2 \leqslant 2Nx \tag{34.2}$$

and the error is again of order $n^{\frac{1}{2}}$.

Errors in similarity transformations

35. Our analysis immediately provides bounds for the errors made in a sequence of N similarity transformations. In fact if we compute $fl_2(\bar{R}_N \bar{R}_{N-1} \dots \bar{R}_1 A_0 \bar{R}_1^T \dots \bar{R}_{N-1}^T \bar{R}_N^T)$ whatever the order of the multiplications, we have

$$\|\bar{A}_N - R_N \dots R_1 A_0 R_1^T \dots R_N^T\|_2 \leqslant 2Nx, \tag{35.1}$$

provided

$$\|A_0\|_2 \leqslant 1 - 2Nx. \tag{35.2}$$

However, the pair of numbers which determines $\bar{R}_p$ will often be elements of $\bar{A}_{p-1}$ and in this case the computations will be performed in the order

$$\left. \begin{array}{ll} \bar{X}_1 = fl_2(\bar{R}_1 A_0), & \bar{A}_1 = fl_2(\bar{X}_1 \bar{R}_1^T) \\ \bar{X}_p = fl_2(\bar{R}_p \bar{A}_{p-1}), & \bar{A}_p = fl_2(\bar{X}_p \bar{R}_p^T) \quad (p = 2, \dots, N) \end{array} \right\}. \tag{35.3}$$

In this case it is advantageous to consider the pre-multiplications and post-multiplications in pairs. We define H_p by the relation

$$H_p = \bar{A}_p - \bar{R}_p \bar{A}_{p-1} \bar{R}_p^T = fl_2(\bar{R}_p \bar{A}_{p-1} \bar{R}_p^T) - \bar{R}_p \bar{A}_{p-1} \bar{R}_p^T, \qquad (35.4)$$

so that

$$\bar{A}_p - R_p \bar{A}_{p-1} R_p^T = H_p + \bar{R}_p \bar{A}_{p-1} \bar{R}_p^T - R_p \bar{A}_{p-1} R_p^T$$

$$= H_p + \delta R_p \bar{A}_{p-1} \bar{R}_p^T + R_p \bar{A}_{p-1} \delta R_p^T + \delta R_p \bar{A}_{p-1} \delta R_p^T. \qquad (35.5)$$

Hence

$$\|\bar{A}_p - R_p \bar{A}_{p-1} R_p^T\|_2 \leqslant \|H_p\|_2 + (2 \|\delta R_p\|_2 + \|\delta R_p\|_2^2) \|\bar{A}_{p-1}\|_2$$

$$\leqslant \|H_p\|_2 + (3 \cdot 43) 2^{-t} \|\bar{A}_{p-1}\|_2 \text{ (say)}, \qquad (35.6)$$

where we have replaced $[2(1 \cdot 71) 2^{-t} + (1 \cdot 71)^2 2^{-2t}]$ by $(3 \cdot 43) 2^{-t}$. Provided $\|\bar{A}_{p-1}\|_2 < 1$ for all relevant p we have then

$$\|\bar{A}_N - R_N \ldots R_1 A_0 R_1^T \ldots R_N^T\|_2 \leqslant \sum_1^N \|H_p\|_2 + N(3 \cdot 43) 2^{-t}. \qquad (35.7)$$

Now H_p is the matrix of errors actually made in computing $\bar{A}_p$ from $\bar{R}_p \bar{A}_{p-1} \bar{R}_p^T$. If the rotation is in the (i, j) plane then the transformation affects only the elements in rows and columns i and j and hence H_p is null elsewhere. Apart from the four elements at the intersections, only rows i and j are affected by the pre-multiplication and only columns i and j by the post-multiplication. We have therefore

$$(H_p)_{ik} = fl_2(\bar{a}_{ik} \bar{c} + \bar{a}_{jk} \bar{s}) - \bar{a}_{ik} \bar{c} - \bar{a}_{jk} \bar{s} = \epsilon_{ik} \quad (|\epsilon_{ik}| \leqslant 2^{-t-1}) \qquad (35.8)$$

for all $k \neq i, j$. Similarly

$$|(H_p)_{jk}|, \; |(H_p)_{ki}|, \; |(H_p)_{kj}| \leqslant 2^{-t-1}. \qquad (35.9)$$

The elements at the four intersections are altered both by the pre-multiplication and the post-multiplication and we now consider them in detail. We assume the pre-multiplication is performed first. The errors made in the pre-multiplication in these four positions are on the same footing as those made in the remainder of rows i and j. We may therefore denote the values obtained by

$$\begin{bmatrix} A + \epsilon_{ii} & B + \epsilon_{ij} \\ C + \epsilon_{ji} & D + \epsilon_{jj} \end{bmatrix} = \begin{bmatrix} \bar{A} & \bar{B} \\ \bar{C} & \bar{D} \end{bmatrix} \text{ (say)}, \qquad (35.10)$$

where A, B, C, D correspond to exact pre-multiplication by $\bar{R}_p$. The values obtained after post-multiplication may therefore be expressed in the form

$$\begin{bmatrix} A\bar{c}+B\bar{s}+\eta_{ii} & -A\bar{s}+B\bar{c}+\eta_{ij} \\ C\bar{c}+D\bar{s}+\eta_{ji} & -C\bar{s}+D\bar{c}+\eta_{jj} \end{bmatrix}, \tag{35.11}$$

where every η_{st} satisfies
$$|\eta_{st}| < 2^{-t-1}. \tag{35.12}$$

If we denote the errors in these positions in the combined operation (relative to $\bar{R}_p \bar{A}_{p-1} \bar{R}_p^T$) by

$$2^{-t-1}\begin{bmatrix} \alpha & \beta \\ \gamma & \delta \end{bmatrix}, \tag{35.13}$$

we have
$$2^{-t-1}\alpha = \epsilon_{ii}\bar{c}+\epsilon_{ij}\bar{s}+\eta_{ii},$$
$$2^{-t-1}|\alpha| < |\epsilon_{ii}|\,|\bar{c}| + |\epsilon_{ij}|\,|\bar{s}| + |\eta_{ii}|$$
$$< (\epsilon_{ii}^2+\epsilon_{ij}^2)^{\frac{1}{2}}(\bar{c}^2+\bar{s}^2)^{\frac{1}{2}} + |\eta_{ii}| < (2{\cdot}42)2^{-t-1}. \tag{35.14}$$

The same bound is obviously adequate for β, γ, δ, and hence typically when $n = 6$, $i = 2$, $j = 4$ we have

$$|H_p| \leqslant 2^{-t-1}\begin{bmatrix} 0 & 1 & 0 & 1 & 0 & 0 \\ 1 & 2{\cdot}42 & 1 & 2{\cdot}42 & 1 & 1 \\ 0 & 1 & 0 & 1 & 0 & 0 \\ 1 & 2{\cdot}42 & 1 & 2{\cdot}42 & 1 & 1 \\ 0 & 1 & 0 & 1 & 0 & 0 \\ 0 & 1 & 0 & 1 & 0 & 0 \end{bmatrix}. \tag{35.15}$$

Now the symmetric matrix formed from the unit elements has the characteristic equation
$$\lambda^n - 2(n-2)\lambda^{n-2} = 0 \tag{35.16}$$

and hence its 2-norm is $(2n-4)^{\frac{1}{2}}$. The 2-norm of the remaining matrix of four elements is not greater than $(4{\cdot}84)2^{-t-1}$. Combining these results we have
$$\|H_p\|_2 < [(2n-4)^{\frac{1}{2}}+4{\cdot}84]2^{-t-1}, \tag{35.17}$$

and therefore from (35.7)

$$\|\bar{A}_N - R_N...R_1A_0R_1^T...R_N^T\|_2 < 2^{-t-1}[(2n-4)^{\frac{1}{2}}+11{\cdot}70]N. \tag{35.18}$$

The corresponding initial scaling requirement is, of course,

$$\|A_0\|_2 < 1 - 2^{-t-1}[(2n-4)^{\frac{1}{2}} + 11 \cdot 70]N. \tag{35.19}$$

General comments on the error bounds

36. The fixed-point bounds we have just obtained appear not to have an exponential factor of the type which we had in the floating-point bounds. However, this difference is illusory since we must remember that our bound holds only under the assumption that (35.19) is satisfied. If n is too large this condition cannot be met since it implies that $\|A_0\|_2$ must be negative. Over their useful ranges the results are essentially of the forms

$$\|\bar{A}_N - R_N \dots R_1 A_0 R_1^T \dots R_N^T\|_E < K_1 n^{\frac{3}{2}} 2^{-t} \|A_0\|_E \text{ (floating-point)}, \tag{36.1}$$

$$\|\bar{A}_N - R_N \dots R_1 A_0 R_1^T \dots R_N^T\|_2 < K_2 n^{\frac{3}{2}} 2^{-t} \|A_0\|_2 \text{ (fixed-point)}, \tag{36.2}$$

$$\|\bar{A}_N - R_N \dots R_1 A_0 R_1^T \dots R_N^T\|_E < K_3 n^{\frac{3}{2}} 2^{-t} \|A_0\|_E \text{ (fixed-point)}. \tag{36.3}$$

There is no doubt that the floating-point result is appreciably stronger for large n.

It is perhaps worth mentioning that it is unlikely that the fixed-point result for a general matrix A_0 could be improved as far as the factor $n^{\frac{3}{2}} 2^{-t}$ is concerned. For consider the following hypothetical experiment in which we compute the matrices $\bar{A}_p$ ($p = 1, \dots, N$) as follows:

At the pth stage the matrices R_p and $R_p \bar{A}_{p-1} R_p^T$ are computed exactly using an infinite number of figures and, after this exact computation, the elements of $R_p \bar{A}_{p-1} R_p^T$ are rounded to give standard fixed-point numbers, the resulting matrix being defined as $\bar{A}_p$. Clearly we have

$$\bar{A}_p = R_p \bar{A}_{p-1} R_p^T + H_p, \tag{36.4}$$

where $|H_p|$ is bounded for $n = 6$, $i = 2$, $j = 4$ by the array

$$2^{-t-1} \begin{bmatrix} 0 & 1 & 0 & 1 & 0 & 0 \\ 1 & 1 & 1 & 1 & 1 & 1 \\ 0 & 1 & 0 & 1 & 0 & 0 \\ 1 & 1 & 1 & 1 & 1 & 1 \\ 0 & 1 & 0 & 1 & 0 & 0 \\ 0 & 1 & 0 & 1 & 0 & 0 \end{bmatrix}. \tag{36.5}$$

(We are assuming the general case here, so that there are no zero elements introduced in the $\bar{A}_p$ by the transformations.) It follows now as in § 35 that even for this experiment the bound obtained for $\|\bar{A}_N - R_N \ldots R_1 A_0 R_1^T \ldots R_N^T\|_2$ has the factor $n^{\frac{1}{2}} 2^{-t}$.

The use of $fi_2(\ )$ rather than $fi(\ )$ from time to time has little effect on the final bound; an additonal factor of 2 is all that is required if $fi(\ )$ is used throughout. In the floating-point analysis we used $fl(\ )$ throughout but the use of $fl_2(\ )$ would have made no essential difference since no high-order inner-products occur anywhere in the whole computation.

Elementary Hermitian matrices in floating-point

37. We now turn our attention to the use of the elementary Hermitian matrices (Chapter 1, §§ 45, 46). It is not practical to give analyses of each of the $fi(\ )$, $fi_2(\ )$, $fl(\ )$ and $fl_2(\ )$ modes and we have selected the $fl_2(\ )$ mode because it is particularly effective and we have given analyses of alternative modes elsewhere. See for example Wilkinson (1961b) and (1962b). The most common situation arising in the relevant algorithms is the following.

We are given a vector x of order n with floating-point components, and we wish to determine an elementary Hermitian matrix P so that pre-multiplication by P leaves the first r components of x unchanged and eliminates elements $r+2$, $r+3, \ldots, n$. We shall assume that x is real and give the relevant mathematical analysis first without regard for rounding errors.

If we write
$$w^T = (0, \ldots, 0, w_{r+1}, \ldots, w_n) \tag{37.1}$$
$$= (O \mid v^T), \tag{37.2}$$

where v is a vector with $(n-r)$ components, then we have

$$P = I - 2ww^T = \left[\begin{array}{c|c} I & O \\ \hline O & I - 2vv^T \end{array}\right]. \tag{37.3}$$

$(I - 2vv^T)$ is an elementary Hermitian matrix of order $(n-r)$ since

$$v^T v = w^T w = 1. \tag{37.4}$$

Partitioning x^T to conform with P, so that

$$x^T = (y^T \mid z^T), \tag{37.5}$$

we have
$$Px = \left[\begin{array}{c|c} I & O \\ \hline O & I - 2vv^T \end{array}\right] \left[\begin{array}{c} y \\ \hline z \end{array}\right] = \left[\begin{array}{c} y \\ \hline (I - 2vv^T)z \end{array}\right]. \tag{37.6}$$

Hence we have only to choose v so that $(I-2vv^T)z$ has zeros everywhere except in its first component.

38. It is evident that our original problem, which had the parameters r and n, is essentially equivalent to a simpler problem with parameters 0 and $n-r$. It will therefore be quite general if we consider for the moment the following problem.

Given a vector x of order n, to construct an elementary Hermitian matrix such that

$$Px = ke_1. \tag{38.1}$$

Since the 2-norm is invariant, if we write

$$S^2 = x_1^2 + \ldots + x_n^2 \tag{38.2}$$

then

$$k = \pm S. \tag{38.3}$$

Equation (38.1) therefore gives

$$x_1 - 2w_1(w^T x) = \pm S, \tag{38.4}$$

$$x_i - 2w_i(w^T x) = 0 \quad (i = 2,\ldots, n), \tag{38.5}$$

and hence

$$2Kw_1 = x_1 \mp S, \qquad 2Kw_i = x_i \quad (i = 2,\ldots, n), \tag{38.6}$$

where

$$K = w^T x. \tag{38.7}$$

Squaring the equations (38.6), adding, and dividing by two we have

$$2K^2 = S^2 \mp x_1 S. \tag{38.8}$$

If we write

$$u^T = (x_1 \mp S, x_2, x_3,\ldots, x_n), \tag{38.9}$$

then

$$w = u/2K, \tag{38.10}$$

and

$$P = I - uu^T/2K^2. \tag{38.11}$$

Only one square root is involved in the whole process, that of S^2 to give S.

Error analysis of the computation of an elementary Hermitian matrix

39. Suppose now that x has floating-point components and we use $fl_2(\ )$ computation to derive a matrix $\bar{P}$. We wish to find a bound for $\|\bar{P}-P\|$, where P is the exact matrix corresponding to the given x. Since x is to be regarded as exact for the purposes of this comparison we omit bars from its components.

There is a choice of signs in the equations of § 38. We shall find that the correct choice for numerical stability is such that

$$2K^2 = S^2 \mp x_1 S = S^2 + |x_1| \, S, \tag{39.1}$$

so that no cancellation takes place in forming $2K^2$. Hence if x_1 is positive we take

$$2K^2 = S^2 + x_1 S \tag{39.2}$$

$$u^T = (x_1 + S, \, x_2, \dots, \, x_n). \tag{39.3}$$

No generality is lost in taking x_1 to be positive; provided we take the stable choice, the error bounds are not dependent on the sign of x_1.

The steps in the computation are as follows.

(i) Calculate the quantity a defined by

$$a = fl_2(x_1^2 + x_2^2 + \dots + x_n^2), \tag{39.4}$$

where we retain the 2t-digit mantissa in a since it is required in steps (ii) and (iii). Hence

$$a \equiv x_1^2(1 + \epsilon_1) + \dots + x_n^2(1 + \epsilon_n), \tag{39.5}$$

where *a fortiori* from (9.4) and (9.5) we have

$$|\epsilon_i| < \tfrac{3}{2} n 2^{-2t_2}. \tag{39.6}$$

Therefore

$$a \equiv (x_1^2 + x_2^2 + \dots + x_n^2)(1 + \epsilon) = S^2(1 + \epsilon) \quad (|\epsilon| < \tfrac{3}{2} n 2^{-2t_2}). \tag{39.7}$$

(ii) Calculate S from a, where

$$S = fl_2(a^{\frac{1}{2}}) \equiv a^{\frac{1}{2}}(1 + \eta) \quad (|\eta| < (1 \cdot 00001)2^{-t}) \tag{39.8}$$

from (10.2). Hence

$$S \equiv S(1 + \epsilon)^{\frac{1}{2}}(1 + \eta) \equiv S(1 + \xi), \tag{39.9}$$

where certainly from (39.7) and (39.8)

$$\xi < (1 \cdot 00002)2^{-t}. \tag{39.10}$$

(iii) Calculate $2K^2$ where

$$2K^2 = fl_2(x_1^2 + x_2^2 + \dots + x_n^2 + x_1 S)$$
$$\equiv [x_1^2(1 + \eta_1) + x_2^2(1 + \eta_2) + \dots + x_n^2(1 + \eta_n) + x_1 S(1 + \eta_{n+1})](1 + \epsilon) \tag{39.11}$$

where *a fortiori* from (9.4) and (9.5)

$$|\eta_i| < \tfrac{3}{2}(n+1)2^{-2t_2}, \qquad |\epsilon| < 2^{-t}. \tag{39.12}$$

In this computation of $2K^2$ most of the accumulation has already been done in (i).

Combining these results and remembering that x_1 is positive we have

$$2\bar{K}^2 \equiv (x_1^2 + x_2^2 + \ldots + x_n^2 + x_1 \bar{S})(1+\zeta)(1+\epsilon) \quad (|\zeta| < \tfrac{3}{2}(n+1)2^{-2t_2})$$

$$\equiv [S^2 + x_1 S(1+\xi)](1+\zeta)(1+\epsilon). \tag{39.13}$$

Now if we write

$$S^2 + x_1 S(1+\xi) \equiv (S^2 + x_1 S)(1+\kappa) \tag{39.14}$$

we have

$$\kappa = \left(\frac{x_1 S}{S^2 + x_1 S}\right)\xi. \tag{39.15}$$

Since $x_1 S$ and S^2 are positive and $x_1 < S$ this gives

$$|\kappa| < \tfrac{1}{2}|\xi|. \tag{39.16}$$

Again our choice of sign is significant; if x_1 is negative we must take $S^2 - x_1 S$ to be sure of a low relative error. We have therefore certainly

$$2\bar{K}^2 \equiv 2K^2(1+\theta) \quad (|\theta| < (1{\cdot}501)2^{-t}). \tag{39.17}$$

(iv) Finally we compute $\bar{u}$. Only one element of $\bar{u}$, the first, involves computation and we have

$$\bar{u}_1 = fl(x_1 + \bar{S}) \equiv (x_1 + \bar{S})(1+\phi) \quad (|\phi| < 2^{-t}) \tag{39.18}$$

$$\equiv [x_1 + S(1+\xi)](1+\phi). \tag{39.19}$$

Here again our choice of sign ensures a low relative error and we obtain certainly

$$\bar{u}_1 \equiv (x_1 + S)(1+\psi) \quad (|\psi| < (1{\cdot}501)2^{-t}) \tag{39.20}$$

$$\equiv u_1(1+\psi). \tag{39.21}$$

We may write $\bar{u}^T \equiv (\bar{u}_1,\, u_2,\, u_3, \ldots,\, u_n) \equiv u^T + \delta u^T,$ \hfill (39.22)

where

$$\|\delta u\|_2 = |u_1|\,|\psi| < (1{\cdot}501)2^{-t}\,\|u\|_2, \tag{39.23}$$

and hence

$$\|\bar{u}\|_2 = \|u + \delta u\|_2 < [1 + (1{\cdot}501)2^{-t}]\,\|u\|_2. \tag{39.24}$$

40. The matrix $\bar{P}$ is now available if we are content to leave it in the factorized form
$$\bar{P} = I - \bar{u}\bar{u}^T/2\bar{K}^2, \tag{40.1}$$

and this is often the most convenient thing to do. No further rounding errors are made until we attempt to use $\bar{P}$ as a multiplier and this involves a separate analysis. From (40.1)

$$\bar{P} - P = uu^T/2K^2 - \bar{u}\bar{u}^T/2\bar{K}^2$$

$$= (uu^T - \bar{u}\bar{u}^T)/2K^2 + \bar{u}\bar{u}^T(1/2K^2 - 1/2\bar{K}^2). \tag{40.2}$$

Since $\|uu^T/2K^2\|_2 = 2$, a simple computation using (39.17), (39.23) and (39.24) shows that certainly

$$\|\bar{P}-P\|_2 < (9{\cdot}01)2^{-t}. \tag{40.3}$$

Note that the bound is independent of n which, since P is as effective as $(n-1)$ plane rotations, is highly satisfactory.

For some applications it is convenient to procede a little further with P and to compute the vector $\bar{v}$ defined by

$$\bar{v}^T = fl(\bar{u}^T/2K^2), \tag{40.4}$$

so that

$$\bar{P} = I - \bar{u}\bar{v}^T. \tag{40.5}$$

From (40.4)

$$\bar{v}_i^T = (\bar{u}_i^T/2\bar{K}^2)(1+\epsilon_i) \quad (|\epsilon_i| < 2^{-t}), \tag{40.6}$$

and hence

$$\bar{v} \equiv \bar{u}/2\bar{K}^2 + \delta v, \tag{40.7}$$

where

$$\|\delta v\|_2 < 2^{-t}\,\|\bar{u}\|_2/2\bar{K}^2. \tag{40.8}$$

For $\bar{P}$ computed in this way we have certainly

$$\|\bar{P}-P\|_2 < (11{\cdot}02)2^{-t}. \tag{40.9}$$

Since $\|ab^T\|_2 = \|ab^T\|_E$ for any vectors a and b, we may replace the 2-norms by Euclidean norms in the above bounds.

Returning to the case when we wish to alter only the last $(n-r)$ components, we produce a matrix of the form

$$\begin{bmatrix} I & O \\ \hline O & \bar{P} \end{bmatrix}$$

where the $\bar{P}$ is of order $(n-r)$. Remember, however, that our error bound was independent of n.

Numerical example

41. The vital importance of the correct choice of sign may be illustrated by the following simple numerical example. Suppose the vector x is given by

$$x = [10^0(0{\cdot}5123);\ 10^{-3}(0{\cdot}6147);\ 10^{-3}(0{\cdot}5135)].$$

Using $fl_2(\ )$ corresponding to 4-digit decimal arithmetic we have

$$a = fl_2(x_1^2+x_2^2+x_3^2) \equiv 10^0(0{\cdot}26245193), \quad \bar{S} = fl_2(a^{\frac{1}{2}}) \equiv 10^0(0{\cdot}5123),$$

and with the stable choice of sign

$$2\bar{K}^2 = fl_2(x_1^2+x_2^2+x_3^2+x_1\bar{S}) = 10^0(0{\cdot}5249),$$

$$\bar{u}_1 = fl_2(x_1+\bar{S}) = 10^1(0{\cdot}1025),$$

$$\bar{u}^T = [10^1(0{\cdot}1025);\ 10^{-3}(0{\cdot}6147);\ 10^{-3}(0{\cdot}5135)].$$

The 'correct' values with this choice of sign are

$$S = 10^0(0.5123006261...), \quad 2K^2 = 10^0(0.5249035...),$$

$$u^T \equiv [10^1(0.10246006...); \ 10^{-3}(0.6147); \ 10^{-3}(0.5135)].$$

It may be verified that the barred values satisfy the inequalities of § 39 where we replace 2^{-t} by $\frac{1}{2}10^{-3}$, the corresponding value for 4-digit decimal computation.

Using the unstable choice of signs we have

$$2\bar{K}^2 = fl_2(x_1^2 + x_2^2 + x_3^2 - x_1 \bar{S}) \equiv 10^{-6}(0.6400),$$

$$\bar{u}_1 = fl_2(x_1 - \bar{S}) \equiv 10^0(0.0000),$$

$$\bar{u}^T = [10^0(0.0000); \ 10^{-3}(0.6147); \ 10^{-3}(0.5135)].$$

The 'correct' values with this choice of sign are

$$2K^2 = 10^{-6}(0.3207...),$$

$$u^T = [10^{-6}(-0.6261); \ 10^{-3}(0.6147); \ 10^{-3}(0.5135)].$$

It will readily be verified that $\bar{P} = I - \bar{u}\bar{u}^T/2\bar{K}^2$ is not even approximately orthogonal in this case and therefore is not close to P. The computed value $2\bar{K}^2$ has a very large relative error.

Pre-multiplication by an approximate elementary Hermitian matrix

42. In analogy with our investigation of plane rotations we now require a bound for

$$fl_2\left(\begin{bmatrix} I & O \\ \hline O & \bar{P} \end{bmatrix} A\right) - \begin{bmatrix} I & O \\ \hline O & P \end{bmatrix} A. \tag{42.1}$$

Clearly the pre-multiplications leave the first r rows of A unaltered and the last $(n-r)$ rows are pre-multiplied by P (or by $\bar{P}$ with rounding errors). The error matrix of (42.1) is therefore null in its first r rows and it is perfectly general to consider the pre-multiplication of a matrix A of m rows and n columns by a P having for its corresponding w a vector of order m with no zero components.

We take $\bar{P}$ in the factorized form $(I - \bar{u}\bar{u}^T/2\bar{K}^2)$ so that the inequalities of § 39 and (40.3) are satisfied. We write

$$fl_2(\bar{P}A) - PA = fl_2(\bar{P}A) - \bar{P}A + \bar{P}A - PA, \tag{42.2}$$

$$\|fl_2(\bar{P}A) - PA\|_E < \|fl_2(\bar{P}A) - \bar{P}A\|_E + \|(\bar{P} - P)A\|_E$$

$$< \|fl_2(\bar{P}A) - \bar{P}A\|_E + (9.01)2^{-t}\|A\|_E. \tag{42.3}$$

We therefore require a bound for $fl_2(\bar{P}A) - \bar{P}A$ which is the matrix of errors actually occurring in the pre-multiplication. Notice that the error bound which we gave in (9.14) does not apply to the calculation of $fl_2(\bar{P}A)$ since $\bar{P}$ is used in factorized form in order to economize on the volume of computation.

In order to emphasize the fact that we are now concerned only with the errors made in the pre-multiplication we introduce y and M defined by

$$y = \bar{u}, \qquad M = 2\bar{K}^2, \tag{42.4}$$

so that

$$\bar{P} = I - yy^T/M \tag{42.5}$$

and

$$\begin{aligned}
\|y\|_2^2/M &= \|\bar{u}\|_2^2/2\bar{K}^2 \\
&< \|u\|_2^2[1 + (1\cdot 501)2^{-t}]^2/2K^2[1 - (1\cdot 501)2^{-t}] \\
&< 2[1 + (4\cdot 51)2^{-t}]. \tag{42.6}
\end{aligned}$$

43. We may express the steps in the exact derivation of $\bar{P}A$ and in the computation of $fl_2(\bar{P}A)$ as follows:

$$\text{Exact} \qquad\qquad\qquad \text{Computed}$$

$$p^T = y^T A/M; \qquad\qquad \bar{p}^T = fl_2(y^T A/M). \tag{43.1}$$

$$B = \bar{P}A = A - yp^T; \qquad \bar{B} = fl_2(\bar{P}A) = fl_2(A - y\bar{p}^T). \tag{43.2}$$

The ith component $\bar{p}_i$ is therefore given by

$$\bar{p}_i \equiv [y_1 a_{1i}(1 + \epsilon_1) + y_2 a_{2i}(1 + \epsilon_2) + \ldots + y_m a_{mi}(1 + \epsilon_m)](1 + \epsilon)/M, \tag{43.3}$$

where *a fortiori* from (9.2) and (9.5) we have

$$|\epsilon_i| < \tfrac{3}{2}(m + 1)2^{-2t_2}, \qquad |\epsilon| < 2^{-t}. \tag{43.4}$$

Hence

$$\bar{p}_i = p_i(1 + \epsilon) + q_i, \tag{43.5}$$

where

$$|q_i| < \tfrac{3}{2}(m + 1)2^{-2t_2}(1 + 2^{-t})[|y_1|\,|a_{1i}| + |y_2|\,|a_{2i}| + \ldots + |y_m|\,|a_{mi}|]/M, \tag{43.6}$$

giving

$$|\bar{p}_i - p_i| < 2^{-t}|p_i| + \tfrac{3}{2}(m + 1)2^{-2t_2}(1 + 2^{-t})[|A|^T\,|y|]_i/M. \tag{43.7}$$

We have therefore

$$\|\bar{p} - p\|_2 < 2^{-t}\|p\|_2 + \tfrac{3}{2}(m + 1)2^{-2t_2}(1 + 2^{-t})\|y\|_2\,\|A\|_E/M, \tag{43.8}$$

and invoking the assumption of (5.2) that $n2^{-t} < 0\cdot 1$, we may write

$$\|\bar{p} - p\|_2 < 2^{-t}\|p\|_2 + (0\cdot 16)2^{-t}\|y\|_2\,\|A\|_E/M. \tag{43.9}$$

For the (i, j) component of $\bar{B}$ we have

$$\bar{b}_{ij} = fl_2(A - y\bar{p}^T)_{ij}$$
$$\equiv [a_{ij}(1+\eta_1) - y_i\bar{p}_j(1+\eta_2)](1+\xi), \qquad (43.10)$$

$$|\eta_1|, |\eta_2| < \tfrac{3}{2}2^{-2t}, \qquad |\xi| < 2^{-t}. \qquad (43.11)$$

If we define δp by the relation

$$\bar{p} = p + \delta p, \qquad (43.12)$$

then

$$\bar{b}_{ij} \equiv [a_{ij}(1+\eta_1) - y_i(p_j+\delta p_j)(1+\eta_2)](1+\xi)$$
$$= b_{ij}(1+\xi) + [a_{ij}\eta_1 - y_ip_j\eta_2 - y_i\,\delta p_j(1+\eta_2)](1+\xi). \qquad (43.13)$$

Hence

$$|\bar{b}_{ij} - b_{ij}| < 2^{-t}|b_{ij}| +$$
$$+[\tfrac{3}{2}2^{-2t}|a_{ij}| + \tfrac{3}{2}2^{-2t}|y_i||p_j| + |y_i||\delta p_j|(1+\tfrac{3}{2}2^{-2t})](1+2^{-t}), \quad (43.14)$$

giving

$$|\bar{B} - B| < 2^{-t}|B| +$$
$$+[\tfrac{3}{2}2^{-2t}|A| + \tfrac{3}{2}2^{-2t}|y||p|^T + |y||\delta p|^T(1+\tfrac{3}{2}2^{-2t})](1+2^{-t}), \quad (43.15)$$

from which we deduce

$$\|\bar{B} - B\|_E < 2^{-t}\|B\|_E +$$
$$+[\tfrac{3}{2}2^{-2t}\|A\|_E + \tfrac{3}{2}2^{-2t}\|y\|_2\|p\|_2 + \|y\|_2\|\delta p\|_2(1+\tfrac{3}{2}2^{-2t})](1+2^{-t}).$$
$$(43.16)$$

44. We now require bounds for each term on the right-hand side in terms of $\|A\|_E$. From (43.1) and (40.3) we have

$$\|B\|_E = \|\bar{P}A\|_E < \|PA\|_E + \|(\bar{P}-P)A\|_E < \|A\|_E + (9{\cdot}01)2^{-t}\|A\|_E.$$
$$(44.1)$$

From (43.1) and (42.6)

$$\|y\|_2\|p\|_2 < \|y\|_2^2\|A\|_E/M < 2\|A\|_E[1+(4{\cdot}51)2^{-t}]. \quad (44.2)$$

From (43.9) and (42.6)

$$\|y\|_2\|\delta p\|_2 < \|y\|_2[2^{-t}\|p\|_2 + (0{\cdot}16)2^{-t}\|y\|_2\|A\|_E/M]$$
$$< 2^{-t}\{2\|A\|_E[1+(4{\cdot}51)2^{-t}] + (0{\cdot}32)\|A\|_E[1+(4{\cdot}51)2^{-t}]\}$$
$$< (2{\cdot}33)2^{-t}\|A\|_E. \qquad (44.3)$$

Combining these results we have certainly

$$\|\bar{B} - B\|_E < (3{\cdot}35)2^{-t}\|A\|_E, \qquad (44.4)$$

and remembering that $\bar{B}-B = fl_2(\bar{P}A)-\bar{P}A$, (42.3) becomes

$$\|fl_2(\bar{P}A)-PA\|_E < (3{\cdot}35)2^{-t}\|A\|_E+(9{\cdot}01)2^{-t}\|A\|_E$$
$$= (12{\cdot}36)2^{-t}\|A\|_E. \qquad (44.5)$$

Returning to the general case we have immediately

$$\left\|fl_2\left\{\left[\begin{array}{c|c}I & O \\ \hline O & \bar{P}\end{array}\right]A\right\}-\left[\begin{array}{c|c}I & O \\ \hline O & P\end{array}\right]A\right\|_E \leqslant (12{\cdot}36)2^{-t}\|A_{n-r}\|_E$$
$$\leqslant (12{\cdot}36)2^{-t}\|A\|_E, \qquad (44.6)$$

where A_{n-r} is the matrix consisting of the last $(n-r)$ rows of A. In the particular case when one of the columns of A is the vector x by means of which P was defined, then we do not actually compute the elements of Px but automatically assign the appropriate number of zero elements and give the remaining element the value $\mp S$. The corresponding values in PA are zero and $\mp S$, so that from (39.9) the contribution to the error matrix is the single element $\mp S\xi$. The special treatment of such a column is therefore covered by our general analysis.

Multiplication by a sequence of approximate elementary Hermitians

45. In a number of algorithms we shall be concerned with a sequence of $(n-1)$ elementary Hermitians $P_0, P_1, \ldots, P_{n-2}$ where the vector w_r corresponding to P_r has zero elements in the first r positions. We consider pre-multiplications first and denote the computed P_r by $\bar{P}_r$ and computed transforms by $\bar{A}_r$. We have then

$$\bar{A}_1-P_0A_0 = F_0, \qquad \|F_0\|_E < (12{\cdot}36)2^{-t}\|A_0\|_E, \qquad (45.1)$$
$$\bar{A}_{r+1}-P_r\bar{A}_r = F_r, \qquad \|F_r\|_E < (12{\cdot}36)2^{-t}\|\bar{A}_r\|_E, \qquad (45.2)$$

and hence if we write $(12{\cdot}36)2^{-t} = x$,

$$\|\bar{A}_{n-1}-P_{n-2}P_{n-3}\ldots P_0A_0\|_E < x[1+(1+x)+(1+x)^2+\ldots+$$
$$+(1+x)^{n-2}]\|A_0\|_E$$
$$\leqslant (n-1)(1+x)^{n-2}x\|A_0\|_E. \qquad (45.3)$$

We shall be interested in the results only if nx is appreciably less than unity, so that although the factor $(1+x)^{n-2}$ is exponentially increasing, it has little effect on the range of applicability. Our bound then, is essentially $(12{\cdot}36)n2^{-t}\|A_0\|_E$, which is remarkably satisfactory, particularly if we remember that we have not taken the statistical distribution of the errors into account. In our experience, computation in the $fl_2(\ )$ mode has proved to be even more accurate than statistical considerations suggest. Note that our argument is weak in some respects since, as in (44.6), we replace the norm of the matrix formed by the modified rows at each stage by the norm of the whole matrix.

If we take $A_0 = I$ we obtain a bound for the error in the computed product of $(n-1)$ approximate elementary Hermitians. This gives

$$\|fl_2(\bar{P}_{n-2}\bar{P}_{n-3}...\bar{P}_0) - P_{n-2}P_{n-3}...P_0\|_E < n^{\frac{3}{2}}x(1+x)^{n-2}$$
$$= (12\cdot36)n^{\frac{3}{2}}2^{-t}[1+(12\cdot36)n2^{-t}]^{n-2}. \quad (45.4)$$

For similarity transformations we have immediately from (45.3), with an obvious notation, the result

$$\|\bar{A}_{n-1} - P_{n-2}P_{n-3}...P_0A_0P_0...P_{n-2}\|_E \leqslant 2(n-1)(1+x)^{2n-4}x \|A_0\|_E.$$
$$(45.5)$$

There are no special difficulties of the type discussed in connexion with plane rotations, since our analysis makes no attempt to establish special relationships between sections of the successive transformed matrices. Clearly the multiplications could take place in any order without affecting the bound.

It is unlikely that the factor $n2^{-t}$ could be improved upon for a general matrix, as we may see by considering the following hypothetical experiment. Suppose we produce a sequence of matrices in the following way. We are given n exact elementary Hermitians P_r $(r = 1,...,n)$ and starting from A_0 we produce n matrices $\bar{A}_r$ $(r = 1,...,n)$ as follows. The matrix $\bar{A}_{r+1}$ is derived from $\bar{A}_r$ by computing $P_{r+1}\bar{A}_rP_{r+1}$ *exactly* and then rounding the elements of the resulting matrix to give t-digit floating-point numbers. If we write

$$\bar{A}_{r+1} \equiv P_{r+1}\bar{A}_rP_{r+1} + G_{r+1}, \qquad P_{r+1}\bar{A}_rP_{r+1} = B_{r+1}, \qquad (45.6)$$

then clearly the optimum bounds for the elements of G_{r+1} are

$$|G_{r+1}|_{ij} \leqslant 2^{-t} |B_{r+1}|_{ij}, \qquad (45.7)$$

and hence

$$\|G_{r+1}\|_E \leqslant 2^{-t} \|B_{r+1}\|_E. \qquad (45.8)$$

Our optimum bound for the error in the final $\bar{A}_n$ is given by

$$\|\bar{A}_n - P_nP_{n-1}...P_1A_0P_1...P_{n-1}P_n\|_E \leqslant$$
$$\leqslant 2^{-t}[1+(1+2^{-t})+...+(1+2^{-t})^{n-1}] \|A_0\|_E, \quad (45.9)$$

and hence is essentially $n2^{-t} \|A_0\|_E$.

If A_0 is symmetric then we may retain symmetry throughout, but it is convenient to reorganize the computation so as to minimize the amount of work. We shall discuss this in Chapter 5 when we describe the relevant transformations.

Non-unitary elementary matrices analogous to plane rotations

46. The great numerical stability of algorithms involving unitary transformations springs from the fact that, apart from rounding errors, the 2-norm and Euclidean norm are invariant, so that there can be no progressive increase in the general size of the elements in the successive transformed matrices. This is important because the current rounding errors at any stage are essentially proportional to the size of the elements of the transformed matrices.

The non-unitary elementary matrices have no natural stability properties of this kind, but we can organize the computation so as to give them a limited form of stability. Let us consider first the problem of § 20. There we had a vector $\bar{x}$, and we reduced its jth element to zero by pre-multiplication with a matrix which affected only the ith and jth elements, both being replaced by linear combinations of their initial values. Now pre-multiplication by the matrix M_{ji} (Chapter 1, § 40, vii) has a similar effect. This leaves all components unaffected except the jth which becomes x'_j defined by

$$x'_j = -m_{ji}\bar{x}_i + \bar{x}_j. \tag{46.1}$$

We can reduce x'_j to zero by taking $m_{ji} = \bar{x}_j/\bar{x}_i$. However, if we do this the value of m_{ji} may be arbitrarily large, possibly infinite. Hence $\|M_{ji}\|$ may be arbitrarily large and any matrix A which is pre-multiplied by M_{ji} may have elements in its jth row arbitrarily augmented. It is obvious from the analysis that we have given that this is unsatisfactory.

We may introduce a measure of stability in the following way.

(i) If $|\bar{x}_j| < |\bar{x}_i|$, then we proceed as above and we have $m_{ji} < 1$.

(ii) If $|\bar{x}_j| > |\bar{x}_i|$, then we first pre-multiply by I_{ij} (Chapter 1, § 40, i), thus *interchanging* the ith and jth elements, and then multiply by the appropriate M_{ji}.

The matrix by which we pre-multiply may therefore be expressed in the form $M_{ji}I'_{ij}$, where I'_{ij} is either I_{ij} or I according as an *interchange* was necessary or not. We have not covered the case when $\bar{x}_i = \bar{x}_j = 0$. This presents no difficulties; we have only to take

$$I_{ij} = I, \qquad m_{ji} = 0, \tag{46.2}$$

so that $M_{ji} I'_{ij} = I$ in this case.

The matrix product $M_{ji} I'_{ij}$ is comparable in effect with a rotation in the (i, j) plane and we might expect it to be reasonably stable. *We shall denote this product by M'_{ji}* and a simple calculation shows that

$$\| M'_{ji} \|_2 < [\tfrac{1}{2}(3 + 5^{\frac{1}{2}})]^{\frac{1}{2}}. \tag{46.3}$$

Corresponding to algorithms in which we use a sequence of rotations in the $(1, 2)$, $(1, 3)$,..., $(1, n)$; $(2, 3)$,..., $(2, n)$;...; $(n-1, n)$ planes we have algorithms in which we use a set of matrices

$$M'_{21}, M'_{31},..., M'_{n1}; M'_{32},..., M'_{n2};...; M'_{n,n-1}.$$

We shall not be able to give any general error analyses leading to bounds of the form $n^k 2^{-t} \| A \|$; usually bounds of the form $2^n n^k 2^{-t} \| A \|$ will be the best attainable for general transformations of general matrices. We shall therefore give only specific analyses at the appropriate point in the later chapters.

There are some algorithms involving sequences of elementary transformations in which interchanges are not permissible because they would destroy an essential pattern of zeros introduced by previous steps. To execute these algorithms we are forced to rely on the M_{ji} matrices rather than the M'_{ji}, and hence situations can occur in which the algorithms are arbitrarily unstable. Such algorithms have no counterparts involving plane rotations since their use, like that of the M'_{ji}, would inevitably destroy the patterns of zeros.

Non-unitary elementary matrices analogous to elementary Hermitian matrices

47. In a similar way, by considering the problem of § 37 we can introduce non-unitary matrices which are analogous to the elementary Hermitian matrices. The problem is, given a vector x, to determine a matrix M such that Mx has zeros in components $(r+1)$ to n while components 1 to r are unaltered. Let us consider the pre-multiplication of x by N_r (Chapter 1, § 40, vi). If we write

$$x' = N_r x \tag{47.1}$$

then

$$\begin{aligned} x'_i &= x_i & (i = 1,..., r) \\ x'_i &= x_i - n_{ir} x_r & (i = r+1,..., n) \end{aligned} \Bigg\}, \tag{47.2}$$

and taking

$$n_{ir} = x_i / x_r \quad (i = r+1,..., n), \tag{47.3}$$

we see that the conditions are satisfied, but the elements of N_r may be arbitrarily large or even infinite and this will lead to numerical instability. Stability is achieved as follows: Suppose the maximum value of $|x_i|$ $(i = r,..., n)$ is attained for $i = r' \geqslant r$. Then the vector y defined by

$$y = I_{r,r'} x \qquad (47.4)$$

is derived from x by interchanging the elements x_r and $x_{r'}$. Hence

$$|y_r| \geqslant |y_i| \quad (i = r+1,..., n), \qquad (47.5)$$

and if we now choose N_r so that $N_r y$ has zero components in positions $(r+1)$ to n, we have

$$n_{ir} = y_i/y_r \qquad (47.6)$$

and

$$|n_{ir}| \leqslant 1. \qquad (47.7)$$

The matrix N_r therefore has elements which are bounded in modulus by unity and we have achieved some measure of numerical stability. We have in fact

$$\|N_r\|_2 \leqslant \{\tfrac{1}{2}[(n-r+2)+(n-r)^{\frac{1}{2}}(n-r+4)^{\frac{1}{2}}]\}^{\frac{1}{2}}. \qquad (47.8)$$

We denote the matrix product $N_r I_{r,r'}$ by N_r'. If $x_i = 0$ $(i = r,..., n)$ then we take $N_r' = I$. Clearly N_r' can be used to achieve the same effect as the product $M_{r+1,r}' M_{r+2,r}'...M_{nr}'$ though starting from the same vector x we will not in general have

$$N_r' = M_{r+1,r}' M_{r+2,r}'...M_{nr}'. \qquad (47.9)$$

Equation (47.9) will be satisfied, however, if no interchanges are required. The process of selecting the largest element from a given set in this context is usually referred to as 'pivoting for size'.

Corresponding to algorithms in which we use a set of elementary Hermitian matrices $(I - 2w_r w_r^T)$ with w_r having zero components in positions 1 to $(r-1)$, we have algorithms in which we use matrices $N_1', N_2',..., N_{n-1}'$. Again we shall not usually be able to give general error analyses leading to bounds comparable with those we gave for transformations based on the Hermitians. We shall refer to M_{ji}' and N_r' as *stabilized elementary matrices*.

As before, we shall find that there are some algorithms based on elementary similarity transformations in which the use of interchanges at each stage is not permissible because it would destroy an essential pattern of zeros produced by preceding transformations. Some algorithms are potentially unstable and always need careful consideration. They have no counterparts based on elementary

Hermitian matrices since their use would inevitably destroy the patterns of zeros in much the same way as the use of interchanges.

Throughout this book we shall be faced repeatedly with a choice between the use of elementary unitary matrices and elementary stabilized matrices. In general the elementary stabilized transformations will be somewhat simpler, but on the other hand we shall not be able to give such good guarantees of numerical stability.

Pre-multiplication by a sequence of non-unitary matrices

48. There is one transformation using matrices of the N'_r type for which we can give a reasonably satisfactory error analysis. This is the pre-multiplication of a matrix A_0 by the sequence of matrices $\bar{N}'_1, \bar{N}'_2, ..., \bar{N}'_{n-1}$. We write

$$\bar{A}_r = fl(\bar{N}'_r \bar{A}_{r-1}) \equiv \bar{N}'_r \bar{A}_{r-1} + H_r, \tag{48.1}$$

so that H_r is the matrix of errors made in the actual multiplication. The pre-multiplication by $I_{r,r'}$ merely interchanges rows r and r' and therefore involves no rounding errors. In the subsequent pre-multiplication by $\bar{N}_r$, rows 1 to r are unaltered, so the H_r is null in these rows. Combining relations (48.1) for $r = 1$ to $n-1$ we have by an argument similar to that of § 15

$$\bar{A}_{n-1} = \bar{N}'_{n-1}\bar{N}'_{n-2}...\bar{N}'_1[A_0+(\bar{N}'_1)^{-1}H_1+(\bar{N}'_1)^{-1}(\bar{N}'_2)^{-1}H_2+...+$$
$$+(\bar{N}'_1)^{-1}...(\bar{N}'_{n-1})^{-1}H_{n-1}]. \tag{48.2}$$

The matrix $\bar{A}_{n-1}$ is therefore exactly equivalent to the matrix in square brackets on the right-hand side of (48.2).

We now show that

$$(\bar{N}'_1)^{-1}(\bar{N}'_2)^{-1}...(\bar{N}'_r)^{-1}H_r = I_{1,1'}I_{2,2'}...I_{r,r'}H_r. \tag{48.3}$$

We have
$$(\bar{N}'_r)^{-1}H_r = (\bar{N}_r I_{r,r'})^{-1}H_r = I_{r,r'}\bar{N}_r^{-1}H_r. \tag{48.4}$$

Now $\bar{N}_r^{-1}$ is the same as the matrix $\bar{N}_r$ but with the signs of the elements below the diagonal changed. Hence

$$\bar{N}_r^{-1}H_r = H_r, \tag{48.5}$$

since pre-multiplication by $\bar{N}_r^{-1}$ merely results in the addition of multiples of the null rth row to rows $(r+1)$ to n. We have therefore

$$(\bar{N}'_r)^{-1}H_r = I_{r,r'}H_r, \tag{48.6}$$

and since $r' \geqslant r$ the matrix $I_{r,r'}H_r$ is null in rows 1 to $(r-1)$. Continuing this argument we obtain (48.3). The matrix on the right of (48.3) is merely H_r with its rows permuted. Equation (48.2) therefore shows that $\bar{A}_{n-1}$ is exactly equivalent to A_0 with the perturbation

$$I_{1,1'}H_1 + I_{1,1'}I_{2,2'}H_2 + \ldots + I_{1,1'}I_{2,2'}\ldots I_{n-1,(n-1)'}H_{n-1}. \qquad (48.7)$$

Hence each set of errors H_r is equivalent in its effect to a perturbation of the original matrix consisting merely of H_r with its rows permuted. No amplification of these errors occurs when they are reflected back into the original matrix. (Note that this result is true, of course, even if no interchanges are used.) There is, naturally, a similar result for post-multiplication by matrices of the form M'_r where

$$M'_r = I_{r,r'}M_r, \qquad (48.8)$$

and post-multiplication by $I_{r,r'}$ now gives column interchanges.

It is immediately apparent from this remarkable result that the danger from instability comes only from the possibility that the H_r may themselves be large. Since the errors made at any stage are, in general, proportional to the size of the current elements of the transformed matrix, we see the vital importance of ensuring that the elements of the $\bar{A}_r$ are kept as small as possible. It is for this reason that interchanges are essential.

The reader should verify that the corresponding analysis for transformations with matrices M'_{ji} does not give such satisfactory results. The equivalent perturbations in the original matrix have a much more complex structure.

A *priori* error bounds

49. We have seen that for some algorithms based on unitary transformations we can expect to be able to find an *a priori* bound for perturbations in the original matrix which are equivalent to the errors made in the computations. This will not, in general, give us *a priori* bounds for the errors in the quantities which we wish to calculate, though the theme of Chapter 2 is that this will be true in the case of similarity transformations of symmetric matrices when the required quantities are the eigenvalues. Even in this very favourable case we cannot obtain *a priori* bounds for the errors in the eigenvectors. However, *a priori* bounds for the equivalent perturbations in the original matrix are of great value in assessing the relative performance of algorithms.

We shall find that the most effective bounds which we can obtain for the equivalent perturbation F are of the form

$$\|F\| < n^k 2^{-t} \|A\| \tag{49.1}$$

in some norm, where k is some constant usually less than three. Such bounds are usually attainable only for algorithms based on unitary transformations. For stabilized elementary non-unitary transformations it is usually quite difficult to obtain bounds even of the form

$$\|F\| < n^k 2^n 2^{-t} \|A\|, \tag{49.2}$$

though often for specific methods grounds exist for believing that the factor 2^n is attainable only in extremely adverse circumstances. For non-stabilized non-unitary transformations even a bound of the form (49.2) is usually unobtainable. Indeed there will usually be circumstances, *not at all related to ill-conditioning of the original problem*, for which such transformations will break down completely through the occurrence of divisions by zero.

Departure from normality

50. In order to obtain bounds for the errors resulting from the use of a given algorithm, we need not only bounds for the equivalent perturbations in the original matrix but also bounds for the sensitivities of the required quantities. It is natural to ask whether there is any simple way of estimating this sensitivity factor in the case of the eigenvalue problem. Now we know from Chapter 2, (31.2) that if A is *normal* then the eigenvalues of $(A+F)$ lie in the discs

$$|\lambda - \lambda_i| < \|F\|_2 < \|F\|_E. \tag{50.1}$$

We might therefore expect that if A is 'almost a normal matrix' then its eigenvalues would be comparatively insensitive. If A is normal then (Chapter 1, § 48) there is a unitary matrix R such that

$$R^H A R = \text{diag}(\lambda_i), \tag{50.2}$$

while (Chapter 1, § 47) for any matrix there is a unitary R such that

$$R^H A R = \text{diag}(\lambda_i) + T = D + T, \tag{50.3}$$

where T is *strictly upper-triangular*, that is

$$t_{ij} = 0 \quad (i > j). \tag{50.4}$$

We might reasonably regard $\|T\|/\|A\|$ as a measure of the 'departure of A from normality.'

Henrici (1962) has used the relation (50.3) to obtain estimates of the effect of perturbations in A on its eigenvalues in the following manner.

Let G be defined by

$$R^H(A+F)R = D+T+G, \tag{50.5}$$

so that

$$G = R^H F R, \quad \|G\|_2 = \|F\|_2, \quad \|G\|_E = \|F\|_E. \tag{50.6}$$

If λ is any eigenvalue of $(A+F)$ then $(D-\lambda I)+T+G$ is singular. We distinguish two cases:

(i) $\lambda = \lambda_i$ for some i.

(ii) $\lambda \neq \lambda_i$ for any i, in which case $(D-\lambda I)$ and $(D+T-\lambda I)$ are non-singular.

Hence

$$0 = \det[D+T-\lambda I+G]$$
$$= \det[D+T-\lambda I] \det[I+(D+T-\lambda I)^{-1}G], \tag{50.7}$$

giving

$$0 = \det[I+(D+T-\lambda I)^{-1}G]. \tag{50.8}$$

Equation (50.8) implies that

$$1 \leqslant \|(D+T-\lambda I)^{-1}G\|_2$$
$$\leqslant \|(D+T-\lambda I)^{-1}\|_2 \, \|G\|_2$$
$$= \|(I+KT)^{-1}K\|_2 \, \|G\|_2 \text{ where } K = (D-\lambda I)^{-1} = \text{diag}[(\lambda_i - \lambda)^{-1}] \tag{50.9}$$

$$= \|[I-KT+(KT)^2-\ldots+(-1)^{n-1}(KT)^{n-1}]K\|_2 \, \|G\|_2, \tag{50.10}$$

since $(KT)^r = 0$ $(r \geqslant n)$ because KT is strictly upper-triangular. Writing

$$\|K\|_2 = a, \quad \|T\|_2 = b, \quad \|T\|_E = c, \tag{50.11}$$

we have

$$1 \leqslant (1+ab+a^2b^2+\ldots+a^{n-1}b^{n-1})a \, \|G\|_2 \tag{50.12}$$

$$\leqslant (1+ac+a^2c^2+\ldots+a^{n-1}c^{n-1})a \, \|G\|_E. \tag{50.13}$$

Notice that the result is still true if we replace c by any larger quantity. Now

$$a = \|K\|_2 = 1/\min |\lambda_i - \lambda|, \tag{50.14}$$

and hence (50.13) shows that there is a λ_i so that

$$\frac{|\lambda_i - \lambda|}{1+ac+a^2c^2+\ldots+a^{n-1}c^{n-1}} \leqslant \|G\|_2 = \|F\|_2 \leqslant \|F\|_E, \tag{50.15}$$

the same result holding when c is replaced by b.

Since in case (i) the relation (50.15) is obviously true, it is true in all cases. When A is normal, $c = 0$ and our result reduces to (50.1).

Simple examples

51. This result is not subject to the limitations which we discussed in Chapter 2, § 26. If we take for example

$$
A = \begin{bmatrix} 1 & \epsilon & & & & \\ & 1 & \epsilon & & & \\ & & \cdot & & & \\ & & & \cdot & & \\ & & & & \cdot & \\ & & & & 1 & \epsilon \\ & & & & & 1 \end{bmatrix}, \qquad \|F\|_E = \eta, \qquad (51.1)
$$

then $b = \epsilon$ and writing $|1 - \lambda| = \xi$ we have

$$
\frac{\xi}{1 + (\epsilon/\xi) + \ldots + (\epsilon/\xi)^{n-1}} \leqslant \eta. \qquad (51.2)
$$

If $n = 10$, $\epsilon = 10^{-10}$, $\eta = 10^{-8}$, then the maximum value of ξ is the solution of

$$
\xi = 10^{-8}[1 + (10^{-10}/\xi) + (10^{-10}/\xi)^2 + \ldots + (10^{-10}/\xi)^9], \qquad (51.3)
$$

and this obviously lies between 10^{-8} and $10^{-8}(1 + 10^{-2})$.

To make any use of (50.15) we require an upper bound for c which can be computed reasonably easily. Now a necessary and sufficient condition for A to be normal is that $(A^H A - A A^H)$ should be zero. We might reasonably expect therefore that c is related to $(A^H A - A A^H)$. Henrici (1962) has shown, in fact, that

$$
c \leqslant \left(\frac{n^3 - n}{12}\right)^{\frac{1}{4}} \|A^H A - A A^H\|_E^{\frac{1}{2}}, \qquad (51.4)
$$

determination of which requires two matrix multiplications. Since there are n^3 multiplications in a matrix multiplication the volume of work is far from negligible.

Unfortunately a matrix may exhibit considerable departure from normality without its eigenvalues being at all ill-conditioned. Consider for example the matrix

$$
A = \begin{bmatrix} 1 & 2 \\ 0 & 3 \end{bmatrix}. \qquad (51.5)
$$

Here we have $b = c = 2$, and hence if λ is an eigenvalue of $(A + F)$ then our result shows that either

$$\frac{|\lambda - 1|}{1 + 2/|\lambda - 1|} \leqslant \|F\|_E = \eta, \tag{51.6}$$

or

$$\frac{|\lambda - 3|}{1 + 2/|\lambda - 3|} \leqslant \|F\|_E = \eta. \tag{51.7}$$

That is, either

or

$$\left.\begin{array}{l} |\lambda - 1| \leqslant \tfrac{1}{2}[\eta + (\eta^2 + 8\eta)^{\frac{1}{2}}] \\ |\lambda - 3| \leqslant \tfrac{1}{2}[\eta + (\eta^2 + 8\eta)^{\frac{1}{2}}] \end{array}\right\}, \tag{51.8}$$

and the radii of these circles are greater than $(2\eta)^{\frac{1}{2}}$. The left-hand and right-hand eigenvectors of this matrix are given in (51.9),

$$\left.\begin{array}{llll} \lambda_1 = 3; & y_1^T = [0, 1] & \text{and} & x_1^T = [1, 1] \\ \lambda_2 = 1; & y_2^T = [1, -1] & \text{and} & x_2^T = [1, 0] \end{array}\right\}. \tag{51.9}$$

Hence (Chapter 2, § 8) $|s_1| = |s_2| = 2^{-\frac{1}{2}}$. If κ is the spectral condition number then from Chapter 2, relation (31.8), $\kappa \leqslant |s_1^{-1}| + |s_2^{-1}| \leqslant 2^{\frac{3}{2}}$; and by (30.7) of Chapter 2, the eigenvalues of $(A + F)$ certainly lie in the circles

$$\left.\begin{array}{l} |\lambda - 1| \leqslant 2^{\frac{3}{2}}\eta \\ |\lambda - 3| \leqslant 2^{\frac{3}{2}}\eta \end{array}\right\}. \tag{51.10}$$

Clearly when η is small the results (51.8) are very weak.

A *posteriori* bounds

52. At present there do not appear to be any useful *a priori* bounds for the perturbations of the eigenvalues of non-normal matrices of general form. Even the result of the previous section, weak though it is, requires $2n^3$ multiplications in order to obtain an estimate for the departure from normality. Now some of the algorithms we shall describe require only a small multiple of n^3 multiplications for the determination of the complete eigensystem. It is not surprising to find that from the approximate computed eigensystem we can obtain much more precise estimates of the eigenvalues. Error bounds determined from a computed eigensystem or partial eigensystem are called *a posteriori* bounds. The case for relying on *a posteriori* bounds for non-normal matrices would appear to be overwhelming.

A *posteriori* bounds for normal matrices

53. Suppose we have an approximate eigenvalue μ and corresponding eigenvector v of a matrix A. Then consider the vector η defined by

$$(A - \mu I)v = Av - \mu v = \eta. \tag{53.1}$$

This is usually called the *residual vector* corresponding to μ and v. If μ and v were exact, η would be zero. Hence if η is 'small' we might expect that μ would be a good approximation to an eigenvalue.

If A has linear divisors then there exists an H such that

$$H \operatorname{diag}(\lambda_i)H^{-1} = A.$$

Hence from (53.1)

$$H \operatorname{diag}(\lambda_i - \mu)H^{-1}v = \eta. \tag{53.2}$$

Two cases arise

(i) $\lambda_i = \mu$ for some i,

(ii) $\lambda_i \neq \mu$ for any i, in which case (53.2) gives

$$v = H \operatorname{diag}[(\lambda_i - \mu)^{-1}]H^{-1}\eta, \tag{53.3}$$

and hence

$$\|v\| < \|H\| \, \|H^{-1}\| \, \|\operatorname{diag}[(\lambda_i - \mu)^{-1}]\| \, \|\eta\|$$

$$= \max_i \frac{1}{|\mu - \lambda_i|} \|H\| \, \|H^{-1}\| \, \|\eta\|, \tag{53.4}$$

for any norm for which the norm of a diagonal matrix is the modulus of its largest element. Hence using the 2-norm

$$\min |\mu - \lambda_i| < \kappa \, \|\eta\|_2 / \|v\|_2 \tag{53.5}$$

where κ is as in § 51. The relation (53.5) is obviously satisfied in case (i) and hence we deduce that there is always at least one eigenvalue of A in a disc with centre μ and radius $\kappa \, \|\eta\|_2/\|v\|_2$. *Clearly there is no loss of generality in assuming that* $\|v\|_2 = 1$ *and we shall do this from now on.* Since, in general, we have no *a priori* estimate of κ this result is not usually of much practical value unless we have more information about the eigensystem. If A is normal, however, we know that $\kappa = 1$ and our result then shows that there is at least one eigenvalue in the disc with centre μ and radius $\|\eta\|_2$.

There is another way of looking at equation (53.1) which relates this result to that of Chapter 2. Equation (53.1) implies that

$$(A - \eta v^H)v = Av - \eta(v^H v) = Av - \eta = \mu v, \tag{53.6}$$

and hence μ and v are an *exact* eigenvalue and eigenvector of the matrix $(A - \eta v^H)$. From Chapter 2, § 30, we know that all the eigenvalues of $(A - \eta v^H)$ lie in the discs

$$|\lambda - \lambda_i| < \kappa \, \|\eta v^H\|_2 = \kappa \, \|\eta\|_2 \quad (i = 1, ..., n), \tag{53.7}$$

and this leads to the same result as we have just obtained.

Rayleigh quotient

54. If we are given any vector v then we may choose a value of μ so that $\|Av - \mu v\|_2$ is a minimum. Writing $k = v^H A v$ and assuming that $\|v\|_2 = 1$ we have

$$\|Av - \mu v\|_2^2 = (Av - \mu v)^H (Av - \mu v)$$
$$= v^H A^H A v - k k^H + (\mu^H - k^H)(\mu - k). \qquad (54.1)$$

Clearly the minimum is achieved for $\mu = k = v^H A v$. This quantity is known as the *Rayleigh quotient* corresponding to v. We denote it by μ_R. The term 'quotient' is used because, if v is not normalized, the appropriate value is $v^H A v / v^H v$. Clearly if

$$Av - \mu_R v = \eta_R, \qquad (54.2)$$

then

$$v^H \eta_R = 0. \qquad (54.3)$$

If for any μ we have

$$Av - \mu v = \eta \quad (\|\eta\|_2 = \epsilon) \qquad (54.4)$$

then

$$Av - \mu_R v = \eta_R \quad (\|\eta_R\|_2 \leqslant \epsilon). \qquad (54.5)$$

The circular disc with centre μ_R has a radius which is at least as small as that corresponding to any other μ.

The Rayleigh quotient is defined for any matrix but it is of special importance for normal matrices as we shall now show. Suppose we have

$$Av - \mu_R v = \eta \quad (\|v\|_2 = 1; \ \|\eta\|_2 = \epsilon) \qquad (54.6)$$

and we know that $n - 1$ of the eigenvalues satisfy the relation

$$|\lambda_i - \mu_R| \geqslant a, \qquad (54.7)$$

where a is some constant. We may take the relevant eigenvalues to be $\lambda_2, \lambda_3, ..., \lambda_n$. Now if A is normal it certainly has an orthonormal system of eigenvectors x_i $(i = 1, ..., n)$ and we may write

$$v = \sum_1^n \alpha_i x_i, \quad 1 = \sum_1^n |\alpha_i|^2. \qquad (54.8)$$

Equation (54.6) therefore gives

$$\eta = \sum_1^n \alpha_i (\lambda_i - \mu_R) x_i,$$

and hence

$$\epsilon^2 = \sum_1^n |\alpha_i|^2 \, |\lambda_i - \mu_R|^2 \geqslant \sum_2^n |\alpha_i|^2 \, |\lambda_i - \mu_R|^2 \qquad (54.9)$$

$$\geqslant a^2 \sum_2^n |\alpha_i|^2 \quad \text{from (54.7).} \qquad (54.10)$$

From this we deduce that

$$|\alpha_1|^2 = 1 - \sum_2^n |\alpha_i|^2 > 1 - \epsilon^2/a^2. \tag{54.11}$$

If we define θ by the relation

$$\alpha_1 = |\alpha_1|\, e^{i\theta}, \tag{54.12}$$

then

$$ve^{-i\theta} = |\alpha_1|\, x_1 + \sum_2^n \alpha_i e^{-i\theta} x_i, \tag{54.13}$$

and hence

$$\|ve^{-i\theta} - x_1\|_2^2 = (|\alpha_1| - 1)^2 + \sum_2^n |\alpha_i|^2$$

$$\leqslant \frac{\epsilon^4/a^4}{(1 + |\alpha_1|)^2} + \frac{\epsilon^2}{a^2} < \frac{\epsilon^4}{a^4} + \frac{\epsilon^2}{a^2}. \tag{54.14}$$

From (54.14) we see that if $a \gg \epsilon$, then $ve^{-i\theta}$ is a good approximation to x_1. The constant factor $e^{-i\theta}$ is of little importance. As a diminishes the result becomes progressively weaker until, when a is of the order of magnitude of ϵ, it no longer gives a useful result. *Notice that the results obtained so far hold whether or not μ_R is a Rayleigh quotient.*

Error in Rayleigh quotient

55. Turning now to μ_R we have from its definition

$$\mu_R = v^H A v = \sum_1^n \lambda_i |\alpha_i|^2. \tag{55.1}$$

Hence

$$\mu_R \sum_1^n |\alpha_i|^2 = \sum_1^n \lambda_i |\alpha_i|^2,$$

giving

$$(\mu_R - \lambda_1) |\alpha_1|^2 = \sum_2^n (\lambda_i - \mu_R) |\alpha_i|^2$$

$$= \sum_2^n |(\lambda_i - \mu_R)|^2 |\alpha_i|^2/(\lambda_i - \mu_R)^H. \tag{55.2}$$

From (54.11), (54.7) and (54.9) we find

$$\left(1 - \frac{\epsilon^2}{a^2}\right) |\mu_R - \lambda_1| < \sum_2^n |\lambda_i - \mu_R|^2 |\alpha_i|^2/|\lambda_i - \mu_R|$$

$$< \sum_2^n |\lambda_i - \mu_R|^2 |\alpha_i|^2/a$$

$$< \epsilon^2/a,$$

$$|\mu_R - \lambda_1| < \frac{\epsilon^2}{a}\bigg/\left(1 - \frac{\epsilon^2}{a^2}\right). \tag{55.3}$$

Relation (55.3) shows that if $a \gg \epsilon$, the Rayleigh quotient has an error which is of order ϵ^2. As a is diminished the result becomes progressively weaker until, when a is of the same order of magnitude as ϵ, it is no stronger than that which we have for an arbitrary μ such that $\|Av-\mu v\|_2 = \epsilon$.

That deterioration in the bounds for the errors in the Rayleigh quotient and the vector is inevitable, when a is of order ϵ, is illustrated by the following example. Consider the matrix A and the alleged eigenvalue and eigenvector given by

$$A = \begin{bmatrix} b & \epsilon \\ \epsilon & b \end{bmatrix}, \qquad \mu_R = b, \qquad v = \begin{bmatrix} 1 \\ 0 \end{bmatrix}. \tag{55.4}$$

We have

$$Av - \mu_R v = \begin{bmatrix} 0 \\ \epsilon \end{bmatrix}, \tag{55.5}$$

and hence $\|\eta\|_2 = \epsilon$ for the normalized vector v. The eigenvalues are $b \pm \epsilon$ so that μ_R has an error of order ϵ. The eigenvectors are

$$\begin{bmatrix} 1 \\ 1 \end{bmatrix} \quad \text{and} \quad \begin{bmatrix} 1 \\ -1 \end{bmatrix}, \tag{55.6}$$

showing that v is in no sense an approximate eigenvector.

Hermitian matrices

56. The only normal matrices with which we shall be much concerned will be Hermitian, usually real and symmetric. The solution of the eigenproblem for a complex Hermitian matrix may be reduced to that of a real symmetric matrix by the following device. Suppose

$$\left. \begin{aligned} A = B+iC \quad & (B \text{ and } C \text{ real}) \\ B^T = B, \qquad & C^T = -C \end{aligned} \right\}. \tag{56.1}$$

Then if λ_i is an eigenvalue of A (necessarily real) and (u_i+iv_i) is the corresponding eigenvector we have

$$(B+iC)(u_i+iv_i) = \lambda_i(u_i+iv_i),$$

i.e.

$$\left. \begin{aligned} Bu_i - Cv_i = \lambda_i u_i \\ Cu_i + Bv_i = \lambda_i v_i \end{aligned} \right\}. \tag{56.2}$$

Hence

$$\begin{bmatrix} B & -C \\ C & B \end{bmatrix} \begin{bmatrix} u_i \\ v_i \end{bmatrix} = \begin{bmatrix} \lambda_i u_i \\ \lambda_i v_i \end{bmatrix}, \quad \begin{bmatrix} B & -C \\ C & B \end{bmatrix} \begin{bmatrix} v_i \\ -u_i \end{bmatrix} = \begin{bmatrix} \lambda_i v_i \\ -\lambda_i u_i \end{bmatrix}, \tag{56.3}$$

and since

$$\begin{bmatrix} u_i \\ v_i \end{bmatrix} \quad \text{and} \quad \begin{bmatrix} v_i \\ -u_i \end{bmatrix} \tag{56.4}$$

are orthogonal, the augmented real matrix has λ_i as a double eigenvalue. The augmented matrix is obviously real and symmetric and its eigenvalues are $\lambda_1, \lambda_1, \lambda_2, \lambda_2, \ldots, \lambda_n, \lambda_n$.

Most of the better algorithms for finding the eigensystems of real symmetric matrices lead to real approximations to the eigenvalues and eigenvectors. The circular discs we have mentioned therefore become *intervals on the real axis*.

Notice that if μ_R is the Rayleigh quotient corresponding to v and

$$Av - \mu_R v = \eta, \tag{56.5}$$

then since $\eta^T v = 0$ we have

$$(A - \eta v^T - v \eta^T)v = \mu_R v \tag{56.6}$$

so that v and μ_R are exact for the matrix $(A - \eta v^T - v \eta^T)$. This matrix is symmetric if A is symmetric and we are now concerned with a symmetric perturbation of a symmetric matrix. It is easy to verify that

$$\|\eta v^T + v \eta^T\|_2 = \|\eta\|_2 \tag{56.7}$$

if $\|v\|_2 = 1$ and $\eta^T v = 0$.

If we obtain n approximate eigenvalues and eigenvectors and the intervals are disjoint, then we know that there is precisely one eigenvalue in each. The result of the last section will enable us to locate the eigenvalues very precisely unless they are poorly separated. Thus if for a matrix of order four we have the Rayleigh quotients μ_i and residuals η_i specified by

i	1	2	3	4	
μ_i	0·1	0·1001	0·2	0·3	(56.8)
$\|\eta_i\|_2$	10^{-8}	$2 . 10^{-8}$	$2 . 10^{-8}$	$3 . 10^{-8}$	

then we know immediately that there is at least one eigenvalue in each of the intervals

$$\left. \begin{array}{ll} |\lambda - 0·1| \leqslant 10^{-8}, & |\lambda - 0·1001| \leqslant 2 . 10^{-8} \\ |\lambda - 0·2| \leqslant 2 . 10^{-8}, & |\lambda - 0·3| \leqslant 3 . 10^{-8} \end{array} \right\}. \tag{56.9}$$

These intervals are clearly disjoint and hence each contains precisely one eigenvalue.

We now illustrate the use of the result of § 55 by obtaining a very precise bound for the eigenvalue λ_2 in the second interval. Clearly we have

$$
\left.\begin{aligned}
|\mu_2 - \lambda_1| &\geqslant |0 \cdot 1001 - 0 \cdot 1 - 10^{-8}| = 10^{-4}(0 \cdot 9999) \\
|\mu_2 - \lambda_3| &\geqslant |0 \cdot 1001 - 0 \cdot 2 + 2 \cdot 10^{-8}| > 0 \cdot 0998 \text{ (say)} \\
|\mu_2 - \lambda_4| &\geqslant |0 \cdot 1001 - 0 \cdot 3 + 3 \cdot 10^{-8}| > 0 \cdot 1998 \text{ (say)}
\end{aligned}\right\}. \qquad (56.10)
$$

Hence we may take $a = 10^{-4}(0 \cdot 9999)$ and (55.3) gives

$$
\begin{aligned}
|\mu_2 - \lambda_2| &\leqslant \frac{4 \cdot 10^{-16}}{10^{-4}(0 \cdot 9999)} \Big/ \left\{ 1 - \left[\frac{2 \cdot 10^{-8}}{10^{-4}(0 \cdot 9999)} \right]^2 \right\} \\
&\doteqdot (4 \cdot 0004)10^{-12}.
\end{aligned}
$$

In spite of the fact that λ_1 and λ_2 might well be regarded as quite close, the Rayleigh quotient is very accurate. The quotients μ_3 and μ_4 are even more accurate. In practice we shall find that the Rayleigh quotient is so accurate (except for pathologically close eigenvalues) that it must be computed with great care if its full potentialities are to be realized.

Pathologically close eigenvalues

57. When the intervals surrounding the Rayleigh quotients are not disjoint the situation is much more obscure and we cannot be sure without further investigation that all the eigenvalues are contained in the union of the intervals. We may illustrate the difficulty by considering an extreme case. Suppose the eigenproblem of a matrix of order 4 with well separated eigenvalues is solved using 4 different algorithms. We denote the approximations to the eigenvalue λ_1 and eigenvector x_1 obtained by the four methods by μ_i and v_i $(i = 1, ..., 4)$. Although these are all approximations to the same eigenvalue and eigenvector, they will not, in general, be identical; in fact the v_i will usually be strictly linearly independent. If we are presented with these four pairs of μ_i and v_i we shall obtain four overlapping intervals, assuming the algorithms were stable, yet there will be only one eigenvalue in their union. Clearly we shall need a good deal more than a mere mathematical independence of vectors if we are to make useful and rigorous claims.

The following considerations shed light on this problem. Suppose v_i $(i = 1, ..., r)$ are orthogonal normalized vectors and μ_i $(i = 1, ..., r)$ are such that

$$
A v_i - \mu_i v_i = \eta_i, \qquad \|\eta_i\|_2 = \epsilon_i. \qquad (57.1)
$$

There exists a set of normalized vectors $v_{r+1}, ..., v_n$ such that the matrix V having the v_i $(i = 1, ..., n)$ as its columns is orthogonal. Consider now the matrix B defined by

$$B = V^T A V. \tag{57.2}$$

B is clearly symmetric and from (57.1) we have

$$
\begin{aligned}
b_{ij} &= v_i^T A v_j \quad &\text{(all } i, j), \\
&= v_i^T (\mu_j v_j + \eta_j) \quad &(j = 1, ..., r).
\end{aligned} \tag{57.3}
$$

From the symmetry we may write

$$
B = \left[
\begin{array}{ccccc|c}
\mu_1 & & & & & \\
& \mu_2 & & & & \\
& & \cdot & & & O \\
& & & \cdot & & \\
& & & & \cdot & \\
& & & & & \mu_r \\
\hline
& & O & & & X
\end{array}
\right]
+
\left[
\begin{array}{c|c}
P & Q \\
\hline
Q^T & O
\end{array}
\right], \tag{57.4}
$$

where X is symmetric, and

$$[P \mid Q]_{ij} = v_i^T \eta_j \quad (i = 1, ..., r; j = 1, ..., n). \tag{57.5}$$

Now B has the eigenvalues $\lambda_1, \lambda_2, ..., \lambda_n$ of A and if $\mu_{r+1}, \mu_{r+2}, ..., \mu_n$ are the eigenvalues of X we have, by the Hoffman-Wielandt theorem (Chapter 2, § 48), for some ordering of the λ_i

$$
\begin{aligned}
\sum_1^r (\lambda_i - \mu_i)^2 &< \sum_1^n (\lambda_i - \mu_i)^2 < \|P\|_E^2 + \|Q\|_E^2 + \|Q^T\|_E^2 \\
&< 2(\|P\|_E^2 + \|Q\|_E^2) \\
&= 2 \sum_1^r \|\eta_i\|^2 \\
&= 2 \sum_1^r \epsilon_i^2 = k^2. \tag{57.6}
\end{aligned}
$$

Hence there are certainly r eigenvalues in the union of the r intervals $\mu_i - k < \lambda < \mu_i + k$. Notice that the μ_i were not necessarily Rayleigh quotients.

As an example we consider the diagonal matrix D of order n with

$$d_{ii} = a \quad (i = 1, 2, ..., n-1), \qquad d_{nn} = a + n^{\frac{1}{2}}\epsilon. \tag{57.7}$$

If we take
$$v = (\pm n^{-\frac{1}{2}}, \pm n^{-\frac{1}{2}}, \ldots, \pm n^{-\frac{1}{2}}), \qquad \mu = a, \qquad (57.8)$$
we have
$$\eta = (0, 0, \ldots, \pm \epsilon) \qquad (57.9)$$

and hence, whatever the distribution of signs, $\|\eta\|_2 = \epsilon$. It is well known that if n is a power of 2, there are n orthogonal vectors of the type (57.8). All the corresponding intervals have the same centre a, and it is evident that the smallest interval which will cover all n eigenvalues is $(a - n^{\frac{1}{2}}\epsilon)$ to $(a + n^{\frac{1}{2}}\epsilon)$. The Rayleigh quotient corresponding to each of the v is $(a + n^{-\frac{1}{2}}\epsilon)$, so that we do not get an appreciably better result.

Suppose then, we are given
$$Av_1 - \mu_1 v_1 = \eta_1, \qquad (57.10)$$
$$Av_2 - \mu_2 v_2 = \eta_2, \qquad (57.11)$$

where $\|\eta_1\|_2$, $\|\eta_2\|_2$, $|\mu_1 - \mu_2|$ are all small and where
$$v_1^T v_2 = \cos \theta. \qquad (57.12)$$

We may write
$$v_2 = v_1 \cos \theta + w_1 \sin \theta, \qquad (57.13)$$

where w_1 is a unit vector in the subspace of v_1 and v_2 and orthogonal to v_1. Multiplying (57.10) by $\cos \theta$ and subtracting from (57.11) we obtain
$$Aw_1 \sin \theta - (\mu_2 - \mu_1)v_1 \cos \theta - \mu_2 w_1 \sin \theta = \eta_2 - \eta_1 \cos \theta,$$
giving
$$Aw_1 - \mu_2 w_1 = \eta_3, \qquad (57.14)$$
where
$$\eta_3 = \frac{\eta_2 - \eta_1 \cos \theta + (\mu_2 - \mu_1)v_1 \cos \theta}{\sin \theta}, \qquad (57.15)$$

$$\|\eta_3\|_2 \leqslant \frac{\|\eta_2\|_2 + \|\eta_1\|_2 + |\mu_2 - \mu_1|}{|\sin \theta|}. \qquad (57.16)$$

Provided $\sin \theta$ is not small, (57.14) and (57.16) now show that the orthogonal vectors v_1 and w_1 have small residuals and we can use our previous result. As θ becomes smaller this result grows progressively weaker.

Non-normal matrices

58. For normal matrices we can often obtain very precise and rigorous bounds for eigenvalues from relatively imprecise information relating to only a part of the eigensystem. For non-normal matrices this is seldom possible. The relation (53.5) is of value only if we have an estimate for κ and this is not obtainable from the relation (53.1). Even if we have a sound estimate for κ the bound (53.5) may be very pessimistic if μ happens to be an approximation to a well-conditioned eigenvalue but A has some ill-conditioned eigenvalues.

However let us consider 'close' approximations u_1 and v_1 to the right-hand and left-hand vectors corresponding to λ_1. For such a u_1 and v_1 we may write

$$u_1 = k(x_1 + f_1), \qquad v_1 = l(y_1 + g_1), \tag{58.1}$$

where f_1 is in the subspace spanned by x_2 to x_n, g_1 is in the subspace spanned by y_2 to y_n and both are small. Hence we have

$$
\begin{aligned}
v_1^T A u_1 / v_1^T u_1 &= (y_1^T + g_1^T)(\lambda_1 x_1 + A f_1)/(y_1^T + g_1^T)(x_1 + f_1) \\
&= (\lambda_1 s_1 + g_1^T A f_1)/(s_1 + g_1^T f_1), \quad \text{where } s_1 = y_1^T x_1, \\
&= \lambda_1 + \frac{g_1^T (A - \lambda_1 I) f_1}{s_1 + g_1^T f_1},
\end{aligned}
\tag{58.2}
$$

giving

$$|v_1^T A u_1 / v_1^T u_1 - \lambda_1| < (\|A\| + |\lambda_1|)\, \|g_1\|_2\, \|f_1\|_2/(|s_1| - \|g_1\|_2\, \|f_1\|_2). \tag{58.3}$$

Provided s_1 is not too small we might therefore expect that the quotient $(v_1^T A u_1 / v_1^T u_1)$ would give a good approximation to λ_1.

For normal matrices the Rayleigh quotient $(u_1^H A u_1 / u_1^H u_1)$ was of particular importance. Since for such matrices we have

$$y_1^T = x_1^H, \tag{58.4}$$

we see that the quotient $(v_1^T A u_1 / v_1^T u_1)$ is a natural extension for non-normal matrices of the Rayleigh quotient for normal matrices. We shall refer to it as the *generalized Rayleigh quotient corresponding to* v_1 *and* u_1.

Suppose then, using a stable algorithm, we have found alleged normalized left-hand and right-hand eigenvectors v_1 and u_1 corresponding to the same eigenvalue, and we verify that for μ defined by

$$\mu = v_1^T A u_1 / v_1^T u_1 \tag{58.5}$$

both $(A u_1 - \mu u_1)$ and $(A^T v_1 - \mu v_1)$ are small, then there is a strong presumption that $v_1^T u_1$ is a close approximation to s_1 and μ is an even better approximation to λ_1. However, without some further information we cannot obtain a rigorous bound for the error in μ or even be absolutely certain that u_1 and v_1 are approximations corresponding to the same eigenvalue.

Error analysis for a complete eigensystem

59. If we are given an approximate solution to the complete eigensystem, then provided the solution is reasonably accurate and the eigensystem is not too ill-conditioned we can obtain rigorous error bounds not merely for the given eigensystem but for an improved eigensystem. Suppose μ_i and u_i $(i = 1,\ldots, n)$ are approximate eigenvalues and eigenvectors of the matrix A. We may write

$$Au_i - \mu_i u_i = r_i, \qquad \|u_i\|_2 = 1, \tag{59.1}$$

or, combining these equations into matrix form,

$$AU - U \operatorname{diag}(\mu_i) = R, \tag{59.2}$$

where U and R are the matrices with columns u_i and r_i respectively. Hence

$$U^{-1}AU = \operatorname{diag}(\mu_i) + U^{-1}R = \operatorname{diag}(\mu_i) + F, \tag{59.3}$$

so that A is exactly similar to $\operatorname{diag}(\mu_i) + F$ whether or not the μ_i and u_i are good approximations. However, if the μ_i and u_i *are* good approximations, then the matrix R will have small components. If the components of the matrix F, which is the solution of the equations

$$UF = R, \tag{59.4}$$

are also small then the right-hand side of (59.3) is a diagonal matrix with a small perturbation. Now in Chapter 2, §§ 14–25, we showed how to reduce the perturbation problem for general matrices to that for diagonal matrices, and a great part of these sections was devoted to the problem of finding bounds for the eigenvalues of a matrix of the form $\operatorname{diag}(\mu_i) + F$ where F was small. The techniques given there may be applied immediately to the matrix on the right of (59.3), and they are usually capable of giving very precise bounds for all except pathologically close eigenvalues.

If their full potentialities are to be realized it is essential that R and $U^{-1}R$ should be computed with some care. The matrix R may be obtained exactly on a fixed-point computer which can accumulate inner-products, but the accurate solution of (59.4) is a difficult problem. It is discussed in detail in Chapter 4, § 69. In general, even if R is given exactly, we will not be able to compute F exactly.

However, if we assume that we can compute an approximate solution $\bar{F}$ of (59.4) *and* give a bound for the elements of G defined by

$$G = F - \bar{F}, \tag{59.5}$$

then we have

$$U^{-1}AU = \operatorname{diag}(\mu_i) + \bar{F} + G. \tag{59.6}$$

Numerical example

60. As an illustration let us assume that for a matrix A of order 3 we have obtained

$$\bar{F} = 10^{-4} \begin{bmatrix} 0\cdot6132 & 0\cdot2157 & -0\cdot4157 \\ 0\cdot2265 & 0\cdot3153 & 0\cdot3123 \\ 0\cdot6157 & 0\cdot2132 & 0\cdot8843 \end{bmatrix}, \qquad (60.1)$$

$$\mu = \begin{bmatrix} 0\cdot8132 \\ 0\cdot6132 \\ 0\cdot2157 \end{bmatrix}, \qquad |g_{ij}| < \tfrac{1}{2}10^{-8}. \qquad (60.2)$$

Then by apply similarity transformations with diagonal matrices we can obtain improved eigenvalues and error bounds for them. In fact, after multiplying the first row of $\mathrm{diag}(\mu_i) + \bar{F} + G$ by 10^{-3} and the first column by 10^3 (cf. Chapter 2, § 15), the Gerschgorin discs are specified by

centre $0\cdot8132 + 10^{-4}(0\cdot6132) + g_{11}$,

radius $10^{-7}(0\cdot2157 + 0\cdot4157) + 10^{-3}(|g_{12}| + |g_{13}|)$

centre $0\cdot6132 + 10^{-4}(0\cdot3153) + g_{22}$,

radius $10^{-1}(0\cdot2265) + 10^{-4}(0\cdot3123) + 10^3 |g_{21}| + |g_{23}|$

centre $0\cdot2157 + 10^{-4}(0\cdot8843) + g_{33}$,

radius $10^{-1}(0\cdot6157) + 10^{-4}(0\cdot2132) + 10^3 |g_{31}| + |g_{32}|$.

Remembering our bound for the g_{ij}, we see that these discs are obviously disjoint and hence there is an eigenvalue in the disc with centre $0\cdot81326132$ and radius $10^{-7}(0\cdot6314) + |g_{11}| + 10^{-3}(|g_{12}| + |g_{13}|)$. This radius is not greater than $10^{-7}(0\cdot6815)$. In a similar way, by operating on the other rows and columns we may improve each of the other eigenvalues and obtain error bounds for the improved values.

Parallel with the analysis in Chapter 2 we find that the error bound for each improved μ_i depends on the size of the elements of F in the ith row and ith column and on the separation of μ_i from the other μ_j.

Conditions limiting attainable accuracy

61. It should be emphasized that $U^{-1}AU$ is exactly similar to A and by computing $\bar{F}$ to sufficient accuracy we can ensure that the eigenvalues of $\mathrm{diag}(\mu_i) + \bar{F}$ are as close to those of A as we please.

In Chapter 4 we describe a method of solving equations (59.4) using t-digit fixed-point arithmetic which normally ensures that $\bar{F}$ is a correctly rounded result so that $|g_{ij}| \leqslant 2^{-t} \max_{k,l} |f_{kl}|$. If we use such a technique, then the bound for g_{ij} will depend directly on the size of F. *We do not need a priori estimates for this because all the information will come from the a posteriori computation itself*, but it is instructive to consider what are the factors governing the size of F.

If U were the matrix of exact normalized eigenvectors, then the rows of U^{-1} would be given by y_i^T/s_i (Chapter 2, § 14). Hence we might well expect that, in general, the ith row of U^{-1}, w_i^T say, is an approximation to y_i^T/s_i. Now the (i, i) element of $U^{-1}AU$ is $w_i^T A u_i$, and since w_i^T is a row of U^{-1} we have

$$w_i^T u_i = 1. \tag{61.1}$$

This (i, i) element may therefore be expressed in the form $w_i^T A u_i / w_i^T u_i$, and hence from § 58 it is the generalized Rayleigh quotient corresponding to w_i and u_i. It is the computed value of $w_i^T A u_i / w_i^T u_i$ which is the improved eigenvalue obtained by the technique of §§ 59, 60. It follows that the basic device involved in that technique then, is equivalent to the determination of a set of approximate left-hand eigenvectors from the given set of approximate right-hand eigenvectors, followed by the calculation of the n generalized Rayleigh quotients. By organizing the work as in §§ 59, 60 we obtain in addition rigorous error bounds.

Since the ith row of U^{-1} will be approximately y_i^T/s_i the ith row of F will be approximately $y_i^T R/s_i$ and we might expect these elements to be large if s_i is small. This in turn will lead to poorer bounds for the g_{ij}. Hence, not surprisingly, small s_i will limit our attainable accuracy for a given precision of computation as will also close eigenvalues.

Notice that when we have the matrix U of approximate eigenvectors the computation of U^{-1} gives us estimates of the s_i immediately and of κ somewhat less directly, from the determination of $\|U^{-1}\|_2$. However at this stage we need attach little importance to an estimate for κ since the technique we have given in § 59 involves less work and gives far more detailed information.

Non-linear elementary divisors

62. So far we have tacitly assumed that the matrix A has linear divisors. When the matrix has non-linear divisors one difficulty immediately presents itself if we use an algorithm which is based on the reduction of the original matrix to one of simpler form. Even if the algorithm is quite stable the computed transform will be exactly similar to $(A+F)$ where, at best, the elements of F will be of order 2^{-t} times those of A. Now we have seen (Chapter 2, § 19) that the eigenvalues corresponding to non-linear divisors may well be very sensitive to perturbations in the matrix elements, and hence the eigenvalues of the transformed matrix might well have quite reasonable separations and independent eigenvectors.

Suppose for example our original matrix A and its Jordan canonical form B are given by

$$ A = \begin{bmatrix} 1 & 1 & -1 \\ -2 & 2 & 1 \\ -2 & 1 & 2 \end{bmatrix}, \qquad B = \begin{bmatrix} 1 & 1 & 0 \\ 0 & 1 & 0 \\ 0 & 0 & 3 \end{bmatrix}, $$

$$ B' = \begin{bmatrix} 1 & 1 & 0 \\ 10^{-10} & 1 & 0 \\ 0 & 0 & 3 \end{bmatrix}. \quad (62.1) $$

The matrix A obviously has the divisors $(\lambda-3)$ and $(\lambda-1)^2$. Using 10-digit decimal computation, even a very stable algorithm might transform A into a matrix which is exactly similar to B' rather than B. The matrix B' has the simple eigenvalues 1 ± 10^{-5} and 3, which are reasonably well separated from the point of view of 10-digit computation, and has three independent eigenvectors. As a result of such a computation we could obtain alleged eigenvalues $1+10^{-5}$, $1-10^{-5}$, $3+10^{-10}$ (say) and corresponding eigenvectors given by

$$ [u_1 \quad u_2 \quad u_3] = \begin{bmatrix} 1 & 1 & 10^{-10} \\ 1+10^{-5} & 1-10^{-5} & 1+10^{-10} \\ 1-10^{-10} & 1-10^{-10} & 1 \end{bmatrix}. \quad (62.2) $$

(We expect errors of order 10^{-10} in the third eigenvalue and eigenvector because these are well-conditioned.)

The corresponding residual matrix is given by

$$ [r_1 \quad r_2 \quad r_3] = \begin{bmatrix} 10^{-10} & 10^{-10} & -10^{-10}-10^{-20} \\ -2.10^{-10} & -2.10^{-10} & -4.10^{-10}-10^{-20} \\ -10^{-10}+10^{-15} & -10^{-10}-10^{-15} & -2.10^{-10} \end{bmatrix}. $$

$$ (62.3) $$

Notice that although the computed eigensystem has errors of order 10^{-5} the residual matrix has components which are of order 10^{-10}. This will generally be true of eigensystems computed by stable methods. Indeed if λ and u are exact for $(A + F)$ the residual vector corresponding to A is $-Fu$, and hence will be dependent on the size of F.

However when we attempt to compute $U^{-1}R$ the inadequacy of the solution reveals itself. We have in fact

$$U^{-1}AU = \begin{bmatrix} 1+10^{-5} & & \\ & 1-10^{-5} & \\ & & 3+10^{-10} \end{bmatrix} + F, \qquad (62.4)$$

where, writing $10^{-5} = x$,

$$F = \begin{bmatrix} -\tfrac{1}{2}x & -\tfrac{1}{2}x^3 & -\tfrac{1}{2}x+x^2-\tfrac{1}{2}x^3 & -x-\tfrac{1}{2}x^2 \\ \tfrac{1}{2}x+x^2+\tfrac{1}{2}x^3 & \tfrac{1}{2}x & +\tfrac{1}{2}x^3 & x-\tfrac{1}{2}x^2 \\ -2x^2+x^3 & -2x^2-x^3 & -x^2 \end{bmatrix} + G;$$

$$(|g_{ij}| \leqslant \tfrac{1}{2}10^{-15}). \quad (62.5)$$

We see that some of the elements of F are of order 10^{-5}. The technique described in Chapter 4, § 69, gives a solution with error bounds of the order of 10^{-15} in such a case; accordingly we have assumed that the elements of $|G|$ are bounded by $\tfrac{1}{2}10^{-15}$.

Gerschgorin's theorem immediately gives us a very precise bound for the third eigenvalue. Corresponding to multiplication of row 3 by 10^{-5} and column 3 by 10^5, the Gerschgorin discs are disjoint and the third has

centre $(3+10^{-10}-10^{-10}+g_{33})$,

radius $10^{-5}(2.10^{-10}-10^{-15}+2.10^{-10}+10^{-15}+|g_{31}| + |g_{32}|)$.

We have therefore located it in a disc of centre 3 and radius less then 6.10^{-15} *in spite of the fact that our eigensystem was largely spurious.* We obviously cannot obtain such precise bounds for the other two eigenvalues without further computation.

Approximate invariant subspaces

63. It is scarcely surprising that the solution of the equation $UF = R$ was much larger than R itself, since two of the columns

of U were in effect approximations to the unique eigenvector corresponding to $\lambda = 1$. The higher the precision to which we work the more nearly singular does U become, but we obtain better and better approximations to the third eigenvalue.

If we knew in advance that the matrix had a quadratic divisor it would be more satisfactory to attempt to compute an *invariant subspace* corresponding to the double eigenvalue. The vectors

$$v_1, v_2, \ldots, v_r$$

are said to span an invariant subspace of A of order r if Av_i $(i = 1, \ldots, r)$ lie in the same subspace. The simplest form of invariant subspace is that spanned by the single eigenvector corresponding to a simple eigenvalue. Obviously the principal vectors corresponding to a non-linear elementary divisor span an invariant subspace. Although the eigenvector corresponding to $\lambda = 1$ is sensitive to perturbations, the invariant subspace of order two is not. This subspace is spanned by e and e_2 since
$$Ae = e, \quad Ae_2 = e + e_2. \qquad (63.1)$$

Each of the vectors u_1 and u_2 of § 62 may be expressed in the form $ae + be_2 + O(10^{-10})$ and hence lies in the subspace to working accuracy. However, the departure of u_1 and u_2 from linear dependence is registered by a vector whose elements are $O(10^{-5})$ only. Since in general we must regard the components of these vectors as being in error in the tenth decimal, u_1 and u_2 do not really determine the subspace to working accuracy.

We digress to mention that the analysis of § 59 may easily be extended to cover the case when we have determined approximate subspaces of orders greater than unity. For example, if for a matrix A of order 4 we have

$$Au_1 = \alpha_{11}u_1 + \alpha_{21}u_2 + r_1, \quad Au_2 = \alpha_{12}u_1 + \alpha_{22}u_2 + r_2 \\ Au_3 = \mu_3 u_3 + r_3, \quad\quad\quad Au_4 = \mu_4 u_4 + r_4 \qquad (63.2)$$

then we have

$$A[u_1\ u_2\ u_3\ u_4] = [u_1\ u_2\ u_3\ u_4]\begin{bmatrix} \alpha_{11} & \alpha_{12} & & \\ \alpha_{21} & \alpha_{22} & & \\ & & \mu_3 & \\ & & & \mu_4 \end{bmatrix} + [r_1\ r_2\ r_3\ r_4], \qquad (63.3)$$

and with an obvious notation

$$U^{-1}AU = D + U^{-1}R. \qquad (63.4)$$

Apart from the case of non-linear elementary divisors we shall frequently be interested in the subspace of order two spanned by a complex conjugate pair of eigenvectors corresponding to a real matrix. If the vectors are $u \pm iv$ where u and v are real, then u and v span the subspace. If $\lambda \pm i\mu$ are the corresponding eigenvalues we have

$$A(u+iv) = (\lambda+i\mu)(u+iv)$$

giving

$$Au = \lambda u - \mu v$$

$$Av = \mu u + \lambda v$$

$$\qquad (63.5)$$

Returning now to the simple problem of § 62, if we attempt to compute an invariant subspace to replace u_1, u_2 of § 62 we might obtain, for example,

$$[u_1 \quad u_2] = \begin{bmatrix} 1+10^{-10} & 1-10^{-10} \\ -10^{-10} & 2+10^{-10} \\ 1-10^{-10} & 1-10^{-10} \end{bmatrix}, \qquad (63.6)$$

and we have

$$A[u_1 \quad u_2] = [u_1 \quad u_2]\begin{bmatrix} \tfrac{1}{2} & \tfrac{1}{2} \\ -\tfrac{1}{2} & 1\tfrac{1}{2} \end{bmatrix} + \begin{bmatrix} 0 & 2.10^{-10} \\ -4.10^{-10} & 2.10^{-10} \\ -5.10^{-10} & 3.10^{-10} \end{bmatrix}. \quad (63.7)$$

The residuals are therefore of the same order of magnitude as in (62.3). If we complete U by adding the same u_3 as before we have

$$A[u_1\,u_2\,u_3] = [u_1\,u_2\,u_3]\begin{bmatrix} \tfrac{1}{2} & \tfrac{1}{2} & \\ -\tfrac{1}{2} & 1\tfrac{1}{2} & \\ \hline & & 3+10^{-10} \end{bmatrix} +$$

$$+ \begin{bmatrix} 0 & 2.10^{-10} & -10^{-10}-10^{-20} \\ -4.10^{-10} & 2.10^{-10} & -4.10^{-10}-10^{-20} \\ -5.10^{-10} & 3.10^{-10} & -2.10^{-10} \end{bmatrix}, \quad (63.8)$$

giving

$$U^{-1}AU = \begin{bmatrix} \tfrac{1}{2} & \tfrac{1}{2} & \\ -\tfrac{1}{2} & 1\tfrac{1}{2} & \\ \hline & & 3+10^{-10} \end{bmatrix} +$$

$$+ \begin{bmatrix} -\tfrac{1}{2}10^{-10}+10^{-20} & 1\tfrac{1}{2}.10^{-10} & \tfrac{1}{2}10^{-10}-2\tfrac{1}{2}.10^{-20} \\ \tfrac{1}{2}10^{-10}+5.10^{-20} & \tfrac{1}{2}10^{-10}-2.10^{-20} & -1\tfrac{1}{2}.10^{-10}+\tfrac{1}{2}10^{-20} \\ -5.10^{-10}-6.10^{-20} & 10^{-10}+4.10^{-20} & -10^{-10}+10^{-20} \end{bmatrix} + G,$$

$$\qquad (63.9)$$

where

$$(|g_{ij}| < \tfrac{1}{2}10^{-20}).$$

We are now able to place the third eigenvalue in a disc with a radius of order 10^{-20} and are in a better position to obtain more accurate estimates of the other eigenvalues.

Almost normal matrices

64. In Chapter 2, § 26, we showed that the error analysis based on the n-condition numbers is unsatisfactory for a matrix which is similar, for example, to A given by

$$A = \begin{bmatrix} 1 & 10^{-10} & 0 \\ 0 & 1 & 0 \\ 0 & 0 & 2 \end{bmatrix}. \qquad (64.1)$$

This matrix has a non-linear divisor but the eigenvalues are comparatively insensitive. It is natural to ask whether we shall expect any special difficulties if we attempt to solve the eigenproblem for such a matrix. In fact such a matrix is surprisingly innocuous in practice. The transformed matrix will be similar to $(A+F)$ and in general this will not have a non-linear divisor, it will have two eigenvalues $(1+a10^{-10})$ and $(1+b10^{-10})$ and two *fully* independent corresponding eigenvectors. Of course it will be extremely difficult to prove *by computation involving rounding errors* that the original matrix has a non-linear divisor. However, it should be appreciated that it is virtually impossible to establish by such methods that a matrix has a multiple eigenvalue as long as rounding errors are made, though often when multiple eigenvalues are exact integers or the roots of factors of the type (x^2+px+q), where p and q are integers, this will be made apparent by numerical methods.

Additional notes

The first detailed analysis of the rounding errors made in the fixed-point arithmetic operations was given by von Neumann and Goldstine (1947) in their fundamental paper on the inversion of matrices. Their notation and general lines of treatment were adopted by several subsequent writers.

The corresponding analysis for the floating-point operations was first given by Wilkinson in 1957 and after a period of development was published in 1960. The choice of the form in which the results were expressed was influenced by the conviction that the effect of the rounding errors made in the course of the solution of matrix problems is best expressed in terms of equivalent perturbations of the original matrix.

The result of § 55 can be sharpened somewhat as follows. For any μ we have

$$Av - \mu v = Av - \mu_R v + (\mu_R - \mu)v = \eta_R + (\mu_R - \mu)v, \tag{1}$$

giving

$$\|Av - \mu v\|_2^2 = \epsilon^2 + |\mu_R - \mu|^2 = y^2 \quad \text{(say)} \tag{2}$$

from the orthogonality of η_R and v. Hence there is at least one eigenvalue in the disc with centre μ and radius y. Consider now the values of μ such that

$$y + |\mu_R - \mu| = a. \tag{3}$$

For such μ the corresponding discs and the disc with centre μ_R and radius a touch internally. Hence the unique eigenvalue in the latter disc lies in each of the relevant discs with centre μ satisfying (3), and therefore in their union. The union is obviously a disc with centre μ_R and radius $y - |\mu_R - \mu|$ and we have

$$y - |\mu_R - \mu| = (y^2 - |\mu_R - \mu|^2)/(y + |\mu_R - \mu|) = \epsilon^2/a. \tag{4}$$

The unique eigenvalue is therefore in a disc with centre μ_R and radius ϵ^2/a. This result is implicit in the paper by Bauer and Householder (1960). The sharper result for Hermitian matrices has been given by Kato (1949) and Temple (1952). In practice the presence of the extra factor $(1 - \epsilon^2/a^2)$ in (55.3) is of no importance since the result is of value only when a is appreciably larger than ϵ.

4

Solution of Linear Algebraic Equations

Introduction

1. A detailed comparison of numerical methods for solving systems of linear equations, inverting matrices and evaluating determinants has been given in a companion volume in this series (Fox, 1964). However, since we shall be concerned with these problems from time to time we shall need to give some consideration to them. Many of the systems we shall have to solve will be very ill-conditioned and it is vital that the significance of this should be fully appreciated. The study of systems of linear equations will also enable us to gain experience of error analyses and stability considerations in the simpler context of one-sided equivalence transformations before passing to the similarity transformations. We therefore consider first the factors which determine the sensitivity of the solution of a system of equations

$$Ax = b \tag{1.1}$$

with respect to changes in the matrix A and the right-hand side b. We shall be concerned only with systems for which the matrix A is square.

Perturbation theory

2. We consider first the sensitivity of the solution of (1.1) to perturbations in b. If b is changed to $b+k$ and

$$A(x+h) = b+k, \tag{2.1}$$

then subtracting (1.1) we have

$$Ah = k, \tag{2.2}$$

that is

$$h = A^{-1}k \tag{2.3}$$

and hence

$$\|h\|/\|k\| \leqslant \|A^{-1}\|. \tag{2.4}$$

We shall usually be interested in the relative error, and for this we have

$$\|h\|/\|x\| \leqslant \|A^{-1}\| \ \|k\|/\|A\|^{-1}\|b\| = \|A\| \ \|A^{-1}\| \ \|k\|/\|b\|, \tag{2.5}$$

showing that $\|A\| \ \|A^{-1}\|$ may be regarded as a condition number for this problem.

When $\|A\|\ \|A^{-1}\|$ is very large this result will be very pessimistic for most right-hand sides b and perturbations k. Most b will then be such that

$$\|x\| \gg \|A\|^{-1}\ \|b\|, \tag{2.6}$$

but there are always some b and k for which (2.5) is realistic.

3. Turning to the effect of perturbations in A we may write

$$(A+F)(x+h) = b, \tag{3.1}$$

giving

$$(A+F)h = -Fx. \tag{3.2}$$

Even if A is not singular (and we naturally assume this), $(A+F)$ may be singular if F is not restricted. Writing

$$A+F = A(I+A^{-1}F), \tag{3.3}$$

we see that $(A+F)$ is certainly non-singular if

$$\|A^{-1}F\| < 1. \tag{3.4}$$

Assuming this condition to be satisfied we have

$$h = -(I+A^{-1}F)^{-1}A^{-1}Fx, \tag{3.5}$$

and hence by (56.3) of Ch. 1

$$\|h\| < \frac{\|A^{-1}F\|\ \|x\|}{1-\|A^{-1}F\|} < \frac{\|A^{-1}\|\ \|F\|\ \|x\|}{1-\|A^{-1}\|\ \|F\|}, \tag{3.6}$$

provided

$$\|A^{-1}\|\ \|F\| < 1. \tag{3.7}$$

For the relative error we are interested in some bound for $\|h\|/\|x\|$ in terms of $\|F\|/\|A\|$. We may write (3.6) in the form

$$\|h\|/\|x\| < \|A\|\ \|A^{-1}\|\frac{\|F\|}{\|A\|}\Big/\Big(1-\|A\|\ \|A^{-1}\|\frac{\|F\|}{\|A\|}\Big), \tag{3.8}$$

and again the quantity $\|A\|\ \|A^{-1}\|$ is the decisive factor. The condition number in most common use is $\|A\|_2\ \|A^{-1}\|_2$; this is usually denoted by $\kappa(A)$ and is called *the spectral condition number of A with respect to inversion.* We see now from Chapter 2, §§ 30, 31, that the spectral condition number of a matrix A with respect to its eigenvalue problem is the spectral condition number of its matrix of eigenvectors with respect to inversion.

In deriving the relations (3.6) and (3.8) we have replaced $\|A^{-1}F\|$ by $\|A^{-1}\|\ \|F\|$. There will always be some perturbations for which this is pessimistic. If, for example, $F = \alpha A$ then from (3.6)

$$\|h\| < \frac{|\alpha|\ \|x\|}{1-|\alpha|}, \tag{3.9}$$

and the condition number is obviously not involved.

Condition numbers

4. The spectral condition number κ is invariant with respect to unitary equivalence transformations. For if R is unitary and

$$B = RA \tag{4.1}$$

we have

$$\|B\|_2 \, \|B^{-1}\|_2 = \|RA\|_2 \, \|A^{-1}R^H\|_2 = \|A\|_2 \, \|A^{-1}\|_2. \tag{4.2}$$

Also if c is any scalar we have obviously

$$\kappa(cA) = \kappa(A). \tag{4.3}$$

This is a reasonable result since we would not expect to affect the condition of a system of equations by multiplying both sides by a constant. From its definition

$$\kappa(A) = \|A\|_2 \, \|A^{-1}\|_2 \geqslant \|AA^{-1}\|_2 = 1 \tag{4.4}$$

for any matrix, equality being achieved if A is a multiple of a unitary matrix.

If σ_1 and σ_n are the largest and smallest singular values of A (Chapter 1, § 53) then

$$\kappa(A) = \sigma_1/\sigma_n, \tag{4.5}$$

while if A is symmetric we have

$$\kappa(A) = |\lambda_1|/|\lambda_n|, \tag{4.6}$$

where λ_1 and λ_n are the eigenvalues of largest and smallest modulus. The emergence of the ratios in equations (4.5) and (4.6) is often found puzzling and it is important to realize that the numerators are merely normalizing factors.

If we take $\|A\|_2 = 1$, which can always be achieved by multiplying the set of equations $Ax = b$ by a suitable constant, then

$$\kappa(A) = \|A\|_2 \, \|A^{-1}\|_2 = 1/\sigma_n \tag{4.7}$$

$$= 1/|\lambda_n| \text{ if } A \text{ is symmetric.} \tag{4.8}$$

A normalized matrix A is therefore ill-conditioned if and only if $\|A^{-1}\|_2$ is large, that is, if σ_n is small or, in the case of a symmetric matrix, if λ_n is small. In using κ as a condition number we are merely implying that a normalized matrix is ill-conditioned if its inverse is 'large', which is in accordance with the popular concept of ill-conditioning since it implies that a small change in the right-hand side can make a large change in the solution.

Equilibrated matrices

5. An orthogonal matrix is not only normalized, but all its rows and columns are of unit length. The same is clearly not true of normalized matrices in general. Suppose we multiply the rth column of an orthogonal matrix A by 10^{-10}, producing the matrix B. The matrix $B^T B$ is the unit matrix apart from its rth diagonal element which is 10^{-20}; hence $\|B\|_2 = 1$. The inverse of B is A^T with its rth row multiplied by 10^{10}. Hence $B^{-1}(B^{-1})^T$ is the unit matrix apart from its rth diagonal element which is 10^{20}. Hence $\|B^{-1}\|_2 = 10^{10}$ and therefore

$$\kappa(B) = \|B\|_2 \, \|B^{-1}\|_2 = 10^{10}. \tag{5.1}$$

We obtain the same result if we multiply any *row* of an orthogonal matrix by 10^{-10}.

A matrix for which every row and column has a length of order unity will be called an *equilibrated* matrix. We shall not use the term in a very precise sense but most commonly it will refer to a matrix in which every element is less than unity and all rows and columns have been scaled so that at least one element is of modulus between $\frac{1}{2}$ and 1. Powers of 2 or 10 are normally used when scaling, so that rounding errors are avoided. If a matrix is symmetric it is convenient to preserve symmetry in the scaling and hence to be satisfied with ensuring that every column contains an element of modulus greater than $\frac{1}{4}$. If any row or column of an equilibrated matrix A is multiplied by 10^{-10} then the resulting matrix B must have a very high condition number. For the largest eigenvalue σ_1^2 of BB^T is greater than its largest diagonal element and its smallest eigenvalue σ_n^2 is smaller than its smallest diagonal element. Hence σ_1^2 is certainly greater than $\frac{1}{4}$ and σ_n^2 less than $10^{-20}n$, so that

$$\kappa(B) \geqslant 10^{10}/(2n^{\frac{1}{4}}). \tag{5.2}$$

Notice that this result is true whatever the condition number of A may be.

6. It is natural to ask whether a matrix B, obtained in this way from a well-conditioned matrix A, is really 'ill-conditioned'. The difficulty in answering this question arises from the fact that the term 'ill-conditioned' is used in a popular sense as well as in a precisely defined mathematical sense. If we *define* κ as the condition number of a matrix, then there is no doubt that the modified orthogonal matrix of § 5 is very 'ill-conditioned'. In the popular sense, however, we say that a set of equations is ill-conditioned if the computed

solutions are of little practical significance. We cannot decide whether equations are ill-conditioned without examining the way in which the coefficients were derived.

Simple practical examples

7. Let us consider a simple example in which the coefficients of a set of equations of order two are determined in the following manner. Each of the four elements of the matrix A is determined by measuring three independent quantities say, each of which lies in the range ± 1, and then adding these three quantities together. Suppose the errors in the individual measurements are of order 10^{-5} and that the resulting matrix is

$$\begin{bmatrix} 1 \cdot 00000 & 0 \cdot 00001 \\ 1 \cdot 00000 & -0 \cdot 00001 \end{bmatrix}, \tag{7.1}$$

cancellation having occurred when computing the elements of the second column. Now this matrix becomes $2^{\frac{1}{2}}$ times an orthogonal matrix if we multiply its second column by 10^5. The set of equations is clearly 'ill-conditioned', however, in the sense that little significance could be attached to computed answers.

Consider now a second situation in which the elements of a matrix of order two are derived from a single measurement with a precision of 1 part in 10^5. Suppose the elements of the second column are measured in units which are much larger than those used for the first column. We might again obtain the matrix (7.1), but now computed solutions may well be fully significant.

Similar considerations apply if the elements of the matrix are computed from mathematical expressions by methods which involve rounding errors. There would be a considerable difference between situations in which a matrix of the type of (7.1) had small elements in its second column as a result of cancellation of nearly equal and opposite numbers, and one in which these elements had a high relative accuracy.

Condition of matrix of eigenvectors

8. In Chapter 2 we were interested in the condition of the matrix with its columns equal to the normalized right-hand eigenvectors of a given matrix. Such a matrix cannot have a *column* of small elements, but it is possible that a whole *row* could be small. If this were true, the eigenvectors would certainly be almost linearly dependent and the associated eigenproblem would be ill-conditioned.

It is evidently 'correct' to regard such a matrix of eigenvectors as ill-conditioned. Even in this case there is much to be said for equilibrating this matrix by multiplying the offending row by a power of 2 (or 10) before inverting it to find the left-hand system of eigenvectors. The computed inverse would then need its corresponding column multiplied by the same power of 2 (or 10) to give the true inverse.

9. We see then that the condition of a system of linear equations in the popular sense is a highly subjective matter. What is important in practice is the condition of the computing problem associated with the solution of the system.

Some of the techniques discussed in this chapter are most appropriate for equilibrated matrices in that decisions are taken from time to time which depend on the relative size of elements of the matrix. Since we almost invariably decrease the condition number of a matrix by equilibration, the case for it would appear very strong. If we are using floating-point arithmetic, normalization, as distinct from equilibration, does not play any essential part. Even so the assumption of normalization simplifies the discussion without really sacrificing generality.

If a matrix is equilibrated and normalized in the strict sense, then, even if we include symmetric equilibration, $\|A\|_2$ lies between $\frac{1}{4}$ and n and will usually be of the order of $n^{\frac{1}{2}}$. Except for very large matrices therefore, $\|A^{-1}\|_2$ (or even the element of A^{-1} of maximum modulus) will be a reasonable condition number. Since the rth column of A^{-1} is the solution of $Ax = e_r$, it is evident that a normalized equilibrated matrix is ill-conditioned if and only if there is a normalized b for which the solution of $Ax = b$ has a large norm.

Explicit solution

10. If A has linear elementary divisors, we can obtain a simple explicit solution of the equation $Ax = b$. For if we write

$$x = \sum \alpha_i x_i, \tag{10.1}$$

where the x_i form a complete set of normalized eigenvectors of A, then

$$\sum \lambda_i \alpha_i x_i = b. \tag{10.2}$$

Hence, if y_i are the corresponding eigenvectors of A^T, we have

$$\alpha_i = y_i^T b (\lambda_i y_i^T x_i)^{-1} = y_i^T b (\lambda_i s_i)^{-1}. \tag{10.3}$$

From (10.1) and (10.3) we have

$$\|x\| < \sum |\alpha_i| \, \|x_i\| = \sum |\alpha_i| < \|b\| \sum |\lambda_i s_i|^{-1}. \qquad (10.4)$$

We cannot have a large solution unless at least one λ_i or s_i (or both) is small. If A is symmetric then all the s_i are unity so that A cannot be ill-conditioned unless it has a small eigenvalue, agreeing with the result obtained previously.

General comments on condition of matrices

11. It is important to appreciate that a normalized unsymmetric matrix may be very ill-conditioned without possessing any small eigenvalues. For example, the set of equations of order 100 given by

$$\begin{bmatrix} 0{\cdot}501 & -1 & & & \\ & 0{\cdot}502 & -1 & & \\ & & \cdots\cdots\cdots & & \\ & & & 0{\cdot}599 & -1 \\ & & & & 0{\cdot}600 \end{bmatrix} x = \begin{bmatrix} 0 \\ 0 \\ \cdots \\ 0 \\ 1 \end{bmatrix}, \qquad (11.1)$$

has for the first component of its solution

$$x_1 = 1/(0{\cdot}600 \times 0{\cdot}599 \times \ldots \times 0{\cdot}501) > (0{\cdot}6)^{-100} > 10^{22}. \qquad (11.2)$$

Hence the norm of the inverse of the matrix of coefficients is certainly greater than 10^{22}. In spite of this, its smallest eigenvalue is $0{\cdot}501$. This matrix must have a singular value of the order of magnitude of 10^{-22}.

On the other hand the possession of a small eigenvalue by a normalized matrix necessarily implies ill-conditioning. For we have

$$\|A^{-1}\|_2 > |\text{maximum eigenvalue of } A^{-1}| = 1/\min |\lambda_i|. \qquad (11.3)$$

Finally the possession of a small s_i does not *necessarily* imply ill-conditioning. This is illustrated by the matrix in Chapter 2, § 26.6. This matrix has an s_i of order 10^{-10} but its inverse is

$$\begin{bmatrix} \tfrac{1}{2} & 0 & 0 \\ 0 & 1 & -1/(1+x) \\ 0 & 0 & 1/(1+x) \end{bmatrix} \quad (x = 10^{-10}). \qquad (11.4)$$

Turning to matrices with non-linear divisors we see from the matrix

$$A = \begin{bmatrix} 1 & 1 \\ 0 & 1 \end{bmatrix} \tag{11.5}$$

that their possession does not necessarily imply ill-conditioning. On the other hand if we replace all the diagonal elements of the matrix in (11.1) by 0·500, it is even more ill-conditioned and it now has an elementary divisor of degree 100.

These considerations show that although the condition of a matrix with respect to its eigenvalue problem and with respect to inversion are entirely different matters they are not completely unrelated.

Relation of ill-conditioning to near-singularity

12. If a matrix is singular it must have at least one zero eigenvalue, since the product of its eigenvalues is equal to its determinant. It is natural to think of singularity as an extreme form of ill-conditioning and therefore tempting to conclude that an ill-conditioned normalized matrix must have a small eigenvalue. We have just seen that this is not true, but it is instructive to see why thinking in terms of the limiting process is so misleading.

We now show that if n is fixed and $\|A\|_2 = 1$, then λ_n, the eigenvalue of A of smallest modulus, does indeed tend to zero as $\|A^{-1}\|_2$ tends to infinity but we can guarantee only a very slow rate of convergence. For any matrix A there is a unitary R such that

$$R^H A R = \mathrm{diag}(\lambda_i) + T = D + T, \tag{12.1}$$

where T is strictly upper triangular (cf. Chapter 3, § 47). Hence we have

$$R^H A^{-1} R = (D+T)^{-1}$$
$$= [I - K + K^2 - \ldots (-)^{n-1} K^{n-1}] D^{-1}$$
$$\text{where } K = D^{-1}T, \tag{12.2}$$

since $K^r = 0$ $(r \geqslant n)$ because K is strictly upper triangular. Quite crude bounds will be adequate for our purpose. We have

$$\|T\|_2 \leqslant \|T\|_E \leqslant \|D+T\|_E = \|A\|_E \leqslant n^{\frac{1}{2}} \|A\|_2 = n^{\frac{1}{2}}, \tag{12.3}$$

and

$$\|K\|_2 \leqslant \|D^{-1}\|_2 \|T\|_2 \leqslant n^{\frac{1}{2}}/|\lambda|, \tag{12.4}$$

where λ is the eigenvalue of smallest modulus. Hence from (12.2)

$$\|A^{-1}\|_2 = \|R^H A^{-1} R\|_2 \leqslant [1 + n^{\frac{1}{2}}/|\lambda| + (n^{\frac{1}{2}}/|\lambda|)^2 + \ldots + (n^{\frac{1}{2}}/|\lambda|)^{n-1}]/|\lambda|$$
$$\leqslant n^{\frac{1}{2}(n+1)}/|\lambda|^n, \tag{12.5}$$

giving
$$|\lambda| \leqslant (n^{\frac{1}{2}(n+1)}/\|A^{-1}\|_2)^{\frac{1}{n}}. \tag{12.6}$$

This certainly shows that, keeping n fixed, λ tends to zero as $\|A^{-1}\|_2$ tends to infinity, but the $1/n$th power is involved. The examples we have discussed show that this bound is quite realistic as far as the factor $\|A^{-1}\|_2^{\frac{1}{n}}$ is concerned and we can construct a normalized matrix of order 100 having a condition number of the order of 2^{100} but with no eigenvalue smaller than 0·5.

Limitations imposed by t-digit arithmetic

13. If the A and b of
$$Ax = b \tag{13.1}$$

require more than t digits for their representation then we must content ourselves with t-digit approximations A' and b'. We have

$$A' = A + F, \qquad b' = b + k, \tag{13.2}$$

where for fixed-point computation

$$|f_{ij}| \leqslant \tfrac{1}{2} 2^{-t}, \qquad |k_i| \leqslant \tfrac{1}{2} 2^{-t}, \tag{13.3}$$

so that
$$\|F\|_2 \leqslant \tfrac{1}{2} n 2^{-t}, \qquad \|k\|_2 \leqslant \tfrac{1}{2} n^{\frac{1}{2}} 2^{-t}. \tag{13.4}$$

It is obviously possible for the upper bounds to be attained. Unless $\tfrac{1}{2} n \|A^{-1}\|_2 2^{-t} < 1$ the matrix $(A + F)$ could be singular from (3.7). If this condition is satisfied a simple manipulation shows that the exact solution x' of $A'x' = b'$ satisfies

$$\|x' - x\|_2 \leqslant (\|A^{-1}\|_2 \|F\|_2 \|x\|_2 + \|A^{-1}\|_2 \|k\|_2)/(1 - \|A^{-1}\|_2 \|F\|_2). \tag{13.5}$$

It is evident that unless $n2^{-t} \|A^{-1}\|_2$ is appreciably smaller than unity the initial errors may well completely destroy the accuracy of the solution. For floating-point computation we have, similarly,

$$|f_{ij}| \leqslant 2^{-t} |a_{ij}|, \qquad |k_i| \leqslant 2^{-t} |b_i|, \tag{13.6}$$

$$\|F\|_2 \leqslant \|F\|_E \leqslant 2^{-t} \|A\|_E \leqslant 2^{-t} n^{\frac{1}{2}} \|A\|_2, \qquad \|k\|_2 \leqslant 2^{-t} \|b\|_2. \tag{13.7}$$

14. We turn now to the rounding errors made during the course of computation. Consider the effect of multiplying the system of equations by a constant c lying between $\tfrac{1}{2}$ and 1 and requiring t digits for its representation, using fixed-point arithmetic. If we round-off the resulting numbers, the transformed set of equations is

$$(cA + F)x = cb + k$$
$$|f_{ij}| < \tfrac{1}{2}2^{-t}, \qquad |k_i| < \tfrac{1}{2}2^{-t} \Big\} .$$

(14.1)

This transformed set is exactly equivalent to

$$(A + F/c)x = b + k/c,$$

(14.2)

so that the rounding errors are equivalent to perturbations F/c and k/c in A and b respectively, having bounds which are $1/c$ times those for the initial roundings. Now a typical step in most direct methods of solving equations consists of adding a multiple of one equation to another. Since even for a matrix of modest order there are many such steps, we might well expect that the sum total of the rounding errors made in these steps would impair the accuracy of the solution much more seriously than the single operation of equation (14.1). It is a matter for some surprise that one of the simplest methods of solution generally leads to an error, the expected value of which is precisely that resulting from random perturbations F and k satisfying equations (13.3). This means that when the original elements are not exactly representable by numbers of t binary digits the errors resulting from any initial rounding that may be necessary are as serious as those arising from all the steps in the solution. Even more surprising is the fact that for some very ill-conditioned matrices the initial roundings have by far the more serious effect. (See § 40 and § 45 respectively.)

Algorithms for solving linear equations

15. We shall be interested only in direct methods of solving linear equations and all the algorithms we use are based on the same basic stratagem. If x is a solution of

$$Ax = b,$$

(15.1)

where A is an $(n \times n)$ matrix, then it is also a solution of

$$PAx = Pb,$$

(15.2)

where P is *any* $(m \times n)$ matrix. However, if P is square and non-singular, then any solution of (15.2) is a solution of (15.1) *and vice versa*. Further, if Q is also non-singular and y is a solution of

$$PAQy = Pb,$$

(15.3)

then Qy is a solution of (15.1) (cf. Chapter 1, § 15). Some of our algorithms will make use of this last result, taking Q to be a permutation matrix (Chapter 1, § 40 (ii)).

The basic stratagem is the determination of a simple non-singular matrix P such that PA is *upper triangular*. The set of equations (15.2) may then be solved immediately, since the nth equation involves only x_n, the $(n-1)$th involves only x_n and x_{n-1}, and so on. The variables may therefore be determined one by one in the order $x_n, x_{n-1}, \ldots, x_1$. This method of solving a set of equations with an upper triangular matrix of coefficients is usually referred to as a *back-substitution*.

The reduction of A to triangular form is also of fundamental importance in contexts in which the back-substitution is not required, so we concentrate for the moment on the triangularization.

The algorithms we shall discuss generally consist of $(n-1)$ major steps, and have the following features in common. We denote the initial set of equations by

$$A_0 x = b_0. \tag{15.4}$$

Each step leads to the production of a new set of equations which is *equivalent* to the original set. We denote the pth equivalent set by

$$A_p x = b_p, \tag{15.5}$$

and an essential feature of the algorithms is that A_p is already upper triangular as far as its first p columns are concerned, so that typically for $n = 6$, $p = 3$, the equations $A_p x = b_p$ have the form

$$
p \overbrace{\qquad\qquad}^{p} \quad \overbrace{\qquad\qquad}^{n-p}
$$

$$
\left.\begin{matrix} p \\ \\ \end{matrix}\right\{
\left.\begin{matrix} n-p \\ \\ \end{matrix}\right\{
\begin{bmatrix}
\times & \times & \times & \times & \times & \times \\
0 & \times & \times & \times & \times & \times \\
0 & 0 & \times & \times & \times & \times \\
0 & 0 & 0 & \times & \times & \times \\
0 & 0 & 0 & \times & \times & \times \\
0 & 0 & 0 & \times & \times & \times
\end{bmatrix}
\begin{bmatrix} x_1 \\ x_2 \\ x_3 \\ x_4 \\ x_5 \\ x_6 \end{bmatrix}
=
\begin{bmatrix} \times \\ \times \\ \times \\ \times \\ \times \\ \times \end{bmatrix}. \tag{15.6}
$$

Here the crosses denote elements which are, in general, non-zero. It is conventional to omit the column of x_i and the equality signs and to present only the matrix $(A_p \mid b_p)$.

The rth major step consists of the determination of an elementary matrix P_r such that A_r, defined by $P_r A_{r-1}$, is upper triangular in its first r columns and is identical with A_{r-1} in its first $(r-1)$ rows and columns. This step therefore leaves the first $(r-1)$ equations completely unaltered. The equations defining the rth transformation are

$$A_r = P_r A_{r-1}, \qquad b_r = P_r b_{r-1}. \tag{15.7}$$

Clearly the rth column of $P_r A_{r-1}$ is

$$P_r \begin{bmatrix} a_{1,r}^{(r-1)} \\ a_{2,r}^{(r-1)} \\ \cdot \\ \cdot \\ \cdot \\ a_{n,r}^{(r-1)} \end{bmatrix} \tag{15.8}$$

and P_r is to be chosen so that $a_{1,r}^{(r-1)}$ to $a_{r-1,r}^{(r-1)}$ are unaltered, while $a_{r+1,r}^{(r-1)}, \ldots, a_{n,r}^{(r-1)}$ are to be reduced to zero. This is precisely the problem we have considered in connexion with our fundamental error analyses in Chapter 3. Most of the rest of this chapter is devoted to a discussion of the relevant algorithms.

Gaussian elimination

16. The simplest algorithm is obtained if we take for the matrix P_r the elementary matrix N_r (Chapter 1, § 40 (vi)) with

$$n_{ir} = a_{ir}^{(r-1)}/a_{rr}^{(r-1)}. \tag{16.1}$$

The divisor $a_{rr}^{(r-1)}$ is called the *r-th pivotal element* or the *r-th pivot*. Pre-multiplication by N_r results in the subtraction of a multiple n_{ir} of the rth row from the ith row for each of the values of i from $(r+1)$ to n, the multiples being chosen so as to eliminate the last $(n-r)$ elements in column r. The same operations are performed on the right-hand side b_{r-1}. Clearly rows 1 to $(r-1)$ are unaltered, and moreover none of the zeros in the first $(r-1)$ columns is affected since each is replaced by a linear combination of zero elements. With this algorithm row r is also unaltered. The reader will recognize this as a systematized version of the method of elimination taught in school algebra. In effect the rth equation is used to eliminate x_r from equations $(r+1)$ to n of $A_{r-1}x = b_{r-1}$. This algorithm is usually referred to as *Gaussian elimination*.

Triangular decomposition

17. Combining equations (15.7) with $P_r = N_r$ for $r = 1$ to $n-1$ we have

$$N_{n-1}...N_2 N_1 A_0 = A_{n-1}, \qquad N_{n-1}...N_2 N_1 b_0 = b_{n-1}. \qquad (17.1)$$

The first of these gives

$$A_0 = N_1^{-1} N_2^{-1}...N_{n-1}^{-1} A_{n-1} = N A_{n-1} \quad \text{(Chapter 1, §§ 40, 41)}, \qquad (17.2)$$

where N is the lower triangular matrix with unit diagonal elements and sub-diagonal elements equal to the n_{ij} defined in (16.1). No products of the n_{ij} occur in the product of the N_i^{-1}. Gaussian elimination therefore produces a unit lower triangular matrix N and an upper triangular matrix A_{n-1} such that their product is equal to A_0.

If A_0 is representable as such a product (cf. § 20) and is non-singular, then the representation is unique. For if

$$A_0 = L_1 U_1 = L_2 U_2, \qquad (17.3)$$

where the L_i are unit lower triangular and the U_i are upper triangular, then since

$$\det(A_0) = \det(L_i) \det(U_i), \qquad (17.4)$$

the U_i cannot be singular. Hence (17.3) gives

$$L_2^{-1} L_1 = U_2 U_1^{-1}. \qquad (17.5)$$

The left-hand matrix is the product of two unit lower triangular matrices and is therefore unit lower triangular, while the right-hand matrix is upper triangular. Hence both sides must be the identity matrix and $L_1 = L_2$, $U_1 = U_2$.

Structure of triangular decomposition matrices

18. For later applications the relations between various sub-matrices of A_0, N, A_{n-1} and the intermediate A_r are of considerable importance. We denote conformal partitions of these matrices by

$$A_0 = \left[\begin{array}{c|c} A_{rr} & A_{r,n-r} \\ \hline A_{n-r,r} & A_{n-r,n-r} \end{array} \right], \quad N = \left[\begin{array}{c|c} L_{rr} & O \\ \hline L_{n-r,r} & L_{n-r,n-r} \end{array} \right], \qquad (18.1)$$

$$A_{n-1} = \left[\begin{array}{c|c} U_{rr} & U_{r,n-r} \\ \hline O & U_{n-r,n-r} \end{array} \right], \quad A_r = \left[\begin{array}{c|c} U_{rr} & U_{r,n-r} \\ \hline O & W_{n-r,n-r} \end{array} \right]. \qquad (18.2)$$

Here a matrix with suffices i, j has i rows and j columns. We shall refer to N and A_{n-1} as L and U respectively when we do not wish to emphasize their association with the execution of the algorithm. The notation in (18.2) reflects the fact that the first r rows of A_r are unaltered in later steps; however, we have seen that row $(r+1)$ of A_r is also unaltered in these steps so that the first rows of $U_{n-r,n-r}$ and $W_{n-r,n-r}$ are also identical.

From the first r major steps we have

$$N_r N_{r-1} \dots N_2 N_1 A_0 = A_r,$$

giving

$$A_0 = N_1^{-1} N_2^{-1} \dots N_r^{-1} A_r = \left[\begin{array}{c|c} L_{rr} & O \\ \hline L_{n-r,r} & I \end{array}\right] A_r, \qquad (18.3)$$

from the properties of the N_i with respect to multiplication (Chapter 1, § 41). Equating partitions in (17.2) we have

$$\left.\begin{array}{ll} A_{rr} = L_{rr} U_{rr}, & A_{r,n-r} = L_{rr} U_{r,n-r} \\ A_{n-r,r} = L_{n-r,r} U_{rr}, & A_{n-r,n-r} = L_{n-r,r} U_{r,n-r} + L_{n-r,n-r} U_{n-r,n-r} \end{array}\right\}, \qquad (18.4)$$

while from the last partitions in (18.3) we find

$$A_{n-r,n-r} = L_{n-r,r} U_{r,n-r} + W_{n-r,n-r}. \qquad (18.5)$$

These last five equations give us a considerable amount of information. The first shows that $L_{rr} U_{rr}$ is the triangular decomposition of A_{rr}, the leading principal submatrix of order r of A_0. In performing Gaussian elimination we therefore obtain the triangular decomposition of each of the leading principal matrices of A_0 in turn. Further, the triangles corresponding to the rth leading principal submatrix of A_0 are the leading principal submatrices of the final triangles N and A_{n-1}. The last equation of (18.4) and (18.5) show that

$$L_{n-r,n-r} U_{n-r,n-r} = W_{n-r,n-r}, \qquad (18.6)$$

so that $L_{n-r,n-r} U_{n-r,n-r}$ is the triangular decomposition of the square matrix of order $(n-r)$ in the bottom right-hand corner of A_r.

Explicit expressions for elements of the triangles

19. Equations (18.4) enable us to obtain explicit expressions for the elements of N and A_{n-1} in terms of minors of A_0 and these too will be required later. If we combine the first and third of the equations we obtain

$$\begin{bmatrix} A_{rr} \\ \hline A_{n-r,r} \end{bmatrix} = \begin{bmatrix} L_{rr} \\ \hline L_{n-r,r} \end{bmatrix} U_{rr}. \tag{19.1}$$

Equating rows 1 to $r-1$ and row i on both sides of this equation, we have

$$A_{ir} = L_{ir}U_{rr} \quad (i = 1,\ldots, n), \tag{19.2}$$

where A_{ir} denotes the matrix formed by the first $(r-1)$ rows and the ith row of the first r columns of A_0, and L_{ir} has a similar definition. Now since L_{ir} is equal to L_{rr} in its first $(r-1)$ rows, it is necessarily triangular, so that

$$\det(L_{ir}) = l_{ir}. \tag{19.3}$$

Hence

$$\det(A_{ir}) = l_{ir}\det(U_{rr}). \tag{19.4}$$

When $i = r$,

$$\det(A_{rr}) = \det(U_{rr}) \tag{19.5}$$

and hence

$$n_{ir} = l_{ir} = \det(A_{ir})/\det(A_{rr}) \quad (i = 1,\ldots, n). \tag{19.6}$$

Notice that $\det(A_{ir}) = 0$ for $i < r$ (since A_{ir} then has two identical rows) agreeing with the triangular nature of N.

Similarly

$$[A_{rr} \mid A_{r,n-r}] = L_{rr}[U_{rr} \mid U_{r,n-r}], \tag{19.7}$$

and with a corresponding notation

$$A_{ri} = L_{rr}U_{ri}, \tag{19.8}$$

giving

$$\det(A_{ri}) = \det(U_{ri}) \quad (i = 1,\ldots, n). \tag{19.9}$$

Now U_{ri} is equal to U_{rr} in its first $(r-1)$ columns and is therefore necessarily triangular. Hence

$$\det(A_{ri}) = \det(U_{r-1,r-1})u_{ri}. \tag{19.10}$$

Combining (19.10) and (19.5) with $r = r-1$

$$u_{ri} = \det(A_{ri})/\det(A_{r-1,r-1}) \quad (i = 1,\ldots, n). \tag{19.11}$$

In particular

$$\left.\begin{array}{l} u_{rr} = \det(A_{rr})/\det(A_{r-1,r-1}) \quad (r > 1) \\ u_{11} = a_{11} \end{array}\right\} . \tag{19.12}$$

We have seen before (18.3) that $u_{r+1,r+1}$ is the $(1, 1)$ element of $W_{n-r,n-r}$ and equation (19.12) shows that it is equal to

$$\det(A_{r+1,r+1})/\det(A_{rr}).$$

We can obtain similar expressions for other elements of the square matrix $W_{n-r,n-r}$. For brevity we denote the (i,j) element of A_r by w_{ij} when $i,j > r$ so that the matrix $W_{n-r,n-r}$ is comprised of such elements. Now in deriving A_r from A_0, multiples of rows 1 to r only are added to other rows. Hence any minor which includes every row from 1 to r is unaltered. Consider then the minor $A_{rr,i,j}$ drawn from rows 1 to r and i and columns 1 to r and j. This minor is unaltered in value and from the structure of A_r it is equal to $\det(U_{rr})w_{ij}$. Hence we have

$$\det(A_{rr,i,j}) = \det(U_{rr})w_{ij}, \tag{19.13}$$

and from (19.5)

$$w_{ij} = \det(A_{rr,i,j})/\det(A_{rr}). \tag{19.14}$$

The value we obtained for $u_{r+1,r+1}$ is merely a special case of (19.14); this is not surprising since, up to the stage when $u_{r+1,r+1}$ is used as a pivot, all elements of $W_{n-r,n-r}$ are on the same footing. We now have explicit expressions for all elements of N and of the A_r.

Breakdown of Gaussian elimination

20. The expressions we have obtained for the elements of N and A_r are based on the assumption that Gaussian elimination can actually be carried out. However, the algorithm cannot fail if all the pivots are non-zero and from (19.13) this will be true if the leading principal minors of orders 1 to $(n-1)$ are all non-zero. Note that we do not require that $\det(A_{nn})$ (that is, $\det(A_0)$) shall be zero.

Suppose that the first pivot to vanish is the $(r+1)$th, so that the first r leading principal minors are non-zero while $\det(A_{r+1,r+1})$ is zero. From (19.6), if $\det(A_{i,r+1}) \neq 0$ for some i greater than $r+1$ the algorithm will fail because $n_{i,r+1}$ is then infinite. However, if $\det(A_{i,r+1}) = 0$ $(i = r+1,\ldots, n)$, then all the $n_{i,r+1}$ are indeterminate. In this case we can take arbitrary values for the $n_{i,r+1}$ and if, in particular, we take them to be zero we have $N_{r+1} = I$ and the $(r+1)$th major step leaves A_r unchanged.

In fact in this case the first column of $W_{n-r,n-r}$ is already null as may be seen from (19.14) if we observe that

$$A_{rr,i,r+1} = A_{i,r+1}. \tag{20.1}$$

Further, the rank of the first $(r+1)$ columns of A_r is obviously r, showing that they are linearly dependent. However, the $(r+1)$ columns are obtained from the same columns of A_0 by pre-multiplication with non-singular matrices and hence these columns of A_0 must also be linearly dependent. A_0 is therefore singular and a solution of $Ax = b$ exists only for b such that rank $(A_0) = $ rank $(A_0 \mid b)$. If the equations are not consistent this is revealed in the back-substitution at the stage when we attempt to compute x_{r+1} by dividing by the zero pivot. If they are consistent then x_{r+1} will be indeterminate.

Notice that breakdowns in the elimination process are not, in general, associated with singularity or ill-condition of the matrix. For example, if A_0 is

$$\begin{bmatrix} 0 & 1 \\ 1 & 0 \end{bmatrix}$$

then the Gaussian elimination fails at the first step, though A_0 is orthogonal and therefore very well-conditioned.

Numerical stability

21. So far we have considered only the possibility of a breakdown in the elimination process. However, if we now think in terms of computation, the process is likely to be numerically unstable if any of the multipliers n_{ir} is large. To overcome this we may replace the N_r by the stabilized elementary matrices N'_r described in Chapter 3, § 47. In the rth major step we multiply by $N_r I_{r,r'}$ where r' is defined by

$$|a_{r,r}^{(r-1)}| = \max_{i=r,\dots,n} |a_{ir}^{(r-1)}|. \tag{21.1}$$

If the r' defined in this way is not unique we take the smallest of its possible values.

The pre-multiplication by $I_{r,r'}$ results in the *interchange* of rows r and r' (where $r' > r$) and the subsequent pre-multiplication by N_r results in the subtraction of a multiple of the new rth row from each of rows $(r+1)$ to n in turn, all the multipliers being bounded in modulus by unity. Note that rows 1 to $(r-1)$ remain unaltered in the rth major step and the zeros introduced by previous steps still persist. For the modified procedure we have

$$A_r = N_r I_{r,r'} \dots N_2 I_{2,2'} N_1 I_{1,1'} A_0, \tag{21.2}$$

where A_r is still of the form (18.2)

This stabilized procedure is uniquely determined at each stage unless at the beginning of the rth major step $a_{ir}^{(r-1)} = 0$ $(i = r,\dots, n)$. Our rule then gives $r' = r$, but the multipliers are indeterminate. We may take $N'_r = I$, in which case the rth step is effectively skipped, this being permissible since A_{r-1} is already triangular in its first r columns. As in the unstabilized procedure, this can happen only if the first r columns of A_0 are linearly dependent. The stabilized elimination process is usually called Gaussian elimination *with interchanges* or *with pivoting for size*. *Note that it cannot break down in any circumstances* and with the conventions we have described it is unique. If the equations are inconsistent this becomes evident during the back-substitution.

Significance of the interchanges

22. The description of the modified process is simpler if we think in terms of actually performing the row interchanges since the final matrix A_{n-1} is then truly upper triangular and the intermediate matrices are of the form exhibited in (15.6). On some computers it may be more convenient to leave the equations in their original places and to keep a record of the positions of the pivotal rows. The final matrix is then given typically for the case $n = 5$ by

$$\begin{bmatrix} 0 & 0 & 0 & \times & \times \\ 0 & \times & \times & \times & \times \\ \times & \times & \times & \times & \times \\ 0 & 0 & \times & \times & \times \\ 0 & 0 & 0 & 0 & \times \end{bmatrix}, \qquad (22.1)$$

where the successive pivotal rows were in the third, second, fourth, first and fifth positions. The final matrix is then a triangular matrix with its rows permuted.

If we actually make the interchanges then we have

$$A_{n-1}x - b_{n-1} = N_{n-1}I_{n-1,(n-1)'} \dots N_2 I_{2.2'} N_1 I_{1,1'}(A_0 x - b_0). \quad (22.2)$$

We now show that pivoting effectively determines a permutation of the rows of A_0 such that, with this permutation, Gaussian elimination may be performed without pivoting, using only multipliers which are bounded in modulus by unity.

In order to avoid obscuring the essential simplicity of the proof we use the case $n = 4$ for illustration. From (22.2)

$$\begin{aligned} A_3 &= N_3 I_{3,3'} N_2 I_{2,2'} N_1 I_{1,1'} A_0 \\ &= N_3 I_{3,3'} N_2 (I_{3,3'} I_{3,3'}) I_{2,2'} N_1 (I_{2,2'} I_{3,3'} I_{3,3'} I_{2,2'}) I_{1,1'} A_0, \end{aligned} \qquad (22.3)$$

since the products in parentheses are equal to the identity matrix. Regrouping the factors we have

$$A_3 = N_3(I_{3,3'}N_2I_{3,3'})(I_{3,3'}I_{2,2'}N_1I_{2,2'}I_{3,3'})(I_{3,3'}I_{2,2'}I_{1,1'}A_0) = N_3\tilde{N}_2\tilde{N}_1\tilde{A}_0,$$

(22.4)

where from Chapter 1, § 41 (iv), $\tilde{N}_2$ and $\tilde{N}_1$ are derived from N_2 and N_1 merely by permuting sub-diagonal elements, while $\tilde{A}_0$ is A_0 with its rows permuted. Suppose now we were to perform Gaussian elimination on $\tilde{A}_0$ without pivoting. We would obtain

$$A_3^* = N_3^* N_2^* N_1^* \tilde{A}_0,$$

(22.5)

where A_3^* is upper triangular and N_3^*, N_2^*, N_1^* are elementary matrices of the usual type. Equations (22.4) and (22.5) give two triangular decompositions of $\tilde{A}_0$ (with unit lower triangular factors) and hence they are identical. This means that $N_3 = N_3^*$, $\tilde{N}_2 = N_2^*$, $\tilde{N}_1 = N_1^*$ and hence the Gaussian elimination of $\tilde{A}_0$ without interchanges leads to multipliers which are bounded in modulus by unity.

Numerical example

23. In Table 1 we show the use of Gaussian elimination with interchanges to solve a set of equations of order 4 using fixed-point arithmetic with 4 decimals. When working with pencil and paper it is inconvenient to carry out the interchanges since this involves rewriting the equations in the correct order. Instead we have underlined the pivotal equation. We have included the equations which have been used as pivots in each of the reduced sets, so that all reduced sets are the equivalents of the original set. The solution has been computed to one decimal place, so that its largest component has 4 figures; in other words we have produced a 4-digit block-floating solution.

We can see immediately that the solution is unlikely to have more than 2 correct significant figures in the largest component, for this is obtained by dividing 0.5453 by -0.0052. Now the presence of only one rounding error in the divisor would mean that even without any 'accumulation' of rounding errors we would be prepared for an error of 1 part in 100 in the computed x_4. In general when there is a loss of k significant figures in any pivot we would expect at least a corresponding loss of k significant figures in some components of the solution.

TABLE 1

$A_0x = b_0$

n	x_1	x_2	x_3	x_4	b	$10^4\epsilon_{ij}^{(1)}$				$10^4\epsilon_i^{(1)}$
(0·2678)	0·2317	0·6123	0·4137	0·6696	0·4753	0·2734	−0·2130	0·3670	−0·4606	0·1870
(0·4950)	0·4283	0·8176	0·4257	0·8312	0·2167	0·2350	−0·3250	0·1750	−0·1150	−0·3250
(0·8461)	0·7321	0·4135	0·3126	0·5163	0·8132	0·3033	−0·1935	0·0165	−0·2297	−0·1065
	0·8653	0·2165	0·8265	0·7123	0·5165	0	0	0	0	0

$A_1x = b_1$

n	x_1	x_2	x_3	x_4	b	$10^4\epsilon_{ij}^{(2)}$				$10^4\epsilon_i^{(2)}$
(0·7803)	0·0000	0·5543	0·1924	0·4788	0·3370	0	0·2512	−0·4702	−0·4842	−0·3170
	0·0000	0·7104	0·0166	0·4786	−0·0390	0	0	0	0	0
(0·3242)	0·0000	0·2303	−0·3867	−0·0864	0·3762	0	0·1168	−0·1828	−0·3788	−0·4380
	0·8653	0·2165	0·8265	0·7123	0·5165	0	0	0	0	0

$A_2x = b_2$

n	x_1	x_2	x_3	x_4	b	$10^4\epsilon_{ij}^{(3)}$				$10^4\epsilon_i^{(3)}$
	0·0000	0·0000	0·1794	0·1053	0·3674	0	0	−0·1425	0·3200	−0·2400
	0·0000	0·7104	0·0166	0·4786	−0·0390	0	0	0	0	0
(−0·4575)	0·0000	0·0000	−0·3921	−0·2416	0·3888	0	0	0	0	0
	0·8653	0·2165	0·8265	0·7123	0·5165	0	0	0	0	0

$A_3x = b_3$

x_1	x_2	x_3	x_4	b	Computed x	A_0x (exact)	r (exact)
0·0000	0·0000	0·0000	−0·0052	0·5453	8·9	0·44234	0·03296
0·0000	0·7104	0·0166	0·4786	−0·0390	69·1	0·18967	0·02703
0·0000	0·0000	−0·3921	−0·2416	0·3888	63·6	0·81003	0·00317
0·8653	0·2165	0·8265	0·7123	0·5165	−104·9	0·50645	0·01005

N

$$\begin{bmatrix} 1\text{·}0000 & & & \\ 0\text{·}4950 & 1\text{·}0000 & & \\ 0\text{·}8461 & 0\text{·}3242 & 1\text{·}0000 & \\ 0\text{·}2678 & 0\text{·}7803 & -0\text{·}4575 & 1\text{·}0000 \end{bmatrix}$$

$(A_3 \mathbin{\vdots} b_3)$ (permuted)

$$\begin{bmatrix} 0\text{·}8653 & 0\text{·}2165 & 0\text{·}8265 & 0\text{·}7123 & 0\text{·}5165 \\ & 0\text{·}7104 & -0\text{·}3921 & -0\text{·}2416 & -0\text{·}3888 \\ & & 0\text{·}0166 & 0\text{·}4786 & -0\text{·}0390 \\ & & & -0\text{·}0052 & 0\text{·}5453 \end{bmatrix}$$

$N(A_3 \mathbin{\vdots} b_3)$ (exact) (Compare with $(A_0 \mathbin{\vdots} b_0)$ with its rows permuted)

0·8653 0000	0·2165 0000	0·8265 0000	0·7123 0000	0·5165 0000
0·4283 2350	0·8175 6750	0·4257 1750	0·8311 8850	0·2166 6750
0·7321 3033	0·4134 9233	0·3125 8337	0·5162 3915	0·8131 6685
0·2317 2734	0·6123 0382	0·4136 7543	0·6695 3752	0·4753 1100

In Table 1 we give the rounding errors made in each stage of the reduction. The significance of these errors is discussed in §§ 24, 25. We also give N and A_3 permuted so as to give the true triangular matrices, and the exact product of $N(A_3 \mid b_3)$ similarly permuted. For exact computation in the triangularization this should have been equal to $(A_0 \mid b_0)$ with rows 4, 2, 3, 1 in the order 1, 2, 3, 4. Note that $N(A_3 \mid b_3)$ is indeed remarkably close to this, the greatest divergence being 0·00006248 in element (1, 4).

Finally we give the exact values of $A_0 x$ for the computed x, and of the vector r defined by $r = b_0 - A_0 x$. This vector is called *the residual vector corresponding to* x and its components are *called the residuals*. It will be seen that the maximum residual is 0·03296. This is surprisingly small since we are expecting an error of order unity in x_4 alone and such an error might be expected to give residuals of order unity. We shall see later that the solutions computed by Gaussian elimination and back-substitution always have errors which are related in such a way as to give residuals which are quite misleadingly small with respect to the errors in the solution. The errors in the components of x are in fact $-0·1$, $-1·0$, $-0·9$, $1·4$ to one decimal place.

Error analysis of Gaussian elimination

24. The steps involved in Gaussian elimination with interchanges are covered by our general analysis of Chapter 3, § 48, except that we have to justify entering the zeros in appropriate positions in the reduced matrices. However, this is the first detailed analysis we have given, and since it is formally the simplest, we propose to give a fairly detailed account so that the reader can fully grasp the essential features.

We give first a rather general analysis which applies to fixed-point or floating-point computation. We assume for the purposes of the analysis that the original matrix has been permuted as described in § 22 so that no interchanges are required. *Our analysis will refer only to elements of the computed* N *and the computed* A_r *and hence no confusion will arise if the usual 'bars' are omitted throughout.*

Consider now the history of the elements in the (i, j) position of the successive matrices A_r.

(i) If $j \geqslant i$, then this element is modified in steps 1, 2,..., $i-1$; it then becomes an element of a pivotal row and is subsequently unaltered. We may write

$$\left.\begin{aligned}
a_{ij}^{(1)} &\equiv a_{ij}^{(0)} \quad -n_{i1}a_{2j}^{(0)} \quad +\epsilon_{ij}^{(1)}\\
a_{ij}^{(2)} &\equiv a_{ij}^{(1)} \quad -n_{i2}a_{2j}^{(1)} \quad +\epsilon_{ij}^{(2)}\\
&\cdots\cdots\cdots\cdots\cdots\cdots\cdots\cdots\\
a_{ij}^{(i-1)} &\equiv a_{ij}^{(i-2)} -n_{i,i-1}a_{i-1,j}^{(i-2)}+\epsilon_{ij}^{(i-1)}
\end{aligned}\right\},\qquad (24.1)$$

where $\epsilon_{ij}^{(k)}$ is the error made in computing $a_{ij}^{(k)}$ from *computed* $a_{ij}^{(k-1)}$, *computed* n_{ik} and *computed* $a_{kj}^{(k-1)}$. In general we will determine only a bound for the $\epsilon_{ij}^{(k)}$ and this will depend on the type of arithmetic used.

(ii) If $i > j$, then this element is modified in steps $1, 2,..., j-1$ and finally, in step j, it is reduced to zero and at the same time used to calculate n_{ij}. The equations are therefore the same as (24.1) except that they terminate with the equation

$$a_{ij}^{(j-1)} \equiv a_{ij}^{(j-2)} -n_{i,j-1}a_{j-1,j}^{(j-2)}+\epsilon_{ij}^{(j-1)}. \qquad (24.2)$$

We have further
$$n_{ij} = a_{ij}^{(j-1)}/a_{jj}^{(j-1)}+\xi_{ij}. \qquad (24.3)$$

Expressing this last equation in the same form as those of (24.1) it becomes
$$0 \equiv a_{ij}^{(j-1)} -n_{ij}a_{jj}^{(j-1)}+\epsilon_{ij}^{(j)} \quad (\epsilon_{ij}^{(j)} = a_{jj}^{(j-1)}\xi_{ij}). \qquad (24.4)$$

Note that the zero on the left-hand side of this equation corresponds to the zero which is inserted in this (i, j) position after the jth reduction. Adding equations (24.1), cancelling common members and rearranging, we have for $j \geqslant i$

$$n_{i1}a_{1j}^{(0)}+n_{i2}a_{2j}^{(1)}+...+n_{i,i-1}a_{i-1,j}^{(i-2)}+a_{ij}^{(i-1)} = a_{ij}^{(0)}+\epsilon_{ij}^{(1)}+\epsilon_{ij}^{(2)}+...+\epsilon_{ij}^{(i-1)}$$
$$(24.5)$$

while for $i > j$, if we include equation (24.4) we have

$$n_{i1}a_{1j}^{(0)}+n_{i2}a_{2j}^{(1)}+...+n_{ij}a_{jj}^{(j-1)} = a_{ij}^{(0)}+\epsilon_{ij}^{(1)}+\epsilon_{ij}^{(2)}+...+\epsilon_{ij}^{(j)}. \qquad (24.6)$$

Similarly for the right-hand side we have

$$b_i^{(k)} \equiv b_i^{(k-1)} -n_{ik}b_k^{(k-1)}+\epsilon_i^{(k)} \quad (k = 1,..., i-1), \qquad (24.7)$$

giving

$$n_{i1}b_1^{(0)}+n_{i2}b_2^{(1)}+...+n_{i,i-1}b_{i-1}^{(i-2)}+b_i^{(i-1)} = b_i^{(0)}+\epsilon_i^{(1)}+...+\epsilon_i^{(i-1)}. \qquad (24.8)$$

Equations (24.5), (24.6), and (24.8) express the fact that

$$N(A_{n-1}\,|\,b_{n-1}) = (A_0+F\,|\,b_0+k),$$

where
$$\left.\begin{aligned}
f_{ij} &= \epsilon_{ij}^{(1)}+...+\epsilon_{ij}^{(i-1)} \quad (j \geqslant i)\\
f_{ij} &= \epsilon_{ij}^{(1)}+...+\epsilon_{ij}^{(j)} \quad (i > j)\\
k_i &= \epsilon_i^{(1)}+...+\epsilon_i^{(i-1)}
\end{aligned}\right\}. \qquad (24.9)$$

In other words, the computed matrices N and A_{n-1} are the matrices which would have been obtained by exact decomposition of $(A+F)$. *This result is exact; there is no multiplication or neglect of second-order quantities.* This is in agreement with our analysis of Chapter 3 but now we are in a better position to obtain upper bounds for the rounding errors.

We draw two main conclusions from these results.

I. The exact determinant of A_{n-1}, that is $a_{11}^{(0)}a_{22}^{(1)}...a_{nn}^{(n-1)}$, is the exact determinant of (A_0+F).

II. The exact solution of $A_{n-1}x = b_{n-1}$ is the exact solution of $(A+F)x = b_0+k$.

Upper bounds for the perturbation matrices using fixed-point arithmetic

25. Let us assume that the original equations were normalized and equilibrated, at least approximately, and that interchanges have been used. We make the further assumption (at the moment completely unwarranted) that all $a_{ij}^{(k)}$ and $b_i^{(k)}$ are bounded in modulus by unity. Now in general $\epsilon_{ij}^{(k)}$ is the error made in computing $a_{ij}^{(k)}$ from the computed values of $a_{ij}^{(k-1)}$, n_{ik} and $a_{kj}^{(k-1)}$. Only one error is made in this process, and that at the stage when the $2t$-digit product $n_{ik}a_{kj}^{(k-1)}$ is rounded to t digits. Hence

$$|\epsilon_{ij}^{(k)}| < \tfrac{1}{2}2^{-t}. \tag{25.1}$$

This holds for all $\epsilon_{ij}^{(k)}$ except those of the form $\epsilon_{ij}^{(j)}$ which arise from the rounding errors in the divisions when the n_{ij} are computed. From (24.3), ξ_{ij} satisfies

$$|\xi_{ij}| < \tfrac{1}{2}2^{-t}, \tag{25.2}$$

since it is the error in a simple division; we are certain that $|n_{ij}| < 1$ because $|a_{jj}^{(j-1)}| \geqslant |a_{i,j}^{(j-1)}|$ and the round-off cannot therefore lead to a value which is greater in modulus than unity. Hence from (24.4) and our assumption about the elements of the reduced matrices we have

$$|\epsilon_{ij}^{(j)}| = |a_{jj}^{(j-1)}\xi_{ij}| < \tfrac{1}{2}2^{-t}. \tag{25.3}$$

Similarly, under our assumption about the elements $b_i^{(k)}$

$$|\epsilon_i^{(k)}| < \tfrac{1}{2}2^{-t}. \tag{25.4}$$

Combining relations (25.1), (25.3), and (25.4) for all relevant k we have

$$|k_i| < \tfrac{1}{2}(i-1)2^{-t}, \qquad |f_{ij}| < \begin{cases} \tfrac{1}{2}(i-1)2^{-t} & (i < j) \\ \tfrac{1}{2}j2^{-t} & (i > j) \end{cases}, \tag{25.5}$$

these results being extreme upper bounds.

The errors in each step of the elimination of our previous example are exhibited in Table 1. At the foot of the table the exact value of $N(A_3 \mid b_3)$ is displayed. It should be compared with $(A_0 \mid b_0)$ with its rows suitably permuted. Note that the difference $N(A_3 \mid b_3) - (A_0 \mid b_0)$ permuted is the exact sum of the error matrices at each stage.

Upper bound for elements of reduced matrices

26. So far the assumption that $|a_{ij}^{(k)}| \leqslant 1$ for all i, j, k would appear to be quite unjustified. Even with pivoting we can guarantee only that $|a_{ij}^{(k)}| < 2^k$. This follows from the inequalities

$$|a_{ij}^{(k)}| = |a_{ij}^{(k-1)} - n_{ik}a_{kj}^{(k-1)}| \leqslant |a_{ij}^{(k-1)}| + |a_{kj}^{(k-1)}|. \tag{26.1}$$

Unfortunately we can construct matrices (admittedly highly artificial) for which this bound is achieved. Examples are matrices of the form

$$\begin{bmatrix} 1 & 0 & 0 & 0 & 1 \\ -1 & 1 & 0 & 0 & 1 \\ -1 & -1 & 1 & 0 & 1 \\ -1 & -1 & -1 & 1 & 1 \\ -1 & -1 & -1 & -1 & 1 \end{bmatrix}. \tag{26.2}$$

Complete pivoting

27. Our analysis has already shown the vital importance of preventing the elements of the successive $a_{ij}^{(k)}$ from increasing rapidly. We now consider an alternative method of pivoting for which we can give appreciably smaller bounds for $|a_{ij}^{(k)}|$.

In the algorithm described in § 21 the variables are eliminated in the natural order and at the rth stage the pivot is selected from the elements in the first column of the square matrix of order $(n-r+1)$ in the bottom right-hand corner of A_{r-1}.

Suppose now we choose the rth pivot to be the element of maximum modulus in the whole of this square array. If this element is in the (r', r'') position $(r', r'' \geqslant r)$ then, thinking in terms of interchanges, we may write

$$A_r = N_r I_{r,r'} A_{r-1} I_{r,r''}. \tag{27.1}$$

In general both row and column interchanges are now required, and the variables are no longer eliminated in the natural order; however, the final A_{n-1} is still upper triangular. We have, typically, when $n = 4$

$$(N_3 I_{3,3'} N_2 I_{2,2'} N_1 I_{1,1'} A_0 I_{1,1'} I_{2,2'} I_{3,3'})(I_{3,3'} I_{2,2'} I_{1,1'} x) =$$
$$= N_3 I_{3,3'} N_2 I_{2,2'} N_1 I_{1,1'} b_0, \quad (27.2)$$

since the product of the factors inserted between A_0 and x is I. The solution of $A_r y = b_r$ is therefore $I_{r,r'} \ldots I_{2,2'} I_{1,1'} x$. We are now using two-sided *equivalence* transformations of A_0, but the right-hand multipliers are the simple permutation matrices.

We refer to this modification as *complete* pivoting in contrast to the earlier form which we call *partial* pivoting. Wilkinson (1961b) has shown that with complete pivoting

$$|a_{ij}^{(r-1)}| < r^{\frac{1}{2}}[2^1 3^{1/2} 4^{1/3} \ldots r^{1/(r-1)}]^{\frac{1}{2}} a = f(r)a, \quad (27.3)$$

where $a = \max |a_{ij}^{(0)}|$. In Table 2 we give values of $f(r)$ for representative values of r.

TABLE 2

r	10	20	50	100
$f(r)$	19	67	530	3300

The function $f(r)$ is *much* smaller than 2^r for large r and the method of proof of (27.3) indicates that the true upper bound is much smaller still. In fact no matrix has yet been discovered for which $f(r) > r$. For matrices of the form (26.2) and of any order, the maximum element with complete pivoting is 2.

One might be tempted to conclude that it would be advisable to use complete pivoting in all cases, but partial pivoting does have considerable advantages. The most important of these are the following.

(I) On a computer with a two-level store it may be easier to organize than complete pivoting.

(II) For matrices having a considerable number of zero elements arranged in a special pattern, this pattern may be preserved by partial pivoting but destroyed by complete pivoting. This is a particularly cogent argument from the point of view of our own applications.

In spite of the existence of matrices of the form shown in (26.2) it is our experience that any substantial increase in size of elements of successive A_r is extremely uncommon even with partial pivoting. If the matrix A_0 is at all ill-conditioned it is common for the elements of successive A_r to show a steady *downward* trend, especially if the

(i, j) element is a smooth function of i and j, in which case the elimination process gives results which are closely akin to differencing. No example which has arisen naturally has in my experience given an increase by a factor as large as 16.

Practical procedure with partial pivoting

28. A recommended procedure with partial pivoting in fixed-point is the following. We scale the matrix initially so that all elements lie in the range $-\frac{1}{2}$ to $\frac{1}{2}$. We can therefore perform one step without any danger of elements exceeding unity. As this step is performed we examine each element as it is computed to see if its modulus exceeds $\frac{1}{2}$. If any element exceeds $\frac{1}{2}$, then on completion the corresponding row is divided through by 2. We therefore start the next step with a matrix having elements lying in the range $-\frac{1}{2}$ to $\frac{1}{2}$.

The corresponding error analysis requires only trivial modifications. At the stage when we divide a row by 2 an extra rounding error, bounded by 2^{-t}, is made in each element. Subsequent rounding errors made in this row are equivalent to perturbations in the original matrix which are bounded by 2^{-t} rather than $\frac{1}{2}2^{-t}$. The rounding errors reflected back into the original matrix are still strictly additive. Similar comments apply if a row is divided on more than one occasion.

Floating-point error analysis

29. The fixed-point procedure described in § 28 is in some respects rather like *ad hoc* floating-point, though, as we pointed out, we expect divisions of rows to be infrequent in practice. The error analysis of true floating-point computation follows much the same lines. We now have

$$a_{ij}^{(k)} = fl(a_{ij}^{(k-1)} - n_{ik}a_{kj}^{(k-1)})$$
$$\equiv [a_{ij}^{(k-1)} - n_{ik}a_{kj}^{(k-1)}(1+\epsilon_1)](1+\epsilon_2) \quad (|\epsilon_i| \leqslant 2^{-t}), \tag{29.1}$$

$$n_{ij} = fl(a_{ij}^{(j-1)}/a_{jj}^{(j-1)}) \equiv a_{ij}^{(j-1)}/a_{jj}^{(j-1)} + \xi_{ij} \quad (|\xi_{ij}| \leqslant \tfrac{1}{2}2^{-t}), \tag{29.2}$$

where we can use an absolute bound for ξ_{ij} since pivoting ensures that $|n_{ij}| \leqslant 1$. If we assume that

$$|a_{ij}^{(k)}| \leqslant a_k \text{ (say)}, \tag{29.3}$$

then using the same notation as in the fixed-point analysis we have

$$|\epsilon_{ij}^{(j)}| = |a_{jj}^{(j-1)}\xi_{ij}| < 2^{-t-1}a_{j-1} \quad (i > j), \tag{29.4}$$

while for all other relevant (i, j) and k a simple manipulation shows that certainly

$$|\epsilon_{ij}^{(k)}| = \left| \frac{a_{ij}^{(k)}\epsilon_2}{1+\epsilon_2} - n_{ik}a_{kj}^{(k-1)}\epsilon_1 \right|$$
$$< 2^{-t}[a_k/(1-2^{-t})+a_{k-1}] < 2^{-t}[(1{\cdot}01)a_k+a_{k-1}]. \qquad (29.5)$$

These bounds show that the computed N and A_{n-1} are such that

$$NA_{n-1} \equiv A_0+F, \qquad (29.6)$$

where

$$|f_{ij}| < \begin{cases} 2^{-t}[a_0+(2{\cdot}01)a_1+...+(2{\cdot}01)a_{i-2}+(1{\cdot}01)a_{i-1}] & (j > i) \\ 2^{-t}[a_0+(2{\cdot}01)a_1+...+(2{\cdot}01)a_{j-2}+(1{\cdot}51)a_{j-1}] & (i > j) \end{cases}. \qquad (29.7)$$

The bounds (29.7) could be sharpened somewhat, but it would not serve much purpose. They show that if $|a_{ij}^{(k)}| < a$ for all relevant i, j, k then certainly

$$|f_{ij}| < \begin{cases} (2{\cdot}01)2^{-t}a(i-1) \\ (2.01)2^{-t}aj. \end{cases} \qquad (29.8)$$

Equations (29.7) show that when the $a_{ij}^{(k)}$ decrease progressively with increasing k, then the bound for F may be far better for floating-point than fixed-point. If, for example,

$$a_k < 2^{-k}, \qquad (29.9)$$

so that roughly one binary figure is lost per stage, then

$$|f_{ij}| < 2^{-t}[1+(2{\cdot}01)(2^{-1}+2^{-2}+...)] < (3{\cdot}01)2^{-t}. \qquad (29.10)$$

A progressive loss of significant figures is quite common with certain types of ill-conditioned equations and our result shows that floating-point computation should then be superior; this has been borne out in practice. (A comparable effect can be obtained in fixed-point if, after each reduction, we multiply each row by the largest power of 2 for which the elements remain in the range ± 1.)

We may obtain results of a similar nature for the rounding errors made in the right-hand sides.

Floating-point decomposition without pivoting

30. In Table 3 we give a simple example which illustrates the catastrophic loss of accuracy which may result from the failure to pivot for size. The matrix involved is very well-conditioned but the failure to pivot is fatal to the accuracy. Floating-point computation has been used because the numbers involved in the A_r span a very wide range.

TABLE 3

n		A_0			b_0
	0·000003	0·213472	0·332147		0·235262
71837·3)	0·215512	0·375623	0·476625		0·127653
57752·3)	0·173257	0·663257	0·625675		0·285321
		A_1			b_1
	0·000003	0·213472	0·332147		0·235262
	0·000000	−15334·9	−23860·0		−16900·5
0·803905)	0·000000	−12327·8	−19181·7		−13586·6
		A_2			b_2
	0·000003	0·213472	0·332147		0·235262
	0·000000	−15334·9	−23860·0		−16900·5
	0·000000	0·000000	−0·500000		−0·200000

Exact equivalent, $Bx = c$, of $A_1 x = b_1$

0·000003	0·213472	0·332147		0·235262
0·2155119	0·3521056	0·5436831		0·0868726
0·1732569	0·6989856	0·5531881		0·3216026

The first pivot is very small and as a result, n_{21} and n_{31} are of order 10^5. The computed elements of A_1 and b_1 are accordingly also very large. If we consider the computation of $a_{23}^{(1)}$ we have

$$a_{23}^{(1)} = fl(a_{23}^{(0)} - n_{21}a_{13}^{(0)}) = fl(0·476625 - 71837·3 \times 0·332147)$$
$$\equiv fl(0·476625 - 23860·5) = -23860·0.$$

The exact value of $a_{23}^{(0)} - n_{21}a_{13}^{(0)}$ is given by

$$0·476625 - 23860·5436831 = -23860·0670581$$

and the quantity $\epsilon_{23}^{(1)}$ is the difference between the two values. Observe that in computing $a_{23}^{(1)}$, nearly all the figures in $a_{23}^{(0)}$ are ignored; we would obtain the same computed value $a_{23}^{(1)}$ for any element $a_{23}^{(0)}$ lying in the range 0·45 to 0·549999. In Table 3 we show the equations obtained by reflecting back the errors into the original equations. It will be seen that the perturbed second and third equations bear no relation to the originals. It may readily be verified that the solution of the computed set of equations is very inaccurate.

We have also exhibited the second major step in the computation. Notice that the large elements which arise in the third equation as a result of the first step disappear in the second step. However, it is obvious that nearly all the figures given in the computed $a_{33}^{(2)}$ are spurious. The value of the determinant given by $a_{11}^{(0)}a_{22}^{(1)}a_{33}^{(2)}$ is also highly inaccurate. This is obvious since $a_{33}^{(2)}$ clearly has at most one correct significant figure.

Loss of significant figures

31. So far we have focused attention on the dangers of techniques which lead to an increase in the size of elements of the successive A_r. It might well be objected that the more serious loss of accuracy occurs in the examples where there is a progressive diminution in the size of the elements. This is indeed true, but there is an important difference between the two phenomena. When, as a result of not pivoting, or of using partial instead of complete pivoting, the elements of a reduced matrix are very large, we suffer a loss of accuracy which is quite unnecessary and could have been avoided by a superior strategy.

On the other hand, the progressive loss of significant figures is a consequence of ill-conditioning, and the loss of accuracy that ensues is more or less inevitable. The argument of § 29 shows that if floating-point is used, then when there is a loss of significant figures during the elimination process, the equivalent perturbations in the original equations are actually smaller than when this loss does not occur.

A popular fallacy

32. There is a widely held belief that the selection of large pivots is, in some respects, harmful. The argument is as follows: 'The product of the pivots is equal to the determinant of A_0 and is therefore independent of pivoting. If we select large pivots initially, then inevitably the later pivots must be smaller and therefore have higher relative errors.'

Now there are matrices for which the last pivot is small with complete pivoting but not with partial pivoting. However, the last pivot is the reciprocal of some element of the inverse matrix whatever strategy may have been used, and hence it cannot be as small as 2^{-k} (say) unless the norm of the inverse is at least as large as 2^k. If the matrix really does have an inverse as large as this the accuracy is bound to suffer accordingly. The emergence of a small pivot is merely one way in which ill-conditioning may make itself felt. In our example of § 30 we saw that the use of an 'unnecessarily' small pivot may not make the last pivot any larger; *it merely leads to a very large pivot in the next stage and a return to 'normal' in the next stage but one.*

We do not claim that selecting the largest elements as pivots is the best strategy and indeed this will often be far from true. We do claim that no alternative practical procedure has yet been proposed. Typical of the alternative proposals is that the pivot should be chosen to be the element nearest in modulus to the nth root of the absolute value of the determinant. There are three objections to this and to related proposals.

(i) The value of the determinant is not usually known until Gaussian elimination is complete.

(ii) There may be no elements near to the prescribed values.

(iii) Even if (i) and (ii) were not true, it is not difficult to invent examples for which it is an extremely poor strategy. Complete pivoting, on the other hand, is never a very poor strategy.

Matrices of special form

33. There are three special classes of matrices which will be of great importance in the eigenproblem, and for which the pivotal strategy is much simpler. The proofs of the following results are only briefly sketched since they are fairly simple and have been given in detail by Wilkinson (1961b).

Class I. Hessenberg matrices. We define a matrix A to be *upper Hessenberg* if

$$a_{ij} = 0 \quad (i > j+2). \tag{33.1}$$

Similarly A is said to be *lower Hessenberg* if

$$a_{ij} = 0 \quad (j > i+2), \tag{33.2}$$

so that if A is upper Hessenberg then A^T is lower Hessenberg. If we perform Gaussian elimination with partial pivoting on an upper Hessenberg matrix, then at the beginning of the rth major step A_{r-1} is of the form given typically when $n = 6$, $r = 3$ by

$$\begin{bmatrix} \times & \times & \times & \times & \times & \times \\ 0 & \times & \times & \times & \times & \times \\ & 0 & \times & \times & \times & \times \\ & & \times & \times & \times & \times \\ & & & \times & \times & \times \\ & & & & \times & \times \end{bmatrix}. \tag{33.3}$$

where we have entered the zeros produced by previous steps. The matrix in the bottom right-hand corner is itself upper Hessenberg. The elements in rows $(r+1)$ to n are the same as those in A_0. There are only two rows in the relevant matrix involving x_r so that r' is either r or $r+1$. We may say that there is, or is not, an interchange at the rth step according as $r' = r+1$, or $r' = r$. From this we can show that if

$$|a_{ij}^{(0)}| < 1 \quad \text{then} \quad |a_{ij}^{(r)}| < r+1, \tag{33.4}$$

and each column of $\tilde{N}$ (the permuted N) has only one sub-diagonal element. If interchanges take place at steps 1, 3, 4, for a matrix of order 6, for example, then $\tilde{N}$ is of the form

$$
\begin{bmatrix}
1 & & & & & \\
\times & 1 & & & & \\
0 & 0 & 1 & & & \\
0 & 0 & 0 & 1 & & \\
0 & \times & \times & \times & 1 & \\
0 & 0 & 0 & 0 & \times & 1
\end{bmatrix}. \tag{33.5}
$$

If Gaussian elimination of a lower Hessenberg matrix is required, then the variables should be eliminated in the order $x_n, x_{n-1}, \ldots, x_1$ so that the final matrix is lower triangular. In either case the use of complete pivoting destroys the simple form of the matrix, and since we now have a much better bound for the $a_{ij}^{(r)}$, complete pivoting has no special merit.

Class II. Tri-diagonal matrices. We have defined tri-diagonal matrices in Chapter 3, § 13; from this definition we now see that tri-diagonal matrices are both upper and lower Hessenberg. The results we obtained for upper Hessenberg matrices therefore hold for tri-diagonal matrices but they admit of further simplification. $\tilde{N}$ is of the same form as before, while A_0 and A_r are given typically for the case $n = 6$, $r = 3$ by

$$
A_0 =
\begin{bmatrix}
\alpha_1 & \beta_1 & & & & \\
\gamma_2 & \alpha_2 & \beta_2 & & & \\
& \gamma_3 & \alpha_3 & \beta_3 & & \\
& & \gamma_4 & \alpha_4 & \beta_4 & \\
& & & \gamma_5 & \alpha_5 & \beta_5 \\
& & & & \gamma_6 & \alpha_6
\end{bmatrix},
\quad
A_3 =
\begin{bmatrix}
u_1 & v_1 & w_1 & & & \\
& u_2 & v_2 & w_2 & & \\
& & u_3 & v_3 & w_3 & \\
& & & p_4 & q_4 & \\
& & & \gamma_5 & \alpha_5 & \beta_5 \\
& & & & \gamma_6 & \alpha_6
\end{bmatrix}.
\tag{33.6}
$$

The element w_i is zero unless there was an interchange in the ith step. It follows by a simple inductive proof that if

$$
|\alpha_i|, \; |\beta_i|, \; |\gamma_i| < 1 \quad (i = 1, \ldots, n) \tag{33.7}
$$

then
$$
|p_i| < 2, \quad |q_i| < 1, \quad |u_i| < 2, \quad |v_i| < 1, \quad |w_i| < 1. \tag{33.8}
$$

Class III. Positive definite symmetric matrices. For matrices of this type Gaussian elimination is quite stable *without any pivotal strategy.* We can show that if, as in § 18, we write

$$A_r = \left[\begin{array}{c|c} U_{rr} & U_{r,n-r} \\ \hline O & W_{n-r,n-r} \end{array}\right], \tag{33.9}$$

then $W_{n-r,n-r}$ is itself positive definite and symmetric at each stage, and if $|a_{ij}^{(0)}| \leqslant 1$ then $|a_{ij}^{(k)}| \leqslant 1$. See for example, Wilkinson (1961b).

We omit the proof because, although the triangularization of positive definite matrices is of fundamental importance in our applications, we shall not use Gaussian elimination to achieve it.

Note that we do not claim that all elements of N are bounded by unity and indeed this may not be true. This is illustrated by the matrix

$$\begin{bmatrix} 0{\cdot}000064 & 0{\cdot}006873 \\ 0{\cdot}006873 & 1{\cdot}000000 \end{bmatrix} \tag{33.10}$$

for which $n_{21} = 107{\cdot}391$. This means we must use floating-point arithmetic, but a large multiplier cannot lead to an increase in the maximum of the elements of the reduced matrix; in this example $a_{22}^{(1)} = 0{\cdot}261902$.

Our error analysis has shown that the errors at each stage are reflected back without amplification into the original matrix. Since all elements of all reduced matrices are bounded by unity, the rounding errors cannot be large.

Gaussian elimination on a high-speed computer

34. We now describe a method of organizing Gaussian elimination with partial pivoting on a computer. At the beginning of the rth major step the configuration in the store is given typically for the case $n = 5$, $r = 3$ by

$$\begin{bmatrix} a_{11} & a_{12} & a_{13} & a_{14} & a_{15} & b_1 \\ n_{21} & a_{22} & a_{23} & a_{24} & a_{25} & b_2 \\ n_{31} & n_{32} & a_{33} & a_{34} & a_{35} & b_3 \\ n_{41} & n_{42} & a_{43} & a_{44} & a_{45} & b_4 \\ n_{51} & n_{52} & a_{53} & a_{54} & a_{55} & b_5 \end{bmatrix}. \tag{34.1}$$

$$1' \quad\ 2' \quad\ 3'$$

The elements $1'$, $2'$,... in the last row describe the positions of the pivotal rows, r' having been determined in the $(r-1)$th step. The method of description is quite close to that used in many automatic programming languages, in that, for example, a_{ij} refers to the current element in the (i, j) position *after all operations up to the current stage* have been performed. We omit the upper suffix which was used in the mathematical discussion. The rth major step is as follows.

(i) Interchange a_{rj} $(j = r,..., n)$ and b_r with $a_{r'j}$ $(j = r,..., n)$ and $b_{r'}$. Note that if $r' = r$ there is no interchange. For each value of s from $(r+1)$ to n in succession perform (ii), (iii), (iv).

(ii) Compute $n_{sr} = a_{sr}/a_{rr}$ and overwrite on a_{sr}.

(iii) For each value of t from $(r+1)$ to n, compute $a_{st}-n_{sr}a_{rt}$ and overwrite on a_{st}.

(iv) Compute $b_s-n_{sr}b_r$ and overwrite on b_s.

As the rth major step progresses, we keep a record of the largest of the quantities $|a_{s,r+1}|$. If, on completion, this is $a_{(r+1)',r+1}$, then $(r+1)'$ is entered in the $(r+1)$th position of the last line.

Obviously a preliminary step is required in which we locate the element of maximum modulus in the first column of A_0.

Solutions corresponding to different right-hand sides

35. We shall often be interested in the solutions of sets of equations $Ax = b$, all having the same matrix A but with a number of different right-hand sides. If the right-hand sides are known *ab initio*, then clearly we may process them all as the elimination proceeds. Note that if we take the n right-hand sides $e_1,..., e_n$ the n solutions are the columns of the inverse of A.

Often though, we shall be using techniques in which each right-hand side is determined from the solution obtained with the previous right-hand side. The separate processing of a right-hand side has $(n-1)$ major steps of which the rth is as follows.

(i) Interchange b_r and $b_{r'}$.

(ii) For each value of s from $(r+1)$ to n compute $b_s-n_{sr}b_r$ and overwrite on b_s.

Direct triangular decomposition

36. We return now for the moment to the elimination process without interchanges. It is evident that the intermediate reduced matrices A_r are of little value in themselves and that the main purpose of the process is to produce the matrices N and A_{n-1} such that

$$NA_{n-1} = A_0. \tag{36.1}$$

If we are given N and A_{n-1}, the solution of $Ax = b$ with any right-hand side b can be reduced to the solution of the two triangular sets of equations

$$Ny = b, \qquad A_{n-1}x = y. \tag{36.2}$$

This suggests that we consider the possibility of producing the matrices N and A_{n-1} directly from A_0 without going through the intermediate stages.

Since the A_r are to be discarded, there is no point in using our old notation. We therefore consider the problem of determining a unit lower triangular matrix L, an upper triangular matrix U and a vector c such that

$$L(U \mathop{:} c) = (A \mathop{:} b). \tag{36.3}$$

Let us assume that the first $(r-1)$ rows of L, U and c can be determined by equating the elements in the first $(r-1)$ rows of both sides of equation (36.3). Then, equating the elements in the rth row, we have

$$\left.\begin{aligned}
l_{r1}u_{11} &= a_{r1} \\
l_{r1}u_{12}+l_{r2}u_{22} &= a_{r2} \\
\dots\dots\dots\dots\dots\dots \\
l_{r1}u_{1r}+l_{r2}u_{2r}+\dots+l_{rr}u_{rr} &= a_{rr}
\end{aligned}\right\}, \tag{36.4}$$

$$\left.\begin{aligned}
l_{r1}u_{1,r+1}+l_{r2}u_{2,r+1}+\dots+l_{rr}u_{r,r+1} &= a_{r,r+1} \\
\dots\dots\dots\dots\dots\dots\dots\dots\dots\dots \\
l_{r1}u_{1n}\ +l_{r2}u_{2n}\ +\dots+l_{rr}u_{rn} &= a_{rn} \\
l_{r1}c_1\ \ +l_{r2}c_2\ \ +\dots+l_{rr}c_r &= b_r
\end{aligned}\right\}. \tag{36.5}$$

From equations 1 to $(r-1)$ of (36.4) l_{r1}, $l_{r2}, \dots, l_{r,r-1}$ are determined necessarily and uniquely. If l_{rr} is chosen arbitrarily, the rth equation determines u_{rr}. The remaining equations then determine $u_{r,r+1}$ to u_{rn} and c_r uniquely. Since the result is obviously true for the elements of the first row, it is therefore true in general.

We see that the product $l_{rr}u_{rr}$ is uniquely determined, but either l_{rr} or u_{rr} may be chosen arbitrarily. If we require L to be unit lower triangular then $l_{rr} = 1$. In this case the only divisions involved are those in which l_{ri} ($i = 1, \dots, r-1$) are determined, and the divisors are the u_{ii}. The decomposition cannot fail if $u_{ii} \neq 0$, and indeed it is then unique. In the above process the elements were determined in the order

first row of U, c: second row of L; second row of U, c: etc.

$$(36.6)$$

It will readily be verified that they also can be determined in the order

first row of U, c; first column of L: second row of U, c;

second column of L: etc.   (36.7)

Relations between Gaussian elimination and direct triangular decomposition

37. We have already shown that if A is non-singular then the triangular decomposition with a unit triangular factor is unique if it exists, and hence L and U must be identical with the N and A_{n-1} of Gaussian elimination. Failure and non-uniqueness of the decomposition occur therefore in just the same circumstances as in Gaussian elimination without pivoting.

Consider now the determination of l_{r4}, for example, from the relation

$$l_{r4} = (a_{r4} - l_{r1}u_{14} - l_{r2}u_{24} - l_{r3}u_{34})/u_{44}. \qquad (37.1)$$

A comparison with Gaussian elimination shows that

$$
\left.
\begin{aligned}
&\text{(i)}\ a_{r4}^{(0)} = a_{r4}, && \text{(ii)}\ a_{r4}^{(1)} = a_{r4} - l_{r1}u_{14}, \\
&\text{(iii)}\ a_{r4}^{(2)} = a_{r4} - l_{r1}u_{14} - l_{r2}u_{24}, && \text{(iv)}\ a_{r4}^{(3)} = a_{r4} - l_{r1}u_{14} - l_{r2}u_{24} - l_{r3}u_{34},
\end{aligned}
\right\}
$$

$$(37.2)$$

and hence in determining the numerator of (37.1) we obtain in turn each of the elements in the $(r, 4)$ positions of A_1, A_2, A_3. Similar remarks apply to the computation of elements of U.

However, if we think in terms of triangular decomposition we do not write down these intermediate results. Even more important, if we have facilities for $fl_2(\ )$ or $fl_2(\ )$ computation, we do not need to round to single-precision after each addition. The numerator may be accumulated and u_{44} divided into the accumulated value. *If, however, we use standard floating-point arithmetic or fixed-point arithmetic without accumulation, even the rounding errors are the same in the two processes.*

Examples of failure and non-uniqueness of decomposition

38. An example of a matrix for which triangular decomposition is impossible is the matrix A given by

$$A = \begin{bmatrix} 1 & 2 & 3 \\ 2 & 4 & 1 \\ 4 & 6 & 7 \end{bmatrix}. \tag{38.1}$$

The elements of L and U obtained up to the stage when the breakdown occurs are given by

$$L = \begin{bmatrix} 1 & 0 & 0 \\ 2 & 1 & 0 \\ 4 & ? & . \end{bmatrix}, \qquad U = \begin{bmatrix} 1 & 2 & 3 \\ 0 & 0 & -5 \\ . & . & . \end{bmatrix}. \tag{38.2}$$

We see that $u_{22} = 0$ and the equation defining l_{32} is

$$4 \times 2 + l_{32} \times 0 = 6.$$

Note that A is not singular. In fact it is well-conditioned.

On the other hand there are an infinite number of decompositions of the matrix A given by

$$A = \begin{bmatrix} 1 & 1 & 1 \\ 2 & 2 & 1 \\ 3 & 3 & 1 \end{bmatrix} \tag{38.3}$$

We have, for example,

$$L = \begin{bmatrix} 1 & 0 & 0 \\ 2 & 1 & 0 \\ 3 & 3 & 1 \end{bmatrix}, \qquad U = \begin{bmatrix} 1 & 1 & 1 \\ 0 & 0 & -1 \\ 0 & 0 & 1 \end{bmatrix}. \tag{38.4}$$

Again $u_{22} = 0$ but when we come to the determination of l_{32} we have

$$3 \times 1 + l_{32} \times 0 = 3$$

and hence l_{32} is arbitrary. When l_{32} has been chosen, u_{33} is uniquely determined. The equivalent triangular set of equations corresponding to any right-hand side is

$$\left. \begin{aligned} x_1 + x_2 + x_3 &= c_1 \\ -x_3 &= c_2 \\ x_3 &= c_3 \end{aligned} \right\}. \tag{38.5}$$

Hence unless $c_2 = -c_3$ there is no solution. Notice that if L is to be unit triangular a *necessary* condition for ambiguity is that at least one u_{rr} should be zero. If, nevertheless, triangular decompositions exist, then A must be singular since $A = LU$ and $\det(U)$ is zero. Hence we know that a solution exists if and only if

$$\text{rank}(A \mid b) = \text{rank}(A).$$

In the case when the only zero element is u_{nn}, then no breakdown or ambiguity arises, but solutions exist only if $c_n = 0$.

Triangular decomposition with row interchanges

39. Since Gaussian elimination and triangular decomposition are identical, except possibly for the rounding errors, it is evident that pivoting is vital for the numerical stability of the decomposition. On an automatic computer it is quite convenient to carry out the equivalent of partial pivoting but it is impractical to do the equivalent of complete pivoting unless one is prepared to use twice as much storage space or, alternatively, to perform nearly all the operations twice.

We describe now the equivalent of partial pivoting in fixed-point arithmetic. It consists of n major steps in the rth of which we determine the rth column of L, the rth row of $(U \mid c)$ and r'. The configuration at the beginning of the rth step for the case $n = 5$, $r = 3$ is given by

$$\begin{bmatrix} u_{11} & u_{12} & u_{13} & u_{14} & u_{15} & c_1 & s_1 \\ l_{21} & u_{22} & u_{23} & u_{24} & u_{25} & c_2 & s_2 \\ l_{31} & l_{32} & a_{33} & a_{34} & a_{35} & b_3 & s_3 \\ l_{41} & l_{42} & a_{43} & a_{44} & a_{45} & b_4 & s_4 \\ l_{51} & l_{52} & a_{53} & a_{54} & a_{55} & b_5 & s_5 \end{bmatrix}. \tag{39.1}$$
$$1' \quad 2'$$

The rth major step is as follows.

(i) For $t = r, r+1, ..., n$:
Calculate $a_{tr} - l_{t1}u_{1r} - l_{t2}u_{2r} - ... - l_{t,r-1}u_{r-1,r}$, storing the result as a double precision number s_t at the end of row t. (Note that if floating-point accumulation is used, little is lost if s_t is rounded on completion.)

(ii) Suppose $|s_{r'}|$ is the maximum of the $|s_t|$ $(t = r, ..., n)$. Then store r' and interchange the whole of rows r and r' including the

l_{ri}, a_{ri}, b_r, s_r. Round the new s_r to single precision to give u_{rr} and overwrite this on a_{rr}.

(iii) For $t = r+1,\dots, n$:

Compute s_t/u_{rr} to give l_{tr} and overwrite on a_{tr}.

(iv) For $t = r+1,\dots, n$:

Compute $a_{rt} - l_{r1}u_{1t} - l_{r2}u_{2t} - \dots - l_{r,r-1}u_{r-1,t}$, accumulating the inner-product and rounding on completion to give u_{rt}. Overwrite u_{rt} on a_{rt}.

(v) Compute $b_r - l_{r1}c_1 - l_{r2}c_2 - \dots - l_{r,r-1}c_{r-1}$, again accumulating the inner-product and rounding on completion to give c_r. Overwrite c_r on b_r.

TABLE 4

Triangular decomposition with interchanges
Original equations

0·7321	0·4135	0·3126	0·5163	0·8132
0·2317	0·6123	0·4137	0·6696	0·4753
0·4283	0·8176	0·4257	0·8312	0·2167
0·8653	0·2165	0·8265	0·7123	0·5165

$$1' = 4$$

Configuration at end of first major step

0·8653	0·2165	0·8265	0·7123	0·5165
0·2678	0·6123	0·4137	0·6696	0·4753
0·4950	0·8176	0·4257	0·8312	0·2167
0·8461	0·4135	0·3126	0·5163	0·8132

(4)

$s_2 = 0·6123 - (0·2678)(0·2165) = 0·55432130$
$s_3 = 0·8176 - (0·4950)(0·2165) = \mathbf{0·71043250}$
$s_4 = 0·4135 - (0·8461)(0·2165) = 0·23031935$

$$2' = 3$$

Configuration at end of second major step

0·8653	0·2165	0·8265	0·7123	0·5165
0·4950	0·7104	0·0166	0·4786	−0·0390
0·2678	**0·7803**	0·4137	0·6696	0·4753
0·8461	**0·3242**	0·3126	0·5163	0·8132

(4)   (3)

$s_3 = 0·4137 - (0·2678)(0·8265) - (0·7803)(0·0166)$
$\quad = 0·17941032$
$s_4 = 0·3126 - (0·8461)(0·8265) - (0·3242)(0·0166)$
$\quad = \mathbf{-0·39208337}$

$$3' = 4$$

Configuration at end of third major step

0·8653	0·2165	0·8265	0·7123	0·5165
0·4950	0·7104	0·0166	0·4786	−0·0390
0·8461	**0·3242**	−0·3921	−0·2415	0·3888
0·2678	**0·7803**	**−0·4576**	0·6696	0·4753

(4)   (3)   (4)

$s_4 = 0·6696 - (0·2678)(0·7123) - (0·7803)(0·4786) - (-0·4576)(-0·2415) = \mathbf{-0·00511592}$

$$4' = 4$$

Configuration at end of fourth major step

					x	r (exact)
0·8653	0·2165	0·8265	0·7123	0·5165	9·1	−0·02332
0·4950	0·7104	0·0166	0·4786	−0·0390	70·4	0·03339
0·8461	**0·3242**	−0·3921	−0·2415	0·3888	64·8	0·03005
0·2678	**0·7803**	**−0·4576**	−0·0051	0·5453	−106·9	−0·01166
(4)	(3)	(4)	(4)			

$L \times (U \mathbin{\vert} c)$ (exact). Compare with $(A \mathbin{\vert} b)$ with its rows permuted

0·86530000	0·21650000	0·82650000	0·71230000	0·51650000
0·42832350	0·81756750	0·42571750	0·83118850	0·21666750
0·73213033	0·41349233	0·31258337	0·51633915	0·81316685
0·23172734	0·61230382	0·41371464	0·66961592	0·47527212

It will be seen that the elements of L and U are found in the order exhibited in (36.7). The final configuration is such that LU is equal to A with its rows permuted, and we have stored sufficient information for the separate processing of a right-hand side. There are n steps in this processing. In the rth step we interchange rows r and r' of the right-hand side and then perform operation (v). One rounding error only is involved in the production of each element c_r.

Note that we could have organized the processing of a right-hand side in the same way for Gaussian elimination. This would have involved interchanging the whole of row r and r' in step (i) of § 34. We could then perform the separate processing of a right-hand side with the accumulation of inner-products.

In Table 4 we exhibit the triangular decomposition with interchanges of the matrix of order 4 of Table 1 with its rows permuted. The configuration at the end of each major step is illustrated; the elements of L are in bold print and the matrix of elements which have not yet been modified (apart from interchanges) is enclosed in dotted lines.

Error analysis of triangular decomposition

40. As in Gaussian elimination with partial pivoting, we would expect it to be rare for the maximum $|u_{ij}|$ obtained by the algorithm of the last section to exceed the maximum $|a_{ij}|$ by any appreciable factor and indeed, for ill-conditioned matrices, the $|u_{ij}|$ will usually be smaller on the whole than the $|a_{ij}|$.

If necessary, additional control on the size of elements may be achieved by a method analogous to that of § 28. We scale A so that $|a_{ij}| < \frac{1}{2}$, and when accumulating the s_t and u_{rt} we examine the inner-product after each addition; if it exceeds $\frac{1}{2}$ in absolute value then we divide s_t or u_{rt} and all elements of A and b in row t or row r by two. Note that more than one such division may be necessary while we are accumulating s_t or u_{rt}. As in Gaussian elimination we expect such divisions to be rare.

Apart from rounding errors we know that $L(U \mathbin{\vert} c)$ is equal to $(A \mathbin{\vert} b)$ with its rows permuted. To simplify the notation, we shall use $(A \mathbin{\vert} b)$ to denote this permuted matrix, so that we now require bounds for $L(U \mathbin{\vert} c) - (A \mathbin{\vert} b)$. Assuming that no divisions by two are necessary we have

$$l_{tr} \equiv s_t/u_{rr} + \xi_{tr} \quad (|\xi_{tr}| < \tfrac{1}{2}2^{-t}), \tag{40.1}$$

where again l_{tr}, s_t, u_{rr} refer to computed values. Hence

$$a_{tr} \equiv l_{t1}u_{1r} + l_{t2}u_{2r} + \ldots + l_{t,r-1}u_{r-1,r} + l_{tr}u_{rr} + u_{rr}\xi_{tr}. \tag{40.2}$$

Similarly for the computed u_{rt} and c_r we have

$$u_{rt} \equiv a_{rt} - l_{r1}u_{1t} - l_{r2}u_{2t} - \ldots - l_{r,r-1}u_{r-1,t} + \epsilon_{tr} \quad (|\epsilon_{tr}| \leqslant \tfrac{1}{2}2^{-t}), \quad (40.3)$$

$$c_r \equiv b_r - l_{r1}c_1 - l_{r2}c_2 - \ldots - l_{r,r-1}c_{r-1} + \epsilon_r \quad (|\epsilon_r| \leqslant \tfrac{1}{2}2^{-t}). \quad (40.4)$$

Collecting terms in L, U and c to one side, relations (40.2), (40.3), (40.4) show that

$$L(U \mathbin{\vdots} c) \equiv (A+F, \; b+k), \quad (40.5)$$

$$|f_{ij}| \leqslant \begin{cases} \tfrac{1}{2}2^{-t} & (i < j) \\ \tfrac{1}{2}|u_{jj}|\,2^{-t} & (i > j) \end{cases}.$$

$$|k_i| \leqslant \tfrac{1}{2}2^{-t} \quad (40.6)$$

Since we are assuming that $|u_{ij}| \leqslant 1$, we see that all f_{ij} satisfy $|f_{ij}| \leqslant \tfrac{1}{2}2^{-t}$, but if some $|u_{jj}|$ are much smaller than unity then many of the elements $|f_{ij}|$ will be much smaller than $\tfrac{1}{2}2^{-t}$.

The relations (40.6) show triangular decomposition with interchanges to be a remarkably accurate process if we can accumulate inner-products. In the case when the elements of A are not exactly representable by t-digit numbers the errors made in rounding the numbers initially to t digits usually have as much effect as all the rounding errors in the triangularization. Note that we are forced to say 'usually', in order to cover those rare cases where, even when using interchanges, some elements of U are appreciably larger than those of A.

At the foot of Table 4 we give the exact products of L and $(A \mathbin{\vdots} c)$ for comparison with $(A \mathbin{\vdots} b)$. It will be seen that the maximum difference is 0·00003915, in the (1, 4) element. This is less than the maximum value of one rounding error.

Evaluation of determinants

41. From the relation $LU = A+F$, where A is now the permuted form of the original matrix we deduce that

$$\det(A+F) = u_{11}u_{22}\ldots u_{nn}. \quad (41.1)$$

For the original ordering we must multiply by $(-1)^k$ where k is the number of values of r for which $r' \neq r$. In practice determinants are invariably evaluated using floating-point arithmetic. The computed value D is therefore given by

$$D = (1+\epsilon)\prod u_{rr} \quad (|\epsilon| < (n-1)2^{-t_1}). \quad (41.2)$$

For any reasonable value of n the factor $(1+\epsilon)$ corresponds to a very low relative error. Apart from this D is the exact determinant of $(A+F)$. If the eigenvalues of $(A+F)$ are λ_i' and those of A are λ_i we have

$$\det(A) = \prod \lambda_i, \qquad D = (1+\epsilon)\prod \lambda_i'. \tag{41.3}$$

There results are of great importance in connexion with the practical eigenvalue problem.

Cholesky decomposition

42. When the matrix A is symmetric and positive definite the diagonal elements of L can be chosen in such a way that L is real and $U = L^T$. The proof is by induction. Let us assume that it is true for matrices of orders up to $(n-1)$. Now if A_n is a positive definite matrix of order n we may write

$$A_n = \left[\begin{array}{c|c} A_{n-1} & b \\ \hline b^T & a_{nn} \end{array}\right], \tag{42.1}$$

where A_{n-1} is the leading principal submatrix of order $(n-1)$ and is therefore positive definite (Chapter 1, § 27). Hence by hypothesis there exists a matrix L_{n-1} such that

$$L_{n-1}L_{n-1}^T = A_{n-1}. \tag{42.2}$$

Clearly L_{n-1} is non-singular, since $[\det(L_{n-1})]^2 = \det(A_{n-1})$; hence there exists a c such that $\qquad L_{n-1}c = b.$ $\tag{42.3}$

Now for any value of x we have

$$\left[\begin{array}{c|c} L_{n-1} & O \\ \hline c^T & x \end{array}\right]\left[\begin{array}{c|c} L_{n-1}^T & c \\ \hline O & x \end{array}\right] = \left[\begin{array}{c|c} A_{n-1} & b \\ \hline b^T & c^T c + x^2 \end{array}\right], \tag{42.4}$$

so that if we define x by the relation

$$c^T c + x^2 = a_{nn} \tag{42.5}$$

we have a triangular decomposition of A_n. To show that x is real, we take determinants of both sides of (42.4), giving

$$[\det(L_{n-1})]^2 x^2 = \det(A_n). \tag{42.6}$$

The right-hand side is positive because A_n is positive definite, and hence x is real and may be taken to be positive. The result is now established since it is obviously true when $n = 1$. Notice that

$$c_1^2 + c_2^2 + \ldots + c_{n-1}^2 + x^2 = a_{nn}. \tag{42.7}$$

Hence if $|a_{ij}| < 1$ then $|l_{ij}| < 1$ and further

$$l_{i1}^2 + l_{i2}^2 + \ldots + l_{ii}^2 = a_{ii} < 1, \tag{42.8}$$

showing that the symmetric decomposition may be carried out in fixed-point arithmetic. It is evident that no interchanges are required. This symmetric decomposition is due to Cholesky.

Symmetric matrices which are not positive definite

43. If A is symmetric but not positive definite there may be no triangular decomposition without interchanges. A simple example is provided by the matrix

$$\begin{bmatrix} 0 & 1 \\ 1 & 0 \end{bmatrix}. \tag{43.1}$$

Even if triangular decompositions exist, there will be none of the form LL^T with *real* non-singular L because LL^T is necessarily positive definite for such an L. However, if triangular decompositions do exist, there will be decompositions of the form LL^T if we are prepared to use complex arithmetic. The interesting thing is that each column of L is either entirely real or entirely imaginary, so that we may write

$$L_n = M_n D_n, \tag{43.2}$$

where M_n is real and D_n is a diagonal matrix having elements each of which is equal to 1 or i. The proof is again by induction and is almost identical with that of § 42. The value of x^2 given by the equivalent of (42.7) is again real, but can be negative.

We may avoid the use of imaginary quantities if we observe that

$$A_n = L_n L_n^T = M_n D_n D_n M_n^T. \tag{43.3}$$

Now the matrix D_n^2 is a diagonal matrix having elements which are all equal to ± 1, and if we write

$$A_n = (M_n D_n D_n)_n M = N_n M_n^T, \tag{43.4}$$

this gives us a triangular decomposition of A_n in which every column of N_n is equal either to the corresponding row of M_n^T or to this row with the sign of each element changed. Alternatively we may associate the signs in D_n with the N_n and the M_n in such a way that the diagonal elements of M_n are all positive. Both matrices are then specified by N_n only. If, for example,

$$N_n = \begin{bmatrix} 1 & 0 & 0 & 0 \\ 2 & -1 & 0 & 0 \\ 1 & 2 & -1 & 0 \\ 1 & 1 & 1 & 1 \end{bmatrix} \quad \text{then } M_n^T = \begin{bmatrix} 1 & 2 & 1 & 1 \\ 0 & 1 & -2 & -1 \\ 0 & 0 & 1 & -1 \\ 0 & 0 & 0 & 1 \end{bmatrix}. \tag{43.5}$$

However, it cannot be emphasized too strongly that *the symmetric decomposition of a matrix which is not positive definite enjoys none of the numerical stability of the positive definite case. To be sure of stability we must use interchanges and this destroys symmetry.* Although selecting the largest *diagonal* element at each stage (this corresponds to identical row and column interchanges) preserves symmetry, it does not *guarantee* stability.

Error analysis of Cholesky decomposition in fixed-point arithmetic

44. We now restrict ourselves to the positive definite case and we show that if

$$\text{(i)} \ |a_{ij}| < 1 - (1 \cdot 00001)2^{-t} \quad \text{and} \quad \text{(ii)} \ \lambda_n = 1/\|A^{-1}\| > \tfrac{1}{2}(n+2)2^{-t},$$
$$(44.1)$$

then the computed Cholesky decomposition using fixed-point arithmetic with accumulation of inner-products gives an L satisfying

$$LL^T = A + F, \qquad |l_{ij}| \leqslant 1, \tag{44.2}$$

$$|f_{rs}| \leqslant \begin{cases} \tfrac{1}{2}l_{rr}2^{-t} \leqslant \tfrac{1}{2}2^{-t} & (r > s) \\ \tfrac{1}{2}l_{ss}2^{-t} \leqslant \tfrac{1}{2}2^{-t} & (r < s) \\ (1 \cdot 00001)l_{rr}2^{-t} \leqslant (1 \cdot 00001)2^{-t} & (r = s) \end{cases}. \tag{44.3}$$

The proof is by induction. Suppose we have computed the elements of L as far as $l_{r,s-1}$ ($s-1 < r$) and that these elements correspond to those of the exact decomposition of $(A + F_{r,s-1})$ where $F_{r,s-1}$ is symmetric, has elements in the (i, j) positions given by

$$i, j \leqslant r-1; \quad i = r, \ j \leqslant s-1; \quad j = r, \ i \leqslant s-1$$

which satisfy relations (44.2) and (44.3) and is null elsewhere. Then we have

$$\lambda_{\min}(A + F_{r,s-1}) > \lambda_{\min}(A) + \lambda_{\min}(F_{r,s-1})$$
$$> \lambda_n - \tfrac{1}{2}(r+2)2^{-t}$$
$$> 0 \quad \text{from (ii)}, \tag{44.4}$$

and $|(A + F_{r,s-1})_{ij}| \leqslant 1$. Hence $(A + F_{r,s-1})$ is positive definite and has elements bounded in modulus by unity. It therefore has an exact triangular decomposition of which we already have the elements up to $l_{r,s-1}$. Now consider the expression

$$\frac{a_{rs} - l_{r1}l_{s1} - l_{r2}l_{s2} - \ldots - l_{r,s-1}l_{s,s-1}}{l_{ss}} = x \quad \text{(say)}, \tag{44.5}$$

where the l_{ij} are the elements which have already been computed. This expression is the (r, s) element of the triangular decomposition of $(A + F_{r,s-1})$ and hence it is bounded in modulus by unity. Now by *definition* $fl_2(x)$ is the correctly rounded value of x and hence it, too, cannot exceed unity in modulus. For the computed l_{rs} we then have

$$l_{rs} \equiv \left(\frac{a_{rs} - l_{r1}l_{s1} - l_{r2}l_{s2} - \ldots - l_{r,s-1}l_{s,s-1}}{l_{ss}} \right) + \xi_{rs} \quad (|\xi_{rs}| < \tfrac{1}{2}2^{-t}). \quad (44.6)$$

This implies

$$(LL^T)_{rs} \equiv a_{rs} + l_{ss}\xi_{rs} \equiv a_{rs} + f_{rs}, \quad (44.7)$$

where

$$|f_{rs}| < \tfrac{1}{2}l_{ss}2^{-t} < \tfrac{1}{2}2^{-t}, \quad (44.8)$$

and we have established the requisite inequalities up to the l_{rs} element. To complete the inductive proof we must consider the computation of the diagonal element l_{rr}. If we write

$$y^2 = a_{rr} - l_{r1}^2 - l_{r2}^2 - \ldots - l_{r,r-1}^2, \quad (44.9)$$

then y is the (r, r) element of the exact triangular decomposition of $(A + F_{r,r-1})$ and therefore lies in the range 0 to 1. Hence

$$l_{rr} = fl_2[(a_{rr} - l_{r1}^2 - l_{r2}^2 - \ldots - l_{r,r-1}^2)^{\tfrac{1}{2}}]$$

$$\equiv y + \epsilon \quad (|\epsilon| < (1{\cdot}00001)2^{-t-1} \quad \text{from Chapter 3, § 10), (44.10)}$$

where we assume that the square root routine always gives an answer which is not greater than unity. Hence

$$l_{rr}^2 \equiv y^2 + 2\epsilon y + \epsilon^2$$

or

$$l_{r1}^2 + l_{r2}^2 + \ldots + l_{rr}^2 \equiv a_{rr} + 2\epsilon y + \epsilon^2 < a_{rr} + 2\epsilon y + 2\epsilon^2 = a_{rr} + 2\epsilon l_{rr},$$

$$(44.11)$$

giving

$$(LL^T)_{rr} \equiv a_{rr} + f_{rr} \quad (|f_{rr}| < (1{\cdot}00001)l_{rr}2^{-t}). \quad (44.12)$$

The result is now fully established.

A similar result can be established if $fl_2(\ )$ is used. Ignoring elements of order 2^{-2t} the corresponding bounds are essentially

$$|f_{rs}| < \begin{cases} |l_{rs}l_{ss}|\, 2^{-t} & (r > s) \\ |l_{rs}l_{rr}|\, 2^{-t} & (r < s) \\ l_{rr}^2 2^{-t} & (r = s) \end{cases}. \quad (44.13)$$

An ill-conditioned matrix

45. If most of the l_{rs} are much smaller than unity, these bounds imply that LL^T differs from A by much less than $\frac{1}{2}2^{-t}$ in most of its elements and *it is quite possible for the errors made in the triangulariza-tion to be far less important than those made in the initial roundings of the elements of the matrix if these were necessary for t-digit repre-sentation.*

In Table 5 we exhibit the triangularization of a very ill-conditioned matrix, H, derived from a submatrix of order five of the Hilbert matrix by multiplying by the factor 1·8144 so as to avoid rounding errors in the original representation. The computation has been performed using $fl_2(\;)$ and 8 decimal digits. The difference between LL^T and H is given; note that most of its elements are far smaller than $\frac{1}{2}10^{-8}$ in magnitude. An interesting feature of this example is that the elements of $(LL^T - H)$ are smallest in those positions in which perturbations have the greatest effect on the solutions. The results may be compared with those obtained by Wilkinson (1961b) using fixed-point arithmetic with accumulation.

It will be seen that the elements of $(LL^T - H)$ are even smaller for the floating-point computation than they are for the fixed. (For discussion of the condition of the Hilbert matrix and related topics see Todd (1950, 1961).)

Triangularization using elementary Hermitian matrices

46. We now turn to the other methods of triangularization, the existence of which was implied in § 15. We describe first a method due to Householder (1958b) based on the use of the elementary Hermitians, and to facilitate comparison with our general analysis in Chapter 3, §§ 37–45, we concentrate on the $(r+1)$th transformation. There are $(n-1)$ major steps and on the completion of the rth step A_r is already upper triangular in its first r columns. We may write

$$A_r = \left[\begin{array}{c|c} U_r & V_r \\ \hline O & W_{n-r} \end{array}\right]^r, \tag{46.1}$$

where U_r is an upper triangular matrix of order r. In the $(r+1)$th step we use a matrix P_r of the form

$$P_r = \left[\begin{array}{c|c} I & O \\ \hline O & I - 2vv^T \end{array}\right]^r, \qquad \|v\|_2 = 1, \tag{46.2}$$

TABLE 5

Upper triangle of H (symmetric)

$$\begin{bmatrix}
0.90720 & 0.60480 & 0.45360 & 0.36288 & 0.30240 & & \\
 & 0.45360 & 0.36288 & 0.30240 & 0.25920 & & \\
 & & 0.36288 & 0.30240 & 0.25920 & 0.22680 & \\
 & & & 0.30240 & 0.25920 & 0.22680 & 0.20160 \\
 & & & & 0.22680 & 0.20160 & 0.18144
\end{bmatrix}$$

L^T

$$\begin{bmatrix}
10^0(0.95247047) & 10^0(0.63498032) & 10^0(0.47623524) & 10^0(0.38098819) & 10^0(0.31749016) \\
 & 10^0(0.22449943) & 10^0(0.26939933) & 10^0(0.26939934) & 10^0(0.25657079) \\
 & & 10^{-1}(0.54990883) & 10^{-1}(0.94270103) & 10^0(0.11783769) \\
 & & & 10^{-1}(0.13606703) & 10^{-1}(0.30237004) \\
 & & & & 10^{-2}(0.33808914)
\end{bmatrix}$$

$10^8(LL^T - H)$

$$\begin{bmatrix}
10^0(-0.37779791) & 10^0(0.38311504) & 10^0(0.28733628) & 10^{-1}(0.3937493) & 10^0(0.19155752) \\
 & 10^{-1}(-0.8576273) & 10^0(0.11178587) & 10^0(0.10747970) & 10^{-1}(-0.4966991) \\
 & & 10^{-2}(0.35426189) & 10^{-3}(0.4728749) & 10^{-2}(-0.9258937) \\
 & & & 10^{-2}(-0.1927482) & 10^{-3}(-0.291118) \\
 & & & & 10^{-4}(0.14040996)
\end{bmatrix}$$

$(L^T)^{-1}$

$$\begin{bmatrix}
10^1(0.10499013) & 10^1(-0.29695696) & 10^1(-0.54554509) & 10^1(-0.83992292) & 10^2(0.11736933) \\
 & 10^1(0.44543543) & 10^2(-0.21821801) & 10^2(0.62994193) & 10^3(-0.14084519) \\
 & & 10^2(0.18184833) & 10^2(-0.12598835) & 10^3(0.49296216) \\
 & & & 10^2(0.73493189) & 10^3(-0.65728637) \\
 & & & & 10^3(0.29577998)
\end{bmatrix}$$

in place of the N_{r+1} used in Gaussian elimination. We have

$$A_{r+1} = P_r A_r = \left[\begin{array}{c|c} U_r & V_r \\ \hline O & (I-2vv^T)W_{n-r} \end{array}\right], \tag{46.3}$$

so that only W_{n-r} is modified. We must choose v so that the first column of $(I-2vv^T)W_{n-r}$ is null apart from its first element. This is precisely the problem we analysed in Chapter 3, §§ 37–45. If we denote the elements of the first column of W_{n-r} by

$$a_{r+1,r+1}, a_{r+2,r+1}, \ldots, a_{n,r+1}, \tag{46.4}$$

omitting the upper suffix for convenience, then we may write

$$I - 2vv^T = I - uu^T/2K^2, \tag{46.5}$$

where $\left.\begin{array}{ll} u_1 = a_{r+1,r+1} \mp S, & u_i = a_{r+i,r+1} \quad (i = 2, 3, \ldots, n-r) \\ S^2 = \sum_{i=1}^{n-r} a_{r+i,r+1}^2, & 2K^2 = S^2 \mp a_{r+1,r+1}S = \mp u_1 S \end{array}\right\}. \tag{46.6}$

The new $(r+1, r+1)$ element is then $\pm S$. Numerical stability is achieved by choosing the signs so that

$$|u_1| = |a_{r+1,r+1}| + S. \tag{46.7}$$

The modified matrix $(I-2vv^T)W_{n-r}$ is computed from

$$(I-uu^T/2K^2)W_{n-r}$$

in the steps

$$p^T = u^T W_{n-r}/2K^2, \tag{46.8}$$

$$(I-2vv^T)W_{n-r} = W_{n-r} - up^T. \tag{46.9}$$

If the right-hand side or sides are known when the triangularization is performed, then in the $(r+1)$th major step we can multiply the current right-hand sides by P_r. However, if we wish to solve the equation with a right-hand side which is determined after the triangularization, then we must store all the relevant information. The most convenient technique is to store the $(n-r)$ elements of u in the $(n-r)$ positions occupied by the first column of W_{n-r}. Since, from (46.6), $u_i = a_{r+i,r+1} \ (i = 2, \ldots, n-r)$, all elements except u_1 are already in position. This means that we must store the diagonal elements of the upper triangular matrix separately; this is not particularly inappropriate since the diagonal elements are used in a different way from the others in the back-substitution. We also need the values taken by the $2K^2$ but these can be determined from the relations $2K^2 = \mp u_1 S$. If the latter policy is adopted, only n extra storage locations are needed in addition to those occupied by A_0. We shall refer to this method as *Householder triangularization*.

Error analysis of Householder triangularization

47. The general analysis we gave in Chapter 3, § 42–45, is more than adequate to cover Householder triangularization and in fact the relevant bound can be reduced somewhat. In § 45 of Chapter 3 we obtained a bound for $\|\bar{A}_{n-1} - P_{n-2}P_{n-3}...P_0 A_0\|_E$ where $\bar{A}_{n-1}$ is the $(n-1)$th *computed* transform and P_r is the exact elementary Hermitian corresponding to the computed $\bar{A}_r$. (We use bars here to facilitate the comparison with our former analysis.) For our present purpose a bound for $\bar{A}_{n-1} - \bar{P}_{n-2}\bar{P}_{n-3}...\bar{P}_0 A_0$ is of greater relevance. For if

$$\bar{A}_{n-1} - \bar{P}_{n-2}\bar{P}_{n-3}...\bar{P}_0 A_0 = F, \qquad (47.1)$$

then

$$\bar{A}_{n-1} = \bar{P}_{n-2}\bar{P}_{n-3}...\bar{P}_0[A_0 + (\bar{P}_{n-2}\bar{P}_{n-3}...\bar{P}_0)^{-1}F]. \qquad (47.2)$$

We see therefore that the closeness of $\bar{P}_r$ to a true elementary Hermitian is important only in so far as it guarantees that $\bar{P}_r^{-1}$ is almost orthogonal, and hence that $\|(\bar{P}_{n-2}\bar{P}_{n-3}...\bar{P}_0)^{-1}F\|_E$ is not appreciably larger than $\|F\|_E$. The term $(9\cdot01)2^{-t}\|A_0\|_E$ of (42.3) of Chapter 3 does not enter into our present analysis and we are left with the term $(3\cdot35)2^{-t}\|A_0\|_E$ of relation (44.4) of Chapter 3. The reader should verify that we have

$$\bar{A}_{n-1} = \bar{P}_{n-2}\bar{P}_{n-3}...\bar{P}_0(A_0 + G) \qquad (47.3)$$

where

$$\|G\|_E \leqslant (3\cdot35)(n-1)[1 + (9\cdot01)2^{-t}]^{n-2}2^{-t}\|A_0\|_E. \qquad (47.4)$$

Similarly

$$\bar{b}_{n-1} = \bar{P}_{n-2}\bar{P}_{n-3}...\bar{P}_0(b_0 + k) \qquad (47.5)$$

$$\|k\|_2 = (3\cdot35)(n-1)[1 + (9\cdot01)2^{-t}]^{n-1}2^{-t}\|b_0\|_2. \qquad (47.6)$$

Triangularization by elementary stabilized matrices of the type M_{ji}

48. We may replace the single elementary stabilized matrix N_r', used in the rth major step of Gaussian elimination with partial pivoting, by the product of the simpler stabilized matrices

$$M_{r+1,r}', \ M_{r+2,r}', ... , \ M_{nr}'.$$

In this case the rth major step is subdivided into $(n-r)$ minor steps in each of which one zero is introduced. The configuration for the case $n = 6$, when two minor steps of the third major step have been completed is

$$
\begin{bmatrix}
u_{11} & u_{12} & u_{13} & u_{14} & u_{15} & u_{16} & c_1 \\
n_{21}^{*} & u_{22} & u_{23} & u_{24} & u_{25} & u_{26} & c_2 \\
n_{31} & n_{32} & a_{33} & a_{34} & a_{35} & a_{36} & b_3 \\
n_{41} & n_{42}^{*} & n_{43} & a_{44} & a_{45} & a_{46} & b_4 \\
n_{51}^{*} & n_{52} & n_{53}^{*} & a_{54} & a_{55} & a_{56} & b_5 \\
n_{61} & n_{62}^{*} & a_{63} & a_{64} & a_{65} & a_{66} & b_6
\end{bmatrix}
. \qquad (48.1)
$$

We have denoted the elements in the first two rows by u_{ij} and c_i since they are final elements. The significance of the asterisks will become apparent when we define the rth step which is as follows.

For each value of i from $(r+1)$ to n:

(i) Compare a_{rr} and a_{ir}. If $|a_{ir}| > |a_{rr}|$ interchange the elements a_{rj} and a_{ij} $(j = r,\dots, n)$ and b_r and b_i.

(ii) Compute a_{ir}/a_{rr} and overwrite as n_{ir} on a_{ir}. If an interchange took place in step (i) then n_{ir} has an asterisk.

(iii) For each value of j from $(r+1)$ to n:
Compute $a_{ij} - n_{ir}a_{rj}$ and overwrite on a_{ij}.

(iv) Compute $b_i - n_{ir}b_r$ and overwrite on b_i.

In practice we may sacrifice the least significant digit of each n_{ir} and store a 'one' if an interchange took place immediately before n_{ir} was computed and a 'zero' otherwise. Clearly we have stored enough information for the separate processing of a right-hand side.

This algorithm is comparable in stability with Gaussian elimination with partial pivoting, although it is not possible to obtain such small upper bounds for the equivalent perturbations in A_0. It does not lend itself to accumulation of inner-products, and as we have just described it the method has almost nothing to recommend its use in the place of Gaussian elimination with partial pivoting.

Evaluation of determinants of leading principal minors

49. However, the steps of the algorithm of § 48 may be rearranged in a manner that has considerable advantages. The rearranged process also has $(n-1)$ major steps but only the first r rows of $(A_0 \,|\, b_0)$ are

involved in the first $(r-1)$ major steps. The configuration at the end
of the first $(r-1)$ major steps for the case $n = 6$, $r = 3$ is given by

$$\begin{bmatrix} a_{11} & a_{12} & a_{13} & a_{14} & a_{15} & a_{16} & b_1 \\ n_{21} & a_{22} & a_{23} & a_{24} & a_{25} & a_{26} & b_2 \\ n_{31} & n_{32} & a_{33} & a_{34} & a_{35} & a_{36} & b_3 \\ a_{41} & a_{42} & a_{43} & a_{44} & a_{45} & a_{46} & b_4 \end{bmatrix}. \tag{49.1}$$

The $(r+1)$th row is the unmodified $(r+1)$th row of A_0; rows 1 to r
have been processed in a way which will become apparent from the
description of the rth major step which is as follows.

For each value of i from 1 to r:

(i) Compare a_{ii} and $a_{r+1,i}$. If $|a_{r+1,i}| > |a_{ii}|$ interchange $a_{r+1,j}$ and
a_{ij} $(j = i,..., n)$ and b_{r+1} and b_i.

(ii) Compute $a_{r+1,i}/a_{ii}$ and overwrite on $a_{r+1,i}$ as $n_{r+1,i}$. If an interchange took place in step (i) then $n_{r+1,i}$ has an asterisk.

(iii) For each value of j from $(i+1)$ to n:
Compute $a_{r+1,j} - n_{r+1,i}a_{ij}$ and overwrite on $a_{r+1,j}$.

(iv) Compute $b_{r+1} - n_{r+1,i}b_i$ and overwrite on b_{r+1}.

This scheme has two advantages.

(I) If we do not wish to retain the n_{ij}, that is, if there are no right-
hand sides to be processed at a later stage, then the maximum storage
required to deal, for example, with one right-hand side, is

$$[(n+1)+n+(n-1)+...+3]+n+1$$

locations, since by the time the last equation is needed the first $(n-1)$
have been reduced almost to triangular form.

(II) More important from our point of view, we can now compute
the leading principal minor of A_0 of order $(r+1)$ during the rth major
step and we can stop at the rth major step if we so desire without
doing any computing on rows $(r+2)$ to n. The $(r+1)$th leading principal
minor p_{r+1} is determined as follows.

If we keep a running total k of the number of interchanges which
have taken place from the beginning of the first major step then at
then end of the rth major step we have

$$p_{r+1} = (-1)^k a_{11}a_{22}...a_{r+1,r+1} \tag{49.2}$$

where the a_{ii} refer to current values.

Often we shall need only the sign of p_{r+1}. This may be obtained in the rth major step as follows:

We assume that we have the sign of p_r initially. Then each time step (i) above is performed we change this sign if *both* an interchange is required *and* a_{ii} and $a_{r+1,i}$ have the same sign. On completion of the rth major step we change the sign again if $a_{r+1,r+1}$ is negative.

The minors can be obtained by the algorithm of § 48 but the method we have described is much more convenient. The leading principal minors cannot in general be obtained if we use Gaussian elimination with partial or complete pivoting or if we use Householder triangularization.

Triangularization by plane rotations

50. Finally we consider triangularization by plane rotations (Givens, 1959). This follows very closely the algorithms based on the use of the M_{ji}. Analogous to the algorithm of § 48 we have an algorithm in which the rth major step consists of a pre-multiplication by rotations in the planes $(r, r+1)$, $(r, r+2), \ldots, (r, n)$ respectively, the zeros in the column r being produced one by one. The rth major step is as follows. (Note the close analogy with § 48.)

For each value of i from $(r+1)$ to n:

(i) Compute $x = (a_{rr}^2 + a_{ir}^2)^{\frac{1}{2}}$.

(ii) Compute $\cos\theta = a_{rr}/x$, $\sin\theta = a_{ir}/x$. If $x = 0$ take $\cos\theta = 1$, $\sin\theta = 0$. Overwrite x on a_{rr}.

(iii) For each value of j from $(r+1)$ to n:
Compute $a_{rj}\cos\theta + a_{ij}\sin\theta$ and $-a_{rj}\sin\theta + a_{ij}\cos\theta$ and overwrite on a_{rj} and a_{ij} respectively.

(iv) Compute $b_r\cos\theta + b_i\sin\theta$ and $-b_r\sin\theta + b_i\cos\theta$ and overwrite on b_r and b_i respectively.

Note that we do not now have room to store the information concerning each rotation in the position of the element which is reduced to zero since both $\cos\theta$ and $\sin\theta$ must be accommodated. It is true that we need only store one of these since the other can then be derived, but this is inefficient, and if numerical stability is to be guaranteed we must store the numerically smaller and remember which of the two is stored!

51. The transformation may be rearranged in a way that is analogous to the rearrangement described in § 49. With this rearrangement rows $(r+1)$ to n are unmodified in the first $(r-1)$ major steps. The rth major step is then as follows.

For each value of i from 1 to r:

(i) Compute $x^2 = (a_{ii}^2 + a_{r+1,i}^2)$.

(ii) Compute $\cos \theta = a_{ii}/x$, $\sin \theta = a_{r+1,i}/x$. Overwrite x on a_{ii}.

(iii) For each value of j from $(i+1)$ to n:

Compute $a_{ij} \cos \theta + a_{r+1,j} \sin \theta$ and $-a_{ij} \sin \theta + a_{r+1,j} \cos \theta$ and overwrite on a_{ij} and $a_{r+1,j}$ respectively.

(iv) Compute $b_i \cos \theta + b_{r+1} \sin \theta$ and $-b_i \sin \theta + b_{r+1} \cos \theta$ and overwrite on b_i and b_{r+1} respectively.

Since each of the rotation matrices has the determinant $+1$, the determinant p_{r+1} of the leading principal minor of order $(r+1)$ is given on completion of the rth major step by

$$p_{r+1} = a_{11}a_{22}\ldots a_{r+1,r+1}. \tag{51.1}$$

We shall refer to triangularization by means of plane rotations as *Givens triangularization* whether it is organized as in § 50 or as we have just described.

Error analysis of Givens reduction

52. Again the error analysis we gave in Chapter 3, § 20–36, is more than adequate to cover Givens triangularization and, as with the Householder triangularization, the bounds can be reduced somewhat.

For standard floating-point computation we can show that if the final computed set of equations is $\bar{A}_N x = \bar{b}_N$ then

$$\bar{A}_N = \bar{R}_N \ldots \bar{R}_2 \bar{R}_1 (A_0 + G) \tag{52.1}$$

$$\bar{b}_N = \bar{R}_N \ldots \bar{R}_2 \bar{R}_1 (b_0 + g) \tag{52.2}$$

where

$$\|G\|_E \leqslant \alpha \|A\|_E, \qquad \|g\|_2 \leqslant \alpha \|b_0\|_2 \tag{52.3}$$

and

$$\alpha \leqslant 3n^{\frac{3}{2}} 2^{-t} \left[\frac{1 + 6 \cdot 2^{-t}}{1 - (4 \cdot 3) 2^{-t}} \right]^{2n-4}. \tag{52.4}$$

The last factor in α is of no importance for any value of n for which α is appreciably less than unity and the Euclidean norm of the equivalent perturbation is effectively bounded by $kn^{\frac{3}{2}} 2^{-t} \|A_0\|_E$.

For fixed-point computation the results of Chapter 3 may be improved somewhat by taking advantage of the fact that the zeros introduced at the successive stages persist throughout. Further, it is a simple deduction from the analysis of Chapter 3 that if every column is scaled initially so that its 2-norm is less than $1 - \frac{5}{2}n^2 2^{-t}$

(say), then no element can exceed capacity during the course of the reduction. Hence, even if we permit scaling only by powers of 2, we can ensure that

$$\tfrac{1}{2}n^{\frac{1}{2}}(1-\tfrac{5}{2}n^2 2^{-t}) \leqslant \|A_0\|_E \leqslant n^{\frac{1}{2}}(1-\tfrac{5}{2}n^2 2^{-t}) \qquad (52.5)$$

without any danger of exceeding capacity during the computation. In fact with this scaling, the 2-norm of each column will remain less than unity at every stage. The bounds for the equivalent perturbation in A_0 may be expressed in many ways but with this scaling we certainly have

$$\|G\|_E < \tfrac{3}{2}n^{\frac{3}{2}}2^{-t} < 3n^2 2^{-t}\|A_0\|_E/(1-\tfrac{5}{2}n^2 2^{-t}). \qquad (52.6)$$

Similarly if b_0 is scaled in the same way we have certainly

$$\|g\|_2 < 0\cdot 4n^2 2^{-t} < 0\cdot 8n^2 2^{-t}\|b_0\|_2. \qquad (52.7)$$

Uniqueness of orthogonal triangularization

53. If we denote by Q^T the product of the orthogonal transformation matrices derived by either the Givens or the Householder triangularization we have

$$Q^T A_0 = U, \qquad (53.1)$$

where Q^T is orthogonal and U is upper triangular. We may write this in the form

$$A_0 = QU. \qquad (53.2)$$

We now show that if A_0 is non-singular and is expressed in the form (53.2), then Q and U are essentially unique. For if

$$Q_1 U_1 = Q_2 U_2 \qquad (53.3)$$

then

$$Q_2^T Q_1 = U_2 U_1^{-1} = V, \qquad (53.4)$$

so that V is also upper triangular. The non-singularity of U_1 follows from that of A_0. Now $Q_2^T Q_1$ is orthogonal and hence

$$V^T V = I. \qquad (53.5)$$

Equating elements on both sides we find that V must be of the form

$$V = D = \operatorname{diag}(d_{ii}), \quad d_{ii} = \pm 1. \qquad (53.6)$$

Hence apart from the signs associated with the columns of Q and the rows of U, the decomposition (53.2) is unique. Similarly if A_0 is complex and non-singular we may show that the unitary triangularization is unique apart from a diagonal multiplier having elements of modulus unity.

The reader may possibly wonder whether the product of the rotation matrices which produce the zeros in the rth column in the Givens triangularization is equal to the elementary Hermitian which produces these zeros in the Householder reduction. That it is not, we can see immediately if we consider the rotations which introduce the zeros in the first column. These are rotations in the

$$(1, 2), (1, 3),\ldots , (1, n)$$

planes. Multiplying the corresponding matrices together we see that the product is of the form illustrated for $n = 6$ by

$$\begin{bmatrix} \times & \times & \times & \times & \times & \times \\ \times & \times & 0 & 0 & 0 & 0 \\ \times & \times & \times & 0 & 0 & 0 \\ \times & \times & \times & \times & 0 & 0 \\ \times & \times & \times & \times & \times & 0 \\ \times & \times & \times & \times & \times & \times \end{bmatrix}. \tag{53.7}$$

There is always a triangle of zero elements in the position shown. On the other hand, the first matrix in the Householder transformation has, in general, no zero elements in any of these positions and, in any case, is symmetric.

Schmidt orthogonalization

54. If we equate corresponding columns in equations (53.2) then we have

$$\left.\begin{aligned} a_1 &= u_{11}q_1 \\ a_2 &= u_{12}q_1 + u_{22}q_2 \\ &\cdots\cdots\cdots\cdots\cdots \\ a_r &= u_{1r}q_1 + u_{2r}q_2 + \ldots + u_{rr}q_r \end{aligned}\right\}, \tag{54.1}$$

where a_r and u_r denote the rth columns of A and U respectively. We may use these equations and the orthonormality condition to determine both the q_i and the u_{ij}. The process consists of n steps in the rth of which we determine $u_{1r}, u_{2r},\ldots , u_{rr}$ and q_r from the previously computed quantities. We have in fact

$$u_{ir} = q_i^T a_r \quad (i = 1,\ldots , r-1), \tag{54.2}$$

$$u_{rr} = \pm \|a_r - u_{1r}q_1 - \ldots - u_{r-1,r}q_{r-1}\|_2, \tag{54.3}$$

the second equation showing the degree of arbitrariness. In the complex case we can take an arbitrary complex multiplier of modulus unity in place of the multiplier ± 1. The determination of the orthonormal set of q_i from the a_i is usually known as the *Schmidt orthogonalization process*.

55. With exact computation the Schmidt process gives matrices Q and U which are identical (apart possibly from the signs) with those given by either the Givens or the Householder process. However, the matrix Q derived in practice using equations (54.1) to (54.3), and those which would be obtained by multiplying together the transposed transformation matrices of the Givens or Householder process may differ almost completely from each other. Incidentally the two product matrices may themselves differ completely. From the error analyses of Chapter 3 we know that the latter matrices must be very close to orthogonal matrices for any reasonable value of n, though we seldom need to form this product in practice. However, the vectors q_i obtained from applying (54.1) to (54.3) may depart from orthogonality to an almost arbitrary extent.

This may at first sight seem very surprising since it would appear that the Schmidt process is specifically designed to produce orthogonal vectors. However, consider the matrix of order 2 given by

$$A = \begin{bmatrix} 0{\cdot}63257 & 0{\cdot}56154 \\ 0{\cdot}31256 & 0{\cdot}27740 \end{bmatrix}. \tag{55.1}$$

The first step in the Schmidt process gives

$$u_{11} = 0{\cdot}70558, \qquad q_1^T = (0{\cdot}89652,\, 0{\cdot}44298), \tag{55.2}$$

and the second step gives $u_{12} = 0{\cdot}62631$. We have therefore

$$u_{22} q_2^T = a_2^T - u_{12} q_1^T = (0{\cdot}00004,\, -0{\cdot}00004). \tag{55.3}$$

In the computation of the right-hand side of (55.3), extensive cancellation takes place, and the resulting components have a high relative error. (This is true whether fixed-point or floating-point is used.) Consequently, when the vector on the right is normalized to give q_2 it is not even approximately orthogonal to q_1! This shows quite convincingly that this is not a phenomenon which is associated with high order matrices and it cannot be attributed to the 'cumulative effect of a large number of rounding errors'.

It is natural to ask whether the computed orthogonal matrix obtained from either the Givens or the Householder process is actually 'close' to that unique orthogonal matrix which corresponds to exact computation with all three methods. The answer is that, in general, we cannot guarantee that this will be so. Our error analysis of Chapter 3 guarantees that the matrices will be 'almost orthogonal' but not that they will be 'close' to the matrix corresponding to exact computation. (In Chapter 5, § 28, we give illustrations of this in connexion with orthogonal similarity transformations.) In some computations the orthogonality of Q is vital. For example, in the QR algorithm for symmetric matrices (Chapter 8), after deriving the QU decomposition of A, the matrix UQ is computed. Since we have $UQ = Q^{-1}(QU)Q = Q^{-1}AQ$, this derived matrix is a similarity transform of A, but it will be symmetric only if Q is orthogonal. The phenomenon we have just discussed in connexion with the Schmidt process will occupy us again when we consider Arnoldi's method for solving the eigenvalue problem (Chapter 6, § 32).

The number of multiplications involved in the Schmidt process is n^3, while the Householder and Givens triangularizations require $\frac{2}{3}n^3$ and $\frac{4}{3}n^3$ respectively. It is true that the Householder and Givens processes give the orthogonal matrix only in factorized form, but this is often more convenient in practice. The computation of the product of the Householder transformation requires $\frac{2}{3}n^3$ further multiplications, so that even if the orthogonal matrix is required explicitly, Householder's method is quite economical. Moreover if we wish to obtain a matrix of 'comparable orthogonality' using the Schmidt process, the vectors must be re-orthogonalized at each stage and this requires a further n^3 multiplications (see Chapter 6).

Comparison of the methods of triangularization

56. We now come for the first time to a comparison of methods based on stabilized elementary transformations and on the elementary orthogonal transformations.

There is one case for which the choice is quite definite and that is the triangularization of a positive definite symmetric matrix. Here the symmetric Cholesky decomposition has all the virtues. No interchanges are needed and the work may conveniently be performed in fixed-point. If accumulation of inner-products is possible the bound for the error, using a given precision of computation, is about as small as can reasonably be expected for any method. It takes full advantage of symmetry and requires only $n^3/6$ multiplications. Note that, ignoring rounding errors, we have

$$A = LL^T \quad \text{and} \quad A^{-1} = (L^T)^{-1}L^{-1}, \tag{56.1}$$

giving

$$\|A\|_2 = \|L\|_2^2 \quad \text{and} \quad \|A^{-1}\|_2 = \|L^{-1}\|_2^2, \tag{56.2}$$

and again this is a most satisfactory state of affairs in that the spectral condition number of L is equal to the square root of that of A. If A is a band matrix, i.e. if $a_{ij} = 0$ ($|i-j| > k$), then $l_{ij} = 0$ ($|i-j| > k$), so that full advantage is taken of this form. Perhaps the only adverse comment we might make is that $n-1$ square roots are required whereas none is needed in ordinary Gaussian elimination.

57. Turning now to the more general case the decisions are not so clear cut. If the matrix is perfectly general then the volume of computation in the four methods we have discussed is approximately as follows.

(I) Gaussian elimination with partial or complete pivoting: $\frac{1}{3}n^3$ multiplications.

(II) Triangularization with matrices of the form M'_{ji} : $\frac{1}{3}n^3$ multiplications.

(III) Householder triangularization: $\frac{2}{3}n^3$ multiplications, n square roots.

(IV) Givens triangularization: $\frac{4}{3}n^3$ multiplications, $\frac{1}{2}n^2$ square roots.

On grounds of speed then, methods I and II are superior to III and IV and they are, moreover, somewhat simpler. On many computers, I and II may be performed in double-precision in much the same time as IV can be performed in single-precision. Unfortunately, as far as numerical stability is concerned I and II are, from a strictly theoretical point of view, rather unsatisfactory. We have seen that if we use partial pivoting, the final pivot *can* be 2^{n-1} times as large as the maximum element in the original matrix and hence the only rigorous *a priori* upper bound we can obtain for the equivalent perturbation of the original matrix contains this factor 2^{n-1}. However, experience suggests that though such a bound is attainable it is quite irrelevant for practical purposes. In fact it is unusual for any growth to take place at all, and if inner-products can be accumulated then the equivalent perturbations in the original matrix are exceptionally small.

For Gaussian elimination with complete pivoting we have a far smaller bound for the maximum growth of the pivotal elements and strong reasons for believing that this bound cannot be approached. Unfortunately it would appear to be impossible to use the accumulation of inner-products to advantage with this method, and hence on a computer with this facility partial pivoting must be considered preferable in spite of the potential dangers. Fortunately for several types of matrix which are important in the eigenproblem, the bound for the pivotal growth with partial pivoting is much lower than for general matrices.

As far as orthogonal triangularization is concerned, Householder's method is superior to Givens' method in speed and also, in general, in accuracy, except possibly for standard floating-point computation. The numerical stability of both methods is guaranteed unconditionally in that we have most satisfactory *a priori* bounds for the equivalent perturbations of the original matrix. Apart from rounding errors, the 2-norm of each column remains constant, so we cannot possibly have any dangerous growth in size of elements during the reduction.

58. If one is of a cautious disposition and has a very fast computer one might feel inclined to use Householder's method for general purpose triangularization. In practice I have preferred to use Gaussian elimination with partial pivoting, risking the remote possibility of extensive pivotal growth; the temptation to do this has been the greater because the computers I have used have had facilities for accumulating inner-products; the results obtained in this way have probably been *more* accurate generally than results which would have been obtained by Householder's method. For Hessenberg matrices, even the most cautious computer might well prefer Gaussian elimination with partial pivoting; while for tri-diagonal matrices there is no longer any reason for preferring Householder's method.

If the object of triangularization is to evaluate the n leading principal minors then I and III are excluded. We are faced with a choice between Givens' method and the use of matrices of the form M'_{ji}. The former has a guaranteed stability but requires four times as many multiplications as the second. Again I have preferred the second in practice since the chances of instability are really rather slight. Note that one cannot take advantage of the facility to accumulate inner-products with either of these methods. The cautious might still prefer Givens' method except possibly for Hessenberg and tri-diagonal matrices, but it is worth mentioning that method II may be performed in double-precision in about the same time as the Givens' in single-precision and would then almost certainly give far higher accuracy.

From time to time throughout this book we shall be faced with a choice of the nature we have just described. It is probably true to say that orthogonal transformations compare rather less favourably with the other elementary transformations for the current problems than in any that arise later.

Back-substitution

59. The reader may have felt surprised that we compared the methods of triangularization before considering the errors made in the back-substitution. We did this for two reasons. The first is that if we are interested in determinant evaluation then there is, in any case, no back-substitution. The second is that the errors made in the back-substitution tend to be rather less important than might be expected.

We shall content ourselves with a fairly brief analysis since we have discussed this problem in considerable detail in Wilkinson (1963b), pp. 99–107. We consider the solution of a set of equations

$$Lx = b \qquad (59.1)$$

having a lower triangular matrix L of coefficients (not necessarily a unit lower triangular matrix). The solution of an upper triangular set is exactly analogous. Suppose $x_1, x_2, \ldots, x_{r-1}$ have been computed from the first $(r-1)$ of the equations (59.1) using floating-point with accumulation. Then x_r is given by

$$x_r = fl_2[(-l_{r1}x_1 - l_{r2}x_2 - \ldots - l_{r.r-1}x_{r-1} + b_r)/l_{rr}]$$
$$\equiv [-l_{r1}x_1(1+\epsilon_1) - l_{r2}x_2(1+\epsilon_2) - \ldots - l_{r.r-1}x_{r-1}(1+\epsilon_{r-1}) + b_r(1+\epsilon_r)] \times$$

where from Chapter 3, § 9, we have certainly $\times (1+\epsilon)/l_{rr}, \quad (59.2)$

$$|\epsilon_i| < \tfrac{3}{2}(r-i+2)2^{-2t_2}, \quad |\epsilon_r| < \tfrac{3}{2}2^{-2t}, \quad |\epsilon| \leqslant 2^{-t}. \qquad (59.3)$$

If we divide the numerator and denominator by $(1+\epsilon_r)$ we may write

$$x_r = [-l_{r1}x_1(1+\eta_1) - l_{r2}x_2(1+\eta_2) - \ldots - l_{r.r-1}x_{r-1}(1+\eta_{r-1}) + b_r]/l_{rr}(1+\eta)$$
$$(59.4)$$

where certainly

$$|\eta_i| < \tfrac{3}{2}(r-i+3)2^{-2t_2}, \quad |\eta| < 2^{-t}(1\cdot00001) \text{ (say)}. \qquad (59.5)$$

Multiplying equation (59.4) by $l_{rr}(1+\eta)$ and rearranging we have

$$l_{r1}(1+\eta_1)x_1+l_{r2}(1+\eta_2)x_2+\ldots+l_{r,r-1}(1+\eta_{r-1})x_{r-1}+l_{rr}(1+\eta)x_r \equiv b_r,$$

$$(59.6)$$

showing that the computed x_i satisfy exactly an equation with co-efficients $l_{ri}(1+\eta_i)$ $(i = 1,\ldots,r-1)$ and $l_{rr}(1+\eta)$. The relative perturbations are of order 2^{-2t_2} in all coefficients except l_{rr} which has a relative perturbation which may be of order 2^{-t}. The computed vector x therefore satisfies

$$(L+\delta L)x = b, \tag{59.7}$$

where δL is a function of b but in any case is uniformly bounded as shown in (59.8) for the case $n = 5$.

$$|\delta L| < 2^{-t}(1\cdot00001)\begin{bmatrix} |l_{11}| & & & & \\ & |l_{22}| & & & \\ & & |l_{33}| & & \\ & & & |l_{44}| & \\ & & & & |l_{55}| \end{bmatrix} +$$

$$+\tfrac{3}{2}2^{-2t_2}\begin{bmatrix} 0 & & & & \\ 4\,|l_{21}| & 0 & & & \\ 5\,|l_{31}| & 4\,|l_{32}| & 0 & & \\ 6\,|l_{41}| & 5\,|l_{42}| & 4\,|l_{43}| & 0 & \\ 7\,|l_{51}| & 6\,|l_{52}| & 5\,|l_{53}| & 4\,|l_{54}| & 0 \end{bmatrix}. \tag{59.8}$$

60. This is one of the most satisfactory error bounds obtained in matrix analysis. Just how remarkable it is will become apparent if we consider the residual vector. We have

$$b-Lx = \delta Lx \tag{60.1}$$

giving

$$\|b-Lx\| \leqslant \|\delta L\|\,\|x\|. \tag{60.2}$$

If the elements of L are bounded by unity then we have certainly

$$\|\delta L\| < 2^{-t}(1\cdot00001)+\tfrac{3}{4}2^{-2t_2}(n+3)^2 \tag{60.3}$$

for the 1, 2 or ∞ norms. If $n^2 2^{-t} \ll 1$ the second term is negligible.

Consider now the vector $\bar{x}$ obtained by rounding the exact solution x_e to t figures in floating-point arithmetic. We may write

$$\bar{x} = x_e + d. \tag{60.4}$$

Clearly

$$\|d\|_\infty \leqslant 2^{-t} \|x_e\|_\infty \tag{60.5}$$

and hence

$$b - L\bar{x} = b - L(x_e + d) = -Ld, \tag{60.6}$$

giving

$$\|b - L\bar{x}\|_\infty = \|Ld\|_\infty \leqslant \|L\|_\infty \|d\|_\infty \leqslant n2^{-t} \|x\|_\infty. \tag{60.7}$$

It would not be difficult to devise examples for which this bound is achieved. *This means that we can expect the residuals corresponding to the computed solution of a triangular set of equations to be smaller than those corresponding to the correctly rounded solution.*

High accuracy of computed solutions of triangular sets of equations

61. In the last paragraph we gave a bound for the residual of the computed solution. Remarkable though this is, it still gives an underestimate of the accuracy of the solution itself when L is very ill-conditioned. We have shown that the computed x satisfies (59.7) and have given the bound (59.8) for δL. All we can deduce from this is that

$$x = (L + \delta L)^{-1} b = (I + L^{-1} \delta L)^{-1} L^{-1} b \tag{61.1}$$

and hence that

$$\|x - L^{-1}b\| / \|L^{-1}b\| \leqslant \|L^{-1} \delta L\| / (1 - \|L^{-1} \delta L\|), \tag{61.2}$$

provided $\|L^{-1} \delta L\| < 1$. Hitherto when we have had an estimate of this kind we have replaced $\|L^{-1} \delta L\|$ by $\|L^{-1}\| \|\delta L\|$ and this is usually by no means a severe overestimate. However for our particular computation this is not so. In fact ignoring the terms in 2^{-2t_2} we have

$$|L^{-1} \delta L| \leqslant |L^{-1}| |\delta L| \leqslant 2^{-t} [|L^{-1}| \operatorname{diag}(|l_{ii}|)]. \tag{61.3}$$

Now the diagonal elements in L^{-1} are $1/l_{ii}$ and hence the matrix in square brackets on the right of (61.3) is *unit* lower triangular. Suppose the elements of L^{-1} satisfy the relations

$$|(L^{-1})_{ij}| \leqslant K |(L^{-1})_{jj}| \quad (i > j) \tag{61.4}$$

for some K. Then (61.3) gives

$$|L^{-1} \delta L| \leqslant 2^{-t} KT, \tag{61.5}$$

where T is a lower triangular matrix every element of which is unity.

Hence
$$\|L^{-1}\,\delta L\|_\infty^{\mathrm{F}} < 2^{-t}nK. \tag{61.6}$$

Similarly if we solve the upper triangular set of equations

$$Ux = b \tag{61.7}$$

the computed solution satisfies

$$(U+\delta U)x = b, \tag{61.8}$$

where, ignoring terms of order 2^{-2t_2}, δU is a diagonal matrix such that
$$|(\delta U)_{ii}| < 2^{-t}(1\!\cdot\!00001)|\,u_{ii}|. \tag{61.9}$$

Again if
$$|(U^{-1})_{ij}| \leqslant K\,|(U^{-1})_{jj}| \quad (i < j), \tag{61.10}$$

then we have
$$\|U^{-1}\,\delta U\|_\infty \leqslant 2^{-t}nK. \tag{61.11}$$

For many very ill-conditioned triangular matrices K is of the order of unity. An example of such a matrix is L^T of § 45. For this upper triangular matrix we may certainly take $K = 2\!\cdot\!5$, so the computed inverse of L^T can be expected to be close to its true inverse in spite of the fact that L^T is quite ill-conditioned. In fact the maximum error in any component is less than one in the least significant figure of the largest element. It is interesting to note that for the corresponding lower triangular matrix L the value of K is considerably larger.

62. In order to emphasize that this rather unexpectedly high accuracy is not restricted to matrices having the property (61.4), we now consider matrices having almost the opposite property. Suppose L is lower triangular and has positive diagonal elements and negative off-diagonal elements. We show that if b is a right-hand side having non-negative elements, then *the computed solution has a low relative error however ill-conditioned L may be.*

Let us denote the exact solution by y and the computed solution by x and define E by the relations

$$x_r = y_r(1+E_r). \tag{62.1}$$

We shall show that

$$(1-2^{-t})^r\,(1-\tfrac{3}{2}2^{-2t})^{r(r+1)/2} \leqslant 1+E_r < (1+2^{-t})^r\,(1+\tfrac{3}{2}2^{-2t})^{r(r+1)/2},$$

$$\tag{62.2}$$

it being convenient to retain this form for the error bounds for the time being. The proof is by induction. Suppose it is true up to x_{r-1}.

Now x_r is given by (59.2) where from Chapter 3, § 8, we may replace the bounds (59.3) by the stricter bounds

$$
\begin{aligned}
(1-\tfrac{3}{2}2^{-2t})^r &\leqslant 1+\epsilon_1 \leqslant (1+\tfrac{3}{2}2^{-2t})^r \\
(1-\tfrac{3}{2}2^{-2t})^{r-i+2} &\leqslant 1+\epsilon_i \leqslant (1+\tfrac{3}{2}2^{-2t})^{r-i+2} \\
1-2^{-t} &\leqslant 1+\epsilon \leqslant 1+2^{-t}
\end{aligned}
\Bigg\}. \tag{62.3}
$$

Remembering that the l_{ri} $(i < r)$ are negative and that the x_i and y_i are obviously non-negative, the expression for x_r is a weighted mean of non-negative terms, and from our inductive hypothesis we see that E_r satisfies the bounds (62.2).

It is easy to prescribe matrices of this kind having arbitrarily high condition numbers. Notice that to invert L we take the n right-hand sides e_1 to e_n and each of these consists entirely of non-negative elements. Hence the elements of the computed inverse all have low relative errors. The reader may well feel that the arguments we have just given are rather tendentious. While admitting that this is true, it is worth stressing that computed solutions of triangular matrices are, in general, much more accurate than can be deduced from the bound obtained from the residual vector, and this high accuracy is often important in practice.

We have analysed only the case when floating-point accumulation is used. For other forms of computation the upper bounds obtained are somewhat poorer but the same broad features persist. Detailed analyses have been given by Wilkinson (1961b) and in Wilkinson (1963b), pp. 99–107.

Solution of a general set of equations

63. We return now to the solution of the general set of equations

$$
Ax = b. \tag{63.1}
$$

Suppose A has been triangularized by some method and we have matrices L and U such that

$$
LU = A + F, \tag{63.2}
$$

where the bound for F depends on the details of the triangularization. If some form of pivoting has been used, then A denotes the original matrix with its rows suitably permuted. To complete the solution we must solve the two sets of triangular equations

$$
Ly = b \quad \text{and} \quad Ux = y. \tag{63.3}
$$

If we use x and y to denote the computed solutions, then for floating-point arithmetic with accumulation we have

$$(L+\delta L)y = b \quad \text{and} \quad (U+\delta U)x = y, \tag{63.4}$$

with bounds for δL as given in (59.8) and corresponding bounds for δU. Hence the computed x satisfies

$$(L+\delta L)(U+\delta U)x = b, \tag{63.5}$$

i.e.
$$(A+G)x = b, \tag{63.6}$$

where
$$G = F+\delta LU+L\,\delta U+\delta L\,\delta U. \tag{63.7}$$

If we assume that $|a_{ij}| \leqslant 1$, that partial pivoting has been used and that all elements of U have remained less than unity in modulus, then for computation in floating-point with accumulation we have for example,

$$\|G\|_\infty \leqslant n2^{-t}+(1\!\cdot\!00001)2^{-t}n+(1\!\cdot\!00001)2^{-t}n+O(n^3 2^{-2t}). \tag{63.8}$$

Hence provided $n^2 2^{-t}$ is appreciably smaller than unity, the bound for $\|G\|_\infty$ is approximately $3.2^{-t}n$. Notice that one third of this bound comes from the error in the triangularization and the other two thirds from the solution of the triangular sets of equations.

From (63.6) we have

$$r = b-Ax = Gx, \quad \|r\|_\infty \leqslant \|G\|_\infty \|x\|_\infty < (3\!\cdot\!1)2^{-t}n\,\|x\|_\infty, \tag{63.9}$$

the last bound certainly holding when $n^2 2^{-t} < 0\!\cdot\!01$ (say). We therefore have a bound for the residual which is dependent only on *the size of the computed solution and not on its accuracy*. If A is very ill-conditioned, then usually the computed solution will be very inaccurate, but if nevertheless $\|x\|_\infty$ is of order unity, the residual will be of order $n2^{-t}$ at most. This means that the computed x will be the exact solution of $Ax = b+\delta b$ for some δb having a norm of order not greater than $n2^{-t}$. Again, as in § 60, it is instructive to compare the residual with that corresponding to the correctly rounded solution. The upper bound for the norm of this residual is $n2^{-t}\|x\|_\infty$ so that the bound of (63.9) corresponding to the computed solution is only three times as large.

Computation of the inverse of a general matrix

64. By taking b to be $e_1, e_2,\ldots, e_n$ in succession we obtain the n computed columns of the inverse of A. For the rth column x_r we have

$$(L+\delta L_r)(U+\delta U_r)x_r = e_r, \tag{64.1}$$

where we have written δL_r and δU_r to emphasize that the equivalent perturbations are different for each right-hand side, though they are uniformly bounded as shown typically in (59.8). Writing

$$AX - I = K \tag{64.2}$$

and using the uniform bounds for δL_r and δU_r we find that certainly

$$\|K\|_\infty < (3 \cdot 1)n2^{-t}\|X\|_\infty \quad \text{if} \quad n^2 2^{-t} < 0 \cdot 01. \tag{64.3}$$

If $\|K\|_\infty$ is less than unity, (64.2) implies that neither A nor X is singular. In this case we have

$$X - A^{-1} = A^{-1}K, \tag{64.4}$$

giving

$$\|X - A^{-1}\|_\infty < (3 \cdot 1)n2^{-t}\|A^{-1}\|_\infty\|X\|_\infty \tag{64.5}$$

and

$$\|X\|_\infty < \|A^{-1}\|_\infty / [1 - (3 \cdot 1)n2^{-t}\|A^{-1}\|_\infty]. \tag{64.6}$$

From (64.5) we then have

$$\|X - A^{-1}\|_\infty / \|A^{-1}\|_\infty < (3 \cdot 1)n2^{-t}\|A^{-1}\|_\infty / [1 - (3 \cdot 1)n2^{-t}\|A^{-1}\|_\infty]. \tag{64.7}$$

From these results we deduce the following: If we compute the inverse of A in floating-point arithmetic with accumulation using triangular decomposition with interchanges, and if no element of U exceeds unity, then provided the computed X is such that $(3 \cdot 1)n2^{-t}\|X\|_\infty$ is less than unity, the relations (64.6) and (64.7) are satisfied. If, in particular, this quantity is small, then (64.5) shows that the computed X has a low relative error. Conversely any normalized matrix for which $(3 \cdot 1)n2^{-t}\|A^{-1}\|_\infty < 1$ must give a computed X satisfying (64.7) provided pivotal growth does not occur.

It is unfortunate that the remark concerning pivotal growth must be inserted, but since such growth is very uncommon this condition is not very restrictive in practice.

For other less accurate modes of computation we have corresponding bounds which are, of course, somewhat weaker, but because of the statistical effect the upper bounds we have just given are not usually exceeded in practice.

Accuracy of computed solutions

65. We notice that for matrix inversion we have obtained much more satisfactory results than were obtained for the solution of a set of linear equations. This is because a small residual matrix $(AX - I)$ necessarily implies that X is a good inverse, whereas a small residual vector $(Ax-b)$ does not necessarily imply that x is a good solution. In practice when we have computed an alleged inverse X by *any technique*, we may compute $(AX-I)$ and therefore obtain a reliable bound for the error in X.

It is natural to ask whether there is any analogous method for estimating the accuracy of a computed solution of $Ax = b$. If we write

$$b - Ax = r \tag{65.1}$$

then the error in x is $A^{-1}r$, and hence if we have a bound for $\|A^{-1}\|$ in some norm we can obtain a bound for the error. However this bound can be very disappointing even when our bound for $\|A^{-1}\|$ is very precise. Suppose, for example, that A is a normalized matrix such that $\|A^{-1}\|_\infty = 10^8$ and that b is a normalized right-hand side such that $\|A^{-1}b\|_\infty = 1$. If x is the vector obtained by rounding $A^{-1}b$ to ten significant decimal places, then we have

$$\|r\|_\infty = \|b - Ax\|_\infty < \tfrac{1}{2}n10^{-10} \tag{65.2}$$

and this bound might well be quite realistic. Hence if we were given such an x and a correct bound 10^8 for $\|A^{-1}\|_\infty$, we would be able to demonstrate only that the norm of the error in x is bounded by $\tfrac{1}{2}n10^{-2}$, a most disappointing result. In order to be able to recognize the accuracy of such a solution we require quite an accurate approximate inverse X of A and a reasonably sharp bound for $\|X-A^{-1}\|$.

Ill-conditioned matrices which give no small pivots

66. It might be thought that if a normalized matrix A were very ill-conditioned, then this must reveal itself in triangular decomposition with pivoting by the emergence of a very small pivot. One might argue that if A were actually singular and if the decomposition were carried out exactly, then one of the pivots would necessarily be zero and hence if A is ill-conditioned one of the pivots should be small.

Unfortunately this is by no means true even when A is symmetric. (This should be a favourable case since all the s_i are then unity.) For consider a matrix having for eigenvalues

$$1 \cdot 00, \ 0 \cdot 95, \ 0 \cdot 90, \dots, \ 0 \cdot 05 \quad \text{and} \quad 10^{-10}.$$

Such a matrix is almost singular in the sense that $(A - 10^{-10}I)$ is exactly singular. Now the product p of the pivots is equal to the product of the eigenvalues and is therefore given by

$$p = 10^{-10}(0 \cdot 05)^{20}20!.$$

If all the pivotal values were equal, then each of them would be approximately 0·1 and we certainly cannot regard this as pathologically small. It might be felt that this is an artificial argument, but we shall see in Chapter 5, § 56, that such examples can actually arise. This phenomenon can occur not only in Gaussian elimination with partial pivoting but also in the Givens and Householder triangularizations.

There are a number of symptoms which might be exhibited during the solution by Gaussian elimination with partial pivoting of the set of equations $Ax = b$ (with normalized A and b) which would immediately indicate their ill-conditioned nature. They are the emergence of

(i) a 'small' pivotal element,
(ii) a 'large' computed solution,
(iii) a 'large' residual vector.

Unfortunately one can invent pathologically ill-conditioned sets of equations which display none of these symptoms. Working with t-digit arithmetic we might well obtain a solution having elements of order unity, and therefore necessarily having residuals of order 2^{-t}, which nevertheless was incorrect even in its most significant figures.

Iterative improvements of approximate solution

67. The considerations of the previous sections show that a reliable bound for the error in a computed solution, which is not pessimistic for accurate solutions of ill-conditioned equations, is by no means a simple requirement. We now show how to improve an approximate solution and at the same time to gain reliable information on the accuracy of the original and the improved solution.

Suppose we have obtained approximate triangular factors L and U of A such that
$$LU = A + F. \tag{67.1}$$

We may use these triangular factors to obtain a sequence of approximate solutions $x^{(r)}$ to the true solution x of $Ax = b$, by the iterative process defined by the relations

$$\left. \begin{array}{lll} x^{(0)} = 0, & r^{(0)} = b \\ LU\, d^{(k-1)} = r^{(k-1)}, & x^{(k)} = x^{(k-1)} + d^{(k-1)}, & r^{(k)} = b - Ax^{(k)} \end{array} \right\}. \tag{67.2}$$

If this process could be performed without rounding errors then we would have
$$x^{(k)} = x^{(k-1)} + (A+F)^{-1}(b - Ax^{(k-1)})$$
$$= x^{(k-1)} + (A+F)^{-1}A(x - x^{(k-1)}), \tag{67.3}$$

giving
$$x - x^{(k)} = [I - (A + F)^{-1}A](x - x^{(k-1)})$$
$$= [I - (A + F)^{-1}A]^k x. \tag{67.4}$$

A sufficient condition for the convergence of the exact iterative process is therefore
$$\| I - (A + F)^{-1}A \| < 1, \tag{67.5}$$

which is certainly satisfied if
$$\| A^{-1} \| \, \| F \| < \tfrac{1}{2}. \tag{67.6}$$

Similarly we have
$$r^{(k)} = A[I - (A + F)^{-1}A]^k x, \tag{67.7}$$

and again this certainly tends to zero if (67.6) is satisfied.

If A has been triangularized by Gaussian elimination with partial pivoting using floating-point with accumulation we have in general
$$\| F \|_\infty \leqslant n2^{-t}, \tag{67.8}$$

and hence the exact iterative process will converge if
$$\| A^{-1} \|_\infty < 2^{t-1}/n. \tag{67.9}$$

If $n2^{-t} \| A^{-1} \|_\infty < 2^{-p}$, then $x^{(k)}$ will gain at least p binary figures per iteration and $r^{(k)}$ will diminish by a factor at least as small as 2^{-p}. If $n2^{-t} \| A^{-1} \|_\infty$ is appreciably greater than $\tfrac{1}{2}$ we cannot expect the process, in general, to converge.

Effect of rounding errors on the iterative process

68. In fact we are not able to perform the iterative process exactly, and at first sight it looks as though the effect of rounding errors could be disastrous. If the iteration is performed using t-digit arithmetic of any kind, the accuracy of $x^{(k)}$ cannot increase indefinitely with k since we are using only t digits to represent the components.

This, of course, is to be expected, but when we turn to the residuals we find a rather disturbing situation. We showed in § 63 that if, for example, we use floating-point arithmetic with accumulation, then the residual corresponding to the first solution is *almost certain to be of the same order of magnitude as that corresponding to the correctly rounded solution*. We can therefore scarcely expect the residual to diminish by the factor 2^{-p} with each iteration and it appears *a priori* improbable that the accuracy of the successive solutions could steadily improve if the residuals remain roughly constant. Nevertheless this is precisely what does happen, and the performance of the practical process using any mode of arithmetic does not fall far short of that corresponding to exact iteration provided only that inner-products are accumulated in the computation of the residuals.

In Wilkinson (1963b), pp. 121–126, I gave a detailed error analysis of the iterative process in fixed-point arithmetic using accumulation at all relevant stages and computing each $x^{(k)}$ and $d^{(k)}$ as block-floating vectors. We shall not repeat the analysis here, but since the technique is of great importance in later chapters we describe it in detail. The relevant changes if floating-point accumulation is used are obvious, and indeed the process is somewhat simpler in this mode.

The iterative procedure in fixed-point computation

69. We assume that A is given with single-precision fixed-point elements and L and U have been computed by triangular decomposition with interchanges, using accumulation of inner-products. We assume further that no element of U is greater than unity in modulus. The right-hand side b also has fixed-point elements, but these may be either single-precision or double-precision numbers.

We express the kth computed solution $x^{(k)}$ in the form $2^{p_k}y^{(k)}$ where $y^{(k)}$ is a normalized block-floating vector. In general, p_k will be constant for all iterations. (At least this is true if A is not too ill-conditioned for the process to work; however, if there is a very close round-off in the largest component, p_k may vary by ± 1 from one iteration to the next.) The general step of the iteration is then as follows.

(i) Compute the exact residual $r^{(k)}$ defined by

$$r^{(k)} = b - 2^{p_k}Ay^{(k)},$$

obtaining this vector in the first instance as a non-normalized double-precision block-floating vector. Clearly we may use a double-precision vector b at this stage without requiring any double-precision multiplication.

(ii) Normalize the residual vector $r^{(k)}$ and then round it to give the single-precision block-floating vector which we may denote by $2^{q_k}s^{(k)}$.

(iii) Solve the equations $LUx = s^{(k)}$, obtaining the solution in the form of a block-floating single-precision vector. Adjust the scale factor to give the computed solution $d^{(k)}$ of $LUx = 2^{q_k}s^{(k)}$.

(iv) Compute the next approximation from the relation

$$x^{(k+1)} = \text{b.f.}(x^{(k)} + d^{(k)}),$$

obtaining $x^{(k+1)}$ as a single-precision block-floating vector.

Simple example of iterative procedure

70. In Table 6 we give an example of the application of this method to a matrix of order 3, the computation being performed using six-digit fixed-point decimal arithmetic with accumulation of inner-products. The matrix has a condition number between 10^4 and 10^5.

TABLE 6

Iterative solution of set of equations with 3 different right-hand sides

A			b_1	b_2	b_3
0·876543	0·617341	0·589973	0·863257	0·8632572136	0·4135762133
0·612314	0·784461	0·827742	0·820647	0·8206474986	0·3712631241
0·317321	0·446779	0·476349	0·450098	0·4500984014	0·5163213142

L			U		
1·000000			0·876543	0·617341	0·589973
0·698556	1·000000			0·353214	0·415613
0·362014	0·632175	1·000000			0·000030

Iterations corresponding to first right-hand side

$x^{(1)}$	$x^{(2)}$	$x^{(3)}$	$x^{(4)}$	$x^{(5)}$
0·639129	0·636280	0·636330	0·636329	0·636329
−0·050678	−0·029140	−0·029513	−0·029507	−0·029507
0·566667	0·548363	0·548680	0·548674	0·548674

$10^6\, r^{(1)}$	$10^6\, r^{(2)}$	$10^6\, r^{(3)}$	$10^6\, r^{(4)}$
0·326160	0·172501	−0·407897	0·304438
0·204138	−0·044726	−0·450687	0·421313
−0·446030	−0·032507	−0·252623	0·242118

Iterations corresponding to second right-hand side

$x^{(1)}$	$x^{(2)}$	$x^{(3)}$	$x^{(4)}$	$x^{(5)}$
0·639129	0·636815	0·636855	0·636855	0·636855
−0·050678	−0·033187	−0·033490	−0·033484	−0·033485
0·566667	0·551803	0·552060	0·552056	0·552056

$10^6\, r^{(1)}$	$10^6\, r^{(2)}$	$10^6\, r^{(3)}$	$10^6\, r^{(4)}$
0·539760	0·307503	0·677045	−0·667109
0·702738	0·147071	0·616500	−0·779298
−0·044630	0·076211	0·335715	−0·439563

Iterations corresponding to third right-hand side

$x^{(1)}$	$x^{(2)}$	$x^{(3)}$	$x^{(4)}$	$x^{(5)}$
1632·2	1603·9	1604·4	1604·4	1604·4
−12336·6	−12123·1	−12126·8	−12126·8	−12126·8
10484·6	10303·2	10306·4	10306·3	10306·3

$10^1\, r^{(1)}$	$10^1\, r^{(2)}$	$10^1\, r^{(3)}$	$10^1\, r^{(4)}$
−0·218436	0·031220	−0·389014	0·200959
−0·098483	−0·113868	−0·638125	0·189617
−0·099289	−0·073525	−0·372475	0·103874

The equations have been solved with three different right-hand sides which illustrate a number of features of special interest.

The first right-hand side has been chosen so that, in spite of the ill-condition of A, the true solution is of order unity. Since the matrix is not too ill-conditioned for the process to succeed, the first solution is also of order unity. Accordingly, the first residual vector has components of order 10^{-6} in spite of the fact that the solution has errors in the second decimal place. The inaccuracy of the first solution is immediately revealed when the second is computed. The changes in the first components of the solution in successive iterations are $-0\cdot002849$, $0\cdot000050$, $-0\cdot000001$, $0\cdot000000$ and the other components behave similarly. The large relative change made by the first correction demonstrates conclusively that the matrix A is ill-conditioned and, in fact, shows that the condition number is probably of the order of 10^4.

In this example this is obvious because the last element of U is $0\cdot000030$, showing that the $(3,3)$ element of the inverse is of the order of 30000. It is not possible to construct a matrix of order 3 which does not reveal its ill-condition in this way, but for high order matrices the ill-conditioning need not be at all obvious. The behaviour of the iterates however, still gives a reliable indication in such cases. We note that all the residual vectors are of much the same size and indeed the final residual corresponding to the correctly rounded solution is one of the largest.

The second right-hand side disagrees with the first only in figures beyond the sixth and the first solution is therefore identical with that for the first right-hand side. However, we may use all the figures of b_2 in forming the residual vector and hence the extra figures make themselves felt after the first iteration. Although the difference between the two right-hand sides is of order 10^{-6}, the two solutions may differ by a vector of order 10^{-2} since $\|A^{-1}\|$ is of order 10^4. This indeed proves to be true and the use of the full right-hand side is therefore of great importance.

The third right-hand side gives a solution having components of the order of 10^4. The emergence of such a solution immediately reveals the ill-condition of the matrix and we are in no danger of being misled. Since the solutions are obtained as block-floating vectors we have only one figure after the decimal place, and even the final correctly rounded solutions may have errors of up to $\pm0\cdot05$. The residuals are now of order $0\cdot01$ at all stages in spite of the fact that

components of the first solution have errors which are greater than
100. *Random* errors of this order of magnitude would give residuals
of order 100.

General comments on the iterative procedure

71. The behaviour exhibited in our example is typical of that for
a matrix with a condition number of order 10^4 and the reader is
advised to study it very carefully. For a matrix of order 3 one would
scarcely think of speaking of 'the cumulative effect' of rounding
errors *and it is important to appreciate that this accumulation plays a
comparatively minor role in general.* We illustrate this by giving rough
bounds for the various errors which arise when we solve a normalized
set of equations with a normalized right-hand side by this technique.
In these bounds the statistical effect has been included. We assume
that the condition number is of order 2^a, the maximum component
of the true solution is of order 2^b and that $n2^a2^{-t}$ is appreciably
smaller than $\frac{1}{2}$.

(a) The maximum component of the first and all subsequent solutions is of order 2^b.

(b) The general level of components of the residual corresponding
to the first solution is of order $(2n)^{\frac{1}{2}}2^{b-t}$.

(c) The general level of errors in components of first solution is of
order $n^{\frac{1}{2}}2^{b-t}2^a$.

(d) The general level of components of the residual corresponding
to the second solution (and all subsequent solutions!) is of order $n^{\frac{1}{2}}2^{b-t}$.

(e) The general level of errors in second solution is of order $n2^{b-2t}2^{2a}$.

Note that the bounds for the residuals contain the factor 2^b, and
hence at all stages are proportional to the size of the computed
solution and are independent of the condition number, 2^a. However,
we know that 2^b cannot be greater than 2^a, so that the condition of
the matrix controls the maximum size attainable by the solution and
hence by the residuals.

Although we have spoken of the process as 'iterative', we expect
to use very few iterations in general. Suppose, for example, we have a
matrix for which $n = 2^{12}$ and $a = 2^{20}$. This corresponds to quite
serious ill-conditioning. Nevertheless, on ACE we would expect to
gain $48 - (\frac{1}{2}.12) - 20$ binary digits per stage, so that after the second
correction the answer should be correct to full working accuracy. If
only a few right-hand sides are involved, then the iterative process

is far more economical than carrying out the whole solution in double-precision arithmetic, besides giving us a virtual guarantee that the final vector is a correctly rounded solution. However if $n2^{-t} \|A^{-1}\| > \frac{1}{2}$, then the iterative process will not in general converge at all and none of the computed solutions will have any correct figures. The precision obtained working in double-precision will have some correct figures provided $n2^{-2t} \|A^{-1}\| < 1$.

Related iterative procedures

72. The procedure we have just described will work using other less accurate modes of arithmetic provided inner-products are accumulated when computing the residuals. If this is not practical, then the residuals must be computed in true double-precision arithmetic. Accumulation of inner-products is not essential at any other stage of the process. In general, if the norm of the equivalent perturbation in A corresponding to the errors made in the solution is bounded by $n^k 2^{-t}$, then the process will work provided $n^k 2^{-t} \|A^{-1}\|$ is less than $\frac{1}{2}$. In practice, because of the statistical distribution of errors, the process will usually work provided $n^{k/2} 2^{-t} \|A^{-1}\|$ is less than unity. (We are assuming here that the pivotal growth is negligible.)

Similarly if we use the Givens or Householder triangularization and store details of the orthogonal transformation we can perform an analogous iterative procedure. Again the only stage at which it is important to accumulate inner-products is when computing the residuals. For the orthogonal methods of triangularization we may leave out the proviso concerning the pivotal growth since this is now precluded. However, I do not think that this justifies the use of the orthogonal transformations.

If inner-products are not accumulated when forming the residuals, the iterative process will not in general converge. In fact the errors made in computing the residuals will then be of the same order of magnitude as the true residuals (except possibly at the first stage, when the true residuals may be somewhat larger for high-order matrices because of the factor n^k). In general, the successive iterates will show *no tendency* to converge; a succession of vectors will be obtained all having errors comparable with that of the first solution.

Limitations of the iterative procedure

73. Analyses similar to that in Wilkinson (1963b), pp. 121–126, show that, provided no pivotal growth occurs, any of the iterative procedures must converge for all matrices satisfying some criterion of the form

$$2^{-t} f(n) \, \|A^{-1}\| < 1, \tag{73.1}$$

where $f(n)$ depends on the method of triangularization and the mode of arithmetic used. If the process does not converge and no pivotal growth has occurred in the triangularization, then we can be absolutely certain that A does not satisfy (73.1). We can say that the matrix is 'too ill-conditioned to be solved by the technique in question' unless higher precision arithmetic is used.

There remains the possibility, however, that in spite of the fact that (73.1) is not satisfied, the process apparently converges and yet gives the wrong answer. Attempts to construct such examples immediately reveal how very improbable this is even for one right-hand side; if we are interested in several different right-hand sides it seems almost inconceivable that we could be misled in this way. Even in the case of the ill-conditioned matrices which were mentioned in § 66, for which no pivot is small, the solution is of order unity and the first residual is of order 2^{-t}, this ill-condition is immediately exposed in full when the first correction is computed.

In any automatic programme based on the iterative technique, one could easily incorporate additional safeguards; pivots of the order of magnitude of $n2^{-t}$ or solutions of the order of magnitude of $2^{-t}/n$ are a sure sign that the condition of the matrix is such that convergence of the iterates, in the unlikely event of its occurring, is not to be taken as guaranteeing a correct solution. Even the failure of two successive corrections to decrease is unmistakable evidence that (73.1) is not satisfied.

Rigorous justification of the iterative method

74. We have always taken the view that if we include the additional safeguards we have mentioned (probably even if we do not), then convergence of the iterates may safely be taken at its face value and that it is pedantic to require further proof.

If we are not prepared to do this, however, we must obtain an upper bound for $\|A^{-1}\|$. Sometimes such a bound is obtainable *a priori* from theoretical considerations. Otherwise we can use the triangular matrices to obtain a computed inverse X and then obtain a bound for $\|A^{-1}\|$ as in § 64 by computing $(AX - I)$. If the bound for $\|A^{-1}\|$ ensures that (73.1) is satisfied then we know that the iterative process must converge and give the correct solution. Otherwise we conclude that A is too ill-conditioned for the precision of computation which is being used.

Notice that having computed X we may use an alternative iterative process in which we compute the correction $d^{(k)}$ from $Xr^{(k)}$ rather than by solving $LUx = r^{(k)}$. I analysed this process in Wilkinson (1963b), pp. 128–131, and showed that it is rather less satisfactory than the process of § 67 and both iterations involve roughly the same number of multiplications. We see that quite a heavy price is paid merely in order to obtain an upper bound for $\|A^{-1}\|$.

75. There is a comparatively economical way of computing an upper bound $\|A^{-1}\|$ though it may be very pessimistic. We write

$$L = D_1 + L_s, \qquad U = D_2 + U_s, \tag{75.1}$$

where D_1 and D_2 are diagonal and L_s and U_s are *strictly* lower-triangular and upper-triangular respectively. Then if x is the solution of

$$(|D_1| - |L_s|)(|D_2| - |U_s|)x = e, \tag{75.2}$$

where e is the **vector** with unit components, then

$$\|x\|_\infty > \|A^{-1}\|_\infty. \tag{75.3}$$

From § 62 we know that the computed solution of (75.2) has components with low relative errors and hence gives us a reliable bound for $\|A^{-1}\|_\infty$.

Additional notes

In this chapter we have covered only those aspects of linear equations with which we shall be directly concerned in this book, so that no reference has been made to the solution by iterative methods. For more general discussions the reader is referred to the books by Fox (1964) and Householder (1964) for both direct and iterative methods and to those by Forsythe and Wasow (1960) and Varga (1962) for iterative methods.

The problem of equilibration has been discussed in papers by Forsythe and Straus (1955) and more recently by Bauer (1963). These papers give a good deal of insight into the factors involved but the problem of how to carry out equilibration in practice before computing the inverse remains intractable. Equilibration is often discussed in connexion with the pivotal strategy in Gaussian elimination and this may give the impression that it is irrelevant when orthogonal triangularization is used. That this is not so becomes apparent if we consider a matrix with elements of the orders of magnitude given by

$$\begin{bmatrix} 1 & 10^{10} & 10^{10} \\ 1 & 1 & 1 \\ 1 & 1 & 1 \end{bmatrix}.$$

The first row should be scaled before orthogonal triangularization or the information in the other rows will be lost.

The first published error analyses of techniques based on Gaussian elimination were those of von Neumann and Goldstine (1947) and Turing (1948). The former may be said to have laid the foundation of modern error analysis, in its rigorous treatment of the inversion of a positive definite matrix. The analyses given in this chapter are based on work done by the author between 1957 and 1962 and are framed everywhere in terms of backward error analysis. Somewhat similar analyses of some of the processes have been given by Gastinel (1960) in his thesis.

5

Hermitian Matrices

Introduction

1. IN THIS chapter we describe techniques for solving the eigen-problem for Hermitian matrices. We shall assume that the matrices are not sparse or special in any other respect; the techniques which are appropriate for sparse matrices will be discussed in later chapters.

We have seen in Chapters 1, 2 and 3 that the eigenproblem is simpler for Hermitian matrices than for matrices in general. A Hermitian matrix A always has linear elementary divisors and there is always a *unitary* matrix R such that

$$R^H A R = \text{diag}(\lambda_i). \tag{1.1}$$

Further, the eigenvalues are real and, as we showed in Chapter 2, § 31, small perturbations in the elements of A necessarily give rise to small perturbations in the eigenvalues. If the matrix of perturbations is Hermitian, then the eigenvalues remain real (Chapter 2, § 38). Although the eigenvectors are not necessarily well-determined, poor determination must be associated with close eigenvalues since $s_i = 1$ ·in (10.2) of Chapter 2.

2. For the greater part of this chapter we shall confine ourselves to real symmetric matrices. For such matrices the matrix R of equation (1.1) can be taken to be real and is therefore an orthogonal matrix. It should be remembered that complex symmetric matrices possess none of the important properties of real symmetric matrices (Chapter 1, § 24).

Most of the methods for solving the eigenproblem for a matrix A of general form depend on the application of a series of similarity transformations which converts A into a matrix of special form. For Hermitian matrices the use of unitary matrices has much to recommend it, since the Hermitian property, with all its attendant advantages, is thereby preserved. In the methods described in this chapter we use the elementary unitary transformations introduced in Chapter 1, §§ 43–46.

The classical Jacobi method for real symmetric matrices

3. In the method of Jacobi (1846) the original matrix is transformed to diagonal form by a sequence of plane rotations. Strictly speaking the number of plane rotations necessary to produce the diagonal form is infinite; this is to be expected since we cannot, in general, solve a polynomial equation in a finite number of steps. In practice the process is terminated when the off-diagonal elements are negligible to working accuracy. For a real symmetric matrix we use real plane rotations.

If we denote the original matrix by A_0 then we may describe Jacobi's process as follows. A sequence of matrices A_k is produced satisfying the relations

$$A_k = R_k A_{k-1} R_k^T, \qquad (3.1)$$

where the matrix R_k is determined by the following rules. (Note that we have used the notation $R_k A_{k-1} R_k^T$ and not $R_k^T A_{k-1} R_k$ in order to conform with our general analysis in Chapter 3.)

Suppose the off-diagonal element of A_{k-1} of maximum modulus is in the (p, q) position, then R_k corresponds to a rotation in the (p, q) plane and the angle θ of the rotation is chosen so as to reduce the (p, q) element of A_{k-1} to zero. We have

$$\left.\begin{aligned} R_{pp} = R_{qq} = \cos\theta, \qquad R_{pq} = -R_{qp} = \sin\theta \\ R_{ii} = 1 \quad (i \neq p, q), \quad R_{ij} = 0 \text{ otherwise} \end{aligned}\right\}, \qquad (3.2)$$

where we assume that p is less than q. A_k differs from A_{k-1} only in rows and columns p and q and all A_k are symmetric. The modified values are defined by

$$\left.\begin{aligned} a_{ip}^{(k)} &= a_{ip}^{(k-1)}\cos\theta + a_{iq}^{(k-1)}\sin\theta = a_{pi}^{(k)} \\ a_{iq}^{(k)} &= -a_{ip}^{(k-1)}\sin\theta + a_{iq}^{(k-1)}\cos\theta = a_{qi}^{(k)} \end{aligned}\right\}, \quad (i \neq p, q) \quad (3.3)$$

$$\left.\begin{aligned} a_{pp}^{(k)} &= a_{pp}^{(k-1)}\cos^2\theta + 2a_{pq}^{(k-1)}\cos\theta\sin\theta + a_{qq}^{(k-1)}\sin^2\theta \\ a_{qq}^{(k)} &= a_{pp}^{(k-1)}\sin^2\theta - 2a_{pq}^{(k-1)}\cos\theta\sin\theta + a_{qq}^{(k-1)}\cos^2\theta \\ a_{pq}^{(k)} &= (a_{qq}^{(k-1)} - a_{pp}^{(k-1)})\cos\theta\sin\theta + a_{pq}^{(k-1)}(\cos^2\theta - \sin^2\theta) = a_{qp}^{(k)} \end{aligned}\right\}. \quad (3.4)$$

Since $a_{pq}^{(k)}$ is to be zero we have

$$\tan 2\theta = 2a_{pq}^{(k-1)}/(a_{pp}^{(k-1)} - a_{qq}^{(k-1)}). \qquad (3.5)$$

We shall always take θ to lie in the range

$$|\theta| < \tfrac{1}{4}\pi, \qquad (3.6)$$

and if $a_{pp}^{(k-1)} = a_{qq}^{(k-1)}$, we take θ to be $\pm\frac{1}{4}\pi$ according to the sign of $a_{pq}^{(k-1)}$. We assume that $a_{pq}^{(k-1)} \neq 0$, since otherwise the diagonalization is complete. The process is essentially iterative since an element which is reduced to zero by one rotation is, in general, made non-zero by subsequent rotations.

Rate of convergence

4. We now prove that if each angle θ is taken in the range defined by (3.6), then

$$A_k \to \text{diag}(\lambda_i) \quad \text{as} \quad k \to \infty, \tag{4.1}$$

where the λ_i are the eigenvalues of A_0, and hence of all A_k, in some order.

We write

$$A_k = \text{diag}(a_{ii}^{(k)}) + E_k, \tag{4.2}$$

where E_k is the symmetric matrix of off-diagonal elements, and we first prove that

$$\|E_k\|_E \to 0 \quad \text{as} \quad k \to \infty. \tag{4.3}$$

For from equations (3.3) we have

$$\sum_{i \neq p,q} [(a_{ip}^{(k)})^2 + (a_{iq}^{(k)})^2] = \sum_{i \neq p,q} [(a_{ip}^{(k-1)})^2 + (a_{iq}^{(k-1)})^2], \tag{4.4}$$

and hence the sum of the squares of the off-diagonal elements excluding the (p, q) and (q, p) elements remains constant. However, these last two elements become zero, and hence

$$\|E_k\|_E^2 = \|E_{k-1}\|_E^2 - 2(a_{pq}^{(k-1)})^2, \tag{4.5}$$

and since $a_{pq}^{(k-1)}$ is the element of E_{k-1} of maximum modulus,

$$\|E_k\|_E^2 < [1 - 2/(n^2 - n)] \|E_{k-1}\|_E^2 < [1 - 2/(n^2 - n)]^k \|E_0\|_E^2. \tag{4.6}$$

The inequality (4.6) provides a rather crude bound for the rate of convergence. If we write

$$N = \frac{1}{2}(n^2 - n) \tag{4.7}$$

then

$$\|E_{rN}\|_E^2 < (1 - 1/N)^{rN} \|E_0\|_E^2 < e^{-r} \|E_0\|_E^2. \tag{4.8}$$

We shall see that this gives a considerable underestimate of the rate of convergence. However, it shows certainly that

$$\|E_{rN}\|_E^2 < \epsilon^2 \|E_0\|^2 \quad \text{when} \quad r > 2\ln(1/\epsilon), \tag{4.9}$$

so that for $\epsilon = 2^{-t}$, for example, this gives

$$r > 2\ln 2^t \doteq 1\cdot39t. \tag{4.10}$$

Convergence to fixed diagonal matrix

5. We have still to show that A_k tends to a fixed diagonal matrix. Suppose we have continued the iteration until

$$\|E_k\|_E < \epsilon. \tag{5.1}$$

Then from (4.2) we see that if the eigenvalues λ_i of A_k (and hence of A_0) and the $a_{ii}^{(k)}$ are arranged in increasing order, corresponding eigenvalues in the two sequences differ by less than ϵ (Chapter 2, § 44). Hence the $a_{ii}^{(k)}$ arranged in some order lie in intervals of width 2ϵ centred on the λ_i. Since ϵ may be made arbitrarily small, we can locate the eigenvalues to any required accuracy.

We now show that each $a_{ii}^{(k)}$ does indeed converge to a specific λ_i. We assume first that the λ_i are distinct and we define ϵ by the relation
$$0 < 4\epsilon = \min_{i \neq j} |\lambda_i - \lambda_j|. \tag{5.2}$$

Let k be chosen so that (5.1) is satisfied; from (4.6) it will be satisfied by all subsequent E_r. With this choice of ϵ the intervals centred on the λ_i are obviously disjoint and there is therefore just one $a_{jj}^{(k)}$ in each interval. We may assume that the λ_i have been numbered so that each $a_{ii}^{(k)}$ lies in the λ_i interval. We show that it must remain in the same interval at all subsequent stages.

For suppose the next iteration is in the (p, q) plane; then the (p, p) and (q, q) elements are the only diagonal elements to change. Hence $a_{pp}^{(k+1)}$ and $a_{qq}^{(k+1)}$ must still be the two diagonal elements in the λ_p and λ_q intervals. We show that it must be $a_{pp}^{(k+1)}$ which lies in the λ_p interval. For we have

$$a_{qq}^{(k+1)} - \lambda_p = a_{pp}^{(k)} \sin^2\theta - 2a_{pq}^{(k)} \cos\theta \sin\theta + a_{qq}^{(k)} \cos^2\theta - \lambda_p$$
$$= (a_{pp}^{(k)} - \lambda_p)\sin^2\theta + (a_{qq}^{(k)} - \lambda_q + \lambda_q - \lambda_p)\cos^2\theta - 2a_{pq}^{(k)}\cos\theta \sin\theta, \tag{5.3}$$

giving

$$|a_{qq}^{(k+1)} - \lambda_p| \geq |\lambda_q - \lambda_p| \cos^2\theta - |a_{pp}^{(k)} - \lambda_p| \sin^2\theta - |a_{qq}^{(k)} - \lambda_q| \cos^2\theta - |a_{pq}^{(k)}|$$
$$\geq 4\epsilon \cos^2\theta - \epsilon \sin^2\theta - \epsilon \cos^2\theta - \epsilon$$
$$= 2\epsilon \cos 2\theta. \tag{5.4}$$

However, we have

$$|\tan 2\theta| = |2a_{pq}^{(k)}/(a_{pp}^{(k)} - a_{qq}^{(k)})| < 2\epsilon/2\epsilon = 1. \tag{5.5}$$

Hence for θ in the range given by (3.6) we have $|2\theta| < \frac{1}{4}\pi$ and hence

$$|a_{qq}^{(k+1)} - \lambda_p| \geq 2^{\frac{1}{2}}\epsilon, \tag{5.6}$$

showing that $a_{qq}^{(k+1)}$ is not in the λ_p interval.

6. The introduction of multiple eigenvalues does not present much difficulty. We now define ϵ so that (5.2) is true for all distinct λ_i and λ_j. We know that if λ_i is a zero of multiplicity s, then precisely s of the $a_{jj}^{(k)}$ lie in the λ_i interval (Chapter 2, § 17). The proof that no $a_{pp}^{(k)}$ and $a_{qq}^{(k)}$ lying in different λ_i intervals can change allegiance remains unaltered. Hence, in all cases, after a certain stage each diagonal element of A_k remains in a fixed λ_i interval as k is increased, and the width of the interval tends to zero as k tends to infinity.

Serial Jacobi method

7. Searching for the largest off-diagonal element is time consuming on an automatic computer and it is usual to select the elements to be eliminated in some simpler manner. Perhaps the simplest scheme is to take them sequentially in the order $(1, 2), (1, 3),\ldots, (1, n)$; $(2, 3), (2, 4),\ldots, (2, n);\ldots; (n-1, n)$ and then return to the $(1, 2)$ element again. This is sometimes referred to as the *special* serial Jacobi method to distinguish it from a generalization of the method in which all off-diagonal elements are eliminated just once in any sequence of N rotations, but not necessarily in the above order. In the serial method (special or general) we shall refer to each sequence of N rotations as a single *sweep*.

Forsythe and Henrici (1960) have shown that the *special* serial Jacobi method does indeed converge if the angles of rotation are suitably restricted, but the proof is quite sophisticated. In § 15 we shall describe a variant of the serial method for which convergence is obviously guaranteed.

The Gerschgorin discs

8. Suppose the eigenvalues λ_i of A_0 are distinct, that

$$\min_{i \neq j} |\lambda_i - \lambda_j| = 2\delta, \tag{8.1}$$

and we have reached a matrix A_k for which

$$|a_{ii}^{(k)} - \lambda_i| < \tfrac{1}{4}\delta, \qquad \max_{i \neq j} |a_{ij}| = \epsilon, \tag{8.2}$$

where

$$(n-1)\epsilon < \tfrac{1}{4}\delta. \tag{8.3}$$

At this stage the Gerschgorin discs are definitely disjoint, since

$$|a_{ii}^{(k)} - a_{jj}^{(k)}| \geqslant |\lambda_i - \lambda_j| - |a_{ii}^{(k)} - \lambda_i| - |a_{jj}^{(k)} - \lambda_j| > \tfrac{3}{2}\delta. \tag{8.4}$$

We have therefore

$$|a_{ii}^{(k)} - \lambda_i| < (n-1)\epsilon. \tag{8.5}$$

Now if we multiply row i by ϵ/δ and column i by δ/ϵ, the ith Gerschgorin disc is certainly contained in the disc

$$|a_{ii}^{(k)} - \lambda| < (n-1)\epsilon^2/\delta, \tag{8.6}$$

while the remainder are contained in the discs

$$|a_{jj}^{(k)} - \lambda| < (n-2)\epsilon + \delta \quad (j \neq i). \tag{8.7}$$

The ith disc is certainly isolated from the others, so from (8.6) we see that when the absolute values of the off-diagonal elements have become less than ϵ, then for sufficiently small ϵ the diagonal elements differ from the eigenvalues by quantities of order ϵ^2. Note that if A had some multiple eigenvalues we could still obtain the same result for any of the simple eigenvalues.

Ultimate quadratic convergence of Jacobi methods

9. We have just shown that we can obtain very good estimates of the eigenvalues when the off-diagonal elements have become small. It is important therefore that the convergence of any Jacobi method should become more rapid in the later stages. It has been shown that for several variants of the Jacobi method convergence does indeed ultimately become quadratic when the eigenvalues are distinct. The first result of this type was obtained by Henrici (1958a). The sharpest results obtained so far, expressed in the notation of §§ 4 and 8, are the following.

(i) For the classical Jacobi method, Schönhage (1961) has shown that if

$$|\lambda_i - \lambda_j| \geqslant 2\delta \quad (i \neq j), \tag{9.1}$$

and we have reached the stage at which

$$\|E_r\|_E < \tfrac{1}{2}\delta, \tag{9.2}$$

then

$$\|E_{r+N}\|_E < (\tfrac{1}{2}n-1)^{\frac{1}{2}} \|E_r\|_E^2/\delta. \tag{9.3}$$

Schönhage has shown further that if (9.1) holds except for p pairs of coincident eigenvalues, then

$$\|E_{r+r'}\|_E < \tfrac{1}{2}n \|E_r\|_E^2/\delta, \tag{9.4}$$

where

$$r' = N + p(n-2). \tag{9.5}$$

(ii) For the general serial Jacobi method Wilkinson (1962c) has shown that, subject to the conditions (9.1) and (9.2),

$$\|E_{r+N}\|_E < \tfrac{1}{2}[n(n-1)]^{\frac{1}{2}} \|E_r\|_E^2/\delta. \tag{9.6}$$

Schönhage has remarked that the same result is true even if A has any number of pairs of coincident eigenvalues, provided that the rotations corresponding to these pairs are performed first.

(iii) For the special serial Jacobi method Wilkinson (1962c) has shown that

$$\|E_{r+N}\|_E < \|E_r\|_E^2/(2^{\frac{1}{2}}\delta). \tag{9.7}$$

We mention that the proofs of these results show only that *if* the processes are convergent, then they are ultimately quadratically convergent. It has never been proved that the general cyclic Jacobi method is convergent. (See, however, § 15.)

Close and multiple eigenvalues

10. The bounds for the convergence rates all contain the factor $1/\delta$ and this would seem to suggest that when δ is small the onset of quadratic convergence is delayed. If δ is zero the proofs break down, though it has been shown that when there are no eigenvalues of multiplicities greater than two, quadratic convergence can be guaranteed if the order in which the elements are eliminated is modified slightly. (See, for example, Schönhage (1961).) Little progress has been made in the investigation of the rate of convergence when there are eigenvalues of multiplicity greater than two.

Our experience with the special serial Jacobi method has suggested that, in practice, very close and multiple eigenvalues usually affect the number of iterations required to reduce the off-diagonal elements below a prescribed level rather less than might be expected in our present state of knowledge. For matrices of orders up to 50 and for word lengths of 32 to 48 binary digits the total number of sweeps has averaged about 5 to 6, and the number of iterations required on matrices having multiple eigenvalues has often proved to be somewhat lower than the average. The following considerations indicate how this may come about.

In the first place if all eigenvalues are equal to λ_1 then the matrix is $\lambda_1 I$ and no iterations are needed! Suppose, to take a less extreme case, that all the eigenvalues are very close so that

$$\lambda_i = a + \delta_i, \tag{10.1}$$

where the δ_i are small compared with a. Then we have

$$A = R^T \operatorname{diag}(a + \delta_i) R \tag{10.2}$$

for some orthogonal matrix R, and hence

$$A = aI + R^T \operatorname{diag}(\delta_i) R. \tag{10.3}$$

Now we have

$$\| R^T \operatorname{diag}(\delta_i) R \|_E = (\textstyle\sum \delta_i^2)^{\frac{1}{2}}, \tag{10.4}$$

showing that the off-diagonal elements must be small initially. Computation with A is identical with computation with $(A - aI)$, and this matrix is no longer special in the original sense. However, since all elements of $(A - aI)$ are small to start with we may expect that the reduction of the off-diagonal elements below a prescribed level will take less time for A than for a general symmetric matrix. In fact if $\max |\delta_i| = O(10^{-r})$, then we would expect to reduce the off-diagonal elements of A below 10^{-t-r} in the time usually taken to reduce them below 10^{-t}.

For example, for the symmetric matrix B defined by

$$B = \begin{bmatrix} 0{\cdot}2512 & 0{\cdot}0014 & 0{\cdot}0017 \\ & 0{\cdot}2511 & 0{\cdot}0026 \\ & & 0{\cdot}2507 \end{bmatrix} \tag{10.5}$$

(we give only the upper triangle), the angles of rotation are clearly the same as for the matrix C defined by

$$C = 100B - 25I = \begin{bmatrix} 0{\cdot}1200 & 0{\cdot}1400 & 0{\cdot}1700 \\ & 0{\cdot}1100 & 0{\cdot}2600 \\ & & 0{\cdot}0700 \end{bmatrix}. \tag{10.6}$$

However, if we think in terms of the matrix C, we must remember that we have only to reduce off-diagonal elements below 10^{-2} in order to give off-diagonal elements of B which are of order 10^{-4}.

Next we may show that if the matrix A has an eigenvalue a of multiplicity $(n-1)$, then it will be reduced to diagonal form in less than one sweep. For denoting the other eigenvalue by b, we know that A may be expressed in the form

$$A = R^T \begin{bmatrix} a & & & & & \\ & a & & & & \\ & & \cdot & & & \\ & & & \cdot & & \\ & & & & a & \\ & & & & & b \end{bmatrix} \qquad R = aI + R^T \begin{bmatrix} O & O \\ \hline O & b-a \end{bmatrix} R. \tag{10.7}$$

The matrix aI does not affect the Jacobi process and the (i, j) element of the second matrix on the right-hand side of (10.7) is given by $(b-a)r_{ni}r_{nj}$. From this we deduce immediately that the $n-1$ rotations required to eliminate the elements of the first row of A do, in fact, eliminate *all* off-diagonal elements.

Numerical examples

11. Tables 1 and 2 illustrate both the process itself and the quadratic convergence of the final stages. Note that we do not need to do both the row and the column transformations except for the four elements which are situated at the intersections of the relevant rows and columns, and two of these become zero. Exact symmetry is preserved by storing and computing only the elements in the upper triangle and assuming the corresponding elements in the lower triangle to be identical with them. There are therefore approximately $4n$ multiplications in each rotation and $2n^3$ multiplications in a complete sweep. The elements which are modified in a matrix of order 6 by a rotation in the (3, 5) plane are indicated by

$$
\begin{bmatrix}
\times & \times & \times & \times & \times & \times \\
 & \times & \times & \times & \times & \times \\
 & & \times & \times & \times & \times \\
 & & & \times & \times & \times \\
 & & & & \times & \times \\
 & & & & & \times
\end{bmatrix}.
\tag{11.1}
$$

The example of Table 1 has been chosen so that, after the first rotation, all off-diagonal elements are small compared with the difference between the diagonal elements. Subsequent rotations are through small angles, and off-diagonal elements which are not currently being reduced to zero are changed so little that they are unaffected to the working accuracy of four decimals.

The example of Table 2 has been chosen to illustrate the fact that the presence of a pair of close diagonal elements does not delay the onset of quadratic convergence. We have taken the first two diagonal elements to be equal and all off-diagonal elements to be small. The first rotation, (in the (1, 2) plane), is through an angle of $\frac{1}{4}\pi$. It reduces the (1, 2) element to zero and leaves the other off-diagonal elements of the same order of magnitude. The angle of the second rotation, (in the (1, 3) plane), is small since the first and third elements are not close. The same applies to the (2, 3) rotation, and at the end of the sweep all off-diagonal elements are zero to working accuracy.

TABLE 1

$$A_0 = \begin{bmatrix} 0 \cdot 6532 & 0 \cdot 2165 & 0 \cdot 0031 \\ & 0 \cdot 4105 & 0 \cdot 0052 \\ & & 0 \cdot 2132 \end{bmatrix}$$

Rotation in (1, 2) plane.
$\tan 2\theta = 2 \times 0 \cdot 2165/(0 \cdot 6532 - 0 \cdot 4105)$
$\cos \theta = 0 \cdot 8628,\ \sin \theta = 0 \cdot 5055$

$$A_1 = \begin{bmatrix} 0 \cdot 7800 & 0 \cdot 0000 & 0 \cdot 0053 \\ & 0 \cdot 2836 & 0 \cdot 0029 \\ & & 0 \cdot 2132 \end{bmatrix}$$

Rotation in (1, 3) plane.
$\tan 2\theta = 2 \times 0 \cdot 0053/(0 \cdot 7800 - 0 \cdot 2123)$
$\cos \theta = 1 \cdot 0000,\ \sin \theta = 0 \cdot 0094$

Note that because $a_{13}^{(1)}$ is small, θ is small.

$$A_2 = \begin{bmatrix} 0 \cdot 7801 & 0 \cdot 0000 & 0 \cdot 0000 \\ & 0 \cdot 2836 & 0 \cdot 0029 \\ & & 0 \cdot 2132 \end{bmatrix}$$

Rotation in (2, 3) plane.
$\tan 2\theta = 2 \times 0 \cdot 0029/(0 \cdot 2836 - 0 \cdot 2132)$
$\cos \theta = 0 \cdot 9991,\ \sin \theta = 0 \cdot 0411$

The (1, 2) element has remained zero to working accuracy. Similarly the (2, 3) element has remained small; in fact it is unchanged to working accuracy.

$$A_3 = \begin{bmatrix} 0 \cdot 7801 & 0 \cdot 0000 & 0 \cdot 0000 \\ & 0 \cdot 2837 & 0 \cdot 0000 \\ & & 0 \cdot 2131 \end{bmatrix}$$

The matrix is now diagonalized to working accuracy.

TABLE 2

$$A_0 = \begin{bmatrix} 0 \cdot 2500 & 0 \cdot 0012 & 0 \cdot 0014 \\ & 0 \cdot 2500 & 0 \cdot 0037 \\ & & 0 \cdot 6123 \end{bmatrix}$$

Rotation in (1, 2) plane.
$\tan 2\theta = 2 \times 0 \cdot 0012/(0 \cdot 2500 - 0 \cdot 2500)$
$\cos \theta = \sin \theta = 0 \cdot 7071$

$$A_1 = \begin{bmatrix} 0 \cdot 2512 & 0 \cdot 0000 & 0 \cdot 0036 \\ & 0 \cdot 2488 & 0 \cdot 0016 \\ & & 0 \cdot 6123 \end{bmatrix}$$

Rotation in (1, 3) plane.
$\tan 2\theta = 2 \times 0 \cdot 0036/(0 \cdot 2512 - 0 \cdot 6123)$
$\cos \theta = 1 \cdot 0000,\ \sin \theta = -0 \cdot 0099$

The next two rotations are both through small angles and the proximity of the (1, 1) and (2, 2) elements is of no consequence.

$$A_2 = \begin{bmatrix} 0 \cdot 2512 & 0 \cdot 0000 & 0 \cdot 0000 \\ & 0 \cdot 2488 & 0 \cdot 0016 \\ & & 0 \cdot 6124 \end{bmatrix}$$

Rotation in (2, 3) plane.
$\tan 2\theta = 2 \times 0 \cdot 0016/(0 \cdot 2488 - 0 \cdot 6124)$
$\cos \theta = 1 \cdot 0000,\ \sin \theta = -0 \cdot 0044$

$$A_3 = \begin{bmatrix} 0 \cdot 2512 & 0 \cdot 0000 & 0 \cdot 0000 \\ & 0 \cdot 2488 & 0 \cdot 0000 \\ & & 0 \cdot 6124 \end{bmatrix}$$

It is obvious that the same argument applies if any number of pairs of diagonal elements are close (or coincident), but not if there is a group of three or more elements.

Calculation of cos θ and sin θ

12. So far the calculation of the angles of rotation has been taken for granted, but it should be noticed that it is not covered by the analysis given in Chapter 3, § 20. In practice considerable attention to detail is necessary if we are to ensure that:

(i) $\cos^2\theta + \sin^2\theta$ is very close to unity, so that the transformation matrix really is almost orthogonal, and the transformed matrix is almost similar to the original.

(ii) $(a_{qq}^{(k)} - a_{pp}^{(k)})\cos\theta\sin\theta + a_{pq}^{(k)}(\cos^2\theta - \sin^2\theta)$ is very small, so that we are justified in replacing $a_{pq}^{(k+1)}$ by zero.

We omit the superfix k and write

$$\tan 2\theta = 2a_{pq}/(a_{pp} - a_{qq}) = x/y, \tag{12.1}$$

where

$$x = \pm 2a_{pq}, \quad y = |a_{pp} - a_{qq}|. \tag{12.2}$$

Then for θ lying in the range $|\theta| < \frac{1}{4}\pi$, we have formally

$$\left.\begin{aligned}
2\cos^2\theta &= [(x^2+y^2)^{\frac{1}{2}} + y]/(x^2+y^2)^{\frac{1}{2}}\\
2\sin^2\theta &= [(x^2+y^2)^{\frac{1}{2}} - y]/(x^2+y^2)^{\frac{1}{2}}
\end{aligned}\right\}, \tag{12.3}$$

where $\cos\theta$ is to be taken to be positive and $\sin\theta$ has the same sign as $\tan 2\theta$.

Now if x is small compared with y, equations (12.3) give a most unsatisfactory determination of $\sin\theta$. For ten-decimal computation with both fixed-point and floating-point, (12.3) gives $\sin\theta = 0$ and $\cos\theta = 1$ for all values of x and y for which $|x| < 10^{-5}|y|$. Hence, although condition (i) is satisfied, condition (ii) is not.

On the other hand, if floating-point is used, the determination of $\cos\theta$ from (12.3) is quite satisfactory. In fact from Chapter 3, §§ 6, 10, we have

$$\left.\begin{aligned}
u &= fl(x^2+y^2) \equiv (x^2+y^2)(1+\epsilon_1) & (|\epsilon_1| < (2 \cdot 00002)2^{-t})\\
v &= fl(u^{\frac{1}{2}}) \equiv u^{\frac{1}{2}}(1+\epsilon_2) & (|\epsilon_2| < (1 \cdot 00001)2^{-t})\\
w &= fl(v+y) \equiv (v+y)(1+\epsilon_3) & (|\epsilon_3| < 2^{-t})\\
z &= fl(w/2v) \equiv w(1+\epsilon_4)/2v & (|\epsilon_4| < 2^{-t})\\
c &= fl(z^{\frac{1}{2}}) \equiv z^{\frac{1}{2}}(1+\epsilon_5) & (|\epsilon_5| < (1 \cdot 00001)2^{-t})
\end{aligned}\right\} \tag{12.4}$$

Combining these results we have certainly

$$c \equiv \cos\theta(1+\epsilon_6) \quad (|\epsilon_6| < (3 \cdot 0003)2^{-t}), \tag{12.5}$$

where $\cos\theta$ is the true value corresponding to x and y. Here and subsequently the bounds we give are quite loose; often it will be possible to improve them by detailed analyses. A value s for $\sin \theta$ may be obtained from the relation

$$2 \sin \theta \cos \theta = x/(x^2+y^2)^{\frac{1}{2}}. \tag{12.6}$$

Using floating-point arithmetic and the computed value c we have

$$s = fl(x/2cv) \equiv x(1+\epsilon_7)/2cv \quad (|\epsilon_7| < (2{\cdot}00002)2^{-t}), \tag{12.7}$$

and from this, (12.4) and (12.5) we may deduce that

$$s \equiv \sin \theta(1+\epsilon_8) \quad (|\epsilon_8| < (6{\cdot}0006)2^{-t}). \tag{12.8}$$

Hence both s and c have low relative errors compared with the exact values corresponding to the given x and y. It is now obvious that condition (ii) is satisfied by the computed values.

13. To obtain comparable results in fixed-point the following technique is very satisfactory if inner-products can be accumulated. We compute x^2+y^2 exactly and then choose k to be the largest integer for which

$$2^{2k}(x^2+y^2) < 1. \tag{13.1}$$

The computation is then performed using $2^k x$ and $2^k y$ in place of x and y in equation (12.3) for $\cos^2\theta$ and again in equation (12.6) for $\sin \theta$.

Alternatively, if inner-products cannot be accumulated, we may write

$$2 \cos^2\theta = \frac{[1+(y/x)^2]^{\frac{1}{2}}+y/x}{[1+(y/x)]^{\frac{1}{2}}}, \tag{13.2}$$

$$2 \cos \theta \sin \theta = 1/[1+(y/x)^2]^{\frac{1}{2}} \tag{13.3}$$

if $|x| > y$, and corresponding expressions if $y > |x|$ (cf. Chapter 3, § 30).

Simpler determination of the angles of rotation

14. On some computers the determination of the trigonometric functions by the methods of §§ 12–13 makes heavier demands on time or on storage than is desirable, and alternative simpler methods of determining the angles of rotation have been developed.

Since any orthogonal transformation leaves the eigenvalues unaltered, we *need* not determine the angle of rotation so as to reduce the current element to zero. Provided we reduce its' value substantially, we can expect convergence since $\|E_k\|_E^2$ is diminished by $2(a_{pq}^{(k)})^2 - 2(a_{pq}^{(k+1)})^2$. However, in order to retain quadratic convergence in the final stages, it is important that the angles of rotation should tend to those computed by the orthodox method when the off-diagonal elements become small.

In one such alternative method, the angle of rotation ϕ is defined by

$$T = \tan \tfrac{1}{2}\phi = a_{pq}/2(a_{pp} - a_{qq}) \qquad (14.1)$$

provided this gives an angle ϕ lying in the range $(-\tfrac{1}{4}\pi, \tfrac{1}{4}\pi)$, and $\phi = \pm \tfrac{1}{4}\pi$ otherwise, the sign being that of the right-hand side of (14.1). Comparing (14.1) and (12.1) we see that

$$\tan 2\theta = 4 \tan \tfrac{1}{2}\phi \qquad (14.2)$$

and hence

$$\phi - \theta \sim \tfrac{5}{4}\theta^3 \quad \text{as} \quad \theta \to 0. \qquad (14.3)$$

We may calculate $\sin \phi$ and $\cos \phi$ from the relation

$$\sin \phi = 2T/(1+T^2), \quad \cos \phi = (1-T^2)/(1+T^2), \qquad (14.4)$$

and since $|T| < 1$ the computation is quite simple in either fixed-point or floating-point. It will readily be verified that with this choice of the angle, the rotation has the effect of replacing $a_{pq}^{(k)}$ by $\alpha a_{pq}^{(k)}$ where α satisfies

$$0 < \alpha < \tfrac{1}{4}(1 + 2^{\frac{1}{2}}) \qquad (14.5)$$

and lies near to zero for quite a large range of the angles defined by (14.1). The speed of convergence with this determination of the angles of rotation has proved to be comparable with that corresponding to the orthodox choice.

The threshold Jacobi method

15. The maximum reduction in the value of $\|E_k\|_E^2$ that can be achieved by a rotation in the (p, q) plane is $2(a_{pq}^{(k)})^2$. If $a_{pq}^{(k)}$ is much smaller than the general level of off-diagonal elements then little purpose is served by a (p, q) rotation. This has led to the development of a variant known as the *threshold serial Jacobi method*.

We associate with each sweep a threshold value, and when making the transformations of that sweep, we omit any rotation based on an off-diagonal element which is below the threshold value. On DEUCE, working in fixed-point, the set of threshold values 2^{-3}, 2^{-6}, 2^{-10}, 2^{-18}, 2^{-30} has been used. All subsequent sweeps have the threshold value 2^{-30}, and since this is the smallest permissible number on DEUCE, only zeros are skipped in the fifth and later sweeps. The process is terminated when $\frac{1}{2}(n^2-n)$ successive elements are skipped. Experience on DEUCE suggests that this modification gives a modest improvement in speed. *Note that convergence is guaranteed with this variant since there can only be a finite number of iterations corresponding to any given threshold.*

A set of threshold values given by k, k^2, k^3,... for some preassigned value of k has been used on SWAC (Pope and Tompkins (1957)). Iteration with each threshold value is continued until all off-diagonals are below that threshold.

Calculation of the eigenvectors

16. If the final rotation is denoted by R_s then we have

$$(R_s...R_2R_1)A_0(R_1^T R_2^T...R_s^T) = \mathrm{diag}(\lambda_i) \qquad (16.1)$$

to working accuracy. The eigenvectors of A_0 are therefore the columns of the matrix P_s defined by

$$P_s = R_1^T R_2^T...R_s^T. \qquad (16.2)$$

If the eigenvectors are required, then at each stage of the process we may store the current product P_k defined by

$$P_k = R_1^T R_2^T...R_k^T. \qquad (16.3)$$

This requires n^2 storage locations in addition to the $\frac{1}{2}(n^2+n)$ needed for the upper triangles of the successive A_k. If we adopt this scheme, we automatically obtain all n eigenvectors.

If only a few of the eigenvectors are required, then we can economize on computing time if we are prepared to remember the $\cos\theta$, $\sin\theta$ and plane of each rotation. Suppose one eigenvector only is required, that corresponding to the jth diagonal element of the final $\mathrm{diag}(\lambda_i)$. Then the eigenvector v_j is given by

$$v_j = R_1^T R_2^T...R_s^T e_j. \qquad (16.4)$$

If we multiply e_j on the left successively by R_s^T, R_{s-1}^T,..., R_1^T we can obtain the jth eigenvector without computing the rest. There are n times as many multiplications in the calculation of P_s as in that of v_j, so that for the calculation of r vectors the number of operations is r/n times the number for the calculation of P_s.

Since, on the average, some five or six sweeps are required, the storage needed to remember the rotations is several times that needed for P_s, though on a computer with a magnetic tape store this is not likely to restrict the order of matrix that can be dealt with.

Numerical example

17. The technique for computing the individual vectors is illustrated in Table 3. There we calculate the eigenvector corresponding to the second eigenvalue of the example in Table 1.

TABLE 3

$$v_2 = \begin{bmatrix} 0\cdot8628 & -0\cdot5055 & 0 \\ 0\cdot5055 & 0\cdot8628 & 0 \\ 0 & 0 & 1\cdot0000 \end{bmatrix} \begin{bmatrix} 1\cdot0000 & 0 & -0\cdot0094 \\ 0 & 1\cdot0000 & 0 \\ 0\cdot0094 & 0 & 1\cdot0000 \end{bmatrix} \times$$

$$\times \begin{bmatrix} 1\cdot0000 & 0 & 0 \\ 0 & 0\cdot9991 & -0\cdot0411 \\ 0 & 0\cdot0411 & 0\cdot9991 \end{bmatrix} \begin{bmatrix} 0 \\ 1 \\ 0 \end{bmatrix}$$

Multiplication by successive matrices gives

$$\begin{bmatrix} 0\cdot0000 \\ 0\cdot9991 \\ 0\cdot0411 \end{bmatrix}, \quad \begin{bmatrix} -0\cdot0004 \\ 0\cdot9991 \\ 0\cdot0411 \end{bmatrix} \quad \text{and finally} \quad \begin{bmatrix} -0\cdot5054 \\ 0\cdot8618 \\ 0\cdot0411 \end{bmatrix}.$$

The final vector already has a Euclidean norm of unity since P_s is orthogonal. This may be used as a check on the computer performance and as a measure of the accumulation of rounding errors.

Error analysis of the Jacobi method

18. Jacobi's method is not quite covered by the general analysis given in Chapter 3, §§ 20–36, since the computation of $\cos \theta$ and $\sin \theta$ is somewhat different from that given there. However, we have shown that the computed $\cos \theta$ and $\sin \theta$ have low relative errors compared with the exact values corresponding to the current computed matrix, and the modifications to the analysis of Chapter 3 are quite simple.

We leave it as an exercise to show that in the result of Chapter 3, equation (21.5), the factor 6.2^{-t} may certainly be replaced by 9.2^{-t} if $\sin \theta$ and $\cos \theta$ are computed as in § 12. If r complete sweeps are used and the final computed diagonal matrix is denoted by $\mathrm{diag}(\mu_i)$ then, corresponding to equation (27.2) of Chapter 3, we have

$$\|\mathrm{diag}(\mu_i) - RA_0 R^T\|_E < 18.2^{-t} rn^{\frac{3}{2}} (1 + 9.2^{-t})^{r(4n-7)} \|A_0\|_E, \quad (18.1)$$

where R is the *exact* orthogonal matrix given by the *exact* product of the *exact* plane rotations corresponding to the *computed* A_k. If we take r equal to six, then we have

$$\|\mathrm{diag}(\mu_i) - RA_0 R^T\|_E < 108.2^{-t} n^{\frac{3}{2}} (1 + 9.2^{-t})^{6(4n-7)} \|A_0\|_E. \quad (18.2)$$

This implies that if λ_i are the true eigenvalues of A_0, and λ_i and μ_i are both arranged in increasing order, then

$$\left[\frac{\sum (\mu_i - \lambda_i)^2}{\sum \lambda_i^2} \right]^{\frac{1}{2}} \leqslant 108.2^{-t} n^{\frac{3}{2}} (1 + 9.2^{-t})^{6(4n-7)}. \quad (18.3)$$

As usual, the final factor is of little importance in practice, since if the right-hand side is to be reasonably small this factor is almost unity. The right-hand side of (18.3) is an extreme upper bound, and we might expect that the statistical distribution of the rounding errors will ensure a bound of order

$$\left[\frac{\sum (\mu_i - \lambda_i)^2}{\sum \lambda_i^2}\right]^{\frac{1}{2}} \leqslant 11.2^{-t} n^{\frac{3}{2}}. \tag{18.4}$$

If we take $t = 40$ and $n = 10^4$, for example, the right-hand side is approximately 10^{-8}. This shows that, in practice, time or storage capacity are more likely to be limiting factors than the accuracy of the computed eigenvalues.

Accuracy of the computed eigenvectors

19. We have been able to obtain satisfactory rigorous *a priori* bounds for the errors in the eigenvalues, but we cannot hope to obtain comparable bounds for the eigenvectors. Inevitably their accuracy will depend, in general, on the separation of the eigenvalues. If we write

$$\text{diag}(\mu_i) - RA_0R^T = F, \tag{19.1}$$

then

$$\text{diag}(\mu_i) = R(A_0 + G)R^T, \tag{19.2}$$

where

$$G = R^T F R, \qquad \|G\|_E = \|F\|_E. \tag{19.3}$$

Even if we could compute R^T exactly, it would give only the eigenvectors of $(A_0 + G)$.

In fact we shall fall somewhat short of this. We shall have only the computed rotations and we shall make errors in multiplying the corresponding matrices together. If we denote the computed product by $\bar{R}^T$, then as in Chapter 3, § 25 with 9.2^{-t} for x we have

$$\|\bar{R}^T - R^T\|_E \leqslant 9.2^{-t} r n^2 (1 + 9.2^{-t})^{r(2n-4)}, \tag{19.4}$$

and again taking $r = 6$ and assuming a reasonable distribution of the errors we might expect something like

$$\|\bar{R}^T - R^T\|_E \leqslant 8n2^{-t}. \tag{19.5}$$

The computed matrix $\bar{R}^T$ will therefore be very close to the exact orthogonal matrix R^T.

We have therefore obtained a bound for the departure of $\bar{R}^T$ from R^T, the exact matrix of eigenvectors of $(A_0 + G)$, and we already have a bound for G. The bounds (19.4) and (19.5) are independent of the separation of the eigenvalues. Even when A_0 has some pathologically close eigenvalues the corresponding eigenvectors are almost exactly orthogonal, though they may be very inaccurate; it is evident that they span the correct subspace and give full digital information. This is perhaps the most satisfactory feature of Jacobi's method. The other techniques we describe in this chapter are superior to Jacobi's method in speed and accuracy, but it is comparatively tiresome to obtain orthogonal eigenvectors corresponding to pathologically close or coincident eigenvalues.

Error bounds for fixed-point computation

20. In a similar way, by methods analogous to those described in Chapter 3, § 29, we can obtain bounds for the errors in the eigenvalues obtained using fixed-point computation. The precise bounds depend on the particular method used to determine $\cos\theta$ and $\sin\theta$ but they are all of the form

$$\|\mathrm{diag}(\mu_i) - RA_0R^T\|_2 \leqslant K_1 rn^{\frac{3}{2}}2^{-t}, \tag{20.1}$$

or

$$\|\mathrm{diag}(\mu_i) - RA_0R^T\|_E \leqslant K_2 rn^{\frac{3}{2}}2^{-t}, \tag{20.2}$$

depending on whether we work with the 2-norm or the Euclidean norm. Here again r is the number of sweeps. As in Chapter 3, §§ 32, 33, the inequality (20.1) is obtained under the assumption that A_0 is scaled initially so that

$$\|A_0\|_2 \leqslant 1 - K_1 rn^{\frac{3}{2}}2^{-t}. \tag{20.3}$$

This is a difficult requirement to meet since we usually have no reliable estimate for $\|A_0\|_2$ until the process is complete. On the other hand (20.2) is derived under the assumption that A_0 is scaled initially so that

$$\|A_0\|_E \leqslant 1 - K_2 rn^{\frac{3}{2}}2^{-t}, \tag{20.4}$$

which is easily met since $\|A_0\|_E$ may be computed quite simply. These scalings guarantee that no elements of any A_k will exceed capacity.

Again by a method similar to that described in Chapter 3, § 34, we find that $\bar{R}^T$, the computed product of the computed rotations, satisfies

$$\|\bar{R}^T - R^T\|_2 < K_3 rn^{\frac{3}{2}}2^{-t}, \tag{20.5}$$

where R^T is the exact matrix of eigenvectors of (A_0+G) and we have either

$$\|G\|_2 \leqslant K_1 rn^{\frac{3}{2}}2^{-t} \quad \text{or} \quad \|G\|_E \leqslant K_2 rn^{\frac{3}{2}}2^{-t}. \tag{20.6}$$

As with floating-point computation, $\bar{R}^T$ will be almost orthogonal for any reasonable value of n, independently of how close its columns may be to the true eigenvectors of A_0.

Organizational problems

21. An inconvenient feature of Jacobi's method is that each transformation requires access to rows and columns of the current matrix. If the matrix can be held in the high-speed store this presents no difficulties, but if the high-speed store is inadequate there seems to be no very satisfactory way of organizing Jacobi's method. However, Chartres (1962) has described *modified* forms of the Jacobi method which are well adapted for use on a computer with magnetic-tape backing store. For details of these techniques we refer the reader to Chartres' paper.

Givens' method

22. The method of Jacobi is essentially iterative in character. The next two methods we shall describe aim at a less complete reduction of the original matrix, producing, by orthogonal similarity transformations, a symmetric tri-diagonal matrix (Chapter 3, § 13). Such a reduction may be achieved by methods which are not of an iterative nature and consequently require very much less computation. Since the computation of the eigenvalues of a symmetric tri-diagonal matrix is comparatively simple, these methods are very effective.

The first such method, which was developed by Givens (1954) is also based on the use of plane rotations. It consists of $(n-2)$ major steps, in the rth of which zeros are introduced in the rth row and rth column without destroying the zeros introduced in the previous $(r-1)$ steps. At the beginning of the rth major step the matrix is tri-diagonal as far as its first $(r-1)$ rows and columns are concerned; for the case $n = 6$, $r = 3$ the configuration is given by

$$
\begin{bmatrix}
\times & \times & 0 & 0 & 0 & 0 \\
\times & \times & \times & 0 & 0 & 0 \\
0 & \times & \times & \times & \underline{\times} & \underline{\times} \\
0 & 0 & \times & \times & \times & \times \\
0 & 0 & \underline{\times} & \times & \times & \times \\
0 & 0 & \underline{\times} & \times & \times & \times
\end{bmatrix}
\qquad (22.1)
$$

The rth major step consists of $(n-r-1)$ minor steps in which zeros are introduced successively in positions $r+2, r+3,..., n$ of the rth row and rth column; the zero in the ith position is produced by rotation in the $(r+1, i)$ plane. In (22.1) we have underlined the elements which are eliminated in the rth major step, and have indicated by arrows the rows and columns which are affected. It is immediately obvious that the first $(r-1)$ rows and columns are entirely unchanged; in so far as elements in these rows and columns are involved at all, some zero elements are replaced by linear combinations of zeros. In the rth major step we are therefore effectively concerned only with the matrix of order $(n-r+1)$ in the bottom right-hand corner of the current matrix.

In order to facilitate comparison with triangularization by plane rotations (Chapter 4, § 50) we shall first give an algorithmic description of the process, assuming that the full matrix is stored at all stages. The current values of $\cos\theta$ and $\sin\theta$ are stored in the positions occupied by the two symmetrically placed elements which are eliminated. The rth major step is then as follows.

For each value of i from $r+2$ to n perform steps (i) to (iv):

(i) Compute $x = (a_{r,r+1}^2 + a_{ri}^2)^{\frac{1}{2}}$.

(ii) Compute $\cos\theta = a_{r,r+1}/x$, $\sin\theta = a_{ri}/x$. (If $x = 0$ we take $\cos\theta = 1$ and $\sin\theta = 0$ and we may omit (iii) and (iv).) Overwrite $\cos\theta$ on a_{ri}, $\sin\theta$ on a_{ir} and x on $a_{r,r+1}$ and $a_{r+1,r}$, x being the correct new value in those positions.

(iii) For each value of j from $r+1$ to n:
Compute $a_{j,r+1}\cos\theta + a_{ji}\sin\theta$ and $-a_{j,r+1}\sin\theta + a_{ji}\cos\theta$ and overwrite on $a_{j,r+1}$ and a_{ji} respectively.

(iv) For each value of j from $r+1$ to n:
Compute $a_{r+1,j}\cos\theta + a_{ij}\sin\theta$ and $-a_{r+1,j}\sin\theta + a_{ij}\cos\theta$ and overwrite on $a_{r+1,j}$ and a_{ij} respectively.

In practice we do not need to do the whole of steps (iii) and (iv) since, because of symmetry, (iv) mainly repeats for the rows the computation which (iii) has performed on the columns. Usually one works only with the upper triangle, and the positions of the elements which are involved in a typical rotation are as shown in (11.1). However, note that if we wish to store the values of $\cos\theta$ and $\sin\theta$ we shall need the full n^2 storage locations.

23. The number of multiplications in the rth step, when full advantage is taken of symmetry, is approximately $4(n-r)^2$ and hence there are approximately $\frac{4}{3}n^3$ multiplications in the complete reduction to tri-diagonal form. We may compare this with $2n^3$ multiplications in *one sweep* of the Jacobi process. In addition there are approximately

$\frac{1}{2}n^2$ square roots in both the Givens reduction and in one sweep of the Jacobi process. Finally, the calculation of the angles of rotation is much simpler for Givens' method. If we assume that there are six sweeps in the Jacobi process then, certainly for large n, the Jacobi reduction involves about nine times as much computation as the Givens reduction. Provided we can find the eigenvalues of the tridiagonal form reasonably quickly, the Givens process is much faster than the Jacobi.

Givens' process on a computer with a two-level store

24. If we think of Givens' process in the terms used in § 22, then it is comparatively inefficient on a computer using a two-level store. Even if we take advantage of symmetry we find that both row and column transfers are involved in each minor step. If, for example, the upper triangle is stored by rows on the backing store, then in the rth major step rows r to n must be transferred from and to the backing store for each minor step, although only two elements are altered in each row except for the two on which the current rotation is based. We now describe a modified procedure due to Rollett and Wilkinson (1961) and Johansen (1961) in which only one transfer to the computing store and back is made for each of rows r to n for the whole of the rth major step.

The first $(r-1)$ rows and columns play no part in the rth major step, which is formally similar to the first but operates on the matrix of reduced order $(n-r+1)$ in the bottom right-hand corner. There is thus no loss of generality in describing the first major step and this will be done for the sake of simplicity. *Four groups of n registers each are required in the high-speed store.* The first major step has five main stages. (Since elements are stored partly on the backing store and partly in the high-speed store, we have departed somewhat from the algorithmic language used previously.)

(1) Transfer the first row of A to the group 1 registers.

(2) For each value of j from 3 to n compute:
$$x_j = [(a_{1,2}^{(j-1)})^2 + a_{1,j}^2]^{\frac{1}{2}}, \quad \cos_{2,j} = a_{1,2}^{(j-1)}/x_j, \quad \sin \theta_{2,j} = a_{1,j}/x_j,$$
where
$$a_{1,2}^{(2)} = a_{1,2} \quad \text{and} \quad a_{1,2}^{(j)} = x_j.$$
Overwrite $\cos \theta_{2,j}$ on $a_{1,j}$ and store $\sin \theta_{2,j}$ in the group 2 registers.

(3) Transfer the second row to the group 3 registers. (Only those elements on and above the diagonal are used in this and all succeeding rows.)

For each value of k from 3 to n perform steps (4) and (5):

(4a) Transfer the kth row to the group 4 registers.

(4b) Perform the row and column operations which involve $\cos \theta_{2,k}$ and $\sin \theta_{2,k}$ on the current elements $a_{2,2}$, $a_{2,k}$ and a_{kk}.

(4c) For each value of l from $k+1$ to n, perform the part of the row transformation involving $\cos \theta_{2,k}$ and $\sin \theta_{2,k}$ on $a_{2,l}$ and a_{kl}.

(4d) For each value of l from $k+1$ to n, perform the part of the column transformation involving $\cos \theta_{2,l}$ and $\sin \theta_{2,l}$ on $a_{2,k}$ and a_{kl}, taking advantage of the fact that $a_{2,k} = a_{k,2}$, by symmetry. When (4a), (4b), (4c) and (4d) have been completed for a given k, all the transformations involved in the first major step have been performed on all the elements of row k and on elements 3, 4,..., k of row 2. Elements 2, $k+1$, $k+2$,..., n of row 2 have been subjected only to the transformations involving $\theta_{2,3}$, $\theta_{2,4}$,..., $\theta_{2,k}$.

(5) Transfer the completed kth row to the backing store and return to (4a) with the next value of k.

25. When stages (4) and (5) have been completed for all appropriate values of k, the whole of the computation on row 2 has also been completed. The $\cos \theta_{2,k}$ and $\sin \theta_{2,k}$ ($k = 3$ to n) and the modified elements of the first row (i.e. the first two elements of the tri-diagonal matrix) may be written on the backing store and row 2 transferred to the group 1 registers. Since the second row plays the same role in the second major step as did the first row in the first major step, everything is ready for stage (2) in the second major step. Stage (1) is not needed in any of the later major steps because the appropriate row is already in the high-speed store.

It should be emphasized that the process we have just described is merely a rearrangement of the Givens process. Although some parts of the later transformations in a major step are done before the earlier transformations have been completed, the elements affected by these later transformations are not subsequently involved in the completion of the earlier transformations. The two schemes are identical as far as the number of arithmetic operations and the rounding errors are concerned but the number of transfers from and to the backing store is substantially reduced.

It is convenient to allocate n^2 locations for the matrix on the backing store, storing each row in full, in spite of the fact that the elements below the diagonal are not used, since the full number of locations is needed for the storage of the sines and cosines. The latter must be stored if the eigenvectors are required. A convenient method of storing the final configuration is shown in (25.1). With this method of storage the 'address' arithmetic is particularly simple.

$$
\begin{bmatrix}
a_{1,1} & a_{1,2} & \cos\theta_{2,3} & \cos\theta_{2,4} & \cos\theta_{2,5} \\
x & a_{2,2} & a_{2,3} & \cos\theta_{3,4} & \cos\theta_{3,5} \\
\sin\theta_{4,5} & x & a_{3,3} & a_{3,4} & \cos\theta_{4,5} \\
\sin\theta_{3,4} & \sin\theta_{3,5} & x & a_{4,4} & a_{4,5} \\
\sin\theta_{2,3} & \sin\theta_{2,4} & \sin\theta_{2,5} & x & a_{5,5}
\end{bmatrix}
\qquad (25.1)
$$

Johansen (1961) has pointed out that if transfers from the backing store take place in units of k words at a time, the high-speed storage requirement may be reduced from $4n$ registers to $3n+k$ registers, since we do not need to have the whole of the current row of A in the high-speed store at one time.

Floating-point error analysis of Givens' process

26. To complete the solution of the original eigenproblem we must solve the eigenproblem for the tri-diagonal matrix. Since our next method also leads to a symmetric tri-diagonal matrix we leave this point for the moment and give error analyses of the reduction to tri-diagonal form.

It is immediately apparent that Givens' reduction is completely covered by the general analyses given in Chapter 3, §§ 20–36, even the computation of the angles of rotation being as described there. If we denote the *computed* tri-diagonal matrix by C and the *exact* product of the $\frac{1}{2}(n-1)(n-2)$ *exact* rotations corresponding to the *computed* A_k by R, then we have from Chapter 3, equation (27.2), the result

$$
\|C - RA_0R^T\|_E \leqslant 12 . 2^{-t}n^{\frac{3}{2}}(1 + 6 . 2^{-t})^{4n-7} \|A_0\|_E \qquad (26.1)
$$

for standard floating-point computation.

Note that the rotations involved in the Givens' reduction do not quite constitute a *complete set* in the sense described in Chapter 3, § 22, since no rotations involving the first plane are required. The bound given there is obviously adequate *a fortiori* for the reduced set of rotations.

The bound (26.1) takes no advantage of the fact that the successive reduced matrices have an increasing number of zero elements, and

indeed it is by no means simple to do this. However, the most that
can be expected from such an analysis is that the factor 12 in (26.1)
might be reduced somewhat. In any case we have certainly

$$\left[\frac{\sum (\mu_i - \lambda_i)^2}{\sum \lambda_i^2}\right]^{\frac{1}{2}} \leqslant 12 \cdot 2^{-t} n^{\frac{3}{2}} (1 + 6 \cdot 2^{-t})^{4n-7}, \qquad (26.2)$$

where the μ_i are the eigenvalues of the computed tri-diagonal matrix.
This may be compared with the bound (18.3) for the Jacobi method.
We show in §§ 40–41 that the errors made in computing the eigenvalues of a tri-diagonal matrix are negligible compared with the
bound of (26.2).

Fixed-point error analysis

27. The general fixed-point analysis of Chapter 3, §§ 29–35, also
applies directly to the Givens reduction, but now it is quite easy to
take advantage of the zeros which are introduced progressively during
the course of the reduction. During the rth major step we are dealing
with a matrix of order $(n-r+1)$ rather than n. Hence in the bound of
Chapter 3, (35.17), we must replace $(2n-4)^{\frac{1}{2}}$ by $(2n-2r-2)^{\frac{1}{2}}$. In-
corporating this modification we have certainly

$$\|C - RA_0 R^T\|_2 \leqslant 2^{-t-1} \sum_r [(2n-2r-2)^{\frac{1}{2}} + 9 \cdot 0](n-r-1)$$

$$< 2^{-t-1}\left(\frac{2^{\frac{3}{2}}}{5} n^{\frac{5}{2}} + 4 \cdot 5n^2\right) = x \qquad \text{(say)}, \qquad (27.1)$$

provided A_0 is scaled so that

$$\|A_0\|_2 < 1 - x. \qquad (27.2)$$

Similarly we can show that

$$\|C - RA_0 R^T\|_E < 2^{-t-1}(\tfrac{4}{5} n^{\frac{5}{2}} + 6n^2) = y \quad \text{(say)}, \qquad (27.3)$$

provided A_0 is scaled so that

$$\|A_0\|_E < 1 - y. \qquad (27.4)$$

The factors associated with the n^2 terms in (27.1) and (27.2) are quite
generous and could probably be reduced considerably by a careful
analysis, but for sufficiently large n the bound is, in any case, dominated by the term in $n^{\frac{5}{2}}$.

Numerical example

28. We now give a simple numerical example of Givens' method which has been specially designed to illustrate the nature of the numerical stability of the process. We have always been at great pains to emphasize that the final tri-diagonal matrix is very close to the matrix which would have been obtained by applying to A_0 the *exact* orthogonal transformations corresponding to the successive *computed* A_r. We do not claim that the tri-diagonal matrix is close to the matrix which would have been obtained by exact computation throughout, but this fact is immaterial provided the eigenvalues and eigenvectors we finally obtain are close to those of A_0.

In Table 4 we give first the reduction of a matrix of order 5 to tri-diagonal form using fixed-point computation and working to 5 decimal places. When the zero is introduced in position $(1, 5)$ the elements in positions $(2, 3)$, $(2, 4)$, $(2, 5)$ all become very small as a result of cancellation. These elements accordingly have a high relative error compared with those which would have been obtained by exact computation. Consequently when we determine the rotations in the $(3, 4)$ and $(3, 5)$ planes, the angles are very different from those corresponding to exact computation.

We have emphasized this by repeating the calculation using six decimal places. The $(2, 3)$, $(2, 4)$, $(2, 5)$ elements are now

$$0 \cdot 000012, \quad 0 \cdot 000007, \quad 0 \cdot 000008 \quad \text{respectively}$$

compared with

$$0 \cdot 00002, \quad 0 \cdot 00001, \quad 0 \cdot 00001 \quad \text{respectively}$$

for 5-decimal computation. For the rotation in the $(3, 4)$ plane, for example, we have $\tan \theta = 7/12$ and $\tan \theta = \frac{1}{2}$ respectively in the two cases. As a result, the two final tri-diagonal matrices differ almost completely as far as later rows and columns are concerned. Our error analysis shows, however, that both tri-diagonal matrices must have eigenvalues which are close to those of A_0.

In Table 4 we have given the eigenvalues of these two tri-diagonal matrices and of A_0 correct to 8 decimals and they confirm our claim. Note that in both cases the effect on each eigenvalue of the rounding errors made in the whole reduction is less than one unit in the last figure retained in the computation. In practice we have found that the actual errors are always well below the upper bounds we have

TABLE 4

$$A_0 = \begin{bmatrix} 0.71235 & 0.33973 & 0.28615 & 0.30388 & 0.29401 \\ & 1.18585 & -0.21846 & -0.06685 & -0.37360 \\ & & 0.18159 & 0.27955 & 0.38898 \\ & & & 0.23195 & 0.20496 \\ & & & & 0.46004 \end{bmatrix}$$

Reduction using 5 decimals

Matrix A_3 obtained after introducing zeros in first row.

$$A_3 = \begin{bmatrix} 0.71235 & 0.61325 & 0 & 0 & 0 \\ & 0.61874 & 0.00002 & 0.00001 & 0.00001 \\ & & 0.81366 & 0.51237 & 0.61325 \\ & & & 0.21436 & 0.21537 \\ & & & & 0.41267 \end{bmatrix}$$

Matrix A_5 obtained after introducing zeros in second row.

$$A_5 = \begin{bmatrix} 0.71235 & 0.61325 & 0 & 0 & 0 \\ & 0.61874 & 0.00002 & 0 & 0 \\ & & 1.48135 & 0.02405 & 0.11049 \\ & & & -0.07568 & -0.10328 \\ & & & & 0.03502 \end{bmatrix}$$

Final tri-diagonal matrix A_6.

$$A_6 = \begin{bmatrix} 0.71235 & 0.61325 & 0 & 0 & 0 \\ & 0.61874 & 0.00002 & 0 & 0 \\ & & 1.48135 & 0.11308 & 0 \\ & & & -0.01292 & 0.11694 \\ & & & & -0.02774 \end{bmatrix}$$

Reduction using 6 decimals

Matrix A_3 obtained after introducing zeros in first row.

$$A_3 = \begin{bmatrix} 0.712350 & 0.613256 & 0 & 0 & 0 \\ & 0.618749 & 0.000012 & 0.000007 & 0.000008 \\ & & 0.813654 & 0.512374 & 0.613255 \\ & & & 0.214360 & 0.215375 \\ & & & & 0.412665 \end{bmatrix}$$

Matrix A_5 obtained after introducing zeros in second row.

$$A_5 = \begin{bmatrix} 0.712350 & 0.613256 & 0 & 0 & 0 \\ & 0.618749 & 0.000016 & 0 & 0 \\ & & 1.486335 & -0.068499 & 0.024712 \\ & & & -0.079492 & -0.102483 \\ & & & & 0.033837 \end{bmatrix}$$

Final tri-diagonal matrix A_6.

$$A_6 = \begin{bmatrix} 0.712350 & 0.613256 & 0 & 0 & 0 \\ & 0.618749 & 0.000016 & 0 & 0 \\ & & 1.486335 & 0.072820 & 0 \\ & & & -0.001012 & -0.115055 \\ & & & & -0.044643 \end{bmatrix}$$

Eigenvalues to eight decimal places

A_0	A_6 (5 decimal computation)	A_6 (6 decimal computation)
1.48991 259	1.48991 000	1.48991 239
1.28058 937	1.28057 855	1.28058 869
−0.14126 880	−0.14126 174	−0.14126 895
0.09203 628	0.09204 175	0.09203 657
0.05051 055	0.05051 145	0.05051 030

obtained and indeed usually much smaller than are indicated by crude estimates of the statistical effect.

Observe that we really have a considerable freedom of choice in the angles of rotation in the (3, 4) and (3, 5) planes. When making the rotation in the (3, 5) plane, for example, we have only to choose $\cos \theta$ and $\sin \theta$ so that they are a true pair to working accuracy and so that, in the six-decimal case for example,

$$|-(0{\cdot}000012) \sin \theta + (0{\cdot}000007)\cos \theta| < \tfrac{1}{2}10^{-6}.$$

The techniques we have described are such that θ corresponds very accurately to that given by $\tan \theta = \dfrac{0{\cdot}000007}{0{\cdot}000012}$, and this is the simplest thing to do. However, any value $\dfrac{x}{y}$ for $\tan \theta$ such that

$$0{\cdot}0000065 \leqslant x \leqslant 0{\cdot}0000075, \qquad 0{\cdot}0000115 \leqslant y \leqslant 0{\cdot}0000125$$

would have worked equally well, though such values correspond to a wide range of final tri-diagonal matrices.

In case this should seem surprising, let us return to exact computation and suppose that complete cancellation occurs so that the (2, 3), (2, 4), (2, 5) elements become zero when a_{15} is eliminated. Now *any* exact plane rotations in the (3, 4) and (3, 5) planes would leave these three zero elements unaltered and could be regarded as appropriate Givens' transformations. Different rotations would give different final tri-diagonal matrices but all such matrices would be exactly similar to A_0. When complete cancellation occurs in the exact process we may choose 'all figures' in the corresponding angles of rotation arbitrarily.

Householder's method

29. For a number of years it seemed that the Givens' process was likely to prove the most efficient method of reducing a matrix to tri-diagonal form, but in 1958 Householder suggested that this reduction could be performed more efficiently using the elementary Hermitian orthogonal matrices rather than plane rotations. In Chapter 6, § 7, we show that the Givens and Householder transformations are essentially the same.

There are $(n-2)$ steps in this reduction, in the rth of which the zeros are introduced in the rth row and rth column without destroying the zeros introduced in the previous steps. The configuration immediately *before* the rth step is illustrated in (29.1) for the case $n = 7$, $r = 4$. We write

$$A_{r-1} = \left. \begin{array}{c} r\{ \\ \\ n-r\{ \end{array} \right. \begin{bmatrix} \times & \times & & & & & \\ \times & \times & \times & & & & \\ & \times & \times & \times & & & \\ & & \times & \times & \times & \times & \times \\ \hline & & & \times & \times & \times & \times \\ & & & \times & \times & \times & \times \\ & & & \times & \times & \times & \times \end{bmatrix} = \begin{bmatrix} C_{r-1} & & O \\ & & \\ \hline & & b_{r-1}^T \\ \hline O & b_{r-1} & B_{r-1} \end{bmatrix},$$

$$\tag{29.1}$$

where b_{r-1} is a vector having $(n-r)$ components, C_{r-1} a tri-diagonal matrix of order r, and B_{r-1} a square matrix of order $(n-r)$. The matrix P_r of the rth transformation may be expressed in the form

$$P_r = \begin{array}{c} r\{ \\ n-r\{ \end{array} \begin{bmatrix} I & O \\ \hline O & Q_r \end{bmatrix} = \begin{bmatrix} I & O \\ \hline O & I - 2v_r v_r^T \end{bmatrix}, \tag{29.2}$$

where v_r is a unit vector having $(n-r)$ components. Hence we have

$$A_r = P_r A_{r-1} P_r = \begin{array}{c} r\{ \\ \\ n-r\{ \end{array} \begin{bmatrix} C_{r-1} & & O \\ & & \\ & & c_r^T \\ \hline O & c_r & Q_r B_{r-1} Q_r \end{bmatrix}, \tag{29.3}$$

where

$$c_r = Q_r b_{r-1}. \tag{29.4}$$

If we choose v_r so that c_r is null except for its first component, then A_r is tri-diagonal as far as its first $(r+1)$ rows and columns are concerned. We have therefore expressed the condition determining P_r in the standard form considered in Chapter 3, §§ 37, 38, but it is convenient to reformulate the results given there in terms of the quantities now involved. We denote the current (i, j) element of A_{r-1} by a_{ij} omitting reference to the suffix r. We shall assume that only the upper triangle of elements is stored and express all formulae in terms of these elements.

The formulae are simplest if we think in terms of the matrix $P_r = I - 2w_r w_r^T$, where w_r is a unit vector having n components of which the first r are zero. We have then

$$P_r = I - 2w_r w_r^T = I - u_r u_r^T / 2K_r^2, \qquad (29.5)$$

where
$$u_{ir} = 0 \qquad (i = 1, 2, ..., r)$$

$$\left. \begin{array}{l} u_{r+1,r} = a_{r,r+1} \mp S_r, \qquad u_{ir} = a_{ri} \quad (i = r+2, ..., n) \\[2mm] S_r = \left(\sum_{i=r+1}^{n} a_{ri}^2 \right)^{\frac{1}{2}}, \qquad 2K_r^2 = S_r^2 \mp a_{r,r+1} S_r \end{array} \right\}. \qquad (29.6)$$

In the equation defining $u_{r+1,r}$, the sign associated with S_r is taken to be that of $a_{r,r+1}$, so that cancellation cannot take place in the addition. The new value of $a_{r,r+1}$ is $\pm S_r$. (We contrast this with Givens' method for which, with our formulation, the super-diagonal elements in the tri-diagonal form are non-negative.) Notice that if all the elements which are to be eliminated in a given step are already zero, the transformation matrix is not the identity matrix as might be expected. (In fact I is not an elementary Hermitian according to our definition.) The transformation matrix differs from the identity matrix in that the $(r+1)$th diagonal element is -1 instead of $+1$. We could arrange to detect this case and omit the corresponding transformation.

Taking advantage of symmetry

30. It is important to take full advantage of symmetry, and to do this the calculation of A_r from

$$A_r = P_r A_{r-1} P_r \qquad (30.1)$$

requires careful formulation. We have in fact

$$A_r = (I - u_r u_r^T / 2K_r^2) A_{r-1} (I - u_r u_r^T / 2K_r^2), \qquad (30.2)$$

and if we write
$$A_{r-1} u_r / 2K_r^2 = p_r, \qquad (30.3)$$

equation (30.2) may be written in the form

$$A_r = A_{r-1} - u_r p_r^T - p_r u_r^T + u_r (u_r^T p_r) u_r^T / 2K_r^2. \qquad (30.4)$$

Hence if we introduce q_r defined by

$$q_r = p_r - \tfrac{1}{2} u_r (u_r^T p_r / 2K_r^2), \qquad (30.5)$$

we have
$$A_r = A_{r-1} - u_r q_r^T - q_r u_r^T, \qquad (30.6)$$

and the right-hand side is in convenient form for taking advantage of symmetry. In fact we are interested only in the matrix of order $(n-r)$ in the bottom right-hand corner of A_r since the elements in the first $(r-1)$ rows and columns are the same as the corresponding elements of A_{r-1} and the transformation P_r is designed to eliminate a_{ri} $(i = r+2,..., n)$ and to convert $a_{r,r+1}$ to $\pm S_r$.

The vector p_r has its first $(r-1)$ components equal to zero and we do not require the rth element. Hence $(n-r)^2$ multiplications are involved in its computation. The number of operations involved in the computation of q_r is of order $(n-r)$ and is relatively negligible. Finally the computation of the relevant elements of A_r from (30.6) involves approximately $(n-r)^2$ multiplications, since we have only to compute the upper triangle of elements. The total number of multiplications in the whole reduction is therefore essentially $\frac{2}{3}n^3$, compared with $\frac{4}{3}n^3$ in Givens' reduction. In the formulation we have given there are essentially n square roots compared with $\frac{1}{2}n^2$ in the Givens process.

Storage considerations

31. If we require the vectors, then details of the transformation matrices must be stored. It is clearly adequate to store the non-zero elements of the u_r and sufficient information for the $2K_r^2$ to be recoverable. Now the vector u_r has $(n-r)$ non-zero elements, while it eliminates only $(n-r-1)$ elements of A_{r-1}. The $(r, r+1)$ element does not become zero and is in fact a super-diagonal element of the final tri-diagonal matrix. However, it is quite convenient to store these super-diagonal elements separately and hence we may overwrite all the non-zero elements of u_r in the space occupied by the original matrix. This is particularly convenient since $u_{ir} = a_{ri}$ $(i = r+2,..., n)$ and hence all elements of u_r except $u_{r+1,r}$ are already in position.

The successive p_r may all be stored in the same set of n registers and we can overwrite each q_r on p_r. Since there are only $(n-r)$ elements in q_r and p_r which are of interest, we can store the super-diagonal elements $\pm S_r$, as they are derived, in the same set of n registers. From $\pm S_r$ and the $u_{r+1,r}$ we can recover the $2K_r^2$ when required, using the relation

$$2K_r^2 = S_r^2 \mp a_{r,r+1}S_r = S_r(S_r \mp a_{r,r+1}) = -(\pm S_r)u_{r+1,r}. \quad (31.1)$$

Hence the whole process requires only $\frac{1}{2}(n^2+3n)$ registers, whereas n^2 are necessary for the Givens process if the details of the transformations are stored.

Householder's process on a computer with a two-level store

32. If Householder's method is performed in this way on a computer having a two-level store then in the rth major step it is necessary to make two transfers of the upper triangle of elements of B_{r-1} (cf. equation (29.1)) to the high-speed store, once in order to compute p_r and then a second time in order to derive the elements of A_r. In the modified form of Givens' method described in § 24 only one such transfer was necessary in each major step, and the same economy can be effected here as follows.

During the computation of A_r, which involves the reading of A_{r-1} from the backing store and its replacement by A_r, u_{r+1} and p_{r+1} might also be computed. A_r differs from A_{r-1} only in rows and columns r to n, and as soon as the $(r+1)$th row of A_r is known u_{r+1} can be computed. As each element of A_r is determined it can be used to compute the corresponding contributions to $A_r u_{r+1}$. Since we assume we are working with the upper triangles only, the computed (i, j) element of A_r must be used to determine the relevant contribution both to $(A_r u_{r+1})_i$ and $(A_r u_{r+1})_j$. When the calculation of A_r and its entry on the backing store is complete, u_{r+1} and $A_r u_{r+1}$ will be determined and hence $2K_{r+1}^2$, p_{r+1} and q_{r+1} may be computed without further reference to the backing store. This technique halves the number of references to the backing store and makes Householder as efficient as Givens in this respect.

If we are to realize the full accuracy attainable by accumulating the inner-products involved in $A_r u_{r+1}$, then each of its elements must be held in double-precision while it is being built up piecemeal as A_r is computed. The working store must therefore be able to accommodate the five vectors u_r, q_r, u_{r+1}, $A_r u_{r+1}$ and the current row of A_{r-1} which is being transformed into a row of A_r. Since $A_r u_{r+1}$ is temporarily in double-precision form, $6n$ registers are required. If transfers to and from the backing store take place in units of k words at a time this may be reduced to $5n+k$ registers since we do not need the whole of the current row of A_{r-1} to be in the store at the same time.

Householder's method in fixed-point arithmetic

33. The technique we described in §§ 29, 30 does not carry over immediately to fixed-point work. In floating-point work the term $(u_r u_r^T) 2K_r^2$ is effectively factorized in the form $u_r (u_r^T / 2K_r^2)$. If those elements which determine the current transformation are very small, then the components of $u_r / 2K_r^2$ will be much greater than unity.

Wilkinson (1960) has described an alternative formulation in which one effectively works with $2(u_r/2K_r)(u_r^T/2K_r)$ which involves one additional square root in each transformation. For high-order matrices the time involved in the additional square roots is negligible compared with the total time taken by the reduction, and moreover the scaling requirements are very easily met in this formulation. There are, however, alternatives in which the extra square roots are avoided, and we now describe one such formulation, partly because it will be required in a later chapter and partly because the error analysis has some instructive features.

Instead of working directly with the vector u_r we use a scaled version defined by

$$y_r = u_r/u_{r+1,r}. \tag{33.1}$$

From (29.5) and (29.6) we have

$$\left. \begin{aligned} y_{ir} &= 0 \quad (i = 1, 2, \dots, r) \\ y_{r+1,r} = 1, \qquad y_{ir} &= a_{ri}/(a_{r,r+1} \mp S_r) \quad (i = r+2, \dots, n) \end{aligned} \right\} \tag{33.2}$$

and

$$P_r = I - u_r u_r^T/2K_r^2 = I - \left(\frac{S_r \mp a_{r,r+1}}{S_r}\right) y_r y_r^T. \tag{33.3}$$

Now if all a_{ri} $(i = r+1, \dots, n)$ are small, the computed S_r will usually have a comparatively high relative error. Nevertheless, all computed elements of y_r will obviously be bounded in modulus by unity, though the computed y_r may differ substantially from that corresponding exactly to the given a_{ri}. In order to make certain that the transformation matrix is accurately orthogonal we should therefore use

$$P_r = I - 2y_r y_r^T/\|y_r\|^2 \tag{33.4}$$

instead of the right-hand side of (33.3).

A_r can now be computed from the sequence

$$p_r = A_{r-1} y_r/\|y_r\|^2, \tag{33.5}$$

$$q_r = p_r - (y_r^T p_r) y_r/\|y_r\|^2, \tag{33.6}$$

$$A_r = A_{r-1} - 2y_r q_r^T - 2q_r y_r^T. \tag{33.7}$$

Since $1 < \|y_r\|^2 < 2$, it will be necessary to introduce factor of $\frac{1}{2}$ at various points in this formulation, but the most convenient way of doing this depends to some extent on the fixed-point facilities which are available and we shall not pursue this further.

We now have a different situation from the one to which we have been accustomed in our error analysis. *The computed transformation matrix may be substantially different from that corresponding to exact computation for the computed* A_{r-1}. However, this proves to be quite unimportant. Because $y_{r+1,r} = 1$, the computed $\|y_r\|^2$ has a low relative error, and hence the matrix computed from (33.4) is almost exactly orthogonal; further, the elements of $P_r A_{r-1} P_r$ in the appropriate positions are zero to working accuracy. These are the only requirements we have to meet.

Numerical example

34. These points are well illustrated in the example of Table 5 which has been computed using the formulation of § 33. The elements of A_0 which determine the first transformation are all very small.

TABLE 5

A_0

$$\begin{bmatrix} 0\cdot23157 & 0\cdot00002 & 0\cdot00003 & 0\cdot00004 \\ 0\cdot00002 & 0\cdot31457 & 0\cdot21321 & 0\cdot31256 \\ 0\cdot00003 & 0\cdot21321 & 0\cdot18175 & 0\cdot21532 \\ 0\cdot00004 & 0\cdot31256 & 0\cdot21532 & 0\cdot41653 \end{bmatrix}$$

$S_1 = 0\cdot00005$

$\|y_1\|^2 = 1\cdot51020$

$y_1^T p_1 / \|y_1\|^2 = 0\cdot49519$

$$\begin{aligned} y_1^T &= (0\cdot00000, \quad 1\cdot00000, \quad 0\cdot42857, \quad 0\cdot57143) \\ p_1^T &= (0\cdot00004, \quad 0\cdot38707, \quad 0\cdot27423, \quad 0\cdot42568) \\ q_1^T &= (0\cdot00004, \quad -0\cdot10812, \quad 0\cdot06201, \quad 0\cdot14271) \end{aligned}$$

A_1

$$\begin{bmatrix} 0\cdot23157 & -0\cdot00005 & 0\cdot00000 & 0\cdot00000 \\ -0\cdot00005 & 0\cdot74705 & 0\cdot18186 & 0\cdot15071 \\ 0\cdot00000 & 0\cdot18186 & 0\cdot07545 & 0\cdot02213 \\ 0\cdot00000 & 0\cdot15071 & 0\cdot02213 & 0\cdot09033 \end{bmatrix}$$

$S_2 = 0\cdot23619$

$\|y_2\|^2 = 1\cdot12997$

$y_2^T p_2 / \|y_2\|^2 = 0\cdot08078$

$$\begin{aligned} y_2^T &= (0\cdot00000, \quad 0\cdot00000, \quad 1\cdot00000, \quad 0\cdot36051) \\ p_2^T &= (0\cdot00000, \quad 0\cdot20903, \quad 0\cdot07383, \quad 0\cdot04840) \\ q_2^T &= (0\cdot00000, \quad 0\cdot20903, \quad -0\cdot00695, \quad 0\cdot01928) \end{aligned}$$

A_2

$$\begin{bmatrix} 0\cdot23157 & -0\cdot00005 & 0\cdot00000 & 0\cdot00000 \\ -0\cdot00005 & 0\cdot74705 & -0\cdot23619 & 0\cdot00000 \\ 0\cdot00000 & -0\cdot23619 & 0\cdot10325 & -0\cdot01142 \\ 0\cdot00000 & 0\cdot00000 & -0\cdot01142 & 0\cdot06253 \end{bmatrix}$$

Accurate computation of first step

$$S_1 = 0\cdot000053852, \quad \|y_1\|^2 = 1\cdot45837, \quad y_1^T p_1 / \|y_1\|^2 = 0\cdot50464$$

$$\begin{aligned} y_1^T &= (0\cdot00000, \quad 1\cdot00000, \quad 0\cdot40622, \quad 0\cdot54162) \\ p_1^T &= (0\cdot00004, \quad 0\cdot39117, \quad 0\cdot27679, \quad 0\cdot42899) \\ q_1^T &= (0\cdot00004, \quad -0\cdot11347, \quad 0\cdot07180, \quad 0\cdot15567) \end{aligned}$$

A_1

$$\begin{bmatrix} 0\cdot23157 & -0\cdot00005 & 0\cdot00000 & 0\cdot00000 \\ & 0\cdot76845 & 0\cdot16180 & 0\cdot12414 \\ & & 0\cdot06508 & 0\cdot01107 \\ & & & 0\cdot07927 \end{bmatrix}$$

As no scaling is used, the computed value of S_1 has a high relative error. In fact the computed value is $0 \cdot 00005$ while the true value is $0 \cdot 000053852\ldots$. If we were to associate the factor $(S_1 + a_{1.2})/S_1$ with $y_1 y_1^T$ as in (33.3), the transformation matrix would not be even approximately orthogonal. Using the formulation (33.4), almost exact orthogonality is assured and we are still fully justified in taking the relevant elements of A_1 to be zeros. However A_1 is quite different from the matrix which would have been obtained using floating-point and the formulation of § 30. That formulation gives an orthogonal matrix which is close to the exact matrix corresponding to A_0.

At the end of the table we give the vector y_1 corresponding to accurate computation, and the corresponding matrix A_1. It will be seen that both the y_1 and the A_1 are substantially different.

Error analyses of Householder's method

35. Quite a number of detailed error analyses of Householder's method have been given. Our general analysis of Chapter 3, § 45, needs only minor modifications to cover the symmetric case. Equation (45.5) there shows that if C is the computed tri-diagonal matrix obtained using floating-point with accumulation, and R is the exact product of the orthogonal matrices corresponding to the computed A_r, then certainly

$$\|C - R A_0 R^T\|_E \leqslant 2K_1 n 2^{-t} (1 + K_1 2^{-t})^{2n} \|A_0\|_E, \qquad (35.1)$$

for some constant K_1. For the floating-point operations used on ACE we have shown that $2K_1$ is certainly less than 40. Hence in the notation of § 26 we have

$$\left[\frac{\sum (\mu_i - \lambda_i)^2}{\sum \lambda_i^2} \right]^{\frac{1}{2}} \leqslant 40 n 2^{-t} (1 + 20 \cdot 2^{-t})^{2n} \qquad (35.2)$$

and as usual, the final factor on the right of (35.2) is unimportant in the useful range of this bound. The first result of this kind was obtained by Ortega (1963) by an argument which was not essentially different from ours. As we pointed out in Chapter 3, § 45, the factor $n2^{-t}$ is scarcely likely to be improved upon, since such a factor is obtained even if we assume that each transformation is performed *exactly* and the resultant A_r then rounded to t binary places before passing to the next step.

Ortega also obtained a bound corresponding to standard floating-point computation. His bound when expressed in our notation is essentially of the form

$$\left[\frac{\sum (\mu_i - \lambda_i)^2}{\sum \lambda_i^2}\right]^{\frac{1}{2}} \leqslant 2K_2 n^2 2^{-t}(1 + K_2 n 2^{-t})^{2n}. \qquad (35.3)$$

The constants K_1 and K_2 depend rather critically on the details of the arithmetic, and careful attention to detail in the analysis may reduce somewhat the values which have been obtained, but comparing (26.2) and (35.3) we see that over the useful range we have bounds of order $n^{\frac{3}{2}}2^{-t}$ and $n^2 2^{-t}$ respectively for Givens and Householder performed in standard floating-point. For large n, the bound for the Givens process is therefore the better. There seems no doubt that the bound (35.3) is rather weaker than most which we have obtained, but there is as yet no conclusive evidence as to which gives the more accurate results in practice. The difficulty is that, because of the statistical distribution of rounding errors, the eigenvalues obtained in practice are very accurate, and significant information on the relative accuracy of the two methods would require the solution of a very large number of high-order matrices.

Although the upper bound (35.2) for Householder's method performed in floating-point with accumulation is, in any case, very encouraging, experience seems to suggest that this is a severe over-estimate and even a value of $n^{\frac{1}{2}}2^{-t}$ has not usually been attained for matrices which I have investigated in detail. If we accept the value $n^{\frac{1}{2}}2^{-t}$ as a guide, then it is obvious that the effect of rounding errors is not likely to be a limiting factor in the foreseeable future. If we take $t = 40$ and $n = 10^4$, for example, then $n^{\frac{1}{2}}2^{-t}$ is approximately 10^{-10}. Hence we might well expect that

$$\left[\frac{\sum (\lambda_i - \mu_i)^2}{\sum \lambda_i^2}\right]^{\frac{1}{2}} < 10^{-10}, \qquad (35.4)$$

a most satisfactory result. However, a full matrix of order 10^4 has 10^8 elements and its reduction involves some $\frac{2}{3}10^{12}$ multiplications. Even with a $1\mu s$ multiplication time the reduction would take about 200 hours.

A detailed error analysis of Householder's method in fixed-point with accumulation has been given by Wilkinson (1962b). This analysis showed that

$$\max |\lambda_i - \mu_i| < K_3 n^2 2^{-t} \qquad (35.5)$$

provided A_0 was scaled so that

$$\|A_0\|_2 < \tfrac{1}{2} - K_3 n^2 2^{-t}. \qquad (35.6)$$

This was for the formulation involving two square roots per major step, but provided inner-products are accumulated, several other formulations give results of the same type as (35.5) with varying values of K_3. Again experience has suggested that even $n2^{-t}$ is a considerable over-estimate in general.

For fixed-point computation without accumulation a bound of the form

$$\max |\lambda_i - \mu_i| < K_4 n^{\frac{3}{2}} 2^{-t} \qquad (35.7)$$

is the best that has so far been obtained.

Eigenvalues of a symmetric tri-diagonal matrix

36. In order to complete the Givens or Householder method we must solve the eigenproblem for a symmetric tri-diagonal matrix. Tri-diagonal matrices commonly arise as primary data in many physical problems and hence are important in their own right. Since there are only $(2n-1)$ independent elements in such a matrix we might expect this problem to be considerably simpler than that for the original matrix.

We consider then a symmetric tri-diagonal matrix C and for convenience we write

$$c_{ii} = \alpha_i, \qquad c_{i,i+1} = c_{i+1,i} = \beta_{i+1}. \qquad (36.1)$$

If r of the β_i are zero, then C is the direct sum of $(r+1)$ tri-diagonal matrices of lower order. We may write

$$
C =
\begin{bmatrix}
C^{(1)} & & & & \\
& C^{(2)} & & & \\
& & \cdot & & \\
& & & \cdot & \\
& & & & \cdot \\
& & & & & C^{(r+1)}
\end{bmatrix}, \qquad (36.2)
$$

where $C^{(s)}$ is of order m_s and $\sum m_i = n$. The eigenvalues of the $C^{(i)}$ together comprise those of C, and if x is an eigenvector of $C^{(s)}$ the corresponding eigenvector y of C is given by

$$y^T = (0, 0, ..., 0, x^T, 0, ..., 0). \qquad (36.3)$$
$$\underbrace{\quad}_{m_1} \underbrace{\quad}_{m_2} \quad \underbrace{\quad}_{m_{s-1}} \underbrace{\quad}_{m_s} \underbrace{\quad}_{m_{s+1}} \underbrace{\quad}_{m_{r+1}}$$

Hence, without loss of generality, we may assume that none of the β_i is zero and this we do from now on.

Sturm sequence property

37. If we denote the leading principal minor of order r of $(C - \lambda I)$ by $p_r(\lambda)$ then, defining $p_0(\lambda)$ to be 1, we have

$$p_1(\lambda) = \alpha_1 - \lambda, \tag{37.1}$$

$$p_i(\lambda) = (\alpha_i - \lambda)p_{i-1}(\lambda) - \beta_i^2 p_{i-2}(\lambda) \quad (i = 2,\dots, n). \tag{37.2}$$

The zeros of $p_r(\lambda)$ are the eigenvalues of the leading principal submatrix of order r of C and hence are all real. We denote this submatrix by C_r.

Now we know that eigenvalues of each C_r separate those of C_{r+1}, at least in the weak sense (Chapter 2, § 47). We show that if none of the β_i is zero, then we must have strict separation. For if μ is a zero of $p_r(\lambda)$ and $p_{r-1}(\lambda)$, then from (37.2) for $i = r$ and the fact that β_r is non-zero, μ is also a zero of $p_{r-2}(\lambda)$. Hence from (37.2) for $i = r-1$, it is also a zero of $p_{r-3}(\lambda)$. Continuing in this way we find that μ is a zero of $p_0(\lambda)$, and since $p_0(\lambda) = 1$ we have a contradiction.

From this we deduce that if a symmetric matrix has an eigenvalue of multiplicity k, then the tri-diagonal matrix obtained by performing the Givens method or Householder method exactly must have at least $(k-1)$ zero super-diagonal elements. On the other hand the presence of a zero super-diagonal element in the Givens or Householder tri-diagonal form *need* not imply the presence of any multiple roots.

38. The strict separation of the zeros for symmetric tri-diagonal matrices with non-zero super-diagonal elements forms the basis of one of the most effective methods of determining the eigenvalues (Givens, 1954). We first prove the following result:

Let the quantities $p_0(\mu), p_1(\mu),\dots, p_n(\mu)$ be evaluated for some value of μ. Then $s(\mu)$, the number of agreements in sign of consecutive members of this sequence, is the number of eigenvalues of C which are strictly greater than μ.

In order that the above statement should be meaningful for all values of μ, we must associate a sign with zero values of $p_r(\mu)$. If $p_r(\mu) = 0$ then $p_r(\mu)$ is taken to have the opposite sign to that of $p_{r-1}(\mu)$. (Note that no two consecutive $p_i(\mu)$ can be zero.)

The proof is by induction. Assume that the number of agreements s in the sequence $p_0(\mu), p_1(\mu),\dots, p_r(\mu)$ is the number of eigenvalues of C_r (i.e. the number of zeros of $p_r(\lambda)$) which are greater than μ. Then we may denote the eigenvalues of C_r by $x_1, x_2,\dots, x_r$ where

$$x_1 > x_2 > \dots > x_s > \mu > x_{s+1} > x_{s+2} > \dots > x_r, \tag{38.1}$$

and there are no coincident eigenvalues since no β_i is zero. Now the eigenvalues $y_1, y_2,..., y_{r+1}$ of C_{r+1} separate those of C_r in the strict sense, and hence

$$y_1 > x_1 > y_2 > x_2 > ... > y_s > x_s > y_{s+1} > x_{s+1} > ...$$
$$... > y_r > x_r > y_{r+1}, \quad (38.2)$$

showing that either s or $s+1$ eigenvalues of C_{r+1} are greater than μ. Clearly we have

$$p_r(\mu) = \prod_{i=1}^{r} (x_i - \mu) \quad \text{and} \quad p_{r+1}(\mu) = \prod_{i=1}^{r+1} (y_i - \mu). \quad (38.3)$$

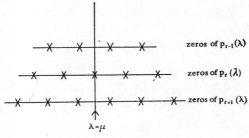

zeros of $p_{r-1}(\lambda)$

zeros of $p_r(\lambda)$

zeros of $p_{r+1}(\lambda)$

$\lambda = \mu$

FIG. 1.

If none of the x_i or y_i is μ then our result is immediately established. For in this case, if $y_{s+1} > \mu$, then $p_r(\mu)$ and $p_{r+1}(\mu)$ have the same sign and hence there is one more agreement in the augmented series, while if $y_{s+1} < \mu$ then $p_r(\mu)$ and $p_{r+1}(\mu)$ have different signs, so that the augmented sequence has the same number of agreements. In either case we have the appropriate correspondence with the number of eigenvalues greater than μ.

If $y_{s+1} = \mu$ then we are to take the sign of $p_{r+1}(\mu)$ to be different from that of $p_r(\mu)$ (which cannot be zero in this case) so that the augmented series has the same number of agreements, corresponding correctly with the same number of eigenvalues strictly greater than μ.

If $x_s = \mu$ then the situation must be as illustrated typically in Fig. 1 for the case $r = 5, s = 2$. The sign associated with the zero $p_r(\mu)$ is to be taken to be the opposite of that associated with $p_{r-1}(\mu)$. It follows from the strict separation of the zeros of $p_{r-1}(\lambda)$, $p_r(\lambda)$ and $p_{r+1}(\lambda)$ that $p_{r+1}(\lambda)$ must have exactly one extra zero greater than μ and one extra zero less than μ when compared with $p_{r-1}(\lambda)$. Hence $p_{r+1}(\mu)$ must be of opposite sign to $p_{r-1}(\mu)$ and therefore of the same sign as we have assumed for $p_r(\mu)$. There is thus one more agreement, corresponding correctly to one more zero greater than μ.

The reader should satisfy himself that if we are interested only in the eigenvalues of C itself, then the signs associated with zero values of $p_r(\mu)$ $(r = 1, ..., n-1)$ are of no importance. We have only to apply the strict rule to $p_n(\mu)$.

We may say that the polynomials $p_0(\lambda), p_1(\lambda), ..., p_n(\lambda)$ have *the Sturm sequence property*.

Method of bisection

39. This Sturm sequence property may be used to locate any individual eigenvalue, the kth in order of decreasing value say, without reference to any of the others. For suppose we have two values a_0 and b_0 such that

$$b_0 > a_0, \qquad s(a_0) \geqslant k, \qquad s(b_0) < k \qquad (39.1)$$

so that we know that λ_k lies in the interval (a_0, b_0). Then we can locate λ_k in an interval (a_p, b_p) of width $(b_0 - a_0)/2^p$ in p steps of the iterative process of which the rth step is as follows. Let

$$c_r = \tfrac{1}{2}(a_{r-1} + b_{r-1})$$

be the mid-point of (a_{r-1}, b_{r-1}). Then compute the sequence

$$p_0(c_r), p_1(c_r), ..., p_n(c_r)$$

and hence determine $s(c_r)$.

If $s(c_r) \geqslant k$, then take $a_r = c_r$ and $b_r = b_{r-1}$.

If $s(c_r) < k$, then take $a_r = a_{r-1}$ and $b_r = c_r$.

In either case we have

$$s(a_r) \geqslant k \quad \text{and} \quad s(b_r) < k, \qquad (39.2)$$

so that λ_k is in the interval (a_r, b_r). Suitable values for a_0 and b_0 are $\pm \|C\|_\infty$, and we have $\|C\|_\infty = \max_i \{|\beta_i| + |\alpha_i| + |\beta_{i+1}|\}$.

Numerical stability of the bisection method

40. With exact computation the method of § 39 could be used to determine any eigenvalue to any prescribed accuracy. In practice rounding errors will limit the attainable accuracy but the method is nevertheless exceptionally stable, as we now show. We give an error analysis corresponding to the use of standard floating-point arithmetic.

We show first that for any value of μ the computed values of the sequence $p_0(\mu)$, $p_1(\mu)$,..., $p_n(\mu)$ are the exact values corresponding to a modified tri-diagonal matrix having elements $\alpha_i + \delta\alpha_i$, $\beta_i + \delta\beta_i$ where $\delta\alpha_i$ and $\delta\beta_i$ satisfy

$$\left. \begin{aligned} |\delta\alpha_i| &< (3{\cdot}01)2^{-t}(|\alpha_i| + |\mu|) \\ |\delta\beta_i| &< (1{\cdot}51)2^{-t}(|\beta_i|) \end{aligned} \right\}. \qquad (40.1)$$

The proof is by induction. We assume that the computed

$$p_0(\mu), p_1(\mu),..., p_{r-1}(\mu)$$

are exact for a matrix having modified elements up to $\alpha_{r-1} + \delta\alpha_{r-1}$, $\beta_{r-1} + \delta\beta_{r-1}$, and then show that the computed $p_r(\mu)$ is the exact value for a matrix having these same modified elements in rows 1, 2,..., $r-1$ and modified elements $\alpha_r + \delta\alpha_r$ and $\beta_r + \delta\beta_r$. We have for the computed $p_r(\mu)$,

$$\begin{aligned} p_r(\mu) &= fl[(\alpha_r - \mu)p_{r-1}(\mu) - \beta_r^2 p_{r-2}(\mu)] \\ &\equiv (\alpha_r - \mu)p_{r-1}(\mu)(1+\epsilon_1) - \beta_r^2 p_{r-2}(\mu)(1+\epsilon_2), \qquad (40.2) \end{aligned}$$

where, from the usual inequalities,

$$(1 - 2^{-t})^3 < 1 + \epsilon_i < (1 + 2^{-t})^3 \quad (i = 1, 2). \qquad (40.3)$$

Hence we can justify the computed $p_r(\mu)$ without modifying previous $\alpha_i + \delta\alpha_i$ and $\beta_i + \delta\beta_i$ if we take

$$\alpha_r + \delta\alpha_r - \mu = (\alpha_r - \mu)(1+\epsilon_1), \qquad (40.4)$$

$$(\beta_r + \delta\beta_r)^2 = \beta_r^2(1+\epsilon_2), \qquad (40.5)$$

giving certainly

$$|\delta\alpha_r| < (3{\cdot}01)2^{-t}(|\alpha_r| + |\mu|), \qquad (40.6)$$

$$|\delta\beta_r| < (1{\cdot}51)2^{-t} |\beta_r|. \qquad (40.7)$$

Our result is now established, since it is obviously true when $r = 1$. Notice that (40.7) implies that if β_r is non-zero then $\beta_r + \delta\beta_r$ is also non-zero and hence the perturbed matrix also has distinct eigenvalues.

We may denote the perturbed tri-diagonal matrix by $(C + \delta C_\mu)$, where we use the suffix μ to emphasize that δC_μ is a function of μ. The number of agreements in sign between consecutive members of the computed sequence is therefore the number of zeros of $(C + \delta C_\mu)$ which are greater than μ.

41. Let us suppose for convenience that α_i and β_i satisfy the relations

$$|\alpha_i|, \ |\beta_i| \leqslant 1. \tag{41.1}$$

Then as in § 39 all the values of μ which are used in the bisection process lie in the interval $(-3, 3)$. Hence we have

$$|\delta\alpha_r| < (12 \cdot 04)2^{-t}, \qquad |\delta\beta_r| \leqslant (1 \cdot 51)2^{-t} \tag{41.2}$$

for any value of μ. The matrices δC_μ are therefore uniformly bounded and, using the infinity norm, we see that the eigenvalues of δC_μ always lie in the intervals of width

$$2[12 \cdot 04 + 2(1 \cdot 51)]2^{-t} = (30 \cdot 12)2^{-t} = 2d \quad \text{(say)}, \tag{41.3}$$

centred on the eigenvalues λ_i of C (Chapter 2, § 44).

Now the vital question that has to be answered in each step of the bisection process when computing the kth eigenvalue is, 'Are there fewer than k eigenvalues of C which are greater than c_r, the mid-point of the current interval?' If the answer to this question is given correctly every time, then λ_k is always located in a correct interval. This is true whether or not the actual number is determined correctly!

We now show that we must obtain the correct answer in practice if c_r is outside the interval (41.3) centred on λ_k, however close the eigenvalues may be and even if some of the other eigenvalues of C lie inside the λ_k interval. For if c_r is less than $\lambda_k - d$, then certainly the first k eigenvalues of $(C + \delta C_\mu)$ must be greater than c_r and possibly some others. The computed sequence gives the correct number of agreements corresponding to $(C + \delta C_\mu)$ and hence it must give the correct answer, 'no', to our question even if it gets the actual number wrong. Similarly if c_r is greater than $\lambda_k + d$ then certainly the kth eigenvalue and all smaller eigenvalues of $(C + \delta C_\mu)$ must be less than c_r. Hence we must obtain the correct answer, 'yes', to our question.

The history of the process is therefore as follows:
Either (i) the correct answer is obtained every time in which case λ_k lies in each of the intervals (a_r, b_r) or (ii) there is a first step at which the process gives an incorrect answer. This cannot happen unless c_r is inside the λ_k interval. (Note that there may have been previous correct decisions corresponding to a c_r inside the λ_k interval.)

We now show that in this case all subsequent intervals (a_i, b_i) have either one or both of their end-points inside the λ_k interval.

By hypothesis this is true for (a_{r+1}, b_{r+1}). Now if both are inside the λ_k interval then obviously all subsequent a_i and b_i are inside. If only one, b_{r+1} say, is inside, then b_{r+1} must be the point at which the incorrect decision was made. The situation is therefore as shown in Fig. 2.

If c_{r+2}, the mid-point of (a_{r+1}, b_{r+1}), is inside the interval, then whatever the decision, c_{r+2} is an end-point of (a_{r+2}, b_{r+2}). If c_{r+2} is outside the interval then the correct decision is made with respect to c_{r+2}

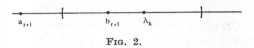

FIG. 2.

and we have $b_{r+2} = b_{r+1}$. Hence in any case at least one end-point of (a_{r+2}, b_{r+2}) is inside the interval. Continuing the argument, our result follows.

Hence, if we stop at the pth stage, the centre c_{p+1} of the final interval (a_p, b_p) is such that

$$|c_{p+1} - \lambda_k| < (15 \cdot 06)2^{-t} + (b_0 - a_0)2^{-p-1} < (15 \cdot 06)2^{-t} + 3.2^{-p}. \qquad (41.4)$$

Numerical example

42. In Table 6 we show the application of the Sturm sequence property to the location of the third eigenvalue of a tri-diagonal matrix of order 5 using 4-decimal floating-point arithmetic. The infinity norm of the matrix is certainly less than $1 \cdot 7$ so that all eigenvalues lie in the interval $(-1 \cdot 7, 1 \cdot 7)$. The first bisection point is

TABLE 6

i	0	1	2	3	4	5	
α_i	—	0·2165	0·3123	0·4175	0·4431	0·6124	
β_i^2	—	—	0·7163	0·3125	0·2014	0·5016	
$p_i(0 \cdot 0000)$	$+1 \cdot 0000$	$+10^0(0 \cdot 2165)$	$-10^0(0 \cdot 6487)$	$-10^0(0 \cdot 3385)$	$-10^{-1}(0 \cdot 1934)$	$+10^0(0 \cdot 1579)$	$s = 3$
$p_i(0 \cdot 8500)$	$+1 \cdot 0000$	$-10^0(0 \cdot 6335)$	$-10^0(0 \cdot 3757)$	$+10^0(0 \cdot 3605)$	$-10^{-1}(0 \cdot 7102)$	$-10^0(0 \cdot 1640)$	$s = 2$
$p_i(0 \cdot 4250)$	$+1 \cdot 0000$	$-10^0(0 \cdot 2085)$	$-10^0(0 \cdot 6928)$	$+10^{-1}(0 \cdot 7035)$	$+10^0(0 \cdot 1408)$	$-10^{-2}(0 \cdot 8902)$	$s = 2$

therefore zero, at which there are 3 agreements, showing that λ_3 is in the interval $(0, 1 \cdot 7)$. The second bisection point is $0 \cdot 85$, and here the Sturm sequence has 2 agreements showing that λ_3 is in the interval $(0, 0 \cdot 85)$. The third bisection point is therefore $0 \cdot 425$ and gives 2 agreements. Hence λ_3 is between 0 and 0·425. Notice that we have also obtained the following information:

$$\lambda_4 \text{ and } \lambda_5 \text{ lie between } -1·7 \text{ and } 0.$$

$$\lambda_1 \text{ and } \lambda_2 \text{ lie between } 0·85 \text{ and } 1·7.$$

General comments on the bisection method

43. If we use t stages of bisection then, from (41.4), the bound for the error in the kth computed eigenvalue is $(18·06)2^{-t}$, and from the analysis it is evident that this will usually be an overestimate. The most remarkable thing about this result is that it is independent of n.

There are essentially $2n$ multiplications and $2n$ subtractions in each computation of the sequence $p_0, p_1, \ldots, p_n$ (the β_i^2 may be computed once and for all), so that if all n eigenvalues are found independently using t bisection steps, then $2n^2t$ multiplications and subtractions are involved. For large n then, the time is of order n^2 while the time taken to perform either a Givens or Householder reduction is of order n^3. For sufficiently large n the computation of the eigenvalues is therefore small compared with either reduction, but because of the factor $2t$ associated with n this is not usually true until n is of order 100 or more. In practice it is uncommon for all the eigenvalues of a large matrix to be required.

The time taken by the bisection method is independent of the separation of the eigenvalues and, as we have shown, the accuracy is quite unaffected even by dense clusters. We are not obliged to use bisection all the time and once an eigenvalue has been isolated in an interval one can switch to an iterative technique which has ultimate quadratic or cubic convergence. We shall discuss such methods in Chapter 7. However, if there are many close eigenvalues, the gain from switching over may be quite slight. Methods which are quadratic or cubic for simple eigenvalues are usually only linearly convergent for multiple eigenvalues. If therefore the matrix has pathologically close eigenvalues (it cannot have true multiple eigenvalues since no β_i is zero) the quadratic or cubic convergence may not come into evidence and the alternative technique may even be slower than bisection.

The bisection method has very great flexibility. We may use it to find specifically selected eigenvalues $\lambda_{r_1}, \lambda_{r_2}, \ldots, \lambda_{r_s}$, the eigenvalues in a given interval (x, y) or a prescribed number of eigenvalues to the

In Table 7 we give the eigenvalues of W_{21}^{+} and W_{21}^{-} to 7 decimal places. The matrix W_{21}^{+} has a number of pairs of eigenvalues which are quite remarkably close. In fact λ_1 and λ_2 agree to 15 significant decimals and λ_3 and λ_4 agree to 11 significant decimals; thereafter the pairs gradually separate. For larger values of n this phenomenon becomes even more marked and it can be shown that $\lambda_1 - \lambda_2$ is roughly of the order of $(n!)^{-2}$. Admittedly the matrix W_{21}^{+} is not normalized, but $\|\frac{1}{11}W_{21}^{+}\|_{\infty} = 1$, so that even after normalization we have $\beta_i = 1/11$

TABLE 7

Eigenvalues of W_{21}^{+}		Eigenvalues of W_{21}^{-}
10·74619 42	5·00024 44	±10·74619 42
10·74619 42	4·99978 25	±9·21067 86
9·21067 86	4·00435 40	±8·03894 11
9·21067 86	3·99604 82	±7·00395 20
8·03894 11	3·04309 93	±6·00022 57
8·03894 11	2·96105 89	±5·00000 82
7·00395 22	2·13020 92	±4·00000 02
7·00395 18	1·78932 14	±3·00000 00
6·00023 40	0·94753 44	±2·00000 00
6·00021 75	0·25380 58	±1·00000 00
	−1·12544 15	0·00000 00

and no β_i can be regarded as pathologically small. Note that a number of the dominant eigenvalues of W_{21}^{-} are remarkably close to some of the eigenvalues of W_{21}^{+}; this too becomes even more marked for larger n. The eigenvalues of W_{21}^{-} are equal and opposite in pairs and it is trivial to show that this is true in general of the W_{2n+1}^{-}.

The eigenvectors of the matrix W_{2n+1}^{+} can be determined indirectly by working with two matrices U_{n+1} and V_n having well-separated eigenvalues. For simplicity we illustrate the reduction for the case $n = 3$, the general case being obvious from this. We write

$$U_4 = \begin{bmatrix} 3 & 1 & & \\ 1 & 2 & 1 & \\ & 1 & 1 & 1 \\ & & 2 & 0 \end{bmatrix}, \qquad V_3 = \begin{bmatrix} 3 & 1 & \\ 1 & 2 & 1 \\ & 1 & 1 \end{bmatrix}. \qquad (45.4)$$

Now if u is an eigenvector of U_4 corresponding to the eigenvalue λ, it is immediately apparent that λ is an eigenvalue of W_7^{+} and the corresponding eigenvector w_1 is given by

$$w_1^{T} = (u_1, u_2, u_3, u_4, u_3, u_2, u_1), \qquad (45.5)$$

and is symmetric about the middle element. On the other hand if v is an eigenvector of V_3 corresponding to the eigenvalue μ, then μ is an eigenvalue of W_7^+ and the corresponding eigenvector w_2 is given by

$$w_2^T = (v_1,\, v_2,\, v_3,\, 0,\, -v_3,\, -v_2,\, -v_1) \qquad (45.6)$$

and is antisymmetric with respect to the middle element.

The eigenvalues of W_{2n+1}^+ are therefore those of U_{n+1} and V_n and each of these last two has well separated eigenvalues. Corresponding to $n = 10$, the largest eigenvalues of U_{11} and V_{10} are

$$10{\cdot}74619\ 41829\ 0339\ldots \quad \text{and} \quad 10{\cdot}74619\ 41829\ 0332\ldots$$

respectively! The corresponding eigenvectors u and v of U_{11} and V_{10} are given in Table 8.

TABLE 8

u	v
1·00000 00000	1·00000 00000
0·74619 41829	0·74619 41829
0·30299 99415	0·30299 99415
0·08590 24939	0·08590 24939
0·01880 74813	0·01880 74813
0·00336 14646	0·00336 14646
0·00050 81471	0·00050 81471
0·00006 65944	0·00006 65942
0·00000 77063	0·00000 77045
0·00000 08061	0·00000 07905
0·00000 01500	

The corresponding eigenvectors w_1 and w_2 of W_{21}^+ are constructed as we have just described. Notice that $(w_1 + w_2)$ is almost null in its lower half and $(w_1 - w_2)$ is almost null in its upper half. The components of the vectors w_1 and w_2 have very pronounced minima.

It is worth remarking that whenever an eigenvector of a tri-diagonal matrix C has a very pronounced minimum at an internal point, then C must have some very close eigenvalues. Suppose, for example, that a normalized matrix of order 17 has an eigenvalue λ corresponding to an eigenvector u with components having logarithms which are approximately as follows:

$$(0, -1, -4, -8,\ -12, -12,\ -8, -3, 0, -2, -8,\ -11, -12,\ -8, -3, -1, 0).$$

Consider now the vectors w_1, w_2, w_3 defined by

$$w_1^T = (u_1, u_2, u_3, u_4, 0,..., 0), \qquad (45.7)$$

$$w_2^T = (0,..., 0, u_7, u_8, u_9, u_{10}, u_{11}, 0,..., 0), \qquad (45.8)$$

$$w_3^T = (0,..., 0, u_{14}, u_{15}, u_{16}, u_{17}). \qquad (45.9)$$

The residual vectors $(Cw_i - \lambda w_i)$ all have components which are either zero or of order 10^{-11}; this follows immediately from the tri-diagonal nature of C. Hence the three *orthogonal* vectors w_1, w_2, w_3 all give very small residuals corresponding to the value λ, and the latter is therefore almost a triple eigenvalue (Chapter 3, § 57).

46. When the original tri-diagonal matrix has some zero β_i we have suggested that it be regarded as the direct sum of a number of smaller tri-diagonal matrices. Suppose, however, we replace all zero β_i by ϵ (a small quantity) and work with the resulting matrix $(C+\delta C)$ instead of C. The matrix δC is symmetric and all eigenvalues λ of δC satisfy

$$|\lambda| < \|\delta C\|_\infty < 2\epsilon. \qquad (46.1)$$

The eigenvalues of $(C+\delta C)$ are therefore all distinct and lie in the intervals of width 4ϵ centred on the eigenvalues of C. If C has components bounded in modulus by unity, and we take $\epsilon = 2^{-t-2}$ (say), then the errors resulting from the addition of the ϵ elements are clearly quite negligible.

We need no longer split $(C+\delta C)$ into smaller submatrices since it has no zero off-diagonal elements. This modification makes for administrative simplicity in the programme but it is obviously somewhat uneconomic in terms of computing time. It has, however, the advantage of dealing with C as a whole, and hence we can find a specific eigenvalue, the rth (say), without dealing separately with the smaller matrices.

If economy of computing time is our main concern we can take a step in the opposite direction and replace any small β_i in the original matrix C by zero, thereby giving a matrix $(C+\delta C)$ which splits into a number of submatrices. If we liquidate all β_i satisfying

$$|\beta_i| < \epsilon, \qquad (46.2)$$

then by the same argument as that used in the last paragraph the maximum error thereby made in any eigenvalue is 2ϵ. For any eigenvalue which is reasonably well separated from the others the error introduced is much smaller than this, as we now show.

To simplify the notation we assume that only β_p and β_q satisfy the relation (46.2). We therefore work with $(C+\delta C)$, and this is the direct sum of three tri-diagonal matrices. We write

$$
C = \begin{bmatrix} C^{(1)} & \beta_p & \\ \hline \beta_p & C^{(2)} & \beta_q \\ \hline & \beta_q & C^{(3)} \end{bmatrix}, \qquad C + \delta C = \begin{bmatrix} C^{(1)} & & \\ \hline & C^{(2)} & \\ \hline & & C^{(3)} \end{bmatrix}. \quad (46.3)
$$

Suppose $C^{(1)}u = \mu u$, so that μ is an eigenvalue of $C^{(1)}$ and therefore of $(C+\delta C)$. If we write

$$
v^T = (u^T \mid 0, 0,..., 0 \mid 0, 0,..., 0), \quad (46.4)
$$

where v is partitioned conformally with C, then defining η by the relation
$$
Cv - \mu v = \eta, \quad (46.5)
$$
we have
$$
\eta^T = (0,..., 0 \mid \beta_p v_{p-1}, 0,..., 0 \mid 0, 0,..., 0). \quad (46.6)
$$

Hence η is orthogonal to v and μ is the Rayleigh quotient of C corresponding to v (Chapter 3, § 54). Assuming v is normalized, we obviously have
$$
\|\eta\|_2 \leqslant \epsilon. \quad (46.7)
$$

Hence, from Chapter 3, § 55, if all but one of the eigenvalues of C are at a distance greater than a from μ, then there is an eigenvalue λ of C such that
$$
|\lambda - \mu| < \frac{\epsilon^2}{a} \Big/ \left(1 - \frac{\epsilon^2}{a^2}\right). \quad (46.8)
$$

If, for example, we know that the minimum separation of the eigenvalues is 10^{-2}, and we are working to 10 decimal places, then we can replace by zero all β_i satisfying

$$
|\beta_i| < 10^{-6}. \quad (46.9)
$$

Fixed-point computation of the eigenvalues

47. Floating-point computation is almost essential for the bisection algorithm as we have described it in §§ 37–39. If fixed-point is used so much scaling is involved that the computation is virtually *ad hoc* floating-point. However, since the essential requirement is the evaluation of the leading principal minors of $(C-\mu I)$ for various values of μ, we may use the technique described in Chapter 4, § 49. Because $(C-\mu I)$ is of tri-diagonal form we find that this technique is identical with Gaussian elimination with partial pivoting. As the process will be required again when we consider the eigenvector problem we shall describe it in detail.

There are $n-1$ major steps, rows $r+1, r+2, ..., n$ being as yet unmodified at the beginning of the rth major step. The configuration at the beginning of this step in the case $n = 6$, $r = 3$ is given by

$$
\begin{bmatrix}
u_1 & v_1 & w_1 & & & \\
m_2 & & u_2 & v_2 & w_2 & & \\
m_3 & & & p_3 & q_3 & & \\
& & & \beta_4 & \alpha_4 - \mu & \beta_5 & \\
& & & & \beta_5 & \alpha_5 - \mu & \beta_6 \\
& & & & & \beta_6 & \alpha_6 - \mu
\end{bmatrix}. \qquad (47.1)
$$

The notation will become clear as we describe the rth step which is as follows.

(i) If $|\beta_{r+1}| > p_r$, then interchange rows r and $r+1$. Denote the elements of the resulting row r by u_r, v_r, w_r and the resulting row $r+1$ by $x_{r+1}, y_{r+1}, z_{r+1}$, so that if an interchange has taken place we have

$$u_r = \beta_{r+1}, \quad v_r = \alpha_{r+1} - \mu, \quad w_r = \beta_{r+2}; \quad x_{r+1} = p_r, \quad y_{r+1} = q_r, \quad z_{r+1} = 0$$

and otherwise

$$u_r = p_r, \quad v_r = q_r, \quad w_r = 0; \quad x_{r+1} = \beta_{r+1}, \quad y_{r+1} = \alpha_{r+1} - \mu, \quad z_{r+1} = \beta_{r+2}.$$

The configuration corresponding to (47.1) is now

$$
\begin{bmatrix}
u_1 & v_1 & w_1 & & & \\
m_2 & & u_2 & v_2 & w_2 & & \\
m_3 & & & u_3 & v_3 & w_3 & \\
& & & x_4 & y_4 & z_4 & \\
& & & & \beta_5 & \alpha_5 - \mu & \beta_6 \\
& & & & & \beta_6 & \alpha_6 - \mu
\end{bmatrix}, \qquad (47.2)
$$

where either w_3 or z_4 is zero.

(ii) Compute $m_{r+1} = x_{r+1}/u_r$ and store it. Replace x_{r+1} by zero.

(iii) Compute $p_{r+1} = y_{r+1} - m_{r+1}v_r$, $q_{r+1} = z_{r+1} - m_{r+1}w_r$, and overwrite on y_{r+1} and z_{r+1} respectively.

The configuration is now

$$\begin{bmatrix} u_1 & v_1 & w_1 & & & \\ m_2 & & u_2 & v_2 & w_2 & \\ m_3 & & & u_3 & v_3 & w_3 \\ m_4 & & & & p_4 & q_4 \\ & & & & \beta_5 & \alpha_5 - \mu & \beta_6 \\ & & & & & \beta_6 & \alpha_6 - \mu \end{bmatrix}.$$

Notice that $p_1 = \alpha_1 - \mu$, $q_1 = \beta_2$. A simple inductive proof shows that if C is scaled so that

$$|\alpha_i|, \ |\beta_i| < \tfrac{1}{5}$$

then, for any permissible value of μ, all elements derived during the reduction are bounded by unity so that fixed-point computation is quite convenient. In order to emphasize that the process really is Gaussian elimination we have displayed $(C - \mu I)$ and the successive reduced matrices in full, but in practice we will, of course, take advantage of the tri-diagonal form and merely store the u_i, v_i, w_i as linear arrays. If we store the m_i, u_i, v_i, w_i and remember whether or not an interchange took place just before m_i was computed, we can subsequently process a right-hand side of the equations

$$(C - \mu I)x = b.$$

The rth principal minor d_r is $(-1)^{k_r} u_1 u_2 \dots u_{r-1} p_r$ where k_r is the total number of interchanges which have taken place up to the end of the $(r-1)$th major step. Of course we do not need to form this product; we merely keep a record of its sign. Since we have

$$\left. \begin{aligned} d_{r+1}/d_r &= (-1)^{k_{r+1}} u_1 u_2 \dots u_r p_{r+1} / (-1)^{k_r} u_1 u_2 \dots u_{r-1} p_r \\ &= (-1)^{k_{r+1} - k_r} u_r p_{r+1} / p_r \end{aligned} \right\} \tag{47.4}$$

we see that d_{r+1} and d_r are of the same sign, that is we have a sign agreement if the right-hand side of (47.4) is positive.

It is not possible to carry out an error analysis similar to that which we gave in § 41. As a result of interchanges the equivalent perturbation of the original matrix can be unsymmetric. If, for example, interchanges take place at every stage, then the matrix which gives precisely the computed multipliers and pivotal rows is, typically, of the form

$$\begin{bmatrix} \alpha_1+\epsilon_1 & \beta_2+\epsilon_2 & \epsilon_3 & \epsilon_4 & \epsilon_5 \\ \beta_2 & \alpha_2 & \beta_3 & & \\ & \beta_3 & \alpha_3 & \beta_4 & \\ & & \beta_4 & \alpha_4 & \beta_5 \\ & & & \beta_5 & \alpha_5 \end{bmatrix}, \qquad (47.5)$$

all perturbations being in the first row. Further, the equivalent perturbation corresponding to the rth step then modifies some of the equivalent perturbations corresponding to the earlier steps. Because the perturbations are unsymmetric, some of the perturbed principal submatrices could even have some complex conjugate zeros.

However it is trivial to show that the 2-norms of all the perturbation matrices are bounded by $Kn^{\frac{1}{2}}2^{-t}$ for some K, and hence the signs of all leading principal minors must be obtained correctly except possibly for values of μ which lie in narrow intervals centred on the $\frac{1}{2}n(n+1)$ eigenvalues of all the leading principal submatrices. Consequently it is not very surprising to find that the bisection process based on this method is very effective. In practice we have found that, for a matrix scaled in the way we have described, the error in a computed eigenvalue seldom attains 2.2^{-t} even for matrices with pathologically close eigenvalues. There is a strong incentive to use this process rather than that of §§ 37–39 since the method we shall recommend for computing the eigenvectors also requires the triangularization of $(C-\mu I)$. On computers on which the number of digits in the mantissa of a floating-point number is appreciably less than that in a fixed-point number the technique we have just described will almost certainly give more accurate results. All that is lacking is a rigorous *a priori* estimate of the error bounds!

Computation of the eigenvectors of a tri-diagonal form

48. We now turn to the computation of the eigenvectors of a symmetric tri-diagonal matrix C. Again we can assume that C has no zero β_i elements since otherwise, as pointed out in § 36, the eigenvector problem for C reduces to the eigenvector problem for a number of smaller tri-diagonal matrices. We assume that very accurate eigenvalues have already been obtained by the bisection technique or some other technique of comparable accuracy.

At first sight the problem appears to be very simple. We know that components of the eigenvector x corresponding to λ satisfy the equations

$$(\alpha_1 - \lambda)x_1 + \beta_2 x_2 = 0, \tag{48.1}$$

$$\beta_i x_{i-1} + (\alpha_i - \lambda)x_i + \beta_{i+1}x_{i+1} = 0 \quad (i = 2, \ldots, n-1), \tag{48.2}$$

$$\beta_n x_{n-1} + (\alpha_n - \lambda)x_n = 0. \tag{48.3}$$

Clearly x_1 cannot be zero, since from equations (48.1) and (48.2), all other x_i would then be zero. Since we are not interested in a normalizing factor we may take $x_1 = 1$. A simple inductive proof shows that

$$x_r = (-1)^{r-1}p_{r-1}(\lambda)/\beta_2\beta_3 \ldots \beta_r \quad (r = 2, \ldots, n). \tag{48.4}$$

To derive this result we require only equations (48.1) and (48.2), but the resulting x_{n-1} and x_n automatically satisfy (48.3) since we have

$$\beta_n x_{n-1} + (\alpha_n - \lambda)x_n = (-1)^{n-1}\left[\frac{(\alpha_n - \lambda)p_{n-1}(\lambda)}{\beta_2\beta_3 \ldots \beta_n} - \frac{\beta_n p_{n-2}(\lambda)}{\beta_2\beta_3 \ldots \beta_{n-1}}\right]$$

$$= (-1)^{n-1}p_n(\lambda)/\beta_2\beta_3 \ldots \beta_n = 0, \tag{48.5}$$

the last equality following because $p_n(\lambda)$ is zero when λ is an eigenvalue.

We therefore have explicit expressions for the components of the eigenvector in terms of the $p_r(\lambda)$. Now we saw that the $p_r(\lambda)$ determined the eigenvalues in an extraordinarily stable manner, and we might therefore imagine that if λ were a very good approximation to an eigenvalue the use of these explicit expressions for the components would give a good approximation to the eigenvector. In fact this is not the case. We shall show that the vector obtained in this way from a very good approximation to an eigenvalue may be catastrophically in error.

Instability of the explicit expression for the eigenvector

49. Let λ be very close to an eigenvalue but not exactly equal to it, and let x_r be the exact values obtained from the relations (48.4) corresponding to this value of λ. By definition then, the x_r satisfy equations (48.1) and (48.2). The components x_{n-1} and x_n cannot satisfy equation (48.3) exactly because, if they did, x would be a non-null vector satisfying $(A - \lambda I)x = 0$, and hence λ would be an exact eigenvalue, contrary to our assumption. We have therefore

$$\beta_n x_{n-1} + (\alpha_n - \lambda)x_n = \delta \neq 0, \tag{49.1}$$

and the vector x satisfies exactly the equations

$$(C - \lambda I)x = \delta e_n, \tag{49.2}$$

where e_n is the nth column of the identity matrix. Since we are not interested in arbitrary non-zero multipliers, x is effectively the solution of the equations

$$(C - \lambda I)x = e_n. \tag{49.3}$$

The matrix $(C - \lambda I)$ is, by assumption, non-singular and hence x is given by

$$x = (C - \lambda I)^{-1} e_n. \tag{49.4}$$

Now in order to emphasize that the phenomenon we are about to exhibit is not associated with ill-conditioning of the eigenvector problem for C, we assume that C has very well separated eigenvalues so that, since C is symmetric, the eigenvectors are not sensitive to perturbations in its elements. We assume that the eigenvalues are ordered so that

$$\lambda_1 > \lambda_2 > ... > \lambda_n, \tag{49.5}$$

and that λ is very close to λ_k. Hence $\lambda - \lambda_k$ is 'small' but $\lambda - \lambda_i$ $(i \neq k)$ is 'not small'.

Let $u_1, u_2, ..., u_n$ be the orthonormal system of eigenvectors of C. Then we may express e_n in the form

$$e_n = \sum_1^n \gamma_i u_i, \qquad \sum \gamma_i^2 = 1. \tag{49.6}$$

Equation (49.4) therefore gives

$$x = \sum_1^n \gamma_i u_i / (\lambda_i - \lambda)$$
$$= \gamma_k u_k / (\lambda_k - \lambda) + \sum_{i \neq k} \gamma_i u_i / (\lambda_i - \lambda). \tag{49.7}$$

If x is to be a good approximation to u_k then it is important that $\gamma_k / (\lambda_k - \lambda)$ should be very large compared with $\gamma_i / (\lambda_i - \lambda)$ $(i \neq k)$. Now we know that $\lambda_k - \lambda$ is very small, but if γ_k is *also very small* x will not be particularly 'rich' in the vector u_k. Suppose for example we are working with ten decimals and $\lambda_k - \lambda = 10^{-10}$. If γ_k happens to be 10^{-18} (say), then x will be almost completely deficient in u_k and therefore almost orthogonal to it.

The reader may feel that this is rather an extravagant argument and that there is no reason for expecting that e_n will ever be so catastrophically deficient in any of the eigenvectors. It is our contention that such a deficiency is extremely common and the main aim of the next few sections is to show how easily this may come about.

50. From equation (49.6) and the orthonormality of the u_i we have

$$\gamma_k = u_k^T e_n = n\text{th component of } u_k. \tag{50.1}$$

If, therefore, the last component of the normalized vector u_k is very small, the eigenvector obtained from the explicit expressions (48.4), corresponding to a very good approximation λ to λ_k, will not be a good approximation to u_k.

Before considering what factors are likely to make the nth component of u_k pathologically small, we return to the set of equations (48.1), (48.2), (48.3). The vector x given by (48.4) is the exact solution of the first $(n-1)$ equations of the system

$$(C - \lambda I)x = 0. \tag{50.2}$$

Since λ is not exactly an eigenvalue, the null vector is the only solution of the complete system. Instead of considering the solution of the first $(n-1)$ of these equations, let us consider the system obtained by omitting the rth. Again we will have a non-null solution which will satisfy all of the equations (50.2) except the rth. Hence this vector x is the exact solution of

$$(C - \lambda I)x = \delta e_r \quad (\delta \neq 0). \tag{50.3}$$

By exactly the same argument as before, this will be a good approximation to u_k provided the rth component of u_k is not small. Hence if the largest component of u_k is the rth, then it is the rth equation which should be omitted when computing u_k. This result is instructive but not particularly useful, since we will not know *a priori* the position of the largest component of u_k. In fact u_k is precisely what we are attempting to compute!

The solution corresponding to the omission of the rth equation may be derived as follows. We take $x_1 = 1$ and compute $x_2, x_3, \ldots, x_r$ using equation (48.1) and equations (48.2) up to $i = r-1$. We then take $x_n' = 1$ and compute $x_{n-1}', x_{n-2}', \ldots, x_r'$ using equation (48.3) and equations (48.2) for $i = n-1, n-2, \ldots, r+1$. We combine these two partial vectors by choosing k so that

$$x_r = kx_r' \tag{50.4}$$

and taking

$$x^T = (x_1, x_2, \ldots, x_r, kx_{r+1}', kx_{r+2}', \ldots, kx_n'). \tag{50.5}$$

If this process is performed *without rounding errors* then obviously we have the exact solution of (50.3) for some non-zero δ.

Numerical examples

51. We consider now a trivial numerical example of order 2. Consider the matrix C defined by

$$C = \begin{bmatrix} 0 \cdot 713263 & 0 \cdot 000984 \\ 0 \cdot 000984 & 0 \cdot 121665 \end{bmatrix}, \quad \left. \begin{matrix} \lambda_1 = 0 \cdot 71326463... \\ \lambda_2 = 0 \cdot 12166336... \end{matrix} \right\}, \tag{51.1}$$

and let us take the approximation $\lambda = 0 \cdot 713265$ to λ_1; this has the error $10^{-6}(0 \cdot 36...)$. Consider now the eigenvector obtained by omitting the second equation of $(C - \lambda I)x = 0$. We have

$$-0 \cdot 000002x_1 + 0 \cdot 000984x_2 = 0 \tag{51.2}$$

giving

$$x^T = (1, 0 \cdot 002033). \tag{51.3}$$

If we omit the first equation we have

$$0 \cdot 000984x_1 - 0 \cdot 591600x_2 = 0 \tag{51.4}$$

giving

$$x^T = (1, 0 \cdot 00166329). \tag{51.5}$$

By considering what would have happened if we had used the exact value of λ_1, it is obvious that it is the second vector which is accurate. The first vector is correct only to 3 decimal places while the second is correct almost to 8 decimal places. This is predicted by our analysis of the previous section. The largest component of u_1 is the first, and hence we should omit the first equation. Turning now to the computation of the eigenvector corresponding to λ_2 we find that it is now the second equation which must be omitted. Notice that this is a very well-conditioned eigenvalue and eigenvector problem. Difficulties which arise cannot therefore be attributed to ill-conditioning of the problem to be solved.

52. This example was, of course, a little artificial, but that is inevitable for a matrix of such a low order. As a second example we consider the matrix W_{21}^- of § 45. This has well separated eigenvalues and hence, since it is symmetric, has a well-conditioned eigenvector problem. Consider now the eigenvector corresponding to the largest eigenvalue λ_1. If we have the exact λ_1 then the eigenvector may be obtained exactly by solving any 20 of the 21 equations

$$(W_{21}^- - \lambda_1 I)x = 0. \tag{52.1}$$

Suppose we use the last 20 equations and take $x_{21} = 1$. We have then

$$x_{20} - (10 + \lambda_1) = 0, \tag{52.2}$$

$$x_s = (\lambda_1 + s - 10)x_{s+1} - x_{s+2} \quad (s = 19, 18, ..., 1). \tag{52.3}$$

Now λ_1 is certainly greater than 10 and hence if we assume that $x_{s+1} > x_{s+2} > 0$ we have

$$x_s > sx_{s+1} - x_{s+2} > (s-1)x_{s+1}. \qquad (52.4)$$

Since x_{20} is obviously greater than x_{21} this shows that the x_i certainly do increase with decreasing i at least as far as x_2, and

$$x_2 > x_3 > 2! \, x_4 > 3! \, x_5 > \ldots > 19! \, x_{21} = 19! > 10^{17} \qquad (52.5)$$

In fact we have also $x_1 > x_2$ since

$$(\lambda_1 - 10)x_1 = x_2 \qquad (52.6)$$

and λ_1 certainly lies between 10 and 11. Hence the last component of the *normalized* eigenvector is less than 10^{-17} and therefore, in the notation of § 49, we have $\gamma_1 < 10^{-17}$. The explicit solution (48.4) corresponding to an approximation to λ which is in error by only 10^{-10} is therefore almost completely deficient in u_1.

This is illustrated in Table 9. There u_1 is the true eigenvector correct to 8 decimal places, while x is the vector obtained by solving *correctly* the first 20 equations of $(A - \lambda I)x = 0$ with $\lambda = 10 \cdot 74619420$. Since this vector is somewhat difficult to compute, we have not given the same number of figures throughout. All figures quoted are correct, however, and they illustrate the behaviour of x. *Note that x is the vector corresponding to the correct use of the explicit formulae* (48.4).

The vector y is the vector we actually obtained from the first 20 equations with this same value of λ, using fixed-point computation with 8 decimal places. Because of the rounding errors (and only because of them) y differs from x, though it shows the same general behaviour in that their components diminish rapidly to a minimum and then increase again. It is evident that this is an unstable process; rounding errors made in the early components are amplified disastrously in the later ones. This has led some to conclude that the divergence of y from the true eigenvector u_1 is a result of these rounding errors and that for a value of λ of the accuracy specified, a good eigenvector would be obtained if the first 20 equations were solved using high-precision arithmetic. The vector x has been given to show the falsity of this assumption.

The fourth vector z in Table 9 is the accurate solution of the first 20 equations with $\lambda = 10 \cdot 74619418$ and again all figures quoted are correct. Like x this diminishes rapidly at first and then increases, though the later components are now negative. *From x and z we see*

TABLE 9

u_1	x	y	z	w
1·00000 000	1·00000 000	1·00000 000	1·00000 000	$1·41440\ 4161 \times 10^{19}$
0·74619 418	0·74619 420	0·74619 420	0·74619 418	$1·05542\ 0141 \times 10^{19}$
0·30299 994	0·30299 998	0·30299 998	0·30299 993	$4·28564\ 3680 \times 10^{18}$
0·08590 249	0·08590 260	0·08590 259	0·08590 248	$1·21500\ 8414 \times 10^{18}$
0·01880 748	0·01880 783	0·01880 780	0·01880 742	$2.66013\ 790 \times 10^{17}$
0·00336 146	0·00336 303	0·00336 288	0·00336 120	$4·75446\ 937 \times 10^{16}$
0·00050 815	0·00051 681	0·00051 596	0·00050 668	$7·18725\ 327 \times 10^{15}$
0·00006 659	0·00012 346	0·00011 789	0·00005 694	$9·41912\ 640 \times 10^{14}$
0·00000 771	0·00043 95.	0·00039 724	$-0·00006\ 562$	$1·08984\ 9597 \times 10^{14}$
0·00000 080	0·00372 1..	0·00335 645	$-0·00063\ 09.$	$1·12909\ 8198 \times 10^{13}$
0·00000 007	0·03582 ...	0·03231 537	$-0·00608\ 3..$	$1·05914\ 3420 \times 10^{12}$
0·00000 001	0·3812 ...	0·34391 079	$-0·06600\ 1..$	$9·07788\ 963 \times 10^{10}$
0·00000 000	$0·4442 \times 10^1$	4·00732 756	$-0·76917 ...$	$7·16312\ 559 \times 10^{9}$
0·00000 000	$0·5624 \times 10^2$	50·73426 451	$-9·738 ...$	$5·23693\ 576 \times 10^{8}$
0·00000 000	$0·7686 \times 10^3$	The components now increase		$3·56680\ 106 \times 10^{7}$
0·00000 000	$0·1128 \times 10^5$	rapidly in magnitude like those		$2·27383\ 533 \times 10^{6}$
0·00000 000	$0·1768 \times 10^6$	of x.		$1·36242\ 037 \times 10^{5}$
0·00000 000	$0·2950 \times 10^7$			$7·70028\ 174 \times 10^{3}$
0·00000 000	$0·5217 \times 10^8$			$4·08658\ 380 \times 10^{2}$
0·00000 000	$0·9751 \times 10^9$			$2·07461\ 942 \times 10^{1}$
0·00000 000	$0·1920 \times 10^{11}$			$1·00000\ 000 \times 10^{0}$

that there is no ten-figure value of λ for which the exact solution of the first 20 equations is anything like u_1!

Finally we give the vector w obtained by solving the last 20 recursions. As our analysis showed the components increase rapidly. When normalized this vector agrees with u to within 1 unit in the eighth decimal place.

Inverse iteration

53. Let us consider now the solution of the inhomogeneous set of equations

$$(A - \lambda I)x = b, \qquad (53.1)$$

where b, for the moment, is an arbitrary normalized vector. If b is expressed in the form

$$b = \sum_{i=1}^{n} \gamma_i u_i \qquad (53.2)$$

then the exact solution of (53.1) is

$$x = \sum_{i=1}^{n} \gamma_i u_i / (\lambda_i - \lambda). \qquad (53.3)$$

Assuming for the moment that we can solve sets of equations such as (53.1) without rounding error, we see that if λ is close to λ_k but not to any other λ_i, then x is much richer in u_k than is b. All that is necessary for x to be a good approximation to u_k is that γ_k should not be small. If we now solve the set of equations

$$(A - \lambda I)y = x \qquad (53.4)$$

without rounding error we obtain

$$y = \sum_{i=1}^{n} \gamma_i u_i / (\lambda_i - \lambda)^2, \qquad (53.5)$$

and y is even richer than x in the vector u_k.

We could obviously repeat this process indefinitely, thereby obtaining vectors which were dominated more and more by u_k, but provided we had some method of choosing b so that it was not pathologically deficient in u_k the second vector y would usually be a very good approximation to u_k. Suppose, for example, that

$$\lambda - \lambda_k = 10^{-10}, \qquad |\lambda - \lambda_i| > 10^{-3} \quad (i \neq k), \qquad \gamma_k = 10^{-4}. \quad (53.6)$$

In such a case we could reasonably say that b was fairly deficient in u_k and that λ_k was not particularly well separated from the other eigenvalues. Yet we have

$$y = \gamma_k u_k / (\lambda_k - \lambda)^2 + \sum_{i \neq k} \gamma_i u_i / (\lambda_i - \lambda)^2$$

$$= 10^{16} u_k + \sum_{i \neq k} \gamma_i u_i / (\lambda_i - \lambda)^2,$$

$$10^{-16} y = u_k + 10^{-16} \sum_{i \neq k} \gamma_i u_i / (\lambda_i - \lambda)^2, \qquad (53.7)$$

and

$$\left\| 10^{-16} \sum_{i \neq k} \gamma_i u_i / (\lambda_i - \lambda)^2 \right\|_2 < 10^{-16} . 10^6 \left[\sum_{i \neq k} \gamma_i^2 \right]^{\frac{1}{2}} \leqslant 10^{-10}. \quad (53.8)$$

Hence even in such an unfavourable case the normalized vector y is almost exactly u_k.

Choice of initial vector b

54. There remains the problem of how to choose b. Notice that we are not requiring b to be chosen particularly well, that is, so that u_k is one of the largest components, but only that it should not be pathologically deficient in u_k. No solution to this problem has been found which can be demonstrated rigorously to be satisfactory in all cases, but the following technique has proved to be extremely effective in practice and there are good reasons for believing that it cannot fail.

Corresponding to the good approximation λ, Gaussian elimination with interchanges is performed on $(C - \lambda I)$ as described in § 47. (It is not relevant now whether fixed-point or floating-point computation is performed.) The pivotal rows u_i, v_i, w_i and the multiplier m_i are stored, together with the information concerning interchanges. Our final stored matrix U (say), is therefore given typically for the case $n = 5$ by

$$U = \begin{bmatrix} u_1 & v_1 & w_1 & & \\ & u_2 & v_2 & w_2 & \\ & & u_3 & v_3 & w_3 \\ & & & u_4 & v_4 \\ & & & & p_5 \end{bmatrix}. \tag{54.1}$$

Note that none of the u_i can be zero, because at each stage of the reduction the pivot is chosen from β_{r+1} and p_r and we are still assuming that no β_i is zero. The final element p_n can, of course be zero, and indeed would always be so if λ were an exact eigenvalue and exact computation were performed. (See, however, § 56.)

We now choose our right-hand side b to be such that when the elimination process is performed it gives as the reduced right-hand side the vector e consisting entirely of unit elements. We do not need to determine the b itself. Notice that b is a function of λ, so that we are effectively using a different right-hand side for each λ. If we now solve
$$Ux = e \tag{54.2}$$
by a back-substitution, we have performed one step of inverse iteration. If p_n is zero we may replace it by a suitably small number $2^{-t} \|C\|_\infty$ (say). As many steps of inverse iteration as are required may be performed, using the multipliers m_i to achieve the reduction and the matrix U for the back-substitution. No further triangularizations are required until we wish to compute the next eigenvector, that is with a different λ.

In practice we have usually found this choice of b to be so effective that after the first iteration, x is already a *good* approximation to the required eigenvector and a third iteration has *never* really been essential.

Error analysis
55. In Chapter 9 we shall discuss inverse iteration in a more general context and moreover we have already examined the

symmetric tri-diagonal case in detail in Wilkinson (1963b) pp. 143–147. For this reason we content ourselves here with a less formal examination.

Notice first that for a given approximate eigenvalue we always use the same triangular decomposition of $(C - \lambda I)$, and this will be an exact triangular decomposition of $(C - \lambda I + F)$ for some small error matrix F. From this it is obvious that the best we can hope to obtain, however many iterations we may use, is an eigenvector of $(C + F)$. We cannot expect to obtain correctly any figures in an eigenvector of C which are modified by small perturbations in C. However, this is true, in general, of any method, and if C has been obtained previously by Givens' or Householder's method, errors of this kind will already be present.

Let us assume, for convenience, that C has been normalized and that we perform the iteration defined by

$$(C - \lambda I)x_{r+1} = y_r, \qquad y_{r+1} = x_{r+1}/\|x_{r+1}\|_\infty, \qquad (55.1)$$

so that x_r is normalized after each iteration. We wish to be able to detect in practice when the iteration should be stopped. Taking advantage of the tri-diagonal nature of C we can apply the argument of Chapter 4, § 63 to show that the computed solution satisfies

$$(C - \lambda I + G)x_{r+1} = y_r, \qquad (55.2)$$

where

$$\|G\|_2 < kn^{\frac{1}{2}}2^{-t}. \qquad (55.3)$$

If we write

$$\lambda = \lambda_k + \epsilon, \qquad (55.4)$$

then (55.2) can be written in the form

$$(C - \lambda_k I)y_{r+1} = \epsilon y_{r+1} - Gy_{r+1} + y_r/\|x_{r+1}\|_\infty. \qquad (55.5)$$

Of the terms on the right, the first is small since λ is assumed to be very close to an eigenvalue, and the second is small from (55.3). The third will be small if $\|x_{r+1}\|_\infty$ is large. Hence the larger the value we obtain for $\|x_{r+1}\|_\infty$ the smaller will be our bound for the right-hand side of (55.5), and this is the residual vector corresponding to y_{r+1} and λ_k. In Wilkinson (1963b), pp. 139–147, we showed that $\|x_{r+1}\|_\infty$ must be large if y_r is not too deficient in u_k.

From these considerations we have derived the following empirical rule for use on ACE, for computation in floating-point with $t = 46$: *Continue iteration until the condition*

$$\|x_r\|_\infty \geqslant 2^t/100n \qquad (55.6)$$

is satisfied and then perform one more step.

In practice this has never involved doing more than three iterations and usually only two iterations are necessary. (Note that the first iteration involves only a back-substitution.) The factor $1/100n$ has no deep significance and merely ensures that we will seldom perform an unnecessary extra iteration. It might be thought that the normalized y_r would tend to a limit, within rounding error, and that some condition based on this, rather than (55.6), would be a more natural terminating criterion. However, this is not so; if C has more than one eigenvalue pathologically close to λ, then ultimately y_r will lie in the subspace spanned by the corresponding eigenvectors, but it sometimes happens that we obtain an entirely different vector lying in this subspace in successive iterations. Successive y_r do not then 'converge' at all.

Numerical example

56. In Table 10 we show the application of this method to the calculation of the dominant eigenvector of W_{21}^{-}. We have taken

TABLE 10

u	v	w	x	Normalized x
1·00000 00	−1·74619 42	1·00000 00	−2·35922 2×10^7	1·00000 00
1·00000 00	−2·74619 42	1·00000 00	−1·76043 8×10^7	0·74619 43
1·00000 00	−3·74619 42	1·00000 00	−7·14844 2×10^6	0·30300 00
1·00000 00	−4·74619 42	1·00000 00	−2·02663 1×10^6	0·08590 25
1·00000 00	−5·74619 42	1·00000 00	−4·43710 5×10^5	0·01880 75
1·00000 00	−6·74619 42	1·00000 00	−7·93046 7×10^4	0·00336 15
1·00000 00	−7·74619 42	1·00000 00	−1·19885 3×10^4	0·00050 82
1·00000 00	−8·74619 42	1·00000 00	−1·57128 3×10^3	0·00006 66
1·00000 00	−9·74619 42	1·00000 00	−1·81935 8×10^2	0·00000 77
1·00000 00	−10·74619 42	1·00000 00	−1·89629 9×10	0·00000 08
1·00000 00	−11·74619 42	1·00000 00	−1·88118 0	0·00000 01
−4·07784 23	0·34996 24	Interchanges	−0·25253 85	0·00000 00
−12·66037 37	1·00000 00	cease at	−0·08518 65	0·00000 00
−13·66720 76	1·00000 00	12th step	−0·07849 27	0·00000 00
−14·67302 64	1·00000 00		−0·07277 54	0·00000 00
−15·67804 19	1·00000 00		−0·06783 52	0·00000 00
−16·68241 07	1·00000 00		−0·06352 36	0·00000 00
−17·68625 08	1·00000 00		−0·05972 74	0·00000 00
−18·68965 31	1·00000 00		−0·05635 39	0·00000 00
−19·69268 87	1·00000 00		−0·05323 40	0·00000 00
−20·69541 39 $= p_{21}$			−0·04831 99	0·00000 00

$\lambda = 10·7461942$ (correct to 9 significant decimals) and we exhibit the matrix of pivotal rows. For ease of presentation we have not given this matrix in the extended form of (54.1) but have written the u_i, v_i and w_i in columns. Interchanges take place at each of the first 11 stages, so the first 11 pivotal rows have three non-zero elements. Thereafter there are no interchanges. Notice that the column u gives the pivotal elements and none of these is at all small (cf. Chapter 4, § 66). For an *exact* eigenvalue exact computation would give a zero value for p_n, the last pivot. However, even if exact computation had been carried out with $\lambda = 10 \cdot 7461942$, the pivots would have been almost exactly as shown (the first 11 would still be unity!) except for the twelfth which would be entirely different. We must have a value of λ of much higher accuracy than the one we have used before the last pivot becomes 'small'. (For a more extended discussion of this example see Wilkinson 1958a.)

The fourth column of Table 10 gives the vector obtained by carrying out a back-substitution with e as the right-hand side. It will be seen that some components of x are very large and, from our remarks of the previous section, this implies that the normalized x and λ give a small residual vector. Hence, since λ_1 is well separated from λ_2, normalized x is a good eigenvector. In fact it is correct almost to 7 decimal places, the working accuracy.

Notice that not all the components of e were of equal importance in the back-substitution; the last 11 components were quite irrelevant and the values they give become negligible when normalization is performed. In fact if the corresponding computation is carried out for each of the eigenvalues of W_{21}^- in turn, we find that for the different eigenvectors it is different components of e which play the vital role. It is a measure of the effectiveness of the technique that working with 46 binary digit arithmetic every eigenvector of W_{21}^- was correct to 40 or more binary places after the first iteration, and the limiting accuracy was obtained in the second iteration.

57. It is natural to ask whether the process can ever fail and a sequence of values be obtained for which (55.6) is never satisfied. This would seem to be almost impossible. It could only occur if the vector b, which is implicit in the use of the technique, contained a component of the required vector which was of order 2^{-t} or less *and the rounding errors were such that this component was not enriched.*

This technique throws an interesting light on the meaning of the term 'ill-conditioned'. The matrices $(C - \lambda I)$ which we use are almost singular, since λ is almost an eigenvalue. The computed solutions we obtain for the sets of equations $(C - \lambda I)x_{r+1} = y_r$ are therefore generally incorrect even in their most significant figures. Yet the process is *well-conditioned* as far as the normalized x_r is concerned unless C has more than one eigenvalue close to λ. Even in this case the errors in the eigenvectors are merely the inevitable errors corresponding to the effect of small perturbations in the elements of C.

Close eigenvalues and small β_i

58. From the arguments of the previous sections we see that the eigenvector of limiting accuracy will be an exact eigenvector of some matrix $(C+F)$ where F is a small perturbation matrix such that

$$\|F\| < f(n)2^{-t} \text{ (say)}, \tag{58.1}$$

$f(n)$ being some quite mild function of n dependent on the type of arithmetic used. The matrix F will, in general, be different for each eigenvalue. Suppose now C has two eigenvalues λ_1 and λ_2 separated by 2^{-p}, while $\lambda_i - \lambda_1$ ($i \neq 1, 2$) is not small. Then we can expect to obtain approximations v_1 and v_2 such that

$$v_1 = u_1 + f(n)2^{-t}\left(2^p h_2 u_2 + \sum_{i=3}^{n} h_i u_i\right), \tag{58.2}$$

$$v_2 = u_2 + f(n)2^{-t}\left(2^p k_1 u_1 + \sum_{i=3}^{n} k_i u_i\right), \tag{58.3}$$

where all h_i and k_i are of order unity. In other words v_1 and v_2 will be contaminated by substantial components of u_2 and u_1 respectively, but by negligible components of the other u_i. We have

$$v_1^T v_2 = (k_1 + h_2)f(n)2^{p-t} + 2^{-2t}(f(n))^2 \sum_{i=3}^{n} h_i k_i. \tag{58.4}$$

In general h_2 and k_1 will not be related, and hence as p tends to t and the two eigenvalues move very close together, v_1 and v_2 will, in general, depart further and further from strict orthogonality.

We may contrast this with the situation which arises when the eigenvectors of the tri-diagonal matrix are found by Jacobi's method. We would then have

$$v_1 = u_1 + 2^{-t}g(n)\left(2^p h_2 u_2 + \sum_{i=3}^{n} h_i u_i\right), \tag{58.5}$$

$$v_2 = u_2 + 2^{-t}g(n)\left(2^p k_1 u_1 + \sum_{i=3}^{n} k_i u_i\right), \tag{58.6}$$

where $g(n)$ is again some mild function of n, but rather larger than $f(n)$, since the effect of rounding errors is somewhat greater. However, we know from § 19 that v_1 and v_2 will be 'almost orthogonal' and hence that h_2 and k_1 will be almost equal and opposite. Although, as p tends to t, v_1 and v_2 inevitably become more contaminated with u_2 and u_1 respectively, the near orthogonality is not in any way affected.

Independent vectors corresponding to coincident eigenvalues

59. Suppose now two eigenvalues are so close that they are identical to working accuracy. This is true, for example, for λ_1 and λ_2 of W_{21}^{+} unless we are working to very high precision. Jacobi's method will give two almost orthogonal eigenvectors spanning the same subspace as u_1 and u_2; these vectors will have very small components of u_i $(i = 3, 4, ..., n)$ but they will be almost arbitrary orthogonal combinations of u_1 and u_2.

The inverse iteration of §§ 53–55, on the other hand produces only one eigenvector for a given value of λ. If the computed λ_1 and and λ_2 happen to be exactly equal, then we will need some method of computing a second vector in the u_1, u_2 subspace. Now when C has pathologically close eigenvalues it will be found that the vector obtained by the technique of § 54 is extremely sensitive to the value of λ which is used. We may illustrate this by means of a simple example.

Consider the tri-diagonal matrix

$$\begin{bmatrix} 3 & 0 & 0 \\ 0 & 2 & 1 \\ 0 & 1 & 2 \end{bmatrix}, \tag{59.1}$$

having $\lambda = 3$ as an eigenvalue of multiplicity 2 with $(1, 0, 0)$ and $(0, 1, 1)$ as independent eigenvectors. Since $\beta_2 = 0$, we replace this first by 10^{-t} so that all β_i are now non-zero. If we now perform one step of inverse iteration with $\lambda = 3$, the vector obtained is

$$\begin{bmatrix} 2 \times 10^t \\ 10^t + 1 \\ 10^t \end{bmatrix} = \begin{bmatrix} 1 \\ \frac{1}{2} \\ \frac{1}{2} \end{bmatrix} \text{ in normalized form.} \tag{59.2}$$

On the other hand if we take $\lambda = 3 - 10^{-t}$, the vector obtained is

$$\begin{bmatrix} 0 \\ 10^t \\ 10^t \end{bmatrix} = \begin{bmatrix} 0 \\ 1 \\ 1 \end{bmatrix} \text{ in normalized form,} \tag{59.3}$$

while $\lambda = 3 - 2 \times 10^{-t}$ gives

$$\begin{bmatrix} -(\frac{1}{7})10^t - \frac{2}{7} \\ (\frac{2}{7})10^t + \frac{4}{7} \\ (\frac{2}{7})10^t \end{bmatrix} = \begin{bmatrix} -\frac{1}{2} \\ 1 \\ 1 \end{bmatrix} \text{ in normalized form.} \tag{59.4}$$

Each of these eigenvectors lies in the correct subspace, though very small changes in λ have produced very different vectors. The third eigenvalue, $\lambda = 1$, is simple and it will be found that any value of λ close to 1 gives the unique eigenvector $(0, 1, -1)$.

The extreme sensitivity of the computed eigenvector to very small changes in the value of λ may be turned to practical advantage and used to obtain independent eigenvectors corresponding to coincident or pathologically close eigenvalues. On ACE, working with normalized matrices, we artificially separate all eigenvalues by at least 3×2^{-t}. This has worked extremely well in practice and has always provided the requisite number of independent eigenvectors. Working with W_{21}^{+}, for example, we obtained completely independent eigenvectors in this way. If we have an eigenvalue λ_i of high multiplicity we use λ_i, $\lambda_i \pm 3 \times 2^{-t}$, $\lambda_i \pm 6 \times 2^{-t}$,.... . It is perhaps worth mentioning that the use of a value $\lambda_i + 9 \times 2^{-t}$ (say) does not have an adverse effect on the accuracy of the computed eigenvector. The closeness of λ to an eigenvalue affects the speed of convergence but not the accuracy attainable by inverse iteration. Of course we cannot expect the computed vectors to be orthogonal, but remember that deterioration of orthogonality sets in well before we reach pathologically close eigenvalues. It must be admitted though, that this solution to the problem is somewhat inelegant. It has the disadvantage that if we require orthogonal vectors we must apply the Schmidt orthogonalization process to the computed vectors. Also, unless the process can be put on a sounder footing, there remains the danger that we shall not have full digital information about the subspace. For example the two computed vectors might conceivably be

$$v_1 = 0 \cdot 6 u_1 + 0 \cdot 4 u_2 + 2^{-t} \sum_{i=3}^{n} h_i u_i, \tag{59.5}$$

$$v_2 = 0 \cdot 61 u_1 + 0 \cdot 39 u_2 + 2^{-t} \sum_{i=3}^{n} k_i u_i, \tag{59.6}$$

where the h_i and k_i are of order unity. These two vectors are mathematically independent, but the direction orthogonal to the first vector is poorly determined. We have in fact

$$v_1 - v_2 = -0 \cdot 01 u_1 + 0 \cdot 01 u_2 + 2^{-t} \sum_{i=3}^{n} (h_i - k_i) u_i, \tag{59.7}$$

and hence two decimals will be lost in the orthogonalization process.

Sometimes, when the original matrix from which C has been determined has multiple eigenvalues, C will have some β_i elements which are almost zero. We can then split C into the direct sum of a number of smaller tri-diagonal matrices and the eigenvectors corresponding to different tri-diagonal matrices are automatically *exactly* orthogonal. Unfortunately experience shows that we cannot rely on the occurrence of small β_i. The matrix W_{21}^+, on the other hand, shows how far we may be from a situation in which we have an obvious decomposition of the tri-diagonal matrix. Nevertheless it is true, even in this case, that if, for example, we separate the matrix into two matrices of order 10 and 11 (say), by putting β_{11} equal to zero, we do obtain independent orthogonal vectors which span the subspaces corresponding to the pathologically close pairs of eigenvalues to a very high accuracy. It is therefore quite possible that such a decomposition is always permissible, and what is needed is some reliable method of deciding when and where to decompose.

Alternative method for computing the eigenvectors

60. It would appear that, at present, the method of inverse iteration we have described is the most satisfactory of known methods for computing the eigenvectors of tri-diagonal matrices. It involves, in any case, considerably less work than the computation of the eigenvalues. Further it gives valuable information on the performance of inverse iteration, which is one of the most important techniques in the more general eigenvector problem.

However, a number of alternative methods have been proposed of which the following, due to Dekker (1962), is perhaps the most successful. It is based essentially on the process described in § 50. We take $x_1 = 1$ and compute $x_2, x_3, \ldots$ in succession using equations (48.1) and (48.2). This process is continued as long as $|x_{r+1}| \geqslant |x_r|$. If we reach a stage at which this condition is not satisfied, then we take $x'_n = 1$ and compute $x'_{n-1}, x'_{n-2}, \ldots$ in succession from equations (48.3) and (48.2). Again we continue until $|x'_{s-1}| < |x'_s|$. When this happens, we return to the forward determination and again continue until a decrease occurs.

If we have $|x_1| > |x_2| > |x_3| > \ldots$ and $|x'_n| > |x'_{n-1}| > |x'_{n-2}| > \ldots$ we shall switch alternately from one end to the other. If on the other hand, we have, for example,

$$|x_1| > |x_2| > |x_3| \leqslant |x_4| \leqslant |x_5| \leqslant \ldots \leqslant |x_n|, \tag{60.1}$$

and

$$|x'_n| > |x'_{n-1}| > |x'_{n-2}|, \tag{60.2}$$

then we shall switch to and fro twice, but on returning to the forward substitution for the second time, the $|x_i|$ increase, and we shall continue with forward substitution for the whole of the rest of the way, discarding the x_i' values. Similarly we may ultimately discard the x_i values. Alternatively we may meet at an internal point, if both the x_i and the x_i' sequences are decreasing when overlap first occurs. In this case the two sequences are combined as described in § 50.

Generally this appears to give good results, but it seems likely that it would be inaccurate when the forward and backward sequences are rapidly decreasing at the time when they intersect. Also it does not appear to be easy to modify the process so as to obtain independent vectors when C has pathologically close eigenvalues, but does not have β_i values which give a clear indication of how to split the matrix. A number of variants of forward and backward substitution have been investigated by the author and others.

Numerical example

61. As an illustration of the behaviour of Dekker's variant we consider the calculation of the eigenvector of W_{21}^+ corresponding to the eigenvalue λ_1, which is pathologically close to λ_2. The matrix is persymmetric (i.e. symmetric about the diagonal from n, 1 to 1, n) and the forward and backward sequences are therefore identical. Both sequences decrease at first, so we switch alternately to the forward and backward sequence. When we reach x_9 the forward sequence begins to increase and continues to do so right up to x_{21}. The backward sequence is therefore discarded. The results are given in Table 11. At first sight they look a little discouraging, since the actual value finally assumed by x_{21} is obviously completely determined by the error in λ and the rounding errors. However, x_{20} and x_{21} satisfy the twenty-first equation almost exactly, and consequently the normalized vector derived from x must give a residual vector with components of order 10^{-8}, and therefore lie in the subspace spanned by the first two eigenvectors to within working accuracy. We give the normalized eigenvector. If u_1 and u_2 are the first two normalized eigenvectors of W_{21}^+, (infinity norm), then we have verified that

$$x = 0 \cdot 50000\ 233u_1 + 0 \cdot 49999\ 767u_2 + O(10^{-8}). \tag{61.1}$$

If we attempt to compute another vector by taking a slightly different value of λ we find that we again accept the forward sequence, but x_{21}

TABLE 11

$$\lambda = 10 \cdot 74619 \ 419$$

$x_1 = 1 \cdot 00000 \ 000$	$x_{12} = 0 \cdot 17133 \ 492$	$x'_{21} = 1 \cdot 00000 \ 000$
$x_2 = 0 \cdot 74619 \ 419$	$x_{13} = 1 \cdot 65376 \ 40$	$x'_{20} = 0 \cdot 74619 \ 419$
$x_3 = 0 \cdot 30299 \ 996$	$x_{14} = 1 \cdot 42928 \ 06 \times 10$	$x'_{19} = 0 \cdot 30299 \ 996$
$x_4 = 0 \cdot 08590 \ 254$	$x_{15} = 1 \cdot 09061 \ 09 \times 10^2$	$x'_{18} = 0 \cdot 08590 \ 254$
$x_5 = 0 \cdot 01880 \ 764$	$x_{16} = 7 \cdot 21454 \ 49 \times 10^2$	$x'_{17} = 0 \cdot 01880 \ 764$
$x_6 = 0 \cdot 00336 \ 217$	$x_{17} = 4 \cdot 03655 \ 65 \times 10^3$	$x'_{16} = 0 \cdot 00336 \ 217$
$x_7 = 0 \cdot 00051 \ 204$	$x_{18} = 1 \cdot 84368 \ 27 \times 10^4$	$x'_{15} = 0 \cdot 00051 \ 204$
$x_8 = 0 \cdot 00009 \ 215$	$x_{19} = 6 \cdot 50313 \ 78 \times 10^4$	$x'_{14} = 0 \cdot 00009 \ 215$
$x_9 = 0 \cdot 00020 \ 177$	$x_{20} = 1 \cdot 60151 \ 97 \times 10^5$	
$x_{10} = 0 \cdot 00167 \ 257$	$x_{21} = 2 \cdot 14625 \ 06 \times 10^5$	
$x_{11} = 0 \cdot 01609 \ 942$		

Normalized vector

$x_1 = 0 \cdot 00000 \ 466$	$x_8 = 0 \cdot 00000 \ 000$	$x_{15} = 0 \cdot 00050 \ 815$
$x_2 = 0 \cdot 00000 \ 348$	$x_9 = 0 \cdot 00000 \ 000$	$x_{16} = 0 \cdot 00336 \ 146$
$x_3 = 0 \cdot 00000 \ 141$	$x_{10} = 0 \cdot 00000 \ 001$	$x_{17} = 0 \cdot 01880 \ 748$
$x_4 = 0 \cdot 00000 \ 040$	$x_{11} = 0 \cdot 00000 \ 007$	$x_{18} = 0 \cdot 08590 \ 249$
$x_5 = 0 \cdot 00000 \ 009$	$x_{12} = 0 \cdot 00000 \ 080$	$x_{19} = 0 \cdot 30299 \ 993$
$x_6 = 0 \cdot 00000 \ 002$	$x_{13} = 0 \cdot 00000 \ 771$	$x_{20} = 0 \cdot 74619 \ 418$
$x_7 = 0 \cdot 00000 \ 000$	$x_{14} = 0 \cdot 00006 \ 659$	$x_{21} = 1 \cdot 00000 \ 000$

is entirely different from its previous value. Unfortunately this new vector $\bar{x}$ is such that

$$\bar{x} = (0 \cdot 5 + \epsilon)u_1 + (0 \cdot 5 - \epsilon)u_2 + O(10^{-8}) \qquad (61.2)$$

where the ϵ is again of the order 10^{-5}. We do not therefore obtain accurate information about the direction orthogonal to x in the subspace of u_1 and u_2. However, it is interesting to note that if, using the original value of λ, we start with the backward sequence instead of the forward, then we ultimately pass over to the backward sequence, and this is identical with the forward sequence. The computed vector is therefore

$$0 \cdot 50000 \ 233u_1 - 0 \cdot 49999 \ 767u_2 + O(10^{-8}), \qquad (61.3)$$

and is almost orthogonal to x of equation (61.1). By this trick we obtain full digital information about the subspace from one value of λ!

Comments on the eigenproblem for tri-diagonal matrices

62. We have discussed the eigenproblem for tri-diagonal matrices at some length, because not only is it important in its own right, but also it provides instructive examples of the differences between classical analysis and numerical analysis. At several points in the techniques we discussed, limiting processes are a poor guide to what can be expected in practice, when the 'smallness' of quantities is strictly limited by the precision of computation. The examples given should not be regarded as mere numerical illustrations; the reader is advised to study them carefully, and should bear them in mind if he attempts to devise a method of computing the eigenvectors which is more aesthetically satisfying than the one we have given.

Completion of the Givens and Householder methods

63. To complete the Givens or Householder methods we must derive the eigenvectors of the original matrix from those of the tri-diagonal matrix. For Givens' method this is almost identical to the computation discussed in § 16 in connexion with Jacobi's method. If x is an eigenvector of C, then z defined by

$$z = R_1^T R_2^T \ldots R_s^T x \qquad (63.1)$$

is an eigenvector of A_0, where the R_i are the matrices of the trans-formations. In general there are $\frac{1}{2}(n-1)(n-2)$ rotations and the computation of each vector z involves $2(n-1)(n-2)$ multiplications. If all the vectors are required this part of the computation therefore involves approximately $2n^3$ multiplications.

For Householder's method we have in the same way

$$z = P_1 P_2 \ldots P_{n-2} x, \qquad (63.2)$$

where the P_r are the matrices of the transformations. Naturally the P_r are used individually in their factorized forms, $I - u_r u_r^T / 2K_r^2$ or $I - 2y_r y_r^T / \|y_r\|^2$, according to the formulations of § 30 or § 33.

If we write $x = x^{(n-2)}$ and $x^{(r-1)} = P_r x^{(r)}$, then we have for the first formulation

$$x^{(r-1)} = x^{(r)} - \left(\frac{u_r^T x^{(r)}}{2K_r^2}\right) u_r, \qquad z = x^{(0)}. \qquad (63.3)$$

There are approximately n^2 multiplications in the recovery of each vector and hence n^3 multiplications if all n are required. Notice that for both Givens' and Householder's methods the recovery of all the vectors involves more computation than the reduction to tri-diagonal form.

64. The recovery of the initial vectors is numerically stable and presents no problems of special interest; almost everything we require has been covered by the analyses we gave in Chapter 3. We may treat the Givens and Householder methods simultaneously.

Let R denote the *exact* product of the *exact* orthogonal matrices corresponding to the *computed* A_r. Then we have

$$C \equiv RA_0R^T + F, \tag{64.1}$$

where bounds for F have been found for the various methods. If x is a computed normalized eigenvector of C, then we have

$$Cx - \lambda x = \eta, \tag{64.2}$$

where the error analysis gives a bound for $\|\eta\|_2$. Hence

$$(C+G)x = \lambda x, \tag{64.3}$$

where G can be taken to be symmetric if λ is the Rayleigh value corresponding to x, and from Chapter 3, § 56, we have

$$\|G\|_2 = \|\eta\|_2. \tag{64.4}$$

Finally the analyses of Chapter 3, §§ 23, 34 and 45, give a bound for f, defined in terms of the computed z by

$$z \equiv R^Tx + f. \tag{64.5}$$

Combining (64.1) and (64.3) we have

$$(A_0 + R^TFR + R^TGR)R^Tx = \lambda R^Tx. \tag{64.6}$$

Hence R^Tx is an exact eigenvector of a symmetric matrix differing from A_0 by a perturbation which is bounded in norm by $\|F\| + \|G\|$, while z differs from R^Tx by f. *The bound for f is not dependent on the condition of the eigenvector problem.*

The main threat to ill-conditioned vectors comes from the possible effect of the symmetric perturbations of A_0. Note that G is dependent only on the technique we used for finding the eigenvector of C while F is dependent on the method of tri-diagonalization, as is also f. Except possibly for Householder's method performed in floating-point with accumulation, the bound for the error in the vector is dominated by the bound for F. For Householder's method, the bound for F is so good that the error introduced by G becomes comparable.

Comparison of methods

65. If we are interested only in eigenvalues, then Jacobi's method has practically nothing to recommend it compared with either Givens' or Householder's method. Assuming 6 sweeps are used, the Jacobi reduction to diagonal form takes about 9 times as long as the Givens tri-diagonalization and about 18 times as long as the Householder tri-diagonalization. If only a few selected eigenvalues are wanted Jacobi compares particularly unfavourably. Since Householder's reduction is about twice as fast as Givens' and information concerning the transformations can be stored in half the space, the case for Householder would appear to be overwhelming.

All three methods are so accurate in practice that accuracy is unlikely to be a decisive point, but in any case the balance is in favour of Householder and Givens. If inner-products can be accumulated Householder's method is extraordinarily accurate.

There is one small point in favour of Givens' method. If an element which is to be eliminated is already zero, then the corresponding rotation is easily omitted; it is true that in this case an element of w_i is zero in the Householder method, but it is not quite so convenient to take advantage of this. If the original matrix is very sparse the number of rotations required is substantially reduced, and although the zeros tend gradually to 'fill up' as the process proceeds, later rotations require less computation than the earlier ones.

If the vectors are required then there is a little more to be said for Jacobi's method. Although it is true that an eigenvector computed by Jacobi's method will, in general, have a larger component orthogonal to the relevant subspace than will the corresponding vector computed by Givens or Householder and the method of §§ 53–55, Jacobi's method has the advantage of giving vectors which are almost orthogonal. The extreme simplicity of Jacobi's method from the programming point of view might also commend its use to some. On the whole though, I have felt that, even when the vectors are required, Householder's method seems preferable and the aesthetic shortcomings of the method of determination of independent vectors corresponding to close eigenvalues are not such as to justify the use of a method which is slower and less accurate.

Quasi-symmetric tri-diagonal matrices

66. The elements of a general tri-diagonal matrix C may be defined by the equations

$$c_{ii} = \alpha_i, \quad c_{i,i+1} = \beta_{i+1}, \quad c_{i+1,i} = \gamma_{i+1}, \quad c_{ij} = 0 \text{ otherwise.}$$

$$(66.1)$$

In general the eigenvalues of such a matrix may be complex even if the elements are real. However if the α_i are real and

$$\beta_i\gamma_i > 0 \quad (i = 2,\ldots, n), \tag{66.2}$$

then C may be transformed into a real symmetric matrix by means of a similarity transformation with diagonal matrix D. If we define D by the relations

$$d_{1,1} = 1, \qquad d_{ii} = (\gamma_2\gamma_3\ldots\gamma_i/\beta_2\beta_3\ldots\beta_i)^{\frac{1}{2}}, \tag{66.3}$$

then

$$D^{-1}CD = T \tag{66.4}$$

where T is tri-diagonal and

$$t_{ii} = \alpha_i, \qquad t_{i,i+1} = t_{i+1,i} = (\beta_{i+1}\gamma_{i+1})^{\frac{1}{2}}. \tag{66.5}$$

If a β_i or γ_i is zero the eigenvalues of C are the eigenvalues of a number of smaller tri-diagonal matrices, so that this case does not cause any difficulties.

We can use the Sturm sequence property to compute the eigenvalues as in §§ 37–39; we merely replace β_i^2 by $\beta_i\gamma_i$. Tri-diagonal matrices having real α_i and with β_i and γ_i satisfying equation (66.2) are sometimes called *quasi-symmetric* tri-diagonal matrices.

Calculation of the eigenvectors

67. The method of calculating the eigenvectors described in §§ 53–55 takes no account of symmetry, and indeed symmetry is usually destroyed in Gaussian elimination with interchanges. It can therefore be applied immediately to quasi-symmetric matrices. However, in later stages in this book where we are concerned with the eigenvectors of tri-diagonal matrices, these will sometimes have been produced by similarity transformations from more general matrices. Thus we shall have

$$X^{-1}AX = C, \tag{67.1}$$

for some non-singular matrix X. Corresponding to an eigenvector y of C we shall be interested in the vector x defined by

$$x = Xy, \tag{67.2}$$

since this will be an eigenvector of A. Now usually when the elements β_i and γ_i are of widely varying orders of magnitude this springs from an X having columns which vary considerably in order of magnitude. If X is replaced by Y defined by

$$Y = XD, \tag{67.3}$$

where D is any non-singular diagonal matrix, then we have

$$Y^{-1}AY = D^{-1}CD = T, \tag{67.4}$$

where T is also tri-diagonal. Variations in the matrix D will cause variations in the interchanges which are used in the Gaussian elimination of $(T - \lambda I)$, and will also influence the effectiveness of the vector e which we used as right-hand side for the back-substitution in § 54. In general it seems advisable to choose D so that the columns of Y are of the same order of magnitude and then to apply the technique of §§ 53–55 to the resulting matrix T.

Equations of the form $Ax = \lambda Bx$ and $ABx = \lambda x$

68. In Chapter 1, §§ 31–33, we considered equations of the forms

$$Ax = \lambda Bx \tag{68.1}$$

and

$$ABx = \lambda x, \tag{68.2}$$

where A and B are real and symmetric and at least one of them is positive definite.

We showed that if, for example, B is positive definite and we have solved its eigenproblem, so that we have

$$R^T BR = \mathrm{diag}(\beta_i^2) = D^2, \tag{68.3}$$

for some orthogonal R, then (68.1) is equivalent to the standard eigenproblem

$$Pz = \lambda z \tag{68.4}$$

where

$$P = D^{-1}R^T ARD^{-1} \quad \text{and} \quad z = DR^T x. \tag{68.5}$$

If we have solved the eigenproblem for B by the method of Jacobi, Givens, or Householder, we shall have R in factorized form, and the real symmetric matrix P may be computed from the factors. Such techniques have in fact been used to solve (68.1) and the corresponding technique to solve (68.2). They are numerically stable and have the advantage that the same programme is used twice for solving the eigenvalue problem for B and P respectively. However, the volume of work is considerable.

69. The following is even more stable numerically and involves considerably less work. By way of contrast we consider the form

(68.2). If B is positive definite then by the Cholesky decomposition we have

$$B = LL^T, \tag{69.1}$$

with a real non-singular lower triangular L. Hence (68.2) may be written in the form

$$ALL^Tx = \lambda x, \tag{69.2}$$

giving

$$(L^TAL)L^Tx = \lambda L^Tx. \tag{69.3}$$

Therefore if λ and y are an eigenvalue and eigenvector of L^TAL, λ and $(L^T)^{-1}y$ are solutions of (68.2).

Now we know that the Cholesky decomposition of B is extremely stable, particularly if we can accumulate inner-products. Further it requires only $\frac{1}{6}n^3$ multiplications. In computing L^TAL from L we can take full advantage of symmetry. For if we write

$$F = AL, \tag{69.4}$$

$$G = L^TF, \tag{69.5}$$

then G is the symmetric matrix L^TAL, and hence we require only its lower triangle. Now L^T is upper triangular, so that only the lower triangular elements of F are used in computing the lower triangle of G. Hence we need compute only the lower triangle of F from (69.4). There are $\frac{1}{3}n^3$ multiplications in the determination of this lower triangle and $\frac{1}{6}n^3$ in the determination of the lower triangle of G. Hence G is obtained from A and B in $\frac{2}{3}n^3$ multiplications! If inner-products are accumulated at all relevant stages the total effect of rounding errors is kept to a very low level.

Turning to the form (68.1) we find that we now require the calculation of $(L^{-1})A(L^{-1})^T$, and the solution of triangular sets of equations is now involved. Again we may accumulate inner-products at all relevant stages and take advantage of the symmetry of the final matrix. Here we benefit considerably from the high accuracy of solutions of triangular sets of equations.

When B is ill-conditioned with respect to inversion this process for $(A - \lambda B)$ will itself be ill-conditioned. If we assume (without essential loss of generality) that $\|B\|_2 = 1$, then in the ill-conditioned case $\|L^{-1}A(L^{-1})^T\|_2$ will be much larger than $\|A\|_2$. In general the original eigenproblem will have some 'large' eigenvalues and others of 'normal size'; the latter will be inaccurately determined *by this process*. In the limit when B has a zero eigenvalue of multiplicity r, $\det(A - \lambda B)$ will be a polynomial of degree $(n - r)$; $(A - \lambda B)$ therefore has r 'infinite' eigenvalues. Nevertheless the finite eigenvalues of $(A - \lambda B)$ will not in general be sensitive to perturbations in A and B and their ill-determination must be regarded as a weakness of the method. A similar difficulty arises if B is diagonalized since D then has r zero elements. (For further comments see p. 344.)

Numerical example

70. In Table 12 we show the calculation of both $L^T A L$ and of $L^{-1} A (L^{-1})^T$ corresponding to the same pair of matrices A and B. Six-digit decimal arithmetic with accumulation of inner-products was used throughout. To find the zeros of $\det(AB - \lambda I)$ and $\det(A - \lambda B)$,

TABLE 12

A (*symmetric*)

$$\begin{bmatrix} 0.935 & 0.613 & 0.217 & 0.413 \\ & 0.216 & 0.317 & 0.323 \\ & & 0.514 & 0.441 \\ & & & 0.315 \end{bmatrix}$$

B (*symmetric and positive definite*)

$$\begin{bmatrix} 0.983 & 0.165 & 0.213 & 0.122 \\ & 0.897 & 0.214 & 0.132 \\ & & 0.903 & 0.213 \\ & & & 0.977 \end{bmatrix}$$

L^T (*where $B = LL^T$*)

$$\begin{bmatrix} 0.991464 & 0.166421 & 0.214834 & 0.123050 \\ & 0.932365 & 0.191177 & 0.119612 \\ & & 0.905703 & 0.180741 \\ & & & 0.956496 \end{bmatrix}$$

$F = AL$ (*lower triangle only*)

$$\begin{bmatrix} 1.126474 & & & \\ 0.751562 & 0.300629 & & \\ 0.432593 & 0.446574 & 0.545238 & \\ 0.596731 & 0.423141 & 0.456348 & 0.301296 \end{bmatrix}$$

$G = L^T F = L^T A L$ (*symmetric*)

$$\begin{bmatrix} 1.408298 & & & \\ 0.854808 & 0.416283 & & \\ 0.499655 & 0.480942 & 0.576304 & \\ 0.570771 & 0.404733 & 0.436495 & 0.288188 \end{bmatrix}$$

$Z^T = L^{-1} A$ (*upper triangle only*)

$$\begin{bmatrix} 0.943050 & 0.618278 & 0.218868 & 0.416556 \\ & 0.121310 & 0.300929 & 0.272078 \\ & & 0.452079 & 0.330676 \\ & & & 0.179229 \end{bmatrix}$$

$W = L^{-1} Z = L^{-1} A (L^{-1})^T$ (*symmetric*)

$$\begin{bmatrix} 0.951169 & 0.493351 & -0.088100 & 0.268090 \\ 0.493351 & 0.042050 & 0.206361 & 0.176732 \\ -0.088100 & 0.206361 & 0.476486 & 0.241206 \\ 0.268090 & 0.176732 & 0.241206 & 0.085213 \end{bmatrix}$$

the eigenvalues of the symmetric matrices $L^T A L$ and $L^{-1} A (L^{-1})^T$ are required. In general we have given only one triangle of symmetric matrices; the exception is the matrix $W = L^{-1} A (L^{-1})^T$. This matrix is obtained by solving $LW = Z$ and when computing the rth column of W we use the previously computed values $w_{r1}, w_{r2}, \ldots, w_{r,r-1}$ as the values $w_{1r}, w_{2r}, \ldots, w_{r-1,r}$. Remarkably little work is involved in the two calculations. For comparison, the matrices $L^T A L$ and $L^{-1} A (L^{-1})^T$ were computed using much higher precision. The maximum error in any element of the presented $L^T A L$ is $10^{-6}(2.04)$, and this occurs in only one element and even there as a result of exceptionally unfortunate roundings; all other elements have errors of less than 10^{-6}.

The maximum error in any element of the computed $L^{-1}A(L^{-1})^T$ is $10^{-6}(0\cdot98)$. These results confirm the remarkable numerical stability of the two processes.

Simultaneous reduction of A and B to diagonal form

71. From the solutions of the eigenproblem

$$Ax = \lambda Bx \tag{71.1}$$

we can derive the non-singular matrix X which is such that

$$X^T AX = \text{diag}(\lambda_i), \qquad X^T BX = I. \tag{71.2}$$

For consider the equivalent symmetric eigenproblem

$$L^{-1}A(L^{-1})^T y = \lambda y, \tag{71.3}$$

and let $y_1, y_2, \ldots, y_n$ be an orthogonal system of eigenvectors. Then we have

$$Y^T Y = I \tag{71.4}$$

and

$$L^{-1}A(L^{-1})^T Y = Y \, \text{diag}(\lambda_i). \tag{71.5}$$

If we define X by the relation

$$Y = L^T X \tag{71.6}$$

then X is non-singular and (71.4) becomes

$$X^T LL^T X = I, \quad \text{i.e. } X^T BX = I. \tag{71.7}$$

On the other hand, pre-multiplying (71.5) by Y^T we have

$$X^T AX = \text{diag}(\lambda_i). \tag{71.8}$$

Tri-diagonal A and B

72. When A and B are both tri-diagonal the method we have discussed for the solution of $Ax = \lambda Bx$ is unattractive because $L^{-1}A\,(L^{-1})^T$ is, in general, a full matrix. This problem may be solved by a method analogous to that used in § 39. We first prove the following simple lemma.

If A and B are symmetric and B is positive definite the zeros of $\det(A_r - \lambda B_r)$ separate those of $\det(A_{r+1} - \lambda B_{r+1})$, where A_i and B_i denote the leading principal submatrices of order i.

This follows if we observe that if

$$y = L^T x \tag{72.1}$$

then

$$\frac{y^T L^{-1} A (L^{-1})^T y}{y^T y} = \frac{x^T A x}{x^T L L^T x} = \frac{x^T A x}{x^T B x}, \tag{72.2}$$

and equation (72.1) gives a one-to-one correspondence between non-null x and y. Hence the zeros of $\det(A - \lambda B)$ are given by the min. max. definition applied to the ratio $x^T A x / x^T B x$ for non-null x (cf. Chapter 2, § 43). We can now prove a separation theorem for the zeros of the $\det(A_r - \lambda B_r)$ in exactly the same way as in Chapter 2, § 47. Hence if we write

$$a_{ii} = \alpha_i, \quad a_{i+1,i} = a_{i,i+1} = \beta_{i+1}; \quad b_{ii} = \alpha'_i, \quad b_{i+1,i} = b_{i,i+1} = \beta'_{i+1}, \tag{72.3}$$

then the polynomials defined by

$$\left. \begin{aligned} p_0(\lambda) &= 1, \qquad p_1(\lambda) = \alpha_1 - \lambda \alpha'_1 \\ p_r(\lambda) &= (\alpha_r - \lambda \alpha'_r) p_{r-1}(\lambda) - (\beta_r - \lambda \beta'_r)^2 p_{r-2}(\lambda) \end{aligned} \right\} \tag{72.4}$$

form a Sturm sequence, and we can find the zeros of $p_n(\lambda)$ by bisection as in § 39. Alternatively we can find the signs of $\det(A_r - \lambda B_r)$ $(r = 1, \ldots, n)$ as in § 47, by performing Gaussian elimination with interchanges on the tri-diagonal matrix $(A - \lambda B)$. As with the simpler problem we do not need to pursue bisection all the time, but can switch to some quadratically convergent process for finding the zeros of $p_n(\lambda) = \det(A - \lambda B)$, when a zero has been isolated,

To compute the eigenvector from a good value of λ, the inverse iteration now requires the solution of

$$(A - \lambda B) x_{r+1} = y_r, \tag{72.5}$$

$$y_{r+1} = B x_{r+1}, \tag{72.6}$$

since these equations imply that

$$[L^{-1} A (L^{-1})^T - \lambda I] L^T x_{r+1} = L^T x_r. \tag{72.7}$$

Again we shall expect two iterations to be adequate in general.

These techniques take full advantage of the tri-diagonal form of A and B.

Complex Hermitian matrices

73. The techniques we have described for real symmetric matrices have exact analogues for complex Hermitian matrices. For the methods of Jacobi and Givens on the one hand, and of Householder on the other hand, we require the complex unitary matrices of Chapter 1, §§ 43–44 and §§ 45–46 respectively. The most natural extension of the formulation used in § 29 gives a method in which two square roots are needed to determine each elementary Hermitian. A slight modification reduces this requirement to one square root. In the notation of § 29 we now have

$$\left.\begin{array}{ll}
P_r = I - u_r u_r^H / 2K_r^2, & u_{ir} = 0 \quad (i = 1, 2, \ldots, r) \\
u_{r+1,r} = a_{r,r+1}(1 + S_r^2/T_r), & u_{ir} = a_{ri} \quad (i = r+2, \ldots, n) \\
S_r^2 = \displaystyle\sum_{i=r+1}^{n} |a_{ri}|^2, & T_r = (|a_{r,r+1}|^2 S_r^2)^{\frac{1}{2}} \\
2K_r^2 = S_r^2 + T_r &
\end{array}\right\}, \quad (73.1)$$

where we have chosen the stable signs.

The matrices remain Hermitian at all stages, whichever of the three methods we use. Jacobi's method gives a real diagonal matrix as the limiting form, while from Chapter 1, § 44, it can be shown that Givens' method gives a real tri-diagonal matrix. Householder's method gives a tri-diagonal Hermitian matrix C which will, in general, have complex off-diagonal elements. This follows from the remark at the beginning of § 46 of Chapter 1. We have

$$c_{ii} = \alpha_i \text{ (real)}, \qquad c_{i,i+1} = \beta_{i+1}, \qquad c_{i+1,i} = \bar{\beta}_{i+1} \qquad (73.2)$$

and hence

$$c_{i,i+1} c_{i+1,i} = |\beta_{i+1}|^2 \text{ (real)}. \qquad (73.3)$$

The fact that the β_i are complex is therefore not of much importance.

Rather surprisingly, in view of their importance in theoretical physics, we have encountered few examples of complex Hermitian matrices and complex versions of our programmes have not been made. As pointed out in Chapter 3, § 56, the eigenvalue problem for a complex Hermitian matrix $(B+iC)$ may be reduced to that of the real symmetric matrix A given by

$$A = \begin{bmatrix} B & -C \\ C & B \end{bmatrix}, \qquad (73.4)$$

and on ACE and DEUCE we have done this. We know that the eigenvalues of A are $\lambda_1, \lambda_1, \lambda_2, \lambda_2, \ldots, \lambda_n, \lambda_n$ respectively, so that we need find only eigenvalues $1, 3, \ldots, 2n-1$ of the tri-diagonal matrix derived from A. The presence of coincident eigenvalues does not affect the accuracy with which they are determined, and only one vector is needed in the subspace of the eigenvectors corresponding to each double eigenvalue. All vectors in this subspace correspond to the same complex eigenvector of A. The tri-diagonal matrix derived by Givens' or Householder's method should split into two equal matrices, but this may not always happen in practice.

However, the matrix A requires twice as much storage as $(B+iC)$ and since the number of multiplications in both the Givens and Householder reductions varies as the cube of the order, there are eight times as many real multiplications in the reduction of A as there are complex multiplications in the reduction of $(B+iC)$. Since one complex multiplication involves four real multiplications the reduction of $(B+iC)$ involves only half as much computation as that of A by the same method. In the Givens reduction two of the four special elements of every plane rotation are real except those in planes $(i, i+1)$ and if we take advantage of this, the number of multiplications in the reduction of $(B+iC)$ is diminished further by the factor $\frac{3}{4}$. If a very large number of such computations were required and if the matrices involved were so large that storage problems became important, then the case for making the more complicated complex versions would become very strong.

Additional notes

The present interest in Jacobi's method dates mainly from its rediscovery in 1949 by Goldstine, Murray, and von Neumann, though it was used on desk computers at the National Physical Laboratory in 1947. The ability to overwrite one number on another makes Jacobi's method much more attractive on high-speed computers. Since 1949 the method has received extensive coverage in the literature; we mention papers by Goldstine, Murray and von Neumann (1959), Gregory (1953), Henrici (1958a), Pope and Tompkins (1957), Schönhage (1961) and Wilkinson (1962c). The paper by Goldstine *et al.* gives an error analysis of the process; although not published until 1959 much of the work was completed by 1951.

The general superiority of the method of Givens to that of Jacobi for the determination of the eigenvalues gained recognition comparatively slowly in spite of the fact that Givens (1954) gave a rigorous *a priori* error analysis which, in my opinion, is one of the landmarks in the history of the subject. Probably the fact that the methods proposed for computing the eigenvectors were found to be unreliable gave rise to unjustified suspicions concerning the accuracy of the eigenvalues.

Householder first suggested the use of the elementary Hermitian matrices for the reduction to tri-diagonal form in a lecture given at Urbana in 1958 and referred again to it briefly in a joint paper with Bauer (1959). Its general superiority to Givens' method both in speed and accuracy was first recognized by

Wilkinson (1960a). Error analyses were given for the fixed-point reduction by Wilkinson (1962b) and for the floating-point reduction by Ortega (1963).

The calculation of the eigenvalues of a tri-diagonal matrix using the Sturm sequence property was described and an error analysis given by Givens (*loc. cit.*). This analysis applied to fixed-point computation with *ad hoc* scaling and was the first in which *explicit* reference was made to what is now known as a 'backward error analysis', though this idea was implicit in the papers by von Neumann and Goldstine (1947) and Turing (1948).

The problem of computing eigenvectors of a tri-diagonal matrix when accurate eigenvalues are known has been considered by Brooker and Sumner (1956), Forsythe (1958), Wilkinson (1958a), and by Lanczos in the paper by Rosser *et al.* (1951). Inverse iteration gives a very satisfactory solution to the problem as far as reasonably well-separated eigenvalues are concerned. The problem of determining reliably full digital information in the subspace spanned by eigenvectors corresponding to coincident or pathologically close eigenvalues has never been satisfactorily solved. The analogous problem is important also in connexion with the determination of eigenvectors of Hessenberg matrices and is discussed in Chapter 9.

The solution of the eigenproblem for $(A - \lambda B)$ is rather unsatisfactory when B is ill-conditioned with respect to inversion. That the finite eigenvalues may be well-determined even when B is singular is illustrated for example by the matrices

$$A = \begin{bmatrix} 2 & 1 \\ 1 & 2 \end{bmatrix}, \qquad B = \begin{bmatrix} 1 & 1 \\ 1 & 1 \end{bmatrix}.$$

We have $\det(A - \lambda B) = 3 - 2\lambda$ so that the eigenvalues are $1\frac{1}{2}$ and ∞. Small changes in A and B make small changes in the finite eigenvalue but the infinite eigenvalue becomes finite, though very large.

If A is well-conditioned and positive or negative definite we can work with $(B - \lambda A)$ or $(-B - \lambda(-A))$ respectively. Unfortunately we will not always know in advance whether A or B is ill-conditioned, though in some problems in statistics and theoretical physics a well-conditioned B is almost guaranteed.

In the ill-conditioned case the method of § 68 has certain advantages in that 'all the ill-condition of B' is concentrated in the small elements of D. The matrix P of (68.5) has a certain number of rows and columns with large elements (corresponding to the small d_{ii}) and eigenvalues of $(A - \lambda B)$ of normal size are more likely to be preserved.

6

Reduction of a General Matrix to Condensed Form

Introduction

1. WE NOW turn to the more difficult problem of computing the eigensystem of a general matrix A_0. We cannot now expect to have *a priori* guarantees of accurate results, but we expect stable transformations to produce matrices which are exactly similar to some matrix $(A_0 + E)$, and to be able to establish satisfactory bounds for E.

In the first half of this chapter we shall consider the reduction of A_0 to Hessenberg form (Chapter 4, § 33) by elementary similarity transformations. Although, in general, a Hessenberg matrix has $\frac{1}{2}n(n+1)$ non-zero elements, a reduction of only one half compared with a full matrix, this reduction effects a remarkable simplification in the eigenproblem. There are two versions of each of the techniques we shall discuss, leading respectively to upper and lower Hessenberg matrices. It is a little more convenient from the point of view of later applications to have upper Hessenberg matrices and accordingly we have described methods appropriate to this form.

This has one slightly unfortunate consequence; it is conventional to describe techniques for symmetric matrices in terms of the upper triangle of elements and therefore in terms of introducing zeros in this upper triangle, and in Chapter 5 we adhered to this convention. (Of course, by symmetry, zeros are simultaneously introduced in the lower triangle.) Our decision therefore makes comparison of techniques for symmetric matrices with corresponding techniques for unsymmetric matrices rather less direct than it might otherwise be, though the modifications required to give lower Hessenberg matrices are quite trivial.

In the second half of the chapter we consider the further reductions from Hessenberg form to tri-diagonal form and to Frobenius form. The solution of the eigenvalue problem for these condensed forms is discussed in Chapter 7.

Givens' method

2. We consider first the analogues of the Givens and Householder methods discussed in Chapter 5. The Givens reduction has $(n-2)$

major steps, and immediately before the rth step A_0 has already been reduced to upper Hessenberg form as far as its first $(r-1)$ columns are concerned, so that typically for $n = 6$, $r = 3$, it has the form

$$\begin{bmatrix} \times & \times & \times & \times & \times & \times \\ \times & \times & \times & \times & \times & \times \\ 0 & \times & \times & \times & \times & \times \\ 0 & 0 & \times & \times & \times & \times \\ 0 & 0 & \underline{\times} & \times & \times & \times \\ 0 & 0 & \underline{\times} & \times & \times & \times \end{bmatrix} \quad . \tag{2.1}$$

In fact, we could say that its leading principal minor of order r is already of Hessenberg form, and this part is unaltered in later steps. The rth major step consists of $(n-r-1)$ minor steps in which zeros are introduced successively in positions $r+2$, $r+3,\ldots, n$ of the rth column; the zero in the ith position is introduced by a rotation in the $(r+1, i)$ plane. In (2.1) we have underlined the elements which are eliminated and indicated by arrows the rows and columns which are affected. The rth major step may be described in the usual algorithmic language as follows.

For each value of i from $r+2$ to n perform steps (i) to (iv):

(i) Compute $x = (a_{r+1,r}^2 + a_{ir}^2)^{\frac{1}{2}}$.

(ii) Compute $\cos\theta = a_{r+1,r}/x$, $\sin\theta = a_{ir}/x$. (If $x = 0$ take $\cos\theta = 1$ and $\sin\theta = 0$ and omit (iii) and (iv).) Overwrite x on $a_{r+1,r}$ and zero on a_{ir}.

(iii) For each value of j from $r+1$ to n:
Compute $a_{r+1,j}\cos\theta + a_{ij}\sin\theta$ and $-a_{r+1,j}\sin\theta + a_{ij}\cos\theta$ and overwrite on $a_{r+1,j}$ and a_{ij} respectively.

(iv) For each value of j from 1 to n:
Compute $a_{j,r+1}\cos\theta + a_{ji}\sin\theta$ and $-a_{j,r+1}\sin\theta + a_{ji}\cos\theta$ and overwrite on $a_{j,r+1}$ and a_{ji} respectively.

Of these, (iii) represents the effect of the pre-multiplication and (iv) the effect of the post-multiplication by the rotation in the $(r+1, i)$ plane. This transformation should be compared with the corresponding symmetric version described in Chapter 5, § 22. Since there is no symmetry, both row and column transfers are now required. The total number of multiplications is essentially

$$\sum 4n(n-r-1)+\sum 4(n-r-1)^2 \doteq 2n^3+\tfrac{4}{3}n^3 = \tfrac{10}{3}n^3. \qquad (2.2)$$

This compares with the $\tfrac{4}{3}n^3$ in the symmetric version of Givens' method. Notice that there is not room to store the values of $\cos\theta$ and $\sin\theta$ in the positions in which the zeros are introduced, since the introduction of a zero in position (i, r) is not now associated with that of a zero in position (r, i). An economical method of storage, consistent with numerical stability, is the following. Of the $\cos\theta$ and $\sin\theta$ we store the one which has the smaller modulus, and we use the two least significant digits to indicate whether or not $\cos\theta$ was stored and the sign of the other one.

Householder's method

3. Similarly we have a natural analogue of the Householder method discussed in Chapter 5. Again there are $(n-2)$ major steps and immediately before the rth step A_0 has been reduced to a form which for $n = 6, r = 3$, is given by

$$A_{r-1} = \begin{array}{c} r\left\{\vphantom{\begin{bmatrix}\times\\\times\\0\end{bmatrix}}\right. \\ \\ n-r\left\{\vphantom{\begin{bmatrix}0\\0\\0\end{bmatrix}}\right. \end{array}\begin{bmatrix} \times & \times & \times & \times & \times & \times \\ \times & \times & \times & \times & \times & \times \\ 0 & \times & \times & \times & \times & \times \\ \hdashline 0 & 0 & \times & \times & \times & \times \\ 0 & 0 & \times & \times & \times & \times \\ 0 & 0 & \times & \times & \times & \times \end{bmatrix} = \left[\begin{array}{c:c} H_{r-1} & C_{r-1} \\ \hline O \;\vdots\; b_{r-1} & B_{r-1} \end{array}\right]. \qquad (3.1)$$

The matrix H_{r-1} is of upper Hessenberg form. This is identical with the form at the beginning of the rth major step of the Givens method, but we have partitioned A_{r-1} in a way which illustrates the effect of the rth transformation. The matrix P_r of this transformation may be expressed in the form

$$P_r = \begin{array}{c} r\{ \\ n-r\{ \end{array}\left[\begin{array}{c:c} I & O \\ \hline O & Q_r \end{array}\right] = \left[\begin{array}{c:c} I & O \\ \hline O & I-2v_rv_r^T \end{array}\right], \qquad (3.2)$$

where v_r is a unit vector with $(n-r)$ components. Hence we have

$$A_r = P_r A_{r-1} P_r = \begin{array}{c} r \\ n-r \end{array} \left\{ \left[\begin{array}{c|cc} H_{r-1} & \multicolumn{2}{c}{C_{r-1}Q_r} \\ \hline O & c_r & Q_r B_{r-1} Q_r \end{array} \right] \right. , \tag{3.3}$$

where $c_r = Q_r b_{r-1}$. If we choose v_r so that c_r is null except for its first element, then the leading principal submatrix of A_r of order $(r+1)$ is of Hessenberg form.

Again, as in Chapter 5, § 29, the formulae are simplest if we think in terms of the matrix $P_r = I - 2w_r w_r^T$, where w_r is a unit vector having n components of which the first r are zero. We have then

$$P_r = I - 2w_r w_r^T = I - u_r u_r^T / 2K_r^2 \tag{3.4}$$

where

$$\left. \begin{array}{l} u_{ir} = 0 \quad (i = 1, \dots, r) \\[2mm] u_{r+1,r} = a_{r+1,r} \mp S, \qquad u_{ir} = a_{ir} \quad (i = r+2, \dots, n) \\[2mm] S_r = \left(\sum_{r+1}^{n} a_{ir}^2 \right)^{\frac{1}{2}}, \qquad 2K_r^2 = S_r^2 \mp a_{r+1,r} S_r \end{array} \right\}. \tag{3.5}$$

In (3.5) a_{ij} denotes the (i, j) element of A_{r-1}. The new $(r+1, r)$ element is $\pm S_r$. The sign is chosen in the usual way.

4. We no longer have symmetry and therefore the pre-multiplication and post-multiplication by P_r are considered separately. For the pre-multiplication we have

$$P_r A_{r-1} = (I - u_r u_r^T / 2K_r^2) A_{r-1} = A_{r-1} - u_r (u_r^T A_{r-1}) / 2K_r^2 = F_r, \tag{4.1}$$

and for the post-multiplication

$$A_r = P_r A_{r-1} P_r = F_r P_r = F_r (I - u_r u_r^T / 2K_r^2) = F_r - (F_r u_r) u_r^T / 2K_r^2. \tag{4.2}$$

Since u_r is null in its first r elements, it is clear from these equations that pre-multiplication by P_r leaves the first r rows of A_{r-1} unchanged while post-multiplication of the resulting F_r by P_r leaves its first r columns unchanged.

We may write

$$u_r^T A_{r-1} = p_r^T, \tag{4.3}$$

where p_r^T has its first $(r-1)$ elements equal to zero because of the zeros in u_r and A_{r-1}. We have then

$$F_r = A_{r-1} - (u_r / 2K_r^2) p_r^T. \tag{4.4}$$

We need not compute the rth element of p_r because this influences only the rth column of F_r, and we know already that the first r elements of this column are those of A_{r-1} and the remainder form the vector c_r; the transformation has been specifically designed so that c_r is null except in its first element which is $\pm S_r$. There are therefore only $(n-r)^2$ multiplications in the calculation of the relevant $(n-r)$ elements of p_r and a further $(n-r)^2$ multiplications in the calculation of F_r from (4.4).

For the post-multiplication, we write

$$F_r u_r = q_r, \tag{4.5}$$

and the vector q_r has no zero components. Because the first r components of u_r are zero, there are $n(n-r)$ multiplications in the computation of q_r. Finally we have

$$A_r = F_r - q_r(u_r/2K_r^2)^T, \tag{4.6}$$

and this requires a further $n(n-r)$ multiplications. Notice that both in (4.4) and (4.6) it is the vector $u_r/2K_r^2$ which is required though u_r itself is still required while F_r is being computed, since we have subsequently to derive $F_r u_r$. This difficulty is avoided if we work with w_r but at the expense of taking the square root of $2K_r^2$.

If the matrix is held on the backing store and economy of transfers is the main requirement, then the following formulation is probably the most convenient. We define the vectors p_r^T, q_r, v_r and the scalar α_r by the relations

$$p_r^T = u_r^T A_{r-1}, \quad q_r = A_{r-1} u_r, \quad v_r = u_r/2K_r^2, \quad p_r^T v_r = \alpha_r. \tag{4.7}$$

Then we compute A_r from the equation

$$\begin{aligned} A_r &= A_{r-1} - v_r p_r^T - q_r v_r^T + (p_r^T v_r) u_r v_r^T \\ &= A_{r-1} - v_r p_r^T - (q_r - \alpha_r u_r) v_r^T. \end{aligned} \tag{4.8}$$

This has more in common with the symmetric formulation in Chapter 5, §§ 29–30, but the term $u_r(u_r^T A_{r-1} u_r) u_r^T/(2K_r^2)^2$ cannot now be shared between the other two terms to such advantage.

The total number of multiplications in the complete reduction, using any of these variants, is essentially

$$\sum 2(n-r)^2 + \sum 2n(n-r) \doteqdot \tfrac{2}{3}n^3 + n^3 = \tfrac{5}{3}n^3. \tag{4.9}$$

This compares with $\tfrac{2}{3}n^3$ in the symmetric version of Householder. Again we find that Householder's method requires half as many multiplications as Givens' method.

Storage considerations

5. The storage of the details of the Householder transformation is quite convenient. There are $(n-r)$ non-zero elements in u_r and the corresponding transformation produces only $(n-r-1)$ zero elements; we cannot therefore store quite all the information in the space occupied by the original matrix. However, we shall find that the sub-diagonal elements of the Hessenberg form play a special role in all subsequent operations on the Hessenberg matrix so that it is quite convenient to store these separately as a linear array of $(n-1)$ elements. Notice that $u_{ir} = a_{ir}$ $(i = r+2,..., n)$, and hence all non-zero elements of u_{ir} except $u_{r+1,r}$ are already stored in the appropriate positions! The allocation of storage at the end of the rth major step for the case $n = 6$, $r = 3$, is given by

$$\begin{bmatrix} h_{11} & h_{12} & h_{13} & h_{14} & a_{15} & a_{16} \\ u_{21} & h_{22} & h_{23} & h_{24} & a_{25} & a_{26} \\ u_{31} & u_{32} & h_{33} & h_{34} & a_{35} & a_{36} \\ u_{41} & u_{42} & u_{43} & h_{44} & a_{45} & a_{46} \\ u_{51} & u_{52} & u_{53} & a_{54} & a_{55} & a_{56} \\ u_{61} & u_{62} & u_{63} & a_{64} & a_{65} & a_{66} \end{bmatrix}. \tag{5.1}$$

The h_{ij} are elements of the final Hessenberg form and the a_{ij} are current elements of A_r. In addition to this we use n storage registers to hold the vector

$$[h_{21}, h_{32},..., h_{r+1,r}; (u_{r+1,r}, u_{r+2,r},..., u_{nr})/2K_r^2]. \tag{5.2}$$

On a computer with a two-level store a further set of n registers is required to hold the vector p_r followed by the vector q_r; these vectors are not required simultaneously. There is no need to store either of these vectors if we have a one-level store since each element of p_r and q_r may be used immediately it is computed, working with the formulations of equations (4.4) and (4.6). Inner-products may be accumulated when computing S_r^2, p_r^T and q_r.

Error analysis

6. The general analyses of Chapter 3 enable us to show that for both the Givens and Householder reductions the final computed Hessenberg matrix H is exactly similar to $(A+E)$, and to give bounds for E. These analyses are no more difficult in the unsymmetric case

than the symmetric; indeed they are often a little simpler. We summarize the results that have been obtained in Table 1. In each case the matrix of the similarity transformation which reduces $(A+E)$ to H is the exact product of the exact orthogonal matrices corresponding to the computed A_r.

TABLE 1

Method	Mode of computation	Initial scaling of A	Bound for E
Givens	fixed-point (fi_2)	$\|A_0\|_2 \leqslant 1 - K_1 2^{-t} n^{\frac{3}{2}}$	$\|E\|_2 \leqslant K_1 2^{-t} n^{\frac{3}{2}}$
Givens	fixed-point (fi_2)	$\|A_0\|_E \leqslant 1 - K_2 2^{-t} n^{\frac{3}{2}}$	$\|E\|_E \leqslant K_2 2^{-t} n^{\frac{3}{2}}$
Householder	fixed-point (fi)	$\|A_0\|_2 \leqslant \frac{1}{2} - K_3 2^{-t} n^{\frac{3}{2}}$	$\|E\|_2 \leqslant K_3 2^{-t} n^{\frac{3}{2}}$
Householder	fixed-point (fi)	$\|A_0\|_E \leqslant \frac{1}{2} - K_4 2^{-t} n^{\frac{3}{2}}$	$\|E\|_E \leqslant K_4 2^{-t} n^{\frac{3}{2}}$
Householder	fixed-point (fi_2)	$\|A_0\|_2 \leqslant \frac{1}{2} - K_5 2^{-t} n^2$	$\|E\|_2 \leqslant K_5 2^{-t} n^2$
Givens	floating-point (fl)	Unnecessary	$\|E\|_E \leqslant K_6 2^{-t} n^{\frac{3}{2}} \|A_0\|_E$
Householder	floating-point (fl)	,,	$\|E\|_E \leqslant K_7 2^{-t} n^2 \|A_0\|_E$
Householder	floating-point (fl_2)	,,	$\|E\|_E \leqslant K_8 2^{-t} n \|A_0\|_E$

The K_r are quantities of order unity. As far as the first two bounds are concerned the use of fi in place of fi_2 affects only the constants K_1 and K_2.

In the fixed-point computations some preliminary scaling is required in order to ensure that all intermediate numbers lie within the permitted range. We have given bounds for $\|E\|_2$ and $\|E\|_E$ where these have been determined. Corresponding to bounds on $\|E\|_E$ we have given bounds on the permissible size of $\|A_0\|_E$, though of course the less restrictive bound on $\|A_0\|_2$ would have been adequate to avoid numbers exceeding capacity. However, the bound on $\|A_0\|_E$ can easily be met in practice, while that on $\|A_0\|_2$ is rather unrealistic.

All the bounds for $\|E\|_E$ corresponding to the use of floating-point arithmetic should contain an extra factor of the form $(1 + f(n)2^{-t})^{g(n)}$ which tends to infinity with n, but since this factor is always close to unity in the useful range of application of the method its omission is not serious.

These results show that when A is reasonably well conditioned the limitation on any of the methods of Table 1 is more likely to be storage capacity or time than accuracy. Of course we can construct an A which is so ill-conditioned that even very small bounds for E are inadequate to ensure the required accuracy in the eigenvalues.

Relationship between the Givens and Householder methods

7. Both the Givens and Householder methods give a reduction of A_0 to an upper Hessenberg matrix by an orthogonal similarity transformation. We now show that the matrices of the two transformations are identical to within multipliers ± 1 of each column. We first prove the following somewhat more general theorem which will be required subsequently.

If $AQ_1 = Q_1H_1$ and $AQ_2 = Q_2H_2$, where the Q_i are unitary and the H_i are upper Hessenberg, and if Q_1 and Q_2 have the same first column, then, in general,

$$Q_2 = Q_1D, \qquad H_2 = D^HH_1D, \tag{7.1}$$

where D is a diagonal matrix with elements of modulus unity. We have only to show that if

$$AQ = QH \tag{7.2}$$

and the first column of Q is given, then Q and H are essentially uniquely determined. The proof is as follows.

Equating the first columns of (7.2) we have

$$Aq_1 = h_{11}q_1 + h_{21}q_2. \tag{7.3}$$

Hence, since Q is unitary,

$$h_{11} = q_1^HAq_1, \qquad h_{21}q_2 = Aq_1 - h_{11}q_1. \tag{7.4}$$

The first equation gives h_{11} and the second gives

$$|h_{21}| = \|Aq_1 - h_{11}q_1\|_2, \tag{7.5}$$

so that q_2 is determined to within a factor $e^{i\theta_2}$ (say).
Equating the second columns of (7.2) we have

$$Aq_2 = h_{12}q_1 + h_{22}q_2 + h_{32}q_3, \tag{7.6}$$

and hence

$$h_{12} = q_1^HAq_2, \quad h_{22} = q_2^HAq_2, \quad h_{32}q_3 = Aq_2 - h_{12}q_1 - h_{22}q_2. \tag{7.7}$$

The first two equations give h_{12} and h_{22} and the third then gives

$$|h_{32}| = \|Aq_2 - h_{12}q_1 - h_{22}q_2\|_2. \tag{7.8}$$

It may readily be verified that $|h_{32}|$ is independent of the factor $e^{i\theta_2}$, so that q_3 is determined to within a factor $e^{i\theta_3}$ (say). Continuing in this way, Q is determined to within a post-multiplying factor D given by $\mathrm{diag}(e^{i\theta_j})$ and the corresponding indeterminacy of H is obviously as shown in (7.1).

Notice that if we insist that $h_{i+1,i}$ be real and positive, then both Q and H are uniquely determined by the first column of Q. The proof breaks down if any of the $h_{i+1,i}$ is zero, since q_{i+1} is then an *arbitrary* unit vector orthogonal to $q_1, q_2, \ldots, q_i$.

The effective equivalence of the Givens and Householder methods now follows if we observe that the matrix of each of the transformations has e_1 for its first column. This is because there are no rotations involving the first plane in the Givens method, and all the vectors w_r in that of Householder have their first elements equal to zero. The equivalence of the *symmetric* Givens and Householder methods is, of course, a special case of the result we have just proved. The argument should be compared with that of Chapter 4, § 53. Again we emphasize that although the matrices of the complete Givens and Householder transformations are essentially the same, the matrix which produces the rth column of zeros is not the same for the two methods. The proof is almost identical with the corresponding proof given in Chapter 4, § 53.

In spite of the formal equivalence of the two transformations and the fact that they are both 'stable', rounding errors may cause them to differ completely in practice. This is not surprising, since we saw in Chapter 5, § 28, that even two applications of Givens' transformation to the same matrix, working to slightly different precisions, may give very different tri-diagonal matrices. We saw there that such a divergence of the two computations would take place if, in any major step, all elements which were to be eliminated were small. Notice that from (7.5) and (7.8) the element $h_{i+1,i}$ is the square root of the sum of the squares of such elements, and we remarked above that the relation $AQ = QH$ does not define Q uniquely when an $h_{i+1,i}$ is zero. The non-uniqueness of the exact transformation when an $h_{i+1,i}$ is zero is matched by a poor determination of the transformation in practice when an $h_{i+1,i}$ is small. We emphasize once again that the accuracy of the final results is in no way influenced by the poor determination of the transformation.

Elementary stabilized transformations

8. For symmetric matrices the advantages of using orthogonal transformations are so great that we did not consider the use of the simpler elementary stabilized matrices, but the position is now very different.

We consider the reduction of A_0 to upper Hessenberg form using stabilized elementary matrices of the type N'_r (Chapter 3, § 47). There are $(n-2)$ major steps and at the beginning of the rth step the matrix

A_0 has already been reduced to A_{r-1} which for the case $n = 6, r = 3$, is of the form

$$
\begin{bmatrix}
h_{11} & h_{12} & h_{13} & a_{14} & a_{15} & a_{16} \\
h_{21} & h_{22} & h_{23} & a_{24} & a_{25} & a_{26} \\
0 & h_{32} & h_{33} & a_{34} & a_{35} & a_{36} \\
0 & 0 & a_{43} & a_{44} & a_{45} & a_{46} \\
0 & 0 & a_{53} & a_{54} & a_{55} & a_{56} \\
0 & 0 & a_{63} & a_{64} & a_{65} & a_{66}
\end{bmatrix}.
\tag{8.1}
$$

We shall see that subsequent steps do not affect the elements which are entered as h_{ij}. The rth major step is then as follows.

(i) Determine the maximum of the quantities $|a_{ir}|$ ($i = r+1,\ldots, n$). (When this is not unique, we take the first of these maximum values.) If this element is $a_{(r+1)',r}$, then interchange rows $r+1$ and $(r+1)'$, i.e. pre-multiply by $I_{r+1,(r+1)'}$. (The element $a_{(r+1)',r}$ may be regarded as the 'pivotal element' with respect to this transformation.)

(ii) For each value of i from $r+2$ to n:
Compute $n_{i,r+1} = a_{ir}/a_{r+1,r}$ so that $|n_{i,r+1}| \leqslant 1$.
Subtract $n_{i,r+1} \times$ row $r+1$ from row i.
The operations in (ii) comprise a pre-multiplication by a matrix of the type N_{r+1}. They produce zeros in positions (i, r) ($i = r+2,\ldots, n$).

(iii) Interchange columns $r+1$ and $(r+1)'$, i.e. post-multiply by $I_{r+1,(r+1)'}$.

(iv) For each value of i from $r+2$ to n:
Add $n_{i,r+1} \times$ column i to column $r+1$.
The operations in (iv) comprise a post-multiplication by N_{r+1}^{-1}. If the pivotal element is zero, the $n_{i,r+1}$ are arbitrary, and it is simplest to take $N_{r+1} I_{r+1,(r+1)'} = I$.

Notice that columns 1 to $r-1$ are unaltered by the rth transformation which is itself determined by the elements in column r in positions $r+1$ to n. The interchange is included in order to avoid numerical instability. Elements 1 to r of column r are also unaltered. There are $(n-r-1)$ quantities $n_{i,r+1}$, and these may be written in the positions previously occupied by the eliminated a_{ir}.

9. The description we gave in § 8 seems the most natural way to introduce the algorithm, but we now describe two different reformulations of the rth major step which make it possible to take

advantage of inner-product accumulation. We revert to the use of the superfix r, because this notation will be required in the error analysis of these reformulations. The interchanges and the multipliers $n_{i,r+1}$ are determined as before and the modified elements are then determined from the following relations.

Reformulation I

$$
\left.
\begin{aligned}
a_{i,r+1}^{(r)} &= a_{i,r+1}^{(r-1)} + \sum_{j=r+2}^{n} a_{ij}^{(r-1)} n_{j,r+1} && (i = 1,\dots, r+1) \\
a_{i,r+1}^{(r)} &= a_{i,r+1}^{(r-1)} + \sum_{j=r+2}^{n} a_{ij}^{(r-1)} n_{j,r+1} - n_{i,r+1} a_{r+1,r+1}^{(r)} && (i = r+2,\dots, n) \\
a_{ij}^{(r)} &= a_{ij}^{(r-1)} - n_{i,r+1} a_{r+1,j}^{(r-1)} && (i,j = r+2,\dots, n)
\end{aligned}
\right\} \text{(9.1)}
$$

Reformulation II

$$
\left.
\begin{aligned}
a_{ij}^{(r)} &= a_{ij}^{(r-1)} - n_{i,r+1} a_{r+1,j}^{(r-1)} && (i,j = r+2,\dots, n) \\
a_{i,r+1}^{(r)} &= a_{i,r+1}^{(r-1)} - n_{i,r+1} a_{r+1,r+1}^{(r-1)} + \sum_{j=r+2}^{n} a_{ij}^{(r)} n_{j,r+1} && (i = r+2,\dots, n) \\
a_{i,r+1}^{(r)} &= a_{i,r+1}^{(r-1)} + \sum_{j=r+2}^{n} a_{ij}^{(r-1)} n_{j,r+1} && (i = 1,\dots, r+1)
\end{aligned}
\right\} \text{(9.2)}
$$

A single rounding error is made when the relevant inner-product has been completed. The two processes are identical mathematically, but not in practice. Since the pivotal element in the next step is chosen from the elements $a_{i,r+1}^{(r)}$ ($i = r+2,\dots, n$), we may determine $(r+2)'$ as these elements are computed.

The approximate number of multiplications in the rth step is $(n-r)^2$ in the pre-multiplication and $n(n-r)$ in the post-multiplication whether we use the original formulation or either of the reformulations. The total number of multiplications in the complete reduction is therefore essentially

$$
\sum (n-r)^2 + \sum n(n-r) \doteq \tfrac{5}{6} n^3, \tag{9.3}
$$

just half the number in Householder's reduction and one quarter the number in Givens' reduction.

Significance of the permutations

10. We may express the complete transformation in the form

$$
N_{n-1} I_{n-1,(n-1)'} \dots N_3 I_{3,3'} N_2 I_{2,2'} A_0 I_{2,2'} N_2^{-1} I_{3,3'} N_3^{-1} \dots I_{n-1,(n-1)'} N_{n-1}^{-1} = H, \tag{10.1}
$$

where H is the final Hessenberg form. We now show that the inter-changes effectively determine a similarity permutation transforma-tion of A_0 such that, with this permutation, pivotal elements arise naturally in the pivotal position. We exhibit the proof for a matrix of order 5, the general case being obvious from this (cf. Chapter 4, § 22). When $n = 5$, equation (10.1) may be written in the form

$$N_4 I_{4,4'} N_3 (I_{4,4'} I_{4,4'}) I_{3,3'} N_2 (I_{3,3'} I_{4,4'} I_{4,4'} I_{3,3'}) I_{2,2'} A_0 \times$$
$$\times I_{2,2'} (I_{3,3'} I_{4,4'} I_{4,4'} I_{3,3'}) N_2^{-1} I_{3,3'} (I_{4,4'} I_{4,4'}) N_3^{-1} I_{4,4'} N_4^{-1} = H, \quad (10.2)$$

since each of the products in parenthesis is equal to the identity matrix. Regrouping the factors, we have

$$(N_4)(I_{4,4'} N_3 I_{4,4'})(I_{4,4'} I_{3,3'} N_2 I_{3,3'} I_{4,4'})(I_{4,4'} I_{3,3'} I_{2,2'} A_0 I_{2,2'} I_{3,3'} I_{4,4'}) \times$$
$$\times (I_{4,4'} I_{3,3'} N_2^{-1} I_{3,3'} I_{4,4'})(I_{4,4'} N_3^{-1} I_{4,4'}) N_4^{-1} = H. \quad (10.3)$$

The central matrix is a similarity permutation transformation of A_0, which we may call $\tilde{A}_0$, and if we denote the pre-multiplying factors in each pair of parentheses by $\tilde{N}_4$, $\tilde{N}_3$, $\tilde{N}_2$, then (10.3) becomes

$$\tilde{N}_4 \tilde{N}_3 \tilde{N}_2 \tilde{A}_0 \tilde{N}_2^{-1} \tilde{N}_3^{-1} \tilde{N}_4^{-1} = H. \quad (10.4)$$

Now since $r' \geqslant r$ $(r = 2,..., n-1)$, each $\tilde{N}_r$ is a matrix of exactly the same form as N_r; it merely has the sub-diagonal elements in the rth column re-ordered (Chapter 1, § 41(iv)). Hence $\tilde{A}_0$, a version of A_0 with its rows and columns similarly permuted, may be reduced to H without pivoting, the multipliers being automatically bounded in modulus by unity. Now we know from Chapter 1, § 41(iii), that in the matrix $\tilde{N}_2^{-1} \tilde{N}_3^{-1} \tilde{N}_4^{-1}$ no products of the n_{ij} arise. We may denote this product by $\tilde{N}$, where $\tilde{N}$ is a unit triangular matrix having its first column equal to e_1 since there is no factor $\tilde{N}_1^{-1}$ in the product. The elements below the diagonal in the rth column of $\tilde{N}$ are equal to the elements n_{ir} suitably permuted. We have therefore

$$\tilde{A}_0 \tilde{N} = \tilde{N} H, \quad (10.5)$$

and if no interchanges were necessary when transforming A_0, this would become

$$A_0 N = N H. \quad (10.6)$$

In Chapter 4, § 5, we referred to the need for equilibration of a matrix before performing Gaussian elimination with pivoting. Sim-ilarly here it is difficult to justify pivoting unless some analogous equilibration is performed. In the case of the eigenvalue problem this may be achieved for example by replacing A_0 by $D A_0 D^{-1}$ where D

is a diagonal matrix chosen so that the elements of maximum modulus in each row and column of the transformed matrix are of comparable magnitude. No completely satisfactory analysis of this problem has yet been given. On ACE we have tried various forms of equilibration and there is no doubt that it is of considerable importance to avoid working with a badly balanced matrix. (For discussions of this problem see Bauer (1963) and Osborne (1960).)

Direct reduction to Hessenberg form

11. In Chapter 4, § 17, we observed that Gaussian elimination without pivoting produced triangular matrices N and A_{n-1} such that

$$NA_{n-1} = A_0, \tag{11.1}$$

and then showed in § 36 that the elements of N and A_{n-1} could be determined directly by equating elements on both sides of equation (11.1). Subsequently we showed that pivoting could be incorporated. In a similar way, ignoring pivoting for the moment, we show that equation (10.6) can be used to find the elements of N and H directly, without going through the intermediate stages of determining each of the transforms A_r. Accordingly we drop the zero suffix in A_0, since this is now irrelevant.

Typically when $n = 5$ equation (10.6) is of the form

$$
\overset{A}{\begin{bmatrix} \times & \times & \times & \times & \times \\ \times & \times & \times & \times & \times \\ \times & \times & \times & \times & \times \\ \times & \times & \times & \times & \times \\ \times & \times & \times & \times & \times \end{bmatrix}}
\overset{N}{\begin{bmatrix} 1 & & & & \\ 0 & 1 & & & \\ 0 & \times & 1 & & \\ 0 & \times & \times & 1 & \\ 0 & \times & \times & \times & 1 \end{bmatrix}}
$$

$$
= \overset{N}{\begin{bmatrix} 1 & & & & \\ 0 & 1 & & & \\ 0 & \times & 1 & & \\ 0 & \times & \times & 1 & \\ 0 & \times & \times & \times & 1 \end{bmatrix}}
\overset{H}{\begin{bmatrix} \times & \times & \times & \times & \times \\ \times & \times & \times & \times & \times \\ & \times & \times & \bullet \times & \times \\ & & \times & \times & \times \\ & & & \times & \times \end{bmatrix}}, \tag{11.2}
$$

and from the structure of the matrices we show that by equating successive columns on both sides we can determine the elements of N and H column by column. In fact by equating the rth columns we determine column r of H and column $(r+1)$ of N. Column 1 of N is, of course, known a priori to be e_1.

The proof is by induction. Suppose that by equating the first $(r-1)$ columns of equation (10.6) we have determined the first $(r-1)$ columns of H and the first r columns of N. Let us now equate the elements of the rth column. The (i, r) elements of the left-hand side are

$$a_{ir}+a_{i,r+1}n_{r+1,r}+a_{i,r+2}n_{r+2,r}+\ldots+a_{in}n_{nr} \quad \text{(all } i\text{)}, \quad\quad (11.3)$$

while the (i, r) elements of the right-hand side are

$$\left.\begin{array}{ll} n_{i1}h_{1r}+n_{i2}h_{2r}+\ldots+n_{i,i-1}h_{i-1,r}+h_{ir} & (i \leqslant r+1) \\ n_{i1}h_{1r}+n_{i2}h_{2r}+\ldots+n_{i,r+1}h_{r+1,r} & (i > r+1) \end{array}\right\} \quad (11.4)$$

where we have included terms in n_{i1}, although these are zero, in order that the first contributions should not be special. Now since the rth column of N is known, the expression (11.3) can be computed for all i. Equating elements we obtain successively $h_{1r}, h_{2r}, \ldots, h_{r+1,r}; n_{r+2,r+1}, n_{r+3,r+1}, \ldots, n_{n,r+1}$. These are the non-zero elements in column r of H and column $(r+1)$ of N. The process is concluded when we find the nth column of H. There is no difficulty in starting, since the first column of N is known.

This formulation of the reduction enables us to take full advantage of inner-product accumulation. Thus we have from (11.3) and (11.4) the formulae

$$\left.\begin{array}{ll} h_{ir} = a_{ir}\times 1+\displaystyle\sum_{k=r+1}^{n} a_{ik}n_{kr}-\sum_{k=1}^{i-1} n_{ik}h_{kr} & (i = 1,\ldots,r+1) \\[4mm] n_{i,r+1} = \left(a_{ir}\times 1+\displaystyle\sum_{k=r+1}^{n} a_{ik}n_{kr}-\sum_{k=1}^{r} n_{ik}h_{kr}\right)\bigg/h_{r+1,r} & (i = r+2,\ldots, n) \end{array}\right\},$$

$$(11.5)$$

so that h_{ir} is given as an inner-product and $n_{i,r+1}$ as an inner-product divided by $h_{r+1,r}$.

Notice that it is because the first column of N is known a priori that we can determine the whole of N and H in this way. It is not particularly relevant that this column is e_1. If we take the first column to be any vector with a unit first element, then again we can determine the whole of N and of H and the equations are unchanged. (The result we have just proved is closely analogous to that of § 7.)

It was partly for this reason that we included the terms in n_{i1} $(i \neq 1)$. *If we think in terms of the formulation of § 8, then taking the first column of N to be other than e_1 is equivalent to computing $N_1 A_0 N_1^{-1}$ before starting the usual process.* This remark will be important later.

Incorporation of interchanges

12. The process of § 11 is formally identical to that of § 8 but without the provision of interchanges. Their incorporation is quite simple and is probably best described in algorithmic language. There are n major steps, in the rth of which we determine the rth column of H and the $(r+1)$th column of N. The configuration immediately before the beginning of the rth step is illustrated for the case $n = 6$, $r = 3$, by the array

$$
\begin{bmatrix}
h_{11} & h_{12} & a_{13} & a_{14} & a_{15} & a_{16} & \sigma_1 \\
h_{21} & h_{22} & a_{23} & a_{24} & a_{25} & a_{26} & \sigma_2 \\
n_{32} & h_{32} & a_{33} & a_{34} & a_{35} & a_{36} & \sigma_3 \\
n_{42} & n_{43} & a_{43} & a_{44} & a_{45} & a_{46} & \sigma_4 \\
n_{52} & n_{53} & a_{53} & a_{54} & a_{55} & a_{56} & \sigma_5 \\
n_{62} & n_{63} & a_{63} & a_{64} & a_{65} & a_{66} & \sigma_6
\end{bmatrix}
. \qquad (12.1)
$$

$$r' = 2' \quad 3'$$

The a_{ij} are elements of the original matrix A though interchanges may have taken place; these interchanges and the significance of the registers σ_i and r' will become apparent when we describe the rth step which is as follows.

(i) For each value of i from 1 to r:

Compute $h_{ir} = a_{ir} + \sum_{k=r+1}^{n} a_{ik} n_{kr} - \sum_{k=2}^{i-1} n_{ik} h_{kr}$, accumulating inner-products if possible. Round the final value to single-precision and overwrite on a_{ir}. (If the upper limit of any sum is smaller than the lower, the sum is ignored; remember that $n_{i1} = 0$ $(i = 2,..., n)$.) If $r = n$ the reduction is now complete.

(ii) For each value of i from $r+1$ to n:

Compute $(a_{ir} + \sum_{k=r+1}^{n} a_{ik} n_{kr} - \sum_{k=2}^{r} n_{ik} h_{kr})$ and store in the register σ_i. Each sum should be accumulated in double-precision if possible and will therefore need two registers. As the successive σ_i are computed, a record is kept of max $|\sigma_i|$. If $\max\limits_{i=r+1}^{n} |\sigma_i| = |\sigma_{(r+1)'}|$ then store $(r+1)'$. When the maximum is assumed by several σ_i we take the first such σ_i.

(iii) Interchange rows $r+1$ and $(r+1)'$ (including the n_{ij} and the σ_i) and columns $r+1$ and $(r+1)'$. Round the current σ_{r+1} to single-precision to give $h_{r+1,r}$ and overwrite on $a_{r+1,r}$. This is the pivotal element.

(iv) For each value of i from $r+2$ to n:
Compute $n_{i,r+1} = \sigma_i/h_{r+1,r}$ and overwrite on a_{ir}.

Several points deserve special comments. If at any stage

$$\max_{i=r+1}^{n} |\sigma_i| = 0$$

then the $n_{i,r+1}$ $(i = r+2,\ldots, n)$ are indeterminate and we could take any values of modulus less than unity and still have numerical stability. The simplest course is to take the relevant $n_{i,r+1}$ to be zero. If we are using floating-point with accumulation little is sacrificed if we round each σ_i to single-precision on completion. Each σ_i can then be stored temporarily on the corresponding a_{ir} until it is used to determine $n_{i,r+1}$. We can then dispense with the registers marked σ_i. If fixed-point accumulation is used, then we must remember the double-precision σ_i if we are to obtain the full benefit. However, since there are only $n-r$ such σ_i in the rth step there is always space in the σ_i registers for the quantities $2', 3',\ldots, r'$. It should be emphasized that the process we have described is the same as that of §§ 8, 9 except for rounding errors. In exact arithmetic, the interchanges, the h_{ij} and the n_{ij} are identical.

Numerical example

13. In Table 2 we give a numerical example of the reduction of a matrix of order 5 by the method of § 12 using 4-decimal fixed-point arithmetic with accumulation of inner-products. Since the algorithm is rather more complex than most which we have described, the example is exhibited in considerable detail. The process is remarkably compact on an automatic computer since all the configurations we give are stored in the same n^2 storage locations. Notice that because the first column of N is e_1 the first step appears to be somewhat different from the others. (If we modify the procedure and allow the first column of N to be any vector with its first element equal to unity

TABLE 2

A

$$\begin{bmatrix} \cdot3200 & \cdot2700 & \cdot2300 & \cdot3200 & \cdot4000 \\ \cdot2500 & \cdot0300 & \cdot7100 & \cdot2100 & \cdot1700 \\ \cdot4300 & \cdot7300 & \cdot1300 & \cdot3700 & \cdot8500 \\ \cdot5300 & \cdot2500 & \cdot5100 & \cdot6200 & \cdot1600 \\ \cdot1600 & \cdot6500 & \cdot4600 & \cdot5600 & \cdot3200 \end{bmatrix}$$

$$\begin{aligned} h_{11} &= \cdot3200\ 0000 \\ \sigma_2 &= \cdot2500\ 0000 \\ \sigma_3 &= \cdot4300\ 0000 \\ \sigma_4 &= \cdot5300\ 0000 \\ \sigma_5 &= \cdot1600\ 0000 \end{aligned}$$

$$\max |\sigma_i| = |\sigma_4|;\ 2' = 4.$$

Configuration at the beginning of second step

$$\begin{bmatrix} \cdot3200 & \cdot3200 & \cdot2300 & \cdot2700 & \cdot4000 \\ \cdot5300 & \cdot6200 & \cdot5100 & \cdot2500 & \cdot1600 \\ (\cdot8113) & \cdot3700 & \cdot1300 & \cdot7300 & \cdot8500 \\ (\cdot4717) & \cdot2100 & \cdot7100 & \cdot0300 & \cdot1700 \\ (\cdot3019) & \cdot5600 & \cdot4600 & \cdot6500 & \cdot3200 \end{bmatrix}$$

$$\begin{aligned} h_{12} &= \cdot3200+\cdot2300\times\cdot8113+\cdot2700\times\cdot4717+\cdot4000\times\cdot3019 && = 0\cdot7547\ 1800 \\ h_{22} &= \cdot6200+\cdot5100\times\cdot8113+\cdot2500\times\cdot4717+\cdot1600\times\cdot3019 && = 1\cdot1999\ 9200 \\ \sigma_3 &= \cdot3700+\cdot1300\times\cdot8113+\cdot7300\times\cdot4717+\cdot8500\times\cdot3019-\cdot8113\times1\cdot2000 = 0\cdot1028\ 6500 \\ \sigma_4 &= \cdot2100+\cdot7100\times\cdot8113+\cdot0300\times\cdot4717+\cdot1700\times\cdot3019-\cdot4717\times1\cdot2000 = 0\cdot2854\ 5700 \\ \sigma_5 &= \cdot5600+\cdot4600\times\cdot8113+\cdot6500\times\cdot4717+\cdot3200\times\cdot3019-\cdot3019\times1\cdot2000 = 0\cdot9741\ 3100 \end{aligned}$$

$$\max |\sigma_i| = |\sigma_5|;\ 3' = 5.$$

Note that the rounded value h_{22} is required immediately in the computation of the σ_i.

Configuration at the beginning of third step

$$\begin{bmatrix} \cdot3200 & \cdot7547 & \cdot4000 & \cdot2700 & \cdot2300 \\ \cdot5300 & 1\cdot2000 & \cdot1600 & \cdot2500 & \cdot5100 \\ (\cdot3019) & \cdot9741 & \cdot3200 & \cdot6500 & \cdot4600 \\ (\cdot4717) & (\cdot2930) & \cdot1700 & \cdot0300 & \cdot7100 \\ (\cdot8113) & (\cdot1056) & \cdot8500 & \cdot7300 & \cdot1300 \end{bmatrix}$$

$$\begin{aligned} h_{13} &= \cdot4000+\cdot2700\times\cdot2930+\cdot2300\times\cdot1056 && = \cdot5033\ 9800 \\ h_{23} &= \cdot1600+\cdot2500\times\cdot2930+\cdot5100\times\cdot1056 && = \cdot2871\ 0600 \\ h_{33} &= \cdot3200+\cdot6500\times\cdot2930+\cdot4600\times\cdot1056-\cdot3019\times\cdot2871 && = \cdot4723\ 5051 \\ \sigma_4 &= \cdot1700+\cdot0300\times\cdot2930+\cdot7100\times\cdot1056-\cdot4717\times\cdot2871-\cdot2930\times\cdot4724 = -\cdot0200\ 7227 \\ \sigma_5 &= \cdot8500+\cdot7300\times\cdot2930+\cdot1300\times\cdot1056-\cdot8113\times\cdot2871-\cdot1056\times\cdot4724 = \cdot7948\ 0833 \end{aligned}$$

$$\max |\sigma_i| = |\sigma_5|,\ 4' = 5.$$

The rounded value h_{23} is used immediately, not only in computing the σ_i but also for computing h_{33}.

Configuration at the beginning of fourth step

$$\begin{bmatrix} \cdot3200 & \cdot7547 & \cdot5034 & \cdot2300 & \cdot2700 \\ \cdot5300 & 1\cdot2000 & \cdot2871 & \cdot5100 & \cdot2500 \\ (\cdot3019) & \cdot9741 & \cdot4724 & \cdot4600 & \cdot6500 \\ (\cdot8113) & (\cdot1056) & \cdot7948 & \cdot1300 & \cdot7300 \\ (\cdot4717) & (\cdot2930) & (-\cdot0253) & \cdot7100 & \cdot0300 \end{bmatrix}$$

$$\begin{aligned} h_{14} &= \cdot2300+(-\cdot0253)\times\cdot2700 && = \cdot2231\ 6900 \\ h_{24} &= \cdot5100+(-\cdot0253)\times\cdot2500 && = \cdot5036\ 7500 \\ h_{34} &= \cdot4600+(-\cdot0253)\times\cdot6500-\cdot3019\times\cdot5037 && = \cdot2914\ 8797 \\ h_{44} &= \cdot1300+(-\cdot0253)\times\cdot7300-\cdot8113\times\cdot5037-\cdot1056\times\cdot2915 && = -\cdot3279\ 0321 \\ \sigma_5 &= \cdot7100+(-\cdot0253)\times\cdot0300-\cdot4717\times\cdot5037-\cdot2930\times\cdot2915-(-\cdot0253)\times(-\cdot3279) = \cdot3779\ 4034 \end{aligned}$$

Of course, σ_5 is now h_{54} since there is only one σ_i.

TABLE 2 CONTINUED

Configuration at the beginning of the fifth step

$$
\begin{bmatrix}
\cdot3200 & \cdot7547 & \cdot5034 & \cdot2232 & \cdot2700 \\
\cdot5300 & 1\cdot2000 & \cdot2871 & \cdot5037 & \cdot2500 \\
(\cdot3019) & \cdot9741 & \cdot4724 & \cdot2915 & \cdot6500 \\
(\cdot8113) & (\cdot1056) & \cdot7948 & -\cdot3279 & \cdot7300 \\
(\cdot4717) & (\cdot2930) & (-\cdot0253) & \cdot3779 & \cdot0300
\end{bmatrix}
$$

$$
\begin{aligned}
h_{15} &= \cdot2700 & &= \cdot2700\ 0000 \\
h_{25} &= \cdot2500 & &= \cdot2500\ 0000 \\
h_{35} &= \cdot6500 - \cdot3019 \times \cdot2500 & &= \cdot5745\ 2500 \\
h_{45} &= \cdot7300 - \cdot8113 \times \cdot2500 - \cdot1056 \times \cdot5745 & &= \cdot4665\ 0780 \\
h_{55} &= \cdot0300 - \cdot4717 \times \cdot2500 - \cdot2930 \times \cdot5745 - (-\cdot0253) \times \cdot4665 &= -\cdot2444\ 5105
\end{aligned}
$$

Final configuration

$$
\begin{bmatrix}
\cdot3200 & \cdot7547 & \cdot5034 & \cdot2232 & \cdot2700 \\
\cdot5300 & 1\cdot2000 & \cdot2871 & \cdot5037 & \cdot2500 \\
(\cdot3019) & \cdot9741 & \cdot4724 & \cdot2915 & \cdot5745 \\
(\cdot8113) & (\cdot1056) & \cdot7948 & -\cdot3279 & \cdot4665 \\
(\cdot4717) & (\cdot2930) & (-\cdot0253) & \cdot3779 & -\cdot2445
\end{bmatrix}
$$

Non-trivial elements of matrix N are in parentheses.

Permutation $\tilde{A}$ of A determined by transformation is given by

$$
\tilde{A} = I_{45} I_{35} I_{24} A I_{24} I_{35} I_{45}
$$

$$
\begin{bmatrix}
\cdot3200 & \cdot3200 & \cdot4000 & \cdot2300 & \cdot2700 \\
\cdot5300 & \cdot6200 & \cdot1600 & \cdot5100 & \cdot2500 \\
\cdot1600 & \cdot5600 & \cdot3200 & \cdot4600 & \cdot6500 \\
\cdot4300 & \cdot3700 & \cdot8500 & \cdot1300 & \cdot7300 \\
\cdot2500 & \cdot2100 & \cdot1700 & \cdot7100 & \cdot0300
\end{bmatrix}
$$

$$
\tilde{A}N =
\begin{bmatrix}
\cdot3200 & \cdot7547\ 18 & \cdot5033\ 98 & \cdot2231\ 69 & \cdot2700 \\
\cdot5300 & 1\cdot1999\ 92 & \cdot2871\ 06 & \cdot5036\ 75 & \cdot2500 \\
\cdot1600 & 1\cdot3364\ 11 & \cdot5590\ 26 & \cdot4435\ 55 & \cdot6500 \\
\cdot4300 & 1\cdot0764\ 25 & 1\cdot0776\ 18 & \cdot1115\ 31 & \cdot7300 \\
\cdot2500 & \cdot8514\ 97 & \cdot2537\ 66 & \cdot7092\ 41 & \cdot0300
\end{bmatrix}
$$

$$
NH =
\begin{bmatrix}
\cdot3200 & \cdot7547 & \cdot5034 & \cdot2232 & \cdot2700 \\
\cdot5300 & 1\cdot2000 & \cdot2871 & \cdot5037 & \cdot2500 \\
\cdot1600\ 07 & 1\cdot3363\ 80 & \cdot5590\ 7549 & \cdot4435\ 6703 & \cdot6499\ 7500 \\
\cdot4299\ 89 & 1\cdot0764\ 2496 & 1\cdot0776\ 0967 & \cdot1115\ 3421 & \cdot7299\ 9220 \\
\cdot2500\ 01 & \cdot8514\ 5130 & \cdot2537\ 2983 & \cdot7092\ 0066 & \cdot0299\ 5105
\end{bmatrix}
$$

and its other elements bounded by unity, then the first step becomes typical.) Special care should be taken in dealing with the first and last steps when programming this algorithm. The arrows indicate the row and column interchanges. The underlined elements are those of the final H and the elements in parentheses are those of N.

At the end of the example we exhibit the matrix $\tilde{A}$ which would have given the final N and H without the need for interchanges. Finally we give the *exact* products of *computed* N and *computed* H and of $\tilde{A}$ and *computed* N. These should be identical apart from rounding errors. Comparison of the two products reveals that they nowhere differ by more than $\frac{1}{2}10^{-4}$, the maximum value of a single rounding error.

Error analysis

14. We now show that the comment we made on the size of the elements of $(\tilde{A}N - NH)$ is true in general, subject to certain assumptions.

Let us assume that pivoting is used in the way we have described, that the computation is in fixed-point arithmetic with inner-product accumulation, and that all elements of A and of the computed H are less than unity. Then from (11.3) and (11.4) when $i < r+1$ we have

$$
\begin{aligned}
h_{ir} &= fi_2(a_{ir} + a_{i,r+1}n_{r+1,r} + \ldots + a_{in}n_{nr} - n_{i1}h_{1r} - \ldots - n_{i,i-1}h_{i-1,r}) \\
&\equiv a_{ir} + a_{i,r+1}n_{r+1,r} + \ldots + a_{in}n_{nr} - n_{i1}h_{1r} - \ldots - n_{i,i-1}h_{i-1,r} + \epsilon_{ir},
\end{aligned}
\tag{14.1}
$$

where
$$
|\epsilon_{ir}| < \tfrac{1}{2}(2^{-t}).
\tag{14.2}
$$

Similarly when $i > r+1$ we have

$$
\begin{aligned}
n_{i,r+1} &= fi_2[(a_{ir} + a_{i,r+1}n_{r+1,r} + \ldots + a_{in}n_{nr} - n_{i1}h_{1r} - \ldots - n_{ir}h_{rr})/h_{r+1,r}] \\
&\equiv (a_{ir} + a_{i,r+1}n_{r+1,r} + \ldots + a_{in}n_{nr} - n_{i1}h_{1r} - \ldots - n_{ir}h_{rr})/h_{r+1,r} + \eta_{ir},
\end{aligned}
\tag{14.3}
$$

where
$$
|\eta_{ir}| < \tfrac{1}{2}(2^{-t}).
\tag{14.4}
$$

Multiplying (14.3) by $h_{r+1,r}$ we have

$$
n_{i,r+1}h_{r+1,r} \equiv a_{ir} + a_{i,r+1}n_{r+1,r} + \ldots + a_{in}n_{nr} - n_{i1}h_{1r} - \ldots - n_{ir}h_{rr} + \epsilon_{ir},
\tag{14.5}
$$

where
$$
\epsilon_{ir} = h_{r+1,r}\eta_{ir}.
\tag{14.6}
$$

Equations (14.1) and (14.5) when rearranged show that

$$
(NH - \tilde{A}N)_{ir} = \epsilon_{ir} \qquad (i, r = 1, 2, \ldots, n),
\tag{14.7}
$$

and from (14.2) and (14.6)

$$
|\epsilon_{ir}| < \tfrac{1}{2}(2^{-t}) \qquad (i, r = 1, 2, \ldots, n),
\tag{14.8}
$$

since we are assuming that all elements of $|H|$ are bounded by unity. Hence we have finally

$$
\tilde{A}N - NH = F,
\tag{14.9}
$$

where
$$
|f_{ij}| < \tfrac{1}{2}(2^{-t}).
\tag{14.10}
$$

This is a most satisfactory result in that we could scarcely have expected $(\tilde{A}N - NH)$ to be smaller! However, it is perhaps not quite so good as it looks. Equation (14.9) implies that

$$H = N^{-1}(\tilde{A} - FN^{-1})N, \tag{14.11}$$

and that
$$N^{-1}\tilde{A}N = H + N^{-1}F. \tag{14.12}$$

Hence the computed H is an exact similarity transformation of $(\tilde{A} - FN^{-1})$ with the computed N as the matrix of the transformation. Alternatively we may say that H differs from an exact similarity transformation of $\tilde{A}$ by the matrix $N^{-1}F$. The fact that $\tilde{A}$ is involved, rather than A, is unimportant since the eigenvalues of $\tilde{A}$ and their individual condition numbers are the same as those of A.

Now we have
$$\|F\|_2 < \|F\|_E < \tfrac{1}{2}n2^{-t}, \tag{14.13}$$

but although the elements n_{ij} are bounded by unity $\|N^{-1}\|_2$ may be quite large. In fact if all the relevant non-zero n_{ij} were equal to -1 then we have, typically for $n = 5$,

$$N^{-1} = \begin{bmatrix} 1 & & & & \\ 0 & 1 & & & \\ 0 & 1 & 1 & & \\ 0 & 2 & 1 & 1 & \\ 0 & 4 & 2 & 1 & 1 \end{bmatrix}, \tag{14.14}$$

and in general n_{n2} is 2^{n-3}. Certainly without some deeper analysis we cannot obtain a bound for $\|N^{-1}F\|_2$ which does not contain the factor 2^n. We may contrast this with the situation which would have existed if N were unitary. We then have immediately

$$\|N^{-1}F\|_2 = \|F\|_2.$$

15. There are then two sources of danger in this process. First the elements of H *may* be very much larger than those of A which invalidates the assumption on which our bound for F was based. If the largest element of H is of order 2^k then the bound we have for F will contain the additional factor 2^k. Secondly, even when we have used pivoting, the norm of N^{-1} can be quite large. The situation is not unlike that which we had in Gaussian elimination (or triangular decomposition) with partial pivoting, though now we have the additional danger from the norm of N^{-1}. No analysis of the pivotal growth factor appears to have been made.

In practice little pivotal growth seems to occur and the matrices N^{-1} have norms of order unity. Consequently $\|N^{-1}F\|_2$ has usually been less than $\frac{1}{2}n2^{-t}$, which is the bound we found for $\|F\|_2$ itself. Again the situation is strongly reminiscent of the situation which we described in connexion with Gaussian elimination.

The net result is that the transformation using stabilized elementary matrices usually proves to be even more accurate than Householder's method. However, there remains the potential danger from pivotal growth and from an N^{-1} of large norm, and anyone of a cautious disposition might well prefer to use Householder's method. On balance there seems now to be slightly more justification for the cautious point of view than there was in the case of triangularization. If one has a computer on which inner-products cannot conveniently be accumulated, then the stabilized elementary matrices in any case are rather less attractive.

Related error analyses

16. We have given the error analysis corresponding to fixed-point accumulation. If we use floating-point accumulation ignoring terms of order 2^{-2t}, we find the bound for F is given typically for the case $n = 5$ by

$$|F| < 2^{-t} \begin{bmatrix} 0 & |h_{12}| & |h_{13}| & |h_{14}| & 0 \\ |h_{21}| & |h_{22}| & |h_{23}| & |h_{24}| & 0 \\ |h_{21}| & |h_{32}| & |h_{33}| & |h_{34}| & |h_{35}| \\ |h_{21}| & |h_{32}| & |h_{43}| & |h_{44}| & |h_{45}| \\ |h_{21}| & |h_{32}| & |h_{43}| & |h_{54}| & |h_{55}| \end{bmatrix}. \tag{16.1}$$

As usual, we gain a little more when there are small elements in the transformed matrix than we do in the case of fixed-point computation, besides dealing automatically with any pivotal growth without the need for scaling.

The analyses of the methods of §§ 8–9 have much in common with that of the direct reduction to Hessenberg form. For the methods of § 9, for example, for reformulation I with fixed-point accumulation of inner-products we have

$$(I_{r+1,(r+1)'}A_{r-1}I_{r+1,(r+1)'})N_{r+1}^{-1} - N_{r+1}^{-1}A_r \equiv F_r \tag{16.2}$$

where from equations (9.1) and the equations determining the $n_{i,r+1}$ we have, typically when $n = 6$, $r = 2$, the result

$$|F_r| < 2^{-t-1} \begin{bmatrix} 0 & 0 & 1 & 0 & 0 & 0 \\ 0 & 0 & 1 & 0 & 0 & 0 \\ 0 & 0 & 1 & 0 & 0 & 0 \\ 0 & 1 & 1 & 1 & 1 & 1 \\ 0 & 1 & 1 & 1 & 1 & 1 \\ 0 & 1 & 1 & 1 & 1 & 1 \end{bmatrix}. \tag{16.3}$$

Again we are assuming that all elements of A_{r-1} and A_r are bounded by unity.

Equation (16.2) may be expressed in the form

$$N_{r+1}(I_{r+1,(r+1)'}A_{r-1}I_{r+1,(r+1)'})N_{r+1}^{-1} = A_r + N_{r+1}F_r = A_r + G_r, \tag{16.4}$$

and we have

$$|G_r| < 2^{-t-1} \begin{bmatrix} 0 & 0 & 1 & 0 & 0 & 0 \\ 0 & 0 & 1 & 0 & 0 & 0 \\ 0 & 0 & 1 & 0 & 0 & 0 \\ 0 & 1 & 2 & 1 & 1 & 1 \\ 0 & 1 & 2 & 1 & 1 & 1 \\ 0 & 1 & 2 & 1 & 1 & 1 \end{bmatrix}. \tag{16.5}$$

Similarly, we can show for reformulation II that

$$N_{r+1}(I_{r+1,(r+1)'}A_{r-1}I_{r+1,(r+1)'} + K_r)N_{r+1}^{-1} \equiv A_r \tag{16.6}$$

and can obtain the same bound for K_r as for G_r. Equation (16.6) gives a perturbation of A_{r-1} which is equivalent to the errors in the rth step.

Unfortunately, with either reformulation, the final bound for the perturbation of A_0 which is equivalent to the rounding errors made in the whole process contains the factor 2^n and, as might be expected, is somewhat poorer than the bound which we obtained for the direct reduction.

17. It is instructive to follow the history of a specific eigenvalue λ of A_0 as we progress from A_0 to A_{n-2}. In general, the condition number (Chapter 2, § 31) of this eigenvalue will change, and if we denote the relevant values by $1/|s^{(i)}|$ $(i = 0, 1, ..., n-2)$, then rough upper bounds for the final errors in λ are given respectively by

$$\sum_{r=1}^{n-2} \|G_r\|_2/|s^{(r)}| \quad \text{and} \quad \sum_{r=1}^{n-2} \|K_r\|_2/|s^{(r-1)}|, \tag{17.1}$$

according as we use reformulation I or II. If we write

$$\alpha = \max |s^{(0)}/s^{(r)}|, \tag{17.2}$$

then the final errors in λ are no greater than those resulting from perturbations in A_0 bounded respectively by

$$\alpha \sum_{r=1}^{n-2} \|G_r\|_2 \quad \text{and} \quad \alpha \sum_{r=1}^{n-2} \|K_r\|_2, \tag{17.3}$$

and hence, from (16.5) using the Euclidean norm as an estimate, certainly by

$$\alpha 2^{-t-1} \sum_{r=1}^{n-2} [r(r+4)+n]^{\frac{1}{2}}. \tag{17.4}$$

The sum tends asymptotically to $\frac{1}{3}n^2$.

If the condition numbers of each eigenvalue did not vary too widely, then α would be of the order unity and the result we have obtained would be very satisfactory. Now if y and x are normalized left-handed and right-handed eigenvectors corresponding to λ the eigenvectors of A_1 are $N_2^T y$ and $N_2^{-1} x$, but these are no longer normalized. We have therefore

$$\left| \frac{s_0}{s_1} \right| = \left| \frac{(y^T x) \|N_2^T y\|_2 \|N_2^{-1} x\|_2}{y^T N_2 N_2^{-1} x} \right| = \|N_2^T y\|_2 \|N_2^{-1} x\|_2. \tag{17.5}$$

Hence the ratio of s_0 to s_1 is dependent upon the lengths of the vectors $N_2^T y$ and $N_2^{-1} x$ and these have components

$$\left. \begin{array}{l} (y_1, \; y_2 + n_{32} y_3 + n_{42} y_4 + ... + n_{n2} y_n, \; y_3, ..., \; y_n) \\ (x_1, \; x_2, \; x_3 - n_{32} x_2, \; x_4 - n_{42} x_2, ..., \; x_n - n_{n2} x_2) \end{array} \right\} \tag{17.6}$$

and

respectively. The length of the transformed eigenvectors can therefore increase progressively even when, as a result of pivoting, we have $|n_{ij}| < 1$.

We have tested this variation for a random selection of matrices, and found that there was surprisingly little variation in the condition numbers. However, since the work involved is substantial, the sample was necessarily small.

We emphasize that if pivoting is not used and we have an $n_{i,r+1}$ of order 2^k, then in general there are elements of A_r which are larger than those of A_{r-1} by a factor of order 2^{2k}. The equivalent perturbations in the original matrix are also larger by this factor 2^{2k}. Failure to pivot can therefore be even more serious than in the case of Gaussian elimination. Our discussion of the variation of the condition numbers of individual eigenvalues shows further that when an $n_{i,r+1}$ is large there is likely to be a sharp deterioration in the condition number as a result of the rth transformation.

Poor determination of the Hessenberg matrix

18. If at any stage in the reduction described in § 12 all the elements σ_i are small as a result of cancellation, then the pivotal element, which is max $|\sigma_i|$, will itself be small. This pivotal element must be rounded before it is divided into the other σ_i to give the multipliers $n_{i,r+1}$, and hence all these multipliers will be poorly determined. In such a case we find that the final computed Hessenberg matrix is very dependent on the precision of computation which is used. It is important to realise that this poor determination of H is in no way associated with a poor determination of the eigenvalues. The situation is exactly analogous to that discussed in Chapter 5, § 28, in connexion with the poor determination of the final tri-diagonal matrix in the symmetric Givens reduction. It is helpful to consider what happens in the exact computation if all the σ_i are zero. The corresponding $n_{i,r+1}$ are then indeterminate and we can give them arbitrary values. Corresponding to this complete arbitrariness in the case of exact computation we have poor determination in practice when the σ_i are small. These considerations will be important later.

Reduction to Hessenberg form using stabilized matrices of the type M'_{ji}

19. In Chapter 4, § 48, we described a method for reducing a matrix to triangular form by pre-multiplication with matrices of the form M'_{ji}. There is a corresponding method for the similarity reduction to Hessenberg form. As in the triangularization, this is closely analogous to the method based on plane rotations; the $(n-r-1)$ zeros in column r are introduced one by one, using $(n-r-1)$ stabilized matrices $M'_{j,r+1}$ instead of one stabilized matrix N'_{r+1}.

In the case of triangularization this technique had much to recommend it because it enabled us to find all the leading principal minors of the original matrix. There appear to be no corresponding advantages associated with the similarity reduction, and since it involves as many multiplications as the direct reduction of § 11 we shall not discuss it in detail.

The method of Krylov

20. We now describe an algorithm (normally used for computing the explicit characteristic polynomial) which will lead us naturally to a second class of methods which are closely related to those we have already presented. This method is due to Krylov (1931).

If b_0 is a vector of grade n, then the vectors b_r defined by

$$b_r = A^r b_0 \qquad (20.1)$$

satisfy the equation

$$b_n + p_{n-1} b_{n-1} + \ldots + p_0 b_0 = 0, \qquad (20.2)$$

where $\lambda^n + p_{n-1}\lambda^{n-1} + \ldots + p_0$ is the characteristic polynomial of A (Chapter 1, §§ 34–39). Equation (20.2) may be written in the form

$$Bp = -b_n, \qquad B = (b_0 \mid b_1 \mid \ldots \mid b_{n-1}), \qquad (20.3)$$

to provide us with a non-singular set of linear equations for determining p and hence for computing the characteristic equation of A. If b_0 is a vector of grade less than n, then the minimum polynomial of b_0 with respect to A will be of degree less than n. We then have

$$b_m + c_{m-1} b_{m-1} + \ldots + c_0 b_0 = 0 \qquad (m < n), \qquad (20.4)$$

and from equation (20.4) we may determine the minimum polynomial $\lambda^m + c_{m-1}\lambda^{m-1} + \ldots + c_0$; this will be a factor of the characteristic polynomial. If A is derogatory every vector will be of grade less than n.

In practice we shall not know in advance whether or not b_0 is of grade less than n. It is desirable that we should use some technique which automatically provides this information. The process we are about to describe is based on the following reorganised version of Gaussian elimination.

Gaussian elimination by columns

21. We denote the matrix on which elimination is to be performed by X and its columns by x_i. The matrix X is reduced to upper triangular form U in n major steps, one for each column separately, columns $x_r, x_{r+1}, \ldots, x_n$ being unaffected by the first $(r-1)$ steps, at the end of which the matrix X has been reduced to the form given typically for the case $n = 5$, $r = 3$, by

$$
\begin{bmatrix}
u_{11} & u_{12} & x_{13} & x_{14} & x_{15} \\
n_{21} & u_{22} & x_{23} & x_{24} & x_{25} \\
n_{31} & n_{32} & x_{33} & x_{34} & x_{35} \\
n_{41} & n_{42} & x_{43} & x_{44} & x_{45} \\
n_{51} & n_{52} & x_{53} & x_{54} & x_{55}
\end{bmatrix}
\begin{matrix}
\sigma_1 \\ \sigma_2 \\ \sigma_3 \\ \sigma_4 \\ \sigma_5
\end{matrix}
\qquad (21.1)
$$
$$
 1' 2'
$$

The rth step is then as follows.

(i) For each value of i from 1 to n multiply x_{ir} by 1 and store the product in the double-precision register σ_i.

For each value of i from 1 to $r-1$ perform steps (ii) and (iii);

(ii) Interchange σ_i and $\sigma_{i'}$, round the new σ_i to single-precision to give u_{ir} and overwrite on x_{ir}.

(iii) For each value of j from $i+1$ to n replace σ_j by $\sigma_j - n_{ji}u_{ir}$.

(iv) Select $\sigma_{r'}$ such that $|\sigma_{r'}| = \max\limits_{i=r}^{n} |\sigma_i|$. If there is more than one such maximum, take the first. Store r' at the foot of the rth column.

(v) Interchange σ_r and $\sigma_{r'}$, round the new σ_r to single-precision to give u_{rr} and overwrite on x_{rr}.

(vi) For each value of i from $r+1$ to n compute $n_{ir} = \sigma_i/u_{rr}$ and overwrite on x_{ir}.

It should be emphasized that this process is identical with triangular decomposition using partial pivoting and accumulation of inner-products (Chapter 4, § 39); even the rounding errors are the same. However, if the rth column of X is linearly dependent on the preceding columns, this is revealed *during the rth major step* of the modified process by the emergence of a set of zero σ_i ($i = r, \ldots, n$) at minor step (iv). At this point no computation has been performed on later columns.

We may apply this process to the matrix B formed by the vectors $b_0, b_1, ..., b_{n-1}$. If, when b_m is processed, we discover that it is linearly dependent on $b_0, b_1, ..., b_{m-1}$, then we do not compute the later vectors. We may write

$$
U_m = \begin{bmatrix} u_{11} & u_{12} & \cdots & u_{1m} \\ & u_{22} & \cdots & u_{2m} \\ & & & \cdot \\ & & & \cdot \\ & & & \cdot \\ & & & u_{mm} \end{bmatrix}, \qquad u_m = \begin{bmatrix} u_{1,m+1} \\ u_{2,m+1} \\ \cdot \\ \cdot \\ \cdot \\ u_{m,m+1} \end{bmatrix}, \qquad (21.2)
$$

so that the solution of the equations

$$
U_m c = -u_m \tag{21.3}
$$

gives the coefficients of $c_0, c_1, ..., c_{m-1}$ of the minimum polynomial of b_0 with respect to A.

Practical difficulties

22. It is not difficult to show that if A is non-derogatory, then almost all vectors are of grade n and therefore yield the full characteristic equation.

In practice this distinction between linear dependence and independence is blurred as the result of rounding errors. Even if b_0 is of grade less than n it will be unusual for the relevant set of σ_i to vanish exactly. Moreover if the matrix B is ill-conditioned this might well reveal itself by the emergence of a set of small σ_i and this will be indistinguishable from true linear dependence. Unfortunately there are a number of factors which contribute to the possibility of B being ill-conditioned, and these we now discuss.

In order to gain insight into these factors let us assume for simplicity that A is the matrix $\operatorname{diag}(\lambda_i)$ and that

$$
|\lambda_1| > |\lambda_2| > ... > |\lambda_n|. \tag{22.1}
$$

If the initial vector b_0 has components $\beta_1, \beta_2, ..., \beta_n$ then we know that if any β_i is zero, the vector b_0 is of grade less than n and hence B is singular. We might therefore expect that a small element β_i

would give an ill-conditioned matrix B but that the choice $\beta_i = 1$ $(i = 1,..., n)$ would be reasonably favourable. The matrix B is then given by

$$B = \begin{bmatrix} 1 & \lambda_1 & \lambda_1^2 & ... & \lambda_1^{n-1} \\ 1 & \lambda_2 & \lambda_2^2 & ... & \lambda_2^{n-1} \\ \cdots\cdots & & ... & \cdots \\ 1 & \lambda_n & \lambda_n^2 & ... & \lambda_n^{n-1} \end{bmatrix}, \tag{22.2}$$

and the equations giving the coefficients of the characteristic polynomial are

$$Bp = -b_n, \qquad b_n^T = (\lambda_1^n, \lambda_2^n,..., \lambda_n^n). \tag{22.3}$$

From the explicit form, B is obviously singular if $\lambda_i = \lambda_j$ (as we already knew) and will be ill-conditioned if any two eigenvalues are very close. We may equilibrate B to give the matrix C defined by

$$C = \begin{bmatrix} 1 & \mu_1 & \mu_1^2 & ... & \mu_1^{n-1} \\ 1 & \mu_2 & \mu_2^2 & ... & \mu_2^{n-1} \\ \cdots\cdots & & ... & ... \\ 1 & \mu_n & \mu_n^2 & ... & \mu_n^{n-1} \end{bmatrix} \text{ where } \mu_i = \lambda_i/\lambda_1, \tag{22.4}$$

and a reasonable estimate of the condition number of C is provided by the largest element of its inverse. The matrix C is a matrix of Vandermonde form and it may be shown that if Z is its inverse, then

$$z_{sr} = p_{rs}/\prod_{i \neq r} (\mu_r - \mu_i) \tag{22.5}$$

where the p_{rs} are defined by

$$\prod_{i \neq r} (y - \mu_i) = \sum_{s=0}^{n-1} p_{r,s+1} y^s. \tag{22.6}$$

Condition of C for some standard distributions of eigenvalues

23. We now estimate the condition numbers of C for a few standard distributions of eigenvalues of matrices of order 20.

(i) Linear distribution $\lambda_r = 21 - r$.
We have

$$\left| 1/\prod_{i \neq r} (\mu_r - \mu_i) \right| = 20^{19}/(r-1)! \, (20-r)! \tag{23.1}$$

and this has its greatest value, approximately 4.10^{12}, when $r = 10$ or 11. Since some of the coefficients $p_{10,s}$ are of order 10^3, we see that Z has some elements which are greater than 10^{15}.

(ii) Linear distribution $\lambda_r = 11 - r$.

We now have

$$\left| 1 \Big/ \prod_{i \neq r} (\mu_r - \mu_i) \right| = 10^{19}/(r-1)! \, (20-r)! \tag{23.2}$$

This is smaller by the factor 2^{-19} than the value obtained for distribution (i) and corresponding quantities $|p_{rs}|$ are also smaller. The condition number is smaller by a factor of at least 10^{-6}.

(iii) Geometric distribution $\lambda_r = 2^{-r}$.

For $r = 20$ we have

$$\left| 1 \Big/ \prod_{i \neq r} (\mu_r - \mu_i) \right| = 2^{-19}/(2^{-1} - 2^{-20})(2^{-2} - 2^{-20})\ldots(2^{-19} - 2^{-20})$$

$$> 2^{-19}/2^{-1}2^{-2}\ldots2^{-19}$$

$$= 2^{171}. \tag{23.3}$$

Now $p_{20,1}$ is almost equal to 1, so that the condition number of C is certainly as great as 2^{171}.

(iv) Circular distribution $\lambda_r = \exp(r\pi i/10)$.

For this distribution

$$\left| 1 \Big/ \prod_{i \neq r} (\mu_r - \mu_i) \right| = 1/20 \qquad \text{(all } r\text{)}. \tag{23.4}$$

Further all p_{rs} are of modulus unity. The matrix is therefore very well conditioned and we may expect to determine the coefficients of the characteristic polynomial to the full working accuracy. (Note that the λ_i are complex and complex arithmetic is therefore needed if we assume that A is in diagonal form.)

If A is not diagonal but

$$A = H^{-1} \operatorname{diag}(\lambda_i) H \tag{23.5}$$

then

$$b_r = A^r b_0 = H^{-1} \operatorname{diag}(\lambda_i^r)(H b_0). \tag{23.6}$$

The sequence corresponding to the matrix A with initial vector $b_0 = H^{-1} c_0$ is H^{-1} times the sequence corresponding to $\operatorname{diag}(\lambda_i)$ with the initial vector c_0. We cannot reasonably expect the matrix corresponding to A to be any better conditioned, in general, than that corresponding to $\operatorname{diag}(\lambda_i)$. Hence, turning to the significance of the use of e as the initial vector, we restrict ourselves again to the use of $\operatorname{diag}(\lambda_i)$.

Now it is evident that when there is a considerable variation in the sizes of the eigenvalues, then the later vectors in the sequence have very small components in the later positions. The following example suggests that the difficulty is fundamental and that any quest for a b_0 which provides a well-conditioned B is doomed to failure.

For the distribution (i) we might accordingly consider the effect of starting with
$$b_0^T = (20^{-10}, 19^{-10}, ..., 1), \tag{23.7}$$
in which the components corresponding to the small eigenvalues are well represented, then we have
$$b_{10}^T = (1, 1, 1, ..., 1) \tag{23.8}$$
and
$$b_{20}^T = 20^{10}[1, (19/20)^{10}, (18/20)^{10}, ..., (1/20)^{10}]. \tag{23.9}$$
in which the components corresponding to large eigenvalues are well represented. The corresponding equilibrated matrix C is given by

$$
\begin{bmatrix}
20^{-10} & 20^{-9} & \cdots & 1 & 1 & \cdots & 1 \\
19^{-10} & 19^{-9} & \cdots & 1 & 19/20 & \cdots & (19/20)^9 \\
\cdots\cdots & & \cdots & \cdots\cdots & & \cdots & \cdots \\
2^{-10} & 2^{-9} & \cdots & 1 & 2/20 & \cdots & (2/20)^9 \\
1 & 1 & \cdots & 1 & 1/20 & \cdots & (1/20)^9
\end{bmatrix} = C' \text{ (say)}. \tag{23.10}
$$

It is evident that if D_1 and D_2 are the diagonal matrices defined by
$$D_1 = [1, (19/20)^{10}, (18/20)^{10}, ..., (1/20)^{10}], \tag{23.11}$$
$$D_2 = (20^{10}, 20^9, ..., 20^1, 1, 1, ..., 1), \tag{23.12}$$
then $D_1 C' D_2$ is the same as the matrix which we obtained corresponding to $b_0 = e$. Hence the new inverse Y is given in terms of the old inverse by the relation
$$Y = D_2 Z D_1 \tag{23.13}$$
and Y is still very ill-conditioned.

Initial vectors of grade less than n

24. It is instructive to consider what will happen in practice when the initial vector is of grade m, less than n. We consider the theoretical position first. We have
$$\sum_{i=0}^{m-1} c_i b_i + b_m = 0 \qquad (m < n), \tag{24.1}$$
and therefore, pre-multiplying (24.1) by A^s,
$$\sum_{i=0}^{m-1} c_i b_{i+s} + b_{m+s} = 0 \qquad (s = 0, ..., n-m), \tag{24.2}$$

showing that any $m+1$ consecutive b_r satisfy the same linear relations. Clearly the equations

$$Bp = -b_n \tag{24.3}$$

are compatible, since $\mathrm{rank}(B) = \mathrm{rank}(B \,|\, b_n) = m$ from relations (24.2). Hence we may choose $p_m, p_{m+1}, \ldots, p_{n-1}$ arbitrarily and $p_0, \ldots, p_{m-1}$ are then determined. The general form of the solution may be derived from equations (24.2). For if we take arbitrary values q_s $(s = 0, \ldots, n-m-1)$ and $q_{n-m} = 1$, multiply corresponding equations of (24.2) by q_s and add, we obtain

$$\sum_{i=0}^{n-1} p_i b_i + b_n = 0, \tag{24.4}$$

where p_i is the coefficient of x^i in the expansion of

$$(c_0 + c_1 x + \ldots + c_{m-1} x^{m-1} + x^m)(q_0 + q_1 x + \ldots + q_{n-m-1} x^{n-m-1} + x^{n-m}).$$

Since we have $n-m$ arbitrary constants, the p_i represent the general solution of (24.3). The corresponding polynomial

$$p_0 + p_1 \lambda + \ldots + p_{n-1} \lambda^{n-1} + \lambda^n$$

is therefore given by

$$(c_0 + c_1 \lambda + \ldots + c_{m-1} \lambda^{m-1} + \lambda^m)(q_0 + q_1 \lambda + \ldots + q_{n-m-1} \lambda^{n-m-1} + \lambda^{n-m}), \tag{24.5}$$

and has the minimum polynomial of b_0 as a factor, the other factor being arbitrary.

In practice if we do not recognize premature linear dependence and continue the process as far as the vector b_n, then, when Gaussian elimination is performed as described in § 21, the triangle of order $(n-m)$ in the bottom right-hand corner of U_n will consist entirely of 'small' elements rather than zeros; the last $(n-m)$ elements of u_n will likewise be small. We therefore obtain more or less random values for $p_{n-1}, p_{n-2}, \ldots, p_m$. However, the corresponding variables

$$p_0, p_1, \ldots, p_{m-1}$$

will be determined correctly from the first m of the triangular equations. Hence the computed polynomial will have the factor

$$c_0 + c_1 \lambda + \ldots + c_{m-1} \lambda^{m-1} + \lambda^m,$$

the other factor of degree $(n-m)$ being quite arbitrary.

Practical experience

25. The results described in the last few sections have been confirmed in practice using a programme written for DEUCE. In order to compensate for the expected ill-condition of the matrix B the process is coded using double-precision fixed-point arithmetic.

The two sequences $b_0, b_1, \ldots, b_n$ and $c_0, c_1, \ldots, c_n$ are calculated from the relations

$$b_0 = c_0, \qquad b_{r+1} = Ac_r, \qquad c_{r+1} = k_{r+1}b_{r+1}, \qquad (25.1)$$

where k_{r+1} is chosen so that $\|c_{r+1}\|_\infty = 1$. The original matrix A has single-precision elements only, and in forming Ac_r inner-products are accumulated exactly and then rounded on completion; this requires only the multiplication of single-precision numbers by double-precision numbers (62 binary digits). Since we work with the original A throughout we can take full advantage of any zero elements it may contain. The equations

$$(c_0 \mid c_1 \mid \ldots \mid c_{n-1})q = -c_n \qquad (25.2)$$

are then solved using true double-precision fixed-point arithmetic, and finally the coefficients p_r of the characteristic polynomial are computed from the relations

$$p_r \, k_{r+1}k_{r+2}\ldots k_n = q_r, \qquad (25.3)$$

using double-precision floating-point arithmetic.

Attempts were made as described in § 21 to recognize linear dependence at the earliest stage at which it occurred by treating as zero all elements which were below a prescribed tolerance in absolute magnitude. (For this it is convenient to normalize the b_r to give the c_r even if double-precision *floating-point* is used throughout.) I also attempted to make the initial vector defective in the eigenvectors corresponding to smaller eigenvalues by a few preliminary multiplications by A, so as to force premature linear dependence. On the whole my experience with these techniques was not very happy and I abandoned their use in favour of the computation of the full set of vectors in all cases. Results obtained in practice were the following.

(i) Using a diagonal matrix $\mathrm{diag}(\lambda_i)$ with $\lambda_r = 21 - r$ and the initial vector e, the characteristic equation had 13 correct binary figures in its most accurate coefficient and 10 in its least. This is roughly what we might expect since the condition number of c is of order 10^{15}. Using the initial vector of equation (23.7) the most accurate coefficient had 15 correct binary figures and the least had 11. Both of the derived equations are quite useless for the determination of the eigenvalues. (See Chapter 7, § 6.)

(ii) For a diagonal matrix of order 21 with $\lambda_r = 11 - r$ and initial vector e, the computed characteristic polynomial had all coefficients correct to at least 30 binary places, some being appreciably more accurate. This was adequate to determine the eigenvalues to about 20 binary places.

(iii) For the distribution $\lambda_r = 2^{-r}$ the computed characteristic equations corresponding to a number of different initial vectors bore no relation to the true equation.

(iv) To experiment with the distribution $\lambda_r = \exp(r\pi i/10)$ without using complex arithmetic, we used the tri-diagonal matrix having 10 blocks of the form

$$\begin{bmatrix} \cos\theta_r & \sin\theta_r \\ -\sin\theta_r & \cos\theta_r \end{bmatrix}, \qquad \theta_r = r\pi/10. \qquad (25.4)$$

With initial vector e this gave the characteristic equation correct to 60 binary places and eigenvalues which were correct to the same accuracy.

For simple distributions with some coincident or very close eigenvalues it was common to obtain some correct eigenvalues and some which were quite arbitrary, thus confirming the analysis of § 24.

Generalized Hessenberg processes

26. Our discussion of Krylov's method shows that it is likely to give good results only when the distribution of eigenvalues is favourable, but perhaps we should mention that favourable distributions are not uncommon for matrices which arise in connexion with damped mechanical and electrical oscillations, for which the eigenvalues are, in general, complex conjugate pairs.

In order to widen the range of satisfactory operation, methods have been sought which 'stabilize' the sequence of Krylov vectors. These may be grouped together under the title 'generalized Hessenberg methods'. The basis of all such methods is the following.

We are given a set of n independent vectors x_i which are the columns of the matrix X. Starting with an arbitrary vector b_1 we form a set of modified Krylov vectors $b_2, b_3, \ldots, b_{n+1}$ defined by the relations

$$k_{r+1}b_{r+1} = Ab_r - \sum_{i=1}^{r} h_{ir}b_i. \qquad (26.1)$$

(Now that we are no longer concerned with the direct determination of the characteristic equations there are advantages in naming the initial vector b_1 rather than b_0.) The k_{r+1} are scalars, inserted for numerical convenience, and are commonly normalizing factors. The h_{ir} are determined so that b_{r+1} is orthogonal to $x_1, \ldots, x_r$. The vector b_{n+1} must be null, if not some earlier b_r, because it is orthogonal to n independent vectors. Equations (26.1) may be assembled into the single matrix equation

$$A[b_1 \mid b_2 \mid \ldots \mid b_n] = [b_1 \mid b_2 \mid \ldots \mid b_n] \begin{bmatrix} h_{11} & h_{12} & h_{13} & \ldots & h_{1n} \\ k_2 & h_{22} & h_{23} & \ldots & h_{2n} \\ 0 & k_3 & h_{33} & \ldots & h_{3n} \\ \cdots\cdots\cdots & & \cdots & \cdots \\ 0 & 0 & \ldots & k_n & h_{nn} \end{bmatrix}, \quad (26.2)$$

i.e.
$$AB = BH \qquad\qquad (26.3)$$

where the matrix H is of upper Hessenberg form.

That equation (26.2) includes the condition that b_{n+1} is null can be seen by equating column n of both sides and comparing with (26.1). The orthogonality condition may be written in the form

$$X^T B = L, \qquad\qquad (26.4)$$

where L is a lower triangular matrix. Provided B is non-singular, equation (26.3) may be written in the form

$$B^{-1} A B = H. \qquad\qquad (26.5)$$

For convenience we have assumed that the vectors x_i are given in advance. There is nothing in the above description that requires this, and we may in fact derive them at the same time as the b_i by any process that may prove expedient.

Failure of the generalized Hessenberg process
27. The process we have just described may break down for two different reasons.

(i) A vector b_{r+1} $(r < n)$ may be null. If this is so, then we may replace the null b_{r+1} by any vector c_{r+1} which is orthogonal to $x_1, \ldots, x_r$.

We may preserve a consistent notation by defining new sets of vectors satisfying the following relations.

$$c_1 = b_1, \quad b_{r+1} = Ac_r - \sum_{i=1}^{r} h_{ir}c_i, \quad k_{r+1}c_{r+1} = b_{r+1} \quad (b_{r+1} \neq 0)$$

c_{r+1} is an arbitrary non-null vector orthogonal to $x_1, ..., x_r$

$$(b_{r+1} = 0)$$

(27.1)

If we take $k_{r+1} = 0$ when b_{r+1} is null, then the relation $k_{r+1}c_{r+1} = b_{r+1}$ is still true.

With these definitions we have

$$AC = CH, \qquad X^T C = L. \tag{27.2}$$

For every null b_r there is a corresponding zero k_r in H. We now see that we should scarcely regard the emergence of a null b_r as a 'breakdown' in the procedure, but rather as providing some choice in the method of continuing. In fact the calculation of the eigenvalues of H is simpler in this case, since when a k_r is zero the eigenvalues of H are those of two smaller matrices. We discuss the process in terms of the c_i from now on.

(ii) The elements of the matrix H are derived from the relation

$$h_{ir} = x_i^T A c_r / x_i^T c_i. \tag{27.3}$$

If an $x_i^T c_i$ is zero, then we cannot determine h_{ir}. It is evident that the vanishing of an $x_i^T c_i$ is a more serious matter than premature vanishing of a b_i. Now if the x_i are not given in advance, we may be able to choose x_i so that $x_i^T c_i$ is not zero. When the x_i are given in advance the vanishing of $x_i^T c_i$ would certainly appear to be fatal and the only solution appears to be to restart with a different vector b_1. An interesting consideration is how to choose b_1 so that this breakdown does not occur.

Notice that from (27.2) we have $l_{ii} = x_i^T c_i$. If no $x_i^T c_i$ is zero, then L is non-singular and hence C is non-singular.

The Hessenberg method

28. An obvious choice for X is the identity matrix and we then have $x_i = e_i$. This gives the method of Hessenberg (1941). The second of equations (27.2) now becomes

$$L = X^T C = C, \tag{28.1}$$

so that C itself is lower triangular, that is the first $(r-1)$ components of c_r are zero. We may choose the k_r so that the rth component of c_r is unity, and we write

$$c_r^T = (0,\dots, 0, 1, n_{r+1,r},\dots, n_{nr}) \qquad (28.2)$$

for reasons which will later become apparent. The rth step in the process is as follows.

Compute Ac_r which will, in general, have no zero elements. Then subtract successively a multiple of c_1 to annihilate the first element, a multiple of c_2 to annihilate the second element,..., and finally a multiple of c_r to annihilate the rth element. This gives the vector b_{r+1}. We then take k_{r+1} to be the $(r+1)$th element of b_{r+1}, so that c_{r+1} is of the form corresponding to equation (28.2). If b_{r+1} is null then we may take c_{r+1} to be an arbitrary vector of the form

$$(0,\dots, 0, 1, n_{r+2,r+1},\dots, n_{n,r+1}) \qquad (28.3)$$

since this is automatically orthogonal to $e_1, e_2,\dots, e_r$. The corresponding value of k_{r+1} is zero. This corresponds to case (i) of § 27.

If b_{r+1} is not null but its $(r+1)$th element is zero, we cannot normalize b_{r+1} to obtain c_{r+1} in the way we described. Whatever normalizing factor we use we have $e_{r+1}^T c_{r+1} = 0$ and hence the process will break down in the next step. This corresponds to case (ii) of § 27.

Practical procedure

29. We can avoid the break-down and also ensure numerical stability if we take the x_i to be the columns of I but not necessarily in the natural order. We take $x_r = e_{r'}$ where $1', 2',\dots, n'$ is some permutation of $1, 2,\dots, n$.

The way in which the r' are determined is probably best explained by describing the rth step. The vectors $c_1,\dots, c_r$ and $1',\dots, r'$ have already been determined, and c_i has zero elements in positions $1',\dots, (i-1)'$. The rth step is then as follows.

Compute Ac_r and subtract multiples of $c_1, c_2,\dots, c_r$ so as to annihilate elements $1', 2',\dots, r'$ respectively. The resulting vector is b_{r+1}. If the maximum element of b_{r+1} is in position $(r+1)'$, then we take this element to be k_{r+1} and x_{r+1} to be $e_{(r+1)'}$. It is clear that elements $1',\dots, r'$ of c_{r+1} are zero and all non-zero elements are bounded by unity. If b_{r+1} is null we may take x_{r+1} to be any column $e_{(r+1)'}$ of I orthogonal to $e_{i'}$ $(i = 1,\dots, r)$ and c_{r+1} to be any vector with unity for its $(r+1)$th component which is also orthogonal to the $e_{i'}$ $(i = 1,\dots, r)$.

Clearly the $e_{i'}$ form the columns of a permutation matrix P (Chapter 1, § 40 (ii)). Equations (27.2) therefore become

$$AC = CH, \quad P^T C = L, \tag{29.1}$$

where L is not merely lower triangular, but *unit* lower triangular. Hence we have

$$P^T A (PP^T) C = P^T CH, \tag{29.2}$$

or

$$(P^T AP)(P^T C) = (P^T C)H, \tag{29.3}$$

that is

$$\tilde{A} L = LH, \tag{29.4}$$

where $\tilde{A}$ is some similarity permutation transform of A and L is C with its rows suitably permuted.

Relation between the Hessenberg method and earlier methods

30. If the e_i are used in the natural order, then we have

$$AC = CH, \tag{30.1}$$

and with the c_r given by equation (28.2) we may write

$$AN = NH, \tag{30.2}$$

where N is a unit triangular matrix. Now if we take the initial vector c_1 to be e_1, then from § 11 we see that the matrices N and H are identical with those which are derived by the direct reduction to Hessenberg form. Taking c_1 in the Hessenberg process to be any other vector with unit first component is equivalent to including the preliminary transformation $N_1 A N_1^{-1}$ if we think in terms of similarity transformations of A.

The practical procedure described in § 29 is the exact equivalent of the direct reduction to Hessenberg form with the inclusion of interchanges. Notice that the r' of § 29 are not the r' of § 12, but the r' of § 12 are such that $I_{2,2'} I_{3,3'} \ldots I_{n-1,(n-1)'} = P$. In fact it is easy to reframe the description of § 29 so that we perform similar row and column interchanges on A and a row interchange of the current matrix of c_i at each stage instead of using the e_i in some permuted order. The two methods are not only mathematically equivalent but, if inner-products are accumulated whenever possible, then even the rounding errors are the same.

We do not therefore need to carry out a separate error analysis for the Hessenberg method. However, we shall describe the results obtained earlier in terms appropriate to the present method. We assume for convenience of description that A and the vectors are actually permuted at each stage.

(i) The use of interchanges is essential in the sense that otherwise we may obtain indefinitely large elements n_{ji} of the vector c_i and this can lead to a great loss of accuracy.

(ii) If at any stage a vector b_i has small elements in all positions from i to n, then no loss of accuracy results from this in spite of the fact that the n_{ji} are calculated as the ratio of small numbers. We merely have a poor determination of the final matrix H. We do not apologize for emphasizing this again; it has been suggested that when this occurs we should start again with a different initial vector in the hope that this situation will be avoided. There is absolutely no justification for such a precaution.

(iii) If a computed b_{r+1} is zero then we may take $c_{r+1} = e_{r+1}$. This corresponds to the fact that when $a_{ir}^{(r-1)} = 0$ $(i = r+1,..., n)$ then the matrix N_{r+1} of § 8 can be taken to be the identity matrix, though of course any N_{r+1} could be used.

The method of Arnoldi

31. Another natural choice for the set of vectors x_i in the generalized Hessenberg process is the set c_i itself. This leads to the method of Arnoldi (1951). The second of equations (27.2) now becomes

$$C^T C = D, \tag{31.1}$$

where D is diagonal, since it is both lower triangular and symmetric. We may choose the normalizing factor so that $C^T C = I$ and the process then gives an orthogonal matrix. (The advantage of doing this is quite slight as compared with normalizing so that the maximum element of each $|c_i|$ lies between $\frac{1}{2}$ and 1 and the latter is more economical; however, we shall see that the normalization $C^T C = I$ reveals an interesting relationship with earlier methods.) We shall assume for the moment that C is not symmetric since the symmetric case is discussed in a different context in § 40.

If we take c_1 to be e_1 then we have

$$AC = CH \tag{31.2}$$

for an orthogonal matrix C having e_1 for its first column. Hence from § 7 we see that C and H are essentially the matrices produced by the Givens and Householder transformations.

Practical considerations

32. Although Arnoldi's process with the normalization $C^T C = I$ is so closely related to Givens' and Householder's methods, the computational process has little in common with either. We now consider the practical process in more detail. We need not assume that c_1 is e_1 but shall take it to be an arbitrary normalized vector. We first consider the vector b_{r+1} defined by

$$b_{r+1} = Ac_r - h_{1r}c_1 - h_{2r}c_2 - \ldots - h_{rr}c_r, \qquad (32.1)$$

where the h_{ir} are chosen so that b_{r+1} is orthogonal to c_i ($i = 1, \ldots, r$). Now it may well happen that when b_{r+1} is computed extensive cancellation takes place so that the components of b_{r+1} are all very small compared with $\|Ac_r\|_2$. In this case b_{r+1} will not even be approximately orthogonal to the c_i. (Compare the similar phenomenon discussed in Chapter 4, § 55, in connexion with Schmidt orthogonalization.)

Now it is essential that the c_i should remain strictly orthogonal to working accuracy or there will be no guarantee that c_{n+1} is null to working accuracy. *The difficulty which arises should not be attributed to the cumulative effect of rounding errors.* It can happen in the first step of the computation with a matrix of order two! This is illustrated in Table 3. Almost complete cancellation takes place when

$$b_2 = Ac_1 - h_{11}c_1$$

is computed. Consequently b_2 is by no means orthogonal to c_1. If we ignore this, normalize b_2 and continue, we find that b_3 is not almost null as it would have been if c_1 and c_2 were almost exactly orthogonal. The matrix H is by no means similar to A. (The computation has been performed using block-floating arithmetic but this is not of any real significance. The same difficulty arises whatever mode of arithmetic is used.) The trace is 0.62888 instead of 0.62697 and $\|b_3\|_2$ is of order 2×10^{-2} instead of 10^{-5}.

Suppose now we re-orthogonalize the computed vector b_2 with respect to c_1. We may write

$$b_2' = b_2 - \epsilon_{11}c_1, \qquad (32.2)$$

and for orthogonality we have

$$\epsilon_{11} = c_1^T b_2. \qquad (32.3)$$

Now because b_2 has already been orthogonalized once with respect to c_1, we can be quite sure that ϵ_{11} is of order 10^{-5}. Hence we have

$$b_2' = b_2 + O(10^{-5}). \qquad (32.4)$$

TABLE 3

A

$$\begin{bmatrix} 0 \cdot 41323 & 0 \cdot 61452 \\ 0 \cdot 54651 & 0 \cdot 21374 \end{bmatrix}$$

Arnoldi's process without re-orthogonalization

c_1	Ac_1	b_2
0·78289	0·70584	0·00004
0·62216	0·56084	−0·00006
$h_{11} = 0 \cdot 90153$	$k_2 = 10^{-4}(0 \cdot 72111)$	

c_2	Ac_2	$Ac_2 - h_{12}c_1$	$b_3 = Ac_2 - h_{12}c_1 - h_{22}c_2$
0·55470	−0·28209	−0·17023	−0·01899
−0·83205	0·12531	0·21420	−0·01266
	$h_{12} = -0 \cdot 14288$	$h_{22} = -0 \cdot 27265$	

H

$$\begin{bmatrix} 0 \cdot 90153 & -0 \cdot 14288 \\ 10^{-4}(0 \cdot 72111) & -0 \cdot 27265 \end{bmatrix}$$

Arnoldi's process with re-orthogonalization

c_1	Ac_1	b_2	$b_2' = b_2 - \epsilon_{11}c_1$
0·78289	0·70584	0·00004	$10^{-4}(\ 0 \cdot 44708)$
0·62216	0·56084	−0·00006	$10^{-4}(-0 \cdot 56258)$
$h_{11} = 0 \cdot 90153$		$k_2 = 10^{-4}(0 \cdot 71859)$	

c_2	Ac_2	$Ac_2 - h_{12}c_1$	$b_3 = Ac_2 - h_{12}c_1 - h_{22}c_2$
0·62216	−0·22401	−0·17082	0·00000
−0·78289	0·17268	0·21495	0·00000
	$h_{12} = -0 \cdot 06794$	$h_{22} = -0 \cdot 27456$	

H

$$\begin{bmatrix} 0 \cdot 90153 & -0 \cdot 06794 \\ 10^{-4}(0 \cdot 71859) & -0 \cdot 27456 \end{bmatrix}$$

However, the original orthogonalization ensured that b_2 satisfied the relation

$$b_2 = Ac_1 - h_{11}c_1 + O(10^{-5}), \tag{32.5}$$

and hence

$$b_2' = Ac_1 - h_{11}c_1 + O(10^{-5}). \tag{32.6}$$

The re-orthogonalization therefore will not make the relation between b_2', Ac_1 and c_1 any less satisfactory, but it will make b_2' orthogonal to c_1 to working accuracy, in the sense that the angle between them will differ from $\dfrac{\pi}{2}$ by an angle of order 10^{-5}.

This is illustrated in Table 3. The normalized vector is computed from b_2' and we have

$$k_2 c_2 = b_2'. \tag{32.7}$$

Since c_1 and c_2 are now almost orthogonal, b_3 really does vanish to working accuracy. Notice that the computed quantities now satisfy

$$k_2c_2 = Ac_1 - h_{11}c_1 + O(10^{-5}), \qquad (32.8)$$

$$b_3 = O(10^{-5}) = Ac_2 - h_{21}c_1 - h_{22}c_2 + O(10^{-5}), \qquad (32.9)$$

so that

$$A[c_1 \mid c_2] = [c_1 \mid c_2]\begin{bmatrix} h_{11} & h_{12} \\ k_2 & h_{22} \end{bmatrix} + O(10^{-5}), \qquad (32.10)$$

which is the best result we can expect. This shows that the computed H is now almost exactly similar to A. Note that the trace is preserved to working accuracy and it may be verified that this is true of the eigenvalues and also of the eigenvectors found via the transformation.

Significance of re-orthogonalization

33. To obtain a vector b_2' which is truly orthogonal to c_1 we applied the Schmidt orthogonalization process to b_2. It is important to realize that the correction to b_2 which is obtained in this way is not the only correction which will give satisfactory results. All that is needed is that we should make a correction which is of the order of 10^{-5} or less in magnitude, and which makes b_2' almost exactly orthogonal to c_1. For example the vector b_2' given by

$$b_2' = 10^{-4}(0{\cdot}43814, \ -0{\cdot}55133) \qquad (33.1)$$

gives just as good results as the b_2' obtained by re-orthogonalization. On normalization it gives a c_2 such that (32.8) is satisfied. The effect of cancellation is to give us some degree of choice in the vector b_2'. In Wilkinson (1963b), pp. 86–91, I have shown that even if we use re-orthogonalization, the final matrices C and H are very dependent on the precision of computation when cancellation takes place.

The simplest procedure, in general, is to compute b_{r+1} given by

$$b_{r+1} = Ac_r - h_{1r}c_1 - h_{2r}c_2 - \ldots - h_{rr}c_r, \qquad (33.2)$$

and then to compute b_{r+1}' such that

$$b_{r+1}' = b_{r+1} - \epsilon_{1r}c_1 - \epsilon_{2r}c_2 - \ldots - \epsilon_{rr}c_r, \qquad (33.3)$$

$$k_{r+1}c_{r+1} = b_{r+1}', \qquad \|c_{r+1}\|_2 = 1, \qquad (33.4)$$

where

$$\epsilon_{ir} = c_i^T b_{r+1}/c_i^T c_i = c_i^T b_{r+1}. \qquad (33.5)$$

We shall then have

$$A(c_1 \mid c_2 \mid ... \mid c_n) = (c_1 \mid c_2 \mid ... \mid c_n)H + O(2^{-t}), \qquad (33.6)$$

if A is a normalized matrix and we are using t-digit computation. The precise meaning of $O(2^{-t})$ has been left deliberately vague. In general it will be a matrix which is bounded in norm by $f(n)2^{-t}$, where $f(n)$ will depend on the type of arithmetic used. The factor $f(n)$ can reasonably be attributed to the cumulative effect of rounding errors; for floating-point computation with accumulation of inner-products a rough estimate shows that $f(n) < Kn^2$, but this result can probably be improved upon.

The need for re-orthogonalization springs from cancellation and not from the accumulation of rounding errors. We have used a matrix of order two for purposes of illustration because it emphasizes this point. If no cancellation takes place at any stage, re-orthogonalization can be safely omitted, but unfortunately this is rare. The inclusion of re-orthogonalization makes the method of Arnoldi far less economical than that of Householder; there are $2n^3$ multiplications without re-orthogonalization and $3n^3$ with it. (In this connexion see our comparison of the Schmidt orthogonalization process and Householder's triangularization in Chapter 4, § 55.) Without re-orthogonalization Arnoldi's method may give arbitrarily poor results even for a well-conditioned matrix. When re-orthogonalization is included however, Arnoldi's method is quite comparable in accuracy with that of Householder; *this accuracy is not in any way impaired if we happen to choose b_1 so that it is deficient or even completely defective in some of the eigenvectors.*

Again we have the situation that, in exact computation, when a b_r vanishes we have complete freedom of choice in 'all figures' of c_r provided only that it is orthogonal to the preceding c_i, while for computation of a restricted precision, when k figures vanish as a result of cancellation we can effectively choose $t - k$ figures arbitrarily. Note that in Arnoldi's method we cannot have the second type of failure described in § 27 because $x_i^T c_i = c_i^T c_i = 1$.

34. Two questions arise quite naturally in connexion with the re-orthogonalization.

(i) Why was there nothing corresponding to this re-orthogonalization in the method of Hessenberg?

(ii) What takes the place of re-orthogonalization in Givens' and Householder's reductions which we know to be mathematically equivalent to that of Arnoldi?

It is imperative that the reader should be quite clear on both of these points, and we now discuss them in some detail. On the first point we notice that if we define c_{r+1} by the equations

$$k_{r+1}c_{r+1} = Ac_r - h_{1r}c_1 - \ldots - h_{rr}c_r, \qquad (34.1)$$

then provided only that $c_{n+1} = 0$ we have

$$AC = CH, \qquad (34.2)$$

however the h_{ir} may have been chosen. Choosing the h_{ir} so that c_{r+1} is orthogonal with respect to the independent vectors x_i is important only because it ensures that the c_i are independent and that c_{n+1} is null. Now when the x_i are the vectors e_i exact orthogonality *is* maintained *merely by taking the appropriate components of c_i to be exactly zero.* We therefore achieve exact orthogonality almost without being aware of it. With most choices of the x_i accurate orthogonality cannot be achieved in such a simple manner, and will normally involve substantial computation. However, when X is any non-singular upper triangular matrix the vectors c_i are exactly the same as those corresponding to $X = I$. For if, in this case, c_{r+1} is to be orthogonal to $x_1, \ldots, x_r$, its first r components must be exactly zero. When computing b_{r+1} from the relations

$$b_{r+1} = Ac_r - h_{1r}c_1 - h_{2r}c_2 - \ldots - h_{rr}c_r \qquad (34.3)$$

we need therefore make no reference to the actual components of the x_i. We first subtract a multiple h_{1r} of c_1 from Ac_r so as to annihilate its first component. We then have orthogonality with respect to x_1. We next subtract a multiple h_{2r} of c_2 so as to annihilate the second component; this leaves the first component still equal to zero. Continuing in this way we obtain a b_{r+1} which is zero in its first r components and is therefore orthogonal to $x_1, \ldots, x_r$.

The second point is a little more subtle. The vectors c_i obtained by the method of Arnoldi are the equivalent of the columns of the matrix obtained by multiplying together the orthogonal transformation matrices in the Givens or Householder methods. We saw in Chapter 3 that comparatively simple devices ensure that these factors are orthogonal to working accuracy. In Arnoldi's method the transformation matrix is not factorized and we determine it directly, column by column. It so happens that the numerical processes used to give orthogonality in this case are much less efficient.

The method of Lanczos

35. There is one further choice of the x_i which is of great practical importance. In this the x_i are determined simultaneously with the b_i and c_i and are derived from A^T in the same way as the c_i are derived from A. We discuss the exact mathematical process first. Since the c_i and x_i play such closely related roles, it is convenient to use a notation which reflects this. Following Lanczos (1950), we refer to the vectors derived from A^T as b_i^* and c_i^*; we wish to emphasize that the asterisk does not denote the conjugate transpose, here or later in this section. The relevant equations are

$$k_{r+1}c_{r+1} = b_{r+1} = Ac_r - \sum_{i=1}^{r} h_{ir}c_i, \tag{35.1}$$

$$k_{r+1}^*c_{r+1}^* = b_{r+1}^* = A^Tc_r^* - \sum_{i=1}^{r} h_{ir}^*c_i^*, \tag{35.2}$$

where the h_{ir} and h_{ir}^* are chosen so that b_{r+1} is orthogonal to $c_1^*, \ldots, c_r^*$ and b_{r+1}^* is orthogonal to $c_1, \ldots, c_r$ respectively.

We may apply the general analysis to both sequences, and hence we have

$$\left. \begin{array}{ll} AC = CH, & (C^*)^TC = L \\ A^TC^* = C^*H^*, & C^TC^* = L^* \end{array} \right\}, \tag{35.3}$$

where H and H^* are of upper Hessenberg form and L and L^* are lower triangular. From the second and fourth of equations (35.3) we find

$$L = (L^*)^T, \tag{35.4}$$

and hence both are diagonal and may be denoted by D. The first and third equations then give

$$H = C^{-1}AC = D^{-1}(H^*)^TD. \tag{35.5}$$

The left-hand side of (35.5) is upper Hessenberg and the right-hand side is lower Hessenberg. *Hence both are tri-diagonal* and

$$h_{ir}^* = h_{ir} = 0 \quad (i = 1, \ldots, r-2), \tag{35.6}$$

showing that if Ac_r is orthogonalized with respect to c_{r-1}^* and c_r^* it is automatically orthogonal with respect to all earlier c_i^* and similarly for $A^Tc_r^*$. We have further, from (35.5), the equations

$$h_{rr} = h_{rr}^*, \qquad h_{r,r+1}k_{r+1} = h_{r,r+1}^*k_{r+1}^*, \tag{35.7}$$

so that if the multiplying factors k_{r+1} and k^*_{r+1} are taken to be unity, $h_{r,r+1} = h^*_{r,r+1}$ and H and H^* are identical. Since the matrices H and H^* are tri-diagonal it is simpler to use the notation

$$\gamma_{r+1}c_{r+1} = b_{r+1} = Ac_r - \alpha_r c_r - \beta_r c_{r-1}, \tag{35.8}$$

$$\gamma^*_{r+1}c^*_{r+1} = b^*_{r+1} = A^T c^*_r - \alpha_r c^*_r - \beta^*_r c^*_{r-1} \tag{35.9}$$

where from (35.7)

$$\gamma_{r+1}\beta_{r+1} = \gamma^*_{r+1}\beta^*_{r+1}. \tag{35.10}$$

It is convenient to give explicit equations for the $\alpha_r, \beta_r, \beta^*_r$. We have

$$\alpha_r = (c^*_r)^T Ac_r/(c^*_r)^T c_r = c^T_r A^T c^*_r/c^T_r c^*_r \tag{35.11}$$

$$\begin{aligned}
\beta_r &= (c^*_{r-1})^T Ac_r/(c^*_{r-1})^T c_{r-1} \\
&= c^T_r A^T c^*_{r-1}/(c^*_{r-1})^T c_{r-1} \\
&= c^T_r(\gamma^*_r c^*_r + \alpha_{r-1}c^*_{r-1} + \beta^*_{r-1}c^*_{r-2})/(c^*_{r-1})^T c_{r-1} \\
&= \gamma^*_r c^T_r c^*_r/(c^*_{r-1})^T c_{r-1} \tag{35.12}
\end{aligned}$$

from the orthogonality and (35.9). Similarly

$$\beta^*_r = \gamma_r(c^*_r)^T c_r/(c_{r-1})^T c^*_{r-1}. \tag{35.13}$$

Failure of procedure

36. Before turning to the practical process we must deal with the case when either b_{r+1} or b^*_{r+1} (or both) is null. These events are quite independent. If b_{r+1} is null, we take c_{r+1} to be any vector orthogonal to $c^*_1, \ldots, c^*_r$ and (35.8) is then satisfied if we take $\gamma_{r+1} = 0$. Similarly if b^*_{r+1} is null, we take c^*_{r+1} to be any vector orthogonal to $c_1, \ldots, c_r$ and $\gamma^*_{r+1} = 0$. Notice from (35.10) that if b_{r+1} is null and b^*_{r+1} is not, then β^*_{r+1} is zero, while if b^*_{r+1} is null and b_{r+1} is not, then β_{r+1} is zero.

The use of the multiplying factors $\gamma_{r+1}, \gamma^*_{r+1}$ makes it quite obvious that the vanishing of a b_{r+1} or a b^*_{r+1} should not really be regarded as a 'failure' of the process. It merely gives us freedom of choice. If we take the multiplying factors equal to unity, then the vanishing of a b_{r+1} or b^*_{r+1} appears to be more exceptional than it really is.

On the other hand if $c^T_r c^*_r$ is zero at any stage the process breaks down completely and we must start again with a fresh vector c_1 or c^*_1 (or both) in the hope that this will not happen again. Unfortunately the breakdown does not seem to indicate any reliable method for re-selecting c_1 or c^*_1.

Let us assume now that c_1 and c_1^* are such that $c_i^T c_i^* \neq 0$ for any value of i relevant to our discussion. We show that, in this case, if c_1 is of grade p with respect to A, then the first b_i to vanish is b_{p+1}.

For, combining equations (35.8) for $r = 1, \ldots, s$, we may write

$$b_{s+1} = (A^s + q_{s-1}^{(s)} A^{s-1} + q_{s-2}^{(s)} A^{s-2} + \ldots + q_0^{(s)} I) c_1 = f_s(A) c_1 \quad (36.1)$$

where $f_s(A)$ is a monic polynomial (Chapter 1, § 18) of degree s in A. Hence b_{s+1} cannot vanish if $s < p$. Now since c_1 is of grade p, $A^p c_1$ may be expressed as a linear combination of $A^i c_1$ ($i = 1, \ldots, p-1$) so that

$$b_{p+1} = f_p(A) c_1 = g_{p-1}(A) c_1 \quad (36.2)$$

where $g_{p-1}(A)$ is a polynomial of degree $(p-1)$ in A. From equations (36.1) for $s = 1, \ldots, p-1$ we see that $g_{p-1}(A) c_1$ may be expressed as a linear combination of $b_p, \ldots, b_1$ and therefore of $c_p, \ldots, c_1$. We may write

$$b_{p+1} = \sum_{i=1}^{p} t_i c_i, \quad (36.3)$$

and since b_{p+1} is orthogonal to $c_p^*, \ldots, c_1^*$, we have

$$0 = t_i (c_i^*)^T c_i \quad \text{i.e.} \quad t_i = 0 \quad (i = 1, \ldots, p) \quad (36.4)$$

and the proof is complete.

We deduce that if A is derogatory, then we must have premature vanishing of the b_i and also, similarly, of the b_i^*.

Numerical example

37. We wish to emphasize that the vanishing of a $c_i^T c_i^*$ is not associated with any shortcoming in the matrix A. It can happen even when the eigenproblem of A is very well conditioned. We are forced

TABLE 4

$$A = \begin{bmatrix} 5 & 1 & -1 \\ -5 & 0 & 1 \\ 1 & 0 & 1 \end{bmatrix} \qquad c_1 = \begin{bmatrix} 0 \cdot 6 \\ -1 \cdot 4 \\ 0 \cdot 3 \end{bmatrix} \qquad c_1^* = \begin{bmatrix} 0 \cdot 6 \\ 0 \cdot 3 \\ -0 \cdot 1 \end{bmatrix}$$

$$c_2 = \tfrac{1}{3} \begin{bmatrix} 1 \cdot 5 \\ -2 \cdot 5 \\ 1 \cdot 5 \end{bmatrix} \qquad c_2^* = \tfrac{1}{3} \begin{bmatrix} 1 \cdot 8 \\ 0 \cdot 6 \\ -0 \cdot 8 \end{bmatrix} \qquad c_2^T c_2^* = 0$$

to regard it as a specific weakness of the Lanczos method itself. In Table 4 we give an example of such a breakdown. Since the use of a normalizing factor introduces rounding errors we have taken all γ_i and γ_i^* to be unity, so that $c_i = b_i$ and $c_i^* = b_i^*$ at all stages. The quantity $c_2^T c_2^*$ is exactly zero; obviously this would still be true whatever scaling factors were used, provided no rounding errors were made. For a matrix which does not have small integer coefficients and if we take initial vectors with random components, the exact vanishing of a $c_i^T c_i^*$ is extremely improbable.

The practical Lanczos process

38. If the Lanczos process is carried out in practice using equations (35.8) to (35.13), then the strict bi-orthogonality of the two sequences c_i and c_i^* is usually soon lost. Whenever considerable cancellation takes place on forming a vector b_i or b_i^*, a catastrophic deterioration in the orthogonality ensues immediately. To avoid this it is essential that re-orthogonalization is performed at each stage. A convenient practical procedure is the following:

At each stage we first compute b_{r+1} and b_{r+1}^* defined by

$$\left.\begin{aligned}
b_{r+1} &= Ac_r - \alpha_r c_r - \beta_r c_{r-1}, & \alpha_r &= (c_r^*)^T Ac_r/(c_r^*)^T c_r, \\
& & \beta_r &= (c_{r-1}^*)^T Ac_r/(c_{r-1}^*)^T c_{r-1} \\
b_{r+1}^* &= A^T c_r^* - \alpha_r c_r^* - \beta_r^* c_{r-1}^*, & \alpha_r &= c_r^T A^T c_r^*/(c_r^*)^T c_r, \\
& & \beta_r^* &= (c_{r-1})^T A^T c_r^*/(c_{r-1}^*)^T c_{r-1}
\end{aligned}\right\}. \quad (38.1)$$

It is instructive to determine the two values of α_r independently, as a check on the computation. We then compute c_{r+1} and c_{r+1}^* from the relations

$$\left.\begin{aligned}
\gamma_{r+1} c_{r+1} &= \bar{b}_{r+1} = b_{r+1} - \epsilon_{r1} c_1 - \epsilon_{r2} c_2 - \ldots - \epsilon_{rr} c_r \\
\gamma_{r+1}^* c_{r+1}^* &= \bar{b}_{r+1}^* = b_{r+1}^* - \epsilon_{r1}^* c_1^* - \epsilon_{r2}^* c_2^* - \ldots - \epsilon_{rr}^* c_r^*
\end{aligned}\right\}, \quad (38.2)$$

where

$$\epsilon_{ri} = (c^*)^T b_{r+1}/(c_i^*)^T c_i, \qquad \epsilon_{ri}^* = c_i^T b_{r+1}^*/c_i^T c_i^*. \quad (38.3)$$

In other words, b_{r+1} is re-orthogonalized with respect to $c_1^*,\ldots, c_r^*$ and b_{r+1}^* with respect to $c_1,\ldots, c_r$. Naturally this is performed progressively, each correction being subtracted immediately it is computed. It cannot be too strongly emphasized that re-orthogonalization does not produce the vectors which would have been obtained by more accurate computation and that this has no deleterious effect.

Notice that it is just as important to re-orthogonalize b_{r+1} with respect to c_{r-1}^* and c_r^* as with respect to $c_1^*,\ldots, c_{r-2}^*$. This is because the need to re-orthogonalize springs from cancellation and not from the accumulation of rounding errors. The factors γ_{r+1} and γ_{r+1}^* may conveniently be chosen to be powers of two such that the maximum elements of $|c_{r+1}|$ and $|c_{r+1}^*|$ lie in the range $(\frac{1}{2}, 1)$, since this involves no rounding errors. If floating-point computation is used, these normalizing factors would appear to be of no significance, but they become important when finding the eigenvectors of the resulting tri-diagonal matrix (cf. Chapter 5, § 67).

Now we know that if $c_i^T c_i^* = 0$ at any stage, then the theoretical process breaks down. We might therefore expect the numerical process to be unstable if $c_i^T c_i^*$ is small. (We are assuming here that c_i and c_i^* are normalized). If all $c_i^T c_i^*$ are of order unity then we may show that, for normalized A, (35.5) produces

$$AC = CT + O(2^{-t}) \qquad (38.4)$$

where T is the tri-diagonal matrix with

$$t_{ii} = \alpha_i, \qquad t_{i,i+1} = \beta_{i+1}, \qquad t_{i+1,i} = \gamma_{i+1}. \qquad (38.5)$$

It might be felt that we should replace T by $(T+F)$, where F is the matrix formed by the ϵ_{ij} and where $(T+F)$ is of upper Hessenberg form with small elements outside its tri-diagonal section. There is no point in doing this since the contribution of F is already covered by the term $O(2^{-t})$. The precise bound for $O(2^{-t})$ depends on the type of arithmetic which is used.

However, when

$$c_i^T c_i^* \doteqdot 2^{-p} \quad (p, \text{ a positive integer}) \qquad (38.6)$$

for some i, then we find that, in general,

$$AC = CT + O(2^{2p-t}), \qquad (38.7)$$

so that we usually lose $2p$ *more* figures of accuracy in such a case than when A is transformed using a stable procedure.

Numerical example

39. The nature of the instability is well illustrated by the example in Table 5, in which for simplicity we have taken c_1 and c_1^* to be almost orthogonal. (We can avoid a breakdown at the first step in practice by taking $c_1 = c_1^*$.) The computation was carried out to high precision, so that the values given are correct in all figures. In order to illustrate our point we have taken the γ_i and γ_i^* to be unity. The β_i and β_i^* are therefore equal.

Notice that if we denote $c_1^T c_1^*$ by ϵ, α_1 and α_2 are of order ϵ^{-1} and are almost equal and opposite, while β_2 is of order ϵ^{-2}. By a rather tedious analysis it can be shown that, in general, if $c_r^T c_r^* = \epsilon$, α_r and α_{r+1}

TABLE 5

$$
A = \begin{bmatrix} 4 & 1 & 3 & 2 \\ 2 & 1 & 2 & 5 \\ 1 & 3 & 3 & 4 \\ 4 & 1 & 2 & 1 \end{bmatrix} \qquad c_1 = \begin{bmatrix} 1 \\ 2^{-10} \\ 0 \\ 0 \end{bmatrix} \qquad c_1^* = \begin{bmatrix} 2^{-10} \\ 1 \\ 0 \\ 0 \end{bmatrix} \qquad c_1^T c_1^* = 2^{-9}
$$

$$
T = \begin{bmatrix} 1026 \cdot 5005 & -1037297 \cdot 7 & & \\ 1 & -1010 \cdot 3982 & -1 \cdot 1047112 & \\ & 1 & -8 \cdot 7263879 & 1 \cdot 0358765 \\ & & 1 & 1 \cdot 6241315 \end{bmatrix}
$$

$$
\tilde{T} = \begin{bmatrix} 1026 \cdot 5005 & -1037 \cdot 2977 & & \\ 1000 & -1010 \cdot 3982 & -1 \cdot 1047112 & \\ & 1 & -8 \cdot 7263879 & 1 \cdot 0358765 \\ & & 1 & 1 \cdot 6241315 \end{bmatrix}
$$

are of order $\|A\|/\epsilon$, β_{r+1} is of order $\|A\|^2/\epsilon^2$ and β_{r+2} is of order $\|A\|^2 \epsilon$. (The last result does not become obvious with the example of Table 5 until we take a much smaller value of ϵ than 2^{-9}.) The upheaval caused by the small value of $c_r^T c_r^*$ then dies out unless some later $c_i^T c_i^*$ is small, but this, if it occurs, is an independent event.

We can now see immediately why a value of $c_r^T c_r^*$ of order 2^{-p} has the effect indicated by equation (38.7). Suppose, for example, we are dealing with a matrix of order 2 only, and correctly rounded values are obtained for α_1, α_2 and β_2. The computed matrix may then be represented by

$$
\begin{bmatrix} \alpha_1(1+\epsilon_1) & \beta_2(1+\epsilon_2) \\ 1 & \alpha_2(1+\epsilon_3) \end{bmatrix}, \quad |\epsilon_i| \leqslant 2^{-t}, \tag{39.1}
$$

where the α_i and β_i are the values corresponding to exact computation. The characteristic polynomial of the matrix in (39.1) is

$$
\lambda^2 - \lambda[\alpha_1(1+\epsilon_1) + \alpha_2(1+\epsilon_3)] + \alpha_1\alpha_2(1+\epsilon_1)(1+\epsilon_3) - \beta_2(1+\epsilon_2). \tag{39.2}
$$

Hence the error in the constant term is

$$
\alpha_1\alpha_2(\epsilon_1 + \epsilon_3 + \epsilon_1\epsilon_3) - \beta_2\epsilon_2, \tag{39.3}
$$

and we have seen that α_1 and α_2 are of order 2^p and β_2 of order 2^{2p}. There is no reason for expecting the terms to cancel in (39.3) and in general they do not. Incidentally this example shows that the trouble does not arise from the amplification of errors in previous steps.

If we use the customary scaling factors in the example of Table 5 we obtain the matrix $\tilde{T}$ instead of T, but our argument is not affected by this. The 'damage' is still proportional to 2^{2p} and not 2^p.

The same matrix was solved using the initial vectors

$$c_1^T = (1, 2^{-p}, 0, 0), \qquad (c_1^*)^T = (2^{-p}, 1, 0, 0)$$

with different values of p, using a double-precision version (60 binary places) of the Lanczos process. As p was increased the eigenvalues of the computed T departed further from those of A, until for $p = 30$ they had no figures in common, thus confirming our analysis. The orders of magnitude of $\alpha_1, \alpha_2, \beta_2, \beta_3$ varied in the manner indicated earlier in this section.

General comments on the unsymmetric Lanczos process

40. Although the Lanczos process has provided us with a tri-diagonal matrix, a very satisfactory compression of the original data, we may well pay a heavy price in terms of numerical stability. Because of this potential instability the Lanczos process has been coded in double-precision arithmetic and with a tolerance set on the acceptable size of $|c_i^T c_i^*|$ for normalized c_i and c_i^*. If this inner-product has a modulus less than $2^{-\frac{1}{2}t}$, then the process is restarted with different initial vectors. Although restarting is quite rare, *moderately* small values of $c_i^T c_i^*$ are quite common, showing that the accuracy attainable with single-precision computation is often inadequate.

The total number of multiplications when re-orthogonalization is included is approximately $4n^3$, and since double-precision work is involved throughout, the volume of computation is substantial. Notice, however, that the original matrix need not be represented to double-precision and hence when forming the Ac_i and $A^T c_i^*$, (involving some $2n^3$ multiplications in all), we require only the multiplication of single-precision elements of A by double-precision elements of the c_i and c_i^*. We show in § 49 that there is an alternative process which is fully the equivalent of the Lanczos process as regards stability, and which involves much less work. In view of this we shall not discuss further the unsymmetric Lanczos process.

The symmetric Lanczos process

41. When A is real and symmetric, and we take $c_1 = c_1^*$, then the two sequences of vectors are identical. If, further, at each stage we choose γ_{r+1} so that $\|c_{r+1}\|_2 = 1$, we see that the symmetric Lanczos process and the symmetric Arnoldi process are the same. Re-orthogonalization is just as important for the symmetric Lanczos process as for the unsymmetric, but only half the computation is required. However, as we pointed out in connexion with Arnoldi's method in the general case, we cannot now have $c_i^T c_i^* = 0$ since $c_i^T c_i^* = \|c_i\|_2^2 = 1$, from the normalization.

If we include re-orthogonalization, the numerical stability of the symmetric Lanczos method is comparable with that of the symmetric Givens and Householder reductions to tri-diagonal form, but of course Householder's reduction is far more economical. No loss of accuracy ensues if the initial vector is completely or almost completely deficient in some eigenvectors, but the re-orthogonalization is particularly important in the latter case. If we take $c_1 = e_1$, then apart from signs, the Lanczos process gives the same tri-diagonal matrix as Givens' and Householder's methods with exact computation. It is difficult to think of any reason why we should use Lanczos' method in preference to Householder's.

We have assumed that the normalizing factors γ_r are chosen so that $\|c_r\|_2^2 = 1$ at each stage since this makes the relationship with other methods particularly close. However, it serves no other useful purpose and necessitates the use of square roots. In practice it is simpler to choose γ_r to be a power of two such that the maximum element of $|c_r|$ lies between $\frac{1}{2}$ and 1. For exact computation, all the β_r are non-negative, since we have, from (35.12)

$$\beta_r = \gamma_r c_r^T c_r / c_{r-1}^T c_{r-1} = \gamma_r \|c_r\|^2 / \|c_{r-1}\|^2. \tag{41.1}$$

However, the β_r are not computed in this way and it is possible for a computed β_r to be a small negative number, in which case T is not quasi-symmetric (Chapter 5, § 66). Such a β_r should be replaced by zero when computing the eigenvalues of T.

Reduction of a Hessenberg matrix to a more compact form

42. We have seen that there are a number of very stable methods of reducing a general matrix to Hessenberg form. The remainder of the chapter is concerned with the further reduction of a Hessenberg matrix to some more compact form. We shall discuss two such forms, the tri-diagonal and the Frobenius. We have already mentioned the production of a tri-diagonal matrix in connexion with Lanczos' unsymmetric process and we saw that it was attained only at the risk of numerical instability. We shall find that this is true, in general, of all the methods which we discuss in the remainder of this chapter.

Reduction of a lower Hessenberg matrix to tri-diagonal form

43. Let us consider first the application of the method of § 11 to a matrix A of lower Hessenberg form. We ignore the question of interchanges for the moment, and hence we require a unit lower triangular matrix N, having e_1 for its first column, such that

$$AN = NH, \qquad (43.1)$$

where H is to be of upper Hessenberg form. We show that in this case, H must be tri-diagonal. We have only to show that $h_{ir} = 0$ $(r > i+1)$ since we know that H is upper Hessenberg.

Let us equate successive elements of the rth columns of both sides of equation (43.1). Since A is of lower Hessenberg form and N is lower triangular, the (i, r) element of AN is zero if $i < r-1$. Hence we have successively

$$\left.\begin{aligned}
0 &= h_{1r} \\
0 &= h_{2r} \\
0 &= n_{32}h_{2r}+h_{3r} = h_{3r} \\
0 &= n_{42}h_{2r}+n_{43}h_{3r}+h_{4r} = h_{4r} \\
&\;\cdots\cdots\cdots\cdots\cdots\cdots\cdots\cdots\cdots\cdots\cdots \\
0 &= n_{r-2,2}h_{2r}+n_{r-2,3}h_{3r}+\ldots+h_{r-2,r} = h_{r-2,r}
\end{aligned}\right\} \qquad (43.2)$$

and the result is proved.

It is not essential for the proof that N have e_1 for its first column. If we specify the first column of N to be any vector with unit first element, then we merely introduce a term $n_{i1}h_{1r}$ in each of equations (43.2) except the first. The whole of N and H is determined and H is still of tri-diagonal form.

For the non-zero elements of column r of H we have

$$\left.\begin{aligned}
a_{r-1,r} &= h_{r-1,r} \\
a_{rr}+a_{r,r+1}n_{r+1,r} &= n_{r,r-1}h_{r-1,r}+h_{rr} \\
a_{r+1,r}+a_{r+1,r+1}n_{r+1,r}+a_{r+1,r+2}n_{r+2,r} &= n_{r+1,r-1}h_{r-1,r}+n_{r+1,r}h_{rr}+h_{r+1,r}
\end{aligned}\right\}, \qquad (43.3)$$

and for the non-zero elements of column $(r+1)$ of N we have

$$\begin{aligned}
a_{ir}+a_{i,r+1}n_{r+1,r}+a_{i,r+2}n_{r+2,r}&+\ldots+a_{i,i+1}n_{i+1,r} \\
&= n_{i,r-1}h_{r-1,r}+n_{ir}h_{rr}+n_{i,r+1}h_{r+1,r} \quad (i = r+2,\ldots, n). \quad (43.4)
\end{aligned}$$

If any suffix is greater than n, the corresponding term is omitted. Inner-products can be accumulated as in § 11, but the relations are simplified as a result of the special forms of A and H. There are approximately $\frac{1}{6}n^3$ multiplications in the whole process.

The use of interchanges

44. We described the process in the spirit of § 11 rather than of § 8 since the former is the more economical as regards rounding errors. For the consideration of interchanges it is a little more convenient to return to the description of § 8. When A_0 is initially of lower Hessenberg form, then provided no interchanges are used, A_{r-1} is of the form given for the case $n = 6$, $r = 3$, by

$$
\begin{bmatrix}
\times & \times & & & & \\
\times & \times & \times & & & \\
0 & \times & \times & \times & & \\
0 & 0 & \times & \times & \times & \\
0 & 0 & \underline{\times} & \times & \times & \times \\
0 & 0 & \underline{\times} & \times & \times & \times
\end{bmatrix}.
\tag{44.1}
$$

The rth step is as follows. For each value of i from $r+2$ to n perform operations (i) and (ii):

(i) Compute $n_{i,r+1} = a_{ir}/a_{r+1,r}$.

(ii) Subtract $n_{i,r+1} \times$ row $r+1$ from row i. (Notice that row $r+1$ has only two non-zero elements besides $a_{r+1,r}$.)

(iii) For each value of i from $r+2$ to n add $n_{i,r+1} \times$ column i to column $r+1$.

It is immediately apparent that step r does not destroy any of the zeros initially in A_0 or any of those introduced by previous steps.

However, if $a_{r+1,r} = 0$ and $a_{ir} \neq 0$ for some i, the process breaks down during the rth step. Now in general we have avoided breakdown of this kind by using interchanges, but if we interchange rows $r+1$ and $(r+1)'$ and columns $r+1$ and $(r+1)'$ in the usual way, we destroy the pattern of zeros above the diagonal. *Interchanges are inconsistent with the preservation of the original lower Hessenberg form.*

The only freedom of choice we have is that we can perform a similarity transformation with any matrix of the type N_1 before we start the reduction to tri-diagonal form. If a breakdown occurs we can go back to the beginning and apply such a transformation in the hope that it will not occur again. In § 51 we show how to improve slightly on this rather haphazard procedure.

Since interchanges are not permitted, if we wish to obtain a tri-diagonal matrix, there is an obvious danger of numerical instability. If at any stage $|a_{r+1,r}|$ is much smaller than some of the $|a_{ir}|$ ($i = r+2,..., n$), then the corresponding $|n_{i,r+1}|$ are much larger than unity.

It is easy to see that the use of plane rotations or elementary Hermitians also destroys the lower Hessenberg pattern, and there is therefore no simple analogue using orthogonal matrices of the method we have just described. In general we find that when 'stabilization' is not permitted in a technique involving the elementary matrices there is *no* corresponding technique involving orthogonal matrices.

Effect of a small pivotal element

45. Let us consider in more detail the effect of the emergence of a small pivotal element $a_{r+1,r}^{(r-1)}$ at the beginning of the rth step. For convenience we denote this element by ϵ and the relevant elements of columns r, $r+1$ and $r+2$ by x_i, y_i, z_i respectively; when $n = 7$, $r = 3$, the relevant array is

$$
\begin{bmatrix}
\times & \times & & & & & \\
\times & \times & \times & & & & \\
0 & \times & \times & \times & & & \\
0 & 0 & \epsilon & y_4 & z_4 & & \\
0 & 0 & x_5 & y_5 & z_5 & \times & \\
0 & 0 & x_6 & y_6 & z_6 & \times & \times \\
0 & 0 & x_7 & y_7 & z_7 & \times & \times
\end{bmatrix} . \tag{45.1}
$$

These are the only columns affected by the rth transformation.
For the asymptotic behaviour of the modified elements we have

$$
\left.
\begin{aligned}
a_{r+1,r+1}^{(r)} &= x_{r+2}z_{r+1}/\epsilon + O(1) \\
a_{i,r+1}^{(r)} &= -x_i x_{r+2} z_{r+1}/\epsilon^2 + O\left(\frac{1}{\epsilon}\right) \quad (i = r+2,..., n) \\
a_{i,r+2}^{(r)} &= -x_i z_{r+1}/\epsilon + O(1) \quad\quad\quad (i = r+2,..., n)
\end{aligned}
\right\} . \tag{45.2}
$$

Notice that the new columns $(r+1)$ and $(r+2)$ have elements which are roughly in a constant ratio. In general the multipliers in the following step are given by

$$n_{i,r+2} = a^{(r)}_{i,r+1}/a^{(r)}_{r+2,r+1} = x_i/x_{r+2} + O(\epsilon), \qquad (45.3)$$

and hence will be of order unity. When $A^{(r)}$ is pre-multiplied by N_{r+2} the ith element of column $(r+2)$ becomes

$$a^{(r)}_{i,r+2} - n_{i,r+2}a^{(r)}_{r+2,r+2} = -x_i z_{r+1}/\epsilon + O(1) - [x_i/x_{r+2} + O(\epsilon)] \times$$

$$\times [-x_{r+2}z_{r+1}/\epsilon + O(1)]$$

$$= O(1). \qquad (45.4)$$

This simple analysis underestimates the degree of cancellation which takes place when we perform the similarity transformation with matrix N_{r+2}. A more detailed analysis shows that

$$a^{(r+1)}_{i,r+2} = O(\epsilon). \qquad (45.5)$$

However, the matrix N_{r+3} is in no way exceptional (unless an independent instability arises) and the instability dies out. Its main effect on the final tri-diagonal matrix is therefore on rows $(r+1)$, $(r+2)$, $(r+3)$ where we have elements of the orders of magnitude given by

$$\left.\begin{array}{ccccc} \text{row } (r+1) & \epsilon & A/\epsilon & B & \\ \text{row } (r+2) & C/\epsilon^2 & -A/\epsilon & D & \\ \text{row } (r+3) & & E\epsilon & F & G \end{array}\right\}. \qquad (45.6)$$

Here $A, B,\ldots$ are of the order of magnitude of $\|A\|$ and

$$A^2 \doteq -BC. \qquad (45.7)$$

Notice that the distribution of exceptional elements in the final matrix is the transpose of that which we observed in the Lanczos method when $c^T_{r+1}c^*_{r+1} = \epsilon$. In § 48 we show why this close relationship should exist.

Error analysis

46. Wilkinson (1962a) has given a very detailed error analysis of this reduction to tri-diagonal form and we content ourselves here with a summary of the results. There is one slight difficulty in such an analysis. We have seen in § 14 that even for the reduction of a general matrix to upper Hessenberg form, *for which interchanges were permissible*, it was not possible to guarantee stability *unconditionally*.

The same point arises in the reduction from upper Hessenberg form to tri-diagonal form even in the most favourable case when all the multipliers happen to be of order unity. Clearly this is not the point at issue at the moment. We are concerned rather with those errors which are peculiar to the emergence of a small pivot coupled with large multipliers. In the analysis in Wilkinson (1962a) we therefore ignored any amplifying effect that *could* result from pre-multiplication and post-multiplication of *normal* rounding errors by a sequence of N_r and N_r^{-1} with elements of order unity. In these circumstances the results are as follows:

When there is an unstable stage which gives rise to multipliers of order 2^k, then the equivalent perturbations in some elements of the original matrix A are of order 2^{2k-t} and, in general, this leads to perturbations in the eigenvalues which are also of order 2^{2k-t}. If we wish to obtain results which are comparable with those which would be obtained by a reduction involving no unstable stages, we must work to $t+2k$ figures instead of t. The backward analysis shows that this extra precision is necessary only for a limited part of the computation, but it is not convenient to take advantage of this on an automatic computer. *It also shows quite clearly that the effect of errors made prior to the unstable stage is not amplified by the large multipliers.*

The general orders of magnitude of the elements of the final tri-diagonal matrix T for the case $n = 6$ with an instability in the second stage are given by

$$|T| = \begin{bmatrix} 1 & 1 & & & & \\ 1 & 1 & 1 & & & \\ & 2^{-k} & 2^k & 1 & & \\ & & 2^{2k} & 2^k & 1 & \\ & & & 2^{-k} & 1 & 1 \\ & & & & 1 & 1 \end{bmatrix}. \tag{46.1}$$

The orders of magnitude of the elements of x and y, the left-hand and right-hand eigenvectors of T corresponding to a well-conditioned eigenvalue of A, are given by

$$(1, 1, 2^k, 1, 1, 1) \tag{46.2}$$

and

$$(1, 1, 1, 2^k, 1, 1) \tag{46.3}$$

respectively, where the vectors are scaled so that $y^T x$ is of order unity. The corresponding condition number is of order 2^{2k} so that a severe worsening has taken place. However, this is a little deceptive, since T is badly balanced. It may be transformed by a diagonal similarity transformation into the matrix $\tilde{T}$ having elements of the orders of magnitude given by

$$|\tilde{T}| = \begin{bmatrix} 1 & 1 & & & & \\ 1 & 1 & 1 & & & \\ & 2^{-k} & 2^k & 2^k & & \\ & & 2^k & 2^k & 2^{-k} & \\ & & & 1 & 1 & 1 \\ & & & & 1 & 1 \end{bmatrix}, \qquad (46.4)$$

and rounding errors may be avoided by using powers of 2 for the elements of the diagonal matrix. The corresponding y and x, still such that $y^T x$ is of order unity, then have components of the orders of magnitude shown respectively by

$$(1, 1, 2^k, 2^k, 1, 1) \qquad (46.5)$$
and $$(1, 1, 1, 1, 1, 1), \qquad (46.6)$$

and the condition number is now of order 2^k. Nevertheless, to compute the eigenvalues of T or $\tilde{T}$ one must work to $2k$ more figures in general than for a tri-diagonal matrix which was obtained without unstable stages.

These results suggest that if we are prepared to work to double-precision then either of the following procedures may be adopted.

(i) We can agree to accept multipliers of orders up to $2^{-\frac{1}{2}t}$ and then, provided multiple instabilities do not occur (see Wilkinson, 1962a), the errors are equivalent to perturbations in A which are no greater than $2^{-2t} \times (2^{\frac{1}{2}t})^2$, that is, 2^{-t}. If a multiplier exceeds this tolerance we return to the matrix A, perform an initial similarity transformation with some matrix N_1 and then repeat the reduction.

(ii) If we are not prepared to return to the beginning, then we must replace all pivots of magnitude less than $2^{-2t/3}$ by $2^{-2t/3}$. The errors involved in this replacement are themselves of order $2^{-2t/3}$, while those resulting from the subsequent reduction are of order $2^{-2t} \times (2^{2t/3})^2$, that is, $2^{-2t/3}$. This is therefore the optimum choice. Notice that if we are prepared to work to triple-precision the corresponding tolerance is 2^{-t}, and the equivalent perturbations in A are of magnitude not greater than 2^{-t}.

The Hessenberg process applied to a lower Hessenberg matrix

47. In § 30 we established that if we take $X = I$ in § 26, the Hessenberg process is identical, even as regards rounding errors, with the direct reduction without interchanges. Since this is true in the general case, it is true in particular when the initial matrix is of lower Hessenberg form. In spite of this relationship it is instructive to describe the reduction again in terms of the Hessenberg process.

At the beginning of the rth step $c_1, c_2, ..., c_r$ have been computed, c_i having its first $(i-1)$ elements equal to zero and its ith element equal to unity. In the rth step we compute c_{r+1} from the relation

$$k_{r+1}c_{r+1} = Ac_r - h_{1r}c_1 - h_{2r}c_2 - ... - h_{rr}c_r, \tag{47.1}$$

where the h_{ir} are determined so that c_{r+1} is orthogonal to $e_1, ..., e_r$. Now when A is of lower Hessenberg form Ac_r has its first $(r-2)$ elements equal to zero, and is therefore already orthogonal to $e_1, ..., e_{r-2}$; hence $h_{ir} = 0$ $(i = 1, ..., r-2)$. Apart from the factor k_{r+1}, c_{r+1} is determined from Ac_r by subtracting multiples $h_{r-1,r}$ and h_{rr} of c_{r-1} and c_r so as to annihilate the $(r-1)$th and rth elements respectively. The factor k_{r+1} is chosen so that the $(r+1)$th element of c_{r+1} is unity. Clearly the matrix H is tri-diagonal. When the vector on the right-hand side of (47.1) is not null, but its $(r+1)$th element is zero, the process breaks down. If the whole vector is null c_{r+1} can be taken to be any vector orthogonal to $e_1, ..., e_r$ with its $(r+1)$th element equal to unity.

If $c_1 = e_1$, the vectors c_i are identical with the columns of the matrix N in the processes of §§ 43, 44. Taking c_1 to be any other vector with a unit first component is equivalent, in terms of § 44, to including the initial transformation $N_1 A N_1^{-1}$, where N_1^{-1} has c_1 as its first column.

Relationship between the Hessenberg process and the Lanczos process

48. We now show further that when A is of lower Hessenberg form there is a close relationship between the Hessenberg process and the Lanczos unsymmetric process.

We first prove that if we take $c_1^* = e_1$ in the Lanczos process then the vectors c_i^* form the columns of an upper triangular matrix, independent of the choice of c_1. For suppose this is true up to c_r^*. We have

$$\gamma_{r+1}^* c_{r+1}^* = A^T c_r^* - \alpha_r c_r^* - \beta_r^* c_{r-1}^*. \tag{48.1}$$

But since A^T is of upper Hessenberg form, $A^T c_r^*$ has zero components in positions $r+2, \ldots, n$ from our inductive hypothesis for c_r^*. Hence the vector c_{r+1}^* also has zero components in these positions whatever values α_r and β_r^* may take. This is the required result for c_{r+1}^*.

We may use this result to show that the vectors c_i are the columns of a lower triangular matrix. This follows from the remark made in § 34 that orthogonalization with respect to the columns of any upper triangular matrix produces the same result as orthogonalization with respect to the columns of I. Further, as we remarked there, this orthogonalization does not require the use of the elements of the upper triangular matrix.

Hence the vectors c_i and the α_i and β_i of the Lanczos process can be obtained without actually computing the c_i^*; since they are the same as the quantities obtained by taking $X = I$ instead of $X = C^*$, we have established the identity of the Lanczos vectors with the Hessenberg vectors. Notice that in this particular use of the Lanczos process no re-orthogonalization is required because exact orthogonality is automatically achieved when X is upper triangular. If, further, $c_1 = e_1$ both the Hessenberg process and the Lanczos process give the same results as the method of § 43.

We now see that it is not at all surprising that the emergence of a small pivot in the process of § 43 led to the same behaviour in the tri-diagonal matrix as we had observed in § 39 for the Lanczos process when $c_i^T c_i^*$ was small. The former is a particular case of the general phenomenon associated with the Lanczos process.

Reduction of a general matrix to tri-diagonal form

49. The methods of § 12 and § 43 may be combined to give a reduction of a general matrix A to tri-diagonal form. We first use the method of § 12 to reduce A to the upper Hessenberg matrix H, using interchanges. This is stable except in very special cases and we might expect to use single-precision arithmetic. Since H^T is of lower Hessenberg form, we can now use the method of § 43 (or the equivalent method of § 47) to obtain a tri-diagonal matrix T, such that

$$H^T N = NT \quad \text{or} \quad N^T H = T^T N^T. \tag{49.1}$$

Hence the matrix T^T, which is also tri-diagonal, is similar to A. This second transformation is appreciably less stable and we should use double-precision arithmetic and the precautions described in § 46. In all, the reduction to tri-diagonal form involves $\frac{5}{6}n^3$ single-precision multiplications and $\frac{1}{6}n^3$ double-precision multiplications.

We have described the second part of the process in terms of H^T for obvious reasons, but since $N^T = M$, a unit upper triangular matrix with its first row equal to e_1^T, we can solve the matrix equation

$$MH = T^T M \qquad (49.2)$$

directly for M and T, by a method analogous to that of § 43 in which the roles of rows and columns are interchanged.

We have been at pains to stress that the reduction to tri-diagonal form is potentially unstable, but we do not wish to be too discouraging. On ACE, for example, it would be extremely rare for a multiplier to approach the tolerance of 2^{-24} ($t = 48$). When the reduction of H is performed in double-precision it is almost always true that the errors arising from this part of the reduction are far smaller than those arising from the reduction from A to H in single-precision. The combined technique is therefore very effective in practice.

Comparison with Lanczos method

50. Now the Lanczos method of § 35 enables us to pass directly from a general matrix A to a tri-diagonal matrix. However, in that method we must calculate both of the sequences c_i and c_i^* and re-orthogonalization is essential. The danger of instability is present at every stage in the transformation, so that as a safeguard we are forced to use double-precision arithmetic the whole time. The total number of multiplications is $4n^3$. If we have too small a value of $c_i^T c_i^*$ at any stage we must return right to the beginning and there is always the possibility, though remote, of a second failure. With the combined method of § 49 we need return only to H. The case for using the combined technique of § 49 in preference to the Lanczos method would appear to be quite overwhelming. In so far as the former has any instability it is precisely the instability associated with any use of the unsymmetric Lanczos process.

Re-examination of reduction to tri-diagonal form

51. In § 44 we remarked that if a breakdown occurs in the rth step of the reduction of a lower Hessenberg matrix to tri-diagonal form we must return to the beginning and compute $N_1 A N_1^{-1}$ for some N_1 in the hope that failure will be avoided with this matrix. If we think in terms of the Hessenberg process, then we can obtain some guidance on the choice of N_1, that is, on the choice of c_1.

We observe that the first $(2r+1)$ elements of c_1 uniquely determine elements $1,\ldots, 2r+2-i$ of c_i $(i = 1,\ldots, r+1)$. This is illustrated for the case $r = 3$ by

$$
\begin{array}{cccc}
c_1 & c_2 & c_3 & c_4 \\[4pt]
\times & 0 & 0 & 0 \\[4pt]
\times & \times & 0 & 0 \\[4pt]
\times & \times & \times & 0 \\[4pt]
\times & \times & \times & \times \\[4pt]
\times & \times & \times \\[4pt]
\times & \times \\[4pt]
\times
\end{array}
\tag{51.1}
$$

Hence we may proceed as follows. We take the first element of c_1 to be unity and then assign its later elements two at a time. Each time we add two extra elements we compute the elements of c_i as far as they are determined, as indicated in (51.1). If considerable cancellation occurs when the $(r+1)$th element of c_{r+1} is computed we change the last two elements of c_1 and hence the last two computed elements of each c_i $(i = 1,\ldots, r+1)$. In this way we avoid a complete re-calculation, but we cannot deal with a breakdown which occurs after we have reached c_{in}. There is a further weakness of the method. It may well happen that all elements of c_{r+1} are small as a result of cancellation. If this is true there is really no need to restart, but unfortunately at the stage when the $(r+1)$th element of c_{r+1} is computed we do not know any of its later elements.

In view of such shortcomings, we must regard these observations as providing insight into the factors influencing the choice of c_1, rather than as forming the basis of a practical method.

Reduction from upper Hessenberg form to Frobenius form

52. Finally we consider the reduction of an upper Hessenberg matrix H to Frobenius form (Chapter 1, §§ 11–14). We shall denote the sub-diagonal elements $h_{r+1,r}$ by k_{r+1} and we may assume that $k_i \neq 0$ $(i = 2,\ldots, n)$. For if $k_{r+1} = 0$ we may write

$$
H = \begin{bmatrix} H_r & B_r \\ O & H_{n-r} \end{bmatrix},
\tag{52.1}
$$

where H_r and H_{n-r} are both of upper Hessenberg form, and hence

$$\det(H - \lambda I) = \det(H_r - \lambda I)\det(H_{n-r} - \lambda I). \qquad (52.2)$$

Notice that if $k_i \neq 0$ $(i = 2,\ldots, n)$, H cannot be derogatory since $(H - \lambda I)$ is then of rank $(n-1)$ for any value of λ. On the other hand an arbitrary number of the k_i may be zero even if H is non-derogatory. In the latter case we shall compute only the Frobenius forms of each of the relevant smaller Hessenberg matrices; this is in any case an advantage since it gives us a factorization of the characteristic polynomial.

We assume first that, apart from being non-zero, the k_i are not otherwise special. The reduction consists of $(n-1)$ major steps; at the beginning of the rth step the current matrix for the case $n = 6$, $r = 4$, is given by

$$
r-1\left\{
\begin{bmatrix}
0 & 0 & 0 & \times & \times & \times \\
k_2 & 0 & 0 & \times & \times & \times \\
 & k_3 & 0 & \times & \times & \times \\
\hline
 & & k_4 & \times & \times & \times \\
 & & & k_5 & \times & \times \\
 & & & & k_6 & \times
\end{bmatrix}
\right.
. \qquad (52.3)
$$

The sub-diagonal elements are unaltered throughout. The rth step is as follows.

For each value of i from 1 to r perform steps (i) and (ii):

(i) Compute $n_{i,r+1} = h_{ir}/k_{r+1}$.

(ii) Subtract $n_{i,r+1} \times$ row $(r+1)$ from row i.

(iii) For each value of i from 1 to r add $k_{i+1}n_{i,r+1}$ to $h_{i,r+1}$.

This completes the similarity transformation.

The final matrix X and its diagonal similarity transformation $F = D^{-1}XD$ are given by

$$
X = \begin{bmatrix}
0 & 0 & 0 & \ldots & x_0 \\
k_2 & 0 & 0 & \ldots & x_1 \\
 & k_3 & 0 & \ldots & x_2 \\
 & & \cdots & & \cdots \\
 & & & k_n & x_{n-1}
\end{bmatrix},
\quad
F = \begin{bmatrix}
0 & 0 & 0 & \ldots & p_0 \\
1 & 0 & 0 & \ldots & p_1 \\
 & 1 & 0 & \ldots & p_2 \\
 & & \cdots & & \cdots \\
 & & & 1 & p_{n-1}
\end{bmatrix},
\qquad (52.4)
$$

where D has the elements

$$1, k_2, k_2 k_3, \ldots, k_2 k_3 \ldots k_n \tag{52.5}$$

and the p_i are given by

$$p_i = k_2 k_3 \ldots k_{n-i} x_i. \tag{52.6}$$

There are approximately $\frac{1}{6} n^3$ multiplications in this reduction.

Effect of small pivot

53. If in the typical step $|k_{r+1}|$ is much smaller than the $|h_{ir}|$ the multipliers $n_{i,r+1}$ will be large and we can expect instability. Unfortunately we cannot use interchanges because this destroys the pattern of zeros. Hence there are no related transformations based on orthogonal matrices. In fact we cannot, in general, obtain the Frobenius canonical form by an orthogonal transformation.

At first sight the effect of a small pivotal value appears to be somewhat less serious than it was in the case of the reduction to tri-diagonal form. For suppose at the rth step k_{r+1} is ϵ and all other elements are of order unity. We see immediately that at the end of this step orders of magnitude of the elements are given typically for the case $n = 5$, $r = 3$, by

$$\begin{bmatrix} 0 & 0 & 0 & \epsilon^{-1} & \epsilon^{-1} \\ 1 & 0 & 0 & \epsilon^{-1} & \epsilon^{-1} \\ & 1 & 0 & \epsilon^{-1} & \epsilon^{-1} \\ & & \epsilon & 1 & 1 \\ & & & 1 & 1 \end{bmatrix}. \tag{53.1}$$

There are no elements of order ϵ^{-2}. In fact even the elements of order ϵ^{-1} would not appear to be harmful. If we carry out the next step for the matrix of (53.1) we see that the final matrix X and the corresponding F have elements of the orders of magnitude given by

$$X = \begin{bmatrix} 0 & 0 & 0 & 0 & \epsilon^{-1} \\ 1 & 0 & 0 & 0 & \epsilon^{-1} \\ & 1 & 0 & 0 & \epsilon^{-1} \\ & & \epsilon & 0 & 1 \\ & & & 1 & 1 \end{bmatrix}, \quad F = \begin{bmatrix} 0 & 0 & 0 & 0 & 1 \\ 1 & 0 & 0 & 0 & 1 \\ & 1 & 0 & 0 & 1 \\ & & 1 & 0 & 1 \\ & & & 1 & 1 \end{bmatrix}. \tag{53.2}$$

All elements of order ϵ^{-1} have vanished in F!

Numerical example

54. The discussion of the last section would seem to indicate that the use of a small pivot is not harmful, in that the computed Frobenius form might not differ much from the matrix which would have been obtained by exact computation. Consider the simple example exhibited in Table 6.

TABLE 6

$$H = \begin{bmatrix} 10^0(0.99995) & 10^{-5}(0.41325) \\ 10^{-5}(0.23125) & 10^1(0.10001) \end{bmatrix}$$

$$X = \begin{bmatrix} 0 & -10^6(0.43245) \\ 10^{-5}(0.23125) & 10^1(0.20000) \end{bmatrix} \qquad F = \begin{bmatrix} 0 & 10^1(-0.10000) \\ 1 & 10^1(\ 0.20000) \end{bmatrix}$$

$$\text{Correct } F = \begin{bmatrix} 0 & 10^1(-0.1000049\ldots) \\ 1 & 10^1(\ 0.200005) \end{bmatrix}$$

We see that the computed F is indeed almost correct to working accuracy, in spite of the use of multipliers of order 10^5 in the computation of X. Nevertheless, the eigenvalues of the computed F do not agree closely with those of the correct F which are, of course, those of H. The cause of the trouble is that the matrix F is very much more ill-conditioned than H, and hence the small 'errors' in the elements of F lead to quite serious errors in the eigenvalues. If we estimate the equivalent perturbations in the original matrix H the harmful effect of the transformation is immediately revealed. In fact if N is the computed matrix of the transformation it may be verified that

$$N^{-1}XN = \begin{bmatrix} 10^0(0.999948125) & 10^2(-0.1756873125) \\ 10^{-5}(0.23125) & 10^1(\ 0.1000051875) \end{bmatrix}. \qquad (54.1)$$

The (1, 2) element has a very large perturbation. In this example it is much better to regard $k_2 = h_{21}$ as zero to working accuracy, so that the eigenvalues of H are obtained from the two smaller Hessenberg matrices of order (1×1).

General comments on the stability

55. In passing to the Frobenius canonical form we find that it is quite common for a serious deterioration in the condition number to take place. In fact all non-derogatory matrices with the same set of eigenvalues, whatever their condition numbers, have the same Frobenius canonical form. Consider for example the symmetric tridiagonal matrix W_{21}^- of Chapter 5, § 45. Since it is symmetric and has

well separated eigenvalues, its eigenvalue and eigenvector problem are well-conditioned. Nevertheless, its corresponding Frobenius form is very ill-conditioned, as we shall see in Chapter 7.

We have described a two-stage reduction of a general matrix A to Frobenius form.

Stage 1. A is reduced to upper Hessenberg form H by stabilized elementary transformations.

Stage 2. H is reduced to Frobenius form F by non-stabilized elementary transformations.

The stability considerations involved in these two stages are very different. In practice we would generally need to work to higher precision in stage 2 to ensure that the errors resulting from this stage were as small as those resulting from stage 1.

In the method of Danilewski (1937) these two stages are combined. This method consists of $(n-1)$ steps, in the rth of which zeros are introduced in positions $1,..., r, r+2,..., n$ of column r by a similarity transformation with a matrix of type S_{r+1} (Chapter 1, § 40(iv)). It is possible by means of interchanges to ensure that the $|s_{i,r+1}|$ $(i = r+2,..., n)$ are bounded by unity but not the remaining $|s_{i,r+1}|$. The usual description of Danilewski's method pays no attention to the difference in the numerical stability of the two halves of the process.

Specialized upper Hessenberg form

56. The reduction to Frobenius form described in § 51 is considerably simplified if the original Hessenberg form has $k_i = 1$ $(i = 2,..., n)$. No divisions are then needed in the reduction and the matrix X is already in the required form F.

If at the time the Hessenberg form H is derived from the general matrix A it is intended that H should subsequently be reduced to Frobenius form, then it is much more convenient to arrange that H should have unit sub-diagonal elements, except where these elements are necessarily zero. This can be arranged in the Hessenberg process of § 29 by taking k_{r+1} to be unity except when b_{r+1} is null, when we *must* take k_{r+1} to be zero.

Similarly if we think in terms of § 11 we solve the matrix equation

$$AN = NH \qquad (56.1)$$

for H and N, but now under the assumption that N is lower tri-angular (with e_1 as its first column) but not otherwise unit triangular, and that H has unit sub-diagonal elements. Just as in § 11 the ele-ments of N and H may be found column by column. Each element of N is now given as an inner-product and each of the elements h_{ij} of H, other than the $h_{i+1,i}$, is given as an inner-product divided by n_{ii}, the reverse of the situation in § 11. We have the same economy of rounding errors as before and interchanges may be incorporated as in § 12. Moreover, the presence of unity sub-diagonal elements considerably reduces the number of rounding errors made in the further reduction to Frobenius form.

The Hessenberg matrix of the present formulation can be obtained from that of §§ 11, 12 by a diagonal similarity transformation. Whereas in the latter case the elements of H are linear in those of A, since the elements of N are homogeneous and of order zero in the a_{ij}, in the present formulation h_{ij} is of degree $(j-i+1)$. This is obviously a disadvantage on a fixed-point computer. Assuming then that we are using floating-point, we shall find that if A is equilibrated and $\|A\|$ is very different from unity, then H will usually be very far from equilibrated. Whereas with the formulation of §§ 11, 12 we regard a sub-diagonal element of H as zero when it is smaller than $2^{-t}\|A\|$ say, and so split H into two smaller Hessenberg matrices, with the present formulation we do not have such a specific criterion to help us.

Direct determination of the characteristic polynominal

57. We have described the determination of the Frobenius form in terms of similarity transformations for the sake of consistency and in order to demonstrate its relation to Danilewski's method. How-ever, since we will usually use higher precision arithmetic in the reduction to Frobenius form than in the reduction to Hessenberg form, the reduced matrices arising in the derivation of the former cannot be overwritten in the registers occupied by the Hessenberg matrix.

It is more straightforward to think in terms of a direct derivation of the characteristic polynomial of H. This polynomial may be obtained by recurrence relations in which we determine successively the characteristic polynomial of each of the leading principal sub-matrices H_r ($r = 1,\ldots, n$) of H. If we denote by $p_r(\lambda)$ the characteristic polynomial of the submatrix of order r, then expanding $\det(H_r-\lambda I)$

in terms of its rth column we have

$$p_0(\lambda) = 1, \qquad p_1(\lambda) = h_{11} - \lambda$$

$$\left.\begin{aligned} p_r(\lambda) = (h_{rr} - \lambda)p_{r-1}(\lambda) - h_{r-1,r}k_r p_{r-2}(\lambda) + h_{r-2,r}k_r k_{r-1} p_{r-3}(\lambda) - \dots \\ \dots + (-1)^{r-1} h_{1r} k_r k_{r-1} \dots k_2 p_0(\lambda) \end{aligned}\right\}. \quad (57.1)$$

To derive the coefficients of $p_r(\lambda)$ we must store those of $p_i(\lambda)$ $(i = 1, \dots, r-1)$. There are essentially $\frac{1}{2}r^2$ multiplications in the computation of $p_r(\lambda)$ and hence $\frac{1}{6}n^3$ in the complete computation of the characteristic polynomial of H. This is the same as when similarity transformations are used. If we decide to compute $p_n(\lambda)$ to double-precision, each of the $p_r(\lambda)$ must be obtained to double-precision. However, if we assume that the elements of H are given only to single-precision, the multiplications involved are of double-precision numbers by single-precision numbers.

No special difficulties arise if some of the k_r are small or even zero, though in the latter case it would be better to deal with the appropriate Hessenberg matrices of lower order. Again the process is simplified somewhat if the k_r are unity.

In general, the reduction to Frobenius form (or the computation of the characteristic equation) is much less satisfactory than the reduction to tri-diagonal form. For many quite harmless looking distributions of eigenvalues the Frobenius form is extremely ill-conditioned, and it is common for double-precision arithmetic to be quite inadequate. We shall cover this point more fully in Chapter 7.

Additional notes

Although pivoting has been widely discussed in connexion with Gaussian elimination its role in similarity transformations has by comparison been neglected; in fact it is even more important here. If A has linear divisors we have $AX = X\Lambda$ for some X and hence

$$(D_1 A D_1^{-1}) D_1 X D_2 = D_1 X D_2 \Lambda$$

for any diagonal matrices D_1 and D_2. (The matrix D_2 gives only a scaling of the eigenvectors.) We wish to choose D_1 so that $\kappa(D_1 X D_2)$ is as small as possible; this problem corresponds to that of equilibration in the case of matrix inversion. Bauer (1963) has given a penetrating discussion of this problem. The difficulty in practice is that one must choose D_1 before X is determined; the problem is therefore in some respects more severe than that of the equilibration of A for matrix inversion. Osborne (1960) has discussed the problem from a more empirical standpoint. Again, as in the matrix inversion problem, it should be stressed that the need for equilibration is not specially related to the use of pivoting; its role can be just as important when orthogonal transformations are used.

Although compact methods of triangular decomposition have long been used, the direct reduction to Hessenberg form discussed in §§ 11–12 seems to be new. It was first used on ACE in 1959 (including the incorporation of interchanges); the computation was performed in double-precision arithmetic with quadruple accumulation of inner-products.

The discussion of the formal aspects of the generalized Hessenberg processes of §§ 26–35 is based on the paper by Householder and Bauer (1959). The method of Lanczos (1950) has been discussed in numerous papers including those by Brooker and Sumner (1956), Causey and Gregory (1961), Gregory (1958), Rosser *et al.* (1951), Rutishauser (1953) and Wilkinson (1958b). A reliable automatic procedure for determining the eigenvalues and eigenvectors has slowly emerged from these successive approaches and the history of the algorithm is particularly instructive for numerical analysts. It is ironic that by the time the practical process was fully understood it should have been superseded by the method of § 43.

A general discussion of methods for reducing a general matrix to tri-diagonal form has been given by Bauer (1959) and the methods described in that paper may well lead to algorithms which are superior to the one we have described.

7

Eigenvalues of Matrices of Condensed Forms

Introduction

1. THIS chapter is concerned with the solution of the eigenvalue problem for matrices of the special forms discussed in the previous chapter. If for the moment we let A denote a matrix of any of these forms, we require the zeros of the function $f(z)$ defined by

$$f(z) = \det(A - zI). \tag{1.1}$$

Since we are computing the zeros of a polynomial, all relevant methods must be essentially iterative in character.

The methods discussed in this chapter depend effectively on the evaluation of $f(z)$, and possibly of some of its derivatives, for sequences of values tending to its zeros. The distinguishing features of the various methods are

  (i) The algorithm used for evaluating $f(z)$ and its derivatives;

  (ii) the process used for deriving the successive members of the sequence;

  (iii) the techniques used to avoid converging to the same zero more than once.

Of these, the first is directly dependent on the form of the matrix A, while the second and third are almost independent of this form.

In all the methods the ultimate location of the zeros is dependent on the computed values obtained in the neighbourhoods of these zeros. Clearly the accuracy to which we can evaluate $f(z)$ in these neighbourhoods is of fundamental importance. We therefore consider the evaluations first.

Explicit polynomial form

2. The Frobenius form is the simplest of those considered in Chapter 6. This gives us either the characteristic polynomial in explicit form or its factorization into two or more explicit polynomials. The coefficient of the highest power of z is unity so we consider the evaluation of $f(z)$ defined by

$$f(z) = z^n + a_{n-1}z^{n-1} + \ldots + a_0. \tag{2.1}$$

The simplest method of evaluation is that usually known as 'nested multiplication'. Corresponding to the value $z = \alpha$ we compute $s_r(\alpha)$ $(r = n, \ldots, 0)$ defined by

$$s_n(\alpha) = 1, \qquad s_r(\alpha) = \alpha s_{r+1}(\alpha) + a_r \quad (r = n-1, \ldots, 0). \qquad (2.2)$$

Clearly we have $f(\alpha) = s_0(\alpha)$. Since $s_0(\alpha)$ satisfies the equation

$$s_0(\alpha) = \prod_{i=1}^{n} (\alpha - \lambda_i) \qquad (2.3)$$

the required values may cover a very wide range and floating-point computation is almost mandatory.

The computed values satisfy the relations

$$\left. \begin{array}{l} s_n(\alpha) = 1 \\[1mm] s_r(\alpha) = fl[\alpha s_{r+1}(\alpha) + a_r] \\[1mm] \qquad \equiv \alpha s_{r+1}(\alpha)(1+\epsilon_1) + a_r(1+\epsilon_2) \quad (r = n-1, \ldots, 0) \end{array} \right\} \qquad (2.4)$$

where
$$|\epsilon_1| < 2 . 2^{-t_1}, \qquad |\epsilon_2| < 2^{-t_1}. \qquad (2.5)$$

Combining these results we have

$$s_0(\alpha) = (1+E_n)\alpha^n + a_{n-1}(1+E_{n-1})\alpha^{n-1} + \ldots + a_0(1+E_0), \qquad (2.6)$$

where certainly
$$|E_r| < (2r+1)2^{-t_1}. \qquad (2.7)$$

The computed value of $s_0(\alpha)$ is therefore the exact value for $z = \alpha$ of the polynomial with coefficients $a_r(1+E_r)$. Hence, however much cancellation may take place in the computation of $s_0(\alpha)$, the errors may be accounted for by small relative perturbations in the coefficients.

Notice, however, that the polynomial with coefficients $a_r(1+E_r)$ does not give the remaining computed $s_r(\alpha)$ $(r = 1, \ldots, n-1)$. In fact from (2.4) we have

$$\begin{aligned} s_r(\alpha) &\equiv \alpha s_{r+1}(\alpha)(1+\epsilon_1) + a_r(1+\epsilon_2) \\ &\equiv \alpha s_{r+1}(\alpha) + [a_r(1+\epsilon_2) + \alpha s_{r+1}(\alpha)\epsilon_1], \end{aligned} \qquad (2.8)$$

showing that the polynomial which gives *all* the computed $s_r(\alpha)$ has coefficients a_r' given by

$$a_r' = a_r(1+\epsilon_2) + \alpha s_{r+1}(\alpha)\epsilon_1. \qquad (2.9)$$

If the change from a_r to a_r' is to correspond to a small relative perturbation, $\alpha s_{r+1}(\alpha)$ must be of the same order of magnitude as a_r, or less, for all r; for many polynomials this is far from true. However, if all the zeros λ_i $(i = 1,..., n)$ are real and of the same sign, these conditions are always satisfied for an α in the neighbourhood of any of the zeros. To prove this we observe that

$$f(z) \equiv (z-\alpha) \sum_1^n s_r(\alpha)z^{r-1} + s_0(\alpha). \qquad (2.10)$$

From (2.1) we have

$$a_r = \pm \sum \lambda_{i_1}\lambda_{i_2}...\lambda_{i_{n-r}}, \qquad (2.11)$$

while taking $\alpha = \lambda_k$ in (2.10) and observing that $s_0(\lambda_k) = 0$ we have

$$s_r(\lambda_k) = \pm \sum \lambda_{i_1}\lambda_{i_2}...\lambda_{i_{n-r}} \quad (\lambda_{i_j} \neq \lambda_k). \qquad (2.12)$$

This shows that $s_r(\lambda_k)$ is of the same sign as a_r and not greater in absolute magnitude, and from (2.2) this implies that $|\lambda_k s_{r+1}(\lambda_k)| < |a_r|$. Hence from (2.9)

$$|a'_r - a_r| < |a_r \epsilon_2| + |\lambda_k s_{r+1}(\lambda_k)\epsilon_1| < |a_r|\, 2^{-t_1} + 2\, |a_r|\, 2^{-t_1} = 3\, |a_r|\, 2^{-t_1}. \qquad (2.13)$$

In general we cannot gain very much from the accumulation of inner-products when evaluating a polynomial. If we are to avoid multiplicands having progressively increasing numbers of digits, it is necessary to round each $s_r(\alpha)$ before computing the next. We have, in fact,

$$s_r(\alpha) = fl_2[\alpha s_{r+1}(\alpha) + a_r] \equiv [\alpha s_{r+1}(\alpha)(1+\epsilon_1) + a_r(1+\epsilon_2)](1+\epsilon), \quad (2.14)$$

where

$$|\epsilon_1| < 3.2^{-2t_2}, \qquad |\epsilon_2| < (1{\cdot}5)2^{-2t_2}, \qquad |\epsilon| \leqslant 2^{-t}. \qquad (2.15)$$

Hence the equivalent perturbation in a_r is of order 2^{-t} unless

$$|s_r(\alpha)| \ll |a_r|, \qquad (2.16)$$

that is, unless severe cancellation takes place when s_r is computed.

The derivatives of $f(z)$ may be computed by recurrence relations similar to those in (2.2). In fact by differentiating these equations we have

$$s'_r(\alpha) = \alpha s'_{r+1}(\alpha) + s_{r+1}(\alpha), \qquad s''_r(\alpha) = \alpha s''_{r+1}(\alpha) + 2s'_{r+1}(\alpha), \quad (2.17)$$

and the first and second derivatives may conveniently be computed at the same time as the function itself.

3. We have shown then that in all cases the computed value $s_0(\alpha)$ of $f(\alpha)$ satisfies the relation

$$s_0(\alpha) = \sum a_r(1+E_r)\alpha^r, \qquad |E_r| < (2r+1)2^{-t_1} \qquad (3.1)$$

and have indicated that for certain distributions of zeros the upper bound for $|E_r|$ may be much smaller. In any case, taking the statistical effect into account we can expect the E_r to satisfy some such bound as

$$|E_r| < (2r+1)^{\frac{1}{2}} 2^{-t_1}. \tag{3.2}$$

If the explicit polynomial has been derived by the reduction of a general matrix to Frobenius form, then the errors in the coefficients which arise from this reduction will often be much larger than those corresponding to (3.2) and the actual evaluations of the polynomial are hardly likely to be the weak link in the chain.

Notice that the perturbations E_r are functions of α. If we denote the zeros of $\sum a_r(1+E_r)z^r$ by $\lambda_i^{(\alpha)}$ ($i = 1,\ldots, n$) then we have

$$s_0(\alpha) = \prod (\alpha - \lambda_i^{(\alpha)}), \tag{3.3}$$

while the true value $f(\alpha)$ is given by

$$f(\alpha) = \prod (\alpha - \lambda_i). \tag{3.4}$$

Condition numbers of explicit polynomials

4. The weakness of eigenvalue techniques which are based on the reduction to Frobenius form arises almost entirely from the extraordinary sensitivities of the zeros of many polynomials with respect to small relative perturbations in their coefficients. This is true not only for polynomials having multiple or pathologically close zeros but also of many having zero distributions which, at first sight, appear to be quite innocuous. This was discussed by Wilkinson (1959a) and also forms the main topic of Wilkinson (1963b), Chapter 2; we content ourselves here with a brief summary of the results.

Let us consider the zeros of the polynomial $g(z)$ defined by

$$g(z) = f(z) + \delta a_k z^k = \sum a_r z^r + \delta a_k z^k. \tag{4.1}$$

Then we have the following simple results.

(i) If λ_i is a simple zero of $f(z)$ then the corresponding zero $(\lambda_i + \delta\lambda_i)$ of $g(z)$ satisfies the relation

$$\delta\lambda_i \sim -\delta a_k \lambda_i^k / f'(\lambda_i) = -\delta a_k \lambda_i^k \Big/ \prod_{j \neq i} (\lambda_i - \lambda_j). \tag{4.2}$$

(ii) If λ_i is a zero of $f(z)$ of degree m, then there are m corresponding zeros $(\lambda_i + \delta\lambda_i)$ of $g(z)$, where the $\delta\lambda_i$ are the m values satisfying

$$\delta\lambda_i \sim [-m!\, \delta a_k \lambda_i^k / f^{(m)}(\lambda_i)]^{\frac{1}{m}} = \Big[-\delta a_k \lambda_i^k \Big/ \prod_{j \neq i} (\lambda_i - \lambda_j) \Big]^{\frac{1}{m}}. \tag{4.3}$$

In the case of a simple zero we see that $\left| \lambda_i^k \middle/ \prod\limits_{j \neq i} (\lambda_i - \lambda_j) \right|$ is effectively a condition number for the zero at λ_i. We may relate this result to the usual matrix perturbation theory applied to the Frobenius form. In Chapter 1, § 11, we saw that the right-hand vector x_i corresponding to one version of the Frobenius form is given by

$$x_i^T = (\lambda_i^{n-1}, \lambda_i^{n-2}, ..., 1). \qquad (4.4)$$

With the substitution of a_r for the p_r of Chapter 1, the corresponding left-hand vector y_i is given by

$$y_i^T = (1;\ \lambda_i + a_{n-1};\ \lambda_i^2 + a_{n-1}\lambda_i + a_{n-2};\ ...;\ \lambda_i^{n-1} + a_{n-1}\lambda_i^{n-2} + ... + a_1). \qquad (4.5)$$

Clearly we have $y_i^T x_i = f'(\lambda_i)$ and hence the change $\delta\lambda_i$ in λ_i corresponding to a perturbation ϵB in the Frobenius matrix satisfies the relation

$$\delta\lambda_i \sim y_i^T(\epsilon B)x_i\ /\ y_i^T x_i. \qquad (4.6)$$

If we take ϵB to be the matrix corresponding to the single perturbation δa_k in the element a_k, the relation (4.6) immediately reduces to (4.2).

Some typical distributions of zeros

5. The relation (4.3) shows immediately that multiple zeros are very sensitive to small perturbations in the coefficients and this is well known. It is perhaps not so widely realized that extreme sensitivity is often associated with zero distributions which appear to be quite innocent. We therefore consider this problem for each of the distributions discussed in Chapter 6, § 23, in connexion with Krylov's method. For purposes of illustration we shall consider the effect of a perturbation $2^{-t}a_k$ in the coefficient a_k.

(i) Linear distribution $\lambda_r = 21 - r$.
We have

$$|\delta\lambda_i| \sim 2^{-t}\ |a_k|\ (21-i)^k/(20-i)!\ (i-1)! \qquad (5.1)$$

and the coefficients are of the orders of magnitude given by

$$x^{20} + 10^3 x^{19} + 10^5 x^{18} + 10^7 x^{17} + 10^8 x^{16} + 10^{10} x^{15} + 10^{11} x^{14} +$$
$$+ 10^{12} x^{13} + 10^{13} x^{12} + 10^{15} x^{11} + 10^{16} x^{10} + 10^{17} x^9 + 10^{17} x^8 + 10^{18} x^7 +$$
$$+ 10^{19} x^6 + 10^{19} x^5 + 10^{19} x^4 + 10^{20} x^3 + 10^{20} x^2 + 10^{19} x + 10^{19}. \qquad (5.2)$$

A trivial but tedious investigation shows that the greatest sensitivity is that of the zero at $x = 16$ ($i = 5$) with respect to perturbations in a_{15}. In fact we have

$$|\delta\lambda_5| \sim 2^{-t}\,10^{10}(16)^{15}/15!\,4! \doteq 2^{-t}(0.38)10^{15}, \qquad (5.3)$$

showing that unless 2^{-t} is appreciably smaller than 10^{-14}, the perturbed zero will bear little relation to the original. Most of the larger zeros are very sensitive to perturbations in the coefficients of the higher powers of x.

This is illustrated in Table 1 where we show the effect of the single perturbation of 2^{-23} in the coefficient x^{19}; this is a change of about one part in 2^{-30}. Ten of the zeros have become complex and have very substantial imaginary parts! On the other hand the smaller zeros are fairly insensitive as may be verified from (5.1).

TABLE 1

Zeros of the polynomial $(x-1)(x-2)\ldots(x-20)-2^{-23}x^{19}$, correct to 9 decimal places.

1·00000 0000	6·00000 6944	10·09526 6145 $\pm$ 0·64350 0904i
2·00000 0000	6·99969 7234	11·79363 3881 $\pm$ 1·65232 9728i
3·00000 0000	8·00726 7603	13·99235 8137 $\pm$ 2·51883 0070i
4·00000 0000	8·91725 0249	16·73073 7466 $\pm$ 2·81262 4894i
4·99999 9928	20·84690 8101	19·50243 9400 $\pm$ 1·94033 0347i

(ii) Linear distribution $\lambda_r = 11-r$.
We now have

$$|\delta\lambda_i| \sim 2^{-t}\,|a_k|\,(11-i)^k/(20-i)!\,(i-1)! \qquad (5.4)$$

and comparing this with (5.1) we see that the factor $(21-i)^k$ has been replaced by $(11-i)^k$. The orders of magnitude of the coefficients are indicated by

$$x^{20}+10x^{19}+10^3x^{18}+10^4x^{17}+10^5x^{16}+10^6x^{15}+10^7x^{14}+$$
$$+10^8x^{13}+10^9x^{12}+10^9x^{11}+10^{10}x^{10}+10^{11}x^9+10^{11}x^8+10^{12}x^7+$$
$$+10^{12}x^6+10^{12}x^5+10^{12}x^4+10^{12}x^3+10^{12}x^2+10^{12}x. \qquad (5.5)$$

The greatest sensitivity is that of λ_8 with respect to a_{16}. We have

$$|\delta\lambda_8| \sim 2^{-t}\,|a_{16}|\,8^{16}/2!\,17! \doteq 2^{-t}(0.4)10^5. \qquad (5.6)$$

The polynomial is therefore very much better conditioned than that corresponding to (i). Notice that if A has the eigenvalues $(21-i)$, $(A-10I)$ has the eigenvalues $(11-i)$. Hence this comparatively trivial change in the original matrix may have a profound effect on the condition of the corresponding Frobenius form. This is not true of the reductions to tri-diagonal and Hessenberg form described in Chapter 6; for these a change of kI in the original matrix produces a change of kI in the derived matrix and hence the condition of the derived matrix is independent of such modifications.

(iii) Geometric distribution $\lambda_r = 2^{-r}$.

The coefficients of the polynomial vary enormously in size, their orders of magnitude being indicated by

$$x^{20} + x^{19} + 2^{-1}x^{18} + 2^{-4}x^{17} + 2^{-8}x^{16} + 2^{-13}x^{15} + 2^{-19}x^{14} + 2^{-26}x^{13} +$$
$$+ 2^{-34}x^{12} + 2^{-43}x^{11} + 2^{-53}x^{10} + 2^{-64}x^9 + 2^{-76}x^8 + 2^{-89}x^7 + 2^{-103}x^6 +$$
$$+ 2^{-118}x^5 + 2^{-134}x^4 + 2^{-151}x^3 + 2^{-169}x^2 + 2^{-189}x + 2^{-209}. \qquad (5.7)$$

For the two previous distributions we considered the *absolute* changes in the zeros. For zeros of such widely varying orders of magnitude it is perhaps more reasonable to consider relative perturbations. We have

$$|\delta\lambda_i/\lambda_i| \sim 2^{-t} |a_k| \, 2^{(1-k)i}/|PQ|, \qquad (5.8)$$

where

$$P = (2^{-i} - 2^{-1})(2^{-i} - 2^{-2})\ldots(2^{-i} - 2^{-i+1}), \qquad (5.9)$$
$$Q = (2^{-i} - 2^{-i-1})(2^{-i} - 2^{-i-2})\ldots(2^{-i} - 2^{-20}). \qquad (5.10)$$

We may write

$$|P| = 2^{-\frac{1}{2}i(i-1)}[(1-2^{-1})(1-2^{-2})\ldots(1-2^{-i+1})], \qquad (5.11)$$
$$|Q| = 2^{-i(20-i)}[(1-2^{-1})(1-2^{-2})\ldots(1-2^{-20+i})], \qquad (5.12)$$

and the expressions in square brackets are both convergents to the infinite product

$$(1-2^{-1})(1-2^{-2})(1-2^{-3})\ldots. \qquad (5.13)$$

Quite a crude inequality shows that they lie between $\frac{1}{2}$ and $\frac{1}{4}$ and hence we have

$$|\delta\lambda_i/\lambda_i| < 2^{-t} |a_k| \, 2^{\theta+4} \quad \text{where } \theta = \tfrac{1}{2}i(41-2k-i). \qquad (5.14)$$

For a fixed value of k this takes its maximum when $i = 20 - k$ so that we have

$$|\delta\lambda_i/\lambda_i| < 2^{-t} |a_k| \, 2^{4+\frac{1}{2}(20-k)(21-k)}, \qquad (5.15)$$

and from (5.7) it may be verified that

$$|a_k| \leqslant 4 . 2^{-\frac{1}{2}(20-k)(21-k)}. \qquad (5.16)$$

Hence we have finally

$$|\delta\lambda_i/\lambda_i| < 2^{6-t} \qquad (5.17)$$

for *relative* perturbations of up to one part in 2^{-t} in any coefficient. The zeros are therefore all quite well-conditioned with respect to such perturbations, which are comparable with those corresponding to the errors made in evaluating the polynomial.

It is natural to ask whether we are likely to obtain the coefficients of the characteristic equation with low relative errors when the eigen-values are of widely varying orders of magnitude. Unfortunately the

the answer is that, in general, we are not. As a very simple illustration consider the matrix

$$A = \begin{bmatrix} 0\cdot63521 & 0\cdot81356 \\ 0\cdot59592 & 0\cdot76325 \end{bmatrix}, \tag{5.18}$$

which has the eigenvalues $\lambda_1 = 10^1(0\cdot13985)$, $\lambda_2 = 10^{-5}(0\cdot52610)$ correct to five significant figures. We may reduce this to Frobenius form by a similarity transformation $M_1^{-1}AM_1$ where

$$M_1 = \begin{bmatrix} 1 & 10^1(-0\cdot12808) \\ 0 & 1 \end{bmatrix}, \tag{5.19}$$

but notwithstanding the use of floating-point arithmetic the computed transform is

$$\begin{bmatrix} 10^1(0\cdot13985) & 10^{-4}(-0\cdot20000) \\ 0\cdot59592 & 0 \end{bmatrix}. \tag{5.20}$$

Severe cancellation takes place when the (1, 2) element is computed; the computed element which gives the constant term in the characteristic polynomial has scarcely one correct significant figure. Generally when there is a great disparity in the size of the eigenvalues the elements in the Frobenius form tend to be obtained with *absolute* errors which are all of much the same orders of magnitude, and this is quite fatal to the relative accuracy of the smaller eigenvalues. It is obvious, for example, that an absolute error of order 10^{-10} in the constant term of the polynomial corresponding to the distribution (iii) has the effect of changing the product of the zeros from 2^{-210} to $(2^{-210}+10^{-10})$; the latter is approximately 10^{53} times the former!

For special types of matrices this deterioration does not take place. A simple example is the matrix

$$\begin{bmatrix} 10^0(0\cdot2632) & 10^{-5}(0\cdot3125) \\ 10^0(0\cdot4315) & 10^{-5}(0\cdot2873) \end{bmatrix}. \tag{5.21}$$

Although this has one eigenvalue of order unity and another of order 10^{-5}, no cancellation takes place when we derive the Frobenius form.

(iv) Circular distribution $\lambda_r = \exp(r\pi i/20)$.
The corresponding polynomial is $z^{20}-1$ and we have

$$|\delta\lambda_i| \sim |\delta a_k| \, |\lambda_i|^k / |f'(\lambda_i)| = \tfrac{1}{20} |\delta a_k|. \tag{5.22}$$

The zeros of this polynomial are all very well-conditioned. (Since most of the coefficients are zero we have not considered relative perturbations in this case.)

Final assessment of Krylov's method

6. The analysis of the previous section confirms the forecasts we made in Chapter 6, § 25, of the degree of effectiveness of Krylov's method applied to matrices having eigenvalues with the various distributions of § 5. For a distribution such as (i) the use of 62 binary places in the normalized Krylov is quite inadequate. The zeros of the computed characteristic equation are unlikely to bear any relation to the eigenvalues of the original matrix. In spite of this, for the closely related distribution (ii) the computed characteristic equation is much more accurate and, moreover, it is much better conditioned.

For a distribution such as (iii) we saw in Chapter 6, § 25, that computation to 62 binary places gave a computed characteristic equation which was quite unrecognizable. The fact that the zeros of the true characteristic equation are comparatively insensitive to small *relative* changes in the coefficients is of no relevance. Krylov's method is quite impractical for such distributions.

Turning now to distribution (iv) we find that not only does Krylov's method give a very accurate determination of the characteristic equation, but also the characteristic equation is very well-conditioned. For matrices having eigenvalues with this type of distribution Krylov's method is quite effective, even in single precision.

We may sum this up by saying that although Krylov's method is effective for matrices having eigenvalues which are 'well-distributed' in the complex plane it has severe limitations as a general-purpose method.

General comments on explicit polynomials

7. It is evident that for a distribution of the type $\lambda_i = 21 - i$ ($i = 1, \ldots, 20$) we shall need to compute the Frobenius form using high-precision arithmetic if its eigenvalues are to be close to those of the original matrix from which it was derived. This is true whatever method of reduction may be used. Further, since the errors made in evaluating $f(\alpha)$ correspond to perturbations in the coefficients of the type shown in equations (2.6) and (2.7), the evaluation of $f(z)$ must also be performed using the same high precision.

It is interesting to note for example that the computed values of $f(z)$ for the distribution (i) of § 5 will usually have no correct figures for values of z in the neighbourhood of $z = 20$ unless high-precision arithmetic is used. As an illustration we consider the evaluation of $f(z)$ when $z = 20 \cdot 00345645$ using 10-decimal floating-point arithmetic.

At the second stage in the nested multiplication we compute $s_{19}(z)$ given by

$$s_{19}(z) = zs_{20}(z) + a_{19} = 20 \cdot 00345645 \times 1 - 210 \cdot 0. \qquad (7.1)$$

The correct value is $-189 \cdot 99654355$, but if we are using only 10 digits this must be replaced by $-189 \cdot 9965436$. Even if no further errors of any kind were made in the nested multiplication, the final error resulting from this rounding would be

$$5 \times 10^{-8}(20 \cdot 00345645)^{19} \doteqdot (2 \cdot 6)10^{17}. \qquad (7.2)$$

Now if, for brevity, we write $x = 0 \cdot 00345645$, the correct value of $f(z)$ is

$$x(1+x)(2+x)...(19+x), \qquad (7.3)$$

and this lies between $19! \, x$ and $20! \, x$, that is between $(4 \cdot 2)10^{14}$ and $(8 \cdot 4)10^{15}$. The very first rounding error affects the computed value by an amount which is far larger than the true value!

In the neighbourhood of an ill-conditioned zero the computed values of $f(z)$ corresponding to a monotonic sequence of values of z may be far from monotonic. This is well illustrated in Table 2 where

TABLE 2

z	Computed $f(z)$		True $f(z)$
$10+2^{-42}$	$+0 \cdot 63811...$		$10^{-6}(+0 \cdot 16501...)$
$10+2 \times 2^{-42}$	$+0 \cdot 57126...$	Dominated by	$10^{-6}(+0 \cdot 33003...)$
$10+3 \times 2^{-42}$	$-0 \cdot 31649...$	the rounding	$10^{-6}(+0 \cdot 49505...)$
$10+2^{-28}+7 \times 2^{-42}$	$+0 \cdot 29389...$	errors.	$10^{-2}(+0 \cdot 27048...)$
$10+2^{-23}+7 \times 2^{-42}$	$+0 \cdot 70396...$		$10^{-1}(+0 \cdot 86518...)$
$10+2^{-18}+7 \times 2^{-42}$	$10^{1}(+0 \cdot 33456...)$		$10^{1}(+0 \cdot 27685...)$
$10+2^{-13}+7 \times 2^{-42}$	$10^{2}(+0 \cdot 89316...)$		$10^{2}(+0 \cdot 88608...)$

we give the computed values obtained for the polynomial $\prod_{r=1}^{12}(z-r)$ for values of z in the neighbourhood of 10. The inclusion of the term 7×2^{-42} in the last four values of z was necessary in order that the rounding errors should be typical. Otherwise z would have terminated with a large number of zeros. The polynomial was evaluated from its exact explicit representation using floating-point arithmetic with a mantissa of 46 binary places. (We did not use $\prod_{r=1}^{20}(z-r)$ because this does not admit of exact representation with this word length.)

For values of z from 10 to approximately $10+2^{-20}$ the computed values are all of the order of unity in absolute value and bear no relation to the true values. This is because the computed values are completely dominated by the rounding errors which, as we saw in (2.13), are equivalent to perturbations of up to $3\,|a_r|\,2^{-t_1}$ in a_r. For values of z which are further from 10 we obtain some correct significant figures. In fact if $z = 10+x$ and $|x|$ is of the order of 2^{-k} then, in general, the computed value of $f(z)$ has approximately $20-k$ correct significant figures, though occasionally exceptionally favourable roundings may give rather better values. One could scarcely expect any iterative technique which is based on values of $f(z)$ computed from the explicit polynomial using a 46-digit mantissa to locate the zero at $z = 10$ with an error of less than 2^{-20}.

Tri-diagonal matrices

8. We now turn to the evaluation of $\det(A-zI)$ where A is a tri-diagonal matrix. If we write

$$a_{ii} = \alpha_i, \qquad a_{i,i+1} = \beta_{i+1}, \qquad a_{i+1,i} = \gamma_{i+1} \qquad (8.1)$$

and let $p_r(z)$ denote the leading principal minors of $(A - zI)$ of order r, then expanding $p_r(z)$ by its last row, we have

$$p_r(z) = (\alpha_r-z)p_{r-1}(z)-\beta_r\gamma_r p_{r-2}(z) \qquad (r = 2,\ldots, n), \qquad (8.2)$$

where

$$p_0(z) = 1, \qquad p_1(z) = \alpha_1-z. \qquad (8.3)$$

The products $\beta_r\gamma_r$ may be computed once and for all. We may assume that $\beta_r\gamma_r \neq 0$, since otherwise we can deal with tri-diagonal matrices of lower order. There are $2n$ multiplications in each evaluation.

If derivatives of $\det(A-zI)$ are required they may be computed at the same time as the function is evaluated. In fact, differentiating (8.2) we have

$$p_r'(z) = (\alpha_r-z)p_{r-1}'(z)-\beta_r\gamma_r p_{r-2}'(z)-p_{r-1}(z) \qquad (8.4)$$

and

$$p_r''(z) = (\alpha_r-z)p_{r-1}''(z)-\beta_r\gamma_r p_{r-2}''(z)-2p_{r-1}'(z). \qquad (8.5)$$

Since it is the accuracy of the computed values of the function itself which is the main limitation on the accuracy attainable in the computed zeros, we content ourselves with an error analysis of the algorithm embodied in equations (8.2). In practice floating-point arithmetic is almost essential. The computed values satisfy the relations

$$p_r(z) = fl[(\alpha_r-z)p_{r-1}(z)-\beta_r\gamma_r p_{r-2}(z)]$$

$$\equiv (\alpha_r-z)(1+\epsilon_1)p_{r-1}(z)-\beta_r\gamma_r(1+\epsilon_2)p_{r-2}(z), \qquad (8.6)$$

where

$$|\epsilon_1| < 3.2^{-t_1}, \qquad |\epsilon_2| < 3.2^{-t_1}. \qquad (8.7)$$

Hence, as in the symmetric case (Chapter 5, § 40), the computed sequence is exact for the tri-diagonal matrix with elements α'_r, β'_r, γ'_r where

$$\alpha'_r - z = (\alpha_r - z)(1 + \epsilon_1), \qquad \beta'_r \gamma'_r = \beta_r \gamma_r (1 + \epsilon_2). \tag{8.8}$$

We may therefore assume that

$$\beta'_r = \beta_r (1 + \epsilon_2)^{\frac{1}{2}}, \qquad \gamma'_r = \gamma_r (1 + \epsilon_2)^{\frac{1}{2}}, \tag{8.9}$$

so that the β'_r and γ'_r correspond to low relative errors. On the other hand we have

$$\alpha'_r - \alpha_r = (\alpha_r - z)\epsilon_1, \tag{8.10}$$

and we cannot guarantee that this corresponds to a low relative error in α_r. However, we can confine ourselves to values of z such that

$$|z| < \|A\|_\infty = \max(|\gamma_i| + |\alpha_i| + |\beta_{i+1}|), \tag{8.11}$$

so that the perturbations in the α_r are small relative to the norm of A.

9. Notice that if $\tilde{A}$ is any matrix obtained from A by a diagonal similarity transformation, the α_r and the products $\beta_r \gamma_r$ are the same for $\tilde{A}$ as for A, though the condition numbers of the eigenvalues will, in general, be different for the two matrices. The computed sequence of $p_r(z)$ is the same for all such $\tilde{A}$. Hence if we have, for example a matrix of the form A given by

$$A = \begin{bmatrix} 0\cdot2532 & 2^{40}(0\cdot2615) & \\ 2^{-38}(0\cdot3125) & 0\cdot3642 & 0\cdot1257 \\ & 0\cdot2135 & 0\cdot4132 \end{bmatrix}, \tag{9.1}$$

for the purposes of the error analysis we can think in terms of the matrix $\tilde{A}$ given by

$$\tilde{A} = \begin{bmatrix} 0\cdot2532 & 2^1(0\cdot2615) & \\ 2^1(0\cdot3125) & 0\cdot3642 & 0\cdot1257 \\ & 0\cdot2135 & 0\cdot4132 \end{bmatrix}. \tag{9.2}$$

Of course we shall also restrict our trial values of z to the region $|z| < \|\tilde{A}\|_\infty$ rather than to the region $|z| < \|A\|_\infty$. In general we may think in terms of an 'equilibrated' version of the matrix A (Chapter 6, § 10).

10. If the tri-diagonal matrix has been derived by a similarity transformation from a matrix of general form then, provided we use the same precision of computation in both cases, it is virtually certain that the errors in the tri-diagonal matrix will be greater than the perturbations which are equivalent to the errors made in an evaluation.

In Chapter 6, § 49, we recommended the use of double-precision arithmetic in the reduction of a Hessenberg matrix to tri-diagonal form, even when the Hessenberg matrix has been derived from a general matrix by a similarity transformation using single-precision arithmetic. This was to guard against the possible emergence of 'unstable' stages in the reduction to tri-diagonal form. It is natural to ask whether we can revert to the use of single-precision arithmetic in the evaluations of the $p_r(z)$. If there have been no unstable stages in the reduction to tri-diagonal form this would indeed be quite justifiable, but if unstable stages have occurred then a reversion to single-precision work would be fatal. Quite an elementary analysis shows why this is true.

We saw in Chapter 6, § 46, that when an instability occurs the corresponding tri-diagonal form has elements of orders of magnitude which can generally be expressed in a form typified by

$$\begin{bmatrix} 1 & 1 & & & & \\ 1 & 1 & 1 & & & \\ & 2^{-k} & 2^k & 2^k & & \\ & & 2^k & 2^k & 2^{-k} & \\ & & & 1 & 1 & 1 \\ & & & & 1 & 1 \end{bmatrix}, \tag{10.1}$$

where the eigenvalues are of order unity and not 2^k as the norm might suggest. We showed that y and x, corresponding left-hand and right-hand eigenvectors normalized such that $y^T x = 1$, have components of the orders of magnitude

$$\begin{aligned} |y^T| &= (1, \quad 1, \quad 2^k, \quad 2^k, \quad 1, \quad 1) \\ |x^T| &= (1, \quad 1, \quad 1, \quad 1, \quad 1, \quad 1) \end{aligned}. \tag{10.2}$$

Now our error analysis of § 8 has shown that the equivalent perturbations corresponding to an evaluation using t-digit arithmetic are of the order of magnitude

$$B = 2^{-t} \begin{bmatrix} 1 & 1 & & & & \\ 1 & 1 & 1 & & & \\ & 2^{-k} & 2^k & 2^k & & \\ & & 2^k & 2^k & 2^{-k} & \\ & & & 1 & 1 & 1 \\ & & & & 1 & 1 \end{bmatrix}. \tag{10.3}$$

Hence from Chapter 2, § 9, the order of magnitude of the perturbation in the eigenvalue will in general be that of $y^T Bx/y^T x$ and we have

$$y^T Bx/y^T x = O(2^{2k-t}).\qquad(10.4)$$

This shows that we should use the same precision in the evaluations as in the reduction, but as we showed in Chapter 6, § 49, we can generally expect double-precision arithmetic to be quite adequate unless the *original* matrix is itself too ill-conditioned. This is a far more satisfactory situation than we found in connexion with the Frobenius form.

Determinants of Hessenberg matrices

11. Finally we consider the evaluation of $\det(A-zI)$ for assigned values of z when A is an upper Hessenberg matrix. Again we find that this can be achieved using simple recurrence relations which are derived as follows (Hyman, 1957). Let us write

$$P = A-zI,\qquad(11.1)$$

and denote the columns of P by $p_1, p_2,..., p_n$. Suppose we can find scalars $x_{n-1}, x_{n-2},..., x_1$ such that

$$x_1 p_1+x_2 p_2+...+x_{n-1}p_{n-1}+p_n = k(z)e_1,\qquad(11.2)$$

that is, such that the left-hand side has zero components in all positions except the first. Then the matrix Q derived from P by replacing its last column by ke_1 clearly has the same determinant as P. Hence, expanding the determinant of Q by its last column, we have

$$\det(A-zI) = \det P = \det Q = \pm k(z)a_{21}a_{32}...a_{n,n-1}.\qquad(11.3)$$

The elements $a_{i+1,i}$ play a special role; we emphasize this with the notation

$$a_{i+1,i} = b_{i+1}.\qquad(11.4)$$

From equation (11.2) we may compute $x_{n-1}, x_{n-2},..., x_1$ in succession by equating elements $n, n-1,..., 2$ on both sides. We have then

$$b_r x_{r-1}+(a_{rr}-z)x_r+a_{r,r+1}x_{r+1}+...+a_{rn}x_n = 0 \quad (r = n,..., 2)\qquad(11.5)$$

where $x_n = 1$. Finally, equating the first components we find

$$(a_{11}-z)x_1+a_{12}x_2+...+a_{1n}x_n = k(z).\qquad(11.6)$$

The determinant of $(A-zI)$ differs from $k(z)$ only by a constant multiplier and hence we may concentrate on finding the zeros of $k(z)$. There are effectively $\tfrac{1}{2}n^2$ multiplications in each evaluation. If the sub-diagonal elements b_r are all unity then the x_{r-1} may be computed without using divisions. We have seen in Chapter 6, § 56, that it is quite convenient to produce upper Hessenberg matrices which are specialized in this way provided we work in floating-point arithmetic. However, we shall make no assumptions about the b_r except that they are non-zero. If any b_r is zero we may work with Hessenberg matrices of lower order.

We may obtain the derivatives of $f(z)$ by similar recurrence relations. Differentiating (11.5) and (11.6) with respect to z we obtain

$$b_r x'_{r-1} + (a_{rr} - z)x'_r + a_{r,r+1}x'_{r+1} + \ldots + a_{rn}x'_n - x_r = 0, \qquad (11.7)$$

$$(a_{11} - z)x'_1 + a_{12}x'_2 + \ldots + a_{1n}x'_n - x_1 = k'(z), \qquad (11.8)$$

and since $x_n = 1$ we have $x'_n = 0$. There are similar relations for the second derivatives. The derivatives of the x_i and $k(z)$ are therefore conveniently evaluated at the same time as the functions themselves.

Effect of rounding errors

12. For the evaluation of the x_i defined by (11.5), floating-point computation is almost essential. The computed x_{r-1} is given in terms of the computed x_i ($i = r, \ldots, n$) by the relation

$$\begin{aligned}
-x_{r-1} &= fl[\{a_{rn}x_n + \ldots + a_{r,r+1}x_{r+1} + (a_{rr} - z)x_r\}/b_r] \\
&\equiv \{a_{rn}x_n(1 + \epsilon_{rn}) + \ldots + a_{r,r+1}x_{r+1}(1 + \epsilon_{r,r+1}) + \\
&\qquad + (a_{rr} - z)x_r(1 + \epsilon_{rr})\}(1 + \epsilon_r)/b_r, \qquad (12.1)
\end{aligned}$$

where $\quad |\epsilon_{ri}| < (i - r + 2)2^{-t_1} \quad (i > r), \quad |\epsilon_{rr}| < 3.2^{-t_1}, \quad |\epsilon_r| < 2^{-t}.$
$$\qquad (12.2)$$

We therefore have an exact evaluation for a matrix with elements a'_{ri} and b'_r given by

$$\left. \begin{aligned}
a'_{ri} &= a_{ri}(1 + \epsilon_{ri}), & |\epsilon_{ri}| &< (i - r + 2)2^{-t_1} \quad (i > r) \\
a'_{rr} &= a_{rr}(1 + \epsilon_{rr}) - z\epsilon_{rr}, & |\epsilon_{rr}| &< 3.2^{-t_1} \\
b'_r &= b_r(1 + \eta_r), & |\eta_r| &< 2^{-t_1}
\end{aligned} \right\}. \qquad (12.3)$$

These elements all have low relative errors except possibly the a'_{rr}; however, we are interested only in values of z satisfying

$$|z| < \|A\| \qquad (12.4)$$

and hence $\qquad |a'_{rr} - a_{rr}| < \{|a_{rr}| + \|A\|\} |\epsilon_{rr}| \qquad (12.5)$

showing that this perturbation is small compared with any norm of A.

As an illustration we consider the case when all elements of A are bounded in modulus by unity, so that we have

$$|z| \leqslant \|A\|_E < [\tfrac{1}{2}(n^2+3n-2)]^{\frac{1}{2}}. \tag{12.6}$$

The equivalent perturbation matrix, which is a function of z, is then uniformly bounded by a matrix F of the form typified for $n = 5$ by

$$F = 2^{-t_1}\begin{bmatrix} 3 & 3 & 4 & 5 & 6 \\ 1 & 3 & 3 & 4 & 5 \\ & 1 & 3 & 3 & 4 \\ & & 1 & 3 & 3 \\ & & & 1 & 3 \end{bmatrix} + 3\,\|A\|_E 2^{-t_1}I, \tag{12.7}$$

and since $\|A\|_E = O(n)$ we have

$$\|F\|_E \sim 2^{-t_1}(n^2/12) \quad \text{as } n \to \infty. \tag{12.8}$$

Once again we find that the errors made in an evaluation can scarcely be serious unless the eigenvalues are extremely sensitive to end-figure perturbations in the elements of the Hessenberg form itself. We have shown in Chapter 6 that there are several very stable methods of reducing a matrix to Hessenberg form. Single-precision computation is therefore normally adequate both for the reduction and subsequent evaluations unless the original matrix is too ill-conditioned.

Floating-point accumulation

13. Although the result of the previous section is by no means unsatisfactory, very much smaller error bounds may be obtained for evaluations of the determinants of Hessenberg matrices using floating-point accumulation. To take full advantage of the accumulation x_{r-1} should be computed from the x_i $(i = r,..., n)$ using the relation

$$-x_{r-1} = fl_2[(a_{rn}x_n + \ldots + a_{r,r+1}x_{r+1} + a_{rr}x_r - zx_r)/b_r]. \tag{13.1}$$

Notice that we have written $a_{rr}x_r - zx_r$ and not $(a_{rr}-z)x_r$ since a single-precision rounding error may be involved in computing $a_{rr}-z$. Equation (13.1) gives

$$-x_{r-1} = [a_{rn}x_n(1+\epsilon_{rn}) + \ldots + a_{rr}x_r(1+\epsilon_{rr}) - zx_r(1+\eta_r)](1+\epsilon_r)/b_r$$

where
$$\tag{13.2}$$

$$|\epsilon_{ri}| < 1{\cdot}5(i-r+3)2^{-2t_2}, \qquad |\eta_r| < 3.2^{-2t_2}, \qquad |\epsilon_r| < 2^{-t}. \tag{13.3}$$

The evaluation is therefore exact for a matrix with elements a'_{ri} and b'_r given by

$$a'_{ri} = a_{ri}(1+\epsilon_{ri}), \qquad |\epsilon_{ri}| < 1 \cdot 5(i-r+3)2^{-2t_2} \quad (i = n,..., r+1)$$
$$a'_{rr} = a_{rr}(1+\epsilon_{rr})-z\eta_r, \quad |\epsilon_{rr}| < (4 \cdot 5)2^{-2t_2}, \quad |\eta_r| < 3.2^{-2t_2}$$
$$b'_r = b_r(1+\xi_r), \qquad |\xi_r| < 2^{-t_1}$$

$$(13.4)$$

Again, for illustration, we consider the case when all elements of A are bounded in modulus by unity. The equivalent perturbations are now bounded by a matrix F of the form typified for $n = 5$ by

$$F = (1 \cdot 5)2^{-2t_2} \begin{bmatrix} 3 & 4 & 5 & 6 & 7 \\ & 3 & 4 & 5 & 6 \\ & & 3 & 4 & 5 \\ & & & 3 & 4 \\ & & & & 3 \end{bmatrix} +$$

$$+3\,\|A\|_E 2^{-2t_2} I + 2^{-t_1} \begin{bmatrix} 0 & 0 & 0 & 0 & 0 \\ 1 & 0 & 0 & 0 & 0 \\ & 1 & 0 & 0 & 0 \\ & & 1 & 0 & 0 \\ & & & 1 & 0 \end{bmatrix}. \quad (13.5)$$

In the general case the three matrices on the right have 2-norms which are bounded respectively by

$$1 \cdot 5(n+2)^2 2^{-2t_2}/12^{\frac{1}{2}}, \qquad 3 \cdot 0(n+2)2^{-2t_2}/2^{\frac{1}{2}} \quad \text{and} \quad 2^{-t_1}.$$

Hence, provided $n^2 2^{-t}$ is appreciably less than unity, $\|F\|_2$ is effectively 2^{-t_1}, a most satisfactory result.

Evaluation by orthogonal transformations

14. Hyman's method of evaluation is based on a post-multiplication of the Hessenberg matrix by an elementary matrix of the type S_n (Chapter 1, § 40(iv)), the strategic elements of S_n being chosen so as to reduce the last column to a multiple of e_1. In previous situations we have found that pivoting was essential for numerical stability, but the error analysis of §§ 12, 13 has shown that it is irrelevant in this case, and the use of large multipliers x_i has no

adverse effect. In the same way we can see that although we divide by the b_r elements no harm results from this, however small they may be. Interchanges can be incorporated but they serve no useful purpose in this case. The way in which they can be introduced will become apparent in the next paragraph when we consider the use of plane rotations.

Fear of numerical instability is therefore misplaced but it has led to the suggestion that plane rotations should be used. (See, for example, White (1958).) The method of evaluation consists of $(n-1)$ major steps, in the rth of which a zero is introduced in position $(n-r+1)$ of the last column. The configuration at the beginning of the rth step is given for the case $n = 5$, $r = 3$ by

$$\begin{bmatrix} \times & \times & \times & \times & \times \\ \times & \times & \times & \times & \times \\ & \times & \times & \times & \times \\ & & \times & \times & 0 \\ & & & \times & 0 \end{bmatrix}. \qquad (14.1)$$

The rth step consists of a post-multiplication by a rotation in the $(n-r, n)$ plane, the angle of the rotation being chosen so that $a_{n-r+1, n}$ is reduced to zero. The steps are:

(i) Compute $x = (b_{n-r+1}^2 + a_{n-r+1, n}^2)$.

(ii) Compute $\cos \theta = b_{n-r+1}/x$ and $\sin \theta = a_{n-r+1, n}/x$.

(iii) For each value of i from 1 to $n-r$:
Compute $a_{j, n-r} \cos \theta + a_{jn} \sin \theta$ and $-a_{j, n-r} \sin \theta + a_{jn} \cos \theta$ and overwrite on $a_{j, n-r}$ and a_{jn} respectively.

(iv) Replace b_{n-r+1} by x and $a_{n-r, n}$ by zero.
Clearly we have replaced columns $n-r$ and n by linear combinations of themselves so that zeros produced by the earlier steps are undisturbed.

The final matrix is of the same form as that corresponding to Hyman's method, and the determinant Δ is given by

$$\Delta = \pm a_{1n} \prod_{r=1}^{n-1} b_{r+1}. \qquad (14.2)$$

Now it is easy to show in the usual way that the computed determinant is the exact determinant of some matrix $(A + \delta A)$ where, for standard floating-point computation for example,

$$\|\delta A\|_E < K n^{\frac{3}{2}} \|A\|_E. \qquad (14.3)$$

The method is therefore stable, but the bound is poorer than for Hyman's method in standard floating-point computation. From relations (12.3) to (12.5) we have certainly

$$\|\delta A\|_E < Kn \|A\|_E, \qquad (14.4)$$

but for Hyman's method using floating-point accumulation we had an even better result.

15. Since there are four times as many multiplications in the method using plane rotations, we can state without any reservations that for this computation the use of elementary transformations *without* interchanges is superior in every respect. Indeed Hyman's method has one further advantage which is not immediately apparent from our analysis. We have shown that the computed determinant is the exact determinant of $(A + \delta A)$, and for each non-zero element of A we have obtained a bound of the form

$$|\delta a_{ij}| < 2^{-t} f(i, j) |a_{ij}|. \qquad (15.1)$$

Now consider any diagonal similarity transformation of $(A + \delta A)$; we may express such a transform in the form $D^{-1}(A + \delta A)D = \tilde{A} + \delta\tilde{A}$, and clearly we have the same relationship between the elements of $\delta\tilde{A}$ and $\tilde{A}$ as we had between those of δA and A. In assessing the effect of the perturbation δA on the eigenvalues of A we can think in terms of the best-conditioned of all diagonal similarity transforms of A. Clearly, therefore, there is no need to equilibrate A before carrying out Hyman's method. None of these comments applies to the use of orthogonal transformations or of the elementary similarity transformations with interchanges.

Evaluation of determinants of general matrices

16. Before taking up the consideration of iterative methods for locating the zeros, we deal with three topics which are related to those we have been discussing. The first is the evaluation of $\det(A - zI)$ when A is of general form. As we showed in Chapter 4, a stable determination is afforded by any of the reductions to triangular form based on stabilized elementary transformations or orthogonal transformations. If we can accumulate inner-products, then direct triangular decomposition with interchanges is particularly favourable. The weakness of such methods lies in the volume of work required and not in their stability. A single evaluation requires at least $\frac{1}{3}n^3$ multiplications, whereas if we reduce first to Hessenberg form an evaluation requires only $\frac{1}{2}n^2$ multiplications.

The generalized eigenvalue problem

17. The second topic concerns the solution of n simultaneous differential equations of order r and in Chapter 1, § 30, we showed that it led to a generalized eigenvalue problem in which we require the zeros of

$$\det[B_0 + B_1 z + \dots + \dots B_{r-1} z^{r-1} - I z^r] = \det[C(z)]. \qquad (17.1)$$

Although this may be reduced to the standard eigenvalue problem the corresponding matrix is of order nr. There are therefore advantages in dealing directly with the form (17.1) in the following manner.

Corresponding to each value of z the elements of the matrix $C(z)$ are evaluated; this involves approximately $n^2 r$ multiplications. The determinant of $C(z)$ may then be evaluated by any of the stable methods of triangularization described in Chapter 4. The most economical of these methods involved $\frac{1}{3}n^3$ multiplications. Since each element of $C(z)$ is obtained by evaluating an explicit polynomial there is a potential danger here, but in practice the order of the original differential equation is usually quite low, orders two and three being by far the most common. Quite a common situation is for the matrices B_i to be functions of a parameter and for the history of specific zeros for a sequence of values of the parameter to be required. We return to this point in § 63.

Indirect determinations of the characteristic polynomial

18. The third point is that because of the compact nature of the explicit polynomial form it is tempting to determine the coefficients of the polynomial from computed values of $\det(A - zI)$. If $\det(A - zI)$, or some constant multiple of it, is evaluated at $z_1, z_2, \dots, z_{n+1}$, then the coefficients of the polynomial are the solutions of the set of equations

$$c_n z_i^n + c_{n-1} z_i^{n-1} + \dots + c_1 z_i + c_0 = v_i \quad (i = 1, \dots, n+1), \qquad (18.1)$$

where v_i are the computed values. Similarly, if the zeros of the expression (17.1) are required we may make $(nr + 1)$ evaluations. The fact that so much effort has been devoted to the solution of polynomial equations has provided a further incentive to reduce the problem to this form.

Our main object here is to show that such methods have severe inherent limitations. To illustrate this, we describe one method of determining the characteristic polynomial and assess its accuracy for one specific distribution of zeros. This method is based on the fact that we can obtain an explicit solution of equations (18.1). In fact we have

$$c_{n-r} = \sum_{i=1}^{n+1} P_i^{(r)} v_i, \tag{18.2}$$

where
$$P_i^{(r)} = (-1)^r \sum_{i_k \neq i} z_{i_1} z_{i_2} \ldots z_{i_r} \Big/ \prod_{j \neq i} (z_i - z_j). \tag{18.3}$$

The choice of the z_i is at our disposal. If we form some estimate of the norm N of the matrix then a reasonable set of values is provided by the $(n+1)$ equidistant points between $\pm N$. With this choice the $P_i^{(r)}$ may be determined once and for all for a given value of n, apart from the simple factor N^r.

Suppose now the matrix A has the eigenvalues $1, 2, \ldots, 20$ and our estimate of the norm happens to be 10. The corresponding values of z_i are then $-10, -9, \ldots, +9, +10$. For the purpose of illustration we consider only the determination of the coefficient of z^{19} and we make the very favourable assumption that all the values v_i are computed with a maximum error of one part in 2^t. In Table 3 we give the orders

TABLE 3

z_i	-10	-9	-8	-7	-6	-5	-4	-3	-2	-1	0
$\|P_i^{(1)}\|$	2^{-58}	2^{-54}	2^{-51}	2^{-48}	2^{-46}	2^{-45}	2^{-44}	2^{-43}	2^{-43}	2^{-44}	0
$\|v_i\|$	2^{86}	2^{85}	2^{83}	2^{81}	2^{79}	2^{77}	2^{75}	2^{72}	2^{69}	2^{66}	2^{62}
$\|P_i^{(1)} v_i\|$	2^{28}	2^{31}	2^{32}	2^{33}	2^{33}	2^{32}	2^{31}	2^{29}	2^{26}	2^{22}	0

of magnitude of the factors $P_i^{(1)}$ and of the values v_i. Since $v_i = 0$ $(i = 11, \ldots, 21)$ we have omitted the corresponding $P_i^{(1)}$ though it is evident from symmetry that $P_i^{(1)} = P_{n-i}^{(1)}$. The true value of c_{19} is, of course, 210 and is therefore of order 2^8, but we see from Table 3 that some of the contributions to the sum $\sum P_i^{(1)} v_i$ are of order 2^{33}. Considerable cancellation therefore takes place when c_{19} is computed. The error in c_{19} arising from the error in $\det(A + 7I)$ alone will be of order 2^{33-t}. Hence, even with our very favourable assumption, the error in c_{19} can scarcely be less than one part in 2^{t-25}, and we have seen in § 5 that this is fatal to the accuracy of the computed eigenvalues unless t is very large.

The reader will find it an instructive exercise to investigate the errors in other coefficients, and to study the effect of varying the estimate of the norm both for this distribution of eigenvalues and for the other distributions discussed in § 5. In general, the results obtainable with this method are most unsatisfactory. Various modifications have been investigated including, in particular, the fitting of an expansion in terms of Chebyshev polynomials, but the results have been unimpressive. Our analysis shows that even when the computed values have errors of only one part in 2^t, the exact zeros of the exact polynomial taking the computed values may well be quite different from the true eigenvalues.

Le Verrier's method

19. An independent method of determining the characteristic equation is based on the observation that the eigenvalues of A^r are λ_i^r and hence the trace of A^r is equal to σ_r where

$$\sigma_r = \sum_{i=1}^{n} \lambda_i^r. \qquad (19.1)$$

If we compute $\sigma_r (r = 1,\dots, n)$ by this means, then the coefficients c_r may be derived using Newton's equations

$$\left.\begin{aligned}
c_{n-1} &= -\sigma_1, \\
rc_{n-r} &= -(\sigma_r + c_{n-1}\sigma_{r-1} + \dots + c_{n-r+1}\sigma_1) \\
(r &= 2,\dots, n). \quad (19.2)
\end{aligned}\right\} \qquad (19.2)$$

Again we find that it is common for severe cancellation to take place when the c_r are computed, as can be verified by estimating the orders of magnitude of the various contributions to c_r. Consider for example the calculation of c_0 for the matrix having the eigenvalues $1, 2,\dots, 20$. We have

$$c_0 = -(\sigma_{20} + c_{19}\sigma_{19} + \dots + c_1\sigma_1)/20. \qquad (19.3)$$

Now we know that c_0 is 20! while the first contribution on the right-hand side is $-\sigma_{20}/20$ which is approximately $-20^{20}/21$. The first contribution is therefore of opposite sign to c_0 and larger by a factor of approximately 2×10^6; some of the later contributions are even larger, so that an error of one part in 2^t in any of the σ_i or c_i produces a much larger relative error in the c_0 computed from it.

An even more striking illustration is provided by the distribution 2^{1-i} $(i = 1,\dots, 20)$. It is clear that all the σ_i satisfy the relation

$$1 < \sigma_i < 2, \qquad (19.4)$$

and the first contribution to c_{n-r} is $-\sigma_r/r$. Now the later c_i are very small, so that cancellation must be very severe; we cannot even guarantee to obtain the correct sign in these coefficients unless we use a very high precision in the computation.

Iterative methods based on interpolation

20. We now turn to the methods used to locate the zeros. We consider first those which are based on computed values of $f(z)$ and not on values of any of the derivatives. These are essentially methods of successive interpolation. (We do not distinguish here between 'interpolation' and 'extrapolation'.) In the most important group of methods in this class the next approximation z_{k+1} to a zero is determined from the $r+1$ previous approximations $z_k, z_{k-1}, ..., z_{k-r}$ as a zero of the polynomial of degree r passing through the points $(z_k, f(z_k)), ..., (z_{k-r}, f(z_{k-r}))$. Of the r such zeros z_{k+1} is taken to be that one which lies closest to z_k. If we take r to be 1 or 2 then the corresponding polynomials are linear and quadratic and the determination of the zeros is trivial. It is convenient to write

$$f(z_i) = f_i. \qquad (20.1)$$

For the case $r = 1$ we have

$$z_{k+1} = z_k + \frac{f_k(z_k - z_{k-1})}{f_{k-1} - f_k} = \frac{z_k f_{k-1} - z_{k-1} f_k}{f_{k-1} - f_k}. \qquad (20.2)$$

For the case $r = 2$, Muller (1956) has suggested a very convenient formulation of the computation. We write

$$h_i = z_i - z_{i-1}, \qquad \lambda_i = h_i/h_{i-1}, \qquad \delta_i = 1 + \lambda_i, \qquad (20.3)$$

and with this notation it may readily be verified that the required λ_{k+1} satisfies the quadratic equation

$$\lambda_{k+1}^2 \lambda_k \delta_k^{-1}(f_{k-2}\lambda_k - f_{k-1}\delta_k + f_k) + \lambda_{k+1}\delta_k^{-1}g_k + f_k = 0 \qquad (20.4)$$

where

$$g_k = f_{k-2}\lambda_k^2 - f_{k-1}\delta_k^2 + f_k(\lambda_k + \delta_k). \qquad (20.5)$$

From this we obtain

$$\lambda_{k+1} = -2f_k\delta_k/\{g_k \pm [g_k^2 - 4f_k\delta_k\lambda_k(f_{k-2}\lambda_k - f_{k-1}\delta_k + f_k)]^{\frac{1}{2}}\}. \qquad (20.6)$$

The sign in the denominator of (20.6) is chosen so as to give the corresponding λ_{k+1} (and hence h_{k+1}) the smaller absolute magnitude.

Clearly successive linear interpolation has the advantage of simplicity, but it suffers from the weakness that if z_k and z_{k-1} are real, and $f(z)$ is a real function, then all subsequent values are real. With successive quadratic interpolation we can move out into the complex plane even if we start with real values.

If we take r to be greater than two, each step of the iteration requires the solution of a polynomial equation of degree three or higher. Hence the use of cubic or higher interpolation could scarcely be justified unless it had superior convergence properties (cf. the end of § 21).

Asymptotic rate of convergence

21. We consider now the ultimate rate of convergence in the neighbourhood of a zero at $z = \alpha$. If we write

$$z - \alpha = w, \qquad z_i - \alpha = w_i, \tag{21.1}$$

then we have

$$f_i = f_i(\alpha + w_i) = f'w_i + \frac{1}{2!}f''w_i^2 + \frac{1}{3!}f'''w_i^3 + ..., \tag{21.2}$$

where $f', f'', ...$ denote the derivatives at $z = \alpha$. For the purposes of this investigation we may express the interpolation polynomial in terms of w. Now the polynomial of order r through the points (w_i, f_i) $(i = k, k-1, ..., k-r)$ is given by

$$\det \begin{bmatrix} 0 & w^r & w^{r-1} & ... & 1 \\ f_k & w_k^r & w_k^{r-1} & ... & 1 \\ f_{k-1} & w_{k-1}^r & w_{k-1}^{r-1} & ... & 1 \\ \cdots\cdots\cdots\cdots\cdots \\ f_{k-r} & w_{k-r}^r & w_{k-r}^{r-1} & ... & 1 \end{bmatrix} = 0. \tag{21.3}$$

Substituting from (21.2) for the f_i and expanding (21.3) by its first row we have

$$-\prod_{i<j}(w_i - w_j)\left[\frac{1}{r!}f^{(r)}w^r + \frac{1}{(r-1)!}f^{(r-1)}w^{r-1} + ... + \frac{1}{1!}f'w + \right.$$

$$\left. + (-1)^r \frac{1}{(r+1)!}f^{(r+1)}w_k w_{k-1}...w_{k-r}\right] = 0, \tag{21.4}$$

where in the coefficient of each w^s we have included only the dominant term. If the w_i are sufficiently small the required solution w_{k+1} is such that

$$w_{k+1} \sim (-1)^{r+1}f^{(r+1)}w_k w_{k-1}...w_{k-r}/[(r+1)!\,f']. \tag{21.5}$$

The nature of this solution justifies *a posteriori* our omission of all but the dominant terms in (21.4).

In order to assess the asymptotic behaviour we express (21.5) in the form

$$|w_{k+1}| = K\,|w_k|\,|w_{k-1}|...|w_{k-r}|, \tag{21.6}$$

that is $\qquad (K^{\frac{1}{r}} |w_{k+1}|) = (K^{\frac{1}{r}} |w_k|)(K^{\frac{1}{r}} |w_{k-1}|)\ldots(K^{\frac{1}{r}} |w_{k-r}|),$ $\qquad$ (21.7)

giving $\qquad u_{k+1} = u_k + u_{k-1} + \ldots + u_{k-r}, \qquad u_i = \log(K^{\frac{1}{r}} |w_i|).$ $\qquad$ (21.8)

From the theory of linear difference equations the solution of (21.8) is

$$u_k = \sum c_i \mu_i^k \qquad (21.9)$$

where the μ_i are the roots of

$$\mu^{r+1} = \mu^r + \mu^{r-1} + \ldots + 1. \qquad (21.10)$$

This equation has one root μ_1 between 1 and 2 which is such that

$$\mu_1 \to 2 \quad \text{as} \quad r \to \infty. \qquad (21.11)$$

The other r roots satisfy $z^{r+2} - 2z^{r+1} + 1 = 0$ and are all inside the unit circle, as can be seen by applying Rouché's theorem to the functions $\qquad\qquad z^{r+2} + 1 \quad \text{and} \quad 2z^{r+1}.$ $\qquad$ (21.12)

Hence $\qquad\qquad u_k \sim c_1 \mu_1^k, \qquad K^{\frac{1}{r}} |w_k| \sim e^{c_1 \mu_1^k}$

giving $\qquad\qquad (K^{\frac{1}{r}} |w_{k+1}|) \sim (K^{\frac{1}{r}} |w_k|)^{\mu_1},$ $\qquad$ (21.13)

$$|w_{k+1}| \sim K^{(\mu_1 - 1)/r} |w_k|^{\mu_1}. \qquad (21.14)$$

When $r = 1$ we have

$$\left.\begin{array}{l} K = |f''/2f'|, \qquad \mu_1 = \tfrac{1}{2}(1 + 5^{\frac{1}{2}}) \doteq 1\cdot 62 \\ |w_{k+1}| \sim K |w_k| |w_{k-1}|, \qquad |w_{k+1}| \sim K^{0\cdot 62} |w_k|^{1\cdot 62} \end{array}\right\}, \qquad (21.15)$$

and when $r = 2$ we have

$$\left.\begin{array}{l} K = |-f'''/6f'|, \qquad \mu_1 \doteq 1\cdot 84 \\ |w_{k+1}| \sim K |w_k| |w_{k-1}| |w_{k-2}|, \qquad |w_{k+1}| \sim K^{0\cdot 42} |w_k|^{1\cdot 84} \end{array}\right\}. \qquad (21.16)$$

The asymptotic rate of convergence is quite satisfactory in either case, but notice that a large value of K delays the onset of the limiting rate of convergence. Since μ_1 is less than 2 for all values of r, there would appear to be little justification for using interpolating polynomials of higher degree as far as the asymptotic convergence rate is concerned.

Multiple zeros

22. In the analysis of the previous paragraph we have assumed that $f' \neq 0$, so that α is a simple zero. We give the corresponding analysis for the cases $r = 1$ and 2 when α is a double zero. In this case we have

$$f_i = \frac{1}{2!} f'' w_i^2 + \frac{1}{3!} f''' w_i^3 + \ldots. \qquad (22.1)$$

When $r = 1$ the linear equation for w_{k+1} is

$$-\tfrac{1}{2}f''(w_k^2-w_{k-1}^2)w+\tfrac{1}{2}f''w_kw_{k-1}(w_k-w_{k-1}) = 0, \qquad (22.2)$$

giving
$$w_{k+1} \sim w_kw_{k-1}/(w_k+w_{k-1}). \qquad (22.3)$$

Writing $u_i = 1/w_i$, (22.3) becomes

$$u_{k+1} = u_k+u_{k-1}. \qquad (22.4)$$

The general solution is
$$u_k = a_1\mu_1^k+a_2\mu_2^k, \qquad (22.5)$$

where μ_1 and μ_2 satisfy the equation

$$\mu^2 = \mu+1. \qquad (22.6)$$

Hence we have finally

$$w_k = 1/u_k \sim 1/(a_1\mu_1^k) \doteq 1/a_1(1{\cdot}62)^k, \qquad (22.7)$$

so that ultimately the error is divided by approximately $1{\cdot}62$ in each step.

When $r = 2$ the quadratic equation for w_{k+1} is

$$-(w_k-w_{k-1})(w_k-w_{k-2})(w_{k-1}-w_{k-2}) \times$$
$$\times[\tfrac{1}{2}f''w^2-\tfrac{1}{6}f'''(w_kw_{k-1}+w_kw_{k-2}+w_{k-1}w_{k-2})w+\tfrac{1}{6}f'''w_kw_{k-1}w_{k-2}] = 0, \qquad (22.8)$$

where we have again included only the dominant term in each coefficient. In the limit the solution is given by

$$w_{k+1}^2 \sim (-f'''/3f'')w_kw_{k-1}w_{k-2}, \qquad (22.9)$$

i.e. $|w_{k+1}|^2 = K\,|w_k|\,|w_{k-1}|\,|w_{k-2}|$ where $K = |-f'''/3f''|$.   (22.10)

We justify this at the end of the section. Writing $u_i = \log K\,|w_i|$ we have
$$2u_{k+1} = u_k+u_{k-1}+u_{k-2}. \qquad (22.11)$$

The general solution is $u_k = c_1\mu_1^k+c_2\mu_2^k+c_3\mu_3^k,$ (22.12)

where μ_1, μ_2, μ_3 are the roots of

$$2\mu^3 = \mu^2+\mu+1. \qquad (22.13)$$

As in the previous paragraph we may show that two roots of this equation are inside the unit circle. The other, μ_1, is approximately $1{\cdot}23$. Hence we have finally

$$|w_{k+1}| \sim K^{\mu_1-1}\,|w_k|^{\mu_1} \doteq |-f'''/3f''|^{0{\cdot}23}\,|w_k|^{1{\cdot}23}. \qquad (22.14)$$

This result justifies *a posteriori* our neglect of the term in w in equation (22.8).

Inversion of the functional relationship

23. If $f(z)$ has a simple zero at $z = \alpha$ then z may be expressed as a convergent power series in f for all sufficiently small f. We write

$$z = g(f) \quad \text{where} \quad z = \alpha \quad \text{when} \quad f = 0. \qquad (23.1)$$

Instead of approximating to f by means of polynomials in z we may approximate to z by means of polynomials in f (see, for example, Ostrowski (1960)). The general case will be sufficiently illustrated by considering the use of polynomials of the second degree in f. For the approximation polynomial $L(f)$ through the points (z_k, f_k), (z_{k-1}, f_{k-1}), (z_{k-2}, f_{k-2}) expressed in Lagrangian form we have

$$L(f) = z_k \frac{(f-f_{k-1})(f-f_{k-2})}{(f_k-f_{k-1})(f_k-f_{k-2})} + z_{k-1} \frac{(f-f_k)(f-f_{k-2})}{(f_{k-1}-f_k)(f_{k-1}-f_{k-2})} +$$
$$+ z_{k-2} \frac{(f-f_k)(f-f_{k-1})}{(f_{k-2}-f_k)(f_{k-2}-f_{k-1})}. \qquad (23.2)$$

The standard theory of Lagrangian interpolation polynomials gives for the error

$$g(f) - L(f) = \frac{g'''(\eta)}{3!}(f-f_k)(f-f_{k-1})(f-f_{k-2}), \qquad (23.3)$$

where η is some point in the interval containing f, f_k, f_{k-1}, f_{k-2}. Taking $f = 0$ we have

$$\alpha - L(0) = -\frac{g'''(\eta)}{3!} f_k f_{k-1} f_{k-2}. \qquad (23.4)$$

Hence if we take z_{k+1} to be $L(0)$, the right-hand side of (23.3) gives us an expression for the error. From (23.2) we have

$$z_{k+1} = L(0) = f_k f_{k-1} f_{k-2} \left[\frac{z_k}{f_k(f_k-f_{k-1})(f_k-f_{k-2})} + \right.$$
$$\left. + \frac{z_{k-1}}{f_{k-1}(f_{k-1}-f_k)(f_{k-1}-f_{k-2})} + \frac{z_{k-2}}{f_{k-2}(f_{k-2}-f_k)(f_{k-2}-f_{k-1})} \right]. \qquad (23.5)$$

It is obvious that whatever the degree of the interpolation polynomial may be, z_{k+1} is unique. Moreover, if f is a real function and we start the iteration with real values of z_i, all iterates are real. This contrasts with our previous technique which involved the solution of a polynomial equation and thus enabled us to move out into the complex plane.

Now we have

$$f_i = f(z_i) \sim (z_i - \alpha) f'(\alpha) \quad \text{as} \quad z_i \to \alpha. \qquad (23.6)$$

Hence (23.4) gives

$$(z_{k+1}-\alpha) \sim \frac{g'''(\eta)}{3!}[f'(\alpha)]^3(z_k-\alpha)(z_{k-1}-\alpha)(z_{k-2}-\alpha), \qquad (23.7)$$

and writing $z_i-\alpha = w_i$ we have

$$w_{k+1} \sim Kw_kw_{k-1}w_{k-2} \quad \text{where} \quad K = -g'''(\eta)[f'(\alpha)]^3/3!. \qquad (23.8)$$

This is a relation of the same form as (21.16) and shows that ultimately

$$|w_{k+1}| \sim K^{0\cdot42}|w_k|^{1\cdot84}. \qquad (23.9)$$

In the general case, when we use an interpolation polynomial of degree r, we have

$$w_{k+1} \sim Kw_kw_{k-1}...w_{k-r} \quad \text{where} \quad K = -g^{r+1}(\eta)[f'(\alpha)]^{r+1}/(r+1)!. \qquad (23.10)$$

Notice that if $r = 1$ this method is identical with the successive linear interpolation discussed in § 20. There is little advantage in taking r greater than two, since as we saw in § 21, corresponding to the relation (23.10), the ultimate rate of convergence is never quite quadratic however large r may be.

If α is a zero of multiplicity m then we have

$$f = c_m(z-\alpha)^m+c_{m+1}(z-\alpha)^{m+1}+... , \qquad (23.11)$$

and $z-\alpha$ is expressible as a power series in $f^{1/m}$. It is obvious that interpolation is far less satisfactory in this case.

In our experience the methods described in this section are less satisfactory in connexion with the eigenvalue problem than those of the previous section. (On the other hand, they have proved very satisfactory for the accurate determination of zeros of Bessel functions *starting from good approximations*.) Hence in the comparisons made in § 26 we do not discuss the methods of this section.

The method of bisection

24. To complete our account of methods which depend on function values alone we mention the method of bisection. We have already discussed this in Chapter 5 in connexion with the eigenvalues of symmetric tri-diagonal matrices, but it is obvious that it can be used in a more general context to find real zeros of a real function. Before we can apply it we must determine by independent means values a and b such that $f(a)$ and $f(b)$ are of opposite signs. Once this has been done a zero is located in an interval of width $(b-a)/2^k$ in k steps. The problem of convergence does not arise.

Newton's method

25. The analysis of §§ 21–23 is concerned only with the ultimate rate of convergence, and we have considered this in detail only for zeros of multiplicities one and two. Furthermore it ignores the effect of rounding errors. In general we shall not be able to start the iterative process with close approximations to the zeros, and the computed values will in any case be subject to rounding errors. We discuss these topics in §§ 59–60 and §§ 35–47 respectively when we have considered other iterative methods.

We turn now to those methods in which values of derivatives are used in addition to functional values. The best known of these is Newton's method, in which the $(k+1)$th approximation is related to the kth by the equation

$$z_{k+1} = z_k - f(z_k)/f'(z_k). \qquad (25.1)$$

In the neighbourhood of a simple zero at $z = \alpha$ we may write

$$w = z - \alpha, \qquad f(z) = f(w+\alpha) = f'w + \tfrac{1}{2}f''w^2 + \dots, \qquad (25.2)$$

where the derivatives are evaluated at $z = \alpha$. Hence we have

$$w_{k+1} = w_k - \frac{f'w_k + \tfrac{1}{2}f''w_k^2 + \tfrac{1}{6}f'''w_k^3 + \dots}{f' + f''w_k + \tfrac{1}{2}f'''w_k^2 + \dots} \qquad (25.3)$$

$$= \frac{\tfrac{1}{2}f''w_k^2 + \tfrac{1}{3}f'''w_k^3 + \dots}{f' + f''w_k + \tfrac{1}{2}f'''w_k^2 + \dots}, \qquad (25.4)$$

giving

$$w_{k+1} \sim (f''/2f')w_k^2 \qquad (f' \neq 0). \qquad (25.5)$$

For a double zero we have $f' = 0, f'' \neq 0$ and hence from (25.3)

$$w_{k+1} = \tfrac{1}{2}w_k + \frac{\tfrac{1}{12}f'''w_k^3 + \dots}{f''w_k + \dots}, \qquad (25.6)$$

giving

$$w_{k+1} \sim \tfrac{1}{2}w_k. \qquad (25.7)$$

Similarly for a zero of multiplicity r we may show that

$$w_{k+1} \sim (1 - 1/r)w_k, \qquad (25.8)$$

so that convergence becomes progressively slower as the multiplicity increases. It is evident that for a zero of multiplicity r the correction made at each stage is too small by a factor $1/r$. This suggests that (25.1) be replaced by

$$z_{k+1} = z_k - rf(z_k)/f'(z_k) \qquad (25.9)$$

for such a zero, and in fact (25.9) gives

$$w_{k+1} \sim \frac{f^{(r+1)}}{r(r+1)f^{(r)}} w_k^2. \qquad (25.10)$$

Unfortunately we do not usually have *a priori* information on the multiplicity of zeros.

Comparison of Newton's method with interpolation

26. For each of the condensed matrix forms the computation of the first derivative involves about the same amount of work as the computation of the function itself. Moreover if the matrix is of reasonably high order we may assume that the work involved in a single application of linear or quadratic interpolation or of Newton's method is negligible compared with the evaluation of a functional value or a derivative.

On this basis a reasonable comparison is between two steps of either linear or quadratic interpolation, and one step of Newton's method. For two steps of linear interpolation in the neighbourhood of a simple zero we have from (21.15)

$$w_{k+2} \sim K^{0\cdot62}(K^{0\cdot62}w_k^{1\cdot62})^{1\cdot62} \doteqdot K^{1\cdot62}w_k^{2\cdot62} \quad \text{where} \quad K = f''/2f', \quad (26.1)$$

while for Newton's method we have

$$w_{k+1} \doteqdot Kw_k^2, \quad (26.2)$$

with the same value of K. Clearly linear interpolation gives the higher asymptotic rate of convergence.

In the neighbourhood of a double zero we have for two steps of linear interpolation

$$w_{k+2} \sim w_k/(1\cdot62)^2 \doteqdot w_k/2\cdot62 \quad (26.3)$$

and once more this compares favourably with Newton's method. Newton's method shares one disadvantage with linear interpolation; if $f(z)$ is a real function and we start with a real value of z_1, then all z are real and we cannot determine a complex zero.

Successive quadratic interpolation emerges even more favourably from such a comparison. For a simple zero, two steps of quadratic interpolation give

$$|w_{k+2}| \sim K^{1\cdot19} |w_k|^{3\cdot39} \quad \text{where} \quad K = |-f'''/6f'|, \quad (26.4)$$

while for a double zero, two steps give

$$|w_{k+2}| \sim K^{0\cdot51} |w_k|^{1\cdot5} \quad \text{where} \quad K = |-f'''/3f''|. \quad (26.5)$$

These comparisons were based on the assumption that the computation of a derivative was strictly comparable with the computation of the function itself. For a matrix A of general form this is certainly not true of the function $\det(A-zI)$; the evaluation of a derivative involves far more work than that of a function value.

The contrast is even more marked for the generalized eigenvalue problem for which the relevant function is of the form

$$\det(A_r z^r + A_{r-1} z^{r-1} + \ldots + A_0).$$

For such problems the balance is tipped even further in the favour of interpolation methods.

Methods giving cubic convergence

27. Methods can be derived having asymptotic convergence of any required order provided we are prepared to evaluate the appropriate number of derivatives. A simple method of deriving such formulae is by reversion of the Taylor-series expansion. We have

$$f(z_k + h) = f(z_k) + hf'(z_k) + \tfrac{1}{2}h^2 f''(z_k) + O(h^3). \tag{27.1}$$

Hence if h is to be chosen so that $f(z_k + h) = 0$ we have

$$z_{k+1} - z_k = h = -f(z_k)/f'(z_k) - h^2 f''(z_k)/2f'(z_k) + O(h^3)$$
$$\sim -f(z_k)/f'(z_k) - [f(z_k)]^2 f''(z_k)/2[f'(z_k)]^3 \tag{27.2}$$

and it may readily be verified that this gives cubic convergence. This formula seems to have been little used.

Laguerre's method

28. A more interesting formula which also gives cubic convergence in the neighbourhood of a simple zero is that due to Laguerre, who developed his method in connexion with polynomials having only real zeros. The motivation is as follows:

Let $f(x)$ have the real zeros $\lambda_1 < \lambda_2 < \ldots < \lambda_n$ and let x lie between λ_m and λ_{m+1}. When x lies outside the (λ_1, λ_n) interval we may regard the real axis as joined from $-\infty$ to $+\infty$ and say that x lies between λ_n and λ_1. Consider now the quadratic $g(X)$ in X defined for any real u by

$$g(X) = (x - X)^2 \left\{ \sum_{i=1}^{n} (u - \lambda_i)^2 / (x - \lambda_i)^2 \right\} - (u - X)^2. \tag{28.1}$$

Clearly if $u \neq x$ we have

$$g(\lambda_i) > 0 \quad (i = 1, \ldots, n), \qquad g(x) < 0. \tag{28.2}$$

Hence for any real $u \neq x$, the function $g(X)$ has two zeros $X'(u)$ and $X''(u)$ such that

$$\lambda_m < X' < x < X'' < \lambda_{m+1}. \tag{28.3}$$

(In the case when x lies outside the (λ_1, λ_n) interval, we have an X' lying 'between' λ_n and x and an X'' lying 'between' x and λ_1.) We may regard X' and X'' as better approximations to the two zeros neighbouring x than was x itself. Consider now the minimum value of X' and the maximum value of X'' attainable for any real u. In general these maxima and minima will be attained for different values of u. For the extreme values we have $\dfrac{\partial X'}{\partial u} = 0 = \dfrac{\partial X''}{\partial u}$. If we write $g(X) \equiv Au^2+2Bu+C$ where A, B, C are functions of X, then for extreme values of the roots we have

$$2Au+2B = 0 \quad \left(\text{since } \frac{\partial A}{\partial u} = \frac{\partial B}{\partial u} = \frac{\partial C}{\partial u} = 0 \quad \text{because } \frac{\partial X}{\partial u} = 0\right),$$

$$(28.4)$$

giving $B^2 = AC$. We may readily verify by an algebraic argument that the vanishing of the discriminant does indeed give the extreme values.

Observing that

$$(f'/f) = \sum 1/(x-\lambda_i) = \Sigma_1, \tag{28.5}$$

$$(f'/f)^2-(f''/f) = \sum 1/(x-\lambda_i)^2 = \Sigma_2, \tag{28.6}$$

so that $\qquad \sum \lambda_i/(x-\lambda_i)^2 = x\,\Sigma_2 - \Sigma_1,$

we see after a little algebraic manipulation that the relation $B^2 = AC$ reduces to

$$n-2\,\Sigma_1\,(x-X)+[\Sigma_1^2 -(n-1)\,\Sigma_2](x-X)^2 = 0. \tag{28.7}$$

Accordingly Laguerre gives the two values

$$X = x - \frac{n}{\Sigma_1 \pm[n(n-1)\,\Sigma_2 -(n-1)\,\Sigma_1^2]^{\frac{1}{2}}}$$

$$= x - \frac{nf}{f' \pm[(n-1)^2(f')^2 -n(n-1)ff'']^{\frac{1}{2}}}, \tag{28.8}$$

one of which is nearer to λ_m and the other to λ_{m+1}. Although the motivation applied only to polynomials with real zeros we may immediately extend the results to the complex plane giving the iterative process

$$z_{k+1} = z_k -nf_k/(f'_k \pm H_k^{\frac{1}{2}}) \quad \text{where} \quad H_k = (n-1)^2(f'_k)^2 -n(n-1)f_kf''_k. \tag{28.9}$$

In the neighbourhood of a simple zero at λ_m (28.9) implies that *if we choose the sign so that the denominator has the larger absolute value* then

$$z_{k+1} - \lambda_m \sim \tfrac{1}{2}(z_k - \lambda_m)^3 [(n-1) \Sigma'_2 - (\Sigma'_1)^2]/(n-1), \qquad (28.10)$$

where

$$\Sigma'_2 = \sum_{i \neq m} 1/(\lambda_m - \lambda_i)^2, \qquad \Sigma'_1 = \sum_{i \neq m} 1/(\lambda_m - \lambda_i). \qquad (28.11)$$

The convergence is therefore cubic in the neighbourhood of a simple zero. It is interesting that in the verification of the cubic nature of the convergence no use is made of the fact that the n occurring in the formulae (28.9) is the degree of f and the process has cubic convergence for any other value of n.

On the other hand if λ_m is a zero of multiplicity r it may be verified that (28.9) implies that

$$z_{k+1} - \lambda_m \sim (r-1)(z_k - \lambda_m)/\{r + [(n-1)r/(n-r)]^{\frac{1}{2}}\} \qquad (28.12)$$

so that the convergence is merely linear.

It is easy to modify (28.9) so as to obtain cubic convergence in the neighbourhood of a zero of multiplicity r. The appropriate formula is

$$z_{k+1} = z_k - nf_k \Big/ \left[f'_k + \left\{ \frac{n-r}{r}[(n-1)(f'_k)^2 - nf_k f''_k] \right\}^{\frac{1}{2}} \right], \qquad (28.13)$$

which reduces to (28.9) when $r = 1$. Unfortunately we must be able to recognize the multiplicity of the zero which is being approached in order to take advantage of this formula.

It is clear that in the case when all the zeros are real the selection of a fixed sign in the formula (28.9) gives a monotonic sequence tending to one or other of the zeros neighbouring the initial approximation. There appears to be no analogous result for polynomials having complex zeros.

Again if we assume that the computation of f'' is comparable with that of f and f', two steps of either of these cubically convergent processes should be compared with three steps of Newton's process. For two steps of the cubically convergent processes we have

$$w_{k+2} \sim A w_k^9, \qquad (28.14)$$

while for three steps of Newton's process we have

$$w_{k+3} \sim B w_k^8. \qquad (28.15)$$

Newton's process is therefore inferior as far as the asymptotic behaviour is concerned.

Complex zeros

29. Up to the present we have made only passing references to complex zeros. Most of the techniques we have discussed, both for the evaluation of $f(z)$ and its derivatives and for the computation of the zeros, carry over immediately to the computation of complex eigenvalues of complex matrices. We merely replace real arithmetic by complex arithmetic; since a complex multiplication involves four real multiplications there is usually about four times as much computation in the complex case.

However, complex matrices are comparatively rare in practice. A more common situation is for the matrix to be real but for some of the eigenvalues to consist of complex conjugate pairs. Naturally we could deal with this situation by treating all quantities as complex, but we would expect to be able to take some advantage of the special circumstances to reduce the volume of computation.

Notice first that if we locate a complex zero we know immediately that the complex conjugate is a zero. Hence, provided the volume of work in one complex iteration is not greater than twice that in a real iteration, we can find a pair of complex conjugate zeros in the same time as two real zeros.

If we evaluate an explicit polynomial for a complex value of z by nested multiplication (cf. § 2) then every value of s_r except s_n is complex, and true complex multiplication is involved at each stage. A single evaluation therefore takes four times as long, and the fact that the a_r are real is of little advantage. We show how to overcome this in the next section.

For real tri-diagonal matrices (cf. § 8) writing

$$z = x+iy, \qquad p_r(z) = s_r(z)+it_r(z), \tag{29.1}$$

equation (8.2) becomes

$$\left.\begin{aligned}
s_r(z) &= (\alpha_r-x)s_{r-1}(z)+yt_{r-1}(z)-\beta_r\gamma_r s_{r-2}(z) \\
t_r(z) &= (\alpha_r-x)t_{r-1}(z)-ys_{r-1}(z)-\beta_r\gamma_r t_{r-2}(z)
\end{aligned}\right\}. \tag{29.2}$$

Thus six multiplications are required compared with two in the real case.

Finally for Hessenberg matrices (cf. § 11), writing

$$x_r = u_r+iv_r, \qquad z = x+iy, \tag{29.3}$$

equation (11.5) is replaced by

$$\left.\begin{aligned}
b_r u_{r-1}+(a_{rr}-x)u_r+a_{r,r+1}u_{r+1}+\ldots+a_{rn}u_n+yv_r &= 0 \\
b_r v_{r-1}+(a_{rr}-x)v_r+a_{r,r+1}v_{r+1}+\ldots+a_{rn}v_n-yu_r &= 0
\end{aligned}\right\}. \tag{29.4}$$

Apart from the presence of the additional terms yv_r and $-yu_r$ there is exactly twice the volume of computation. The same comment applies to the differentiated equation (11.7). Hence in this case we can find complex conjugate pairs of zeros with about the same efficiency as two real zeros without effectively modifying our procedure.

30. It might be thought that the error analysis which we gave in §§ 12, 13 does not carry over to the complex case, since it would appear that each of the equations in (29.4) determines independently an equivalent perturbation in each of the a_{rs}, and that these will not in general be the same. However, this difficulty proves to be illusory and is met by permitting a complex perturbation in each a_{rs}. To avoid a multiplicity of suffices we concentrate our attention on a specific element a_{rs}, and we assume that the equivalent perturbations corresponding to the rounding errors are respectively $a_{rs}\epsilon_1$ and $a_{rs}\epsilon_2$ in the first and second of equations (29.4). We wish to show that there exist satisfactory η_1 and η_2 so that

$$a_{rs}(\eta_1+i\eta_2)(u_s+iv_s) = a_{rs}u_s\epsilon_1+ia_{rs}v_s\epsilon_2. \tag{30.1}$$

Clearly we have
$$(\eta_1^2+\eta_2^2)(u_s^2+v_s^2) = u_s^2\epsilon_1^2+v_s^2\epsilon_2^2, \tag{30.2}$$

giving
$$\eta_1^2+\eta_2^2 < \epsilon_1^2+\epsilon_2^2. \tag{30.3}$$

It can be shown that the error bounds which we gave for floating-point operations with real numbers carry over to complex work without any essential modifications. The corresponding results are

$$\left.\begin{array}{ll} fl(z_1\pm z_2) \equiv (z_1+z_2)(1+\epsilon_1), & |\epsilon_1| < 2^{-t} \\[4pt] fl(z_1z_2) \equiv z_1z_2(1+\epsilon_2), & |\epsilon_2| < 2.2^{\frac{1}{2}}2^{-t_1} \\[4pt] fl(z_1/z_2) \equiv z_1(1+\epsilon_3)/z_2, & |\epsilon_3| < 5.2^{\frac{1}{2}}2^{-t_1} \end{array}\right\} \tag{30.4}$$

where the ϵ_i are now, in general, complex. In these results we have assumed that the rounding errors in the real arithmetic which these complex computations involve are of the type discussed in Chapter 3. The proof of these results is left as an exercise.

Complex conjugate zeros

31. We now show that we can evaluate a polynomial with real coefficients for a complex value of the argument in approximately $2n$ multiplications. We observe first that $f(\alpha)$ is the remainder when

$f(z)$ is divided by $z-\alpha$. This suggests that $f(\alpha+i\beta)$ may be obtained from the remainder when $f(z)$ is divided by z^2-pz-l where

$$[z-(\alpha+i\beta)][z-(\alpha-i\beta)] = z^2-pz-l. \tag{31.1}$$

In fact if

$$f(z) \equiv (z^2-pz-l)q(z)+rz+s \tag{31.2}$$

then

$$f(\alpha+i\beta) = (r\alpha+s)+i(r\beta). \tag{31.3}$$

If we write

$$\left. \begin{aligned} f(z) &= a_n z^n + a_{n-1} z^{n-1} + \ldots + a_0 \\ q(z) &= q_n z^{n-2} + q_{n-1} z^{n-3} + \ldots + q_2 \end{aligned} \right\}, \tag{31.4}$$

then, equating powers of z on both sides of (31.2) and rearranging, we have

$$\left. \begin{aligned} q_n &= a_n, \qquad q_{n-1} = a_{n-1}+pq_n \\ q_s &= a_s+pq_{s+1}+lq_{s+2} \quad (s = n-2,\ldots, 2) \\ r &= a_1+pq_2+lq_3, \qquad s = a_0+lq_2 \end{aligned} \right\}. \tag{31.5}$$

These equations may be used to determine successively q_n, $q_{n-1},\ldots,$ q_2, r, s. Notice that if we continue the recurrence relation to give

$$q_1 = a_1+pq_2+lq_3 \tag{31.6}$$

$$q_0 = a_0+pq_1+lq_2 \tag{31.7}$$

then

$$r = q_1, \quad \text{and} \quad s = q_0-pq_1 = q_0' \ (\text{say}). \tag{31.8}$$

It is then convenient to write (31.2) in the form

$$f(z) \equiv (z^2-pz-l)q(z)+q_1 z+q_0'. \tag{31.9}$$

By means of the recurrence relations and equations (31.3) we may evaluate $f(\alpha+i\beta)$ in approximately $2n$ real multiplications.

32. A similar device may be used to evaluate $f'(\alpha+i\beta)$. To do this we first divide $q(z)$ by z^2-pz-l. If we write

$$q(z) \equiv (z^2-pz-l)T(z)+T_1 z+T_0', \tag{32.1}$$

$$T(z) \equiv T_{n-2} z^{n-4}+T_{n-3} z^{n-5}+\ldots+T_2, \tag{32.2}$$

then we have by a similar argument

$$\left. \begin{aligned} T_{n-2} &= q_n, \qquad T_{n-3} = q_{n-1}+pT_{n-2} \\ T_s &= q_{s+2}+pT_{s+1}+lT_{s+2} \quad (s = n-4,\ldots, 0) \\ T_0' &= T_0-pT_1. \end{aligned} \right\} \tag{32.3}$$

If the two sets of recurrence relations are carried out *pari passu* then we do not need to store the intermediate q_s. Differentiating (31.9) we have

$$f'(z) = (z^2 - pz - l)q'(z) + (2z - p)q(z) + q_1, \qquad (32.4)$$

$$f'(\alpha + i\beta) = (2\alpha + 2i\beta - p)q(\alpha + i\beta) + q_1, \qquad (32.5)$$

and from (32.1) we find

$$q(\alpha + i\beta) = T_1\alpha + T_0' + iT_1\beta. \qquad (32.6)$$

Hence remembering that $2\alpha = p$ we have

$$f'(\alpha + i\beta) = 2i\beta(T_1\alpha + T_0' + iT_1\beta) + q_1. \qquad (32.7)$$

Similarly we may obtain $f''(\alpha + i\beta)$ if we divide $T(z)$ by $z^2 - pz - l$ and so on.

Bairstow's method

33. In the last section we introduced division by $z^2 - pz - l$ as a means of evaluating $f(z)$ for a complex value of the argument. Notice though, that the recurrence relations for the q_r enable us to evaluate $f(z)$ at z_1 and z_2, the zeros of $z^2 - pz - l$, independently of whether they are a complex conjugate pair or real.

In Bairstow's method (1914) we iterate directly for real quadratic factors of $f(z)$. Differentiating (31.9) with respect to l and p respectively we have

$$0 = -q(z) + (z^2 - pz - l)\frac{\partial}{\partial l}q(z) + \frac{\partial q_1}{\partial l}z + \frac{\partial q_0'}{\partial l} \qquad (33.1)$$

$$0 = -zq(z) + (z^2 - pz - l)\frac{\partial}{\partial p}q(z) + \frac{\partial q_1}{\partial p}z + \frac{\partial q_0'}{\partial p} \qquad (33.2)$$

which may be written in the form

$$\frac{\partial q_1}{\partial l}z + \frac{\partial q_0'}{\partial l} \equiv q(z) \qquad \mathrm{mod}(z^2 - pz - l), \qquad (33.3)$$

$$\frac{\partial q_1}{\partial p}z + \frac{\partial q_0'}{\partial p} \equiv zq(z) \qquad \mathrm{mod}(z^2 - pz - l). \qquad (33.4)$$

Hence from (32.1) we obtain

$$\frac{\partial q_1}{\partial l}z + \frac{\partial q_0'}{\partial l} \equiv T_1z + T_0', \qquad (33.5)$$

$$\frac{\partial q_1}{\partial p}z + \frac{\partial q_0'}{\partial p} \equiv z(T_1z + T_0') \equiv z(pT_1 + T_0') + lT_1. \qquad (33.6)$$

Using Newton's method to improve an approximate zero of $q_1 z + q_0'$ we choose δp and δl so that

$$(q_1 z + q_0') + \frac{\partial}{\partial p}(q_1 z + q_0')\,\delta p + \frac{\partial}{\partial l}(q_1 z + q_0')\,\partial l = 0 \qquad (33.7)$$

for all z, and hence

$$q_1 + (pT_1 + T_0')\,\delta p + T_1\,\delta l = 0, \qquad (33.8)$$

$$q_0' + l T_1\,\delta p + T_0'\,\delta l = 0. \qquad (33.9)$$

The solution may be expressed in the form

$$\left.\begin{aligned} D\,\delta p &= T_1 q_0 - T_0 q_1, \quad D\,\delta l = M q_1 - T_0 q_0 \\ M &= l T_1 + p T_0, \qquad\qquad D = T_0^2 - M T_1 \end{aligned}\right\}. \qquad (33.10)$$

where

Since terms in δp^2, $\delta p\,\delta l$, and δl^2 and higher have been omitted in (33.7) this method gives quadratic convergence to a quadratic factor. Even when it converges to a quadratic factor with complex zeros, Bairstow's method is not identical with that of Newton. In particular if we apply it to a polynomial of degree two, then it converges to the correct factor from an arbitrary start in one iteration. Newton's method merely gives quadratic convergence.

The generalized Bairstow method

34. Bairstow's method is usually discussed in connexion with explicit polynomials. However, we can apply it to a polynomial expressed in any form provided we can compute the remainders corresponding to two successive divisions by $z^2 - pz - l$.

For a tri-diagonal matrix we can obtain simple recurrence relations for the remainders obtained when the leading principal minors of $(A - zI)$ are divided by $z^2 - pz - l$. If we write

$$\left.\begin{aligned} p_r(z) &\equiv (z^2 - pz - l)q_r(z) + A_r z + B_r \\ q_r(z) &\equiv (z^2 - pz - l)T_r(z) + C_r z + D_r \end{aligned}\right\}, \qquad (34.1)$$

then from the relation

$$p_r(z) = (\alpha_r - z)p_{r-1}(z) - \beta_r \gamma_r p_{r-2}(z) \qquad (34.2)$$

we have on substituting for the $p_i(z)$ from (34.1) and equating coefficients of $z^2 - pz - l$

$$q_r(z) = (\alpha_r - z)q_{r-1}(z) - \beta_r \gamma_r q_{r-2}(z) - A_{r-1} \qquad (34.3)$$

and similarly

$$T_r(z) = (\alpha_r - z)T_{r-1}(z) - \beta_r \gamma_r T_{r-2}(z) - C_{r-1}. \qquad (34.4)$$

These equations give

$$
\left.
\begin{aligned}
A_r &= (\alpha_r - p)A_{r-1} - \beta_r\gamma_r A_{r-2} - B_{r-1} \\
B_r &= \alpha_r B_{r-1} - \beta_r\gamma_r B_{r-2} - lA_{r-1} \\
C_r &= (\alpha_r - p)C_{r-1} - \beta_r\gamma_r C_{r-2} - D_{r-1} \\
D_r &= \alpha_r D_{r-1} - \beta_r\gamma_r D_{r-2} - lC_{r-1} - A_{r-1}
\end{aligned}
\right\}. \qquad (34.5)
$$

Since $p_0(z) = 1$, $q_0(z) = 0$, $p_1(z) = \alpha_1 - z$, $q_1(z) = 0$ we have

$$
A_0 = 0, \; B_0 = 1, \; A_1 = -1, \; B_1 = \alpha_1; \; C_0 = D_0 = C_1 = D_1 = 0.
$$
$$(34.6)$$

Notice that the first two of equations (34.5) enable us to determine $p_n(\alpha + i\beta)$ in $5n$ multiplications as compared with the $6n$ required if we used equations (29.2). The quantities A_n, B_n, C_n, D_n are respectively the q_1, q_0', T_1, T_0' of the previous section.

We may obtain corresponding relations for the leading principal minors of $(A - zI)$ when A is upper Hessenberg. To simplify the notation we assume that the sub-diagonal elements are unity. We have then

$$
p_r(z) = (a_{rr} - z)p_{r-1}(z) - a_{r-1,r}p_{r-2}(z) +
$$
$$
+ a_{r-2,r}p_{r-3}(z) + \ldots + (-1)^{r-1}a_{1r}p_0(z), \quad (34.7)
$$

where $p_0(z) = 1$.

Defining $q_r(z)$, A_r, B_r, C_r, D_r as in (34.1), the $q_r(z)$ satisfy the relations

$$
q_r(z) = (a_{rr} - z)q_{r-1}(z) - a_{r-1,r}q_{r-2}(z) +
$$
$$
+ a_{r-2,r}q_{r-3}(z) + \ldots + (-1)^{r-1}a_{1r}q_0(z) - A_{r-1}, \quad (34.8)
$$

and the A_r, B_r, C_r, D_r satisfy the relations

$$
\left.
\begin{aligned}
A_r &= (a_{rr} - p)A_{r-1} - a_{r-1,r}A_{r-2} + a_{r-2,r}A_{r-3} + \ldots + (-1)^{r-1}a_{1r}A_0 - B_{r-1} \\
B_r &= a_{rr}B_{r-1} - a_{r-1,r}B_{r-2} + a_{r-2,r}B_{r-3} + \ldots + (-1)^{r-1}a_{1r}B_0 - lA_{r-1} \\
C_r &= (a_{rr} - p)C_{r-1} - a_{r-1,r}C_{r-2} + a_{r-2,r}C_{r-3} + \ldots + (-1)^{r-1}a_{1r}C_0 - D_{r-1} \\
D_r &= a_{rr}D_{r-1} - a_{r-1,r}D_{r-2} + a_{r-2,r}D_{r-3} + \ldots \\
&\qquad \ldots + (-1)^{r-1}a_{1r}D_0 - lC_{r-1} - A_{r-1}
\end{aligned}
\right\}
$$
$$(34.9)$$

where

$$
A_0 = 0, \; B_0 = 1, \; A_1 = -1, \; B_1 = a_{11}; \; C_0 = D_0 = C_1 = D_1 = 0.
$$
$$(34.10)$$

There are approximately $2n^2$ multiplications in an iteration.

Handscomb (1962) has given a slightly different formulation of the recurrence relations.

Practical considerations

35. Up to the present our discussion of the iterative methods has made no concessions to the exigencies of numerical computation. Our discussion of the asymptotic behaviour of the convergence would be realistic only if it were customary to compute eigenvalues to some prodigious number of significant figures. In practice we are seldom interested in more than 10 correct significant decimals, though we may have to work to much higher precision in order to attain them.

A common way of expressing the behaviour of methods having quadratic or cubic convergence is to say that 'ultimately the number of figures is doubled or trebled with each iteration'. Such a comment might well be made by a 'classical' analyst but is inappropriate in the field of practical numerical analysis. Suppose, for example, that using a cubically convergent process the ultimate behaviour of the error w_i is described by the relation

$$w_{k+1} \sim 10^4 w_k^3. \tag{35.1}$$

If we take $w_k = 10^{-3}$ then we have

$$w_{k+1} \doteqdot 10^{-5}, \quad w_{k+2} \doteqdot 10^{-11}, \quad w_{k+3} \doteqdot 10^{-29}, \tag{35.2}$$

assuming that the asymptotic relation is substantially true when $w_k = 10^{-3}$. Now in practice we cannot expect to have $w_{k+3} = 10^{-29}$ unless we are using more than 29 decimal places in our function values; the iteration is quite likely to be stopped at w_{k+2}, and if we proceed to w_{k+3} the accuracy will probably be limited by the precision of the computation. This is true even if the computed values of $f(z)$ and the derivatives are correct to working accuracy. Almost as soon as we reach a value of w_k which is small enough for the asymptotic behaviour to come into force we find that the precision of the computation limits our ability to take advantage of it.

The attainable accuracy is further limited by our ability to evaluate $f(z)$ itself in the neighbourhood of a zero. An examination of the formulae we have given reveals that in every case the 'correction', $z_{k+1} - z_k$, is directly proportional to the current computed value of $f(z_k)$. The computed correction will therefore not bear any relation to the true correction unless $f(z_k)$ has some correct significant figures.

Effect of rounding errors on asymptotic convergence

36. For each of the special matrix forms we have found bounds for the perturbation δA in A which is equivalent to the errors made in evaluating $\det(A - zI)$. In each case the bound for $\|\delta A\|_E$ is proportional to 2^{-t} and hence the perturbations may be made as small as possible by working to a sufficiently high precision. It must be emphasized that δA is a function of z *and* of the algorithm used to evaluate $\det(A - zI)$.

If the eigenvalues of A are denoted by λ_i and those of $(A + \delta A)$ by λ_i', then from the continuity of the eigenvalues we know that the λ_i' can be arranged so that

$$\lambda_i' \to \lambda_i \quad \text{as} \quad t \to \infty.$$

However, unless t is sufficiently large, the eigenvalues of $(A + \delta A)$ may not have much in common with those of A. The Frobenius form with eigenvalues $1, 2, \ldots, 20$ provides a good example of this; we have seen (equation (5.3)) that unless 2^{-t} is smaller than 10^{-14}, the eigenvalues of the perturbed matrix can scarcely be regarded as approximations to those of the original matrix.

The method of bisection

37. We consider first the limitations of the method of bisection when used to locate a simple real eigenvalue λ_p of a real matrix. We assume that we have located two points a and b at which the computed values of $\det(A - zI)$ are of opposite signs. Corresponding to our evaluation of $\det(A - zI)$ for a specific real value of z we obtain $\det(A + \delta A - zI)$, where δA is real and

$$\det(A + \delta A - zI) = \prod_{i=1}^{n} (\lambda_i' - z). \tag{37.1}$$

For each of the compact forms we have given bounds for δA, and for any permissible δA each corresponding λ_i' will lie in a connected domain containing λ_i. We may call this domain the *domain of indeterminacy* associated with λ_i. Notice that this domain is dependent on t *and the algorithm used for evaluating* $\det(A - zI)$. We assume that t is sufficiently large for the domain containing λ_p to be isolated from the other domains. In this case all λ_p' must, of course, be real and the domain containing λ_p is an interval on the real axis.

A typical situation is illustrated in Fig. 1. The eigenvalue in question is λ_5, and this remains real for all perturbations δA satisfying the relevant bounds. The other eigenvalues behave as follows.

λ_1 is a double eigenvalue which becomes a complex conjugate pair for some δA.

λ_3 and λ_4 are close eigenvalues which become a complex conjugate pair for some δA.

λ_6 and λ_7 are real eigenvalues which remain real.

λ_8 and λ_9 are a complex conjugate pair which remain a complex conjugate pair. The diameter of the domain containing λ_i is a measure of its condition.

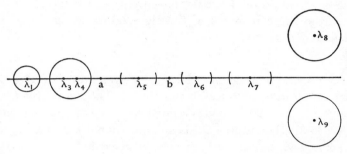

FIG. 1.

Now if $\det(A-zI)$ is evaluated at the points marked a and b we obtain a positive and a negative value respectively, irrespective of the positions of the λ_i' in their domains. If we now bisect the interval (a, b) repeatedly we must obtain the correct sign of $\det(A-zI)$ for any bisection point which is not in the domain containing λ_5. Hence, exactly as in Chapter 5, § 41, we may show that either all the bisection intervals really do contain λ_5, or from some stage onwards at least one of the end points of the bisection intervals lies in the domain containing λ_5. If this domain is the interval $(\lambda_5-x, \lambda_5+y)$, then clearly after k steps of bisection the centre of the final bisection interval lies in the interval

$$. \; [\lambda_p-x-2^{-k-1}(b-a), \; \lambda_p+y+2^{-k-1}(b-a)]. \tag{37.2}$$

Since we can use as many steps as we please, the ultimate limitation on the method is that imposed by the size of the interval $(\lambda_5-x, \lambda_5+y)$.

Notice that in general we cannot expect computed values of small eigenvalues to have a low relative error. There is no reason to expect that the domain containing a small zero will be any smaller than those containing the larger zeros. In general the best we can expect is that the domain will be of the order of magnitude of $2^{-t}\|A\|_E$ when the eigenvalue is well-conditioned.

Successive linear interpolation

38. With the method of bisection there are no convergence problems and it might be felt that the success of the method is strictly dependent on the fact that we make no quantitative use of the function values. For example, if a bisection point z lies just outside the domain containing some eigenvalue λ_p, then the sign of the computed $\det(A-zI)$ will certainly be correct, but its relative error may be arbitrarily large as can be seen from (37.1) and the fact that λ'_p may be anywhere in the λ_p domain.

On the other hand if we are using successive linear interpolation we have
$$z_{k+1} = z_k + f_k(z_k - z_{k-1})/(f_{k-1} - f_k), \tag{38.1}$$
and at first sight a high relative error in f_k appears to be very serious. However, this is illusory. If the precision of computation is such that the possibility of locating the eigenvalue at λ_p with reasonable precision may be seriously considered, then at the stage when we reach a z_k which is close to the λ_p domain, z_k and z_{k-1} will be on the same side of λ_p, and, from § 21, $|z_k - \lambda_p|$ will be much smaller than $|z_{k-1} - \lambda_p|$. If λ'_i and λ''_i be the eigenvalues of the perturbed matrix $(A + \delta A)$ corresponding to the evaluations at z_k and z_{k-1} respectively, we have

$$z_{k+1} = z_k + (z_k - z_{k-1}) \prod_{i=1}^{n} (\lambda'_i - z_k) \bigg/ \bigg[\prod_{i=1}^{n} (\lambda''_i - z_{k-1}) - \prod_{i=1}^{n} (\lambda'_i - z_k) \bigg]$$

$$\doteqdot z_k + (z_k - z_{k-1})(\lambda'_p - z_k)/[(\lambda''_p - z_{k-1}) - (\lambda'_p - z_k)], \tag{38.2}$$

since
$$\lambda'_i - z_k \doteqdot \lambda''_i - z_k \qquad (i \neq p). \tag{38.3}$$

The relation (38.2) may be expressed in the form

$$z_{k+1} \doteqdot z_k + (\lambda'_p - z_k) \bigg/ \bigg[1 + \frac{\lambda''_p - \lambda'_p}{z_k - z_{k-1}} \bigg], \tag{38.4}$$

showing that z_{k+1} is very close to λ'_p. In general, z_{k+1} will be a considerably less accurate value than that which would have been obtained by exact computation; the latter is such that $z_{k+1} - \lambda_p$ is of the order of magnitude of $(z_k - \lambda_p)(z_{k-1} - \lambda_p)$, as we have shown in (21.15). Nevertheless, our argument shows that we can expect the iterates to improve until we have obtained a value which is inside the domain of indeterminacy, and even with the method of bisection we cannot do better than this.

Having reached the domain of indeterminacy, the situation is less satisfactory. The computed values of f are almost exactly proportional to $z_k - \lambda'_k$ and λ'_k will vary with z_k. Hence for z_k inside the domain of indeterminacy the computed values of $\det(A - z_k I)$ will fluctuate according to the vagaries of the rounding errors, the 'expected' value being more or less constant for any such z_k. This is well illustrated in Table 2 of § 7 where each of the first five values of z_k lies in the domain of indeterminacy surrounding $\lambda_p = 10$. In spite of the fact that the distance of the fifth value from the zero is 2^{19} times as great as that of the first value, all computed function values are of the same order of magnitude. Now from equation (38.1) we see that the modification resulting from an iteration is

$$(z_k - z_{k-1})[f_k/(f_{k-1} - f_k)].$$

For values of z_k and z_{k-1} inside the domain of indeterminacy the quantity in the square brackets is entirely dependent on the rounding errors. In general its value will be of order unity (this is true, for example, for any pair of z_i taken from the first five entries in Table 2), but if it happens that f_{k-1} and f_k are exceptionally close, then its value will be very large.

39. Accordingly the behaviour of iterates may be described as follows. At a sufficient distance from an eigenvalue the computed function values have a low relative error, and the correction made in a typical step is much the same as that corresponding to exact computation. As we approach the domain of indeterminacy the relative error in the computed values increases, though as long as the current value is outside the domain of indeterminacy the move is in the correct direction. The ordered progress ceases when we reach a point inside the domain of indeterminacy. Thereafter the successive approximations tend to remain inside or near the domain of indeterminacy, though an unfortunate combination of rounding errors in two successive evaluations may produce a value which is well outside this domain.

It is instructive to carry out the relevant computation for pairs of values of z_k taken from Table 2 using both the computed values and the 'true' values. Remember that even these so-called true values are not exact, since they are subject to a single rounding error inherent in their representation to a small number of significant decimal places. It is important to realize that if, say, the eigenvalue at $z = 10$ is required correct to s binary places then when $|z_k - 10|$ is of the order of 2^{-p}, it is adequate to compute $f(z_k)$ correct to $s - p$ binary places. The correction made at each stage will then agree with the correction corresponding to the exact computation in all $s - p$ figures which are relevant.

Thus in Table 2 the value at $z = 10 + 2^{-28} + 7 \times 2^{-42}$ is given to about 15 significant binary figures. If we take this value and perform one step of linear interpolation using $z = 10 + 2^{-18} + 7 \times 2^{-42}$ as our other interpolation point, the interpolated value has 43 correct binary places, that is $28 + 15$.

Multiple and pathologically close eigenvalues

40. The method of bisection cannot be used to locate a multiple zero of even order, but we can still use successive linear interpolation. It may be thought that the accuracy attainable in a multiple zero by such means is necessarily severely limited but this is not so. Although convergence is slow the ultimate attainable accuracy is still dependent only on the size of the domain of indeterminacy. The situation is rather less favourable now than it was for a simple zero. For a double eigenvalue, for example, the relation (38.2) becomes

$$z_{k+1} \doteq z_k + (z_k - z_{k-1})(\lambda_p' - z_k)^2 / [(\lambda_p'' - z_{k-1})^2 - (\lambda_p' - z_k)^2] \qquad (40.1)$$

but we cannot claim that $z_k - \lambda_p$ is much smaller than $z_{k-1} - \lambda_p$ since, even with exact computation, the error is reduced only by the factor $1 \cdot 62$ in each iteration (cf. (22.7)). Instability can therefore set in when $z_k - \lambda_p$ is several times the diameter of the domain of indeterminacy if λ_p' and λ_p'' happen to be on opposite sides of λ_p.

It is true, though, that if the multiple eigenvalue is ill-conditioned, or even if it is well-conditioned but the method of evaluation is ill-conditioned, the domain of indeterminacy will be comparatively large. For example, if we use the explicit form of the function $z^3 - 3z^2 + 3z - 1$ and evaluate $f(z)$ by nested multiplication, then we know that each evaluation gives the exact value for some polynomial

$$(1 + \epsilon_3)z^3 - 3(1 + \epsilon_2)z^2 + 3(1 + \epsilon_3)z - (1 + \epsilon_4), \qquad (40.2)$$

where the bounds on the ϵ_i are of the order of 2^{-t}. Such perturbations in the coefficients may cause perturbations of order $2^{-t/3}$ in the zeros (Chapter 2, § 3). The domain of indeterminacy is therefore of width $k2^{-t/3}$ for some k of order unity. Accordingly, linear interpolation ceases to give a steady improvement once we reach a z having an error of the order of magnitude of $2^{-t/3}$. However if we evaluate the determinant of

$$\begin{bmatrix} 1 & & \\ & 1 & \\ & & 1 \end{bmatrix} - zI, \qquad (40.3)$$

treating the matrix as a tri-diagonal form, the domain of indeterminacy containing $z = 1$ is of order 2^{-t} and a zero at $z = 1$ can be accurately determined. (For the determination of the multiplicity see § 57.)

41. This example may seem a little artificial and it is unfortunately true that a tri-diagonal matrix with linear elementary divisors and an eigenvalue of multiplicity r must have $r-1$ zero off-diagonal elements; this follows from simple considerations of rank. However, if we are willing to consider eigenvalues which are equal to working accuracy rather than truly multiple, then we can give much more satisfactory examples. The matrix W_{21}^+ of Chapter 5 provides a good illustration. Using a mantissa of 30 binary places, the dominant zero was found correct to the last digit using successive linear interpolation, in spite of the fact that the second eigenvalue is identical with it to working accuracy. Similar examples having eigenvalues of high multiplicity to working accuracy have been solved, and although the convergence is extremely slow the ultimate attainable accuracy has been very high.

Other interpolation methods

42. Similar results may be established for the other interpolation methods which we have discussed. In each case the z_k of limiting accuracy is in the domain of indeterminacy or its immediate neighbourhood and again there is a danger of instability after reaching the limiting accuracy. This phenomenon has been very marked with Muller's method, which we have used more frequently than any of the other interpolation methods. With this method the most common experience has been that if iteration is continued after reaching a value in the domain of indeterminacy, a succession of values of much the same accuracy is obtained, but sooner or later a large jump occurs. In experimental runs this jump has been sufficiently violent for later iterations to converge to a different eigenvalue! In library programmes of course this is prevented by limiting the current modification (§ 62).

There is one point that deserves special mention. It might well be thought that Muller's method cannot locate accurately a double eigenvalue even if the method of evaluation is such that the domain of indeterminacy is very small. One might argue as follows.

Consider for example the polynomial $(z-0.98762\ 3416)^2$. If we compute $f(z)$ exactly at any three points and derive the exact polynomial passing through these points we shall obtain

$$z^2 - 1.97524\ 6832z + 0.97540\ 00118\ 31509\ 056. \tag{42.1}$$

Now if we are prepared to use only 9 decimal digits we cannot possibly obtain these coefficients exactly. At best we are bound to have a rounding in the 9th significant figure and in general this will lead to errors in the zeros of order $(10^{-9})^{\frac{1}{2}}$. Hence even if we evaluate $f(z)$ directly from the expression $(z-0.98762\ 3416)^2$, so that computed values have a low relative error, the quadratic fitted to these values must give comparatively poor results. In a more practical situation we are hardly likely to obtain function values in the neighbourhood of a double-zero with the accuracy attainable in this very special example, so that the usual situation would appear to be even less favourable.

The fallacy in this argument is that the computation is not carried out in this way. If we use Muller's method (§ 20) we derive the *modification* to be made in z_k in each iteration. Consider the expression for the increment $z_{k+1}-z_k$ derived from equation (20.6). We have

$$z_{k+1}-z_k = \lambda_{k+1}(z_k-z_{k-1})$$

$$= -2f_k\delta_k(z_k-z_{k-1})/\{g_k\pm[g_k^2\pm4f_k\delta_k\lambda_k(f_{k-2}\lambda_k-f_{k-1}\delta_k+f_k)]^{\frac{1}{2}}\}.$$

$$(42.2)$$

If the computation were exact the quantity in the square brackets would be zero for any z_{k-2}, z_{k-1}, z_k. In practice this will not be true, but if the maximum error in f_k, f_{k-1}, f_{k-2} is one part in 2^s the denominator will differ from its true value by at most one part in $2^{s/2}$. The *correction* made will therefore differ from that corresponding to exact computation by about one part in $2^{s/2}$. Hence even if the accuracy of computed values of $f(z)$ diminishes as we approach the eigenvalue λ_p in such a way that when $|z_k-\lambda_p| = O(2^{-s})$ the computed value has an error of 1 part in 2^{t-s}, the successive iteratives will continue to improve in practice until a value is reached having a relative error of a few parts in 2^t. This is exactly the situation for a well-conditioned double eigenvalue using a method of evaluation which gives a domain of indeterminacy having a diameter of order 2^{-t}.

Methods involving the use of a derivative

43. Methods which make use of one or more of the derivatives behave rather differently in general from interpolation methods. For the study of the limiting behaviour near a simple eigenvalue λ_p, we first make the very reasonable assumption that in the neighbourhood of a zero the relative error in the computed derivatives is smaller than in the function values.

For the purpose of illustration we may consider Newton's method (§ 25), since the effect of rounding errors on the limiting behaviour is much the same for all such methods. If we denote the computed values of $f(z_k)$ and $f'(z_k)$ by $\bar{f}_k$ and $\bar{f}'_k$ respectively, then the correction obtained in a typical step is $-\bar{f}_k/\bar{f}'_k$ instead of $-f_k/f'_k$. We may write

$$\bar{f}_k = f_k(1+\epsilon_k), \qquad \bar{f}'_k = f'_k(1+\eta_k), \tag{43.1}$$

and our assumption is that in the neighbourhood of λ_p we have

$$|\eta_k| \ll |\epsilon_k|. \tag{43.2}$$

The ratio between the computed correction and the true correction is therefore essentially that between $\bar{f}_k$ and f_k, and in the notation of § 38 we find

$$\bar{f}_k/f_k = \prod_{i=1}^{n} (\lambda'_i - z_k) \Big/ \prod_{i=1}^{n} (\lambda_i - z_k)$$

$$\doteq (\lambda'_p - z_k)/(\lambda_p - z_k). \tag{43.3}$$

As long as $|\lambda_p - z_k|$ is appreciably greater than $|\lambda_p - \lambda'_p|$, that is as long as z_k is well outside the domain of indeterminacy, the computed correction has a low relative error, but when these two quantities approach the same order of magnitude this is no longer true. If at this stage z_k is $\lambda_p + O(\theta)$ where θ is small, the true correction would produce a z_{k+1} given by $\lambda_p + O(\theta^2)$ where we are assuming that the precision of computation is such that the domain of indeterminacy is small compared with the distance of λ_p from neighbouring eigenvalues. Hence the computed correction will be $\bar{f}_k[\lambda_p - z_k + O(\theta^2)]/f_k$ so that the computed z_{k+1} is approximately λ'_p. Again we find that the computed iterates improve right up to the stage when the domain of indeterminacy is reached.

Thereafter the behaviour differs from that exhibited by interpolation methods. For points z_k inside the domain of indeterminacy the computed values of f' will still have a comparatively low relative error. Hence the ratio of the computed correction to the true correction will be given almost exactly by the right-hand side of (43.3), and since the true correction will give $z_{k+1} = \lambda_p$ almost exactly, the

computed correction gives λ'_p almost exactly. The iterates will therefore move about, more or less at random inside the domain of
indeterminacy.

44. Our argument has been based on the assumption that the
relative error in the computed derivative is much lower than that in
the function value. This will generally be true, even for multiple or
very close eigenvalues, for values of z outside the domain of indeterminacy, but inside this domain it will, in general, no longer hold for
such eigenvalues. Indeed the computed value of $f'(z)$ may even turn
out to be zero and $f(z)$ non-zero.

However, the comments we made in § 40 apply with equal force
to iterative methods in which the derivative is used. Even when a
matrix has a multiple zero or a very severe cluster of zeros, provided
the corresponding domains of indeterminacy are small the limiting
accuracy attainable in the zero will be high; only the rate of convergence will be affected. Furthermore, in the unlikely event of our
knowing the multiplicity of a zero, then the use of formula (25.9) or
(28.12) will give a high rate of convergence.

Criterion for acceptance of a zero

45. The practical problem of determining when a value attained
by an iterative procedure should be accepted as a zero presents
considerable difficulties. We consider first the simplified problem in
which it is assumed that no rounding errors are made in the computation, so that any prescribed accuracy may be attained in a zero
α by continuing the iteration for a sufficient number of steps.

Now in all the methods we have described we have ultimately

$$|z_{i+1}-\alpha| = O(|z_{i+1}-z_i|), \tag{45.1}$$

and indeed for simple zeros we have the stronger result

$$|z_{i+1}-\alpha| = o(|z_{i+1}-z_i|). \tag{45.2}$$

Hence an obvious criterion to use for stopping the iteration is

$$|z_{i+1}-z_i| < \epsilon, \tag{45.3}$$

where ϵ is some pre-assigned 'small' quantity. Notice, however, that
if this criterion is to be adequate to cover multiple zeros it is necessary
to take ϵ to be appreciably smaller than the maximum error that is

regarded as acceptable in a zero. For example if α is a zero of multiplicity r and we are using Newton's method, then we have ultimately

$$\alpha \sim z_{i+1} + (r-1)(z_{i+1} - z_i), \tag{45.4}$$

and we must take ϵ to be smaller than our required tolerance by the factor $(r-1)^{-1}$ at least.

It is tempting to use some criterion based on the relative change made by the current iteration, for example,

$$|(z_{i+1} - z_i)/z_i| < \epsilon. \tag{45.5}$$

For convergence to a zero at $z = \alpha \neq 0$ such a criterion is indeed satisfactory, but if $\alpha = 0$ we have

$$|(z_{i+1} - z_i)/z_i| \to 1 \tag{45.6}$$

for a simple zero. If we are dealing with the explicit polynomial form, then zero eigenvalues can be detected quite simply and discarded. Hence in this case if all the non-zero eigenvalues are simple, the relative criterion can be used.

Effect of rounding errors

46. There is little point in pursuing the analysis of the previous section any further since in practice the effect of rounding errors entirely dominates the practical problem. In practice the attainable accuracy in any zero with a given precision of computation is limited by the domain of indeterminacy associated with that zero and, in general, we shall have no *a priori* information on the size of these domains. If we rely on the criterion (45.3), then if we choose too small a value of ϵ the criterion may never be satisfied however long the iteration is continued. On the other hand if we take ϵ to be too large then we may accept a much less accurate value than may ultimately be attainable. This will be true, for example, in the case of *well-determined multiple* zeros since for these the passage to the limit is slow but persistent.

If the domain of indeterminacy surrounding a zero is comparatively large then we may not be able to meet even the most modest demands. For example, if we use the Frobenius form and our matrix has the eigenvalues 1, 2,..., 20, then working to single-precision on ACE (almost 14 decimals) a value of ϵ of 10^{-2} will not be met when converging to the zeros 10, 11,..., 19. In fact these eigenvalues have such large domains of indeterminacy that it is impossible to detect when iterates are in their neighbourhoods unless we work to an appreciably higher precision than 14 decimals.

An ideal programme would be able to detect when higher-precision computation is essential and to distinguish this from the disordered progress which precedes 'homing' on a zero. Further it should be designed to move progressively to the precision of computation which is required in order to determine the eigenvalues to the prescribed accuracy. In practice the aim is usually rather more modest and the criterion (45.3) with some compromise value for ϵ is used.

47. Garwick has proposed a simple device which reduces the risk of iterating indefinitely for ill-conditioned zeros without sacrificing the accuracy attainable in well-conditioned zeros. This is based on the observation that when we are moving towards a zero, then from a certain stage onwards the differences $|z_{i+1} - z_i|$ progressively decrease until we effectively reach the domain of indeterminacy. Thereafter the differences may behave erratically in the way we have described. Garwick suggested that a relatively modest criterion of the form

$$|z_{i+1} - z_i| < \epsilon \qquad (47.1)$$

should be used, and when this criterion has been satisfied iteration should continue as long as the differences between consecutive z_i are diminishing. When finally we reach values for which

$$|z_{r+2} - z_{r+1}| > |z_{r+1} - z_r|, \qquad (47.2)$$

then the value z_{r+1} is accepted.

As an example of its application Newton's method was used on the explicit polynomial with zeros 1, 2,..., 20, working to double-precision on ACE (about 28 decimals) and with $\epsilon = 10^{-6}$. It was found that for each of the zeros iteration continued until a value inside the corresponding domain of indeterminacy was reached. Thus the error in the accepted approximation to the zero at $z = 1$ was about 10^{-27} while that in the accepted approximation to the ill-conditioned zero at $z = 15$ was about 10^{-14}, so that optimum efficiency was effectively achieved.

Little seems to have been done in the direction of designing programmes which detect automatically when the limiting precision has been reached. On a computer employing non-normalized floating-point arithmetic this should not be too difficult, since it is obvious from our earlier discussion that iteration should continue as long as the current computed value of the function has some correct significant figures. As soon as we reach a value of z_i for which this is not true, z_i is indistinguishable from a zero of the function with the precision of computation and the algorithm which is being used.

Suppression of computed zeros

48. When a zero of $f(z)$ has been accepted we wish to avoid converging to this zero on subsequent occasions. If we are working with the explicit characteristic polynomial, then when α has been accepted we can compute explicitly the quotient polynomial $f(z)/(z-\alpha)$. Similarly if z^2-pz-l has been accepted as a quadratic factor we can compute the quotient polynomial $f(z)/(z^2-pz-l)$. This process is usually referred to as *deflation*. Notice from (2.10) that

$$f(z)/(z-\alpha) = \sum_{r=1}^{n} s_r(\alpha)z^{r-1} \qquad (48.1)$$

where the $s_r(\alpha)$ are the values obtained using nested multiplication with $z = \alpha$, while in the notation of § 31

$$f(z)/(z^2-pz-l) = \sum_{r=2}^{n} q_r z^{r-2}. \qquad (48.2)$$

In practice, an accepted zero or quadratic factor will usually be in error and further errors will occur in the deflation process itself. There is thus a danger that successive computed zeros may suffer a progressive loss of accuracy and, indeed, the deflation process is generally regarded as being extremely dangerous in this respect.

In Wilkinson (1963b), Chapter 2, I gave a very detailed analysis of this problem and I shall content myself here with a summary of the main results:

If before each deflation *iteration is continued until the limiting accuracy for the precision of computation is attained*, then provided the zeros are found roughly in order of increasing absolute magnitude, little progressive deterioration takes place. In general the limiting accuracy attainable in a zero by iterating in a computed deflated polynomial falls only a little short of that attainable by iteration in the original polynomial. On the other hand if a polynomial has zeros which vary considerably in order of magnitude, an approximation to one of the larger zeros is accepted and deflation carried out before finding a much smaller zero, then a catastrophic loss of accuracy may take place. *This is likely to be true even when all the zeros of the original polynomial are very well-conditioned in that they are insensitive to small relative changes in the coefficients.*

Although deflation is somewhat less dangerous than is commonly supposed, there seems to be no reliable method of ensuring in general that zeros are found roughly in order of increasing absolute magnitude, and in § 55 we describe techniques for suppressing computed zeros without explicit deflation.

Deflation for Hessenberg matrices

49. For the explicit polynomial form, after accepting a zero we are able to derive an explicit polynomial having the remaining $(n-1)$ zeros. It is natural to ask whether there is an analogous process for each of the other condensed forms, and if so whether it is stable. The problem can be posed explicitly as follows:

If we are given an eigenvalue of a Hessenberg (tri-diagonal) matrix of order n, can we produce a Hessenberg (tri-diagonal) matrix of order $(n-1)$ having the remaining $(n-1)$ eigenvalues?

We consider the Hessenberg case first. Let λ_1 be an eigenvalue of the upper Hessenberg matrix A. As always we may assume without loss of generality that the sub-diagonal elements of A are non-zero. Now taking $z = \lambda_1$ in Hyman's method as described in § 11, we find that this method effectively determines a matrix M of the form

$$
M = \begin{bmatrix}
1 & & & & -x_1 \\
& 1 & & & -x_2 \\
& & \cdot & & \\
& & & \cdot & \\
& & & & \cdot & -x_{n-1} \\
& & & & & 1
\end{bmatrix}, \qquad (49.1)
$$

such that $B = (A - \lambda_1 I)M$ has all elements in its last column equal to zero except the first which is proportional to $\det(A - \lambda_1 I)$. Now since λ_1 is an eigenvalue of A, $\det(A - \lambda_1 I)$ is zero and hence the whole of the last column is zero. Notice that apart from its last column, B is the same as $(A - \lambda_1 I)$, so that when $n = 5$, B is typically of the form

$$B = \begin{bmatrix} \times & \times & \times & \times & 0 \\ \times & \times & \times & \times & 0 \\ & \times & \times & \times & 0 \\ & & \times & \times & 0 \\ & & & \times & 0 \end{bmatrix}. \qquad (49.2)$$

We may complete a similarity transformation by pre-multiplying B by M^{-1}, and this has the effect of adding x_i times row n to row i for each value of i from 1 to $n-1$. This clearly does not disturb the Hessenberg form or the zeros in the last column, so that $M^{-1}(A - \lambda_1 I)M$ is also of the form shown in (49.2). Hence $M^{-1}(A - \lambda_1 I)M + \lambda_1 I$ is of the form

$$\begin{bmatrix} \times & \times & \times & \times & 0 \\ \times & \times & \times & \times & 0 \\ & \times & \times & \times & 0 \\ & & \times & \times & 0 \\ \hline & & & \times & \lambda_1 \end{bmatrix}. \qquad (49.3)$$

Clearly the remaining eigenvalues are those of the Hessenberg matrix of order $(n-1)$ in the top left-hand corner.

50. Since the tri-diagonal form is a special case of the Hessenberg form we may apply this process to it. The post-multiplication by M leaves the first $(n-1)$ columns unaltered and the form (49.2) now becomes

$$\begin{bmatrix} \times & \times & & & 0 \\ \times & \times & \times & & 0 \\ & \times & \times & \times & 0 \\ & & \times & \times & 0 \\ & & & \times & 0 \end{bmatrix}. \qquad (50.1)$$

Pre-multiplication by M^{-1} merely alters column $(n-1)$ so that the final matrix is of the form

$$
\begin{bmatrix}
\times & \times &   & \times & 0 \\
\times & \times & \times & \times & 0 \\
       & \times & \times & \times & 0 \\
       &   & \times & \times & 0 \\
       &   &   & \times & 0
\end{bmatrix}. \qquad (50.2)
$$

The matrix of order $(n-1)$ in the top left-hand corner is no longer tri-diagonal; it has additional non-zero elements in the whole of its last column. The tri-diagonal form is therefore not invariant with respect to our deflation process.

It is convenient to refer to a matrix obtained from a tri-diagonal matrix by the addition of non-zero elements in its last column as an *augmented tri-diagonal matrix*. It is immediately obvious that the augmented tri-diagonal matrix *is* invariant under our deflation process. More specifically the deflation of a matrix A_n leads to a matrix C_n where A_n and C_n are of the forms

$$
A_n =
\begin{bmatrix}
\times & \times &   &   &   &   &   & \times \\
\times & \times & \times &   &   &   &   & \times \\
       & \times & \times & \times &   &   &   & \times \\
       &   & \times & \times & \times &   &   & \times \\
       &   &   & \times & \times & \times & \times \\
       &   &   &   & \times & \times & \times \\
       &   &   &   &   & \times & \times
\end{bmatrix},
$$

$$
C_n =
\left[
\begin{array}{ccccccc:c}
\times & \times &   &   &   & \times & 0 \\
\times & \times & \times &   &   & \times & 0 \\
       & \times & \times & \times &   & \times & 0 \\
       &   & \times & \times & \times & \times & 0 \\
       &   &   & \times & \times & \times & 0 \\
       &   &   &   & \times & \times & 0 \\
\hdashline
       &   &   &   &   & \times & \lambda_1
\end{array}
\right]. \qquad (50.3)
$$

Hence although the tri-diagonal form is not preserved the deflated matrices do not become progressively more complex in structure.

Deflation of tri-diagonal matrices

51. We have introduced the deflation of tri-diagonal matrices as a special case of that of *upper* Hessenberg matrices, but the process has a closer relationship with that of § 8 if we introduce the non-zero elements in the first column of the deflated matrix rather than the last. We illustrate the process for a matrix of order five. Suppose we take our original tri-diagonal matrix to be of the form A_n given by

$$A_n = \begin{bmatrix} \alpha_1 & 1 & & & \\ \beta_2 & \alpha_2 & 1 & & \\ & \beta_3 & \alpha_3 & 1 & \\ & & \beta_4 & \alpha_4 & 1 \\ & & & \beta_5 & \alpha_5 \end{bmatrix}. \tag{51.1}$$

If λ_1 is an eigenvalue of A_n we define $p_r(\lambda_1)$ by the relations

$$p_0(\lambda_1) = 1, \quad p_1(\lambda_1) = \alpha_1 - \lambda_1, \quad p_r(\lambda_1) = (\alpha_r - \lambda_1)p_{r-1}(\lambda_1) - \beta_r p_{r-2}(\lambda_1)$$

$$(r = 2, ..., n). \tag{51.2}$$

Then if we take M to be the matrix

$$M = \begin{bmatrix} 1 & & & & \\ -p_1(\lambda_1) & 1 & & & \\ p_2(\lambda_1) & & 1 & & \\ -p_3(\lambda_1) & & & 1 & \\ p_4(\lambda_1) & & & & 1 \end{bmatrix} \tag{51.3}$$

we have

$$(A_n - \lambda_1 I)M = \begin{bmatrix} 0 & 1 & & & \\ 0 & (\alpha_2 - \lambda_1) & 1 & & \\ 0 & \beta_3 & (\alpha_3 - \lambda_1) & 1 & \\ 0 & & \beta_4 & (\alpha_4 - \lambda_1) & 1 \\ 0 & & & \beta_5 & (\alpha_5 - \lambda_1) \end{bmatrix}, \tag{51.4}$$

the $(n, 1)$ element being zero because $p_n(\lambda_1) = 0$. Hence we have

$$M^{-1}(A_n - \lambda_1 I)M + \lambda_1 I = \begin{bmatrix} \lambda_1 & 1 & & & \\ 0 & \alpha_2 + p_1(\lambda_1) & 1 & & \\ 0 & \beta_3 - p_2(\lambda_1) & \alpha_3 & 1 & \\ 0 & p_3(\lambda_1) & \beta_4 & \alpha_4 & 1 \\ 0 & -p_4(\lambda_1) & & \beta_5 & \alpha_5 \end{bmatrix}. \quad (51.5)$$

The deflated matrix A_{n-1} of order $(n-1)$ in the bottom right-hand corner therefore has the extra elements $\pm p_r(\lambda_1)$ in its first column.

Let us denote the leading principal minor of order r of $(A_{n-1} - zI)$ by $q_{r+1}(z)$. We then have

$$\left.\begin{aligned} q_1(z) &= 1, \qquad q_2(z) = (\alpha_2 - z)q_1(z) + p_1(\lambda_1) \\ q_r(z) &= (\alpha_r - z)q_{r-1}(z) - \beta_r q_{r-2}(z) + p_{r-1}(\lambda_1) \quad (r = 3, ..., n) \end{aligned}\right\}, \quad (51.6)$$

and we see that the column of elements $p_r(\lambda_1)$ has scarcely added to the complexity of the recurrence relations.

To carry out a second deflation after finding an eigenvalue λ_2 of A_{n-1} we post-multiply by a matrix of the form

$$\begin{bmatrix} 1 & & & \\ -q_2(\lambda_2) & 1 & & \\ q_3(\lambda_2) & & 1 & \\ -q_4(\lambda_2) & & & 1 \end{bmatrix}, \quad (51.7)$$

and the final deflated matrix of order $(n-2)$ is of the form

$$\begin{bmatrix} \alpha_3 + q_2(\lambda_2) & 1 & \\ -q_3(\lambda_2) & \alpha_4 & 1 \\ q_4(\lambda_2) & \beta_5 & \alpha_5 \end{bmatrix}. \quad (51.8)$$

Hence the recurrence relations connecting the principal minors of the deflated matrices never become more complex than those of (51.6)

Deflation by rotations or stabilized elementary transformations

52. The deflation process for Hessenberg matrices has a close formal relationship with Hyman's method, and we have seen that the latter has remarkable numerical stability. Similarly the deflation of tri-diagonal matrices is associated with the simple recurrence relations used in § 8 and these too are very stable.

Nevertheless the deflation processes do not enjoy the same numerical stability. It might be felt that this is because we have abandoned the use of interchanges, and that if they were incorporated or alternatively if plane rotations were used, we might obtain a stable form of deflation. We now consider this.

We have already shown in § 14 how to reduce the last $(n-1)$ elements of the last column of the Hessenberg matrix $(A-zI)$ to zero by post-multiplying successively by rotations in the planes $(n-1, n)$, $(n-2, n), ..., (1, n)$. In the particular case when $z = \lambda_1$, an eigenvalue of A, the first element of the last column of the transformed matrix will also be zero (if the computation is exact) and hence the transformed matrix will be as in (49.2). We complete the similarity transformation by pre-multiplying successively by the rotations in planes $(n-1, n)$, $(n-2, n), ..., (1, n)$. Clearly each pre-multiplication preserves the zeros in the last column and the Hessenberg nature of the matrix of order $(n-1)$ in the top left-hand corner, but the rotation in the (r, n) plane $(r = n-1, ..., 2)$ introduces a non-zero element into position $(n, r-1)$. The final matrix obtained after adding back $\lambda_1 I$ is therefore of the form

$$\begin{bmatrix} \times & \times & \times & \times & 0 \\ \times & \times & \times & \times & 0 \\ & \times & \times & \times & 0 \\ & & \times & \times & 0 \\ \hline \times & \times & \times & \times & \lambda_1 \end{bmatrix}. \qquad (52.1)$$

The presence of the non-zero elements in the last row is of no importance and the Hessenberg matrix of order $(n-1)$ in the top left-hand corner again has the remaining $(n-1)$ eigenvalues. The Hessenberg form is therefore invariant with respect to our deflation technique even when rotations are used. An exactly equivalent deflation may be performed using stabilized non-unitary elementary transformations, and again it leads to a matrix of the form (52.1).

Before discussing the effect of rounding errors we consider the same reductions applied to a tri-diagonal matrix. At the end of the post-multiplications the transformed matrix is clearly of the form (50.1). We shall show that the pre-multiplication leads to a matrix of the form (52.1), that is the tri-diagonal matrix is replaced, in general, by a full Hessenberg matrix. The proof is by induction. Suppose that after the pre-multiplication by a rotation in the (r, n)

plane the matrix is of the form given typically for the case $n = 7$, $r = 4$, by

$$
\begin{bmatrix}
\times & \times & & & & & 0 \\
\times & \times & \times & & & & 0 \\
 & \times & \times & \times & & & 0 \\
 & & \times & \times & \times & \times & 0 \\
 & & & \times & \times & \times & 0 \\
 & & & & \times & \times & 0 \\
0 & 0 & \times & \times & \times & \times & 0
\end{bmatrix} . \qquad (52.2)
$$

The next rotation is in the $(r-1, n)$ plane and replaces rows $r-1$ and n by linear combinations of themselves. Hence in general row $r-1$ gains non-zero elements in positions $r+1$ to $n-1$ and row n gains a non-zero element in position $r-2$. Again elementary stabilized non-unitary transformations have a similar effect, though the Hessenberg matrix will have quite a high percentage of zero elements. However, if the original tri-diagonal matrix is symmetric, symmetry is preserved by the orthogonal deflation and knowing this we can show that the tri-diagonal form is also preserved. The argument is as follows.

Since the final deflated matrix has a null last column it must also have a null last row. This in turn implies that after the pre-multiplication by the rotation in the (r, n) plane the elements in positions $(n, r+1)$, $(n, r+2)$,..., $(n, n-1)$ must already be zero. For the only operations which have still to be carried out are the successive replacements of row n by

$$
(\text{row } n)\cos \theta_i - (\text{row } i)\sin \theta_i \quad (i = r-1,..., 1), \qquad (52.3)
$$

(using an obvious notation), and from the non-vanishing nature of the subdiagonal elements of the original matrix we cannot have $\cos \theta_i = 0$. Since from (52.2) the elements in positions $r+1$, ..., $(n-1)$ of each of rows $r-1$,..., 1 are zero, it is clear that the specified elements of row n must already be zero. From this it now follows that we do not introduce non-zero elements in positions above the diagonal, so that the final matrix is tri-diagonal. Deflation using elementary stabilized non-unitary matrices does not preserve symmetry or the tri-diagonal form.

Stability of the deflation

53. We now consider the effect of rounding errors both in the deflation itself and in the value of λ_1; we discuss first the deflation of a Hessenberg matrix using plane rotations.

As far as the post-multiplications are concerned our general analysis in Chapter 3, §§ 20–26, immediately shows that if $R_{n-1,n}, R_{n-2,n}, \ldots, R_{1n}$ are the exact plane rotations corresponding to the successive computed matrices, then the final computed matrix is very close to $A R_{n-1,n} R_{n-2,n} \ldots R_{1n}$. This analysis includes a justification for entering zero elements in positions $(n, n), (n-1, n), \ldots, (2, n)$. The emergence of a zero in position $(1, n)$, which is guaranteed in the case of exact computation, is not however a direct consequence of a specific rotation, and our general analysis does not show that the computed element will be small even when λ_1 is correct to working accuracy. It is easy to construct matrices having well-conditioned eigenproblems for which the computed $(1, n)$ element on the completion of the post-multiplications is quite large, and consequently the final computed form of $R_{1n}^T R_{2n}^T \ldots R_{n-1,n}^T (A - \lambda_1 I) R_{n-1,n} \ldots R_{2n} R_{1n}$ is of the form

$$
\begin{bmatrix}
\times & \times & \times & \times & \times & \times \\
\times & \times & \times & \times & \times & 0 \\
& \times & \times & \times & \times & 0 \\
& & \times & \times & \times & 0 \\
& & & \times & \times & 0 \\
\hline
\times & \times & \times & \times & \times & \times
\end{bmatrix}, \tag{53.1}
$$

neither the $(1, n)$ element nor the (n, n) element being small.

Similarly if we use plane rotations to deflate a symmetric tridiagonal matrix the final computed matrix may well be of the form

$$
\begin{bmatrix}
\times & \times & & & & \underline{\times} \\
\times & \times & \times & & & 0 \\
& \times & \times & \times & & 0 \\
& & \times & \times & \times & 0 \\
& & & \times & \times & 0 \\
\hline
\underline{\times} & 0 & 0 & 0 & 0 & \underline{\times}
\end{bmatrix} \tag{53.2}
$$

where the three underlined elements, which would be zero with exact computation, may be far from negligible.

Much the same remarks apply to deflation using elementary stabilized non-unitary transformations. We now show that for such deflations the failure of the $(1, n)$ element to vanish on completion of the post-multiplication is closely related to the phenomenon which we discussed in Chapter 5, § 56.

Consider the matrix A of order 21 given by

$$A = \begin{bmatrix} -10 & 1 & & & & & & \\ 1 & -9 & 1 & & & & & \\ & 1 & -8 & & & & & \\ & & & \cdot & & & & \\ & & & & \cdot & & & \\ & & & & & \cdot & & \\ & & & & & 1 & 8 & 1 \\ & & & & & & 1 & 9 & 1 \\ & & & & & & & 1 & 10 \end{bmatrix}, \qquad (53.3)$$

which obviously has the same eigenvalues as W_{21}^-. Now

$$\lambda_1 = 10 \cdot 74619\ 42$$

is an eigenvalue correct to 9 significant decimals. Let us perform the deflation process on $(A - \lambda_1 I)$ using elementary stabilized non-unitary transformations.

The first step is to compare absolute values of elements $(21, 20)$ and $(21, 21)$, and if $(21, 21)$ has the larger modulus we interchange columns 21 and 20. We then take a multiple of column 20 from column 21 to produce a zero in position $(21, 21)$. We immediately see that the arithmetic is exactly the same as it was in the first step of Gaussian elimination with interchanges for the matrix $(W_{21}^- - \lambda_1 I)$. This identity between the two processes persists throughout, but notice that the presence of an interchange at any stage in the Gaussian elimination corresponds to the absence of an interchange in the deflation process and vice versa. When the post-multiplication has been completed columns $20, 19, \ldots, 1$ are identical with the pivotal rows $1, 2, \ldots, 20$ obtained in the Gaussian elimination, while the twenty-first column of the reduced $(A - \lambda_1 I)$ is equal to the final pivotal row obtained in Gaussian elimination, so that the element $(1, 21)$ which should have been zero is in fact $-20 \cdot 69541\ 39$. The elements in column $(21 - i)$ are the w_i, v_i and u_i of Table 10, Chapter 5.

General comments on deflation

54. In Chapter 8 we describe some methods of deflation which are both economical and stable, but since they are rather intimately connected with the topics which are presented there we defer the discussion of them.

Each of the techniques of deflation which we have described in this chapter is unsatisfactory in some respect or other. In general we are forced to compute eigenvalues to an unnecessarily high accuracy and to use arithmetic of high precision in the deflation process, in order to preserve the eigenvalues in the deflated matrices. This can be avoided in the case of the explicit polynomial only if we can find the zeros in order of increasing absolute magnitude and in general there appears to be no simple method of guaranteeing this. It is natural to ask whether there are alternative methods of suppressing computed zeros without an explicit deflation.

Suppression of computed zeros

55. Let $f(z)$ be a polynomial defined in some manner, not necessarily explicitly, and let $\lambda_1, \lambda_2, \dots, \lambda_n$ be its zeros. If approximations $\lambda_1', \lambda_2', \dots, \lambda_r'$ to the first r zeros have been accepted then we may define a function $g_r(z)$ by the relation

$$g_r(z) \equiv f(z) \Big/ \prod_{i=1}^{r} (z - \lambda_i'). \tag{55.1}$$

If each λ_i' were exactly equal to the corresponding λ_i then $g_r(z)$ would be a polynomial of degree $(n - r)$ having the zeros $\lambda_{r+1}, \dots, \lambda_n$. However, even if the λ_i' are not exact the function $g_r(z)$ will, in general, have zeros at $\lambda_{r+1}, \dots, \lambda_n$. Indeed $g_r(z)$ has these zeros whatever the values of the λ_i' unless some of these quantities happen to coincide with some of the $\lambda_{r+1}, \dots, \lambda_n$. Provided we can evaluate $f(z)$ for an assigned value of z we can obviously evaluate $g_r(z)$ merely by dividing the computed $f(z)$ by the computed value of $\prod (z - \lambda_i')$. Hence we can immediately apply any of the methods depending on function values alone to find the zeros of $g_r(z)$.

To apply Newton's method to the function $g_r(z)$ we require the function $g_r(z)/g'_r(z)$, but since we have

$$g'_r(z)/g_r(z) = f'(z)/f(z) - \sum_{i=1}^{r} 1/(z - \lambda'_i) \qquad (55.2)$$

we can evaluate this function provided we can evaluate $f(z)$ and $f'(z)$. Finally for methods requiring $g''_r(z)$ we have

$$[g'_r(z)/g_r(z)]^2 - g''_r(z)/g_r(z) = [f'(z)/f(z)]^2 - f''(z)/f(z) - \sum_{i=1}^{r} 1/(z - \lambda'_i)^2.$$
$$(55.3)$$

Hence we can compute everything that is required by the iterative methods.

Suppression of computed quadratic factors

56. There is an analogous technique for suppressing accepted quadratic factors which can be used in connexion with methods of the Bairstow type. Suppose we have accepted the quadratic factor $z^2 - Pz - L$. If this were an exact factor and we could compute exactly the explicit polynomial $g(z) = f(z)/(z^2 - Pz - L)$, then the essential requirement in a Bairstow step is the computation of the remainders when $g(z)$ is divided twice in succession by a trial factor $z^2 - pz - l$. Formulae are required therefore which obtain these remainders by working directly with $f(z)$ itself. Suppose that

$$f(z) \equiv T(z)(z^2 - pz - l)^2 + (cz + d)(z^2 - pz - l) + az + b, \qquad (56.1)$$

so that a, b, c and d are the quantities q_1, q'_0, T_1, T'_0 of §§ 31, 32. Now if

$$g(z) \equiv V(z)(z^2 - pz - l)^2 + (c''z + d'')(z^2 - pz - l) + a''z + b'', \qquad (56.2)$$

then

$$(z^2 - Pz - L)[V(z)(z^2 - pz - l)^2 + (c''z + d'')(z^2 - pz - l) + a''z + b'']$$

$$\equiv T(z)(z^2 - pz - l)^2 + (cz + d)(z^2 - pz - l) + az + b. \qquad (56.3)$$

This identity enables us to express a'', b'', c'', d'' in terms of a, b, c, d and hence to obtain the remainders corresponding to $g(z)$ from those corresponding to $f(z)$.

Following Handscomb (1962), but with a slight change in notation, we may express the solution in the form

$$\left.\begin{aligned}
p' &= p - P, & l' &= l - L \\
f &= pp' + l, & e &= fl' - lp'^2 \\
a' &= al' - bp', & b' &= bf - alp' \\
c' &= ce - a', & d' &= de - b' - a'p' \\
a''e &= a', & b''e &= b' \\
c''e^2 &= c'l' - d'p', & d''e^2 &= d'f - c'lp'
\end{aligned}\right\} . \qquad (56.4)$$

Notice that as a and b tend to zero the same is true of a'' and b''; moreover since the convergence of a and b is ultimately quadratic the same is true of a'' and b''. We therefore get true convergence to the zeros of $f(z)$ even if the accepted factor $z^2 - Pz - L$ is quite poor. Since we are not concerned with an arbitrary constant multiplier of a, b, c and d we may take

$$\left.\begin{aligned}
a'' &= a'e, & b'' &= b'e \\
c'' &= c'l' - d'p', & d'' &= d'f - c'lp'
\end{aligned}\right\} . \qquad (56.5)$$

Obviously starting with a'', b'', c'', d'' we can suppress a second quadratic factor in the same way, and so on.

General comments on the methods of suppression

57. At first sight these processes appear to be rather dangerous, since for example the function $g_r(z)$ will, in general, still have all the zeros of $f(z)$ together with a number of poles at the points λ'_i ($i = 1, ..., r$), these poles being very close to the corresponding zeros. However, in practice we work with computed values of $g_r(z)$ and these in turn are obtained from computed values of $f(z)$. An essential feature of the practical application of the process is that the same precision of computation should be used throughout.

Consider then the computation of a typical value of $g_r(z)$. The computed value of $f(z)$ is $\prod_{i=1}^{n} (z - \bar{\lambda}_i)$, where the $\bar{\lambda}_i$ are values inside the domain of indeterminacy of the corresponding λ_i, each being, of course, a function of z. The accepted zeros λ'_i will also be in the domains of indeterminacy, or very close to them, and hence the computed $g_r(z)$ is given by $\prod_{i=1}^{n} (z - \bar{\lambda}_i) \Big/ \prod_{i=1}^{r} (z - \lambda'_i)$. For values of z which are not too close to any of the λ_i $(i = 1,..., r)$ the computed value of $g(z)$ is therefore essentially $\prod\limits_{i=r+1}^{n} (z - \bar{\lambda}_i)$. Hence we can expect to obtain any of the remaining zeros just as accurately using values of $g_r(z)$ as values of $f(z)$. If z is in the neighbourhood of any of the λ_i $(i = 1,..., r)$ the computed values will be so erratic that the question of converging again to one of these zeros can scarcely arise. Naturally we must avoid attempting an evaluation at any of the points $z = \lambda_i'$, but this can easily be arranged.

In this argument we have assumed that the λ_i are simple zeros. However, multiple or pathologically close zeros do not give rise to any special difficulties. They are determined to lower accuracy only if the domains of indeterminacy are large, as indeed they will be, for example, for the explicit polynomial form. Thus if $f(z)$ has a double zero at $z = \lambda_1 = \lambda_2$, and we have accepted a zero at λ_1', then the computed values of $f(z)/(z - \lambda_1')$ behave as though the function has a simple zero in the neighbourhood of $z = \lambda_1$, provided z is appreciably outside the domain of indeterminacy. Zeros of multiplicities as high as five have been computed on ACE using this method of suppression, and all of them were obtained with errors which were no greater than were to be expected having regard to the size of the domains of indeterminacy.

We notice that the accidental acceptance of a zero which is seriously in error does not affect the accuracy of the later computed zeros, but if a λ_i' is accepted which is not of the limiting accuracy we are likely to converge again to a true limiting approximation to λ_i. Since programmes will normally be designed to find precisely n zeros of a polynomial of degree n, the acceptance of a spurious zero will naturally imply that a true zero is missed.

An important feature of the method of suppression is that each zero is in effect determined independently. The agreement of the sum of the computed zeros with the trace of the original matrix therefore constitutes quite a good check on their accuracy. This is not true of many methods of deflation. These are often of such a nature that the sum of the computed zeros is correct even if the zeros have little relation to the eigenvalues. With explicit polynomial deflation, for example, even if we accept random values as zeros at each stage except the last (at this stage the polynomial is of degree unity and hence the root is given without iteration!) it is easy to see that the sum is correct.

Asymptotic rates of convergence

58. It is not immediately obvious that Newton's method (§ 25) and Laguerre's method (§ 28) remain quadratically and cubically convergent when applied to $g_r(z)$. We now show that this is indeed true. Consider first the use of Newton's method and suppose that $z = \lambda_{r+1} + h$; then we have

$$g'_r(z)/g_r(z) = \sum_{i=1}^{n} 1/(z - \lambda_i) - \sum_{i=1}^{r} 1/(z - \lambda'_i)$$
$$= 1/h + \sum_{i=1}^{n}{}' 1/(z - \lambda_i) - \sum_{i=1}^{r} 1/(z - \lambda'_i), \qquad (58.1)$$

where $\sum'$ denotes that $i = r+1$ is omitted from the sum. Hence

$$g'_r(z)/g_r(z) = 1/h + A + O(h), \qquad (58.2)$$

where

$$A = \sum_{i=1}^{n}{}' 1/(\lambda_{r+1} - \lambda_i) - \sum_{i=1}^{r} 1/(\lambda_{r+1} - \lambda'_i), \qquad (58.3)$$

giving

$$z - g_r(z)/g'_r(z) = \lambda_{r+1} + Ah^2 + O(h^3). \qquad (58.4)$$

With Laguerre's method we use the formula

$$z_{k+1} = z_k - \frac{(n-r)g_r(z_k)}{g'_r(z_k) \pm \{(n-r-1)^2[g'_r(z_k)]^2 - (n-r)(n-r-1)g_r(z_k)g''_r(z_k)\}^{\frac{1}{2}}} \qquad (58.5)$$

and again taking $z_k = \lambda_{r+1} + h$ and using (55.2) and (55.3) it can be verified that

$$z_{k+1} - \lambda_{r+1} \sim \tfrac{1}{2}h^3[(n-r-1)B - A^2]/(n-r-1), \qquad (58.6)$$

where A is as defined in (58.3) and B is defined by

$$B = \sum_{i=1}^{n}{}' 1/(\lambda_{r+1} - \lambda_i)^2 - \sum_{i=1}^{r} 1/(\lambda_{r+1} - \lambda'_i)^2. \qquad (58.7)$$

Convergence in the large

59. So far we have avoided any reference to the problem of obtaining initial approximations to the eigenvalues. Naturally if we have approximations from some independent source the methods of this chapter are very satisfactory, but in practice they are most frequently used as the sole means of locating eigenvalues.

When it is known that all the eigenvalues are real then the situation is quite satisfactory. As far as Laguerre's method is concerned we know that convergence to neighbouring zeros is guaranteed starting from any initial approximation. A simple procedure is to use $\|A\|_\infty$ as the starting value for each zero and the method of suppression described in § 55. The convergence in the large is so remarkable that this simple procedure is perfectly satisfactory unless A has some multiple or pathologically close zeros. For such zeros (28.12) implies that

$$z_{k+1} - \lambda_m \sim C(z_k - \lambda_m), \tag{59.1}$$

and

$$z_{k+2} - \lambda_m \sim C(z_{k+1} - \lambda_m), \tag{59.2}$$

giving

$$(z_{k+2} - z_{k+1}) \sim C(z_{k+1} - z_k). \tag{59.3}$$

Writing $z_{k+1} - z_k = \Delta z_k$ we have

$$\Delta z_{k+1} / \Delta z_k \sim C. \tag{59.4}$$

Parlett (1964) suggested that convergence to a simple zero was usually so rapid that if from a certain point on

$$|\Delta z_{k+1} / \Delta z_k| \div |\Delta z_k / \Delta z_{k-1}| > 0 \cdot 8, \tag{59.5}$$

then it could be assumed that the zero which was being approached was at least of multiplicity two and (28·13) could be used with $r = 2$. Parlett found that it was reasonably safe to use this criterion after three iterations. This process could be continued in an attempt to recognize zeros of higher multiplicity.

If we use Newton's method and $\|A\|_\infty$ as the starting value for each zero with the method of suppression described in § 55, obviously we obtain the eigenvalues in monotonically decreasing order. Finally if we use successive linear interpolation and, say, $(1 \cdot 1)\|A\|_\infty$ and $\|A\|_\infty$ as the starting values we again obtain the eigenvalues in decreasing order. The convergence in the large of these last two methods is not as satisfactory as that of Laguerre's method, and it seems worth while to try to obtain somewhat better starting values. Since the eigenvalues are found in decreasing order the last computed zero λ_m is an upper bound of the remaining zeros but the method of suppression precludes its use. Instead we could use $\frac{1}{2}(\lambda_m + \lambda_{m-1})$ at all stages after the second for Newton's method and $\frac{1}{3}(\lambda_{m-1} + 2\lambda_m)$ and $\frac{1}{3}(2\lambda_{m-1} + \lambda_m)$ for successive interpolation. When λ_{m-1} is equal to λ_m it should be replaced by the first of $\lambda_{m-2}, ..., \lambda_1$ which is greater than λ_{m-1} or, if none such exists, by $\|A\|_\infty$. Notice that instead of $\|A\|_\infty$ we may use any known upper bound for the eigenvalues. If A has been obtained from an original matrix A_0 by a similarity transformation it may well happen that $\|A\|_\infty \gg \|A_0\|_\infty$ in which case it is much better to use $\|A_0\|_\infty$. With each of the last two methods not only are the zeros found in monotonic decreasing order but the convergence to each zero is monotonic.

When the starting value is much larger than any of the eigenvalues, both Newton's method and successive linear interpolation give rather slow convergence. The use of the following (unpublished) result (discovered independently by Kahan and Maehly) can lead to a considerable reduction in the required number of iterations. Suppose first that we have

$$x > \lambda_1 > \lambda_2 > \lambda_3 > \ldots > \lambda_n, \tag{59.6}$$

and that we apply twice the Newton correction to give as our next approximation the value ξ defined by

$$\xi = x - 2f(x)/f'(x). \tag{59.7}$$

Let the zeros of $f'(x)$ be denoted by μ_i $(i = 1, \ldots, n-1)$ so that

$$x > \lambda_1 > \mu_1 > \lambda_2 > \mu_2 > \lambda_3 > \ldots > \lambda_{n-1} > \mu_{n-1} > \lambda_n. \tag{59.8}$$

Then we shall show that $\xi > \mu_1$. Since $f'(x) = \prod (x - \mu_i)$ it is clear that all derivatives of f are non-negative for $x > \mu_1$. Hence we may write

$$f(x) = f(\mu_1) + \sum_{2}^{n} b_r (x - \mu_1)^r \quad (b_r > 0), \tag{59.9}$$

and we have

$$\xi - \mu_1 = (x - \mu_1) - 2f(x)/f'(x)$$

$$= (x - \mu_1) - 2\left[f(\mu_1) + \sum_{2}^{n} b_r (x - \mu_1)^r \right] \bigg/ \sum_{2}^{n} r b_r (x - \mu_1)^{r-1}$$

$$= \left[\sum_{2}^{n} (r-2) b_r (x - \mu_1)^r - 2f(\mu_1) \right] \bigg/ \sum_{2}^{n} r b_r (x - \mu_1)^{r-1}. \tag{59.10}$$

Since $f(\mu_1) < 0$ and $b_r > 0$, both the numerator and the denominator consist entirely of non-negative terms, showing that $\xi > \mu_1$. Notice that we required only that λ_1 and μ_1 were real and that all derivatives of f were non-negative for $x > \mu_1$. The proof therefore applies for example to $f(x) = x^{20} - 1$.

Similarly with successive linear interpolation starting with an x_1 and x_2 both greater than λ_1 we can apply double the shift given by linear interpolation without obtaining a value less than μ_1. This follows since if $x_1 > x_2$ the correction corresponding to linear interpolation is smaller than that corresponding to Newton's method applied at the point x_2.

Hence we may safely apply the double-shift technique and if it is a simple eigenvalue we shall not progress beyond μ_1. The iterates will be monotonically decreasing until a value is obtained which is less than λ_1. The next shift will be positive and at this point the double shift can be abandoned. Notice that the proof continues to be valid for eigenvalues of any multiplicity. If in particular λ_1 is a multiple eigenvalue the double-shift technique never gives a value below λ_1 and convergence is monotonic throughout until rounding errors dominate the computation. Kahan has extended these ideas considerably and has further results in connexion with Muller's method.

When A is a symmetric or quasi-symmetric tri-diagonal matrix we can obtain information on the distribution of the zeros from the Sturm sequence property and bisection (Chapter 5, § 39). We may therefore combine the methods of this chapter with the method of bisection. A particularly elegant combination of the methods of bisection and of successive linear interpolation has been made by Dekker (1962). This may be used to locate a real zero of any real function once an interval has been found containing such a zero.

Complex zeros

60. When we turn to complex eigenvalues of either real or complex matrices we no longer have any satisfactory methods of ensuring convergence. In general Laguerre's method appears to have by far the best general convergence properties, but Parlett (1964) has shown that there exist simple polynomials for which some starting values give a repeated cycle of iterates. He considered the function $z(z^2+b^2)$ with $b > 0$, for which

$$z_{k+1} = z_k - 3z_k(z_k^2+b^2)/[3z_k^2+b^2\pm 2b(b^2-3z_k^2)^{\frac{1}{2}}]. \qquad (60.1)$$

If $z_1 = 3^{-\frac{1}{2}}b$ then $z_2 = -z_1$ and $z_3 = z_1$. Cycling is also possible with each of the other methods, and in particular only the methods of Muller and Laguerre are capable of producing complex iterates starting from a real value for a real function. There is the additional complication that cycling may result from ill-conditioning and not from a fundamental failure of the method which is being used. In such a case the use of sufficiently high precision in the computation would remove the apparent cycling.

It is idle to deny that these comments constitute a serious criticism of the methods of this chapter. The purist may feel disinclined to use methods the success of which would appear to be largely fortuitous, but it is salutary to remember that there are as yet no alternative methods for the general eigenvalue problem which can be *guaranteed* to give results *in an acceptable time*. Our experience has been that programmes based on the methods of this chapter are very powerful as well as being among the most accurate.

Recommendations

61. Of the methods I have used for real matrices the most effective has been that based on the following combination of the methods of Laguerre and Bairstow. For each eigenvalue Laguerre's method is tried first, but if convergence is not obtained in a prescribed number, k, of iterations I switch to Bairstow's method. In practice the value 32 has been used for k, though most commonly Laguerre's method gives convergence in from 4 to 10 iterations. The methods of suppression described in §§ 55–56 are used and not deflation.

If the original matrix is of general form then my preferred method has been to reduce it to Hessenberg form and to use this form for evaluating the function and its derivatives. Single-precision computation has almost invariably given correctly all those figures which are unaltered by perturbations of up to a few parts in 2^t in the elements of the original matrix. For general purposes this method has proved superior to all others I have used except the method of Francis described in the next chapter.

I have also experimented with the further reduction to tri-diagonal form, but recommend that in this case both the reduction to tri-diagonal form and the evaluations of the function and its derivatives should be performed in double-precision in order to guard against the possible deterioration in the condition of the eigenvalues. For matrices of orders greater than about sixteen this has proved faster than the use of the Hessenberg matrix in single-precision, and except in rare cases where the deterioration in condition has been exceptionally severe the tri-diagonal matrix has given the more accurate results. My preference for the Hessenberg form might therefore be regarded by some as rather too cautious.

Although we have made it clear that we regard the use of the Frobenius form as dangerous, in that it may well be catastrophically worse-conditioned than the original matrix, we have found the programme based on its use surprisingly satisfactory in general for matrices arising from damped mechanical or electrical systems. It is common for the corresponding characteristic polynomial to be well-conditioned. When this is true methods based on the use of the explicit characteristic polynomial are both fast and accurate.

Complex matrices

62. Surprisingly few general matrices with complex elements appear to have been solved. For almost all those solved at the

National Physical Laboratory by the methods of this chapter, Muller's method has been used. Where the original matrix was of general form the evaluations were again performed in single-precision using the reduced Hessenberg form. However a number of examples have been encountered in which the original matrix was itself of tri-diagonal form and for these a single-precision evaluation based on the method of § 8 has been used. As an indication of the accuracy attained we may mention that for a group of 8 such complex tri-diagonal matrices of order 105 (the highest order of complex matrix with which we have dealt) subsequent analysis showed that the maximum error in any eigenvalue was less than $2^{-43} \|A\|_\infty$, this accuracy being achieved using a 46 binary digit mantissa. The average number of iterations per eigenvalue was 10.

In all the iterative eigenvalue programmes an upper bound for the eigenvalues is computed before starting the iterations. Where the original matrix was of a different form from that used for the evaluation, the smaller of the infinity norms of the original matrix and of the reduced matrix is used. This norm gives an upper bound for the modulus of all eigenvalues; when an iterate exceeds this norm it is replaced by some suitable value. The way in which this is done is not critical, but unless some such device is used an excessive number of iterations may be required when an iterate is accepted which is far outside the domain containing the eigenvalues. An example of such an excursion arises if we take $z = \frac{1}{2}$ for the polynomial $z^{20} - 1$ and use Newton's method. The next approximation is roughly $2^{19}/20$, and a return to the relevant domain would take place very slowly.

Matrices containing an independent parameter

63. As we have remarked, iterative methods appear to best advantage when approximations to the eigenvalues are available from an independent source. There is one important situation when such approximations are indeed available. This is when the matrix elements are functions of an independent parameter v, and the history of some or all of the eigenvalues is required as v varies.

This situation is quite common in connexion with the generalized eigenvalue problem

$$\det[B_0(v) + \lambda B_1(v) + \ldots + \lambda^r B_r(v)] = 0. \tag{63.1}$$

An important example is provided by the 'flutter' problem in aerodynamics; for this problem we have $r = 2$ and v is the velocity of the aircraft. The flutter velocity is the value of v for which an eigenvalue has a zero real part.

The eigenvalue problem is solved completely for an initial velocity v_0, and the history of some or all of these eigenvalues is traced for a series of values v_0, v_1,.... If the difference between successive values of v_i is kept sufficiently small each eigenvalue at v_i will serve as a good approximation to the corresponding eigenvalue at v_{i+1}. The programme on ACE based on Muller's method in which the function values are obtained directly from $B_0(v) + \lambda B_1(v) + \ldots + \lambda^r B_r(v)$ by Gaussian elimination with interchanges, has proved particularly successful. Even better initial approximations could have been obtained by differencing $\lambda_s(v_i)$, but since convergence is usually immediate for the intervals in v which are required in practice, there has been little incentive to use such a refinement. No doubt successive linear interpolation would have been equally satisfactory but we have no experience of its use in this connexion. The ACE programme has been used very successfully to deal with the generalized problem in which the B_i were complex matrices of order 6 and r was equal to 5, so that there were 30 eigenvalues for each value of v.

Additional notes

The current interest in Laguerre's method was stimulated by Maehly (1954). Probably the most careful analysis of its effectiveness in practice is that given by Parlett (1964) and the reader is strongly advised to study this paper carefully. Practical experience of iterative methods soon reveals that attention to detail, particularly as regards the determination of the attainment of limiting accuracy and the recognition of multiple roots and of ill-conditioning, is vital to the success of any automatic procedure.

Since going to press Traub (1964) has published an excellent treatise on iterative methods which includes an improved formulation of Muller's method based on Newtonian interpolation polynomials instead of Lagrangian. For a quadratic we have in the usual notation for divided differences

$$f(w) = f(w_i) + (w - w_i)f[w_i, w_{i-1}] + (w - w_i)(w - w_{i-1})f[w_i, w_{i-1}, w_{i-2}]$$
$$= f(w_i) + (w - w_i)\{f[w_i, w_{i-1}] + (w_i - w_{i-1})f[w_i, w_{i-1}, w_{i-2}]\} + $$
$$+ (w - w_i)^2 f[w_i, w_{i-1}, w_{i-2}]$$
$$= f(w_i) + (w - w_i)p + (w - w_i)^2 q \quad \text{(say)},$$

giving
$$w = w_i - 2f(w_i)/\{p \pm [p^2 - 4f(w_i)q]^{\frac{1}{2}}\}.$$

Since $f[w_i, w_{i-1}, w_{i-2}] = \{f[w_i, w_{i-1}] - f[w_{i-1}, w_{i-2}]\}/(w_i - w_{i-2})$ we have only to compute $f[w_i, w_{i-1}]$ at each stage.

8

The *LR* and *QR* Algorithms

Introduction

1. IN CHAPTER 1, § 42, we showed that any matrix may be reduced to triangular form by a similarity transformation and in § 47 we showed further that the matrix of the transformation may be chosen to be unitary. Obviously any algorithm which gives such a reduction must be essentially iterative since it gives the zeros of a polynomial.

As we showed in Chapter 1, § 48, a unitary transformation which reduces a normal matrix to *triangular* form must, in fact, reduce it to *diagonal* form. The method of Jacobi (Chapter 5, § 3) gives such a reduction in the case of real symmetric matrices. The basic stratagem in that method is the successive reduction of the norm of the matrix of super-diagonal elements by a sequence of elementary orthogonal transformations. Since symmetry is retained this automatically ensures a simultaneous reduction in the norm of the matrix of sub-diagonal elements.

Jacobi's method may be extended in a fairly simple way to cover normal matrices in general (Goldstine and Horwitz, 1959), but it is natural to attempt to develop a method based on a similar stratagem for the reduction of a general matrix to triangular form. Unfortunately, although a number of attempts have been made, no algorithm has yet emerged which is comparable in practice with the more successful of those described in Chapters 6 and 7.

The methods of deflation described in Chapter 7, §§ 49–54, effectively lead to a reduction of a matrix of Hessenberg form to one of triangular form, but these were numerically unstable. In Chapter 9 we describe stable methods of deflation which provide a similarity reduction of a general matrix to triangular form.

In this chapter we shall be concerned with two algorithms known respectively as the *LR* and *QR* algorithms. The first of these, which was developed by Rutishauser (1958), gives a reduction of a general matrix to triangular form by means of non-unitary transformations. In my opinion its development is the most significant advance which has been made in connexion with the eigenvalue problem since the advent of automatic computers. The *QR* algorithm, which was developed later by Francis, is closely related to the *LR* algorithm but is based on the use of unitary transformations. In many respects this has proved to be the most effective of known methods for the solution of the general algebraic eigenvalue problem.

Real matrices with complex eigenvalues

2. A disadvantage of the triangular form is that for real matrices having some complex conjugate eigenvalues it takes us into the field of complex numbers. We introduce a simple modification of the triangular form which enables us to remain in the real field. We show that if A is a real matrix having the eigenvalues $\lambda_1 \pm i\mu_1$, $\lambda_2 \pm i\mu_2$,..., $\lambda_s \pm i\mu_s$, λ_{2s+1},..., λ_n, where the λ_i and μ_i are real, then there exists a real matrix H such that $B = HAH^{-1}$ is of the form

$$
B = \begin{bmatrix}
X_1 & & & & \\
 & X_2 & & P & \\
 & & X_s & & \\
 & & & \lambda_{2s+1} & \\
 & & & & \ddots \\
 & & & & & \lambda_n
\end{bmatrix}, \tag{2.1}
$$

where X_r is a 2×2 matrix, on the diagonal of B, having the eigenvalues $\lambda_r \pm i\mu_r$. The matrix B is otherwise null below the diagonal and hence differs from a triangular matrix only in possessing s real sub-diagonal elements, one associated with each X_r. We shall show further that H can be chosen to be orthogonal.

The proofs follow closely those given in Chapter 1, § 42 and § 47, for the non-unitary and unitary reductions to triangular form. The only new feature is that introduced by the pairs of complex conjugate eigenvalues. Reference to the proofs in Chapter 1 shows that it is sufficient to demonstrate that there exists a matrix H_1 such that

$$
H_1 A H_1^{-1} = \begin{bmatrix}
X_1 & P_1 \\
\hline
O & A_2
\end{bmatrix}, \tag{2.2}
$$

since the result then follows by induction. We may show this as follows.

Let $x_1 \pm iy_1$ be eigenvectors corresponding to $\lambda_1 \pm i\mu_1$; clearly x_1 and y_1 must be independent since otherwise the corresponding eigenvalue would be real, and we have

$$A[x_1 \mid y_1] = [x_1 \mid y_1]\begin{bmatrix} \lambda_1 & \mu_1 \\ -\mu_1 & \lambda_1 \end{bmatrix} = [x_1 \mid y_1]\Lambda_1. \tag{2.3}$$

Suppose we can find a non-singular H_1 such that

$$H_1[x_1 \mid y_1] = \begin{bmatrix} V_1 \\ \overline{O} \end{bmatrix}, \tag{2.4}$$

where V_1 is a non-singular 2×2 matrix. Then from (2.3) we have

$$H_1 A H_1^{-1} H_1[x_1 \mid y_1] = H_1[x_1 \mid y_1]\Lambda_1, \tag{2.5}$$

giving

$$H_1 A H_1^{-1}\begin{bmatrix} V_1 \\ \overline{O} \end{bmatrix} = \begin{bmatrix} V_1 \\ \overline{O} \end{bmatrix}\Lambda_1. \tag{2.6}$$

Hence if we write

$$H_1 A H_1^{-1} = \begin{bmatrix} X_1 & P_1 \\ \hline Q_1 & A_2 \end{bmatrix}, \tag{2.7}$$

equation (2.6) gives

$$X_1 V_1 = V_1 \Lambda_1, \qquad Q_1 V_1 = O, \tag{2.8}$$

showing that Q_1 is null and that X_1 is similar to Λ_1 and therefore has the eigenvalues $\lambda_1 \pm i\mu_1$. Any H_1 for which equation (2.4) is true therefore meets our requirements.

In fact we can build up such an H_1 either as the product of real stabilized elementary non-unitary matrices, or of real plane rotations or real elementary Hermitians. This is obvious if we think of x_1 and y_1 as the first two columns of an $n \times n$ matrix, and consider the triangularization of that matrix by Gaussian elimination with partial pivoting or by Givens' or Householder's method. In each of these cases V_1 is of the form

$$\begin{bmatrix} \times & \times \\ 0 & \times \end{bmatrix}, \tag{2.9}$$

and its non-singularity follows from the independence of x_1 and y_1.

The *LR* algorithm

3. Rutishauser's algorithm is based on the triangular decomposition of a matrix (Chapter 4, § 36); Rutishauser (1958) writes

$$A = LR, \tag{3.1}$$

where L is unit lower-triangular and R is upper-triangular. In order to facilitate comparison with Rutishauser's work we shall use R instead of U in this chapter.

Suppose now we form the similarity transform $L^{-1}AL$ of the matrix A. We have

$$L^{-1}AL = L^{-1}(LR)L = RL. \tag{3.2}$$

Hence if we decompose A and then multiply the factors in the reverse order we obtain a matrix similar to A. In the LR algorithm this process is repeated indefinitely. If we rename the original matrix A_1 then the algorithm is defined by the equations

$$A_{s-1} = L_{s-1}R_{s-1}, \qquad R_{s-1}L_{s-1} = A_s. \tag{3.3}$$

Clearly A_s is similar to A_{s-1} and hence by induction to A_1. Rutishauser showed that under certain restrictions

$$L_s \to I \quad \text{and} \quad R_s \to A_s \to \begin{bmatrix} \lambda_1 & & & & \\ & \lambda_2 & & X & \\ & & \cdot & & \\ & & & \cdot & \\ & O & & & \cdot \\ & & & & \lambda_n \end{bmatrix} \quad \text{as } s \to \infty. \tag{3.4}$$

4. Before proving this result we establish relations between the successive iterates which will be required repeatedly. From (3.3) we have

$$A_s = L_{s-1}^{-1}A_{s-1}L_{s-1}, \tag{4.1}$$

and repeated application of this result gives

$$A_s = L_{s-1}^{-1}L_{s-2}^{-1}...L_2^{-1}L_1^{-1}A_1L_1L_2...L_{s-1}, \tag{4.2}$$

or

$$L_1L_2...L_{s-1}A_s = A_1L_1L_2...L_{s-1}. \tag{4.3}$$

Now the matrices T_s and U_s defined by

$$T_s = L_1L_2...L_s \quad \text{and} \quad U_s = R_sR_{s-1}...R_1 \tag{4.4}$$

are *unit lower-triangular* and *upper-triangular* respectively. Consider the product T_sU_s. We have

$$\begin{aligned}
T_sU_s &= L_1L_2...L_{s-1}(L_sR_s)R_{s-1}...R_2R_1 \\
&= L_1L_2...L_{s-1}A_sR_{s-1}...R_2R_1 \\
&= A_1L_1L_2...L_{s-1}R_{s-1}...R_2R_1 \quad \text{(from (4.3))} \\
&= A_1T_{s-1}U_{s-1}. \tag{4.5}
\end{aligned}$$

Repeated application of this result shows that

$$T_s U_s = A_1^s, \tag{4.6}$$

so that $T_s U_s$ gives the triangular decomposition of A_1^s.

Proof of the convergence of the A_s

5. Equation (4.6) is the fundamental result on which the analysis depends. We shall use it to prove that if the eigenvalues λ_i of A_1 satisfy the relations

$$|\lambda_1| > |\lambda_2| > ... > |\lambda_n| \tag{5.1}$$

then, in general, the results (3.4) are true. Before giving the formal proof we consider the simple case of a matrix of order three. If for simplicity we express the matrix X of right-hand eigenvectors in the form

$$X = \begin{bmatrix} x_1 & y_1 & z_1 \\ x_2 & y_2 & z_2 \\ x_3 & y_3 & z_3 \end{bmatrix}, \tag{5.2}$$

and its inverse Y by

$$X^{-1} = Y = \begin{bmatrix} a_1 & b_1 & c_1 \\ a_2 & b_2 & c_2 \\ a_3 & b_3 & c_3 \end{bmatrix}, \tag{5.3}$$

then we have

$$F = A_1^s = X \begin{bmatrix} \lambda_1^s & & \\ & \lambda_2^s & \\ & & \lambda_3^s \end{bmatrix} X^{-1} \tag{5.4}$$

$$= \begin{bmatrix} \lambda_1^s x_1 a_1 + \lambda_2^s y_1 a_2 + \lambda_3^s z_1 a_3 & \lambda_1^s x_1 b_1 + \lambda_2^s y_1 b_2 + \lambda_3^s z_1 b_3 & \lambda_1^s x_1 c_1 + \lambda_2^s y_1 c_2 + \lambda_3^s z_1 c_3 \\ \lambda_1^s x_2 a_1 + \lambda_2^s y_2 a_2 + \lambda_3^s z_2 a_3 & \lambda_1^s x_2 b_1 + \lambda_2^s y_2 b_2 + \lambda_3^s z_2 b_3 & \lambda_1^s x_2 c_1 + \lambda_2^s y_2 c_2 + \lambda_3^s z_2 c_3 \\ \lambda_1^s x_3 a_1 + \lambda_2^s y_3 a_2 + \lambda_3^s z_3 a_3 & \lambda_1^s x_3 b_1 + \lambda_2^s y_3 b_2 + \lambda_3^s z_3 b_3 & \lambda_1^s x_3 c_1 + \lambda_2^s y_3 c_2 + \lambda_3^s z_3 c_3 \end{bmatrix}. \tag{5.5}$$

Now $T_s U_s$ is the triangular decomposition of A_1^s and hence the elements of the first column of T_s are given by

$$t_{11}^{(s)} = 1$$
$$\left. \begin{aligned} t_{21}^{(s)} &= (\lambda_1^s x_2 a_1 + \lambda_2^s y_2 a_2 + \lambda_3^s z_2 a_3)/(\lambda_1^s x_1 a_1 + \lambda_2^s y_1 a_2 + \lambda_3^s z_1 a_3) \\ t_{31}^{(s)} &= (\lambda_1^s x_3 a_1 + \lambda_2^s y_3 a_2 + \lambda_3^s z_3 a_3)/(\lambda_1^s x_1 a_1 + \lambda_2^s y_1 a_2 + \lambda_3^s z_1 a_3) \end{aligned} \right\}. \tag{5.6}$$

Clearly if $x_1 a_1 \neq 0$ then

$$t_{21}^{(s)} = x_2/x_1 + O(\lambda_2/\lambda_1)^s \quad \text{and} \quad t_{31}^{(s)} = x_3/x_1 + O(\lambda_2/\lambda_1)^s, \tag{5.7}$$

and elements of T_s tend, in general, to the corresponding elements obtained in the triangular decomposition of X.

For the elements of the second column of T_s we have similarly

$$\left.\begin{aligned}
t_{22}^{(s)} &= 1 \\
t_{32}^{(s)} &= (f_{11}f_{32}-f_{12}f_{31})/(f_{11}f_{22}-f_{12}f_{21}) \\
&= \frac{(\lambda_1\lambda_2)^s(x_1y_3-x_3y_1)(a_1b_2-a_2b_1)+\dots}{(\lambda_1\lambda_2)^s(x_1y_2-x_2y_1)(a_1b_2-a_2b_1)+\dots} \\
&= \frac{x_1y_3-x_3y_1}{x_1y_2-x_2y_1}+O\left(\frac{\lambda_3}{\lambda_2}\right)^s+\dots
\end{aligned}\right\}, \qquad (5.8)$$

provided

$$(a_1b_2-a_2b_1)(x_1y_2-x_2y_1) \neq 0. \qquad (5.9)$$

From (5.8) we see that the limiting value of $t_{32}^{(s)}$ is equal to the corresponding element obtained in the triangular decomposition of X. We have therefore established that if

$$X = TU \qquad (5.10)$$

then provided $x_1a_1 \neq 0$ and $(x_1y_2-x_2y_1)(a_1b_2-a_2b_1) \neq 0$ then

$$T_s \to T. \qquad (5.11)$$

6. We have shown in this simple case that the matrix T_s tends to the unit lower-triangular matrix obtained from the triangular decomposition of the matrix X of eigenvectors, provided the leading principal minors of X and Y are non-zero. We now show that this result is true in general for a matrix with distinct eigenvalues. If we write

$$T_sU_s = A^s = B, \qquad (6.1)$$

then from the explicit expressions obtained in Chapter 4, § 19, for the elements of the triangular factors of a matrix, we have

$$t_{ji}^{(s)} = \det(B_{ji})/\det(B_{ii}). \qquad (6.2)$$

Here B_{ii} denotes the leading principal submatrix of B and B_{ji} the matrix obtained from B_{ii} by replacing its ith row by the corresponding elements of the jth row of B. From the relation

$$B = A^s = X \operatorname{diag}(\lambda_i^s)Y \qquad (6.3)$$

we have

$$
B_{ji} =
\begin{bmatrix}
x_{11} & x_{12} & \cdots & x_{1n} \\
x_{21} & x_{22} & \cdots & x_{2n} \\
\cdots & \cdots & \cdots & \cdots \\
x_{i-1,1} & x_{i-1,2} & \cdots & x_{i-1,n} \\
x_{j1} & x_{j2} & \cdots & x_{jn}
\end{bmatrix}
\times
\begin{bmatrix}
\lambda_1^s y_{11} & \lambda_1^s y_{12} & \cdots & \lambda_1^s y_{1i} \\
\lambda_2^s y_{21} & \lambda_2^s y_{22} & \cdots & \lambda_2^s y_{2i} \\
\cdots & \cdots & \cdots & \cdots \\
\lambda_n^s y_{n1} & \lambda_n^s y_{n2} & \cdots & \lambda_n^s y_{ni}
\end{bmatrix}.
$$

$$(6.4)$$

By the Theorem of Corresponding Matrices (Chapter 1, § 15) $\det(B_{ji})$ is equal to the sum of the products of corresponding i-rowed minors of the two matrices on the right of (6.4). Hence we may write

$$
t_{ji}^{(s)} = \frac{\sum x_{p_1 p_2 \ldots p_i}^{(j)} y_{p_1 p_2 \ldots p_i}^{(i)} (\lambda_{p_1} \lambda_{p_2} \ldots \lambda_{p_i})^s}{\sum x_{p_1 p_2 \ldots p_i}^{(i)} y_{p_1 p_2 \ldots p_i}^{(i)} (\lambda_{p_1} \lambda_{p_2} \ldots \lambda_{p_i})^s},
\tag{6.5}
$$

where $x_{p_1 p_2 \ldots p_i}^{(j)}$ is the i-rowed minor consisting of rows $1, 2, \ldots, i-1$ and j and columns $p_1, p_2, \ldots, p_i$ of X, and $y_{p_1 p_2 \ldots p_i}^{(i)}$ is the i-rowed minor consisting of rows $p_1, p_2, \ldots, p_i$ and columns $1, 2, \ldots, i$ of Y.

The dominant terms in the numerator and denominator of (6.5) are those associated with $(\lambda_1 \lambda_2 \ldots \lambda_i)^s$ provided the corresponding coefficients are non-zero. The appropriate term in the denominator is

$$
\det(X_{ii})\det(Y_{ii})(\lambda_1 \lambda_2 \ldots \lambda_i)^s,
\tag{6.6}
$$

where X_{ii} and Y_{ii} are the leading principal submatrices of order i. Provided $\det(X_{ii})\det(Y_{ii})$ is non-zero, we have

$$
t_{ji}^{(s)} \to \frac{\det(X_{ji})\det(Y_{ii})}{\det(X_{ii})\det(Y_{ii})} = \frac{\det(X_{ji})}{\det(X_{ii})},
\tag{6.7}
$$

showing that $T_s \to T$ where
$$
X = TU.
\tag{6.8}
$$

From (4.3) and (4.4) we have

$$
A_s = T_{s-1}^{-1} A_1 T_{s-1} \to T^{-1} A_1 T = U X^{-1} A X U^{-1} = U \operatorname{diag}(\lambda_i) U^{-1},
$$
$$(6.9)$$

showing that the limiting A_s is upper-triangular with diagonal elements λ_i.

From the relation $L_s = T_{s-1}^{-1} T_s$ and equation (6.5) it can be proved by an elementary but rather tedious argument that

$$
l_{ij}^{(s)} = O(\lambda_i/\lambda_j)^s \quad (s \to \infty).
\tag{6.10}
$$

From this we may deduce, using the relation $A_s = L_s R_s$ and the fact that R_s tends to a limit, that

$$a_{ij}^{(s)} = O(\lambda_i/\lambda_j)^s \quad (s \to \infty) \quad (i > j). \tag{6.11}$$

Hence if the separation of some of the eigenvalues is poor, the convergence of A_s to triangular form may be slow. An alternative (and simpler) proof is given in § 33 but the method of proof used here is instructive.

7. In establishing these results the following assumptions have been made either explicitly or implicitly.

(i) The eigenvalues are all of different absolute magnitude. Note that A may be complex, but not a real matrix having some complex conjugate eigenvalues (the latter case is discussed in § 9).

(ii) The triangular decomposition exists at every stage. It is easy to construct matrices, which are not otherwise exceptional, for which this condition is not satisfied, for example

$$A = \begin{bmatrix} 0 & 1 \\ -3 & 4 \end{bmatrix}. \tag{7.1}$$

This matrix has eigenvalues 1 and 3 but it has no triangular decomposition. Notice though, that $(A+I)$, for example, has a triangular decomposition and we could well work with this modified matrix.

(iii) The leading principal minors of X and Y are all non-zero. There are quite simple matrices for which this requirement is not met. The most obvious examples are provided by matrices of the form

$$\begin{bmatrix} A_1 & O \\ \hline O & A_2 \end{bmatrix} \tag{7.2}$$

which clearly retain this form at each stage. The matrices A_1 and A_2 are effectively treated independently and the limiting matrix has the eigenvalues of A_1 in the upper left-hand corner and those of A_2 in the lower right-hand corner *independent of their relative magnitudes*. It is instructive to consider the 2×2 matrix

$$\begin{bmatrix} a & \epsilon_1 \\ \epsilon_2 & b \end{bmatrix}, \tag{7.3}$$

where $|b| > |a|$. If the ϵ_i are zero the matrix remains unchanged by the *LR* algorithm and the limiting matrix does not have the eigenvalues a and b in the correct order. The matrix of eigenvectors is

$$\begin{bmatrix} 0 & 1 \\ 1 & 0 \end{bmatrix}$$ and its first principal minor is zero. If the ϵ_i are non-zero but small, then the eigenvalues are very close to a and b; the limiting matrix now has the eigenvalues in the correct order but, not surprisingly, the convergence may be slow even when a and b are not particularly close together.

In this example it is obvious that the two halves of the matrix are completely 'decoupled' when $\epsilon_1 = \epsilon_2 = 0$, and the introduction of small ϵ_1 and ϵ_2 effects a 'weak coupling'. In more complicated examples we may have complete decoupling initially, that is for some i we have

$$\det(X_{ii})\det(Y_{ii}) = 0, \tag{7.4}$$

but rounding errors made in the *LR* process may have the effect of introducing weak coupling.

Positive definite Hermitian matrices

8. When A_1 is a positive definite Hermitian matrix we can remove the restrictions discussed in § 7. We now have

$$A_1 = X \operatorname{diag}(\lambda_i) X^H, \tag{8.1}$$

where X is unitary and the λ_i are real and positive. Hence equation (6.5) becomes

$$t_{ji}^{(s)} = \frac{\sum x_{p_1 p_2 \ldots p_i}^{(j)} \, \bar{x}_{p_1 p_2 \ldots p_i}^{(i)} (\lambda_{p_1} \lambda_{p_2} \ldots \lambda_{p_i})^s}{\sum |x_{p_1 p_2 \ldots p_i}^{(i)}|^2 (\lambda_{p_1} \lambda_{p_2} \ldots \lambda_{p_i})^s} . \tag{8.2}$$

Now consider the class of aggregates $p_1 p_2 \ldots p_i$ for which $x_{p_1 p_2 \ldots p_i}^{(i)} \neq 0$; let $q_1 q_2 \ldots q_i$ be a member of this class such that $\lambda_{q_1} \lambda_{q_2} \ldots \lambda_{q_i}$ takes a value which is not exceeded for any other member. The denominator is clearly dominated by the term (or terms) in $(\lambda_{q_1} \lambda_{q_2} \ldots \lambda_{q_i})^s$. As far as the numerator is concerned we know from our definition of $q_1, q_2, \ldots, q_i$ that there cannot be a term which is of greater order of magnitude than $(\lambda_{q_1} \lambda_{q_2} \ldots \lambda_{q_i})^s$ and indeed this term may have a zero coefficient. Hence $t_{ji}^{(s)}$ tends to a limit which can be zero.

We wish here to make no assumptions about the principal minors of X and so we cannot assert that the limiting T_s is the matrix obtained from the triangular decomposition of X. Accordingly we do not proceed as in equation (6.9), but we write

$$T_s \to T_\infty, \tag{8.3}$$

so that

$$L_s = T_{s-1}^{-1} T_s \to T_\infty^{-1} T_\infty = I. \tag{8.4}$$

Further we have

$$A_s = T_{s-1}^{-1} A_1 T_{s-1} \to T_\infty^{-1} A_1 T_\infty, \tag{8.5}$$

so that A_s tends to a limit, A_∞ say. Now

$$R_s = L_s^{-1} A_s \to I T_\infty^{-1} A_1 T_\infty, \tag{8.6}$$

and hence R_s tends to the same limit as A_s. Since R_s is triangular for all s this limit is necessarily triangular. It must have the eigenvalues of A_1 in some order on its diagonal since from (8.5) it is similar to A_1.

Notice that the proof is unaffected by the presence of multiple eigenvalues or by the vanishing of some of the leading principal minors of X, though the λ_i may not necessarily be in order of decreasing magnitude on the diagonal of A. Again there is the danger that in the case when exact computation would lead to a limiting matrix with 'disordered' eigenvalues, the incidence of rounding errors will destroy this condition and may result in a *very slow* convergence to a triangular matrix with the eigenvalues correctly ordered.

Complex conjugate eigenvalues

9. When A_1 is real but has one or more pairs of complex conjugate eigenvalues it is obvious that A_s cannot tend to an upper triangular matrix having these eigenvalues on the diagonal since all A_s are real. We consider first the case when there is one pair of complex conjugate eigenvalues λ_{m-1} and $\bar{\lambda}_{m-1}$ but otherwise $|\lambda_i| > |\lambda_j|$ if $i < j$. Since this is to be an extension of § 6 we shall also assume that none of the leading principal minors of X or Y is zero. From (6.5) we see that

$$t_{ji}^{(s)} \to \det X_{ji}/\det X_{ii} \quad (i \neq m-1) \tag{9.1}$$

just as in the previous case. For $i = m-1$ however, there are terms in both $(\lambda_1 \lambda_2 \ldots \lambda_{m-2} \lambda_{m-1})^s$ and $(\lambda_1 \lambda_2 \ldots \lambda_{m-2} \bar{\lambda}_{m-1})^s$ in the numerator and denominator. In order to study the asymptotic behaviour of these elements it is convenient to introduce a simplified notation for the minors of order $(m-1)$ of X and Y. We write

$$p_j = \det \begin{bmatrix} x_{11} & x_{12} & \cdots & x_{1,m-1} \\ x_{21} & x_{22} & \cdots & x_{2,m-1} \\ \cdots\cdots\cdots\cdots & & \cdots & \cdots\cdots\cdots \\ x_{m-2,1} & x_{m-2,2} & \cdots & x_{m-2,m-1} \\ x_{j1} & x_{j2} & \cdots & x_{j,m-1} \end{bmatrix},$$

$$q_{m-1} = \det \begin{bmatrix} y_{11} & y_{12} & \cdots & y_{1,m-1} \\ y_{21} & y_{22} & \cdots & y_{2,m-1} \\ \cdots\cdots\cdots\cdots & & \cdots & \cdots\cdots\cdots \\ y_{m-1,1} & y_{m-1,2} & \cdots & y_{m-1,m-1} \end{bmatrix}, \qquad (9.2)$$

so that, remembering that columns $(m-1)$ and m of X are complex conjugates as are also rows $(m-1)$ and m of Y, we have

$$\bar{p}_j = \det \begin{bmatrix} x_{11} & x_{12} & \cdots & x_{1,m-2} & x_{1m} \\ x_{21} & x_{22} & \cdots & x_{2,m-2} & x_{2m} \\ \cdots\cdots\cdots\cdots & & \cdots & \cdots\cdots\cdots\cdots \\ x_{m-2,1} & x_{m-2,2} & \cdots & x_{m-2,m-2} & x_{m-2,m} \\ x_{j1} & x_{j2} & \cdots & x_{j,m-2} & x_{jm} \end{bmatrix},$$

$$\bar{q}_{m-1} = \det \begin{bmatrix} y_{11} & y_{12} & \cdots & y_{1,m-1} \\ y_{21} & y_{22} & \cdots & y_{2,m-1} \\ \cdots\cdots\cdots\cdots & & \cdots & \cdots\cdots\cdots \\ y_{m-2,1} & y_{m-2,2} & \cdots & y_{m-2,m-1} \\ y_{m1} & y_{m2} & \cdots & y_{m,m-1} \end{bmatrix}. \qquad (9.3)$$

Equation (6.5) now gives

$$t_{j,m-1}^{(s)} \sim \frac{p_j q_{m-1}(\lambda_1\lambda_2\ldots\lambda_{m-1})^s + \bar{p}_j \bar{q}_{m-1}(\lambda_1\lambda_2\ldots\lambda_{m-2}\bar{\lambda}_{m-1})^s}{p_{m-1}q_{m-1}(\lambda_1\lambda_2\ldots\lambda_{m-1})^s + \bar{p}_{m-1}\bar{q}_{m-1}(\lambda_1\lambda_2\ldots\lambda_{m-2}\bar{\lambda}_{m-1})^s}, \qquad (9.4)$$

which does not converge as $s \to \infty$. We may express (9.4) in the form

$$t_{j,m-1}^{(s)} \sim \frac{a_s p_j + \bar{a}_s \bar{p}_j}{a_s p_{m-1} + \bar{a}_s \bar{p}_{m-1}}, \qquad (9.5)$$

and from this we see that

$$t_{j,m-1}^{(s+1)} - t_{j,m-1}^{(s)} \sim \frac{(p_{m-1}\bar{p}_j - \bar{p}_{m-1}p_j)(\bar{a}_{s+1}a_s - a_{s+1}\bar{a}_s)}{(a_{s+1}p_{m-1} + \bar{a}_{s+1}\bar{p}_{m-1})(a_s p_{m-1} + \bar{a}_s \bar{p}_{m-1})}$$

$$= P_s(p_{m-1}\bar{p}_j - \bar{p}_{m-1}p_j). \tag{9.6}$$

The difference between the $(m-1)$th column of T_{s+1} and T_s is therefore parallel to the fixed vector with components $p_{m-1}\bar{p}_j - \bar{p}_{m-1}p_j$. From the determinantal expressions for p_j, $\bar{p}_j$, p_{m-1} and $\bar{p}_{m-1}$, Sylvester's theorem (see for example Gantmacher, 1959a, Vol. 1) gives

$$p_{m-1}\bar{p}_j - \bar{p}_{m-1}p_j = (x_{12\ldots m-2})(x_{12\ldots m}^{(j)}). \tag{9.7}$$

The first factor is independent of j and since we have

$$\lim t_{jm}^{(s)} = x_{12\ldots m}^{(j)} / x_{12\ldots m}^{(m)}, \tag{9.8}$$

we may write

$$t_{j,m-1}^{(s+1)} - t_{j,m-1}^{(s)} \sim Q_s t_{jm}^{(s)}. \tag{9.9}$$

10. We have now proved that all columns of T_s tend to a limit except column $(m-1)$; ultimately columns $(m-1)$ of T_{s+1} and T_s differ by a multiple of column m of T_s. Hence since $T_{s+1} = T_s L_{s+1}$ we have

$$L_{s+1} \sim \begin{bmatrix} 1 & & & & & & \\ & 1 & & & & & \\ & & \cdot & & & & \\ & & & \cdot & & & \\ & & & & \cdot & & \\ & & & & 1 & & \\ & & & & Q_s & 1 & \\ & & & & & 1 & \\ & & & & & & \cdot \\ & & & & & & & \cdot \\ & & & & & & & & 1 \end{bmatrix} \tag{10.1}$$

where the element Q_s is in position $(m, m-1)$. From the relation $T_{s+1}^{-1} = L_{s+1}^{-1} T_s^{-1}$ we see that all rows of T_s^{-1} tend to a limit except row m; ultimately row m of T_{s+1}^{-1} differs from row m of T_s^{-1} by a multiple $-Q_s$ of row $(m-1)$. Hence $A_{s+1} = T_s^{-1}A_1T_s$ tends to a limit except in row m and column $(m-1)$. Now L_{s+1} is obtained from the triangular decomposition of A_{s+1}, and from the limiting form of L_{s+1} we see that either all sub-diagonal elements of A_{s+1} except that in position $(m, m-1)$ tend to zero, or some of the diagonal elements of A_{s+1} tend to infinity. Now from (9.4) $t_{j,m-1}^{(s)}$ may be expressed in the form

$$t_{j,m-1}^{(s)} \sim R \cos(s\theta+\alpha_j+\beta_{m-1})/\cos(s\theta+\alpha_{m-1}+\beta_{m-1}). \qquad (10.2)$$

Unless θ is expressible in the form $d\pi/e$, where d and e are integers, the expression on the right will take arbitrarily large values; however it will also take values of order R for an infinite number of values of s. Hence we can exclude the possibility of elements of A_{s+1} tending to infinity.

11. Our analysis can be extended in an obvious manner to give the following result.

Let A be a real matrix and let the eigenvalues be arranged so that $|\lambda_1| \geqslant |\lambda_2| \geqslant \ldots \geqslant |\lambda_n|$. Let us denote a typical real eigenvalue by λ_r and a typical complex pair by λ_{c-1} and λ_c so that $\lambda_c = \bar{\lambda}_{c-1}$. If

(i) $|\lambda_i| > |\lambda_{i+1}|$ except for the conjugate pairs,
(ii) $\det(X_{ii})\det(Y_{ii}) \neq 0 \quad (i = 1,\ldots, n)$,
(iii) at each stage A_r has a triangular decomposition,

then

(i) $a_{ij}^{(s)} \to 0 \ (i > j)$ except for the $a_{c,c-1}^{(s)}$,
(ii) $a_{rr}^{(s)} \to \lambda_r$,
(iii) all elements $a_{ij}^{(s)} \ (j > i)$ tend to a limit except for those in rows c and columns $(c-1)$,
(iv) the 2×2 matrices $\begin{bmatrix} a_{c-1,c-1}^{(s)} & a_{c-1,c}^{(s)} \\ a_{c,c-1}^{(s)} & a_{c,c}^{(s)} \end{bmatrix}$

do not converge but their eigenvalues converge to the corresponding λ_{c-1} and λ_c,

(v) the elements $t_{ij}^{(s)}$ tend to the limits given in (6.7) except for those in the columns $(c-1)$.

Notice that the LR transformation effectively gives a reduction of a real matrix to the form discussed in § 2.

12. We can now attempt to assess the value of the LR algorithm as a practical technique. At first sight it does not appear to be very promising for the following reasons.

(i) Matrices exist which have no triangular decomposition, in spite of the fact that their eigenproblem is well-conditioned. Without some modification of the process such matrices cannot be treated by the LR algorithm. Further there is a much larger class of matrices for which triangular decomposition is numerically unstable: numerical instability can arise at any step in the iterative process and may lead to a severe loss in accuracy in the computed eigenvalues.

(ii) The volume of computation is very high. There are $\frac{2}{3}n^3$ multiplications in each step of the iteration, half of these being in the triangularization and the remainder in the post-multiplication.

(iii) The convergence of the sub-diagonal elements to zero depends on the ratios $(\lambda_{r+1}/\lambda_r)$ and will be very slow if separation of the eigenvalues is poor.

Clearly if the LR algorithm is to compete with the better of the methods described in Chapters 6 and 7 it must be modified to meet these criticisms.

Introduction of interchanges

13. In triangular decomposition numerical stability was maintained by the introduction of interchanges where necessary, and the same device was used in the similarity reduction to Hessenberg form. We now consider the analogous modification for the LR algorithm.

If A is *any* matrix, we saw in Chapter 4, § 21, that there exist matrices $I_{r,r'}$ and N_r such that

$$N_{n-1}I_{n-1,(n-1)'}\ldots N_2I_{2,2'}N_1I_{1,1'}A = R \tag{13.1}$$

where all elements of $|N_r|$ are bounded by unity. We can complete a similarity transformation of A by post-multiplying R by

$$I_{1,1'}N_1^{-1}I_{2,2'}N_2^{-1}\ldots I_{n-1,(n-1)'}N_{n-1}^{-1}$$

and we have

$$N_{n-1}I_{n-1,(n-1)'}\ldots N_2I_{2,2'}N_1I_{1,1'}AI_{1,1'}N_1^{-1}I_{2,2'}N_2^{-1}\ldots I_{n-1,(n-1)'}N_{n-1}^{-1}$$
$$= RI_{1,1'}N_1^{-1}I_{2,2'}N_2^{-1}\ldots I_{n-1,(n-1)'}N_{n-1}^{-1}. \tag{13.2}$$

In the particular case when no interchanges are required so that $I_{r,r'} = I$ ($r = 1,\ldots, n-1$), the matrix product preceding A in (13.1) is the matrix L^{-1} for which $L^{-1}A = R$ or, equivalently, such that $A = LR$; in this case the matrix on the right of equation (13.2) is the matrix RL. Accordingly we propose the following modification to the LR algorithm.

At each stage we reduce the matrix A_s to an upper-triangular matrix R_s using Gaussian elimination with interchanges. We then post-multiply R_s by the inverses of the factors used in the reduction. This gives us the matrix A_{s+1}. The volume of computation is the same as in the orthodox *LR* process.

To avoid a tedious complexity we analyse the relationship between A_1 and A_2 a little more closely for the case when $n = 4$. The analysis extends immediately to the general case. We have

$$N_3 I_{3,3'} N_2 I_{2,2'} N_1 I_{1,1'} A_1 = R_1$$

$$A_2 = \{N_3 I_{3,3'} N_2 I_{2,2'} N_1 I_{1,1'}\} A_1 \{I_{1,1'} N_1^{-1} I_{2,2'} N_2^{-1} I_{3,3'} N_3^{-1}\}$$

$$= \{N_3 I_{3,3'} N_2 (I_{3,3'} I_{3,3'}) I_{2,2'} N_1 (I_{2,2'} I_{3,3'} I_{3,3'} I_{2,2'}) I_{1,1'}\} A_1 \times$$

$$\times \{I_{1,1'} (I_{2,2'} I_{3,3'} I_{3,3'} I_{2,2'}) N_1^{-1} I_{2,2'} (I_{3,3'} I_{3,3'}) N_2^{-1} I_{3,3'} N_3^{-1}\}, \quad (13.3)$$

where the terms in parenthesis are each equal to I. Hence remembering that

$$I_{3,3'} I_{2,2'} N_1 I_{2,2'} I_{3,3'} = \tilde{N}_1, \qquad I_{3,3'} N_2 I_{3,3'} = \tilde{N}_2 \qquad (13.4)$$

where $\tilde{N}_1$ and $\tilde{N}_2$ are of the same form as N_1 and N_2 but with sub-diagonal elements reordered, (13.3) gives on regrouping the factors

$$A_2 = \tilde{L}_1^{-1} \tilde{A}_1 \tilde{L}_1, \quad \tilde{L}_1 = \tilde{N}_1^{-1} \tilde{N}_2^{-1} N_3^{-1}, \quad \tilde{A}_1 = I_{3,3'} I_{2,2'} I_{1,1'} A_1 I_{1,1'} I_{2,2'} I_{3,3'}.$$
$$(13.5)$$

Clearly $\tilde{L}_1$ is a unit lower-triangular matrix and $\tilde{A}_1$ is A_1 with its rows and columns suitably permuted. If we write

$$I_{3,3'} I_{2,2'} I_{1,1'} = P_1 \qquad (13.6)$$

where P_1 is a permutation matrix then

$$\tilde{A}_1 = P_1 A_1 P_1^T, \quad P_1 A_1 = \tilde{L}_1 R_1, \quad A_2 = R_1 P_1^T \tilde{L}_1 = \tilde{L}_1^{-1} (P_1 A P_1^T) \tilde{L}_1.$$
$$(13.7)$$

Numerical example

14. As a simple numerical example of the modified process we have taken a 3×3 matrix which was used by Rutishauser (1958) to illustrate the non-convergence of the orthodox *LR* process. The matrix A_1, the corresponding λ_i, and the X and Y matrices are given by

$$A_1 = \begin{bmatrix} 1 & -1 & 1 \\ 4 & 6 & -1 \\ 4 & 4 & 1 \end{bmatrix}, \quad \begin{matrix} \lambda_1 = 5, \\ \lambda_2 = 2, \\ \lambda_3 = 1, \end{matrix}$$

$$X = \begin{bmatrix} 0 & -1 & -1 \\ 1 & 1 & 1 \\ 1 & 0 & 1 \end{bmatrix}, \quad Y = \begin{bmatrix} 1 & 1 & 0 \\ 0 & 1 & -1 \\ -1 & -1 & 1 \end{bmatrix}. \quad (14.1)$$

The leading principal minor of X of order 1 is zero and hence we cannot be sure that the orthodox process converges or, if it does, that the eigenvalues will not be in the usual order in the limiting matrix. In fact in this case the elements of T_s diverge, as may be verified using the simple analysis of § 5. In Table 1 we show the results obtained from the first 3 steps of the orthodox *LR* process and from these the form of A_s is immediately obvious.

TABLE 1

Orthodox LR algorithm

L_1 L_2 L_3

$$\begin{bmatrix} 1 & & \\ 4 & 1 & \\ 4 & 0\cdot8 & 1 \end{bmatrix} \quad \begin{bmatrix} 1 & & \\ 20 & 1 & \\ 4 & 0\cdot16 & 1 \end{bmatrix} \quad \begin{bmatrix} 1 & & \\ 100 & 1 & \\ 4 & 0\cdot032 & 1 \end{bmatrix}$$

A_2 A_3 A_4

$$\begin{bmatrix} 1 & -0\cdot2 & 1 \\ 20 & 6 & -5 \\ 4 & 0\cdot8 & 1 \end{bmatrix} \quad \begin{bmatrix} 1 & -0\cdot04 & 1 \\ 100 & 6 & -25 \\ 4 & 0\cdot16 & 1 \end{bmatrix} \quad \begin{bmatrix} 1 & -0\cdot008 & 1 \\ 500 & 6 & -125 \\ 4 & 0\cdot032 & 1 \end{bmatrix}$$

LR with interchanges

Step 1 in detail

$$1' = 2; \; N_1 = \begin{bmatrix} 1 & & \\ -0\cdot25 & 1 & \\ 1 & 0 & 1 \end{bmatrix}; \quad 2' = 2; \; N_2 = \begin{bmatrix} 1 & & \\ 0 & 1 & \\ 0 & -0\cdot8 & 1 \end{bmatrix}; \quad R_1 = \begin{bmatrix} 4 & 6 & 1 \\ & -2\cdot5 & 1\cdot25 \\ & & 1 \end{bmatrix}$$

$$A_2 = R_1 I_{12} N_1^{-1} N_2^{-1} = \begin{bmatrix} 6 & 3\cdot2 & -1 \\ -1\cdot25 & 1 & 1\cdot25 \\ 1 & 0\cdot8 & 1 \end{bmatrix}$$

Steps 2 and 3

A_3 A_4

$$\begin{bmatrix} 5\cdot167 & 3\cdot040 & -1\cdot000 \\ -0\cdot174 & 1\cdot833 & 1\cdot042 \\ 0\cdot167 & 0\cdot160 & 1\cdot000 \end{bmatrix} \quad \begin{bmatrix} 5\cdot032 & 3\cdot008 & -1\cdot000 \\ -0\cdot033 & 1\cdot968 & 1\cdot008 \\ 0\cdot032 & 0\cdot032 & 1\cdot000 \end{bmatrix}$$

$$A_\infty = \begin{bmatrix} 5 & 3 & -1 \\ 0 & 2 & 1 \\ 0 & 0 & 1 \end{bmatrix}$$

Table 1 also gives the results obtained using 3 steps of the modified process. The first step is shown in detail. In the reduction of A_1 to triangular form, row 1 is interchanged with row 2. This is the only interchange. Hence to obtain A_2 from R_1 we first interchange columns 1 and 2 and then multiply successively by N_1^{-1} and N_2^{-1}. In subsequent steps no interchanges are necessary so that we are effectively using the orthodox *LR* algorithm. The matrices A_s converge quite rapidly to upper-triangular form and it is easy to see that A_∞ is the matrix given at the end of the table.

The use of interchanges has not only given convergence, but also the eigenvalues are in decreasing order on the diagonal of the limiting matrix.

Convergence of the modified process

15. The proof we have given of the convergence of the *LR* algorithm does not carry over to the modified procedure. However, if convergence to upper-triangular form *does* take place and if none of the eigenvalues is zero, it is obvious that interchanges must ultimately cease because the sub-diagonal elements are tending to zero. This is certainly true of the example in § 14.

Unfortunately it is easy to construct simple examples for which the orthodox *LR* process converges but the modified process does not. Consider for example the matrix

$$A_1 = \begin{bmatrix} 1 & 3 \\ 2 & 0 \end{bmatrix}, \quad \lambda_1 = 3, \quad \lambda_2 = -2. \tag{15.1}$$

We have

$$A_2 = \begin{bmatrix} 1 & 2 \\ 3 & 0 \end{bmatrix}, \quad A_3 = \begin{bmatrix} 1 & 3 \\ 2 & 0 \end{bmatrix} = A_1, \tag{15.2}$$

an interchange being needed both in the reduction of A_1 and of A_2. Since $A_3 = A_1$ the process is cyclic and we have no convergence. On the other hand the orthodox *LR* algorithm gives convergence with the eigenvalues in the correct order.

We shall return to this problem again in § 24 and shall show that this example of non-convergence is not as damaging as it appears. At this stage we content ourselves with the remark that in our opinion the requirement of numerical stability is paramount and the modified algorithm must be used.

Preliminary reduction of original matrix

16. With regard to the second of the criticisms made in § 12, it would appear that the volume of work involved in the *LR* algorithm is prohibitive when A_1 is a matrix with few zero elements. If A_1 is of some condensed form *which is invariant with respect to the LR algorithm* then the volume of work might well be considerably reduced.

There are two main forms which meet these requirements. The first is the upper Hessenberg form which is invariant with respect to the modified *LR* algorithm (and therefore *a fortiori* with respect to the orthodox *LR* algorithm). The second consists of those matrices in which the only non-zero elements are those in a band centred on the main diagonal. These are invariant with respect to the orthodox algorithm but not with respect to the modified version (see §§ 59–68). In my experience it is only for matrices of these two types that the *LR* algorithm is a practical proposition. Since there are several stable methods of reducing a matrix to upper Hessenberg form this is not a severe restriction.

Invariance of upper Hessenberg form

17. We first prove that the upper Hessenberg form is indeed invariant with respect to the modified algorithm. We have already discussed the reduction of an upper Hessenberg matrix to triangular form by Gaussian elimination with interchanges (Chapter 4, § 33). At the rth step of the reduction the choice of pivotal element lies between rows r and $(r+1)$ only, and the only non-zero element in the relevant matrix N_r (apart from the unit diagonal elements) is that in position $(r+1, r)$. The matrix N_r is therefore a matrix of the form $M_{r+1,r}$ (Chapter 1, § 40 (vi)). We are required to show that the matrix A_2 given by

$$A_2 = R_1 I'_{12} M_{21}^{-1} I'_{23} M_{32}^{-1} \ldots I'_{n-1,n} M_{n,n-1}^{-1} \tag{17.1}$$

is of upper Hessenberg form. Here $I'_{r,r+1}$ denotes either $I_{r,r+1}$ or I according as an interchange was or was not necessary in the rth step of the reduction to triangular form.

The proof is by induction. Let us assume that the matrix

$$R I'_{12} M_{21}^{-1} I'_{23} M_{32}^{-1} \ldots I'_{r-1,r} M_{r,r-1}^{-1}$$

is of upper Hessenberg form in its first $(r-1)$ columns and of triangular form in the remainder, so that typically for $n = 6$, $r = 4$, it is of the form

$$
\left. r-1 \left\{ \quad \right. \right.
\left. n-r+1 \left\{ \quad \right. \right.
\begin{bmatrix}
\times & \times & \times & \times & \times & \times \\
\times & \times & \times & \times & \times & \times \\
 & \times & \times & \times & \times & \times \\
 & & & \times & \times & \times & \times \\
 & & & & \times & \times \\
 & & & & & \times
\end{bmatrix} . \tag{17.2}
$$

The next step is the post-multiplication by $I'_{r,r+1}$. If this factor is $I_{r,r+1}$ we interchange columns r and $(r+1)$; otherwise it has no effect.

The resulting matrix is therefore of the form (a) or (b) given by

$$(a) = \;\; r-1 \left\{ \begin{array}{ccc|ccc} \times & \times & \times & \times & \times & \times \\ \times & \times & \times & \times & \times & \times \\ & \times & \times & \times & \times & \times \\ \hline & & \times & \times & \times & \times \\ & & & \times & 0 & \times \\ & & & & & \times \end{array} \right] ,$$

$$(b) = \;\; r-1 \left\{ \begin{array}{ccc|ccc} \times & \times & \times & \times & \times & \times \\ \times & \times & \times & \times & \times & \times \\ & \times & \times & \times & \times & \times \\ \hline & & & \times & \times & \times & \times \\ & & & & \times & \times \\ & & & & & \times \end{array} \right] . \qquad (17.3)$$

The post-multiplication by $M_{r+1,r}^{-1}$ results in the addition of a multiple of column $(r+1)$ to column r, and hence the resulting matrix is of the form (c) or (d) given by

$$(c) = \;\; r \left\{ \begin{array}{cccc|cc} \times & \times & \times & \times & \times & \times \\ \times & \times & \times & \times & \times & \times \\ & \times & \times & \times & \times & \times \\ & & \times & \times & \times & \times \\ \hline & & & \times & 0 & \times \\ & & & & & \times \end{array} \right] ,$$

$$(d) = \;\; r \left\{ \begin{array}{cccc|cc} \times & \times & \times & \times & \times & \times \\ \times & \times & \times & \times & \times & \times \\ & \times & \times & \times & \times & \times \\ & & \times & \times & \times & \times \\ \hline & & & \times & \times & \times \\ & & & & & \times \end{array} \right] . \qquad (17.4)$$

In either case the matrix is of Hessenberg form in its first r columns and triangular otherwise, the extra zero element in (c) being of no special significance. The result is therefore established.

There are essentially $\frac{1}{2}n^2$ multiplications in the reduction to triangular form and a further $\frac{1}{2}n^2$ in the post-multiplication, giving n^2 in a complete step of the LR iteration compared with $\frac{2}{3}n^3$ for a full matrix.

Simultaneous row and column operations

18. A further advantage of the Hessenberg form is that the reduction of A_s and its post-multiplication can be combined in such a way that on a computer with a two-level store, the sth iteration requires only one transfer of A_s from the backing store and its replacement by A_{s+1}. In order to do this it is necessary to store A_s by columns.

We observe that only $(n-1)$ multipliers $m_{r+1,r}$ $(r = 1, \ldots, n-1)$ are required in the reduction, and to determine the first r of these multipliers we require only columns 1 to r of the matrix A_s. We may describe the proposed technique by considering the typical major step. At the beginning of this step columns 1 to $(r-1)$ of A_{s+1} have been computed and overwritten on the corresponding columns of A_s on the backing store. Column r of A_{s+1} has not yet been completed and the partly processed column is in the high-speed store. The elements $m_{21}, m_{32}, \ldots, m_{r,r-1}$ and information indicating whether or not an interchange took place at the stage when each of these was computed are also in the high-speed store. The typical step may then be described as follows, using the usual conventions.

(i) Read column $(r+1)$ into the high-speed store.
For each value of i from 1 to r perform steps (ii) and (iii):

(ii) Interchange $a_{i,r+1}$ and $a_{i+1,r+1}$ or not according as an interchange took place or not just before $m_{i+1,i}$ was computed.

(iii) Compute $a_{i+1,r+1} - m_{i+1,i}a_{i,r+1}$ and overwrite on $a_{i+1,r+1}$.
On completion, column $(r+1)$ has been subjected to the operations involved in reducing the first r rows of A_s.

(iv) If $|a_{r+2,r+1}| > |a_{r+1,r+1}|$ interchange these two elements and then compute $m_{r+2,r+1} = a_{r+2,r+1}/a_{r+1,r+1}$. Store $m_{r+2,r+1}$ and record whether an interchange took place. Replace $a_{r+2,r+1}$ by zero.
Column $(r+1)$ is now the $(r+1)$th column of the triangular matrix

which would have been obtained by completing the reduction independently.

(v) Interchange the modified columns r and $(r+1)$ (which are both in the high-speed store) if there was an interchange just before computing $m_{r+1,r}$.

(vi) Add $m_{r+1,r} \times$column $(r+1)$ to column r.

The current column r is now the rth column of A_{s+1} and may be written on the backing store.

The combined process gives identical results with the original, even as regards rounding errors, and involves no more arithmetic operations. Notice that we can assume that none of the subdiagonal elements $a_{r+1,r}$ of A_s is zero because otherwise we could deal with two or more smaller Hessenberg matrices. Hence in step (iv) the denominator $a_{r+1,r+1}$ is never zero, since at the stage when the division is performed $|a_{r+1,r+1}|$ is the larger of the original $|a_{r+2,r+1}|$ and some other number.

Acceleration of convergence

19. Although a preliminary reduction to Hessenberg form reduces the volume of computation very substantially the modified *LR* algorithm will still be uneconomical compared with earlier methods unless we can improve the rate of convergence.

We have seen that for a general matrix the elements in position (i, j) $(i > j)$ tend to zero roughly as $(\lambda_i/\lambda_j)^s$. For Hessenberg matrices the only non-zero sub-diagonal elements are those in positions $(r+1, r)$. Consider now the matrix $(A - pI)$. This matrix has the eigenvalues $(\lambda_i - p)$ and, for the orthodox *LR* technique at least, $a_{n,n-1}^{(s)}$ tends to zero as $\{(\lambda_n - p)/(\lambda_{n-1} - p)\}^s$. If p is a good approximation to λ_n the element $a_{n,n-1}^{(s)}$ will decrease rapidly. Hence it would be advantageous to work with $(A_s - pI)$ rather than A_s.

Notice, in particular, that if p is exactly equal to λ_n then, if the modified process were performed using exact computation, the $(n, n-1)$ element would be zero after one iteration. For consider the triangularization process. None of the pivots (the diagonal elements of the triangle R) can be zero except the last, because at each stage in the reduction the pivot is either $a_{r+1,r}$ or some other number and we are assuming that no $a_{r+1,r}$ is zero originally. Hence since the determinant of $(A - \lambda_n I)$ is zero the final pivot must be zero, and the triangle R is of the form

$$\begin{bmatrix} \times & \times & \times & \times & \times \\ & \times & \times & \times & \times \\ & & \times & \times & \times \\ & & & \times & \times \\ & & & & 0 \end{bmatrix}, \qquad (19.1)$$

the whole of its last row being null. The subsequent post-multiplication merely combines columns and hence at the end of the iteration the matrix is of the form

$$\begin{bmatrix} \times & \times & \times & \times & \times \\ \times & \times & \times & \times & \times \\ & \times & \times & \times & \times \\ & & \times & \times & \times \\ & & & 0 & 0 \end{bmatrix}. \qquad (19.2)$$

Incorporation of shifts of origin

20. The discussion of the previous section provides the motivation for the following modification of the LR algorithm. At the sth stage we perform a triangular decomposition of $(A_s - k_s I)$, where k_s is some suitable value, rather than of A_s. Accordingly we produce the sequence of matrices defined by

$$A_s - k_s I = L_s R_s, \qquad R_s L_s + k_s I = A_{s+1}. \qquad (20.1)$$

We have

$$A_{s+1} = R_s L_s + k_s I = L_s^{-1}(A_s - k_s I)L_s + k_s I = L_s^{-1}A_s L_s, \qquad (20.2)$$

and hence the matrices A_s are again all similar to A_1. In fact (20.2) shows that

$$A_{s+1} = L_s^{-1}A_s L_s = L_s^{-1}L_{s-1}^{-1}A_{s-1}L_{s-1}L_s = L_s^{-1}...L_2^{-1}L_1^{-1}A_1 L_1 L_2...L_s, \qquad (20.3)$$

or

$$L_1 L_2 ... L\, A_{s+1} = A_1 L_1 L_2...L_s. \qquad (20.4)$$

This formulation of the modification is usually described as LR with shifts of origin and 'restoring', because the shift is *added back* at each stage. For programming convenience it is sometimes advantageous to use the 'non-restoring' process defined by

$$A_s - z_s I = L_s R_s, \qquad R_s L_s = A_{s+1}, \qquad (20.5)$$

so that

$$A_{s+1} = R_s L_s = L_s^{-1}(A_s - z_s I)L_s = L_s^{-1}A_s L_s - z_s I$$
$$= L_s^{-1}(L_{s-1}^{-1}A_{s-1}L_{s-1} - z_{s-1}I)L_s - z_s I$$
$$= L_s^{-1}L_{s-1}^{-1}A_{s-1}L_{s-1}L_s - (z_{s-1} + z_s)I.$$
$$(20.6)$$

Continuing this argument we find

$$A_{s+1} = L_s^{-1}...L_2^{-1}L_1^{-1}A_1 L_1 L_2...L_s - (z_1 + z_2 + ... + z_s)I$$
$$= L_s^{-1}...L_2^{-1}L_1^{-1}[A_1 - (z_1 + z_2 + ... + z_s)I]L_1 L_2...L_s, \quad (20.7)$$

so that the eigenvalues of A_{s+1} differ from those of A_1 by $\sum_{i=1}^{s} z_i$.

Returning to the restoring version we have

$$L_1 L_2...L_{s-1}(L_s R_s)R_{s-1}...R_2 R_1$$
$$= L_1 L_2...L_{s-1}(A_s - k_s I)R_{s-1}...R_2 R_1$$
$$= (A_1 - k_s I)L_1 L_2...L_{s-1}R_{s-1}...R_2 R_1$$
$$= (A_1 - k_s I)(A_1 - k_{s-1}I)L_1 L_2...L_{s-2}R_{s-2}...R_2 R_1$$
$$= (A_1 - k_s I)(A_1 - k_{s-1}I)...(A_1 - k_1 I). \qquad (20.8)$$

Hence writing

$$T_s = L_1 L_2...L_s, \qquad U_s = R_s...R_2 R_1, \qquad (20.9)$$

we see that $T_s U_s$ gives the triangular decomposition of $\prod_{i=1}^{s} (A_1 - k_i I)$, the order of the factors being immaterial. This result is required later.

Clearly we can combine the use of shifts of origin with the use of interchanges. Each member of the series of matrices A_s is again similar to A_1 but we do not have any simple analogue of equation (20.8).

Choice of shift of origin

21. In practice we are faced with the problem of choosing the sequence of k_s so as to give rapid convergence. If the eigenvalues all have different moduli, then, whether they are real or complex, we are expecting that $a_{n,n-1}^{(s)}$ and $a_{nn}^{(s)}$ will tend to zero and λ_n respectively. Hence it is attractive to take $k_s = a_{nn}^{(s)}$ as soon as $a_{n,n-1}^{(s)}$ becomes 'small' or $a_{nn}^{(s)}$ shows signs of convergence. In fact it is easy to show that when $a_{n,n-1}^{(s)}$ is of order ϵ then, if we choose $k_s = a_{nn}^{(s)}$ we have

$$a_{n,n-1}^{(s+1)} = O(\epsilon^2). \tag{21.1}$$

For $(A_s - a_{nn}^{(s)}I)$ is typically, when $n = 6$, of the form shown at (a) in the arrays

$$
\begin{array}{cc}
\text{(a)} & \text{(b)}
\end{array}
$$

$$
\begin{bmatrix}
\times & \times & \times & \times & \times & \times \\
\times & \times & \times & \times & \times & \times \\
 & \times & \times & \times & \times & \times \\
 & & \times & \times & \times & \times \\
 & & & \times & \times & \times \\
 & & & & \epsilon & 0
\end{bmatrix},
\begin{bmatrix}
\times & \times & \times & \times & \times & \times \\
 & \times & \times & \times & \times & \times \\
 & & \times & \times & \times & \times \\
 & & & \times & \times & \times \\
 & & & & a & b \\
 & & & & \epsilon & 0
\end{bmatrix},
$$

$$
\begin{array}{c}
\text{(c)}
\end{array}
$$

$$
\begin{bmatrix}
\times & \times & \times & \times & \times & \times \\
 & \times & \times & \times & \times & \times \\
 & & \times & \times & \times & \times \\
 & & & \times & \times & \times \\
 & & & & a & b \\
 & & & & & -b\epsilon/a
\end{bmatrix}. \tag{21.2}
$$

Consider now the reduction of $(A_s - a_{nn}^{(s)}I)$ to triangular form by Gaussian elimination with interchanges. When the reduction is almost complete and only the last row remains to be reduced, the current matrix will be of the form shown at (b) of (21.2). The element a in row $(n-1)$ will not be small unless $a_{nn}^{(s)}$ happened to bear some special relationship to the leading principal minor of order $(n-1)$ of A_s, for example, if the shift was an eigenvalue of this minor. *Hence no interchange will be necessary in the last step* and we have

$$m_{n,n-1} = \epsilon/a; \tag{21.3}$$

the triangular matrix is therefore of the form shown at (c) in (21.2).

Turning now to the post-multiplication, when we have post-multiplied by all factors except $I'_{n-1,n}M^{-1}_{n,n-1}$ the current matrix will be of the form (a) or (b) given by

(a)   (b)

$$
\begin{bmatrix}
\times & \times & \times & \times & \times & & \times \\
\times & \times & \times & \times & \times & & \times \\
 & \times & \times & \times & \times & & \times \\
 & & \times & \times & \times & & \times \\
 & & & & a & 0 & b \\
 & & & & & & -b\epsilon/a
\end{bmatrix},
\qquad
\begin{bmatrix}
\times & \times & \times & \times & \times & & \times \\
\times & \times & \times & \times & \times & & \times \\
 & \times & \times & \times & \times & & \times \\
 & & \times & \times & \times & & \times \\
 & & & \times & a & & b \\
 & & & & & & -b\epsilon/a
\end{bmatrix},
$$
$$(21.4)$$

according as an interchange did or did not take place just before computing $m_{n-1,n-2}$ during the reduction. To complete the post-multiplication we add $m_{n,n-1}$ times column n to column $(n-1)$, no interchange being necessary in general. The final matrix is therefore of the form (a) or (b) in the arrays

(a)   (b)

$$
\begin{bmatrix}
\times & \times & \times & \times & & \times & & \times \\
\times & \times & \times & \times & & \times & & \times \\
 & \times & \times & \times & & \times & & \times \\
 & & \times & \times & & \times & & \times \\
 & & & \times & & b\epsilon/a & & b \\
 & & & & & -b\epsilon^2/a^2 & & -b\epsilon/a
\end{bmatrix},
\quad
\begin{bmatrix}
\times & \times & \times & \times & & \times & & \times \\
\times & \times & \times & \times & & \times & & \times \\
 & \times & \times & \times & & \times & & \times \\
 & & \times & \times & & \times & & \times \\
 & & & \times & & (a+b\epsilon/a) & & b \\
 & & & & & -b\epsilon^2/a^2 & & -b\epsilon/a
\end{bmatrix}.
$$
$$(21.5)$$

Hence after restoring the shift we have

$$a^{(s+1)}_{nn} = a^{(s)}_{nn} - b\epsilon/a, \qquad a^{(s+1)}_{n,n-1} = -b\epsilon^2/a^2 \qquad (21.6)$$

so that $a^{(s)}_{nn}$ is indeed converging and $a^{(s+1)}_{n,n-1}$ is of order ϵ^2, as we wished to show. Notice that *any interchanges which may take place in the other steps of the reduction are of little consequence.*

Deflation of the matrix

22. In general, once $a^{(s)}_{n,n-1}$ has become small it will diminish rapidly in value. When it is negligible to working accuracy we can treat it as zero and the current value of $a_{nn}^{(s)}$ is then an eigenvalue. The remaining eigenvalues are those of the leading principal submatrix of order $(n-1)$; this matrix is itself of Hessenberg form so that we can continue with the same method, working with a matrix of order one less than the original. Since we are expecting all sub-diagonal elements to tend to zero $a_{n-1,n-2}^{(s)}$ may already be fairly small, and in this case we can immediately use $a_{n-1,n-1}^{(s)}$ as the next value of k_s.

Continuing in this way we may find the eigenvalues one by one, working with matrices of progressively decreasing order. The later stages of the convergence to each eigenvalue will generally be quadratic, and moreover we can expect that when finding the later eigenvalues in the sequence we shall have a good start.

This method of deflation is very stable; in fact eigenvalues can only be seriously affected if they are sensitive to the sub-diagonal values. Since we are assuming that in general the original Hessenberg matrix A_1 has been obtained by the reduction of a full matrix A, if the eigenvalues are sensitive to small errors in the sub-diagonal elements of A_1 we shall already have lost accuracy before starting on the LR process. The only additional danger is that the condition of the A_s may degenerate progressively. We comment on this in § 47 when we compare the LR and QR algorithms.

Practical experience of convergence

23. We have experimented with the LR algorithm (with interchanges) on Hessenberg matrices using shifts of origin determined by the current value of $a_{nn}^{(s)}$. Naturally it was used only on real matrices which were known to have real eigenvalues or on matrices with complex elements. It is not at all obvious when we should start using this value of the shift and we have investigated the effects of the following procedures.

(i) using the shift $a_{nn}^{(s)}$ at every stage,

(ii) using the shift only when $|1-a_{n,n}^{(s)}/a_{nn}^{(s-1)}| < \epsilon$, where ϵ is some tolerance (i.e. the shift is used when $a_{nn}^{(s)}$ has begun to 'settle down'),

(iii) using the shift only when $|a_{n,n-1}^{(s)}| < \eta \|A_1\|_\infty$, where η is again some tolerance.

In general, convergence using (i) has been remarkably satisfactory, and this suggests that when using (ii) and (iii) the tolerances should not be too stringent. On the whole, method (ii) with $\epsilon = \frac{1}{3}$ has been marginally the most effective. For the majority of the matrices which have been tried (of orders between 5 and 32) the average number of iterations per eigenvalue has been of the order of 5, and usually several of the later eigenvalues have been obtained after only 1 or 2 iterations. However, matrices were encountered which did not give convergence using (i), (ii) or (iii). This is obviously true, for example, of the 2×2 matrices

$$\begin{bmatrix} a & b \\ c & 0 \end{bmatrix}, \quad |b|, |c| > |a|, \tag{23.1}$$

an example of which was discussed in § 15. These become

$$\begin{bmatrix} a & c \\ b & 0 \end{bmatrix} \quad \text{and} \quad \begin{bmatrix} a & b \\ c & 0 \end{bmatrix} \tag{23.2}$$

successively if we use *LR* with interchanges. Clearly $a_{nn}^{(s)} = 0$ for all s and the choice $k_s = a_{nn}^{(s)}$ has no effect.

Since, if we have convergence, $a_{n,n-1}^{(s)}$ tends to zero, shifts given by

$$k_s = \tfrac{1}{2} a_{n,n-1}^{(s)} + a_{nn}^{(s)}, \tag{23.3}$$

for which $k_s \to a_{nn}^{(s)}$ were also tried. This simple empirical device gave convergence in nearly all cases when the choice $k_s = a_{nn}^{(s)}$ had failed. Nevertheless, there were some matrices with distinct eigenvalues which did not give convergence even with this modified shift. In the next section we describe a strategy for choosing the shift which has proved much more effective.

Improved shift strategy

24. The motivation for the improved shift strategy is provided by considering the case when the matrix is real but has some complex conjugate eigenvalues. We know that in this case the limiting matrix will not be strictly upper-triangular. If the eigenvalues of smallest modulus are a complex conjugate pair $\lambda \pm i\mu$, then the iterates will ultimately become of the form shown at (a), in the arrays where the eigenvalues of the 2×2 matrix in the bottom right-hand corner tend to the relevant values. On the other hand if the eigenvalue of smallest modulus has a real value λ, the iterates will assume the form shown at (b) in (24.1). Suppose we take the eigenvalues of the 2×2 matrix in both cases, then in case (a) these will tend to $\lambda \pm i\mu$, and in case (b) the smaller of the two will tend to λ. This point being established we leave the complex conjugate case for the moment.

For real matrices with real eigenvalues we propose the following choice for k_s. We compute the eigenvalues of the 2×2 matrix in the bottom right-hand corner of A_s. If these are complex and equal to $p_s \pm iq_s$ then we take $k_s = p_s$. If they are real and are equal to p_s and q_s, we take k_s equal to p_s or q_s according as $|p_s - a_{nn}^{(s)}|$ is less than or greater than $|q_s - a_{nn}^{(s)}|$.

If the original matrix is complex then, in general, the relevant 2×2 matrix will have two complex eigenvalues p_s and q_s and again the decision is made by comparing $|p_s - a_{nn}^{(s)}|$ and $|q_s - a_{nn}^{(s)}|$.

We can use the shift determined in this way at every stage or we can wait until the suggested shift has shown some indication of convergence. In practice the behaviour has not proved very sensitive to this decision. The best results were obtained by accepting the indicated k_s as a shift as soon as the criterion

$$|k_s/k_{s-1} - 1| < \tfrac{1}{2} \tag{24.2}$$

was satisfied.

On average the rate of convergence using this technique has proved to be remarkably high. It has been used most extensively on matrices with complex elements. It is worth noting that with this choice of the shift the matrices of (23.1) give convergence in one iteration. Since the relevant 2×2 matrix is now the full matrix this is not surprising, but it suggests that this choice of shift might well give a better general performance.

Complex conjugate eigenvalues

25. Returning now to the case when the smallest eigenvalues are a complex conjugate pair we will expect the 2×2 matrix ultimately to have eigenvalues which are approximations to this pair. Suppose at a given stage the eigenvalues of the 2×2 matrix are $p \pm iq$. If we perform two steps of the orthodox LR algorithm using first a shift of $p - iq$ and then of $p + iq$, this will certainly involve us in complex arithmetic, but we now show that if the computation is performed exactly the

second of the derived matrices is real. Denoting the three relevant matrices by A_1, A_2 and A_3 we have

$$A_1-(p-iq)I = L_1R_1, \qquad R_1L_1+(p-iq)I = A_2 \Big\} \tag{25.1}$$
$$A_2-(p+iq)I = L_2R_2, \qquad R_2L_2+(p+iq)I = A_3 \Big\}$$

Applying the result of (20.8) we have

$$L_1L_2R_2R_1 = [A_1-(p-iq)I][A_1-(p+iq)I], \tag{25.2}$$

and obviously the right-hand side is real. Now provided this real matrix is non-singular, that is, provided $p\pm iq$ are not exactly eigen-values of A_1, its triangular decomposition, if it exists, is unique and the triangles are obviously real. Equation (25.2) shows that L_1L_2 and R_2R_1 are indeed the triangular factors and hence L_1L_2 must be real. Since we have

$$A_3 = (L_1L_2)^{-1}A_1L_1L_2 \tag{25.3}$$

we see that A_3 is also real.

We might expect that if we carry out the two steps of the *LR* algorithm the computed matrix A_3 will have a negligible imaginary component and can be taken to be exactly real. Unfortunately this is not true, and it is by no means uncommon to find that the final matrix has quite a substantial imaginary part. The process is numerically stable in that the computed A_3 has eigenvalues which are close to those of A_1 (unless A_1 is ill-conditioned with respect to its eigenvalue problem) but as we have often found on previous occasions *there is no guarantee that the final computed matrix will be close to the matrix which would be obtained by exact computation.*

26. To see how this may come about we consider the case when A_1 is the 2×2 matrix shown in (26.1) having eigenvalues λ and $\bar{\lambda}$.

$$A_1 = \begin{bmatrix} a & b \\ c & d \end{bmatrix}, \quad A_2 = \begin{bmatrix} \bar{\lambda} & b \\ 0 & \lambda \end{bmatrix}, \quad (A_2-\bar{\lambda}I) = \begin{bmatrix} 0 & b \\ 0 & \lambda-\bar{\lambda} \end{bmatrix}. \tag{26.1}$$

If we perform one step of the *LR* algorithm using a shift of exactly λ, then A_2 is as shown and $(A_2-\bar{\lambda}I)$ has zero elements in both positions in its leading column. Its triangular decomposition is therefore not uniquely determined. Consequently if we use shifts $(\lambda+\epsilon)$ and $(\bar{\lambda}+\bar{\epsilon})$ where ϵ is small then we have

$$A_2-(\bar{\lambda}+\bar{\epsilon})I = \begin{bmatrix} \epsilon_1 & b \\ \epsilon_2 & (\lambda-\bar{\lambda})+\epsilon_3 \end{bmatrix}, \tag{26.2}$$

where ϵ_1, ϵ_2 and ϵ_3 are of the same order of magnitude as ϵ. If the computation is performed exactly then A_3 will be real, but quite small errors in ϵ_1 and ϵ_2 will alter A_3 substantially. In general, using t-digit computation, A_3 will have imaginary components as large as $2^{-t}/\epsilon$ and these cannot be neglected, though provided we retain the imaginary components the eigenvalues are fully preserved. Here again the use of the exact λ and $\bar{\lambda}$ is instructive. In the triangular decomposition of $(A_2-\bar{\lambda}I)$ of (26.1) the matrix L_2 may be taken to be

$$\begin{bmatrix} 1 & 0 \\ m & 1 \end{bmatrix}, \tag{26.3}$$

where the m is arbitrary. The final matrix A_3 is given by

$$\begin{bmatrix} \bar{\lambda}+mb & b \\ m[(\lambda-\bar{\lambda})-mb] & \lambda-mb \end{bmatrix}, \tag{26.4}$$

which is *exactly similar* to A_1 for all values of m, but it is not a real matrix unless m is chosen so that $\lambda-mb$ is real. The indeterminacy in the exact computation with exact eigenvalue shifts is matched by the poor determination of m in the practical process. This is illustrated in Table 2.

TABLE 2

$$A_1 = \begin{bmatrix} 0 \cdot 31257 & 0 \cdot 61425 \\ -0 \cdot 51773 & 0 \cdot 41631 \end{bmatrix} \quad \begin{array}{l} \text{Eigenvalues } 0 \cdot 36444 \pm 0 \cdot 56154i \\ \text{Shifts } \lambda, \bar{\lambda} = 0 \cdot 36442 \pm 0 \cdot 56156i \end{array}$$

$$\bar{A}_2 - \bar{\lambda}I = \begin{bmatrix} 0 \cdot 00000+0 \cdot 00004i & 0 \cdot 61425 \\ -0 \cdot 00003-0 \cdot 00004i & 0 \cdot 00004+1 \cdot 12308i \end{bmatrix}$$

$$\bar{A}_3 = \begin{bmatrix} -0 \cdot 24983-0 \cdot 10083i & 0 \cdot 61425 \\ -1 \cdot 11108-0 \cdot 20167i & 0 \cdot 97871+0 \cdot 10083i \end{bmatrix}$$

The eigenvalues of A_1 are $0 \cdot 36444 \pm 0 \cdot 56154i$, and shifts of $0 \cdot 36442 \pm 0 \cdot 56156i$ were used. The elements of $(\bar{A}_2-\bar{\lambda}I)$ in the first column are both very small and have very high relative errors. The matrix $\bar{A}_3$, which would have been real with exact computation, has an imaginary part of the same order of magnitude as the real part. Nevertheless its eigenvalues are the same as those of A_1 to working accuracy. Further if we compute the eigenvectors of $\bar{A}_3$, and use them to determine the eigenvectors of A_1 making use of the computed transformation, the vectors obtained are accurate vectors of A_1.

Criticisms of the modified *LR* algorithm

27. In discussing the use of complex conjugate shifts we have ignored the question of interchanges. Exact computation does not guarantee, in general, that A_3 is real if we include interchanges. However if we use interchanges only in passing from A_1 to A_2, then in the notation of §13 we have

$$P_1(A_1-\lambda I) = \tilde{L}_1 R_1, \quad R_1 P_1^T \tilde{L}_1 + \lambda I = A_2, \quad A_2 = \tilde{L}_1^{-1}(P_1 A_1 P_1^T)\tilde{L}_1 \Bigg\}$$
$$A_2 - \bar{\lambda}I = L_2 R_2, \qquad R_2 L_2 + \bar{\lambda}I = A_3, \quad A_3 = L_2^{-1} A_2 L_2$$
(27.1)

giving

$$\tilde{L}_1 L_2 R_2 R_1 = \tilde{L}_1(A_2 - \bar{\lambda}I)R_1 = (P_1 A_1 P_1^T - \bar{\lambda}I)\tilde{L}_1 R_1$$
$$= (P_1 A_1 P_1^T - \bar{\lambda}I)P_1(A_1 - \lambda I), \quad (27.2)$$

and post-multiplication on the right of P^T shows that this last product is real. Unfortunately we cannot guarantee that interchanges are not required in passing from A_2 to A_3.

Even if we restrict ourselves to cases where there are no complex conjugate zeros (and the situations in which one can guarantee this *a priori* are rare), the modified algorithm is still open to criticism on the grounds of numerical stability. As we pointed out in Chapter 4, even when interchanges are used the stability of triangular decomposition cannot be guaranteed unconditionally. Even assuming that no more than four iterations per eigenvalue are required, this implies some 200 iterations in dealing with a matrix of order 50. The danger of a progressive growth in size of super-diagonal elements cannot be ignored. (The diagonal elements are tending to the eigenvalues and the sub-diagonal elements to zero.) Experience suggests that loss of accuracy from this cause is not too uncommon.

The *QR* algorithm

28. In line with the general treatment in this book it seems natural to attempt to replace the stabilized elementary transformations used in the modified *LR* algorithm by elementary unitary transformations. This leads directly to the *QR* algorithm of Francis (1961). In place of triangular decomposition Francis uses a factorization into the product of a unitary matrix Q and an upper-triangular matrix R.

The algorithm is defined by the relations

$$A_s = Q_s R_s, \qquad A_{s+1} = Q_s^H A_s Q_s = Q_s^H Q_s R_s Q_s = R_s Q_s, \quad (28.1)$$

so that at each stage we now use a unitary similarity transformation. The required factorization has already been discussed in Chapter 4, §§ 46–55, and we have shown that if A_s is non-singular this factorization is essentially unique and indeed *is* unique if we take the diagonal elements of R_s to be real and positive. If A_s is real both Q_s and R_s are real. This factorization has the advantage that the vanishing of a leading principal minor of A_s does not cause a breakdown as it does in the orthodox LR decomposition.

The successive iterates satisfy relations similar to those derived for the LR transformation. We have

$$A_{s+1} = Q_s^H A_s Q_s = Q_s^H Q_{s-1}^H A_{s-1} Q_{s-1} Q_s = (Q_s^H \ldots Q_2^H Q_1^H A_1 Q_1 Q_2 \ldots Q_s),$$

$$(28.2)$$

giving
$$Q_1 Q_2 \ldots Q_s A_{s+1} = A_1 Q_1 Q_2 \ldots Q_s. \tag{28.3}$$

All A_s are therefore unitarily similar to A_1, and if we write

$$Q_1 Q_2 \ldots Q_s = P_s, \qquad R_s R_{s-1} \ldots R_1 = U_s, \tag{28.4}$$

then $P_s U_s = Q_1 \ldots Q_{s-1} (Q_s R_s) R_{s-1} \ldots R_1$

$$= Q_1 \ldots Q_{s-1} A_s R_{s-1} \ldots R_1 = A_1 Q_1 \ldots Q_{s-1} R_{s-1} \ldots R_1 \text{ from (28.3)}$$

$$= A_1 P_{s-1} U_{s-1}. \tag{28.5}$$

Hence
$$P_s U_s = A_1^s, \tag{28.6}$$

and we have shown that P_s and U_s give the corresponding factorization of A_1^s. This factorization is unique if the diagonal elements of the upper triangle are positive, and this will be true of U_s if it is true of the R_s.

Convergence of the QR algorithm

29. In general the matrix A_s tends to upper-triangular form under somewhat less stringent conditions than were necessary for the convergence of the LR algorithm. Before discussing the convergence in detail we mention one minor difference between the convergence of the two techniques.

If A_1 is an upper-triangular matrix then with the LR algorithm we have $L_1 = I$, $R_1 = A_1$ and hence $A_s = A_1$ for all s. This is not true of the QR algorithm if we insist that R_s have positive diagonal elements. In fact writing

$$a_{ii} = |a_{ii}| \exp(i\theta_i), \qquad D = \text{diag}[\exp(i\theta_i)], \tag{29.1}$$

we have
$$A_1 = D(D^{-1} A_1), \qquad A_2 = D^{-1} A_1 D. \tag{29.2}$$

Hence although A_2 has the same diagonal elements as A_1, the super-diagonal elements are multiplied by complex factors of modulus unity. Obviously therefore we cannot expect A_s to tend to a strict limit unless all λ_i are real and positive. The factors of modulus unity are of little importance and we shall say that A_s is 'essentially' convergent if asymptotically $A_{s+1} \sim D^{-1}A_s D$ for some unitary diagonal matrix D.

As in § 5 we consider first the case of a 3×3 matrix with

$$|\lambda_1| > |\lambda_2| > |\lambda_3|$$

so that A_1^s has the form of (5.5). It is convenient to think of the factorization in terms of the Schmidt orthogonalization process with normalization (Chapter 4, § 54). The first column of P_s is therefore that of A_1^s scaled so that its Euclidean norm is unity so that its components are in the ratios

$$\lambda_1^s x_1 a_1 + \lambda_2^s y_1 a_2 + \lambda_3^s z_1 a_3 \ : \ \lambda_1^s x_2 a_1 + \lambda_2^s y_2 a_2 + \lambda_3^s z_2 a_3 \ :$$
$$: \ \lambda_1^s x_3 a_1 + \lambda_2^s y_3 a_2 + \lambda_3^s z_3 a_3. \tag{29.3}$$

Provided a_1 is not zero these elements are ultimately in the ratios $x_1:x_2:x_3$. If we insist that all diagonal elements of all R_i be positive then the first column of P_s assumes asymptotically the form

$$(a_1/|a_1|)(\lambda_1/|\lambda_1|)^s(x_1, y_1, z_1)/(|x_1|^2 + |y_1|^2 + |z_1|^2)^{\frac{1}{2}}, \tag{29.4}$$

and is therefore 'essentially' convergent to the first column of the unitary matrix obtained by the factorization of X. If a_1 is zero but a_2 is not, then the components of the first column of P_s are ultimately in the ratios $y_1:y_2:y_3$. A similar phenomenon occurs in connexion with the LR technique.

Notice that the vanishing of x_1 is irrelevant for the QR algorithm though in general it causes divergence in the LR algorithm. The 'accidental' failure of the LR process for a specific value of s, as a result of the vanishing of $\lambda_1^s x_1 a_1 + \lambda_2^s y_1 a_2 + \lambda_3^s z_1 a_3$, has no counterpart in the QR technique.

We can continue this elementary analysis to find the asymptotic behaviour of the second and third columns of P_s, but this is rather more tedious than it was for the LR algorithm. Clearly we can expect that P_s will converge essentially to the unitary factor of X.

Formal proof of convergence

30. We now give a formal proof of the essential convergence of P_s and in the first instance we assume that the eigenvalues of A satisfy

$$|\lambda_1| > |\lambda_2| > \ldots > |\lambda_n|. \tag{30.1}$$

Since such an A_1 necessarily has linear divisors we may write

$$A_1^s = X \operatorname{diag}(\lambda_i^s) X^{-1} = X D^s Y. \tag{30.2}$$

We define matrices Q, R, L, U by the relations

$$X = QR, \quad Y = LU, \tag{30.3}$$

where R and U are upper-triangular, L is unit lower-triangular and Q is unitary. All four matrices are independent of s. The non-singularity of R follows from that of X. Note that the QR decomposition always exists, but a triangular decomposition of Y exists only if all its leading principal minors are non-zero. We have

$$A_1^s = QRD^sLU = QR(D^sLD^{-s})D^sU, \tag{30.4}$$

and clearly D^sLD^{-s} is a unit lower-triangular matrix. Its (i, j) element is given by $l_{ij}(\lambda_i/\lambda_j)^s$ when $i > j$, and hence we may write

$$D^sLD^{-s} = I + E_s \quad \text{where} \quad E_s \to 0 \quad \text{as} \quad s \to \infty. \tag{30.5}$$

Equation (30.4) therefore gives

$$\begin{aligned}
A_1^s &= QR(I + E_s)D^sU \\
&= Q(I + RE_sR^{-1})RD^sU \\
&= Q(I + F_s)RD^sU \quad \text{where} \quad F_s \to 0 \quad \text{as} \quad s \to \infty. \tag{30.6}
\end{aligned}$$

Now $(I + F_s)$ may be factorized into the product of a unitary matrix $\tilde{Q}_s$ and an upper-triangular matrix $\tilde{R}_s$ and since $F_s \to 0$, $\tilde{Q}_s$ and $\tilde{R}_s$ both tend to I. Hence we have finally

$$A_1^s = (Q\tilde{Q}_s)(\tilde{R}_sRD^sU). \tag{30.7}$$

The first factor in parenthesis is unitary and the second is upper-triangular. Provided A_1^s is non-singular its factorization into such a product is essentially unique and therefore P_s is equal to $Q\tilde{Q}_s$ apart possibly from a post-multiplying diagonal unitary matrix. Hence P_s converges essentially to Q. If we insist that all R_s have positive diagonal elements we can find the unitary diagonal factor from (30.7). Writing

$$D = |D|D_1, \quad U = D_2(D_2^{-1}U), \tag{30.8}$$

where D_1 and D_2 are unitary diagonal matrices and $D_2^{-1}U$ has positive diagonal elements ($\tilde{R}_s$ and R already have positive diagonal elements), we obtain from (30.7)

$$A_1^s = Q\tilde{Q}_sD_2D_1^s[(D_2D_1^s)^{-1}\tilde{R}_sR(D_2D_1^s)\,|D|^s(D_2^{-1}U)]. \tag{30.9}$$

The matrix in braces is upper-triangular with positive diagonal elements and hence $P_s \sim QD_2D_1^s$ showing that ultimately Q_s becomes D_1.

Disorder of the eigenvalues

31. The proof shows that provided all leading principal minors of Y are non-zero we not only have essential convergence but the λ_i are correctly ordered on the diagonal. No corresponding restriction is placed on the matrix X. It is obvious from the 3×3 matrix that if a_1 (i.e. the first leading principal minor of Y) is zero, P_s still tends to a limit. We now establish this formally for the general case when a number of the principal minors of Y vanish.

Although Y will not have a triangular decomposition when a principal minor of Y vanishes, there always exists a permutation matrix P such that PY has a triangular decomposition. In Chapter 4, § 21, we showed how to determine such a P by using a pivotal strategy. In terms of Gaussian elimination, at each stage we took as pivot the element of largest modulus in the leading column. Suppose instead at the rth stage we choose as pivot the first non-zero element of those in positions (r, r), $(r+1, r)$,..., (n, r), and if this is the (r', r) element we rearrange the rows in the order r', r, $r+1$,..., $r'-1$, $r'+1$,..., n. Then we have finally

$$PY = LU, \tag{31.1}$$

where L has a number of zero elements below the diagonal in positions which are related to the permutation matrix P. Since Y is non-singular there will certainly be a non-zero pivot at each stage. We may therefore write

$$A_1^s = XD^sY = XD^sP^TLU = XP^T(PD^sP^T)LU. \tag{31.2}$$

The matrix PD^sP^T is a diagonal matrix with the elements λ_i^s in a different order, while XP^T is obtained from X by permuting its columns. If we write

$$XP^T = QR, \qquad PD^sP^T = D_3^s \tag{31.3}$$

then

$$A_1^s = QRD_3^sLU = QR(D_3^sLD_3^{-s})D_3^sU. \tag{31.4}$$

Since the elements λ_i^s are not in the natural order in D_3 it appears at first sight that $D_3^sLD_3^{-s}$ no longer tends to I. However, if we denote the diagonal elements of D_3 by λ_{i_1}, λ_{i_2},..., λ_{i_n}, then the (p, q) element $(p > q)$ of $D_3^sLD_3^{-s}$ is given by $(\lambda_{i_p}/\lambda_{i_q})^s l_{pq}$ and the pivotal strategy for Y assures that $l_{pq} = 0$ whenever $i_p < i_q$. Hence we may write

$D_3^s L D_3^{-s} = I + E_s$ where $E_s \to 0$ as $s \to \infty$. The matrix P_s therefore tends essentially to the matrix Q obtained by the factorization of $X P^T$.

Eigenvalues of equal modulus

32. We now turn to the case when A_1 has some eigenvalues of equal modulus, but all its elementary divisors are linear. We shall assume that all leading principal minors of Y are non-zero since we have already examined the effect of their vanishing. We have then

$$A_1^s = X D^s L U. \tag{32.1}$$

Suppose $|\lambda_r| = |\lambda_{r+1}| = \ldots = |\lambda_t|$, all other eigenvalues having distinct moduli. The element of $D^s L D^{-s}$ in position (i, j) below the diagonal is $(\lambda_i/\lambda_j)^s l_{ij}$ and therefore tends to zero unless

$$t \geqslant i > j \geqslant r, \tag{32.2}$$

in which case it remains equal to l_{ij} in modulus.

When all the eigenvalues of equal modulus are actually equal we may write
$$D^s L D^{-s} = \tilde{L} + E_s \quad (E_s \to 0) \tag{32.3}$$

where $\tilde{L}$ is a fixed unit lower-triangular matrix which is equal to I except for the elements in position (i, j) satisfying relations (32.2), where it is equal to l_{ij}. If we write $X \tilde{L} = QR$ then we have

$$\begin{aligned}
A_1^s &= QR(I + \tilde{L}^{-1} E_s) D^s U \\
&= Q(I + R \tilde{L}^{-1} E_s R^{-1}) R D^s U \\
&= Q(I + F_s) R D^s U \quad (F_s \to 0) \\
&= (Q \tilde{Q}_s)(\tilde{R}_s R D^s U), \tag{32.4}
\end{aligned}$$

where $\tilde{Q}_s \tilde{R}_s$ is the factorization of $(I + F_s)$. Hence apart from the usual unitary diagonal post-multiplier, P_s tends to Q, the matrix obtained by factorizing $X \tilde{L}$. Notice that the columns of $X \tilde{L}$ are a set of independent eigenvectors of A_1 since it differs from X only in that columns r to s are replaced by combinations of themselves. Multiple eigenvalues corresponding to linear divisors do not therefore prevent convergence.

When the eigenvalues of equal modulus are not equal then the matrix $\tilde{L}$ is of the same form as before but the non-zero sub-diagonal elements are not now fixed. If $\lambda_i = |\lambda_i| \exp(i\theta_i)$ we have

$$l_{ij} = l_{ij} \exp[is(\theta_i - \theta_j)]. \tag{32.5}$$

The matrix $X\tilde{L}$ is fixed apart from columns r to t. For each value of s these columns consist of a linear combination of the corresponding columns of X. Therefore in the QR decomposition of $X\tilde{L}$ all columns of Q are fixed except columns r to t which consist of linear combinations of columns r to t of the unitary factor of X. Hence apart from columns r to t, P_s is essentially convergent.

The most important case is that of a real matrix with some real eigenvalues and some complex conjugate eigenvalues. Clearly just as in §§ 9, 10 we may show that A_s tends essentially to an upper-triangular matrix apart from a single sub-diagonal element associated with each complex conjugate pair. Each such element is associated with a 2×2 matrix centred on the diagonal and having eigenvalues which converge to the relevant conjugate pair.

When we have r complex eigenvalues of equal modulus, then in the limit A_s has a matrix of order r centred on the diagonal with eigenvalues which tend to these r values. As a simple example we consider the matrix A_1 given by

$$A_1 = \begin{bmatrix} 0 & 0 & 0 & 1 \\ 1 & 0 & 0 & 0 \\ 0 & 1 & 0 & 0 \\ 0 & 0 & 1 & 0 \end{bmatrix}, \quad \begin{aligned} Q_1 &= A_1, \quad R_1 = I, \\ A_2 &= R_1 Q_1 = A_1, \end{aligned} \tag{32.6}$$

which has the eigenvalues $e^{\frac{1}{2}i\pi r}$ $(r = 0, 1, 2, 3)$. This is invariant with respect to the QR algorithm and therefore remains a 4×4 matrix centred on the diagonal. Such blocks may be difficult to detect by an automatic procedure, particularly when their size is not known in advance. (See § 47, however, for the effect of shifts of origin.)

Alternative proof for the *LR* technique

33. We sketch the adaptation of the above method of proof to the *LR* algorithm. When the λ_i are distinct we have

$$A_1^s = XD^sY = L_X U_X D^s L_Y U_Y, \tag{33.1}$$

where $L_X U_X$ and $L_Y U_Y$ are the triangular decompositions of X and Y, the existence of which requires that the leading principal minors of both X and Y are non-zero. Hence

$$\begin{aligned} A_1^s &= L_X U_X (D^s L_Y D^{-s}) D^s U_Y = L_X U_X (I + E_s) D^s U_Y \ (E_s \to 0) \\ &= L_X (I + F_s) U_X D^s U_Y \ (F_s \to 0). \end{aligned} \tag{33.2}$$

The matrix $(I+F_s)$ certainly has a triangular decomposition for all sufficiently large s, each of the factors tending to I, but for early values of s this decomposition may not exist. This corresponds to the case of 'accidental' failure which arises when a principal minor of some A_s is zero. It has no counterpart in the QR algorithm. Ignoring this possibility we see that in the notation of § 4, $T_s \rightarrow L_X$. From this it follows that A_s tends to upper-triangular form with the λ_i in the correct order on the diagonal.

If one (or more) of the leading principal minors of Y vanishes there will nevertheless be a permutation matrix such that PY has a triangular decomposition. Denoting this again by $L_Y U_Y$ we have

$$A_1^s = X D^s P^T L_Y U_Y = (X P^T)(P D^s P^T) L_Y U_Y. \qquad (33.3)$$

If $X P^T$ has a triangular decomposition $L_X U_X$, that is if all its leading principal minors are non-zero, then we can show in the usual way that $T_s \rightarrow L_X$ and A_s tends to a triangular matrix with $D P^T$ as its diagonal.

34. The treatment of multiple eigenvalues (corresponding to linear divisors) and eigenvalues of the same modulus is analogous to that we have just given for the QR algorithm. In the former case we have convergence if $X P^T \tilde{L}$ has a triangular decomposition, where $\tilde{L}$ is a fixed matrix derived from the lower triangle of the factorization of Y in the way we have described. However, the case when A_1 is symmetric is of special interest. We now have $X = Y^T$, and if P is chosen so that PY has a triangular decomposition, $X P^T$ also has a decomposition because its leading principal minors have the same values as those of PY and are therefore non-zero.

If there are no eigenvalues of equal modulus then it is precisely the matrix $X P^T$ which is required to have a triangular decomposition, and at first sight it looks as though the LR process must always converge. However we must not forget that $(I+F_s)$ may fail to have a decomposition for some specific value of s even though we know that it must have such a decomposition for all sufficiently large s. Hence for a symmetric matrix having eigenvalues with distinct moduli, exact computation must certainly give convergence apart from the possibility of an accidental failure at an early stage.

If A_1 is symmetric and positive definite we can show that convergence must take place even when there are multiple eigenvalues. (Since all eigenvalues are positive, eigenvalues of equal modulus are necessarily equal.) It is easy to show that in this case accidental breakdown cannot occur at any stage. If we write

$$PY = L_1 U_1, \qquad X P^T = U_1^T L_1^T, \qquad (34.1)$$

then, as in § 32, for convergence of the *LR* algorithm we require that $X P^T \tilde{L}$ should have a triangular decomposition, where $\tilde{L}$ is derived from L by deleting all elements below the diagonal except those in rows and columns corresponding to the equal diagonal elements in $P D^s P^T$. Now we have

$$(X P^T) \tilde{L} = U_1^T L_1^T \tilde{L}, \qquad (34.2)$$

and hence all that is necessary is that $L_1^T \tilde{L}$ should have a triangular decomposition, that is, all its leading principal minors should be non-zero. From the relationship of $\tilde{L}$ to L, it follows, using the theorem of corresponding matrices, that every leading principal minor of $L_1^T \tilde{L}$ is greater than or equal to unity. The result is therefore established.

Practical application of the *QR* algorithm

35. The typical step of the QR algorithm requires the factorization of A_s into $Q_s R_s$. For theoretical purposes it was convenient to think of this in terms of the Schmidt orthonormalization of the columns of A_s but as we have remarked in Chapter 4, § 55, in practice this commonly gives rise to a Q_s which is by no means orthogonal. In such a case the numerical stability normally associated with orthogonal similarity transformations is likely to be lost. In practice we usually determine the matrix Q_s^T such that

$$Q_s^T A_s = R_s, \qquad (35.1)$$

rather than Q_s itself. The matrix Q_s^T is not determined explicitly but rather in factorized form, either as the product of plane rotations or of elementary Hermitians using the Givens or Householder triangularization respectively. The matrix $R_s Q_s$ is then computed by successive post-multiplication of R_s with the transposes of the factors of Q_s^T. We have shown in Chapter 3, §§ 26, 35, 45, that the product of the computed factors of Q_s^T will then be almost exactly orthogonal and that the computed $R_s Q_s$ will be close to the matrix obtained by multiplying exactly by the computed factors. Hence we can show that the computed A_{s+1} is close to an exact orthogonal transformation of the computed A_s. (We cannot show that the computed A_{s+1} is close to that transform of the computed A_s which would be obtained by performing the sth step exactly, and indeed this is not true in general.)

Clearly the volume of work is even greater than that for the LR transformation. The QR algorithm is of practical value only when A_1 is of upper Hessenberg form or symmetric and of band form. We deal with the upper Hessenberg form first, and we assume that none of the sub-diagonal elements is zero since otherwise we can work with Hessenberg matrices of lower order.

The Hessenberg form is invariant with respect to the QR transformation, as we now show. Consider first the triangularization of the Hessenberg matrix using Givens' process. It is easy to see that this is achieved by rotations in planes $(1, 2)$, $(2, 3),\ldots, (n-1, n)$ respectively, and that only $2n^2$ multiplications are required. No advantage is gained by using Householder's reduction instead of Givens', since in each major step of the triangularization there is only one element to be reduced to zero. On post-multiplying the upper triangle by the transposes of the rotations, the Hessenberg form is regained just as in the LR algorithm with interchanges. There are $4n^2$ multiplications in one complete step of the QR algorithm compared with n^2 in LR with interchanges, but the convergence properties are on a much sounder footing with the QR algorithm.

Just as with the LR algorithm with interchanges, we find that if we store each A_s by columns then the decomposition and post-multiplication can be combined in such a way that in each step A_s is read once from the backing store and replaced by A_{s+1}. Accommodation in the high-speed store is required for two columns of A_s and for the cosines and sines of the $(n-1)$ angles of rotation, amounting to approximately $4n$ words at the stage requiring maximum storage.

Shifts of origin

36. Shifts of origin with or without restoring may be introduced in the QR algorithm just as in the LR algorithm. The process with restoring is defined by

$$A_s - k_s I = Q_s R_s, \qquad R_s Q_s + k_s I = A_{s+1}, \tag{36.1}$$

and we may show as before that

$$A_s = (Q_1 Q_2 \ldots Q_{s-1})^T A_1 Q_1 Q_2 \ldots Q_{s-1}, \tag{36.2}$$

and

$$(Q_1 Q_2 \ldots Q_s)(R_s \ldots R_2 R_1) = (A_1 - k_1 I)(A_1 - k_2 I)\ldots(A_1 - k_s I). \tag{36.3}$$

These shifts of origin are essential in practice. When A_1 is real and is known to have real eigenvalues or if A_1 is complex we can choose k_s

exactly as in § 24. In the former case all the arithmetic is real. In general the number of iterations has proved to be considerably less for the QR transformation than with the modified LR process using shifts determined on the same basis in each case.

When A_1 is real but has some complex conjugate eigenvalues then, in general, the 2×2 matrix in the bottom right-hand corner will have complex conjugate eigenvalues λ, $\bar{\lambda}$ at some stage in the iteration. If we perform two steps of QR using shifts λ and $\bar{\lambda}$, then exact computation would give a real matrix on completion though the intermediate matrix would be complex. Although we do not now have the difficulty that was noticed in § 27 in connexion with interchanges we may still find in practice that the final matrix has a substantial imaginary part although the eigenvalues have been accurately preserved. Francis has described an elegant method of combining the two steps in such a way that complex arithmetic is avoided. The final matrix is necessarily real and the process is as stable as the standard QR algorithm. Before discussing this technique however, we shall describe two other stratagems which give a considerable economy in the volume of computation. They can be used in connexion with both the QR and the modified LR algorithm and are most simply considered in connexion with the latter.

Decomposition of A_s

37. The first point is that although we have assumed that A_1 has non-zero sub-diagonal elements, we know that, in general, some of the sub-diagonal elements of A_s are tending to zero. It may well happen during the course of the iteration that a matrix A_s has one or more very small sub-diagonal elements. If any such element is sufficiently small we may regard it as zero and deal with the smaller Hessenberg matrices. Thus in the matrix

$$A_s = \begin{bmatrix} \times & \times & \times & \times & \times & \times \\ \times & \times & \times & \times & \times & \times \\ & \epsilon & \times & \times & \times & \times \\ & & \times & \times & \times & \times \\ & & & \times & \times & \times \\ & & & & \times & \times \end{bmatrix} \doteq \begin{bmatrix} B & C \\ O & D \end{bmatrix}, \qquad (37.1)$$

if ϵ is sufficiently small we may find the eigenvalues of D and then of B. Note that the elements C then serve no further purpose. Remember that in general the original Hessenberg matrix will itself have been obtained by a previous computation, and hence when ϵ is smaller than the errors in the original elements of A_1, little can be lost by replacing ϵ by zero. A number of the order of magnitude of $2^{-t}\|A\|_\infty$ is a suitable tolerance.

Hence at each stage of the iteration we can examine the sub-diagonal elements and if the latest of the sub-diagonal elements which are negligible is in position $(r+1, r)$ we can continue iterating with the matrix of order $(n-r)$ in the bottom right-hand corner, returning to the upper matrix when all the eigenvalues of the lower matrix are found. If $r = n-1$ the lower matrix is of order unity and its eigenvalue is known. At this stage we can deflate by omitting the last row and column. If $r = n-2$ the lower matrix is of order 2 and we can find its two eigenvalues by solving a quadratic. Again we can deflate by omitting the last two rows and columns.

38. The second point concerns an extension of this idea by Francis (1961). Consider a Hessenberg matrix A having small elements ϵ_1 and ϵ_2 in positions $(r+1, r)$ and $(r+2, r+1)$. This is illustrated for the case $n = 6$, $r = 2$ by the matrix given by

$$
A = \left. r \left\{ \begin{array}{}
\left[\begin{array}{cc|cccc}
\times & \times & \times & \times & \times & \times \\
\times & \times & \times & \times & \times & \times \\
\hline
\epsilon_1 & \times & \times & \times & \times & \\
 & \epsilon_2 & \times & \times & \times & \\
 & & \times & \times & \times & \\
 & & & \times & \times &
\end{array}\right]
\end{array} \right. \right. = \left[\begin{array}{c|c} X & Y \\ \hline E & W \end{array}\right]. \tag{38.1}
$$

Let A be partitioned with respect to these elements in the manner illustrated in (38.1), where E is null apart from the element ϵ_1. We show that any eigenvalue μ of W is also an eigenvalue of a matrix A' which differs from A only by an element $\epsilon_1\epsilon_2/(a_{r+1,r+1}-\mu)$ in position $(r+2, r)$.

For since $(W-\mu I)$ is singular there exists a non-null vector x such that

$$
x^T(W-\mu I) = 0. \tag{38.2}
$$

The first two components of x satisfy the relations

$$(a_{r+1,r+1}-\mu)x_1+\epsilon_2 x_2 = 0 \tag{38.3}$$

giving

$$\epsilon_1 x_1+[\epsilon_1\epsilon_2/(a_{r+1,r+1}-\mu)]x_2 = 0. \tag{38.4}$$

Equations (38.2) and (38.4) together imply that the last $(n-r)$ rows of the modified matrix $(A'-\mu I)$ are linearly dependent and hence that μ is an eigenvalue of A'. If $\epsilon_1\epsilon_2/(a_{r+1,r+1}-\mu)$ is negligible then μ is effectively an eigenvalue of A. Provided $a_{r+1,r+1}-\mu$ is not small, the small terms ϵ_1 and ϵ_2 serve to decouple W from the rest of the matrix.

Numerical example

39. This effect is illustrated in Table 3. The matrix of order 6 has elements 10^{-5} in positions (3, 2) and (4, 3). The eigenvalues μ_1, μ_2, μ_3, μ_4 of the matrix of order 4 in the bottom right-hand corner are given, and only μ_4 is close to a_{33}. Hence we can expect that the full matrix has eigenvalues λ_1, λ_2, λ_3 differing from μ_1, μ_2, μ_3 by quantities of order 10^{-10}. The eigenvalue λ_4 differs from μ_4 by a quantity of order 10^{-5} approximately. Eigenvalues λ_5 and λ_6 have no counterparts though of course they are very close to eigenvalues of the leading principal matrix of order 3.

TABLE 3

$$A_1 = \begin{bmatrix} X & Y \\ \hline E & W \end{bmatrix} = \left[\begin{array}{cc|cccc} 4 & 3 & 2 & 1 & 5 & 6 \\ 3 & 1 & 4 & 2 & 1 & 7 \\ \hline & 10^{-5} & 6 & 1 & 5 & 3 \\ & & 10^{-5} & 2 & 5 & 7 \\ & & & 1 & 2 & 3 \\ & & & & 4 & 1 \end{array}\right]$$

Eigenvalues of A_1	*Eigenvalues of W*
$\lambda_1 = -0{\cdot}46410\ 31621$	$\mu_1 = -0{\cdot}46410\ 31621$
$\lambda_2 = -0{\cdot}99999\ 85712$	$\mu_2 = -0{\cdot}99999\ 85714$
$\lambda_3 = 6{\cdot}46412\ 31631$	$\mu_3 = 6{\cdot}46412\ 31611$
$\lambda_4 = 6{\cdot}00011\ 84534$	$\mu_4 = 5{\cdot}99997\ 85724$
$\lambda_5 = -0{\cdot}85410\ 48844$	
$\lambda_6 = 5{\cdot}85396\ 50012$	

Practical procedure

40. To take advantage of this phenomenon in the use of the *LR* algorithm we proceed as follows. At the end of each iteration we determine the shift of origin k_s to be used in the next stage. We then test whether the condition

$$a^{(s)}_{r+1,r}a^{(s)}_{r+2,r+1}/(a^{(s)}_{r+1,r+1}-k_s) < 2^{-t}\|A_1\| \tag{40.1}$$

is satisfied for $r = 1, 2, ..., n-1$ and denote by p the largest value of r for which this is true. We next perform triangularization with interchanges of $(A_s - k_s I)$ *starting at row* $(p+1)$ *instead of the first row*. This introduces an element $a_{p+1,p}^{(s)} a_{p+2,p+1}^{(s)}/(a_{p+1,p+1}^{(s)} - k_s)$ in position $(p+2, p)$, no interchange being required in the first step, and under our assumption this element is negligible. At the end of the partial triangularization the matrix is of the form given for the case $n = 6$, $p = 2$ by

$$\begin{bmatrix} \times & \times & \times & \times & \times & \times \\ \times & \times & \times & \times & \times & \times \\ & \epsilon_1 & \times & \times & \times & \times \\ & — & 0 & \times & \times & \times \\ & & & 0 & \times & \times \\ & & & & 0 & \times \end{bmatrix}. \tag{40.2}$$

The first p rows are unmodified; the bar denotes the negligible element introduced by the reductions. We now complete the similarity transformation. Obviously this involves only operations on columns $(p+1)$ to n including the elements in the matrix in the upper right-hand corner. Usually if the condition (40.1) has been satisfied in the sth step it will be satisfied in subsequent steps.

We have described the process in terms of the LR technique with interchanges but it carries over immediately to the QR algorithm. In fact the same criterion is adequate, since when it is satisfied we can start the unitary triangularization at row $(p+1)$; the angle of the first rotation is small and the negligible element introduced in position $(p+2, p)$ is $\epsilon_1 \epsilon_2 / [|a_{p+1,p+1}^{(s)} - k_s|^2 + \epsilon_2^2]^{\frac{1}{2}}$.

Clearly we can combine the two techniques we have described, taking advantage both of individually negligible $a_{p+1,p}^{(s)}$ and of negligible $a_{p+1,p}^{(s)} a_{p+2,p+1}^{(s)}/(a_{p+1,p+1}^{(s)} - k_s)$.

Avoiding complex conjugate shifts

41. We are now in a position to deal with the case when the matrix is real but may have some complex conjugate eigenvalues. For the moment we consider the QR algorithm only. Suppose at the sth stage the eigenvalues of the 2×2 matrix in the bottom right-hand corner are k_1 and k_2; these may be real or a complex conjugate pair. In either case let us consider the effect of performing two steps of QR

with shifts k_1 and k_2 respectively. It is convenient to rename the sth matrix A_1 for the purpose of this investigation. We have then

$$A_1 - k_1 I = Q_1 R_1, \qquad R_1 Q_1 + k_1 I = A_2 \qquad (41.1)$$

$$A_2 - k_2 I = Q_2 R_2, \qquad R_2 Q_2 + k_2 I = A_3 \qquad (41.2)$$

$$(Q_1 Q_2)(R_2 R_1) = (A_1 - k_1 I)(A_1 - k_2 I) \qquad (41.3)$$

$$A_3 = (Q_1 Q_2)^H A_1 Q_1 Q_2. \qquad (41.4)$$

Provided k_1 and k_2 are not exactly eigenvalues, the QR decompositions are unique if we insist that the upper triangular matrices have positive diagonal elements.

Now the matrix on the right of (41.3) is real in any case, and hence $Q_1 Q_2$ is real and is obtained by the orthogonal triangularization of $(A_1 - k_1 I)(A_1 - k_2 I)$. We shall write

$$Q_1 Q_2 = Q, \qquad R_2 R_1 = R, \qquad QR = (A_1 - k_1 I)(A_1 - k_2 I). \quad (41.5)$$

Equation (41.4) implies that the real matrix Q is such that

$$A_1 Q = Q A_3. \qquad (41.6)$$

Now assume that by any other method we can determine an orthogonal matrix $\tilde{Q}$ having the same first column as Q and such that

$$A_1 \tilde{Q} = \tilde{Q} B, \qquad (41.7)$$

where B is a Hessenberg matrix with positive sub-diagonal elements. From Chapter 6, § 7, we know that such a $\tilde{Q}$ must indeed be Q and B must be A_3. We propose to determine such $\tilde{Q}$ and B without using complex arithmetic at any stage. In effect we shall do this by computing a $\tilde{Q}^T$ having the same first row as Q^T and such that

$$\tilde{Q}^T A_1 = B \tilde{Q}^T. \qquad (41.8)$$

Now from (41.5) we have

$$Q^T (A_1 - k_1 I)(A_1 - k_2 I) = R \qquad (41.9)$$

and hence Q^T is the orthogonal matrix that reduces $(A_1 - k_1 I)(A_1 - k_2 I)$ to upper triangular form. In terms of Givens' method for the triangularization, Q^T is determined as the product of $\frac{1}{2} n(n-1)$ plane rotations R_{ij}. Hence we may write

$$Q^T = R_{n-1,n} \cdots R_{2,n} \cdots R_{2,3} R_{1,n} \cdots R_{1,3} R_{1,2}. \qquad (41.10)$$

If we build up the product on the right starting with $R_{1,2}$ we find that the first row of Q^T is the first row of $R_{1,n}\ldots R_{1,3}R_{1,2}$, the subsequent multiplication by the other rotations leaving this row unaltered. Now $R_{1,2}$, $R_{1,3},\ldots$, $R_{1,n}$ are determined entirely by the first column of $(A_1-k_1I)(A_1-k_2I)$ and hence can be computed without knowing the rest of this matrix. This first column consists of three elements only and if we denote these by x_1, y_1, z_1 then

$$\left.\begin{aligned} x_1 &= (a_{11}-k_1)(a_{11}-k_2)+a_{12}a_{21} = a_{11}^2+a_{12}a_{21}-a_{11}(k_1+k_2)+k_1k_2 \\ y_1 &= a_{21}(a_{11}-k_2)+(a_{22}-k_1)a_{21} = a_{21}(a_{11}+a_{22}-k_1-k_2) \\ z_1 &= a_{32}a_{21} \end{aligned}\right\}.$$

$$(41.11)$$

Consequently $R_{1,4}$, $R_{1,5},\ldots$, $R_{1,n}$ are each the identity matrix, while $R_{1,2}$ and $R_{1,3}$ are determined by x_1, y_1, z_1 which are obviously real. Suppose $R_{1,2}$ and $R_{1,3}$ are computed from x_1, y_1 and z_1, and we then form the matrix C_1 defined by

$$R_{1,3}R_{1,2}A_1R_{1,2}^TR_{1,3}^T = C_1. \tag{41.12}$$

In Chapter 6, § 7, we showed that corresponding to *any* real matrix C_1 there is an orthogonal matrix S (say) *having its first column equal to e_1* and such that

$$S_1^TC_1S_1 = B, \tag{41.13}$$

where B is an upper Hessenberg matrix. This matrix S_1 can be determined either by Givens' or Householder's process. Hence

$$(S_1^TR_{1,3}R_{1,2})A_1(R_{1,2}^TR_{1,3}^TS_1) = B, \tag{41.14}$$

and if we write $(S_1^TR_{1,3}R_{1,2}) = \tilde{Q}^T$, then from (41.14)

$$\tilde{Q}^TA_1 = B\tilde{Q}^T. \tag{41.15}$$

Since the first row of S_1^T is e_1^T, the first row of $\tilde{Q}^T$ is the same as the first row of $R_{1,3}R_{1,2}$ which we have already shown to be the same as the first row of Q^T. Hence B must be the matrix A_3 which would be obtained by performing two steps of QR with shifts of k_1 and k_2. We therefore have a method of deriving A_3 without using complex arithmetic even when k_1 and k_2 are complex.

42. This derivation is remarkably efficient, as we now show. The number of multiplications involved in computing the rotations $R_{1,2}$

and $R_{1,3}$ is independent of n. Having obtained them, C_1 can be computed from equation (41.12) and again the number of multiplications is independent of n. For the case $n = 8$, C_1 has the form

$$C_1$$

$$
\begin{bmatrix}
\times & \times & \times & \times & \times & \times & \times & \times \\
\times & \times & \times & \times & \times & \times & \times & \times \\
\times & \times & \times & \times & \times & \times & \times & \times \\
\times & 0 & \times & \times & \times & \times & \times & \times \\
 & & & \times & \times & \times & \times & \times \\
 & & & & \times & \times & \times & \times \\
 & & & & & \times & \times & \times \\
 & & & & & & \times & \times
\end{bmatrix},
$$

$$C_3$$

$$
\begin{bmatrix}
\times & \times & \times & \times & \times & \times & \times & \times \\
\times & \times & \times & \times & \times & \times & \times & \times \\
 & \times & \times & \times & \times & \times & \times & \times \\
 & & \times & \times & \times & \times & \times & \times \\
 & & \underline{\times} & \times & \times & \times & \times & \times \\
 & & \underline{\times} & 0 & \times & \times & \times & \times \\
 & & & & & \times & \times & \times \\
 & & & & & & \times & \times
\end{bmatrix}
\begin{matrix} \\ \\ \\ \leftarrow \\ \leftarrow \\ \leftarrow \\ \\ \end{matrix}
\qquad . \qquad (42.1)
$$

$$\uparrow \quad \uparrow \quad \uparrow$$

No advantage results from the vanishing of the element denoted by zero. C_1 is not quite of Hessenberg form, extra non-zero elements having been introduced in columns 1 and 2 by the pre-multiplications and post-multiplications. We now consider the similarity reduction of C_1 to Hessenberg form using Givens' method. Because C_1 is already almost of Hessenberg form the work involved in this is much less than if C_1 had been a full matrix.

As in the general case (Chapter 6, § 2), there are $(n-2)$ major steps, in the rth of which the zeros are produced in the rth column. We now show that at the beginning of the rth step C_r is of Hessenberg form apart from non-zero elements in positions $(r+2, r)$ and $(r+3, r)$.

The element $(r+3, r+1)$ is zero but this is of no advantage. This is illustrated for the case $n = 8$, $r = 3$, in (42.1). Assuming such a configuration in the rth step we pre-multiply by rotations in the planes $(r+1, r+2)$ and $(r+1, r+3)$ to eliminate those two elements underlined in (42.1). The pre-multiplication destroys the zero in position $(r+3, r+1)$ but otherwise introduces no new non-zero elements. We next post-multiply by the transposes of these two rotations. This replaces columns $(r+1)$ and $(r+2)$ by linear combinations of themselves, and then columns $(r+1)$ and $(r+3)$ by linear combinations of themselves. In (42.1) we have indicated by arrows the rows and columns of C_r which are modified in the rth step. It is now obvious that the form of C_{r+1} is exactly analogous to that of C_r.

We have described the preliminary transformation of A_1 to give C_1 as though it were distinct from the operations required to reduce C_1. In fact if we augment the matrix A_1 by adding at the front an extra column consisting of the elements x_1, y_1, z_1 as illustrated by the array

$$
C_0 =
\begin{bmatrix}
x_1 & \times & \times & \times & \times & \times & \times & \times & \times \\
y_1 & \times & \times & \times & \times & \times & \times & \times & \times \\
z_1 & 0 & \times & \times & \times & \times & \times & \times & \times \\
 & & & \times & \times & \times & \times & \times & \times \\
 & & & & \times & \times & \times & \times & \times \\
 & & & & & \times & \times & \times & \times \\
 & & & & & & \times & \times & \times \\
 & & & & & & & \times & \times
\end{bmatrix}
=
\begin{bmatrix}
x_1 \\
y_1 \\
z_1 \\
0 \\
\cdot \\
\cdot \\
\cdot \\
0
\end{bmatrix}
A_1
\Bigg| \quad , \quad (42.2)
$$

and denote the augmented matrix by C_0, then C_1 is derived from C_0 by steps which are exactly analogous to those used to derive C_{r+1} from C_r. There are essentially $8n^2$ multiplications involved in passing from A_1 to A_3, the same number as in two real steps of the QR algorithm. If we actually perform two steps using complex shifts k_1 and k_2, two steps of QR in complex arithmetic are required, involving $16n^2$ multiplications.

Double QR step using elementary Hermitians

43. We have described the double QR step in terms of Givens' transformation since it is conceptually simpler. For the single-shift technique Householder's method is in no way superior, but for the double-shift technique it is appreciably more economical.

We shall use the notation $P_r = I - 2w_r w_r^T$, where

$$w_r^T = \underbrace{(0, 0,\ldots, 0}_{r-1}, \times, \times, \times,\ldots, \times).\qquad(43.1)$$

We show that A_3 can be derived in the following way. We first produce an elementary Hermitian P_1 such that

$$P_1 x = k e_1 \quad \text{where } x^T = (x_1, y_1, z_1, 0,\ldots, 0).\qquad(43.2)$$

Clearly only the first three components of the corresponding w_1 are non-zero.

We then compute the matrix $C_1 = P_1 A_1 P_1$, which is of the form illustrated by

$$C_1$$

$$\begin{bmatrix}
\times & \times & \times & \times & \times & \times & \times & \times \\
\times & \times & \times & \times & \times & \times & \times & \times \\
\times & \times & \times & \times & \times & \times & \times & \times \\
\times & \times & \times & \times & \times & \times & \times & \times \\
 & & & & \times & \times & \times & \times & \times \\
 & & & & & \times & \times & \times & \times \\
 & & & & & & \times & \times & \times \\
 & & & & & & & \times & \times
\end{bmatrix},$$

$$C_3$$

$$\begin{bmatrix}
\times & \times & \times & \times & \times & \times & \times & \times \\
\times & \times & \times & \times & \times & \times & \times & \times \\
 & \times & \times & \times & \times & \times & \times & \times \\
 & & \times & \times & \times & \times & \times & \times \\
 & & \times & \times & \times & \times & \times & \times \\
 & & \times & \times & \times & \times & \times & \times \\
 & & & & \times & \times & \times \\
 & & & & & \times & \times
\end{bmatrix}.\qquad(43.3)$$

It is of almost the same form as the matrix C_1 obtained using plane rotations (cf. (42.1)); the only difference is that there is no longer a zero in the (4, 2) position and this is of no particular significance.

The matrix C_1 is next reduced to Hessenberg form using Householder's method (Chapter 6, § 3). This involves successive pre-multiplication and post-multiplication by matrices $P_2, P_3, ..., P_{n-2}$. The resulting Hessenberg matrix B is such that

$$P_{n-2}...P_2P_1A_1P_1P_2...P_{n-2} = B, \tag{43.4}$$

and $P_{n-2}...P_1$ has the same first row as P_1 which itself has the same first row as the orthogonal matrix which triangularizes

$$(A-k_1I)(A-k_2I).$$

Hence B is A_3.

We now show that just before the pre-multiplication by P_r, the current matrix C_{r-1} has the form illustrated in (43.3) for the case $n = 8$, $r = 4$. For, assuming this is true of C_{r-1}, then there are only two elements in column $(r-1)$ to be eliminated by P_r and hence the corresponding w_r has non-zero components only in positions r, $r+1$, $r+2$. Pre-multiplication and post-multiplication by P_r therefore replace rows and columns r, $r+1$, $r+2$ by linear combinations of themselves and thereby eliminate elements $(r, r+1)$ and $(r, r+2)$. The matrix C_r is of the same form as C_{r-1} and since C_1 is of the required form the result is established. As in § 42 we can include the initial transformation in the general scheme, by augmenting the matrix A_1 with an extra first column consisting of the elements x_1, y_1, z_1.

Computational details

44. The non-zero elements of P_r are determined by the elements $c_{r,r-1}^{(r-1)}$, $c_{r+1,r-1}^{(r-1)}$ and $c_{r+2,r-1}^{(r-1)}$. We shall denote these elements by x_r, y_r, z_r, which is consistent with our previous use of x_1, y_1, z_1. The most economical representation of P_r is that described in Chapter 5, § 33. This gives

$$\left.\begin{aligned}
&P_r = I - 2p_rp_r^T/\|p_r\|_2^2 \\
&p_r^T = (0,..., 0, 1, u_r, v_r, 0,..., 0) \\
&S_r^2 = x_r^2+y_r^2+z_r^2, \qquad u_r = y_r/(x_r\mp S_r), \qquad v_r = z_r/(x_r\mp S_r) \\
&2/\|p_r\|_2^2 = 2/(1+u_r^2+v_r^2) = \alpha_r \quad \text{(say)}
\end{aligned}\right\}, \tag{44.1}$$

where the sign is chosen so that

$$|x_r\mp S_r| = |x_r|+S_r. \tag{44.2}$$

This method of computation ensures that P_r is orthogonal to working accuracy. We use P_r in the form

$$P_r = I - p_r \, (\alpha_r \, p_r^T), \qquad (44.3)$$

since this takes advantage of the fact that one of the three components of p_r is unity. There are only $5n^2$ multiplications involved in passing from A_1 to A_3 in this way.

Decomposition of A_s

45. We can still take advantage of the two stratagems described in §§ 37, 38 while using the double-step technique. If a sub-diagonal element is zero to working accuracy at any stage this is obvious. In the case when two consecutive sub-diagonal elements of an A_s are 'small', the modification is not quite trivial. If $a_{r+1,r}^{(s)}$ and $a_{r+2,r+1}^{(s)}$ are both sufficiently small we hope to start the transformation with P_{r+1} rather than P_1. Accordingly by analogy with (41.11) we calculate elements x_{r+1}, y_{r+1}, z_{r+1} defined by

$$\left. \begin{aligned} x_{r+1} &= a_{r+1,r+1}^2 + a_{r+1,r+2} a_{r+2,r+1} - a_{r+1,r+1}(k_1 + k_2) + k_1 k_2 \\ y_{r+1} &= a_{r+2,r+1}(a_{r+1,r+1} + a_{r+2,r+2} - k_1 - k_2) \\ z_{r+1} &= a_{r+3,r+2} a_{r+2,r+1} \end{aligned} \right\}, \qquad (45.1)$$

and from them we determine P_{r+1} in the usual way. In the case when $n = 7$, $r = 2$, the matrices P_{r+1} and A_s are given by

$$P_3 = \begin{bmatrix} I & \\ \hline & I - 2w_3 w_3^T \end{bmatrix},$$

$$A_s = \begin{bmatrix} \times & \times & \times & \times & \times & \times & \times \\ \times & \times & \times & \times & \times & \times & \times \\ \hline & & \epsilon_1 & \times & \times & \times & \times & \times \\ & & & \epsilon_2 & \times & \times & \times & \times \\ & & & & \times & \times & \times & \times \\ & & & & & \times & \times & \times \\ & & & & & & \times & \times \end{bmatrix} = \begin{bmatrix} X & Y \\ \hline E & Z \end{bmatrix}, \qquad (45.2)$$

where we have denoted $a_{r+1,r}$ and $a_{r+2,r+1}$ by ϵ_1 and ϵ_2 to emphasize that they are small. Pre-multiplication by P_{r+1} leaves the first r rows unaffected. The part marked E in (45.2) has only one non-zero element and in general after pre-multiplication it will have three non-zero elements as illustrated by

$$
\begin{bmatrix}
\times & \times & \times & \times & \times & \times & \times \\
\times & \times & \times & \times & \times & \times & \times \\
\eta_1 & & \times & \times & \times & \times & \times \\
\eta_2 & & \times & \times & \times & \times & \times \\
\eta_3 & & \times & \times & \times & \times & \times \\
& & & & \times & \times & \times \\
& & & & & \times & \times
\end{bmatrix},
\qquad (45.3)
$$

Z having been modified in rows $r+1$, $r+2$, $r+3$ in the usual way. We require that η_2 and η_3 should be negligible. In the notation of § 44 we have

$$
\left.\begin{aligned}
\eta_1 &= \epsilon_1 - 2\epsilon_1/(1+u_{r+1}^2+v_{r+1}^2) \\
\eta_2 &= \quad -2\epsilon_1 u_{r+1}/(1+u_{r+1}^2+v_{r+1}^2) \\
\eta_3 &= \quad -2\epsilon_1 v_{r+1}/(1+u_{r+1}^2+v_{r+1}^2)
\end{aligned}\right\}. \qquad (45.4)
$$

From (45.1) we see that y_{r+1} and z_{r+1} have the factor $a_{r+2,r+1} = \epsilon_2$, but x_{r+1} is not, in general, small. Hence u_{r+1} and v_{r+1} have the factor ϵ_2, and η_2 and η_3 have the factor $\epsilon_1\epsilon_2$, while η_1 will be $-\epsilon_1$ if ϵ_2 is small enough. There is no need to compute η_2 and η_3 exactly and an adequate criterion for starting at row $(r+1)$ is

$$
|\epsilon_1(|y_{r+1}| + |z_{r+1}|)/x_{r+1}| < 2^{-t}\|A_1\|_E. \qquad (45.5)
$$

Notice that the sign of ϵ_1 must be changed to give η_1.

The total saving resulting from taking advantage of small sub-diagonal elements in the two ways we have described is often very substantial. However, it is worth pointing out that we should not think of these devices merely in terms of the saving of computation in those iterations in which they are used. The reduction to Hessenberg form is unique only if sub-diagonal elements are non-zero and therefore it is important to use only the lower submatrix when a sub-diagonal element is negligible. When they are relevant it means that the top and bottom halves of the matrix are weakly coupled. If the upper half has some eigenvalues which are smaller than any corresponding to the lower half, then certainly if we iterate for long

enough with a zero shift of origin the small eigenvalue in the top half
will be 'fed down' to the bottom right-hand corner, but because of
the weak coupling this will be a slow process. Applying the stratagems
we have described avoids the waste of time but unfortunately makes
it less likely that eigenvalues will be found in order of increasing
absolute magnitude.

Double-shift technique for *LR*

46. Ignoring interchanges we have an analogous double-shift technique for the *LR* algorithm. The equations analogous to those of
(41.1) to (41.4) now give

$$L_1 L_2 R_2 R_1 = (A_1 - k_1 I)(A_1 - k_2 I), \qquad A_1 L_1 L_2 = L_1 L_2 A_3,$$
(46.1)

and writing $L_1 L_2 = L$, $R_2 R_1 = R$, we have

$$LR = (A_1 - k_1 I)(A_2 - k_2 I), \qquad A_1 L = LA_3. \tag{46.2}$$

We now make use of the fact that if $\tilde{L}$ is a unit lower-triangular
matrix having the same first column as L, and such that

$$A_1 \tilde{L} = \tilde{L} B, \tag{46.3}$$

where B is upper Hessenberg, then $\tilde{L}$ is L and B is A_3 (Chapter 6,
§ 11). Analogously with the QR case we effectively find the factors
of $\tilde{L}^{-1}$ rather than $\tilde{L}$ itself.

We calculate the first column of $(A_1 - k_1 I)(A_1 - k_2 I)$ and, as before,
it has only the three non-zero elements x_1, y_1, z_1 given in (41.11). We
then determine an elementary matrix of type N_1 which reduces this
first column to a multiple of e_1. The first column of N_1^{-1} is obviously
the first column of L. Finally we compute C_1 given by $C_1 = N_1 A_1 N_1^{-1}$,
and reduce it to Hessenberg form B using elementary non-unitary
matrices of types $N_2, N_3, ..., N_{n-2}$ (Chapter 6, § 8). Hence

$$N_{n-2}...N_2 N_1 A_1 N_1^{-1} N_2^{-1}...N_{n-2}^{-1} = B, \tag{46.4}$$

and the matrix $N_1^{-1} N_2^{-1}...N_{n-2}^{-1}$ is a unit lower-triangular matrix
having the same first column as N_1^{-1}, which itself has the same first
column as L. B is therefore the matrix A_3. At the rth stage of the
reduction the current matrix is of exactly the same form as in the
corresponding stage of the Givens reduction illustrated in (42.1).
Hence each N_r except the last has only two non-zero elements below

the diagonal, those in positions $(r+1, r)$ and $(r+2, r)$. As with QR the initial step can be combined with the others. The total number of multiplications is of the order of $2n^2$.

We can introduce interchanges at all stages by replacing each N_i, including N_1, by $N_r I_{r,r'}$ but the process is now on a rather shaky theoretical basis. In general interchanges continue to take place in earlier rows and columns even when some of the other eigenvalues are almost determined. Nevertheless we have found in practice that the double-shift technique works quite satisfactorily though it is absolutely essential that negligible sub-diagonal elements and small consecutive sub-diagonal elements be detected in the way we have described.

Assessment of *LR* and *QR* algorithms

47. We are now in a position to compare the relative merits of the LR and QR algorithms as applied to Hessenberg matrices. In my opinion the case for preferring the unitary transformations is very much stronger than it has been on previous occasions when faced with such a choice. The convergence properties of the LR technique with interchanges are far from satisfactory, and there is the further point that a considerable number of transformations are required so that there may well be a progressive deterioration of the condition of the matrix and a steady increase in the size of off-diagonal elements. For the QR algorithm on the other hand, it is easy to prove by the methods of Chapter 3, § 45, that A_s is exactly unitarily similar to a matrix $(A_1 + E_s)$, where E_s is small. For example it can be shown, using floating-point with accumulation of inner-products wherever possible, that $\|E_s\|_E < Ksn2^{-t}\|A_1\|_E$. Moreover the number of iterations required using QR has proved to be consistently less than with LR.

The procedure based on the double-shift QR technique, with shifts determined by the bottom 2×2 matrix and including the detection of small sub-diagonal elements, has proved to be the most powerful general-purpose programme of any I have used. In view of its importance it seems worth while to deal with one or two common misconceptions concerning its convergence.

In § 32 we pointed out that when the matrix A_1 has r eigenvalues of equal modulus then, if no shifts are used, there may be a corresponding square block of order r centred on the diagonal at every stage in the iteration. This may give the impression that the detection of such a group always gives difficulties. In fact when shifts of origin are incorporated this is rarely true. For example, each of the matrices

(i)

$$\begin{bmatrix} -1 & -1 & -1 & -1 & -1 \\ 1 & 0 & 0 & 0 & 0 \\ & 1 & 0 & 0 & 0 \\ & & 1 & 0 & 0 \\ & & & 1 & 0 \end{bmatrix},$$

(ii)

$$\begin{bmatrix} 0 & -1 & -1 & 0 & -1 \\ 1 & 0 & 0 & 0 & 0 \\ & 1 & 0 & 0 & 0 \\ & & 1 & 0 & 0 \\ & & & 1 & 0 \end{bmatrix},$$

(iii)

$$\begin{bmatrix} 0 & 0 & 0 & 0 & -1 \\ 1 & 0 & 0 & 0 & 0 \\ & 1 & 0 & 0 & -1 \\ & & 1 & 0 & -1 \\ & & & 1 & 0 \end{bmatrix},$$

(iv)

$$\begin{bmatrix} 0 & 0 & 0 & 0 & -1 \\ 1 & 0 & 0 & 0 & -\frac{1}{2} \\ & 1 & 0 & 0 & -1 \\ & & 1 & 0 & -1 \\ & & & 1 & -\frac{1}{2} \end{bmatrix}$$

(v)

and

$$\begin{bmatrix} -\frac{1}{2} & -1 & -1 & -\frac{1}{2} & -1 \\ 1 & 0 & 0 & 0 & 0 \\ & 1 & 0 & 0 & 0 \\ & & 1 & 0 & 0 \\ & & & 1 & 0 \end{bmatrix} \qquad (47.1)$$

was solved using the standard double-shift QR procedure. Although the modulus of every eigenvalue of matrices (i), (ii) and (iii) is unity the average number of iterations per eigenvalue was less than three. The shifts of origin determined by the standard process are such that the modified matrices do not have eigenvalues of equal modulus and the difficulty disappears. This was not true of the matrix (32.6) which has the characteristic equation $\lambda^4 + 1 = 0$. This matrix is invariant with respect to QR and the shifts of origin determined by the usual process are both zero. However, if we perform one iteration on this matrix with an arbitrary shift of origin then convergence takes place.

Further if $(\lambda^r + 1)$ is a factor of the characteristic equation of a matrix, this does not mean that the zeros of this factor will not be isolated by the QR algorithm. Provided the other factor is such that non-zero shifts are used, convergence will take place. The second matrix provides an example of this. The characteristic equation is $(\lambda^3 + 1)(\lambda^2 + 1)$ but after the first iteration the usual process determines non-zero shifts. All five eigenvalues were determined to 12 decimals in a total of 12 iterations.

Multiple eigenvalues

48. We now turn to multiple eigenvalues. The argument of § 32 shows that multiple eigenvalues corresponding to linear elementary divisors do not adversely affect the speed of convergence. This is sometimes found puzzling and gives rise to the following type of argument.

Let the matrix A_1 have the elementary divisors $(\lambda - a)$, $(\lambda - a)$ and $(\lambda - b)$ where a and b are approximately 1 and 3, and suppose that at some stage A_r is given by

$$A_r = \begin{bmatrix} 3 & 1 & 2 \\ \epsilon_1 & 1 & 2 \\ \epsilon_2 & \epsilon_3 & 1 \end{bmatrix}. \tag{48.1}$$

It may be verified that one step of LR produces a matrix A_{r+1} given by

$$A_{r+1} \doteq \begin{bmatrix} 3 & 1 & 2 \\ \tfrac{1}{3}\epsilon_1 + \tfrac{2}{3}\epsilon_2 & 1 & 2 \\ \tfrac{1}{3}\epsilon_3 & \epsilon_3 + \tfrac{1}{3}\epsilon_2 & 1 \end{bmatrix}. \tag{48.2}$$

(The QR algorithm gives much the same results.) Clearly the (3, 2) element will not converge satisfactorily.

The fallacy in this argument is that if A_1 has linear divisors it cannot give rise to such an A_r. The rank of $(A - aI)$ must be unity and hence, for example, the (2, 3) element must satisfy the relation $(3 - a)a_{23} - 2\epsilon_1 = 0$, so that a_{23} must be approximately equal to ϵ_1. If one allows for the relations that must exist between the elements of A_r, then one finds that sub-diagonal elements are indeed diminished by a factor of about $\tfrac{1}{3}$ in each iteration. The use of shifts gives a much higher rate of convergence (cf. § 54 for symmetric matrices).

If A_1 has non-linear divisors then the *LR* algorithm does not in general give convergence to upper-triangular form. Thus the matrix

$$A_1 = \begin{bmatrix} a & 0 & 0 \\ 1 & a & 0 \\ 0 & 1 & a \end{bmatrix} \tag{48.3}$$

is invariant with respect to the orthodox *LR* technique, and when $|a| > 1$ this is true even if interchanges are used. (The fact that it is of lower-triangular form is of no value with a procedure which is designed to detect the emergence of upper-triangular form.)

The QR algorithm nevertheless gives convergence. In fact we may prove that convergence takes place for any matrix having the Jordan form A_1 of (48.3). For we have

$$A_1 = X \begin{bmatrix} a & 0 & 0 \\ 1 & a & 0 \\ 0 & 1 & a \end{bmatrix} Y = XJY, \quad A_1^s = X \begin{bmatrix} a^s & 0 & 0 \\ \binom{s}{1} a^{s-1} & a^s & 0 \\ \binom{s}{2} a^{s-2} & \binom{s}{1} a^{s-1} & a^s \end{bmatrix} Y, \tag{48.4}$$

and hence if $Y = LU$ we have

$$A_1^s = X \begin{bmatrix} 0 & 0 & 1 \\ 0 & 1 & 0 \\ 1 & 0 & 0 \end{bmatrix} \times$$

$$\times \begin{bmatrix} \binom{s}{2} a^{s-2} + \binom{s}{1} l_{21} a^{s-1} + l_{31} a^s & \binom{s}{1} a^{s-1} + l_{32} a^s & a^s \\ \binom{s}{1} a^{s-1} + l_{21} a^s & a^s & 0 \\ a^s & 0 & 0 \end{bmatrix} U$$

$$= XPK_s U \quad \text{(say).} \tag{48.5}$$

Now if we write

$$K_s = L_s U_s, \tag{48.6}$$

it is obvious from the form of K_s that $L_s \to I$. Hence

$$A_1^s = XPL_sU_sU = XP(I+E_s)U_sU \quad \text{where } E_s \to 0, \quad (48.7)$$

showing, in the usual way, that in the QR algorithm the product $Q_1Q_2...Q_s$ tends essentially to the matrix obtained by the QR factorization of XP. If we write

$$XP = QR, \tag{48.8}$$

then

$$A_s \to Q^TA_1Q = Q^TXJX^{-1}Q = RP^TJPR^{-1}, \tag{48.9}$$

and we have

$$P^TJP = \begin{bmatrix} a & 1 & 0 \\ 0 & a & 1 \\ 0 & 0 & a \end{bmatrix}, \tag{48.10}$$

showing that A_s tends to an upper-triangular matrix.

Obviously the argument may be extended to elementary divisors of any degree and then, as in §§ 30–32, to cover the case when A_1 has a composite Jordan form.

The speed of convergence is governed by the rate at which a/s tends to zero and clearly if a is small the limit is reached more quickly. When a is sufficiently small two iterations are adequate for the matrix of (48.3). We might expect that in practice the use of shifts will mean that surprisingly few iterations are needed. As illustration we took the matrix A_1 given by

$$A_1 = \begin{bmatrix} -5 & -9 & -7 & -2 \\ 1 & 0 & 0 & 0 \\ & 1 & 0 & 0 \\ & & 1 & 0 \end{bmatrix}, \quad \det(A_1 - \lambda I) = (\lambda+1)^3(\lambda+2);$$

$$(48.11)$$

this has a cubic elementary divisor. Only eleven iterations were needed using the double-shift QR technique and working to 12 significant decimals. Of course this matrix is ill-conditioned, changes of order ϵ in elements of A_1 making changes of order $\epsilon^{\frac{1}{3}}$ in the eigenvalues. The multiple eigenvalues had errors of order 10^{-4} and the simple eigenvalue an error of order 10^{-12}. We must regard these as 'best possible' results for this precision of computation. For large matrices with non-linear elementary divisors the double-shift programme may be rather slow to home on accurate values of the shift and convergence can be fairly slow at first.

Strictly speaking our upper Hessenberg matrices will never have multiple eigenvalues corresponding to linear divisors because we are assuming that all sub-diagonal elements are non-zero and hence the rank of $(A - \lambda I)$ is $n - 1$ for any value of λ. An eigenvalue of multiplicity r must therefore correspond to a single divisor of degree r. However, as we showed in Chapter 5, § 45, matrices exist having pathologically close eigenvalues for which no sub-diagonal element is small. Provided the matrix is not such that a small perturbation would give rise to a non-linear divisor, the convergence behaviour will be much as for the case of a multiple root corresponding to linear divisors.

It is clear from this discussion that the incorporation of the shifts might be expected to make the QR remarkably effective for nearly all matrices, and this has certainly been my experience.

Special use of the deflation process

49. Our next point is best introduced by considering the QR process primarily as a method of deflation. If λ_1 were an exact eigenvalue then one exact step of the QR process applied to $(A_1 - \lambda_1 I)$ would produce a matrix having a null last row and we could deflate. Suppose now we have a value of λ_1 which is correct to working accuracy, can we perform one step of QR and deflate immediately?

The answer to this question is an emphatic 'no'. In fact we have constructed very well-conditioned matrices for which the (n, n) element of the transformed matrix before the shift is restored, so far from being zero, is the largest element in the whole matrix! The matrix W_{21}^- is a good example of such a matrix. We have already discussed the triangularization of $W_{21}^- - (\lambda_1 + \epsilon)I$ in Chapter 5, § 56, and have seen that the last element of the triangle is greater than 21 when ϵ is of order 10^{-7}.

At first sight this is rather alarming and one might feel that it throws doubt on the numerical stability of the process. However, any such fears are unfounded. The computed transform is always exactly similar to a matrix which is very close to the original and hence the eigenvalues are preserved. If we continue to iterate using the same shift we shall ultimately be able to deflate. In practice it is rare to meet an example for which two iterations are not sufficient. With the matrix W_{21}^-, for example, two iterations gave a matrix in which the elements (21, 20) and (21, 21) were zero to working accuracy.

We may contrast this with the deflation discussed in Chapter 7, § 53, where the failure to vanish of quantities which should have been exactly zero was absolutely fatal.

Symmetric matrices

50. The other class of matrices for which the LR and QR algorithms are relevant (cf. § 35) is that of the band symmetric matrices. Before concentrating on these we consider symmetric matrices in general. It is immediately obvious that the QR algorithm preserves symmetry since

$$A_2 = Q_1^T A_1 Q_1, \tag{50.1}$$

but in general this is not true of the LR algorithm. This is unfortunate because symmetry should lead to considerable economies in the computation.

If A_1 is positive definite we can modify the LR algorithm by using the Cholesky symmetric decomposition (Chapter 4, § 42). Denoting the matrices thus obtained by $\tilde{A}_s$ we have

$$A_1 = L_1 L_1^T, \qquad L_1^T L_1 = \tilde{A}_2 = L_1^{-1} A_1 L_1 = L_1^T A_1 (L_1^{-1})^T. \tag{50.2}$$

Obviously $\tilde{A}_2$ is symmetric and since it is similar to A_1 it is also positive definite. Hence this process may be continued and we can prove that

$$\tilde{A}_s = L_{s-1}^{-1}...L_1^{-1} A_1 L_1...L_{s-1} = L_{s-1}^T...L_1^T A_1 (L_1^{-1})^T...(L_{s-1}^{-1})^T, \tag{50.3}$$

and

$$L_1 L_2...L_s L_s^T...L_1^T = A_1^s, \tag{50.4}$$

or

$$(L_1 L_2...L_s)(L_1 L_2...L_s)^T = A_1^s. \tag{50.5}$$

Not only is the volume of work halved but, as we saw in Chapter 4, §§ 43, 44, the Cholesky decomposition enjoys great numerical stability and no interchanges are required. If A_1 is not positive definite the Cholesky decomposition leads to complex numbers and numerical stability is no longer assured. In general A_2 is not even Hermitian (merely complex symmetric) but of course it still has real eigenvalues since it is similar to A_1. Thus if

$$A_1 = \begin{bmatrix} 1 & 1 \\ 1 & 0 \end{bmatrix} \quad \text{then} \quad \tilde{A}_2 = \begin{bmatrix} 2 & i \\ i & -1 \end{bmatrix}. \tag{50.6}$$

Relationship between *LR* and *QR* algorithms

51. The QR algorithm and the Cholesky LR algorithm are very closely related. In fact we have

$$(Q_1 Q_2 \ldots Q_s)(R_s \ldots R_2 R_1) = A_1^s, \tag{51.1}$$

and hence

$$(R_s \ldots R_2 R_1)^T (Q_1 Q_2 \ldots Q_s)^T (Q_1 Q_2 \ldots Q_s)(R_s \ldots R_2 R_1) = (A_1^s)^T A_1^s = A_1^{2s} \tag{51.2}$$

or

$$(R_s \ldots R_2 R_1)^T (R_s \ldots R_2 R_1) = A_1^{2s}. $$

But from (50.5) there follows

$$(L_1 L_2 \ldots L_{2s})(L_1 L_2 \ldots L_{2s})^T = A_1^{2s}. \tag{51.3}$$

Hence we have two Cholesky decompositions of A_1^{2s}, and since these must be the same we find

$$L_1 L_2 \ldots L_{2s} = (R_s \ldots R_2 R_1)^T. \tag{51.4}$$

Now (50.3) gives

$$\tilde{A}_{2s+1} = (L_1 L_2 \ldots L_{2s})^T A_1 [(L_1 L_2 \ldots L_{2s})^T]^{-1}, \tag{51.5}$$

while from the equations defining the QR algorithm we have

$$\begin{aligned} A_{s+1} &= R_s \ldots R_1 A_1 (R_s \ldots R_1)^{-1} \\ &= (L_1 L_2 \ldots L_{2s})^T A_1 [(L_1 L_2 \ldots L_{2s})^T]^{-1} = \tilde{A}_{2s+1}. \end{aligned} \tag{51.6}$$

Hence the $(2s+1)$th matrix obtained using the Cholesky LR is equal to the $(s+1)$th matrix obtained using QR. (This result does not hold for unsymmetric matrices but it is perhaps indicative of the greater power of QR.)

The proof applies whether A is positive definite or not (provided it is not singular, this being the case for which the triangular decomposition of A_1^s is not unique). Now the QR algorithm always gives real iterates if A_1 is real, and hence the Cholesky LR algorithm must give real matrices for the $\tilde{A}_{2s+1}$ even though the $\tilde{A}_{2s}$ are complex. For example for the matrix of (50.6) we have

$$\tilde{A}_3 = \begin{bmatrix} \frac{3}{2} & -\frac{1}{2} \\ -\frac{1}{2} & -\frac{1}{2} \end{bmatrix}. \tag{51.7}$$

A more important point is that the 2-norm and the Euclidean norm are preserved by the QR algorithm and no elements of any A_s can become large. This is not true of the Cholesky LR algorithm when A_1

is not positive definite. $\tilde{A}_2$ can have indefinitely large elements but we know that there must be a return to normal size when $\tilde{A}_3$ is reached. An example of an instability is given in Table 4. There $\|A_2\|_E$ is much larger than $\|A_1\|_E$. Although the computed $\tilde{A}_3$ is real the eigenvalues have suffered a fatal loss of accuracy. An accurate $\tilde{A}_3$ is given for comparison.

TABLE 4

Computed transformations

$$A_1 = \begin{bmatrix} 10^{-3}(0\cdot1000) & 10^0(0\cdot9877) \\ 10^0(0\cdot9877) & 10^0(0\cdot1471) \end{bmatrix} \quad L_1 = \begin{bmatrix} 10^{-1}(0\cdot1000) & 0 \\ 10^2(0\cdot9877) & 10^2(0\cdot9877)i \end{bmatrix}$$

$$\tilde{A}_2 = \begin{bmatrix} 10^4(0\cdot9756) & 10^4(0\cdot9756)i \\ 10^4(0\cdot9756)i & 10^4(0\cdot9756) \end{bmatrix} \quad L_2 = \begin{bmatrix} 10^2(0\cdot9877) & 0 \\ 10^2(0\cdot9877)i & 10^0(0\cdot6979)i \end{bmatrix}$$

$$\tilde{A}_3 = \begin{bmatrix} 0 & 10^2(-0\cdot6893) \\ 10^2(-0\cdot6893) & 10^0(-0\cdot4871) \end{bmatrix}$$

$$\text{Accurate } \tilde{A}_3 = \begin{bmatrix} 0\cdot1473 & 0\cdot9877 \\ 0\cdot9877 & -0\cdot0001 \end{bmatrix}$$

Convergence of the Cholesky LR algorithm

52. When A_1 is positive definite $\tilde{A}_s$ tends to a diagonal matrix whatever the nature of the eigenvalues. We have already proved that the QR algorithm always converges for positive definite matrices and in view of the relationship established between the Cholesky LR and QR the convergence of the former is assured. The following proof is independent of this relationship.

Let A_{rs} and L_{rs} denote the leading principal minors of $\tilde{A}_s$ and L_s respectively. From the relation $\tilde{A}_s = L_s L_s^T$ it follows that

$$A_{rs} = (L_{rs})^2. \tag{52.1}$$

On the other hand, from the relation $\tilde{A}_{s+1} = L_s^T L_s$ and the theorem of corresponding matrices, $A_{r,s+1}$ is equal to the sum of the products of corresponding minors of order r formed from the first r rows of L_s^T and the first r columns of L_s. Now the corresponding minors are equal since the corresponding matrices are transposes of each other, and hence $A_{r,s+1}$ is given as the sum of squares. One contribution, however, is L_{rs}^2 and hence

$$A_{r,s+1} > A_{rs}. \tag{52.2}$$

Each leading principal minor of $\tilde{A}_s$ therefore increases with s and since they are obviously bounded they tend to a limit. From this it follows immediately that all off-diagonal elements of A_s tend to zero. By arguments similar to those in §§ 33, 34, it may be shown that when the eigenvalues are distinct and the limiting diagonal matrix has the eigenvalues in decreasing order, then $\tilde{a}_{ij}^{(s)}$ tends to zero as $(\lambda_j/\lambda_i)^{\frac{1}{2}s}$ when $j > i$ and as $(\lambda_i/\lambda_j)^{\frac{1}{2}s}$ when $i > j$.

It is interesting to compare the standard LR and the Cholesky LR when applied to the same symmetric matrix. If $A_1 = \tilde{L}_1 \tilde{L}_1^T$ and $A_1 = L_1 R_1$, then provided A_1 is non-singular, we must have

$$L_1 = \tilde{L}_1 D_1, \qquad R_1 = D_1^{-1} \tilde{L}_1^T \tag{52.3}$$

for some diagonal matrix D_1. Hence

$$\tilde{A}_2 = \tilde{L}_1^T \tilde{L}_1, \qquad A_2 = R_1 L_1 = D_1^{-1} \tilde{L}_1^T \tilde{L}_1 D_1 = D_1^{-1} \tilde{A}_2 D_1. \tag{52.4}$$

Similarly if
$$\tilde{A}_2 = \tilde{L}_2 \tilde{L}_2^T, \qquad A_2 = L_2 R_2 = D_1^{-1} \tilde{L}_2 \tilde{L}_2^T D_1, \tag{52.5}$$

then from the essential uniqueness of the triangular decomposition we must have
$$L_2 = D_1^{-1} \tilde{L}_2 D_2, \qquad R_2 = D_2^{-1} \tilde{L}_2^T D_1 \tag{52.6}$$

for some diagonal matrix D_2. Hence

$$\tilde{A}_3 = \tilde{L}_2^T \tilde{L}_2, \qquad A_3 = R_2 L_2 = D_2^{-1} \tilde{L}_2^T D_1 D_1^{-1} \tilde{L}_2 D_2 = D_2^{-1} \tilde{A}_3 D_2. \tag{52.7}$$

We see that at each stage A_s is obtained from $\tilde{A}_s$ by a diagonal similarity transformation. We may write

$$A_s = D_{s-1}^{-1} \tilde{A}_s D_{s-1}. \tag{52.8}$$

Notice that this is true even if A_1 is not positive definite, though in this case the matrices D_{s-1} will be complex for even values of s, and we may reach a stage at which A_s has no triangular decomposition.

In the positive definite case we know that $\tilde{A}_s$ ultimately becomes diagonal while A_s merely becomes upper-triangular. The elements of D_s are ultimately of the order of magnitude of $(\lambda_i)^{-\frac{1}{2}s}$.

53. We may incorporate shifts of origin in the Cholesky LR algorithm, but if we are to ensure that all elements remain real and that the process is numerically stable, these shifts of origin k_s should be chosen so that $(A_s - k_s I)$ is positive definite at each stage. On the other hand, in order that convergence should be rapid, it is desirable to choose k_s as close as possible to λ_n, the smallest eigenvalue. The choice of k_s is thus a more delicate problem than it is for Hessenberg matrices. With the QR algorithm there are no such difficulties and we need not be concerned with positive definiteness at all.

Cubic convergence of the QR algorithm

54. The simplest choice of the shift of origin is $k_s = a_{nn}^{(s)}$. We shall show that, in general, for the QR algorithm this ultimately gives cubic convergence to the eigenvalue of minimum modulus *whatever its multiplicity*. We assume that

$$\left.\begin{array}{l} \lambda_n = \lambda_{n-1} = \ldots = \lambda_{n-r+1} \\[4pt] |\lambda_1| > |\lambda_2| \geqslant \ldots \geqslant |\lambda_{n-r}| > |\lambda_{n-r+1}| \end{array}\right\}. \tag{54.1}$$

If we write

$$A_s = \begin{array}{c} \\ n-r \end{array}\left\{\left[\begin{array}{c|c} F_s & G_s \\ \hline G_s^T & H_s \end{array}\right], \tag{54.2}$$

then we know from the general theory of § 32 that if no shifts are used G_s tends to the null matrix, the eigenvalues λ_i' $(i = 1,\ldots, n-r)$ of F_s tend to the corresponding λ_i while the eigenvalues

$$\lambda_i'' \quad (i = n-r+1,\ldots, n)$$

of H_s all tend to λ_n. Suppose we have reached the stage at which

$$A_s = \left[\begin{array}{c|c} F_s & \epsilon K_s \\ \hline \epsilon K_s^T & H_s \end{array}\right], \qquad \|K_s\|_E = 1, \tag{54.3}$$

$$|\lambda_i' - \lambda_n| > \tfrac{2}{3}|\lambda_{n-r} - \lambda_n| \quad (i = 1,\ldots, n-r), \qquad \epsilon < \tfrac{1}{3}|\lambda_{n-r} - \lambda_n|. \tag{54.4}$$

Now we know that the rank of $(A_s - \lambda_n I)$ is $n-r$ at all stages, while from (54.4) the rank of $(F_s - \lambda_n I)$ is also $n-r$. Hence we must have

$$H_s - \lambda_n I - \epsilon^2 K_s^T (F_s - \lambda_n I)^{-1} K_s = 0 \tag{54.5}$$

giving $\quad H_s = \lambda_n I + \epsilon^2 K_s^T (F_s - \lambda_n I)^{-1} K_s = \lambda_n I + M_s \quad$ (say). (54.6)

A simple exercise in matrix norms shows that

$$\|M_s\|_E < \epsilon^2 \, \|K_s\|_E^2 \, \|(F_s - \lambda_n I)^{-1}\|_2$$

$$< \epsilon^2 / \tfrac{2}{3} |\lambda_{n-r} - \lambda_n| = \tfrac{3}{2}\epsilon^2 / |\lambda_{n-r} - \lambda_n| < \tfrac{1}{6}|\lambda_{n-r} - \lambda_n|. \tag{54.7}$$

These results show that the norm of the off-diagonal elements of H_s is of order ϵ^2 and that every diagonal element of H_s differs from λ_n by a quantity of order ϵ^2. In particular

$$|a_{nn}^{(s)} - \lambda_n| < \tfrac{3}{2}\epsilon^2 / |\lambda_{n-r} - \lambda_n| < \tfrac{1}{6}|\lambda_{n-r} - \lambda_n|, \tag{54.8}$$

and we deduce that

$$|\lambda_i' - a_{nn}^{(s)}| > |\lambda_i' - \lambda_n| - |\lambda_n - a_{nn}^{(s)}| > \tfrac{1}{2}|\lambda_{n-r} - \lambda_n| \quad (i = 1,\ldots, n-r) \tag{54.9}$$

$$|\lambda_i'' - a_{nn}^{(s)}| < |\lambda_i'' - \lambda_n| + |\lambda_n - a_{nn}^{(s)}| < 3\epsilon^2 / |\lambda_{n-r} - \lambda_n|$$

$$(i = n-r+1,\ldots, n). \tag{54.10}$$

Consider now the effect of performing one step of QR with a shift of $a_{nn}^{(s)}$. Since $\lambda_n - a_{nn}^{(s)}$ is of order ϵ^2 while $\lambda_i' - a_{nn}^{(s)}$ $(i = 1, ..., n-r)$ do not depend on ϵ, from the general theory of the convergence (cf. § 19) we can expect the elements of ϵK_s^T to be multiplied by a factor of order ϵ^2, but it is instructive to give an explicit proof. Let P_s be the orthogonal matrix such that $P_s(A_s - a_{nn}^{(s)} I)$ is upper-triangular. Then writing

$$P_s = \begin{bmatrix} Q_s & R_s \\ \hline S_s & T_s \end{bmatrix}, \qquad A_{s+1} = P_s(A_s - a_{nn}^{(s)} I)P_s^T + a_{nn}^{(s)} I = \begin{bmatrix} F_{s+1} & G_{s+1} \\ \hline G_{s+1}^T & H_{s+1} \end{bmatrix},$$

(54.11)

we have

$$O = S_s(F_s - a_{nn}^{(s)} I) + \epsilon T_s K_s,$$ (54.12)

and

$$G_{s+1}^T = [\epsilon S_s K_s + T_s(H_s - a_{nn}^{(s)} I)] R_s^T.$$ (54.13)

Hence $\quad \|G_{s+1}\|_E < \epsilon \|S_s\|_E^2 + \|T_s\|_E \|S_s\|_E \|H_s - a_{nn}^{(s)} I\|_2$

$$\text{(since } \|R_s\|_E = \|S_s\|_E), \quad (54.14)$$

while from (54.12) and (54.10)

$$\|S_s\|_E = \epsilon \|T_s K_s(F_s - a_{nn}^{(s)} I)^{-1}\|_E$$
$$< \epsilon \|T_s\|_E \max |(\lambda_i' - a_{nn}^{(s)})^{-1}| < 2\epsilon \|T_s\|_E /|\lambda_{n-r} - \lambda_n|. \quad (54.15)$$

Hence finally

$$\|G_{s+1}\|_E < 4\epsilon^3 \|T_s\|_E^2 /|\lambda_{n-r} - \lambda_n|^2 + 2\epsilon \|T_s\|_E^2 \max |\lambda_i'' - a_{nn}^{(s)}|/|\lambda_{n-r} - \lambda_n|$$
$$< 10\epsilon^3 \|T_s\|_E^2 /|\lambda_{n-r} - \lambda_n|^2$$
$$< 10r\epsilon^3 /|\lambda_{n-r} - \lambda_n|^2. \quad (54.16)$$

The cubic convergence is therefore established. Notice that it means that all r coincident eigenvalues will be found simultaneously and that coincident eigenvalues are an advantage rather than a disadvantage.

In practice we shall start to apply the shifts much earlier than our proof implies, usually as soon as one binary digit of $a_{nn}^{(s)}$ has settled.

Shift of origin in Cholesky LR

55. Turning now to the Cholesky LR, we notice that if

$$(A_s - k_s I) = L_s L_s^T,$$ (55.1)

then by (50.3)

$$\det(A_1 - k_s I) = \det(A_s - k_s I) = (\prod l_{ii}^{(s)})^2.$$ (55.2)

Hence each time a decomposition is performed we can compute $\det(A_1 - \lambda I)$ for the value $\lambda = k_s$. If both k_s and k_{s-1} are less than λ_n then since

$$\det(A_1 - \lambda I) = f(\lambda) = \prod(\lambda_i - \lambda), \qquad (55.3)$$

it is evident that if

$$k_{s+1} = [f(k_s)k_{s-1} - f(k_{s-1})k_s]/[f(k_s) - f(k_{s-1})], \qquad (55.4)$$

then $k_{s+1} < \lambda_n$ and $\lambda_n - k_{s+1} < (\lambda_n - k_s)$ and $(\lambda_n - k_{s-1})$. (55.5)

Once we have succeeded in choosing k_1 and k_2 both less than λ_n, then we can produce a sequence of such values all less than λ_n and converging to it. If A_1 is positive definite we may take $k_1 = -2^{-\frac{1}{2}t}\|A_1\|$, $k_2 = 0$. All subsequent k_i will be positive. Notice that we do not need to continue iterating until $k_s = \lambda_n$ to working accuracy. We can stop as soon as the off-diagonal elements in row n and column n are negligible and then deflate. If we keep a record of the last two values of $\prod_{i=1}^{n-1} l_{ii}^{(s)}$ at all stages we can compute a good initial value for the shift immediately on deflation.

This technique is quite simple and is very effective provided there are no multiple or pathologically close eigenvalues.

Failure of the Cholesky decomposition

56. Rutishauser (1960) has suggested a method of choosing the shifts of origin which ultimately gives cubic convergence. It is based on an analysis of the breakdown which occurs when a k_s is taken which is greater than λ_n. We dispense with the suffix s for the purpose of this investigation. If we write

$$A - kI = LL^T, \qquad A = \begin{array}{c} p \\ \\ \end{array}\left\{\begin{bmatrix} F & G \\ \hline G & H \end{bmatrix}, \qquad L = \begin{bmatrix} M & O \\ \hline N & P \end{bmatrix} \right. \qquad (56.1)$$

then

$$MM^T = F - kI, \qquad MN^T = G, \qquad NN^T + PP^T = H - kI. \quad (56.2)$$

Suppose the decomposition has proceeded up to the stage where M and N have been determined, but that when we attempt to determine the first element of P the square root of a negative number is required. From (56.2) we have

$$PP^T = H - kI - G^T(F - kI)^{-1}G = X \quad \text{(say)}, \qquad (56.3)$$

and our assumption is that x_{11} is negative. Suppose τ is the algebraically smallest eigenvalue of X; from our assumption about X this is obviously *negative*. We assert that $(A-kI-\tau I)$ is positive definite; in other words if we start again using the shift $(k+\tau)$ instead of k we shall be able to complete the reduction.

Obviously since τ is negative we can proceed at least as far as before and the relevant matrix P now satisfies the relation

$$PP^T = H-kI-\tau I-G^T(F-kI-\tau I)^{-1}G$$

$$= (X-\tau I)+G^T(F-kI)^{-1}G-G^T(F-kI-\tau I)^{-1}G = Y \quad \text{(say).}$$

$$(56.4)$$

We must show that Y is positive definite. Now since τ is the smallest eigenvalue of X, $(X-\tau I)$ is non-negative definite. Let Q be an orthogonal matrix such that

$$F = Q^T \text{diag}(\lambda_i')Q. \quad (56.5)$$

Since we have assumed that $(F-kI)$ has a decomposition, $\lambda_i'-k > 0$ $(i=1,...,p)$. Hence we have

$$G^T(F-kI)^{-1}G-G^T(F-kI-\tau I)^{-1}G$$

$$= G^TQ^T[\text{diag}(\lambda'-k)^{-1}-\text{diag}(\lambda_i'-k-\tau)^{-1}]QG$$

$$= S^T \text{diag}[-\tau/(\lambda_i'-k)(\lambda_i'-k-\tau)]S. \quad (56.6)$$

The elements of the diagonal matrix are all positive and hence the matrix on the right of (56.6) is certainly non-negative definite, and is positive definite unless S, and therefore G, has linearly dependent rows. Y is therefore always non-negative definite and in general is positive definite.

To turn this result to advantage it is essential first to ensure that the breakdown occurs at a very late stage so that X is a matrix of low order (preferably 1 or 2) and its eigenvalues can be readily computed, and secondly that $k+\tau$ is very close to λ_n. We consider this problem in the next section.

Cubically convergent *LR* process

57. Let A_1 be a matrix of the type discussed in § 54 except that we now assume that it is positive definite, and suppose we have iterated using the Cholesky *LR* algorithm until A_s satisfies the conditions of (54.3), (54.4). We know from (54.6) that at this stage all diagonal elements of H_s differ from λ_n by a quantity of order ϵ^2.

Suppose we use $a_{nn}^{(s)} = \lambda_n + m_{nn}^{(s)}$ as our shift of origin. The decomposition must then fail since any diagonal element of a symmetric matrix exceeds its smallest eigenvalue. However, from (54.9) the failure will not occur until stage $(n-r+1)$. The relevant matrix X at this stage is given by

$$
\begin{aligned}
X &= H_s - (\lambda_n + m_{nn}^{(s)})I - \epsilon^2 K_s^T (F_s - a_{nn}^{(s)} I)^{-1} K_s \\
&= \lambda_n I + \epsilon^2 K_s^T (F_s - \lambda_n I)^{-1} K_s - (\lambda_n + m_{nn}^{(s)} I) - \epsilon^2 K_s^T (F_s - a_{nn}^{(s)} I)^{-1} K_s
\end{aligned}
$$

from (54.6)

$$
= -m_{nn}^{(s)} I + \epsilon^2 K_s^T [(F_s - \lambda_n I)^{-1} - (F_s - a_{nn}^{(s)} I)^{-1}] K_s, \tag{57.1}
$$

and it is a matrix of order r. If τ is the smallest eigenvalue of X the argument of § 56 shows that $a_{nn}^{(s)} + \tau$ is a safe shift. We now show that it is remarkably close to λ_n. Writing $X = -m_{nn}^{(s)} I + Y$ we have

$$
\|Y\|_E < \epsilon^2 \|(F_s - \lambda_n I)^{-1} - (F_s - a_{nn}^{(s)} I)^{-1}\|_2. \tag{57.2}
$$

Since the two matrices on the right have the same system of eigenvectors we have

$$
\begin{aligned}
\|Y\|_E &< \epsilon^2 \max |1/(\lambda_i' - \lambda_n) - 1/(\lambda_i' - a_{nn}^{(s)})| \\
&= \epsilon^2 \max |(a_{nn}^{(s)} - \lambda_n)/(\lambda_i' - \lambda_n)(\lambda_i' - a_{nn}^{(s)})| \\
&< \epsilon^2 [\tfrac{3}{2}\epsilon^2/(\lambda_{n-r} - \lambda_n)][3/(\lambda_{n-r} - \lambda_n)^2] \text{ by (54.8), (54.9) and (54.4)} \\
&= \tfrac{9}{2}\epsilon^4/(\lambda_{n-r} - \lambda_n)^3. \tag{57.3}
\end{aligned}
$$

Hence since $\tau = -m_{nn}^{(s)} + $ smallest eigenvalue of Y
we have

$$
a_{nn}^{(s)} + \tau = \lambda_n + m_{nn}^{(s)} - m_{nn}^{(s)} + \text{smallest eigenvalue of } Y, \tag{57.4}
$$

giving

$$
|a_{nn}^{(s)} + \tau - \lambda_n| < \|Y\|_E < \tfrac{9}{2}\epsilon^4/(\lambda_{n-r} - \lambda_n)^3. \tag{57.5}
$$

We could now carry out a direct analysis similar to that of § 54 to show that if we perform one iteration on $A_s - (a_{nn}^{(s)} + \tau)I$, then $\|G_{s+1}\|_E$ is of order ϵ^3. We leave this as an exercise. The general theory shows that G_s will be diminished by a factor of order

$$
|[\lambda_n - (a_{nn}^{(s)} + \tau)]/[\lambda_{n-r} - (a_{nn}^{(s)} + \tau)]|^{\frac{1}{2}},
$$

that is, of order ϵ^2. Since G_s is of order ϵ the result is established.

58. We have treated the case when the minimum eigenvalue is of arbitrary multiplicity, but in practice the method is most satisfactory when r is 1 or 2 since only then is the smallest eigenvalue τ readily available.

It is not easy to be certain when we have reached the stage at which it is advantageous to apply this procedure and in practice the decision must be taken on empirical grounds. The broad strategy for a simple eigenvalue is as follows.

We determine positive shifts in any appropriate manner, such as the method of § 55, until $(a_{nn}^{(s)} - k_{s-1})$ is considerably smaller than all the other diagonal elements of A_s and the off-diagonal elements in the last row and column are small. We then take $a_{nn}^{(s)}$ as the next shift and the decomposition fails. If the remaining matrix at the failure point is of order greater than two, we have changed methods too early (notice that if the eigenvalue is really of multiplicity greater than two the breakdown will always occur too early). Rutishauser and Schwarz (1963) have developed very ingenious stratagems for applying this process and the reader is strongly advised to refer to their paper.

It is perhaps worth remarking that it is conceptually simpler to use the 'non-restoring' method with the Cholesky *LR*. The element $a_{nn}^{(s)}$ then gives us important information on the stage of convergence. Since A_s is formed by computing $L_{s-1}^T L_{s-1}$ it is always at least non-negative definite. If $a_{nn}^{(s)} < \epsilon$ then deflation at this stage gives an error of less than ϵ in every eigenvalue. For if the eigenvalues of the leading principal submatrix of order $(n-1)$ are λ_i', then

$$\left.\begin{aligned}
\lambda_1 > \lambda_1' > \ldots > \lambda_{n-1} > \lambda_{n-1}' \\
\lambda_1' + \ldots + \lambda_{n-1}' + \epsilon = \lambda_1 + \ldots + \lambda_n \\
(\lambda_1 - \lambda') + \ldots + (\lambda_{n-1} - \lambda_{n-1}') = \epsilon - \lambda_n > 0
\end{aligned}\right\}, \qquad (58.1)$$

from which the result follows. Notice though that if the off-diagonal elements of the last row and column are small enough we may deflate whether or not $a_{nn}^{(s)}$ is small; the latter will, in fact, be small only if the shifts have been well chosen.

Band matrices

59. For general symmetric matrices the volume of work is prohibitive but for band matrices such that

$$a_{ij} = 0 \quad (|i-j| > m), \qquad 2m+1 \ll n \qquad (59.1)$$

this is no longer true. We consider first the Cholesky *LR* algorithm. It is obvious that this preserves the band form (for unsymmetric matrices cf. § 66) and that at each stage L_s is such that

$$l_{ij}^{(s)} = 0 \quad (i-j > m). \tag{59.2}$$

Hence although A_s has $2m+1$ non-zero elements in a typical row, L_s has only $m+1$.

There are a number of ways of performing the decomposition and organizing the storage; the following is perhaps the most convenient if one wishes to take full advantage of the accumulation of inner-products.

It is convenient to name the elements of A in an unconventional manner so as to take full advantage of the band form and the symmetry. This is illustrated for the case $n = 8$, $m = 3$, by the arrays

Upper triangle of A   *Storage array*

$$\begin{bmatrix} a_{10} & a_{11} & a_{12} & a_{13} & & & & \\ & a_{20} & a_{21} & a_{22} & a_{23} & & & \\ & & a_{30} & a_{31} & a_{32} & a_{33} & & \\ & & & a_{40} & a_{41} & a_{42} & a_{43} & \\ & & & & a_{50} & a_{51} & a_{52} & a_{53} \\ & & & & & a_{60} & a_{61} & a_{62} \\ & & & & & & a_{70} & a_{71} \\ & & & & & & & a_{80} \end{bmatrix}, \quad \begin{bmatrix} a_{10} & a_{11} & a_{12} & a_{13} \\ a_{20} & a_{21} & a_{22} & a_{23} \\ a_{30} & a_{31} & a_{32} & a_{33} \\ a_{40} & a_{41} & a_{42} & a_{43} \\ a_{50} & a_{51} & a_{52} & a_{53} \\ a_{60} & a_{61} & a_{62} & 0 \\ a_{70} & a_{71} & 0 & 0 \\ a_{80} & 0 & 0 & 0 \end{bmatrix}.$$

Notice that A is stored as a rectangular $n \times (m+1)$ array. The bottom right-hand corner is filled up with zeros in the illustration but these elements may be arbitrary if the decomposition is performed as we describe it. The dotted lines indicate the elements of a column of A. The upper triangle $U = L^T$ is produced in the same form as the upper half of A. It could be overwritten on A as it is produced but if we are considering using shifts which are greater than λ_n we cannot do this, because such a decomposition will fail and we shall need to start again.

60. The elements of U are produced row by row, those in the ith row being computed as follows.

(i) Determine $p = \min(n-i, m)$.
Then for each value of j from 0 to p perform steps (ii) and (iii):
(ii) Determine $q = \min(m-j, i-1)$.

(iii) Compute $x = a_{ij} - \sum_{k=1}^{q} (u_{i-k,k})(u_{i-k,j+k})$, where the sum is zero if $q = 0$.
If $j = 0$ then $u_{i0} = (x-k_s)^{\frac{1}{2}}$ (If $x-k_s$ is negative, k_s is too large).
If $j \neq 0$ then $u_{ij} = x/u_{i0}$.

As an example for $i = 5$, $m = 3$, $n = 8$, we have

$$\left.\begin{array}{l} u_{50}^2 = a_{50} - u_{41}u_{41} - u_{32}u_{32} - u_{23}u_{23} - k_s \\[2mm] u_{50}u_{51} = a_{51} - u_{41}u_{42} - u_{32}u_{33} \\[2mm] u_{50}u_{52} = a_{52} - u_{41}u_{43} \\[2mm] u_{50}u_{53} = a_{53} \end{array}\right\} . \tag{60.1}$$

The quantities p and q are introduced to deal with the end effects.

We now have to compute $(UU^T + k_s I)$. The computed matrix will be over-written on the original A. If U has already been overwritten on the original A this does not give rise to any difficulties since by the time the new a_{ij} is computed u_{ij} is no longer required. The elements of the ith row of A are computed as follows.

(i) Determine $p = \min(n-i, m)$.
Then for each value of j from 0 to p perform steps (ii) and (iii):
(ii) Determine $q = \min(m-j, n-i-j)$.

(iii) Compute $x = \sum_{k=0}^{q} u_{i,j+k} u_{i+j,k}$.

If $j = 0$ then $a_{i0} = x + k_s$.
If $j \neq 0$ then $a_{ij} = x$.
As an example when $i = 5$, $m = 3$

$$\left.\begin{array}{l} a_{50} = u_{50}u_{50} + u_{51}u_{51} + u_{52}u_{52} + u_{53}u_{53} + k_s \\[2mm] a_{51} = \qquad\qquad u_{51}u_{60} + u_{52}u_{61} + u_{53}u_{62} \\[2mm] a_{52} = \qquad\qquad\qquad\qquad u_{52}u_{70} + u_{53}u_{71} \\[2mm] a_{53} = \qquad\qquad\qquad\qquad\qquad\qquad u_{53}u_{80} \end{array}\right\} . \tag{60.2}$$

On a computer with a two-level store the decomposition and the recombination can be combined. When computing row i of U_s, rows $i-1$, $i-2$,..., $i-m$ of U_s and row i of A_s must be in the high-speed store; as soon as row i of U_s is complete we can compute row $i-m$ of A_{s+1}. Again we cannot do this if we are considering using values of k_s which will cause a breakdown.

61. If we are using Rutishauser's technique of § 56 then we must examine each breakdown. If a breakdown occurs when $i = n$, then the current value of x is the next shift. If the breakdown occurs when $i = n-1$ we must calculate the 2×2 matrix X of equation (56.3). We have

$$\left.\begin{aligned}
x_{11} &= \text{current } x \\
x_{12} &= a_{n-1,1} - u_{n-2,1}u_{n-2,2} - u_{n-3,2}u_{n-3,3} - \cdots - u_{n-m,m-1}u_{n-m,m} \\
x_{22} &= a_{n,0} - u_{n-2,2}^2 - u_{n-3,3}^2 - \cdots - u_{n-m,m}^2
\end{aligned}\right\},$$

$$(61.1)$$

and we require the smaller eigenvalue of this 2×2 matrix. If the breakdown occurs at any earlier stage then it is not convenient to take advantage of the remaining matrix.

There are approximately $\frac{1}{2}nm^2$ multiplications in the decomposition and the same number in the recombination; if $m \ll n$ this is far less than is required for a full matrix. We have seen in Chapter 4, § 40, that the error in the Cholesky decomposition is very small. For example, using fixed-point arithmetic with accumulation of inner-products we have

$$L_s L_s^T = A_s + E_s, \qquad |e_{ij}^{(s)}| < \tfrac{1}{2}2^{-t}, \qquad (61.2)$$

where E_s is of the same band form.

Similarly

$$A_{s+1} = L_s^T L_s + F_s, \qquad |f_{ij}^{(s)}| < \tfrac{1}{2}2^{-t}, \qquad (61.3)$$

where F_s is again of band form. From these results it is easy to see that the maximum error introduced in any eigenvalue by a complete transformation is $(2m+1)2^{-t}$. If A_1 is scaled so that $\|A_1\|_\infty < 1$, then all elements of all $|A_s|$ are bounded by unity. Notice that if we compare the computed A_s with the exact A_s we might be misled into thinking that the eigenvalues will be of much lower accuracy (see for example Rutishauser, 1958, p. 80).

QR decomposition of a band matrix

62. It is not quite so obvious that the QR algorithm preserves the band form. The matrix R produced by the QR decomposition is of width $2m+1$ and not $m+1$ as was the case with the Cholesky LR decomposition. If we use Householder's method for the triangularization (Chapter 4, § 46), $n-1$ elementary Hermitians $P_1, P_2, ..., P_{n-1}$ are required, and the vector w_i corresponding to P_i has non-zero elements only in positions $i, i+1, ..., i+m$, (except of course for the later w_i). The matrices $P_2 P_1 A_s$ and R_s for the case $n = 8$, $m = 2$, are given by

$$P_2 P_1 A_s$$

$$\begin{bmatrix} \times & \times & \times & \times & \times & & & \\ 0 & \times & \times & \times & \times & \times & & \\ 0 & 0 & \times & \times & \times & \times & & \\ & 0 & \times & \times & \times & \times & & \\ & & \times & \times & \times & \times & \times & \\ & & & \times & \times & \times & \times & \times \\ & & & & \times & \times & \times & \times \\ & & & & & \times & \times & \times \end{bmatrix},$$

$$P_{n-1}...P_1 A_s = R_s$$

$$\begin{bmatrix} \times & \times & \times & \times & \times & & & \\ & \times & \times & \times & \times & \times & & \\ & & \times & \times & \times & \times & \times & \\ & & & \times & \times & \times & \times & \times \\ & & & & \times & \times & \times & \times \\ & & & & & \times & \times & \times \\ & & & & & & \times & \times \\ & & & & & & & \times \end{bmatrix}. \qquad (62.1)$$

In the former the zeros introduced by P_1 and P_2 are indicated. Consider now the post-multiplication by P_1. This replaces columns 1 to

$m+1$ by combinations of themselves. Hence R_sP_1 is given typically by

$$R_sP_1$$

$$\begin{bmatrix} \times & \times & \times & \times & \times & & & \\ \times & \times & \times & \times & \times & \times & & \\ \times & \times & \times & \times & \times & \times & \times & \\ & & \times & \times & \times & \times & \times & \\ & & & \times & \times & \times & \times & \\ & & & & \times & \times & \times & \\ & & & & & \times & \times & \\ & & & & & & \times & \end{bmatrix},$$

$$R_sP_1P_2P_3$$

$$\begin{bmatrix} \times & \times & \times & \times & \times & & & \\ \times & \times & \times & \times & \times & \times & & \\ \times & \times & \times & \times & \times & \times & \times & \\ & \times & \times & \times & \times & \times & \times & \times \\ & \times & \times & \times & \times & \times & \times & \\ & & & \times & \times & \times & \times & \\ & & & & \times & \times & \times & \\ & & & & & \times & \end{bmatrix}. \qquad (62.2)$$

Subsequent post-multiplications by $P_2,\ldots,P_{n-1}$ leave the first column unaltered and hence A_{s+1} has $m+1$ elements in this column. Continuing in this way we see that A_{s+1} has only m elements below the diagonal in each column (except of course for the last m columns). However, we know that the final matrix is symmetric and hence those elements in i, j positions for which $j > i+m$ must vanish.

It is evident that to compute the *lower triangle* of A_{s+1} we use only those elements in R_s in positions (i, j) satisfying $j = i, i+1,\ldots, i+m$. Hence when computing R_s we need produce only the first $m+1$ non-zero elements in each row.

63. To take advantage of this and of symmetry we require some auxiliary storage. For simplicity we shall assume first that the computation of R_s is completed before we start to compute A_{s+1}. The triangularization is probably most simply described by considering a specific step. We therefore concentrate on the fourth major step in the triangularization of a matrix A_s for which $n = 8$, $m = 2$. The configuration at the beginning of this step is illustrated by the arrays

$$A_s \text{ and } R_s$$

$$
\begin{bmatrix}
r_{10} & r_{11} & r_{12} \\
r_{20} & r_{21} & r_{22} \\
r_{30} & r_{31} & r_{32} \\
a_{40} & a_{41} & a_{42} \\
a_{50} & a_{51} & a_{52} \\
a_{60} & a_{61} & a_{62} \\
a_{70} & a_{71} & \times \\
a_{80} & \times & \times
\end{bmatrix},
\qquad
\begin{array}{c}
W_s \\[4pt]
\begin{bmatrix}
w_{10} & w_{11} & w_{12} \\
w_{20} & w_{21} & w_{22} \\
w_{30} & w_{31} & w_{32}
\end{bmatrix},
\end{array}
$$

$$(63.1)$$

$$\text{Auxiliary store}$$

$$
\begin{bmatrix}
y_{40} & y_{41} & y_{42} & y_{43} & 0 \\
z_{50} & z_{51} & z_{52} & z_{53} & 0 \\
a_{42} & a_{51} & (a_{60}-k_s) & a_{61} & a_{62}
\end{bmatrix}.
$$

The matrix W_s consists of the vectors w_i of the matrices P_i. The auxiliary store in the general case holds $m+1$ vectors of $2m+1$ elements each. In our case, at the beginning of the fourth major step it already contains the row vectors denoted by $y_{40},...,y_{43}$ and $z_{50},...,z_{53}$ which are the partially processed rows 4 and 5. The arrangement displayed in (63.1) is that which would obtain in the non-triangularized bottom right-hand corner of the complete matrix. Thus the suffices of the y and z vectors signify different locations relative to the diagonal. The fourth step is as follows.

(i) Transfer the *complete* row 6 of A_s to the auxiliary store and incorporate the shift as shown. (Note that all relevant rows of A_s are still available.)

(ii) Compute the vector w_4 from the elements y_{40}, z_{50}, a_{42}. (In the general case w_4 has $m+1$ non-zero elements.) Store w_4 as the fourth row of W_s.

(iii) Pre-multiply by P_4. The only elements affected by this transformation are those in the auxiliary store. The effect of this pre-multiplication (but not the storage) is indicated in (63.2). Only the first 3 elements of row 4 of R_s need be computed.

$$P_4 \times \begin{bmatrix} y_{40} & y_{41} & y_{42} & y_{43} & 0 \\ z_{50} & z_{51} & z_{52} & z_{53} & 0 \\ a_{42} & a_{51} & (a_{60}-k_s) & a_{61} & a_{62} \end{bmatrix} = \begin{bmatrix} r_{40} & r_{41} & r_{42} & - & - \\ 0 & y_{50} & y_{51} & y_{52} & y_{53} \\ 0 & z_{60} & z_{61} & z_{62} & z_{63} \end{bmatrix}.$$

$$(63.2)$$

As the fourth row of R_s is produced it is overwritten on the fourth row of A_s, (note that only the first 3 components are produced), and as the elements y_{5i} and z_{6i} are produced they are written in the y and z positions of the auxiliary store. Hence on the completion of (iii) the stage is set for the fifth major step. Of course, there are the usual end effects. Since we do not contemplate using abortive decompositions with the QR algorithm we can always overwrite R_s on A_s.

64. Turning now to the recombination, A_{s+1} is produced row by row and overwritten on R_s. Actually we produce the elements below the diagonal of each column of A_{s+1} and then enter them as the elements above the diagonal of the corresponding row of A_{s+1}, thereby taking advantage of the known symmetry of A_{s+1} and economizing considerably on the computation.

We describe the typical step in which the fourth row of A_{s+1} is computed. The auxiliary store in the general case holds an $(m+1) \times (m+1)$ array. In our case, at the beginning of the fourth major step the column vectors denoted by $\tilde{y}_{40}$, $\tilde{y}_{41}$ and $\tilde{z}_{50}$, $\tilde{z}_{51}$ (the tildes indicate that these quantities are unrelated to those of (63.1)) are already in the auxiliary store as represented by the arrays

$$\begin{bmatrix} \tilde{y}_{40} & z_{50} & r_{42} \\ \tilde{y}_{41} & \tilde{z}_{51} & r_{51} \\ 0 & 0 & r_{60} \end{bmatrix} \times {}^{'}P_4{}^{'} = \begin{bmatrix} a_{40} & - & - \\ a_{41} & \tilde{y}_{50} & \tilde{z}_{60} \\ a_{42} & \tilde{y}_{51} & \tilde{z}_{61} \end{bmatrix}. \qquad (64.1)$$

The fourth major step is as follows.

(i) Transfer the elements r_{42}, r_{51}, r_{60} of the sixth column of R_s to the auxiliary store as shown.

(ii) Post-multiply the 3×3 matrix in the auxiliary store by 'P_4', the submatrix of P_4 which is a function of w_4. The first column of the resulting matrix gives the relevant elements of the fourth row of A_{s+1}. The other two columns provide the $\tilde{y}$ and $\tilde{z}$ vectors for the next major step. Notice that only part of the computation involved in the post-multiplication by P_4 is actually performed.

It is now evident that the decomposition and the recombination can be combined. In our example as soon as the first three rows of R_s are computed we have sufficient information to determine the first row of A_{s+1}. In general the recombination can be pursued m rows behind the decomposition and if we do this we do not need to store the whole of W_s but merely the last m vectors w_i. Since we recommend the use of QR (and LR) techniques only on narrow bands, the auxiliary storage requirement is slight compared with that for A_s. It is interesting that if we combine the decomposition and recombination the storage required for QR is actually less than in the Cholesky LR in which abortive decompositions are used!

Error analysis

65. In taking advantage of symmetry we are effectively assuming that the elements a_{ij} $(m < j - i \leqslant 2m)$, which would have been zero if exact computation had been used, will in fact be negligible. It is important to justify this assumption and fortunately it is almost covered by our general analysis of Chapter 3, § 45. If $\bar{R}_s$ is the computed matrix (with exact zeros below the diagonal) and $P_1, P_2, ..., P_{n-1}$ are the *exact* elementary Hermitians corresponding to the successive *computed* reduced matrices, then we have

$$\bar{R}_s = (P_{n-1}P_{n-2}...P_1)A_s + E_s, \qquad (65.1)$$

where E_s is small. In fact we compute only part of $\bar{R}_s$ but we use exactly the same arithmetic operations as would have been used if we were computing the whole of it.

We now multiply the computed $\bar{R}_s$ successively by the computed elementary Hermitians $\bar{P}_i$. In spite of any rounding errors which might be made the resulting matrix $\bar{A}_{s+1}$ certainly has only m sub-diagonal elements in each column. If we were to do the whole computation our general analysis shows that the computed $\bar{A}_{s+1}$ would satisfy the relation

$$\begin{aligned}
\bar{A}_{s+1} &= \bar{R}_s P_1 ... P_{n-1} + F_s \\
&= P_{n-1} ... P_1 A_s P_1 ... P_{n-1} + E_s P_1 ... P_{n-1} + F_s, \qquad (65.2)
\end{aligned}$$

where F_s is small. In fact we compute only the lower half of $\bar{A}_{s+1}$ in this way and complete our matrix $\bar{\bar{A}}_{s+1}$ by symmetry. Clearly we have

$$\begin{aligned}
\|G_{s+1}\|_E &\equiv \|\bar{\bar{A}}_{s+1} - P_{n-1}...P_1 A_s P_1...P_{n-1}\|_E \\
&\leqslant 2^{\frac{1}{2}}[\|E_s P_1...P_{n-1}\|_E + \|F_s\|_E] \\
&= 2^{\frac{1}{2}}[\|E_s\|_E + \|F_s\|_E]. \qquad (65.3)
\end{aligned}$$

For floating-point computation with accumulation it can be shown that

$$\|G_{s+1}\|_E < Km2^{-t}\, \|A_1\|_E \qquad (65.4)$$

for some constant K, so that very high accuracy can be expected.

There is about three times as much work in one QR step as in one Cholesky LR step though we have seen (§ 51) that when no shifts are used a QR step is equivalent to two steps of Cholesky LR. The choice of shifts of origin is much simpler with QR; if we take $k_s = a_{nn}^{(s)}$ then convergence is ultimately cubic. Choosing k_s equal to the smaller eigenvalue of the 2×2 matrix in the bottom right-hand corner gives even better convergence. Usually remarkably few iterations are required in practice. There is no guarantee that eigenvalues will be found in increasing order, though if the original matrix is positive definite and one or two preliminary steps with zero shifts are performed this is usually true.

The choice of the shifts of origin with Cholesky LR is rather complex if one uses Rutishauser's technique of abortive decompositions, but the scheme is nonetheless very effective. Eigenvalues are always found in increasing order, and this is important since band matrices often arise from finite-difference approximations to differential equations and only the smaller eigenvalues are significant. On the whole Rutishauser's technique seems preferable, but there is much to be said on both sides.

Unsymmetric band matrices

66. If we use the orthodox LR algorithm without interchanges then band form is preserved. In fact the band need not even have the same number of elements on each side of the diagonal. However, numerical stability is no longer assured and in line with our usual policy we cannot, in general, recommend this use of the LR algorithm. If the LR algorithm with interchanges or the QR algorithm is used, then in general the band form *above the diagonal* is gradually destroyed.

The case when A is tri-diagonal is of special interest. Let us compare two sequences of tri-diagonal matrices A_s and $\tilde{A}_s$ produced as follows. In the first we start with A_1 and use LR decompositions (without interchanges) in which each L_s is unit lower-triangular. In the second we start with DA_1D^{-1} for some diagonal matrix D and use any other type of LR decomposition (without interchanges). The argument of § 52 shows that at corresponding stages we have

$$\tilde{A}_s = D_s A_s D_s^{-1} \tag{66.1}$$

for some diagonal matrix D_s. Now if we write

$$A_s = \begin{bmatrix} \alpha_1^{(s)} & \beta_2^{(s)} & & \\ \gamma_2^{(s)} & \alpha_2^{(s)} & \beta_3^{(s)} & \\ & \cdots\cdots\cdots\cdots & \\ & & \gamma_n^{(s)} & \alpha_n^{(s)} \end{bmatrix}, \quad \tilde{A}_s = \begin{bmatrix} a_1^{(s)} & b_2^{(s)} & & \\ c_2^{(s)} & a_2^{(s)} & b_3^{(s)} & \\ & \cdots\cdots\cdots\cdots & \\ & & c_n^{(s)} & a_n^{(s)} \end{bmatrix},$$

$$\tag{66.2}$$

then from (66.1) we find

$$\alpha_i^{(s)} = a_i^{(s)}, \qquad \beta_i^{(s)}\gamma_i^{(s)} = b_i^{(s)}c_i^{(s)}. \tag{66.3}$$

Hence whatever triangular decomposition we may use, the diagonal elements and the product of opposing off-diagonal elements are the same, and it is these quantities which dominate the eigenvalue problem for A_1. There is therefore no essential loss of generality if we choose D so that

$$DA_1D^{-1} = \begin{bmatrix} \alpha_1 & 1 & & \\ \beta_2 & \alpha_2 & 1 & \\ & \cdots\cdots\cdots\cdots & \\ & & \beta_n & \alpha_n \end{bmatrix}, \tag{66.4}$$

and then use the LR decomposition with unit lower-triangular L_s at every stage. If we write

$$\begin{bmatrix} \alpha_1^{(s)} & 1 & & & \\ \beta_2^{(s)} & \alpha_2^{(s)} & 1 & & \\ & \beta_3^{(s)} & \alpha_3^{(s)} & 1 & \\ & & \cdots\cdots\cdots\cdots & \\ & & & \beta_n^{(s)} & \alpha_n^{(s)} \end{bmatrix}$$

$$= \begin{bmatrix} 1 & & & & \\ l_2^{(s)} & 1 & & & \\ & l_3^{(s)} & 1 & & \\ & & \cdots\cdots\cdots & \\ & & & l_n^{(s)} & 1 \end{bmatrix} \begin{bmatrix} u_1^{(s)} & 1 & & & \\ & u_2^{(s)} & 1 & & \\ & & u_3^{(s)} & 1 & \\ & & & \cdots\cdots\cdots & \\ & & & & u_n^{(s)} \end{bmatrix}, \tag{66.5}$$

then we have

$$u_1^{(s)} = \alpha_1^{(s)}, \qquad l_i^{(s)} u_{i-1}^{(s)} = \beta_i^{(s)}, \qquad l_i^{(s)} + u_i^{(s)} = \alpha_i^{(s)} \quad (i = 2,\ldots, n)$$

$$u_i^{(s)} + l_{i+1}^{(s)} = \alpha_i^{(s+1)} \quad (i = 1,\ldots, n), \qquad \beta_i^{(s+1)} = l_i^{(s)} u_i^{(s)} \quad (i = 2,\ldots, n)$$

$$\tag{66.6}$$

and hence

$$u_i^{(s)} + l_{i+1}^{(s)} = \alpha_i^{(s+1)} = l_i^{(s+1)} + u_i^{(s+1)}, \qquad l_i^{(s)} u_i^{(s)} = \beta_i^{(s+1)} = l_i^{(s+1)} u_{i-1}^{(s+1)}.$$

$$\tag{66.7}$$

These equations show that after computing the $u_i^{(1)}$, $l_i^{(1)}$ all later $u_i^{(s)}$ and $l_i^{(s)}$ may be determined without computing the $\alpha_i^{(s)}$ and $\beta_i^{(s)}$. In fact if we regard $l_1^{(s)}$ as zero for all s then we have the scheme

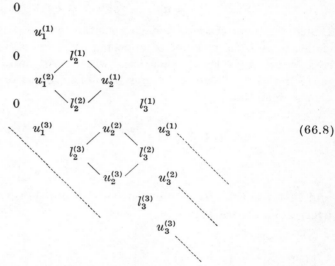

$$\tag{66.8}$$

in which each sloping line with the upper suffix s can be derived from the sloping line above it using equations (66.7), the related elements being those situated at the vertices of the rhombi of the type we have indicated. Readers who are familiar with the QD algorithm will recognize the $l_i^{(s)}$ and $u_i^{(s)}$ arrays as Rutishauser's $q_i^{(s)}$ and $e_i^{(s)}$ arrays. The QD algorithm is of much wider significance in the general problem of computing the zeros of meromorphic functions and a full discussion of it is beyond the scope of this book.

From the point of view of our present problem the QD algorithm is seen to be a little unsatisfactory since it corresponds to elimination without interchanges. However when A_1 is positive definite we know that if we use Cholesky decompositions throughout then numerical stability is guaranteed. If we write

$$\tilde{A}_s = \begin{bmatrix} a_1^{(s)} & b_2^{(s)} & & \\ b_2^{(s)} & a_2^{(s)} & b_3^{(s)} & \\ & \cdots\cdots\cdots\cdots & \\ & & b_n^{(s)} & a_n^{(s)} \end{bmatrix}, \tag{66.9}$$

then in this case our analysis shows that the $a_i^{(s)}$ and the $(b_i^{(s)})^2$ are exactly the same quantities as the $\alpha_i^{(s)}$ and the $\beta_i^{(s)}$ obtained using the form (66.3). It seems more appropriate to work in terms of the $\alpha_i^{(s)}$ and $\beta_i^{(s)}$. This means effectively that we can obtain the full advantages of the Cholesky LR without computing any square roots. It is probably simplest to think of passing directly from the $\alpha_i^{(s)}$, $\beta_i^{(s)}$ to the $\alpha_i^{(s+1)}$, $\beta_i^{(s+1)}$, and if we incorporate a shift k_s at each stage and use the non-restoring method the equations become

$$\left. \begin{array}{ll} & u_1^{(s)} = \alpha_1^{(s)} - k_s \\ l_i^{(s)} = \beta_i^{(s)}/u_{i-1}^{(s)}, & \alpha_{i-1}^{(s+1)} = u_{i-1}^{(s)} + l_i^{(s)} \\ u_i^{(s)} = \alpha_i^{(s)} - k_s - l_i^{(s)}, & \beta_i^{(s+1)} = l_i^{(s)} u_i^{(s)} \\ & \alpha_n^{(s+1)} = u_n^{(s+1)} \end{array} \right\} \ (i = 2,...,n) \Bigg\}. \tag{66.10}$$

Provided the total shift at all stages is less than the smallest eigenvalues, all $\alpha_i^{(s)}$ and $\beta_i^{(s)}$ will be positive and the process will be stable. Only $n-1$ divisions and $n-1$ multiplications are involved in each step.

Simultaneous decomposition and recombination in *QR* algorithm

67. Ortega and Kaiser (1963) have pointed out that a similar process may be used to pass from A_s to A_{s+1} by the QR technique without using square roots. With this transformation symmetry is required but not positive definiteness.

To avoid the upper suffix s we consider the passage from A to $\tilde{A}$ via the relations

$$A = Q_1 R_1, \qquad R_1 Q_1 = \tilde{A}, \tag{67.1}$$

and omit the use of a shift. The reduction of A to R is achieved by premultiplication by $n-1$ rotations in planes $(i, i+1)$ $(i = 1,..., n-1)$.

If we write

$$A = \begin{bmatrix} a_1 & b_2 & & & \\ b_2 & a_2 & b_3 & & \\ & b_3 & a_3 & b_4 & \\ & & \cdots\cdots\cdots & & \\ & & & b_n & a_n \end{bmatrix}, \qquad \bar{A} = \begin{bmatrix} \bar{a}_1 & \bar{b}_2 & & & \\ \bar{b}_2 & \bar{a}_2 & \bar{b}_3 & & \\ & \bar{b}_3 & \bar{a}_3 & \bar{b}_4 & \\ & & \cdots\cdots\cdots & & \\ & & & \bar{b}_n & \bar{a}_n \end{bmatrix},$$

$$R = \begin{bmatrix} r_1 & q_1 & t_1 & & \\ & r_2 & q_2 & t_2 & \\ & & r_3 & q_3 & t_3 \\ & & & \cdots\cdots\cdots & \\ & & & & r_n \end{bmatrix}, \qquad (67.2)$$

and denote the cosine and sine of the jth rotation by c_j and s_j, then

$$s_j = b_{j+1}/(p_j^2+b_{j+1}^2)^{\frac{1}{2}}, \quad c_j = p_j/(p_j^2+b_{j+1}^2)^{\frac{1}{2}} \quad (j = 1,\dots, n-1), \quad (67.3)$$

where

$$\left. \begin{aligned} p_1 &= a_1, \qquad p_2 = c_1 a_2 - s_1 b_2, \\ p_j &= c_{j-1} a_j - s_{j-1} c_{j-2} b_j \quad (j = 3,\dots, n) \end{aligned} \right\}, \qquad (67.4)$$

and

$$\left. \begin{aligned} r_j &= c_j p_j + s_j b_{j+1} \quad (j = 1,\dots, n-1), \qquad r_n = p_n \\ q_1 &= c_1 b_2 + s_1 a_2, \qquad q_j = c_j c_{j-1} b_{j+1} + s_j a_{j+1} \quad (j = 2,\dots, n-1) \\ t_j &= s_j b_{j+2} \quad (j = 1,\dots, n-2) \end{aligned} \right\}. \quad (67.5)$$

On the other hand from the recombination we have

$$\left. \begin{aligned} \bar{a}_1 &= c_1 r_1 + s_1 q_1, \qquad \bar{a}_j = c_{j-1} c_j r_j + s_j q_j \quad (j = 2,\dots, n-1) \\ \bar{a}_n &= c_{n-1} r_n, \qquad \bar{b}_{j+1} = s_j r_{j+1} \quad (j = 1,\dots, n-1) \end{aligned} \right\}. \quad (67.6)$$

If we introduce the quantities γ_j defined by

$$\gamma_1 = p_1, \qquad \gamma_j = c_{j-1} p_j \quad (j = 2,\dots, n), \qquad (67.7)$$

then from (67.6) and (67.5) we find

$$\left. \begin{aligned} \bar{a}_j &= (1+s_j^2)\gamma_j + s_j^2 a_{j+1}, \qquad \bar{a}_n = \gamma_n \\ \bar{b}_{j+1}^2 &= s_j^2(p_{j+1}^2 + b_{j+2}^2) \quad (j = 1,\dots, n-2) \\ \bar{b}_n^2 &= s_{n-1}^2 p_n^2 \end{aligned} \right\}. \qquad (67.8)$$

while for the γ_j and p_j^2 we have

$$\left.\begin{array}{lll} \gamma_1 = p_1 = a_1, & \gamma_j = a_j - s_{j-1}^2(a_j + \gamma_{j-1}) & (j = 2, \ldots, n) \\ p_1^2 = a_1^2, & p_j^2 = \gamma_j^2/c_{j-1}^2 \quad \text{if } c_{j-1} \neq 0 \\ & p_j^2 = c_{j-2}^2 b_j^2 \quad \text{if } c_{j-1} = 0 \end{array}\right\}. \qquad (67.9)$$

The final algorithm is therefore described by the equations

$$u_0 = 0, \qquad c_0 = 1, \qquad b_{n+1} = 0, \qquad a_{n+1} = 0, \qquad (67.10)$$

$$\left.\begin{array}{l} \gamma_i = a_i - u_{i-1} \\ p_i^2 = \gamma_i^2/c_{i-1}^2 \quad (\text{if } c_{i-1} \neq 0) \\ \quad = c_{i-2}^2 b_i^2 \quad (\text{if } c_{i-1} = 0) \\ \bar{b}_i^2 = s_{i-1}^2(p_i^2 + b_{i+1}^2) \quad (i \neq 1) \\ s_i^2 = b_{i+1}^2/(p_i^2 + b_{i+1}^2) \\ c_i^2 = p_i^2/(p_i^2 + b_{i+1}^2) \\ u_i = s_i^2(\gamma_i + a_{i+1}) \\ \bar{a}_i = \gamma_i + u_i \end{array}\right\} \quad (i = 1, \ldots, n). \qquad (67.11)$$

Note that we compute s_i^2 and c_i^2 because we cannot use $(1 - s_i^2)$ for c_i^2 when c_i is small. Each iteration requires $3n$ divisions and $3n$ multiplications but no square roots. The $\bar{a}_i$ and $\bar{b}_i^2$ are overwritten on a_i and b_i^2.

As with general symmetric band matrices both the Cholesky LR and QR have points of superiority and the choice between them is difficult.

Reduction of band width

68. When only a few eigenvalues of a general band symmetric matrix are required the techniques of §§ 59–64 are very efficient. To calculate a larger number, however, it may well be worth while to reduce the matrix to tri-diagonal form by a preliminary transformation. This can be done using Givens' or Householder's method but the band width increases at intermediate stages in the reduction.

Rutishauser (1963) has described two interesting reductions using plane rotations and elementary Hermitians respectively in which the band symmetric form is preserved at all stages. We shall give only the reduction based on plane rotations and in our description we refer only to elements in the upper triangle.

Suppose the band is of width $(2m+1)$. The first move is to eliminate the element $(1, m+1)$ while still retaining a band of width $(2m+1)$

in rows and columns 2 to n. Altogether this requires $p = [(n-1)/m]$ rotations where $[x]$ denotes the integral part of x, because in eliminating the $(1, m+1)$ element the rotation introduces a non-zero element just outside the band lower down the matrix. This must be eliminated in turn and the progress of the computation may be represented by the table

Element eliminated	Plane of rotation	Extra element introduced
1, $m+1$	m, $m+1$	m, $2m+1$
m, $2m+1$	$2m$, $2m+1$	$2m$, $3m+1$
$2m$, $3m+1$	$3m$, $3m+1$	$3m$, $4m+1$
....	....	....
$(p-1)m$, $pm+1$	pm, $pm+1$	none.

As far as the first row and column are concerned the matrix is now a band of width $(2m-1)$ while the remainder is a band of width $(2m+1.)$ The next move is to eliminate element $(2, m+2)$. This is performed in the same way and requires $[n-2/m]$ rotations. Continuing this process the whole matrix is gradually reduced to a band of width $(2m-1)$. An exactly analogous process reduces it to a band of width $(2m-3)$ and hence the matrix ultimately becomes tri-diagonal.

The number of rotations needed to reduce the matrix from a width of $(2m+1)$ to $(2m-1)$ is $[(n-1)/m]+[(n-2)/m]+[(n-3)/m]+...$, which is approximately $n^2/2m$, and the number of multiplications is approximately $4(m+1)n^2/m$. The process is most likely to be used when $m = 2$ or 3.

Additional notes

The suggestion that methods analogous to that of Jacobi should be used for the reduction of a general matrix to triangular form was made by von Neumann, and Causey (1958), Greenstadt (1955) and Lotkin (1956), have described methods of this type. At the matrix conference at Gatlinburg in 1961 Greenstadt gave a summary of the progress which had been made up to that time and concluded that no satisfactory procedure of this kind had yet been developed.

Eberlein (1962) described a modification of this method based on the observation that for any matrix A there exists a similarity transform $B = P^{-1}AP$ which is arbitrarily close to a normal matrix. In Eberlein's algorithm P is constructed as the product of a sequence of matrices which are generalizations of plane rotations but are no longer unitary. Iteration is continued until B is normal to working accuracy and a feature of the method is that in general the limiting B is the direct sum of a number of 1×1 and 2×2 matrices, so that the eigenvalues are available.

Eberlein has also considered the reduction of a general matrix to normal form using a combination of plane rotations and diagonal similarity transformations, and similar ideas have been developed independently by Rutishauser. Generally the guiding strategy at each stage is the reduction of the Henrici departure from normality (Chapter 3, § 50). Development on these lines may yet give rise to methods which are superior to any which I have described and would seem to be one of the most promising lines of research.

Methods of this class are not covered by any of the general error analyses I have given. However, one would expect them to be stable since the successive reduced matrices are tending to a normal matrix and the latter has a perfectly conditioned eigenvalue problem.

Rutishauser's first account of the *LR* algorithm was given in 1955. Since then he has steadily developed and extended the theory in a series of papers. I have attempted here to give an account based on the most recent information available in Rutishauser's papers and on our own experience at the National Physical Laboratory. I have included the proofs of convergence given by Rutishauser in his original papers since they are typical of proofs that have been used by others. The *LR* algorithm with shifts of origin and interchanges applied to Hessenberg matrices was first used by the author in 1959, mainly on matrices with complex elements.

Francis' work on the *QR* algorithm dates from 1959 but was not published until 1961. Virtually the same algorithm was discovered independently by Kublanovskaya (1961). In the light of our parallel treatment of orthogonal and non-orthogonal transformations in this book it is a natural analogue of the *LR* algorithm, but Francis' papers contain much more than the mere description of the *QR* algorithm. Both the technique for combining complex conjugate shifts of origin and that described in § 38 for dealing with consecutive small sub-diagonal elements are vital contributions to the effectiveness of the *QR* algorithm.

The proofs of convergence of the *QR* algorithm given in §§ 29–32 were discovered by the author while this chapter was being written. Proofs depending on more sophisticated determinantal theory have been given by Kublanovskaya (1961) and Householder (1964).

The reduction of the band width of a symmetric matrix both by plane rotations and elementary Hermitians was described by Rutishauser (1963) in a paper which includes a number of other interesting transformations based on the elementary orthogonal matrices.

9

Iterative Methods

Introduction

1. IN THIS final chapter we shall be concerned with what are usually called iterative methods for solving the eigenvalue problem. This is something of a misnomer since all eigenvalue techniques are essentially iterative. A distinguishing feature of the methods of this chapter is that the eigenvalues are found by a technique which is primarily concerned with the determination of an eigenvector or, possibly, of several eigenvectors. The reader may have noticed that so far we have said little about the computation of eigenvectors of non-Hermitian matrices; this is because most of the stable techniques for computing eigenvectors are of the 'iterative' type. It is interesting that stable methods for computing eigenvalues do not necessarily lead to stable methods for the eigenvectors. We have already drawn attention to this in connexion with the computation of eigenvectors of tri-diagonal matrices (Chapter 5, §§ 48–52).

The methods described in this chapter are rather less precisely defined than the methods of previous chapters. Although they are mostly quite simple conceptually they do not readily lend themselves to the design of automatic procedures. This is particularly true of those techniques which are designed to recognize the presence of non-linear elementary divisors and to determine the corresponding invariant subspaces. In practice it is much easier to deal with non-linear divisors if they are known in advance to be present.

An exception to this general criticism is the method of inverse iteration described in §§ 47–60. This is by far the most useful method for computing eigenvectors.

The power method

2. Until specific reference is made to the contrary we shall confine ourselves to matrices having linear elementary divisors. For any such matrix A we have

$$A = X \operatorname{diag}(\lambda_i) X^{-1} = X \operatorname{diag}(\lambda_i) Y^T = \sum_1^n \lambda_i x_i y_i^T, \qquad (2.1)$$

where the columns x_i of X and rows y_i^T of Y^T are the right-hand and left-hand eigenvectors of A normalized so that

$$y_i^T x_i = 1. \tag{2.2}$$

Hence

$$A^s = X \operatorname{diag}(\lambda_i^s) Y^T = \sum_1^n \lambda_i^s x_i y_i^T \tag{2.3}$$

and if

$$|\lambda_1| = |\lambda_2| = \ldots = |\lambda_r| > |\lambda_{r+1}| \geqslant \ldots \geqslant |\lambda_n| \tag{2.4}$$

the expression on the right of (2.3) is ultimately dominated by the terms $\sum_1^r \lambda_i^s x_i y_i^T$. This is the fundamental result on which the methods of this chapter are based. (We have already seen that it is also the basis of the LR and QR algorithms of Chapter 8.)

We shall refer to the eigenvalues $\lambda_1, \ldots, \lambda_r$ as the *dominant eigenvalues* and to the corresponding eigenvectors as the *dominant eigenvectors*. Most commonly $r = 1$ and in this case A^s is ultimately dominated by $\lambda_1^s x_1 y_1^T$.

Most of the practical techniques based on this so-called *power method* incorporate some device for increasing the dominance of the eigenvalue of maximum modulus. They make use of the fact that if $p(A)$ and $q(A)$ are polynomials in A then

$$p(A)\{q(A)\}^{-1} = X \operatorname{diag}\{p(\lambda_i)/q(\lambda_i)\} Y^T \tag{2.5}$$

provided $q(A)$ is non-singular. If we write

$$\mu_i = p(\lambda_i)/q(\lambda_i) \tag{2.6}$$

the object is to choose the polynomials $p(A)$ and $q(A)$ if possible so that one of the $|\mu_i|$ is much larger than the others.

Direct iteration with a single vector

3. The simplest application of the power method is the following. Let u_0 be an arbitrary vector and let the sequences v_s and u_s be defined by the equations

$$v_{s+1} = A u_s, \qquad u_{s+1} = v_{s+1}/\max(v_{s+1}), \tag{3.1}$$

where here and later we use the notation $\max(x)$ to denote the element of maximum modulus of the vector x. Clearly we have

$$u_s = A^s u_0/\max(A^s u_0) \tag{3.2}$$

and if we write

$$u_0 = \sum_1^n \alpha_i x_i, \tag{3.3}$$

then apart from the normalizing factor, u_s is given by

$$\sum_1^n \alpha_i \lambda_i^s x_i = \lambda_1^s \left[\alpha_1 x_1 + \sum_2^n \alpha_i (\lambda_i/\lambda_1)^s x_i \right]. \tag{3.4}$$

If $|\lambda_1| > |\lambda_2| \geqslant |\lambda_3| \geqslant \dots \geqslant |\lambda_n|$, then provided $\alpha_1 \neq 0$ we have

$$u_s \to x_1 / \max(x_1) \quad \text{and} \quad \max(v_s) \to \lambda_1. \tag{3.5}$$

Hence this process provides simultaneously the dominant eigenvalue and the corresponding eigenvector. If $|\lambda_1/\lambda_2|$ is close to unity the convergence is very slow.

If there are a number of independent eigenvectors corresponding to the dominant eigenvalue this does not affect the convergence. Thus if
$$\lambda_1 = \lambda_2 = \dots = \lambda_r \quad \text{and} \quad |\lambda_1| > |\lambda_{r+1}| \geqslant \dots \geqslant |\lambda_n|, \tag{3.6}$$
we have

$$A^s u_0 = \lambda_1^s \left[\sum_1^r \alpha_i x_i + \sum_{r+1}^n \alpha_i (\lambda_i/\lambda_1)^s x_i \right]$$

$$\sim \lambda_1^s \sum_1^r \alpha_i x_i. \tag{3.7}$$

The iterates therefore tend to some vector lying in the subspace spanned by the eigenvectors $x_1, \dots, x_r$, the limit depending upon the initial vector u_0.

Shift of origin

4. The simplest polynomials in A are those of the form $(A - pI)$. The corresponding eigenvalues are $\lambda_i - p$, and if p is chosen appropriately convergence to an eigenvector may be accelerated. We consider first the case when A and all λ_i are real. Clearly if $\lambda_1 > \lambda_2 \geqslant \dots \geqslant \lambda_{n-1} > \lambda_n$ then however we choose p, either $\lambda_1 - p$ or $\lambda_n - p$ is the dominant eigenvalue. For convergence to x_1 the optimum value of p is $\frac{1}{2}(\lambda_2 + \lambda_n)$ and the convergence is determined by the speed at which

$$\{(\lambda_2 - \lambda_n)/(2\lambda_1 - \lambda_2 - \lambda_n)\}^s$$

tends to zero. Similarly the optimum choice for convergence to x_n is $p = \frac{1}{2}(\lambda_1 + \lambda_{n-1})$, the convergence being governed by the quantity $\{(\lambda_1 - \lambda_{n-1})/(\lambda_1 + \lambda_{n-1} - 2\lambda_n)\}^s$. We shall analyse first the effect of using specific values of p and leave until later the problem of determining such values.

For some distributions of eigenvalues this simple device can be quite effective. For example, for a matrix of order 6 with $\lambda_i = 21 - i$, iteration without shift of origin gives convergence to x_1 at a rate governed by $(19/20)^s$. If we take $p = 17$ then the eigenvalues become 3, ± 2, ± 1, 0 and we still have convergence to x_1 but at a rate governed by $(2/3)^s$. To obtain a prescribed precision in x_1 the un-accelerated iteration requires about eight times as many iterations as the accelerated version. Similarly if we take $p = 18$ the eigenvalues become ± 2, ± 1, 0, -3 and we have convergence to x_6 at a speed governed by $(2/3)^s$.

Unfortunately distributions quite commonly arise for which the maximum attainable rate of convergence with this device is not very satisfactory. Suppose for example

$$\lambda_i = 1/i \quad (i = 1,\dots, 20) \tag{4.1}$$

and we wish to choose p so as to obtain x_{20}. The optimum choice of p is $\frac{1}{2}(1+1/19)$ and the rate of convergence is governed by $(180/181)^s$. Approximately 181 iterations are required to reduce the relative error in the vector by a factor e^{-1}. It is quite common to encounter distributions for which $|\lambda_{n-1}-\lambda_n| \ll |\lambda_1-\lambda_n|$ and with such distributions convergence to x_n is slow even with the optimum choice of p.

Effect of rounding errors

5. It is important before considering more complicated acceleration techniques to assess the effect of rounding errors. It might well be felt that since we are working directly with A itself the attainable accuracy will always be high even when A has an ill-conditioned eigenvector problem but this is not so. If we include the effect of rounding errors in the process of § 3, then denoting $\max(v_{s+1})$ by k_{s+1} we have

$$k_{s+1}u_{s+1} = Au_s+f_s, \tag{5.1}$$

where in general f_s will be a 'small' non-null vector. It is quite possible for u_{s+1} to be equal to u_s when it is still far from an accurate eigen-vector.

Consider, for example, the matrix

$$A = \begin{bmatrix} 0\cdot9901 & 10^{-3}(0\cdot2000) \\ 10^{-3}(-0\cdot1000) & 0\cdot9904 \end{bmatrix} \tag{5.2}$$

for which

$$\left.\begin{aligned} \lambda_1 &= 0\cdot9903, \quad x_1^T = (1, 1) \\ \lambda_2 &= 0\cdot9902, \quad x_2^T = (1, 0\cdot5) \end{aligned}\right\}. \tag{5.3}$$

If we take $u_s^T = (1, 0\cdot9)$, exact computation gives

$$v_{s+1}^T = (0\cdot99028, 0\cdot89126), \quad u_{s+1}^T = (1\cdot0, 0\cdot900008\dots), \quad k_{s+1} = 0\cdot99028. \tag{5.4}$$

The differences between the components of u_{s+1} and of u_s are less than 5×10^{-5} and hence if we are working with 4-decimal floating-point arithmetic it is quite likely that the computed u_{s+1} will be exactly equal to u_s and the process will become stationary. It may be verified that widely different vectors u_s show this behaviour.

6. If the component of u_s of maximum modulus is the ith, then (5.1) may be written in the form

$$k_{s+1}u_{s+1} = (A + f_s e_i^T)u_s, \qquad (6.1)$$

and hence we have effectively performed an exact iteration with $A + f_s e_i^T$ rather than with A. As far as this step is concerned we are progressing towards an eigenvector of $A + f_s e_i^T$. (There are of course many other perturbations F of A such that $(A + F)u_s = k_{s+1}u_{s+1}$.) Corresponding to a given technique for evaluating Au_s there will, in effect, be a domain of indeterminacy associated with each eigenvector, and ordered progress towards the true vector will, in general, cease once a vector is attained which lies in this domain.

If, for example, Au_s is computed using standard floating-point arithmetic we have from Chapter 3, § 6(v),

$$fl(Au_s) = (A + F_s)u_s, \qquad |F_s| < 2^{-t_1}|A|\operatorname{diag}(n+1-i). \qquad (6.2)$$

We cannot guarantee that progress will continue after we have reached a vector u_s which is an exact eigenvector of some matrix $(A + G)$ with

$$|G| < 2^{-t_1}|A|\operatorname{diag}(n+1-i), \qquad (6.3)$$

though usually the error in the u_s of limiting accuracy is much less than that corresponding to the maximum that can be caused by perturbations satisfying (6.2). Notice that the bound for each $|f_{ij}|$ in (6.2) is given by $2^{-t_1}(n+1-j)|a_{ij}|$ and since this is directly proportional to $|a_{ij}|$ it is clear as in Chapter 7, § 15, that the accuracy attainable in the dominant eigenvalue by iterating with $D^{-1}AD$ (where D is any diagonal matrix) is the same as that attainable by iterating with A.

In the example of § 5 the k_{s+1} of limiting accuracy was a very good approximation to λ_1 in spite of the inaccuracy of the vector from which it was determined. This is because the relevant matrix has well-conditioned *eigenvalues* but ill-conditioned *eigenvectors* because the eigenvalues are close. When λ_1 is an ill-conditioned eigenvalue, that is when s_1 is small (Chapter 2, § 31), the limiting accuracy attainable in the eigenvalue is also low.

Consider for example the matrix

$$\begin{bmatrix} 1 & 1 & 0 \\ -1+10^{-8} & 3 & 0 \\ 0 & 1 & 1 \end{bmatrix}, \tag{6.4}$$

which has the eigenvalues 2 ± 10^{-4} and 1. If we take $u_s^T = (1, 1, 1)$ then

$$(Au_s)^T = (2, 2+10^{-8}, 2) \tag{6.5}$$

so that using floating-point arithmetic with an 8-digit mantissa the process is stationary and gives $\lambda = 2$ as an eigenvalue.

7. We see then that the effect of rounding errors made in the iterative process is not essentially of a different nature from that corresponding to errors made in successive similarity transformations. Indeed it is our experience that if, for example, we iterate for the dominant eigenvector using *standard* floating-point arithmetic the error introduced is usually greater than that corresponding to errors made in the reduction to Hessenberg form using elementary Hermitians with *floating-point accumulation*.

With the matrix of (5.2) it is possible to obtain the eigenvectors accurately by using suitable shifts of origin. We have, for example,

$$A-0.99I = \begin{bmatrix} 10^{-3}(0.1000) & 10^{-3}(0.2000) \\ 10^{-3}(0.1000) & 10^{-3}(0.4000) \end{bmatrix} \tag{7.1}$$

and this matrix has the eigenvalues $10^{-3}(0.3)$ and $10^{-3}(0.2)$. This is rather a special case. In general when $|\lambda_1-\lambda_2| \ll |\lambda_1-\lambda_n|$ the difficulty involved in computing x_1 and x_2 accurately is quite fundamental and cannot be overcome by such a simple device.

8. There is one other respect in which rounding errors have a profound influence. Suppose, for example, we have a matrix A of order 3 with eigenvalues of the orders of magnitude of 1.0, 0.9 and 10^{-4}. If we iterate directly with A then with exact computation the component of x_3 will be reduced by a factor of 10^{-4s} in s iterations. However, the effect of rounding each element of u_s to t binary places will be to introduce a component of x_3 which will, in general, be of the order of magnitude $2^{-t}x_3$. In fact there will not normally be any t-digit approximation to x_1 which contains a component of x_3 which is appreciably smaller than this. The importance of this comment will become apparent in the next section.

Variation of p

9. The value of p may be varied in successive iterations to considerable advantage. If we denote the current vector by $\sum \alpha_i x_i$ then after one iteration the vector becomes $\sum \alpha_i (\lambda_i - p) x_i$ apart from the normalizing factor. If $p = \lambda_k$ the component of x_k is completely eliminated, while if p is close to λ_k the relative size of the component x_k is considerably reduced. Suppose we have reached a stage at which

$$u_s = \alpha_1 x_1 + \alpha_2 x_2 + \sum_{i=3}^{n} \epsilon_i x_i, \qquad |\epsilon_i| \ll |\alpha_1|, |\alpha_2|, \qquad (9.1)$$

and we have some estimate λ_2' of λ_2. Taking $p = \lambda_2'$ we have

$$(A - pI)u_s = \alpha_1(\lambda_1 - \lambda_2')x_1 + \alpha_2(\lambda_2 - \lambda_2')x_2 + \sum_{3}^{n} \epsilon_i(\lambda_i - \lambda_2')x_i. \qquad (9.2)$$

If $|\lambda_2 - \lambda_2'| \ll |\lambda_1 - \lambda_2'|$ the component of x_2 is greatly reduced relative to that of x_1, though if $|\lambda_i - \lambda_2'| > |\lambda_1 - \lambda_2'|$ for some $i > 2$ the corresponding components will increase relative to that of x_1. If the ϵ_i are very small we may perform a number of iterations with $p = \lambda_2'$, but we must not continue to do so indefinitely or ultimately the component of x_n will dominate.

By way of illustration suppose a matrix of order 4 has the eigenvalues $1 \cdot 0$, $0 \cdot 9$, $0 \cdot 2$, $0 \cdot 1$. If we iterate without shift of origin, the components of x_3 and x_4 will rapidly die out but rounding errors will prevent their being diminished indefinitely. Suppose, working to 10 decimal places, we have reached a u_s given by

$$x_1 + 10^{-2}x_2 + 10^{-10}x_3 + 10^{-10}x_4 \qquad (9.3)$$

and that at this stage we have an estimate $0 \cdot 901$ for λ_2. If we perform r iterations with $p = 0 \cdot 901$ then we have

$$(A - pI)^r u_s = (0 \cdot 099)^r x_1 + 10^{-2}(-0 \cdot 001)^r x_2 +$$
$$+ (-0 \cdot 701)^r 10^{-10} x_3 + (-0 \cdot 801)^r 10^{-10} x_4. \qquad (9.4)$$

The offending component of x_2 is rapidly attenuated, but we must remember that in the normalized u_i, rounding errors will always maintain a component of x_2 of the order of magnitude of 10^{-10}. Taking $r = 4$ we see that u_{s+4} is roughly proportional to

$$10^{-4}(0 \cdot 96)x_1 + 10^{-14}x_2 + 10^{-10}(0 \cdot 24)x_3 + 10^{-10}(0 \cdot 41)x_4,$$

that is to
$$x_1 + 10^{-10}(1 \cdot 04)x_2 + 10^{-6}(0 \cdot 25)x_3 + 10^{-6}(0 \cdot 43)x_4. \qquad (9.5)$$

Further iteration with $p = 0 \cdot 901$ would not reduce the component of x_2 and the components of x_3 and x_4 are rapidly increasing. If we return now to $p = 0$ the components of x_3 and x_4 will rapidly die out again and in 5 or 6 iterations the vector will be essentially

$$x_1 + 10^{-10}x_2 + 10^{-10}x_3 + 10^{-10}x_4. \tag{9.6}$$

It would take about 200 iterations to achieve this result starting from (9.3) and using $p = 0$ throughout. When λ_1 and λ_2 are even closer the advantages to be gained from such a technique are even greater but in order to reap the benefit a more accurate approximation to λ_2 is required.

Ad hoc choice of *p*

10. On the computers at the National Physical Laboratory *ad hoc* versions of this technique have been developed in which the choice of the value of p is made by the operator. This value of p is set up on hand switches and read by the computer at the beginning of each iteration, the value of the current estimate $[\max(v_s) + p]$ of the eigenvalue is displayed on output lights at the end of each iteration and the successive vectors u_s are displayed in binary form on a cathode ray tube.

The use of these programmes is probably best illustrated by discussing a specific field of application. Consider the determination of the dominant eigenvalue and eigenvector of a matrix known to have all its eigenvalues real and positive. Iteration is begun using the value $p = 0$. If convergence is very rapid then there is no need to use any other value of p, but if not, an estimate can be made of the speed at which u_s is converging. Since the programme uses fixed-point arithmetic this can quite easily be done by observing how many iterations are needed for the digits in each binary place of u_s to become stationary. Suppose it takes k iterations per digit then we may assume that

$$(\lambda_2/\lambda_1)^k \doteqdot \tfrac{1}{2} \tag{10.1}$$

and taking the current value of $[\max(v_s) + p]$ as λ_1 we have

$$\lambda_2 = [\max(v_s) + p](\tfrac{1}{2})^{1/k}. \tag{10.2}$$

This approximation to λ_2 is set up on the input switches. Normally the vectors u_i will then show a much higher rate of convergence for the next few iterations but gradually the components of $x_n, x_{n-1}, \ldots$ assume greater importance and digits in the u_i which have settled down begin to be disturbed.

In practice it is usually safe to continue until almost all digits of u_i are changing; at this stage the component of x_n will probably be somewhat larger than that of x_1. We then revert to the value $p = 0$ and the components of x_n and x_{n-1} are rapidly reduced again. If λ_1 and λ_2 are very close it is sometimes necessary to repeat this process several times in order to obtain x_1 to the required accuracy.

This technique has proved remarkably successful on matrices of quite high order provided λ_1 and λ_2 are not pathologically close. In practice it is advisable to use a value of p which is somewhat smaller than that given by (10.2) since it is safer to have p less than λ_2 rather than greater.

Aitken's acceleration technique

11. Aitken (1937) has described a method for speeding the convergence of the vectors which is sometimes of great value. Suppose u_s, u_{s+1}, u_{s+2} are three successive iterates and that we have reached the stage at which u_s is essentially of the form $(x_1 + \epsilon x_2)$ where ϵ is small. If x_1 and x_2 are both normalized so that $\max(x_i) = 1$, then for sufficiently small ϵ the largest element of u_s will usually be in the same position as that of x_1. Suppose the element of x_2 in this position is k, so that we have $|k| < 1$, then u_s, u_{s+1} and u_{s+2} are given by

$$(x_1 + \epsilon x_2)/(1 + \epsilon k), \quad (\lambda_1 x_1 + \epsilon \lambda_2 x_2)/(\lambda_1 + \epsilon k \lambda_2), \quad (\lambda_1^2 x_1 + \epsilon \lambda_2^2 x_2)/(\lambda_1^2 + \epsilon k \lambda_2^2).$$

$$\tag{11.1}$$

Consider now the vector a with components defined by

$$a_i = [u_i^{(s)} u_i^{(s+2)} - (u_i^{(s+1)})^2]/[u_i^{(s)} - 2u_i^{(s+1)} + u_i^{(s+2)}]$$

$$= u_i^{(s)} - [u_i^{(s)} - u_i^{(s+1)}]^2 / [u_i^{(s)} - 2u_i^{(s+1)} + u_i^{(s+2)}], \tag{11.2}$$

where $u_i^{(s)}$ denotes the ith component of $u^{(s)}$. Substituting from (11.1) and writing $\lambda_2/\lambda_1 = r$ we have after some simplification

$$a_i = \frac{\epsilon(1-r)^2(x_i^{(2)} - kx_i^{(1)})(x_i^{(1)} - \epsilon^2 r^2 kx_i^{(2)})}{\epsilon(1-r)^2(x_i^{(2)} - kx_i^{(1)})(1 - \epsilon^2 k^2 r^2)} \tag{11.3}$$

$$= [x_i^{(1)} - \epsilon^2 r^2 kx_i^{(2)}]/[1 - \epsilon^2 k^2 r^2], \tag{11.4}$$

giving $$a = x_1 + O(\epsilon^2). \tag{11.5}$$

It might well be thought that since the denominator and numerator of a_i both have the factor ϵ the process will be seriously affected by rounding errors, but this is not so. If u_s, u_{s+1} and u_{s+2} have errors which are denoted by $2^{-t}\alpha$, $2^{-t}\beta$ and $2^{-t}\gamma$ then ignoring $2^{-t}\epsilon$ etc. equation (11.3) becomes

$$a_i = \frac{\epsilon(1-r)^2(x_i^{(2)} - kx_i^{(1)})(x_i^{(1)} - \epsilon^2 r^2 kx_i^{(2)}) + x_i^{(1)} 2^{-t}(\alpha_i + \gamma_i - 2\beta_i)}{\epsilon(1-r)^2(x_i^{(2)} - kx_i^{(1)})(1 - \epsilon^2 k^2 r^2) + 2^{-t}(\alpha_i + \gamma_i - 2\beta_i)}.$$

$$\tag{11.6}$$

We may write this in the form

$$a_i = \frac{px_i^{(1)} - \epsilon^2 q x_i^{(2)}}{p - \epsilon^2 q} \tag{11.7}$$

showing that the rounding errors are comparatively unimportant.

Hence when we have reached the stage at which all eigenvectors other than x_1 and x_2 have been eliminated, the vector given by Aitken's process is much closer to x_1 than those from which it is derived. However, if λ_3 and λ_2 are not well separated it may require many iterations before the component of x_3 is negligible compared with that of x_2.

It is surprisingly difficult to design an automatic programme which uses Aitken's process efficiently. In the *ad hoc* programme designed at the National Physical Laboratory the use of Aitken's process was controlled by the operator. By setting an input switch he could request the use of the process at the end of the current iteration and by observing the behaviour of the vector iterates he could assess whether the process had been beneficial. When no gain was observed the use of Aitken's process was usually abandoned as far as the determination of the current vector was concerned. In spite of the difficulty of designing an automatic procedure, the use of Aitken's process in this way was found to be of very considerable value.

Complex conjugate eigenvalues

12. If the dominant eigenvalues of a real matrix are a complex conjugate pair λ_1 and $\bar{\lambda}_1$ the iterated vectors will not converge. In fact if x_1 and $\bar{x}_1$ are the corresponding eigenvectors an arbitrary real vector u_0 will be expressible in the form

$$u_0 = \alpha_1 x_1 + \bar{\alpha}_1 \bar{x}_1 + \sum_3^n \alpha_i x_i. \tag{12.1}$$

Hence we have

$$A^s u_0 = r_1^s [\rho_1 e^{i(\alpha+s\theta)} x_1 + \rho_1 e^{-i(\alpha+s\theta)} \bar{x}_1 + \sum_3^n \alpha_i (\lambda_i/r_1)^s x_i] \tag{12.2}$$

where

$$\lambda_1 = r_1 e^{i\theta}, \qquad \alpha_1 = \rho_1 e^{i\alpha}. \tag{12.3}$$

The components of $x_3, \ldots, x_n$ ultimately die out, but if we write

$$v_{s+1} = A u_s, \qquad \max(v_{s+1}) = k_{s+1}, \qquad u_{s+1} = v_{s+1}/k_{s+1}, \tag{12.4}$$

it is clear from (12.2) that neither k_{s+1} nor u_{s+1} tend to a limit. If we denote the jth component of x_1 by $\xi_j e^{i\varphi_j}$, equation (12.2) gives

$$(A^s u_0)_j \sim 2\rho_1 r_1^s \xi_j \cos(\alpha + \varphi_j + s\theta) \tag{12.5}$$

and hence the components of u_s oscillate in sign.

If λ_1 and $\bar{\lambda}_1$ are the roots of $\lambda^2 - p\lambda - q = 0$ we have

$$(A^{s+2} - pA^{s+1} - qA^s)u_0 \to 0 \quad (s \to \infty) \qquad (12.6)$$

or

$$k_{s+1}k_{s+2}u_{s+2} - pk_{s+1}u_{s+1} - qu_s \to 0 \qquad (12.7)$$

and hence ultimately any three successive iterates are linearly dependent. By the method of least squares we can determine successive approximations p_s and q_s to p and q. We have

$$k_{s+1}k_{s+2}\begin{bmatrix} u_{s+1}^T u_{s+2} \\ u_s^T u_{s+2} \end{bmatrix} = \begin{bmatrix} u_{s+1}^T u_{s+1} & u_{s+1}^T u_s \\ u_s^T u_{s+1} & u_s^T u_s \end{bmatrix}\begin{bmatrix} p_s k_{s+1} \\ q_s \end{bmatrix}. \qquad (12.8)$$

When p_s and q_s have tended to limits p and q, λ_1 and $\bar{\lambda}_1$ may be computed from the relations

$$\mathscr{R}(\lambda_1) = \tfrac{1}{2}p, \qquad \mathscr{I}(\lambda_1) = \tfrac{1}{2}(p^2 + 4q)^{\frac{1}{2}}. \qquad (12.9)$$

13. Since they are obtained from an explicit polynomial equation, λ_1 and $\bar{\lambda}_1$ will be poorly determined from the coefficients p and q when the roots are close, that is, when $\mathscr{I}(\lambda_1)$ is small. It might be felt that the loss of accuracy in the eigenvalues is attributable to ill-condition of the original problem but this is not necessarily true. Consider for example the matrix A given by

$$\begin{bmatrix} 0.4812 & 0.0023 \\ -0.0024 & 0.4810 \end{bmatrix}. \qquad (13.1)$$

The eigenvalues are $0.4811 \pm 0.0023i$ and it may readily be verified that they are relatively insensitive to small perturbations in the elements of A. However, the characteristic equation is

$$\lambda^2 - 0.9622\lambda + 0.23146272 = 0$$

and the roots of this equation are very sensitive to *independent* changes in the coefficients. Hence even a comparatively accurate determination of p and q by the method of § 12 will give a poor determination of the imaginary parts of the eigenvalue. Unfortunately it is precisely in this case that the equations (12.8) are ill-conditioned.

This is well illustrated by the matrix of (13.1). Since this is of order two we can take any vector as u_s. If we take $u_s^T = (1, 1)$ then we have

$$(Au_s)^T = (0.4835, 0.4786), \qquad (A^2u_s)^T = (0.2338, 0.2290), \qquad (13.2)$$

working to 4 decimals. Since there are only two components, p and q are uniquely determined and we have

$$0 \cdot 2338 = 0 \cdot 4835p + q$$

$$0 \cdot 2290 = 0 \cdot 4786p + q$$

giving

$$p = 0 \cdot 9796, \quad q = -0 \cdot 2398. \tag{13.3}$$

Exact computation, of course, gives p and q exactly. The values of p and q in (13.3) are quite useless for the determination of λ_1.

Calculation of the complex eigenvector

14. When the components of $x_3, \ldots, x_n$ have been eliminated we have

$$u_s = \alpha_1 x_1 + \bar{\alpha}_1 \bar{x}_1, \qquad v_{s+1} = A u_s = \alpha_1 \lambda_1 x_1 + \bar{\alpha}_1 \bar{\lambda}_1 \bar{x}_1. \tag{14.1}$$

If we write

$$\alpha_1 x_1 = z_1 + i w_1, \qquad \lambda_1 = \xi_1 + i \eta_1 \tag{14.2}$$

then

$$u_s = 2 z_1, \qquad v_{s+1} = 2 \xi_1 z_1 - 2 \eta_1 w_1, \tag{14.3}$$

$$z_1 + i w_1 = \tfrac{1}{2} [u_s + i(\xi_1 u_s - v_{s+1})/\eta_1]. \tag{14.4}$$

Apart from a normalizing factor we have therefore

$$x_1 = \eta_1 u_s + i(\xi_1 u_s - v_{s+1}). \tag{14.5}$$

When η_1 is small, u_s and v_{s+1} may well be almost parallel and hence the vector x_1 is poorly determined. This point also is well illustrated by the example of § 13. Notice that $(z_1 \mid w_1)$ is an invariant subspace (Chapter 3, § 63) since the equation

$$A(z_1 + i w_1) = (\xi_1 + i \eta_1)(z_1 + i w_1) \tag{14.6}$$

implies

$$A z_1 = \xi_1 z_1 - \eta_1 w_1, \qquad A w_1 = \eta_1 z_1 + \xi_1 w_1. \tag{14.7}$$

Since a real eigenvalue has a real eigenvector, that is, a vector with zero imaginary part, it might be thought that if η_1 is small the eigenvector x_1 when normalized so that $\max(x_1) = 1 + i0$ will have a comparatively small imaginary part. However, this will not be true unless there is a matrix close to A having a real quadratic divisor close to $(\lambda - \lambda_1)^2$. For example, the normalized eigenvector x_1 corresponding to the matrix of (13.1) is given by

$$x_1^T = (-0 \cdot 0417 - i0 \cdot 9782, \ 1 + i0). \tag{14.8}$$

The poor results we obtained with this matrix are not caused by our choice $(1, 1)$ for u_s^T. Almost any vector would have given equally poor results.

This becomes clear if we consider the case of a real matrix having a real dominant eigenvalue with two corresponding eigenvectors. It is clear that we cannot obtain both of these eigenvectors from one simple sequence $u_0, u_1, u_2, \ldots$; we can only obtain one vector in the appropriate subspace. This is the limiting situation corresponding to a complex pair of eigenvalues $(\xi_1 \pm i\eta_1)$, as $\eta_1 \to 0$ and the elementary divisors remain linear.

Shift of origin

15. It is evident that if we iterate with $(A - pI)$ using any real value of p we cannot 'separate' a complex conjugate pair of eigenvectors. If, however, we take complex values of p and use complex vector iterates, $(\lambda_1 - p)$ and $(\bar{\lambda}_1 - p)$ are of different modulus and it is possible to obtain convergence to x_1. There is now about twice as much work in one iteration but this is quite reasonable since when x_1 and λ_1 are found we immediately have $\bar{x}_1$ and $\bar{\lambda}_1$. In this connexion it is worth remarking that in the method of § 12 we are attempting to determine a complex conjugate pair of eigenvectors with little more computation than is involved in computing one real eigenvector.

My experience with using complex shifts has not been a very happy one and suggests that it would be better to use the method of § 12 incorporating only real shifts but retaining double-precision vectors u_s. Since A has only single-precision elements this involves only twice as much arithmetic as the use of single-precision vectors, quite a reasonable ratio in view of the fact that we obtain two eigenvalues. It may be verified that provided $|\eta_1| > 2^{-\frac{1}{2}t} \|A\|$ the double-precision vectors give full protection against the ill-conditioning we have discussed, though to gain the full benefit we must continue iteration until the components of the remaining eigenvectors are of order 2^{-2t} rather than 2^{-t}.

Non-linear divisors

16. The behaviour of the simple power method when A has non-linear elementary divisors is illustrated by considering the case when A is a simple Jordan submatrix $C_r(a)$ (Chapter 1, § 7). We have

$$[C_r(a)]^s e_i = \binom{s}{i-1} a^{s-i+1} e_1 + \binom{s}{i-2} a^{s-i+2} e_2 + \ldots + \binom{s}{1} a^{s-1} e_{i-1} + a^s e_i.$$

$$(16.1)$$

Clearly as $s \to \infty$ the term in e_1 becomes predominant but convergence is slow since the right-hand side of (16.1) may be written in the form

$$\binom{s}{i-1} a^{s-i+1}\left[e_1 + \frac{i-1}{s-i+2}\, ae_2 + \ldots\right], \qquad (16.2)$$

and the term in brackets is asymptotic to

$$e_1 + \frac{i-1}{s}\, ae_2. \qquad (16.3)$$

Starting with any vector u_0 we do ultimately have convergence to the unique eigenvector e_1, but the asymptotic convergence is slower than that corresponding to distinct eigenvalues however poorly they may be separated. The asymptotic rate of convergence, however, may be a little misleading; if a is small we see that the term in e_2 in (16.3) is small. This apparently trivial comment is important in connexion with inverse iteration (§ 53).

Returning now to the general case we have

$$A^s = XC^sX^{-1}, \qquad (16.4)$$

where C is the Jordan canonical form. If the dominant eigenvalue of A is λ_1 and this corresponds to a single divisor of degree r, then if $C_r(\lambda_1)$ is the first submatrix in the Jordan form C, $C^sX^{-1}u_0 \sim k_s e_1$ for an arbitrary u_0 provided not all of the first r components of $X^{-1}u_0$ are zero. Hence

$$A^s u_0 \sim k_s x_1. \qquad (16.5)$$

The convergence is usually too slow to be of practical value but we notice that the components of invariant subspaces corresponding to the eigenvalues λ_i of smaller modulus are diminished by the factor $(\lambda_i/\lambda_1)^s$ after s iterations. We are therefore left with a vector lying in the subspace corresponding to λ_1 at a comparatively early stage. We return to non-linear divisors in §§ 32, 41, 53.

Simultaneous determination of several eigenvalues

17. An essential feature of the method of § 12 is the determination of two eigenvalues from a single sequence of iterates. The fact that the eigenvalues were complex conjugates is really quite incidental and the method may be extended to cover the determination of several real or complex eigenvalues. Suppose for example

$$|\lambda_1| > |\lambda_2| > |\lambda_3| \gg |\lambda_4| > \ldots > |\lambda_n|. \qquad (17.1)$$

The components of x_4 to x_n will die out rapidly in the iterated vectors and we shall soon reach a stage at which u_s is effectively given by $\alpha_1 x_1 + \alpha_2 x_2 + \alpha_3 x_3$. If we define quantities p_2, p_1, p_0 by the equation

$$(\lambda - \lambda_1)(\lambda - \lambda_2)(\lambda - \lambda_3) \equiv \lambda^3 + p_2 \lambda^2 + p_1 \lambda + p_0, \qquad (17.2)$$

then
$$(A^3 + p_2 A^2 + p_1 A + p_0 I) u_s = 0, \qquad (17.3)$$

giving

$$-[A^3 u_s] = [A^2 u_s \;\vdots\; A u_s \;\vdots\; u_s] \begin{bmatrix} p_2 \\ p_1 \\ p_0 \end{bmatrix}. \qquad (17.4)$$

The coefficients p_i may therefore be obtained by least squares from any four consecutive iterates following u_s.

The use of such a technique would have most to recommend it when $|\lambda_1|$, $|\lambda_2|$, $|\lambda_3|$ were close, since it is in such circumstances that the iterates are slow to converge to the dominant eigenvector. Unfortunately if λ_1, λ_2, λ_3 are of the same sign and close together, not only is the relevant polynomial ill-conditioned but the equations determining the p_i are also ill-conditioned. For these reasons the method is of little practical value. The method of Krylov (Chapter 6, § 20) is precisely this method carried to the full limit. In general to determine r eigenvalues in a well-conditioned manner we require r independent sets of iterated vectors. Such methods are discussed in §§ 36–39.

Complex matrices

18. If the matrix A is complex then we may use the simple iterative procedure working in complex arithmetic throughout, but there is now four times as much work in each iteration. Complex shifts of origin may be used but in general the choice of such values presents considerable difficulties. In general to obtain the dominant eigenvector of a complex matrix it is unsatisfactory to rely entirely on iteration with matrices of the type $(A - pI)$, and the procedure described at the end of § 62 is recommended.

Deflation

19. We have seen that when A is real and has real eigenvalues we can employ real shifts of origin to give convergence either to λ_1 or to λ_n. In the case of complex matrices by using appropriate shifts of origin we could in principle make quite a number of the eigenvalues dominant in turn but this would usually be prohibitively difficult to do.

It is natural to ask whether we can use our knowledge of λ_1 and x_1 in such a way that another eigenvector may be found without danger of converging again to x_1. One class of such methods depends essentially on replacing A by a matrix which possesses only the remaining eigenvalues. We shall refer to such methods as methods of *deflation*.

Probably the simplest is that due to Hotelling (1933) which can be applied when an eigenvalue λ_1 and vector x_1 of a symmetric matrix A_1 are known. If we define A_2 by the relation

$$x_1^T x_1 = 1, \qquad A_2 = A_1 - \lambda_1 x_1 x_1^T \qquad (19.1)$$

then from the orthogonality of the x_i we have

$$
\begin{aligned}
A_2 x_i = A_1 x_i - \lambda_1 x_1 x_1^T x_i &= 0 \qquad (i = 1) \\
&= \lambda_i x_i \quad (i \neq 1)
\end{aligned} \Bigg\} . \qquad (19.2)
$$

Hence the eigenvalues of A_2 are $0, \lambda_2, \ldots, \lambda_n$ corresponding to eigenvectors $x_1, x_2, \ldots, x_n$, and the dominant eigenvalue λ_1 has been reduced to zero.

When A_1 is unsymmetric there is a corresponding deflation technique, also due to Hotelling (1933), but it requires the determination of the left-hand eigenvector y_1 as well as x_1. If both are determined and normalized so that $y_1^T x_1 = 1$ then defining A_2 by

$$A_2 = A_1 - \lambda_1 x_1 y_1^T, \qquad (19.3)$$

we have from the bi-orthogonality of the x_i and y_i

$$
\begin{aligned}
A_2 x_i = A_1 x_i - \lambda_1 x_1 y_1^T x_i &= 0 \qquad (i = 1) \\
&= \lambda_i x_i \quad (i \neq 1)
\end{aligned} \Bigg\} . \qquad (19.4)
$$

These two methods of deflation have rather poor numerical stability and I do not recommend their use.

Deflation based on similarity transformations

20. We consider now deflations which depend on similarity transformations of A_1. Suppose H_1 is any non-singular matrix such that

$$H_1 x_1 = k_1 e_1, \qquad (20.1)$$

where clearly $k_1 \neq 0$. We have

$$A_1 x_1 = \lambda_1 x_1 \quad \text{giving} \quad H_1 A_1 (H_1^{-1} H_1) x_1 = \lambda_1 H_1 x_1. \qquad (20.2)$$

Equations (20.1) and (20.2) give

$$H_1 A_1 H_1^{-1} e_1 = \lambda_1 e_1 \tag{20.3}$$

and hence the first column of $H_1 A_1 H_1^{-1}$ must be $\lambda_1 e_1$. We may write

$$A_2 = H_1 A_1 H_1^{-1} = \left[\begin{array}{c|c} \lambda_1 & b_1^T \\ \hline O & B_2 \end{array}\right], \tag{20.4}$$

where B_2 is a matrix of order $(n-1)$ which obviously has the eigenvalues $\lambda_2, \ldots, \lambda_n$. To find the next eigenvalue λ_2 we can work with the matrix B_2. If the corresponding eigenvector of B_2 is denoted by $x_2^{(2)}$, then writing

$$\left[\begin{array}{c|c} \lambda_1 & b_1^T \\ \hline O & B_2 \end{array}\right] \left[\begin{array}{c} \alpha \\ \hline x_2^{(2)} \end{array}\right] = \lambda_2 \left[\begin{array}{c} \alpha \\ \hline x_2^{(2)} \end{array}\right] = \lambda_2 x_2^{(1)} \quad \text{(say)} \tag{20.5}$$

we have

$$(\lambda_1 - \lambda_2)\alpha + b_1^T x_2^{(2)} = 0, \tag{20.6}$$

and hence finally from (20.4)

$$x_2 = H_1^{-1} x_2^{(1)}. \tag{20.7}$$

From an eigenvector of B_2 the corresponding eigenvector of A_1 may therefore readily be determined.

In a similar way after finding an eigenvector $x_2^{(2)}$ of B_2, if we determine H_2 so that

$$H_2 x_2^{(2)} = k_2 e_1, \tag{20.8}$$

we have

$$H_2 B_2 H_2^{-1} = \left[\begin{array}{c|c} \lambda_2 & b_2^T \\ \hline O & B_3 \end{array}\right] \tag{20.9}$$

where B_3 is of order $(n-2)$ and has eigenvalues $\lambda_3, \ldots, \lambda_n$. From an eigenvector $x_3^{(3)}$ of B_3 we may determine an eigenvector $x_3^{(2)}$ of B_2, then a vector $x_3^{(1)}$ of $H_1 A_1 H_1^{-1}$ and hence finally a vector x_3 of A_1. Continuing this process we may find all eigenvalues and eigenvectors of A_1. If we write

$$K_r = \left[\begin{array}{c|c} I_{r-1} & O \\ \hline O & H_r \end{array}\right] \tag{20.10}$$

then we have

$$K_r \ldots K_2 K_1 A_1 K_1^{-1} K_2^{-1} \ldots K_r^{-1} = \left[\begin{array}{c|c} T_r & C_r \\ \hline O & B_{r+1} \end{array}\right] \tag{20.11}$$

where T_r is an upper-triangular matrix with diagonal elements $\lambda_1, \ldots, \lambda_r$ and B_{r+1} has the eigenvalues $\lambda_{r+1}, \ldots, \lambda_n$. Successive deflation therefore gives a similarity reduction to triangular form (Chapter 1, §§ 42, 47).

Deflation using invariant subspaces

21. Before considering details of the deflation process we discuss the case when we have determined an invariant subspace of A (Chapter 3, § 63) but not necessarily individual eigenvectors. Suppose we have a matrix X having m independent columns and an $m \times m$ matrix M such that

$$AX = XM. \tag{21.1}$$

Suppose further that H is a non-singular matrix such that

$$HX = \begin{bmatrix} T \\ \hline O \end{bmatrix} \tag{21.2}$$

where T is a non-singular $m \times m$ matrix (usually upper-triangular in practice). If we write

$$HAH^{-1} = \begin{bmatrix} B & C \\ \hline D & E \end{bmatrix} \tag{21.3}$$

then from the relation

$$HAH^{-1}HX = HXM \tag{21.4}$$

we have

$$\begin{bmatrix} B & C \\ \hline D & E \end{bmatrix} \begin{bmatrix} T \\ \hline O \end{bmatrix} = \begin{bmatrix} T \\ \hline O \end{bmatrix} M \tag{21.5}$$

giving

$$BT = TM \quad \text{and} \quad DT = 0. \tag{21.6}$$

From the second of these relations $D = O$ and hence the eigenvalues of A are those of B and E. From the first of equations (21.6) the eigenvalues of B are those of M.

Deflation using stabilized elementary transformations

22. Returning now to the case when we have a single eigenvector x_1 we require a matrix H_1 satisfying (20.1). We can determine H_1 in a stable manner as a product of the form $N_1 I_{1,1'}$ where the element of x_1 of maximum modulus is in position $1'$. Assuming x_1 is normalized so that $\max(x_1) = 1$ the deflation takes the following form.

(i) Interchange rows 1 and 1' and columns 1 and 1' of A_1.

(ii) Interchange elements 1 and 1' of x_1 calling the resulting vector y_1.

(iii) For each value of i from 2 to n compute

$$\text{row } i = \text{row } i - (y_i \times \text{row } 1).$$

(iv) For each value of i from 2 to n compute

$$\text{column } 1 = \text{column } 1 + (y_i \times \text{column } i).$$

In fact step (iv) need not be performed since we know that column 1 must become $(\lambda_1, 0, \dots, 0)$. There are therefore only $(n-1)^2$ multiplications in the deflation. Notice that with this deflation, b_1^T of equation (20.4) consists of the elements of row 1' of A_1 other than $a_{1',1'}$ permuted in the obvious way. Even when A_1 is symmetric this will not in general be true of A_2.

23. When we have an invariant subspace X as in § 21 we wish to determine an H satisfying (21.2). Again this can be performed quite conveniently using stabilized elementary transformations. A suitable H may be determined as the product

$$N_m I_{m,m'} \dots N_2 I_{2,2'} N_1 I_{1,1'} \qquad (23.1)$$

where these matrices are precisely those which would be obtained by performing Gaussian elimination with interchanges on the matrix X. (The fact that it has only m columns instead of n is irrelevant.) Obviously we have

$$N_m I_{m,m'} \dots N_2 I_{2,2'} N_1 I_{1,1'} X = \begin{bmatrix} T \\ \hline O \end{bmatrix}, \qquad (23.2)$$

where T is upper-triangular. When interchanges are unnecessary H has the form $\begin{bmatrix} L & O \\ \hline M & I_{n-m} \end{bmatrix}$ and if X^T is $[U \mid V]$, (23.2) is equivalent to

$$LU = T \quad \text{and} \quad MU + V = O. \qquad (23.3)$$

The computation of HAH^{-1} takes place in m steps in the rth of which we perform a similarity transformation using the matrix $N_r I_{r,r'}$. Notice that we need not compute that part of the deflated matrix denoted by D in (21.5) because we know that this should be null. Since we have full information on the interchanges before starting the deflation we know which rows and columns of A will give rise to the submatrix D. It is quite easy to organize the computation so that inner-products are accumulated both in determining H (as in triangular decomposition with interchanges) and in computing the deflated matrix.

24. The most common example of the use of invariant subspaces arises in association with a complex pair of eigenvalues. If we are using the method of § 12, then after convergence has taken place any two consecutive u_s and u_{s+1} constitute an invariant subspace.

When the complex eigenvector $(z_1 + iw_1)$ itself has been determined the deflation process has a particularly interesting feature. Let us assume that $(z_1 + iw_1)$ has been normalized so that $\max(x_1) = 1 + 0i$. We then have typically when $n = 5$ and $1' = 3$ the array

$$\left.\begin{aligned} z_1^T &= a_1 \quad a_2 \quad 1 \quad a_4 \quad a_5 \\ w_1^T &= b_1 \quad b_2 \quad 0 \quad b_4 \quad b_5 \end{aligned}\right\} . \tag{24.1}$$

The first interchange gives

$$\begin{bmatrix} 1 & a_2 & a_1 & a_4 & a_5 \\ 0 & b_2 & b_1 & b_4 & b_5 \end{bmatrix} = \begin{bmatrix} 1 & a_2' & a_3' & a_4' & a_5' \\ 0 & b_2' & b_3' & b_4' & b_5' \end{bmatrix} \tag{24.2}$$

and in N_1 we have $n_{j1} = a_j'$. The reduction of the first column of x leaves the second unaffected; this is then reduced independently by the pre-multiplication by $N_2 I_{2,2'}$. The net effect is that we first treat z_1 as a real eigenvector and perform one deflation; we then treat w_1 (with its zero element omitted) as though it were a real eigenvector of the deflated matrix and perform the next deflation accordingly.

Deflation using unitary transformations

25. In a similar way we can use unitary transformations to effect a deflation. In the case of a single vector the matrix H_1 of equation (20.1) can be chosen either to be a single elementary Hermitian $(I - 2ww^T)$, where in general w has no zero components, or the product of rotations in planes $(1, 2), (1, 3), \ldots, (1, n)$. If x_1 is real the matrices are orthogonal.

There are approximately $8n^2$ multiplications in the deflation if we use plane rotations, and $4n^2$ if we use elementary Hermitians. This gives the usual two to one ratio and there is little to be said for the plane rotations. However, we notice that with elementary Hermitians there are four times as many multiplications as with stabilized non-unitary transformations; in previous situations the ratio has been two to one. This is because in the case of deflation by stabilized non-unitary transformations we do not actually have to perform the post-multiplication by H_1^{-1}. It leaves the last $(n-1)$ columns unaltered in any case, and we set the first column equal to $\lambda_1 e_1$ without computation. Although with the unitary transformations we can take the final first column to be $\lambda_1 e_1$ this does not save much computation since all other columns are modified in the post-multiplication.

If the original matrix is Hermitian the deflated matrix is also Hermitian and is therefore of the form

$$
\begin{bmatrix}
\lambda_1 & O \\
\hline
O & B_2
\end{bmatrix},
\tag{25.1}
$$

where B_2 is Hermitian. The deflation can be carried out in almost the same way as the first step of Householder's method (except that w no longer has a zero first element) including the device described in Chapter 5, § 30, for simplifying the symmetric case. It is interesting that this method of deflation using elementary Hermitians was first described by Feller and Forsythe (1951) though in rather different terms. It is unfortunate that it did not attract more attention (probably because it was only one of a very large number of deflation techniques described in the paper) since otherwise it might have led to the general use of the elementary Hermitians at a much earlier stage in the history of matrix work in numerical analysis.

Similarly we can deal with an invariant subspace consisting of m columns by treating them in the same manner as the first m columns of a square matrix are treated in orthogonal triangularization (Chapter 4, § 46). This requires m elementary Hermitians with vectors w_i having respectively $0, 1, 2, \ldots, m-1$ zero elements. The corresponding matrix T is again upper-triangular. Corresponding to a complex eigenvector $(z_1 + iw_1)$ of a real matrix we use the real invariant subspace $(z_1 \mid w_1)$ so that real (orthogonal) matrices are adequate and no complex arithmetic is involved.

Numerical stability

26. The methods of deflation of § 22 and § 25 are remarkably stable. Suppose the vector u_s has been accepted as an eigenvector and $\lambda = \max(Au_s)$ as an eigenvalue. We may drop the suffix s for the purpose of this investigation and it will simplify the notation

without any real loss of generality if we assume that the element of u_s of maximum modulus is in the first position. We write

$$A_1 u - \lambda u = \eta \qquad (26.1)$$

and we assume that $\|\eta\|$ is small compared with $\|A\|$. However, (53.6) of Chapter 3 and § 10 of Chapter 2 show that this does not mean that u is necessarily accurate. Indeed if there is more than one eigenvalue close to λ our u might well be a very inaccurate eigenvector. If we write

$$A_1 = \left[\begin{array}{c|c} a_{11} & a^T \\ \hline c & B_1 \end{array}\right], \quad u^T = (1 \mid v^T) \quad \text{where } |v_i| \leqslant 1, \qquad (26.2)$$

then

$$H_1 = \left[\begin{array}{c|c} 1 & O \\ \hline -v & I \end{array}\right]. \qquad (26.3)$$

Using A_2 now to denote the *computed* deflated matrix we have

$$A_2 = \left[\begin{array}{c|c} \lambda & a^T \\ \hline O & B_2 \end{array}\right], \qquad (26.4)$$

where only the elements of B_2 are actually computed. In fixed-point computation for example we have

$$a_{ij}^{(2)} = a_{ij}^{(1)} - u_i a_{1j}^{(1)} + \epsilon_i, \qquad |\epsilon_{ij}| < \tfrac{1}{2} 2^{-t} \qquad (1 < i, j < n), \qquad (26.5)$$

giving

$$B_2 = B_1 - va^T + F, \qquad (26.6)$$

where F is the matrix formed by the ϵ_{ij}. Consider now the matrix $\tilde{A}_1$ defined in terms of the computed A_2 by the relation

$$\tilde{A}_1 = H_1^{-1} A_2 H_1 = \left[\begin{array}{c|c} \lambda - a^T v & a^T \\ \hline \lambda v - B_2 v - va^T v & B_2 + va^T \end{array}\right]$$

$$= \left[\begin{array}{c|c} \lambda - a^T v & a^T \\ \hline (\lambda I - B_1 - F)v & B_1 + F \end{array}\right]. \qquad (26.7)$$

The computed A_2 has exactly the same eigenvalues as $\tilde{A}_1$ and it is obvious that $\tilde{A}_1$ has λ and u as *exact* eigenvalue and eigenvector. We may therefore say that the computed A_2 is the matrix obtained by finding an exact eigenvalue and eigenvector of $\tilde{A}_1$ and then performing an exact deflation. We wish to relate $\tilde{A}_1$ and A_1. Writing $\eta^T = (\eta_1 \mid \zeta)$ equation (26.1) gives

$$a_{11}+a^Tv-\lambda = \eta_1, \qquad c+B_1v-\lambda v = \zeta, \qquad (26.8)$$

and hence

$$\tilde{A}_1 = \left[\begin{array}{c|c} a_{11}-\eta_1 & a^T \\ \hline c-\zeta-Fv & B_1+F \end{array}\right] \qquad (26.9)$$

$$\tilde{A}_1-A_1 = \left[\begin{array}{c|c} -\eta_1 & O \\ \hline -\zeta-Fv & F \end{array}\right], \qquad (26.10)$$

giving, for example,

$$\|\tilde{A}_1-A_1\|_\infty \leqslant \|\eta\|_\infty + (n-1)2^{-t}. \qquad (26.11)$$

Provided the residual η is small we can be certain that the eigenvalues of A_2 will be close to those of A_1 (if they are well-conditioned) even though A_2 may be very different from the matrix which would have been derived if u had been exact.

If we consider now a sequence of deflations there is a danger that elements of the successive A_i may steadily increase in magnitude and they may become steadily more ill-conditioned. We have drawn attention to this possibility on earlier occasions when a sequence of stabilized non-unitary transformations has been applied. In this particular case these dangers seem to be more remote than usual. Indeed at successive stages in the deflation we work only with the matrices B_r in the bottom right-hand corner of the A_r and it is usual for the conditions of B_r to improve with increasing r even when this is not true of the A_r. An extreme example of this occurs when A_r has a quadratic divisor. If rounding errors are ignored every A_r must have a quadratic divisor, but after one of the corresponding eigenvalues has been found B_r no longer has a quadratic divisor. Hence the condition of the remaining eigenvalue of the pair must have improved!

Numerical example

27. In Table 1 we show the deflation of a matrix of order three. From the eigenvalues and eigenvectors given there we observe that λ_1 and λ_2 are close. The vector u_s is accepted as an eigenvector and gives an accurate eigenvalue and a small residual vector in spite of the fact that it has errors in the second decimal place. The computed deflated matrix has substantial disagreements with the matrix which would have been obtained by exact computation throughout but

nevertheless both λ_2 and λ_3 are correct to working accuracy. The vector u is an exact eigenvector of $\tilde{A}_1$ and exact deflation of $\tilde{A}_1$ gives the computed A_2. Hence the eigenvalues of the computed A_2 are exactly those of $\tilde{A}_1$; we observe that $\tilde{A}_1$ is very close to A_1.

TABLE 1

$$A_1 \qquad\qquad x_1 \quad x_2 \quad x_3$$

$$\begin{bmatrix} 0\cdot987 & 0\cdot400 & -0\cdot487 \\ -0\cdot079 & 0\cdot500 & 0\cdot479 \\ 0\cdot082 & 0\cdot400 & 0\cdot418 \end{bmatrix} \quad \begin{array}{l} \lambda_1 = 0\cdot905 \\ \lambda_2 = 0\cdot900 \\ \lambda_3 = 0\cdot100 \end{array} \quad \begin{array}{ccc} 1\cdot000 & 1\cdot0 & 1\cdot0 \\ 0\cdot143 & 1\cdot0 & -1\cdot0 \\ 0\cdot286 & 1\cdot0 & 1\cdot0 \end{array}$$

$$u_s^T = [\ 1\cdot000 \quad 0\cdot200 \quad 0\cdot333] = \text{computed } x_1.$$
$$\eta_s^T = 10^{-3}[-0\cdot171 \quad -0\cdot498 \quad -0\cdot171] = \text{residual vector.}$$

$$\text{Computed } A_2 = \begin{bmatrix} 0\cdot905 & 0\cdot400 & -0\cdot487 \\ 0 & 0\cdot420 & 0\cdot576 \\ 0 & 0\cdot267 & 0\cdot580 \end{bmatrix} \quad \begin{array}{l} \lambda_2' = 0\cdot900 \\ \lambda_1' = 0\cdot100 \end{array}$$

$$\text{Accurate } A_2 = \begin{bmatrix} 0\cdot905 & 0\cdot400 & -0\cdot487 \\ 0 & 0\cdot443 & 0\cdot549 \\ 0 & 0\cdot286 & 0\cdot557 \end{bmatrix}$$

$$\tilde{A}_1 = \begin{bmatrix} 0\cdot987\ 171 & 0\cdot400\ 000 & -0\cdot487\ 000 \\ -0\cdot078\ 373\ 800 & 0\cdot500\ 000 & 0\cdot478\ 600 \\ 0\cdot082\ 187\ 943 & 0\cdot400\ 200 & 0\cdot417\ 829 \end{bmatrix}$$

Transpose of eigenvector of computed $B_2 = [1\cdot000,\ 0\cdot833] = [x_2^{(2)}]^T$.

$$\text{Computed } A_3 = \begin{bmatrix} 0\cdot905 & 0\cdot400 & -0\cdot487 \\ 0 & 0\cdot900 & 0\cdot576 \\ 0 & 0 & 0\cdot100 \end{bmatrix}$$

$$\begin{array}{cccc} \text{Computed} & x_1 & x_2 & x_3 \\ & 1\cdot000 & 0\cdot750 & 1\cdot000 \\ & 0\cdot200 & 1\cdot000 & -1\cdot000 \\ & 0\cdot333 & 0\cdot958 & 1\cdot000 \end{array}$$

The second eigenvalue is determined from the 2×2 submatrix B_2 of the computed A_2. Since this has two well separated eigenvalues its eigenvector problem *is* well determined and hence iteration does lead to an accurate eigenvector of B_2. The computed A_3 is displayed and finally the three computed eigenvectors of A_1 are given. The computed x_2 is inaccurate but this is to be expected; its inaccuracy cannot be fairly attributed to the deflation. It is no less accurate than the computed x_1 and this was obtained using the original matrix. However, the computed x_1 and x_2 span the subspace accurately. In fact the computed $x_2^{(2)}$ agrees with the correct $x_2^{(2)}$ to working accuracy. This is because the vector in the x_1, x_2 subspace having a zero first component *is* well-determined. On the other hand the x_3 which was obtained using 'highly inaccurate' deflations is correct to working accuracy. This is to be expected since x_3 is a well-conditioned eigenvector and the errors made in the first deflation are merely equivalent to working with $\tilde{A}_1$ instead of A_1.

In recovering x_2 from the eigenvector of B_2 we have

$$x_2 = \begin{bmatrix} 1\cdot000 \\ 0\cdot200 \\ 0\cdot333 \end{bmatrix} + \alpha \begin{bmatrix} 0 \\ 1\cdot000 \\ 0\cdot833 \end{bmatrix}, \qquad (27.1)$$

where α is to be determined by equating the first elements of Ax_2 and $\lambda_2 x_2$. This gives

$$\lambda_2 = \lambda_1 + \alpha(-0\cdot005\ 671), \quad \text{or} \quad -0\cdot005 = \alpha(-0\cdot005\ 671).$$

Observe that α is determined as the ratio of two small quantities. This is invariably the case when A_1 has close eigenvalues but is not close to a matrix having a non-linear divisor. In the limiting case when λ_1 and λ_2 are equal but A_1 has linear divisors, the coefficient of α (that is $a^T x_2^{(2)}$) is exactly zero, showing that α is arbitrary as we might expect. When A_1 has a quadratic divisor, $a^T x_2^{(2)}$ is non-zero and hence effectively $x_2 = x_1$ and we obtain only one eigenvector. The vector $x_2^{(2)}$ gives an independent vector in the subspace of A_1 corresponding to the quadratic divisor.

Stability of unitary transformations

28. As might be expected the unitary deflation of § 25 is also very stable and does not even suffer from the dangers that the elements of the successive deflated matrices might progressively increase in size and the matrices themselves become more ill-conditioned. There are no new points in the analysis. We again observe that λ and u are an exact eigenvalue and eigenvector of $(A_1 - \eta u^T)$, and the general analyses of Chapter 3 then apply immediately. The analysis in the case of fixed-point computation with accumulation of inner-products has been given in detail by Wilkinson (1962b).

For deflation using elementary Hermitians with floating-point accumulation the analysis of Chapter 3, § 45, shows that the computed A_r is exactly similar to $(A_1 + F_r)$ where

$$\|F_r\|_E \leqslant \|\eta_1\|_2 + \|\eta_2\|_2 + \ldots + \|\eta_{r-1}\|_2 + 2(r-1)(1+x)^{2r-4}x\,\|A_1\|_E.$$
$$(28.1)$$

Here $x = (12\cdot36)2^{-t}$ and η_i is the residual vector at the ith stage.

Hence provided we iterate until the residuals are small the eigen-values are preserved (in so far as they are well-conditioned) even if the accepted eigenvectors are poor.

Obviously symmetry in A_1 is preserved. An apparent difficulty in the analysis arises from the fact that $(A_1 - \eta u^T)$, of which u^T is an exact eigenvector, will not in general be symmetric. However, if we take λ to be the Rayleigh quotient (Chapter 3, § 54) corresponding to u we have $\eta^T u = 0$ and hence

$$(A_1 - \eta u^T - u \eta^T) u = \lambda u. \tag{28.2}$$

The matrix in parenthesis *is* exactly symmetric.

Ignoring rounding errors we know that the individual condition numbers of the eigenvalues of the A_i are invariant under the unitary deflations, but this is a little misleading. At each stage we do not work with the full matrix A_i but only with its submatrix B_i where

$$A_i = \left[\begin{array}{c|c} T_{i-1} & C_{i-1} \\ \hline O & B_i \end{array} \right], \tag{28.3}$$

and T_{i-1} is an upper-triangular matrix having the computed eigen-values for its diagonal elements. If λ_k is an eigenvalue of B_i it is easy to show that the individual condition number $\tilde{s}_k$ of λ_k with respect to B_i is not less in modulus than the condition number s_k of λ_k with respect to A_i and therefore with respect to A_1. For if

$$\left[\begin{array}{c|c} T_{i-1} & C_{i-1} \\ \hline O & B_i \end{array} \right] \left[\begin{array}{c} u \\ \hline v \end{array} \right] = \lambda_k \left[\begin{array}{c} u \\ \hline v \end{array} \right], \quad \left[\begin{array}{c|c} T_{i-1}^T & O \\ \hline C_{i-1}^T & B_i^T \end{array} \right] \left[\begin{array}{c} O \\ \hline w \end{array} \right] = \lambda_k \left[\begin{array}{c} O \\ \hline w \end{array} \right] \tag{28.4}$$

then
$$B_i v = \lambda_k v, \qquad B_i^T w = \lambda_k w, \tag{28.5}$$

showing that $|\tilde{s}_k| \geqslant |s_k|$. Hence in general the eigenvalues of B_i are less sensitive than those of A_i; the extra sensitivity of those of A_i arises because of possible perturbations in the remaining three sub-matrices. We did not have quite such a good result in the case of non-unitary deflations, though experiments on ACE have shown that the condition numbers do *usually* show a progressive improvement.

The number of multiplications in the non-unitary deflation is approximately n^2 compared with $4n^2$ in the unitary deflation, while if A_1 is a full matrix there are approximately n^2 multiplications in one iteration. If a considerable number of iterations are used to find each eigenvector the work involved in the deflations may well be quite a small fraction of the total. In this case the argument for using the unitary transformations with their unconditional guarantee of stability is very strong.

Deflation by non-similarity transformations

29. For the case when we have determined a single eigenvector x_1, Wielandt (1944b) has described a general form of deflation which is not a similarity transformation. Suppose u_1 is any vector such that

$$u_1^T x_1 = \lambda_1. \tag{29.1}$$

Consider the matrix A_2 defined by

$$A_2 = A - x_1 u_1^T. \tag{29.2}$$

We have
$$A_2 x_1 = A x_1 - x_1 u_1^T x_1 = \lambda_1 x_1 - \lambda_1 x_1 = 0, \tag{29.3}$$

and hence x_1 is still an eigenvector but the corresponding eigenvalue has become zero. On the other hand

$$\begin{aligned}
A_2(x_i - \alpha_i x_1) &= A x_i - \alpha_i A x_1 - x_1 u_1^T x_i + \alpha_i x_1 u_1^T x_1 \\
&= \lambda_i x_i - x_1 u_1^T x_i \\
&= \lambda_i [x_i - (u_1^T x_i / \lambda_i) x_1] \quad (i = 2, \ldots, n).
\end{aligned} \tag{29.4}$$

Hence taking $\alpha_i = (u_1^T x_i / \lambda_i)$ we see that λ_i has remained an eigenvalue and the corresponding eigenvector is given by

$$x_i^{(2)} = x_i - (u_1^T x_i / \lambda_i) x_1. \tag{29.5}$$

Notice that the n vectors x_1, $x_i^{(2)}$ $(i \geqslant 2)$ are independent. For if

$$\beta_1 x_1 + \sum_{i=2}^{n} \beta_i [x_i - (u_1^T x_i / \lambda_i) x_1] = 0 \tag{29.6}$$

we must have $\beta_i = 0$ $(i = 2, \ldots, n)$ because the x_i are independent, and hence β_1 is also zero.

Hotelling's deflation (§ 19) is a special case of this in which u_1 is taken to be $\lambda_1 y_1$, where y_1 is the left-hand eigenvector normalized so that $y_1^T x_1 = 1$. Since $y_1^T x_i = 0$ $(i \neq 1)$ the eigenvectors are invariant in this case.

30. In the general process we have tacitly assumed that $\lambda_i \neq 0$ and, although this can always be assured by a shift of origin, it is an unsatisfactory feature of the method. Consider the matrix given by

$$A = \begin{bmatrix} 2 & 1 \\ 4 & 2 \end{bmatrix}, \quad \lambda_1 = 4, \quad \lambda_2 = 0, \quad x_1 = \begin{bmatrix} 1 \\ 2 \end{bmatrix}, \quad x_2 \begin{bmatrix} 1 \\ -2 \end{bmatrix}. \tag{30.1}$$

An appropriate vector u is given by $u^T = [4, 0]$ and we have

$$A_2 = \begin{bmatrix} 2 & 1 \\ 4 & 2 \end{bmatrix} - \begin{bmatrix} 1 \\ 2 \end{bmatrix} [4, 0] = \begin{bmatrix} -2 & 1 \\ -4 & 2 \end{bmatrix}. \tag{30.2}$$

The matrix A_2 has the *quadratic divisor* λ^2 and is much more ill-conditioned than A.

Suppose now that v_1 is any vector such that

$$v_1^T x_1 = 1. \tag{30.3}$$

We have $v_1^T A x_1 = v_1^T \lambda_1 x_1 = \lambda_1$, and $v_1^T A$ is an acceptable vector for the u_1^T in relation (29.1). Taking this vector, Wielandt's process gives

$$A_2 = A - x_1 v_1^T A. \tag{30.4}$$

The eigenvectors are now

$$x_i - (v_1^T A x_i / \lambda_i) x_1 = x_i - (v_1^T x_i) x_1 \quad (i = 2, \dots, n), \tag{30.5}$$

and zero values of λ_i are no longer of special significance. It may be verified directly that $x_i - (v_1^T x_i) x_1$ $(i = 2, \dots, n)$ are indeed eigenvectors of A_2. This is clearly not a similarity transformation since in general λ_1 is changed.

If x_1 is normalized so that $\max(x_1) = 1$ and if the unit element is in position $1'$ an obvious choice for v_1 is the vector $e_{1'}$. We then have

$$A_2 = A - x_1 e_{1'}^T A. \tag{30.6}$$

Clearly row $1'$ of A_2 is null and row i of A_2 is obtained from row i of A by subtracting x_i' times row $1'$. Hence $I_{1,1'} A_2 I_{1,1'}$ has the same elements in the matrix of order $(n-1)$ in the bottom right-hand corner as the matrix given by the deflation of § 22. Since A_2 has a null row $1'$ the eigenvectors corresponding to λ_i $(i = 2, \dots, n)$ all have zero components in position $1'$. This is otherwise evident since the relevant eigenvectors of A_2 are given by $x_i - (e_1^T x_i) x_1$. We can clearly omit column $1'$ and row $1'$ when iterating with A_2 and deal only with a matrix of order $(n-1)$.

It will simplify the subsequent discussion if we assume that $1' = 1$; the appropriate modifications when $1' \neq 1$ are trivial. The deflated matrix A_2 is then given by

$$A_2 = A - x_1 e_1^T A = (I - x_1 e_1^T) A. \tag{30.7}$$

We denote the eigenvectors of A_2 by x_1, $x_2^{(2)}$, $x_3^{(2)}, \dots, x_n^{(2)}$, where $x_i^{(2)}$ $(i > 2)$ has a zero first component. When $x_2^{(2)}$ has been found we may perform a second deflation and again assume for simplicity of notation that $2' = 2$. Hence we have

$$A_3 = A_2 - x_2^{(2)} e_2^T A_2 = (I - x_2^{(2)} e_2^T)(I - x_1 e_1^T) A, \tag{30.8}$$

and A_3 has null rows 1 and 2.

In general it is convenient to perform the deflations explicitly to obtain A_2, A_3,... or at least the relevant submatrices of respective orders $n-1$, $n-2$,... which are used in later iterations. If A is very sparse it may be uneconomic to do this since A_2 and subsequent matrices will not in general be so sparse. If just a few eigenvectors of a large sparse matrix are required it will usually be better to work with the deflated matrices in the factorized form of which (30.8) is typical. When iterating with A_3 for example we know that every relevant eigenvector has its first two components equal to zero. Hence we may take our initial vector u_0 to be of this form and u_s will remain of this form throughout. If we write $Au_s = p_{s+1}$ then we compute first q_{s+1} defined by

$$q_{s+1} = (I - x_1 e_1^T)p_{s+1} = p_{s+1} - (e_1^T p_{s+1})x_1. \tag{30.9}$$

Clearly q_{s+1} is obtained from p_{s+1} by subtracting the appropriate multiple of x_1 so as to eliminate the first element. We then have

$$(I - x_2^{(2)} e_2^T)q_{s+1} = q_{s+1} - (e_2^T q_{s+1})x_2^T = v_{s+1} \quad \text{(say)}, \tag{30.10}$$

so that v_{s+1} is the unnormalized vector obtained from q_{s+1} by subtracting a multiple of $x_2^{(2)}$ so as to eliminate the second element. This process may be used at the general stage when we have A_r. When $n = 6$ and $r = 4$ for example, we have

$$[x_1 \mid x_2^{(2)} \mid x_3^{(3)}] = \begin{bmatrix} 1 & 0 & 0 \\ \times & 1 & 0 \\ \times & \times & 1 \\ \times & \times & \times \\ \times & \times & \times \\ \times & \times & \times \end{bmatrix}, \quad u_s = \begin{bmatrix} 0 \\ 0 \\ 0 \\ 1 \\ \times \\ \times \end{bmatrix},$$

$$Au_s = \begin{bmatrix} \times \\ \times \\ \times \\ \times \\ \times \\ \times \end{bmatrix}, \quad v_{s+1} = \begin{bmatrix} 0 \\ 0 \\ 0 \\ \times \\ \times \\ \times \end{bmatrix}, \tag{30.11}$$

and Au_s must be reduced to the form v_{s+1} by an elimination process which operates column-wise rather than row-wise. The vector v_{s+1} is then normalized to give u_{s+1}. We shall see in § 34 that this is almost exactly the process of treppen-iteration introduced by F. L. Bauer.

General reduction using invariant subspaces

31. If X is an invariant subspace with m columns such that

$$AX = XM, \tag{31.1}$$

where M is an $m \times m$ matrix and if P is any $n \times m$ matrix such that

$$P^T X = M, \tag{31.2}$$

then we have a generalized form of Wielandt's deflation in which

$$A_2 = A - X P^T. \tag{31.3}$$

Clearly $\quad A_2 X = AX - X P^T X = XM - XM = 0, \tag{31.4}$

so that the m independent columns of X are all eigenvectors of A_2 corresponding to the eigenvalue zero. (This is true even if X corresponds to an elementary divisor of degree m.) The eigenvalues of A_2 are otherwise the same as those of A, as we now show. The proof depends on the simple identity

$$\det(I_n - RS) = \det(I_m - SR) \tag{31.5}$$

which is true when R and S are arbitrary $n \times m$ and $m \times n$ matrices respectively. Equation (31.5) is an immediate consequence of the fact that the non-zero eigenvalues of RS are the same as those of SR (Chapter 1, § 51). We see further from (31.1) that

$$\lambda X - AX = \lambda X - XM, \tag{31.6}$$

giving

$$(\lambda I_n - A)X = X(\lambda I_m - M), \quad \text{or} \quad X(\lambda I_m - M)^{-1} = (\lambda I_n - A)^{-1} X. \tag{31.7}$$

We now have

$$\begin{aligned}
\det(\lambda I_n - A_2) &= \det(\lambda I_n - A + X P^T) \\
&= \det(\lambda I_n - A) \det[I_n + (\lambda I_n - A)^{-1} X P^T] \\
&= \det(\lambda I_n - A) \det[I_n + X(\lambda I_m - M)^{-1} P^T] \quad \text{from (31.7)} \\
&= \det(\lambda I_n - A) \det[I_m + P^T X(\lambda I_m - M)^{-1}] \quad \text{from (31.5)} \\
&= \det(\lambda I_n - A) \det[I_m + M(\lambda I_m - M)^{-1}] \quad \text{from (31.2)} \\
&= \det(\lambda I_n - A) \lambda^m \det(\lambda I_m - M)^{-1}, \tag{31.8}
\end{aligned}$$

showing that A_2 has the eigenvalues of A except that those belonging to M are replaced by zeros.

32. Householder (1961) discusses the relations between the invariant subspaces of A_2 and A. These are the generalizations of equation (29.5) and they suffer defects comparable with the presence of the divisor λ_i in that equation. We shall consider only the case which is used in practice, for which somewhat stronger results may be obtained more simply than in the general case.

As a natural extension of equation (30.3) let Q be any matrix such that
$$Q^T X = I_m. \tag{32.1}$$
Then from (31.1) we have
$$Q^T A X = Q^T X M = M, \tag{32.2}$$
and hence $Q^T A$ is an appropriate matrix P^T in equation (31.2). Accordingly we have in this case
$$A_2 = A - X Q^T A = (I - X Q^T) A, \qquad A_2 X = 0. \tag{32.3}$$
Now suppose that Y consists of a set of p vectors independent of those of X but such that
$$A Y = [X \mid Y] \left[\frac{S}{N} \right] = [X \mid Y] W \quad \text{(say)}. \tag{32.4}$$
(Notice that Y need not itself be an invariant subspace of A but (31.1) and (32.4) imply that $[X \mid Y]$ *is* an invariant subspace.) Consider now $A_2(Y - XZ)$ where Z is any matrix such that $(Y - XZ)$ exists. Since $A_2 X = 0$ we have
$$\begin{aligned}
A_2(Y - XZ) = A_2 Y &= A Y - X Q^T A Y \\
&= [X \mid Y] W - X Q^T [X \mid Y] W \\
&= [X \mid Y] W - [X \mid X Q^T Y] W \\
&= (Y - X Q^T Y) N. \tag{32.5}
\end{aligned}$$
If we take $Z = Q^T Y$, equation (32.5) shows that $Y - X(Q^T Y)$ is an invariant subspace of A_2. The independence of X and $Y - X(Q^T Y)$ follows immediately from that of X and Y. The significance of this result becomes more apparent if we consider a particular application. Suppose A has a quartic elementary divisor $(\lambda - \lambda_1)^4$ and that we have found vectors x_1 and x_2 of grades one and two but not x_3 and x_4. If we deflate, by (31.4) the matrix A_2 now has two linear divisors λ and λ corresponding to x_1 and x_2, and a quadratic divisor $(\lambda - \lambda_1)^2$ having vectors of grades one and two respectively corresponding to the reduced forms of x_3 and x_4. In practice, however, iteration with A is more likely to provide the whole space spanned by x_1, x_2, x_3, x_4 than the subspace spanned by x_1 and x_2 only.

Practical application

33. The only restriction on the matrix Q is that it should satisfy equation (32.1). In theory this gives considerable freedom of choice but in practice one chooses Q in such a way as to minimize the computation. If we write $Q^T = [J \mid K]$ and $X^T = [U^T \mid V^T]$ where J and U are $m \times m$ matrices, then probably the simplest choice of Q is given by

$$J = U^{-1}, \qquad K = 0. \tag{33.1}$$

If we write

$$A = \begin{bmatrix} B & C \\ \hline D & E \end{bmatrix} \quad \text{then} \quad A_2 = \begin{bmatrix} B & C \\ \hline D & E \end{bmatrix} - \begin{bmatrix} U \\ \hline V \end{bmatrix} [U^{-1} \mid O] \begin{bmatrix} B & C \\ \hline D & E \end{bmatrix},$$

giving $\tag{33.2}$

$$A_2 = \begin{bmatrix} O & O \\ \hline D - VU^{-1}B & E - VU^{-1}C \end{bmatrix}, \tag{33.3}$$

so that A_2 is null in rows 1 to m. The remaining eigenvalues are those of $(E - VU^{-1}C)$.

In practice we would not take the first m rows of X but would use pivoting to select the relevant m rows. When this is done it may be verified that the matrix $(E - VU^{-1}C)$ (which is the part of A_2 which is actually used in subsequent iteration) is the same as the corresponding part of the deflated matrix obtained by the technique of § 23.

However, it is more convenient to reduce the invariant subspace X to *standard trapezoidal form* before carrying out the deflation. This form is obtained by performing the relevant part of a triangular decomposition with partial pivoting. Thus if we ignore the pivoting for convenience of notation, we may write

$$X = TR, \tag{33.4}$$

where typically X, T and R are of the forms

$$X = \begin{bmatrix} \times & \times & \times \\ \times & \times & \times \\ \times & \times & \times \\ \times & \times & \times \\ \times & \times & \times \\ \times & \times & \times \end{bmatrix}, \quad T = \begin{bmatrix} 1 & 0 & 0 \\ \times & 1 & 0 \\ \times & \times & 1 \\ \times & \times & \times \\ \times & \times & \times \\ \times & \times & \times \end{bmatrix}, \quad R = \begin{bmatrix} \times & \times & \times \\ & \times & \times \\ & & \times \end{bmatrix}.$$

$$\tag{33.5}$$

T is a *unit lower trapezoidal* $n \times m$ matrix and R is an upper-triangular $m \times m$ matrix. Obviously if X is an invariant subspace the same is true of T, so that T may be used in place of X throughout. Since the submatrix of T corresponding to U is now a unit lower-triangular $m \times m$ matrix the computation of $(E - VU^{-1}C)$ is particularly simple.

Again if A is sparse we may use (33.2) in the form

$$A_2 = \left\{ I - \left[\begin{array}{c|c} I & O \\ \hline VU^{-1} & O \end{array} \right] \right\} A, \qquad (33.6)$$

if it is more economical. We know from (33.3) that relevant invariant subspaces Y of A_2 have their first m rows null and hence when iterating with A_2 we may start with our initial Y_0 in this form and all Y_s will retain this form. If we write

$$AY_s = \left[\begin{array}{c} K_s \\ M_s \end{array} \right], \qquad (33.7)$$

then $A_2 Y_s$ is obtained from AY_s by subtracting multiples of the column of T so as to eliminate the first m rows. This is particularly easy to do because T is of unit trapezoidal form.

Treppen-iteration

34. One difficulty in the use of iteration and deflation is that in general we shall not know at each stage whether the current dominant eigenvalue is real or complex or whether it corresponds to a non-linear divisor. We can overcome this difficulty by working with a complete set of n vectors simultaneously and can ensure that they are independent by taking the n vectors to be the columns of a unit lower-triangular matrix. If at each stage we denote the matrix formed by the set of n vectors by L_s, the process becomes

$$X_{s+1} = AL_s, \qquad X_{s+1} = L_{s+1}R_{s+1}, \qquad (34.1)$$

where each L_s is unit lower-triangular and each R_s is upper-triangular. If we take L_0 to be I then we have

$$L_s R_s = X_s = AL_{s-1}, \qquad (34.2)$$

$$L_s R_s R_{s-1} = AL_{s-1}R_{s-1} = AAL_{s-2} = A^2 L_{s-2}, \qquad (34.3)$$

$$L_s(R_s R_{s-1} \ldots R_1) = A^s L_0 = A^s. \qquad (34.4)$$

Hence L_s and $R_s R_{s-1} \ldots R_1$ are the matrices obtained by the triangular decomposition of A^s so that from Chapter 8, § 4, L_s is equal to the product of the first s lower-triangular matrices obtained in the LR algorithm while the R_s are identical with the individual upper-triangular matrices in the same algorithm. We know therefore (Chapter 8, § 6) that if, for example, all eigenvalues are of distinct moduli, then $L_s \rightarrow L$, the matrix obtained by the triangular decomposition of the matrix of right-hand eigenvectors of A. One should not assume that this formal relationship with LR implies that this technique will be as useful in practice as the LR algorithm.

The practical LR technique depends critically for its success on a number of factors. Of these the invariance of the upper Hessenberg form (even when interchanges are used), the effectiveness of the shifts of origin and the stability of the deflation process are the most important. If in treppen-iteration we take A to be upper Hessenberg and L_0 to be I, then although L_1 has only $(n-1)$ non-zero sub-diagonal elements, the number of such elements in successive L_i progressively increases until with L_{n-1} we have a full lower-triangular matrix, and from then on $A L_i$ is a full matrix.

If on the other hand we take A to be lower Hessenberg, $A L_i$ is also of lower Hessenberg form at each stage, but although there are only $\frac{1}{2} n^2$ multiplications in the triangular decomposition of $A L_i$, the computation of $A L_i$ requires $\frac{1}{6} n^3$ multiplications. The introduction of row interchanges in the decomposition of $A L_i$ destroys the advantages of the Hessenberg form but column interchanges do not, though as with the LR algorithm the introduction of interchanges destroys the theoretical justification of the process. Acceleration of convergence may be achieved by operating with $(A - pI)$ instead of A, and p may be varied from iteration to iteration, but the absence of a really effective deflation process makes this much less satisfactory than in the LR technique. When a small eigenvalue λ_i of A has been determined we cannot choose p freely since this may make $(\lambda_i - p)$ a dominant eigenvalue of A.

Obviously we need not use a complete set of n vectors. We can work with a block of m vectors in unit lower-trapezoidal form. Denoting these by T_s the process becomes

$$AT_s = X_{s+1}, \qquad X_{s+1} = T_{s+1} R_{s+1}, \qquad (34.5)$$

where R_{s+1} is an $m \times m$ upper-triangular matrix. If the dominant m eigenvalues of A have distinct moduli, $T_s \rightarrow T$ where T is obtained by trapezoidal decomposition of the m dominant eigenvectors. More generally if

$$|\lambda_1| > |\lambda_2| > \ldots > |\lambda_m| > |\lambda_{m+1}|, \tag{34.6}$$

T_s does not necessarily tend to a limit but it does tend to an invariant subspace (Chapter 8, §§ 32, 34). This process was first described by F. L. Bauer (1957) and was called *treppen-iteration* or *staircase iteration*.

If $|\lambda_1| \gg |\lambda_2|$ the first column of T_s will converge to x_1 in a few iterations and at this stage there is little point in including x_1 in the matrix T_s when computing X_{s+1}. In general if the first k vectors of T_s have converged we need not multiply these vectors by A in subsequent steps. We may write

$$T_s = [T_s^{(1)} \mid T_s^{(2)}], \tag{34.7}$$

where $T_s^{(1)}$ consists of the first k vectors, which have converged, and $T_s^{(2)}$ consists of the remaining $(m-k)$ which have not. We now define X_{s+1} by the relation

$$X_{s+1} = [T_s^{(1)} \mid A T_s^{(2)}], \tag{34.8}$$

of which $T_s^{(1)}$ is already in trapezoidal form but $A T_s^{(2)}$ in general consists of $(m-k)$ full vectors. X_{s+1} is then reduced to trapezoidal form in the obvious way. It is immediately apparent that the process we are using is identical with that discussed in § 30.

It should be emphasized that it is not necessarily the earlier columns which converge first. For example if the four dominant eigenvalues of A are 1, 0·99, 10^{-2}, 10^{-3} the first two columns will soon span the subspace corresponding to x_1 and x_2 and hence columns 3 and 4 will converge. The separation of columns 1 and 2 into the appropriate eigenvectors will take place more slowly. If the dominant eigenvalue corresponds to a non-linear divisor this effect may be even more marked. A numerical example is given in § 41.

Accurate determination of complex conjugate eigenvalues

35. The discussion of the last few sections suggests a method for the determination of complex conjugate eigenvalues of a real matrix A which does not suffer from any loss of accuracy when the corresponding eigenvalues have small imaginary parts (except, of course, when A is close to a matrix having a quadratic elementary divisor). We describe the process first without the inclusion of a pivotal strategy. At each stage we iterate with a pair of real vectors a_s and b_s which are in standard trapezoidal form, so that

$$a_s^T = (1, \alpha_s, \times, \ldots, \times, \times) \\ b_s^T = (0, 1, \times, \ldots, \times, \times) \Big\}. \tag{35.1}$$

We compute c_{s+1} and d_{s+1} given by

$$c_{s+1} = A a_s, \qquad d_{s+1} = A b_s, \tag{35.2}$$

and write

$$c_{s+1}^T = (\beta_{s+1}, \gamma_{s+1}, \times, \ldots, \times, \times) \\ d_{s+1}^T = (\delta_{s+1}, \epsilon_{s+1}, \times, \ldots, \times, \times) \Big\}. \tag{35.3}$$

The next vectors a_{s+1}, b_{s+1} are obtained by reducing (c_{s+1}, d_{s+1}) to standard trapezoidal form.

If the dominant eigenvalues of A are a complex conjugate pair $(\xi_1 \pm i\eta_1,)$ then ultimately

$$[c_{s+1} \,|\, d_{s+1}] = [a_s \,|\, b_s] M_s, \tag{35.4}$$

where M_s is a 2×2 matrix which varies with s but has fixed eigen-values $(\xi_1 \pm i\eta_1)$. At each stage we determine the matrix M_s such that

$$\begin{bmatrix} 1 & 0 \\ \alpha_s & 1 \end{bmatrix} M_s = \begin{bmatrix} \beta_{s+1} & \delta_{s+1} \\ \gamma_{s+1} & \epsilon_{s+1} \end{bmatrix}, \tag{35.5}$$

and obtain the current approximations to $\xi_1 \pm i\eta_1$ from the eigen-values of M_s.

It might be thought that if η_1 is small we shall still lose accuracy when we compute the eigenvalues of M_s, but this will not happen if $(\xi_1 \pm i\eta_1)$ are well-conditioned eigenvalues of A. In this case it will be found that $m_{11}^{(s)}$ and $m_{22}^{(s)}$ will be close to ξ_1 and $m_{12}^{(s)}$ and $m_{21}^{(s)}$ will be small. In fact in the limit when η_1 tends to zero both $m_{12}^{(s)}$ and $m_{21}^{(s)}$ become zero provided A does not have the quadratic divisor $(\lambda - \xi_1)^2$. The vectors a_s and b_s or c_{s+1} and d_{s+1} ultimately determine the invari-ant subspace accurately but we have seen in § 14 that the determina-tion of the two individual eigenvectors *is* an ill-conditioned problem. If the eigenvector u_s of M_s corresponding to $\xi_1 + i\eta_1$ is given by

$$u_s^T = (\varphi_s, \psi_s), \tag{35.6}$$

then the corresponding eigenvector of A is given by $(\varphi_s a_s + \psi_s b_s)$. Naturally u_s should be normalized so that the larger of $|\varphi_s|$ and $|\psi_s|$ is unity.

For numerical stability pivoting is essential in the reduction of $[c_{s+1} \,|\, d_{s+1}]$ to trapezoidal form. The unit elements in a_s and b_s will not in general be in the first two positions and moreover the positions will vary with s. Notice, however, that the matrix M_s is determined from the *four elements of* c_{s+1} *and* d_{s+1} *which are in the same positions as the unit elements in* a_s *and* b_s.

This technique involves about twice as much computation as that of § 12 in which we were attempting to determine $(\xi_1 \pm i\eta_1)$ from a single sequence of iterated vectors.

Very close eigenvalues

36. An exactly analogous technique may be used to determine accurately a group of poorly separated dominant eigenvalues provided, of course, they are well-conditioned. Suppose, for example, a real matrix A has dominant eigenvalues λ_1, $\lambda_1 - \epsilon_1$, $\lambda_1 - \epsilon_2$, where the ϵ_i are small. If we iterate using a trapezoidal set of vectors a_s, b_s, c_s and write

$$[d_{s+1} \mid e_{s+1} \mid f_{s+1}] = A[a_s \mid b_s \mid c_s], \qquad (36.1)$$

then as in the treatment in § 35 we can determine a 3×3 matrix M_s from the appropriate elements of a_s, b_s, c_s and d_{s+1}, e_{s+1}, f_{s+1}. The eigenvalues of M_s will tend to λ_1, $\lambda_1 - \epsilon_1$, $\lambda_1 - \epsilon_2$ respectively. If the eigenvalues of A are not ill-conditioned the eigenvalues of M_s will be determined accurately. In fact ultimately the diagonal elements of M will be close to λ_1 and all off-diagonal elements will be of the same order of magnitude as the ϵ_i.

Orthogonalization techniques

37. We now turn to an alternative technique for suppressing the dominant eigenvector (or eigenvectors). The simplest example of its use is in connexion with real symmetric matrices. Suppose λ_1 and x_1 have been determined so that $Ax_1 = \lambda_1 x_1$. We know that the remaining eigenvectors are orthogonal to x_1 and hence we may suppress the component of x_1 in any vector by orthogonalizing it with respect to x_1. This gives us the iterative procedure defined by the relations

$$v_{s+1} = Au_s, \qquad w_{s+1} = v_{s+1} - (v_{s+1}^T x_1) x_1, \qquad u_{s+1} = w_{s+1} / \max(w_{s+1}), \qquad (37.1)$$

where it is assumed that $\|x_1\|_2 = 1$. Obviously u_s tends to the subdominant eigenvector or, if A has a second eigenvalue equal to λ_1, it tends to an eigenvector corresponding to λ_1 which is orthogonal to x_1.

Obviously unless $|\lambda_1| \gg |\lambda_2|$ it will not strictly be necessary to orthogonalize with respect to x_1 in every iteration but if A is of high order the work involved in the orthogonalization is in any case slight.

We may generalize this process to find x_{r+1} when $x_1, x_2, ..., x_r$ have been determined. The corresponding iteration is defined by

$$v_{s+1} = Au_s, \qquad w_{s+1} = v_{s+1} - \sum_1^r (v_{s+1}^T x_i)x_i, \qquad u_{s+1} = w_{s+1}/\max(w_{s+1}).$$
$$(37.2)$$

Note that apart from rounding errors this method gives results identical with those of Hotelling (§ 19) since from (37.1) we have

$$\begin{aligned} w_{s+1} = v_{s+1} - (v_{s+1}^T x_1)x_1 &= Au_s - x_1(u_s^T A x_1) \\ &= Au_s - \lambda_1 x_1(u_s^T x_1) \\ &= Au_s - \lambda_1 x_1(x_1^T u_s) \\ &= (A - \lambda_1 x_1 x_1^T)u_s. \end{aligned} \qquad (37.3)$$

There is an analogous process which may be used when A is unsymmetric, but this requires the computation of both the left-hand eigenvector y_1 and the right-hand eigenvector x_1. If these are normalized so that $x_1^T y_1 = 1$ we may use the iterative procedure

$$v_{s+1} = Au_s, \quad w_{s+1} = v_{s+1} - (v_{s+1}^T y_1)x_1, \quad u_{s+1} = w_{s+1}/\max(w_{s+1}). \quad (37.4)$$

Because of the bi-orthogonality of the left-hand and right-hand eigenvectors the component of x_1 in w_{s+1} is suppressed. Again we have the obvious generalization when r eigenvectors have been determined.

Analogue of treppen-iteration using orthogonalization

38. In treppen-iteration we iterate simultaneously on a number of vectors and their independence is maintained by reducing them to standard trapezoidal form at each stage. There is an analogous procedure in which independence of the vectors is maintained by using the Schmidt orthogonalization procedure (Chapter 4, § 54) or its equivalent. We consider first the case when we iterate with n vectors simultaneously. If we denote the matrix formed by these vectors at each stage by Q_s then the process becomes

$$AQ_s = V_{s+1}, \qquad V_{s+1} = Q_{s+1}R_{s+1}, \qquad (38.1)$$

where the columns of Q_{s+1} are orthogonal and R_{s+1} is upper-triangular. Hence

$$AQ_s = Q_{s+1}R_{s+1}, \qquad (38.2)$$

and if we take $Q_0 = I$ we have

$$A^s = Q_s(R_s R_{s-1} ... R_1). \qquad (38.3)$$

Equation (38.3) implies that Q_s and $(R_s R_{s-1} \dots R_1)$ are the matrices obtained by the orthogonal triangularization of A^s. Comparison with the QR algorithm (Chapter 8, § 28) shows that Q_s is equal to the product of the first s orthogonal matrices determined by the QR algorithm. Hence, for example, if the $|\lambda_i|$ are all different, Q_s essentially tends to a limit, this being the matrix obtained from the triangular orthogonalization of X, the matrix of eigenvectors. The diagonal elements of R_s tend to $|\lambda_i|$. In the more general case, where some of the eigenvalues are of the same modulus, the corresponding columns of Q_s may not tend to a limit but they will ultimately span the appropriate subspaces. As in § 34 we warn the reader that the formal equivalence of the method of this section with the QR algorithm does not imply that it leads to an equally valuable practical procedure.

The invariance of the upper Hessenberg form and an automatic and stable method of deflation were essential to the success of the QR algorithm and we do not have anything of comparable convenience in the orthogonalization technique. (Compare the corresponding discussion of treppen-iteration at the end of § 34.)

39. We can use Householder's process for the orthogonal triangularization (Chapter 4, § 46) in which case the transpose Q_s^T rather than Q_s itself is obtained at each stage as the product $P_{n-1}^{(s)} P_{n-2}^{(s)} \dots P_1^{(s)}$ of $(n-1)$ elementary Hermitians. In these terms the process becomes

$$A P_1^{(s)} P_2^{(s)} \dots P_{n-1}^{(s)} = V_{s+1}, \qquad P_{n-1}^{(s+1)} \dots P_2^{(s+1)} P_1^{s+1} V_{s+1} = R_{s+1}. \quad (39.1)$$

If all the $|\lambda_i|$ are distinct the diagonal elements of $|R_s|$ converge and at this stage we can compute the current Q_s from the $P_i^{(s)}$; we do not need to compute Q_s at the earlier stages. When some of the $|\lambda_i|$ are equal the corresponding elements of $|R_s|$ may not converge and it will not be possible to recognize when the 'limit' has been reached in the sense that groups of columns of Q_s span the appropriate subspaces.

We have described the case when Q_s has the full number of columns in order to emphasize the relationship with the QR algorithm. In fact there is no reason why we should not work with a smaller number of columns. If Q_s has r columns the process is still described by equation (37.1) but R_s is now an upper-triangular $r \times r$ matrix. A case of special interest arises when $r = 2$ and the dominant eigenvalues are a complex conjugate pair $(\xi_1 \pm i\eta_1)$. The columns of Q_s then ultimately consist of two orthogonal vectors spanning the relevant subspace. The method can be used to determine accurately both

$(\xi_1 + i\eta_1)$ and the invariant subspace even when η_1 is small. At each stage we write

$$V_{s+1} = Q_s M_s, \tag{39.2}$$

where M_s is a 2×2 matrix. From the orthogonality of the columns of Q_s we have

$$M_s = Q_s^T V_{s+1}. \tag{39.3}$$

The eigenvalues of M_s tend to $(\xi_1 \pm i\eta_1)$ and when η_1 is small but the eigenvalues are well-conditioned it will be found that $m_{12}^{(s)}$ and $m_{21}^{(s)}$ are of order η_1 so that $(\xi_1 \pm i\eta_1)$ will be accurately determined.

Bi-iteration

40. Bauer (1958) has suggested a generalization of the method of § 36 which he has called *bi-iteration*; in this all the x_i and y_i may be found simultaneously. At each stage two systems of n vectors are retained comprising the columns of two matrices X_s and Y_s respectively. The two matrices P_{s+1} and Q_{s+1} defined by

$$P_{s+1} = AX_s, \qquad Q_{s+1} = A^T Y_s \tag{40.1}$$

are formed, and from them two matrices X_{s+1} and Y_{s+1} are derived; the columns of these matrices are chosen so as to be bi-orthogonal. The equations defining the ith columns of X_{s+1} and Y_{s+1} are therefore

$$\left. \begin{array}{l} x_i^{(s+1)} = p_i^{(s+1)} - r_{1i} x_1^{(s+1)} - r_{2i} x_2^{(s+1)} - \ldots - r_{i-1,i} x_{i-1}^{(s+1)} \\ y_i^{(s+1)} = q_i^{(s+1)} - u_{1i} y_1^{(s+1)} - u_{2i} y_2^{(s+1)} - \ldots - u_{i-1,i} y_{i-1}^{(s+1)} \end{array} \right\}, \tag{40.2}$$

where the r_{ki} and u_{ki} are chosen so as to make

$$(y_k^{(s+1)})^T x_i^{(s+1)} = 0, \qquad (x_k^{(s+1)})^T y_i^{(s+1)} = 0 \quad (k = 1, \ldots, i-1). \tag{40.3}$$

Clearly we have

$$P_{s+1} = X_{s+1} R_{s+1}, \qquad Q_{s+1} = Y_{s+1} U_{s+1}, \tag{40.4}$$

where R_{s+1} and U_{s+1} are the unit upper-triangular matrices formed from the r_{ki} and u_{ki}. From the bi-orthogonality we have

$$Y_{s+1}^T X_{s+1} = D_{s+1}, \tag{40.5}$$

where D_{s+1} is a diagonal matrix. If we take $X_0 = Y_0 = I$ we have

$$\left. \begin{array}{l} A = X_1 R_1, \qquad AX_s = X_{s+1} R_{s+1} \\ A^T = Y_1 U_1, \qquad A^T Y_s = Y_{s+1} U_{s+1} \end{array} \right\}, \tag{40.6}$$

giving

$$A^s = X_s R_s \ldots R_2 R_1, \qquad (A^T)^s = Y_s U_s \ldots U_2 U_1. \tag{40.7}$$

Hence
$$A^{2s} = (Y_s U_s \dots U_2 U_1)^T X_s R_s \dots R_2 R_1$$
$$= (U_1^T U_2^T \dots U_s^T) D_s (R_s \dots R_2 R_1), \tag{40.8}$$

showing that the unit lower-triangular matrix $U_1^T U_2^T \dots U_s^T$ is that corresponding to the triangular decomposition of A^{2s}. Now we have seen that in the LR algorithm, $L_1 L_2 \dots L_{2s}$ is the matrix obtained by the triangular decomposition of A^{2s} (Chapter 8, § 4) and hence

$$U_s^T = L_{2s-1} L_{2s}. \tag{40.9}$$

Results from the LR algorithm therefore carry over immediately to bi-iteration. In particular if the $|\lambda_i|$ are distinct $U_1^T U_2^T \dots U_s^T$ tends to the matrix given by the triangular decomposition of X and hence $U_s \to I$. Similarly $R_s \to I$. In practice the columns of X and Y are normalized at each stage, usually so that $\max(x_i^{(s)}) = \max(y_i^{(s)}) = 1$. Hence in the case of distinct $|\lambda_i|$, $X_s \to X$ and $Y_s \to Y$. In one step of bi-iteration convergence proceeds as far as in two steps of LR. When some of the $|\lambda_i|$ are equal the corresponding columns of X_s and Y_s tend to the relevant invariant subspaces. If A is symmetric the two systems X_s and Y_s are identical and the method becomes essentially that of § 37, since the columns of X_s itself are made orthogonal at each stage. This corresponds to the fact that two steps of the LR algorithm become equivalent to one of the QR algorithm for symmetric matrices.

We have dealt with the case when the X_s and Y_s consist of complete systems of n vectors in order to show the relationship with the LR and QR algorithms, but the process can obviously be used when X_s and Y_s have any number r of columns from 1 to n. The same equations apply but now R_s and U_s are $r \times r$ matrices. In general the process will provide the r dominant left-hand and right-hand eigenvectors (or invariant subspaces).

Numerical example

41. In order to draw attention to some of the problems which arise in the use of the methods of the previous section we give a simple numerical example. In Table 2 we show the application of treppen-iteration (§ 34) and of the orthogonalization technique of § 38. We have taken A to be a 3×3 matrix having the elementary divisors $(\lambda + 1)^2$ and $(\lambda + 0.1)$.

TABLE 2

A

$$\begin{bmatrix} -2\cdot1 & -1\cdot2 & -0\cdot1 \\ 1 & 0 & 0 \\ 0 & 1 & 0 \end{bmatrix} \quad \begin{array}{l} \lambda_1 = \lambda_2 = -1 \\ \lambda_3 = -0\cdot1 \end{array} \quad \text{Elementary divisors } (\lambda+1)^2, (\lambda+0\cdot1)$$

	x_1	x_2	x_3	y_1	y_2	y_3
	1	-2	0·01	1·0	1	1
	-1	1	$-0\cdot1$	1·1	0·1	2
	1	0	1	0·1	0	1

x_1 and x_2 form an invariant subspace of A. x_1 is the eigenvector.
y_1 and y_2 form an invariant subspace of A^T. y_1 is the eigenvector.

Treppen-iteration

$$\qquad AI \qquad\qquad\qquad L_1 \qquad\qquad\qquad R_1$$

$$\begin{bmatrix} -2\cdot1 & -1\cdot2 & -0\cdot1 \\ 1 & 0 & 0 \\ 0 & 1 & 0 \end{bmatrix} \rightarrow \begin{bmatrix} 1 & \cdot & \cdot \\ -0\cdot476 & -0\cdot571 & 1 \\ 0 & 1 & \cdot \end{bmatrix} \begin{bmatrix} -2\cdot1 & -1\cdot2 & -0\cdot1 \\ \cdot & \cdot & -0\cdot048 \\ \cdot & 1 & 0 \end{bmatrix}$$

Note that pivotal rows are in positions 1, 3, 2 respectively and remain so.

$$\qquad AL_1 \qquad\qquad\qquad L_2 \qquad\qquad\qquad R_2$$

$$\begin{bmatrix} -1\cdot529 & 0\cdot585 & -1\cdot200 \\ 1\cdot000 & 0 & 0 \\ -0\cdot476 & -0\cdot571 & 1\cdot000 \end{bmatrix} \rightarrow \begin{bmatrix} 1 & \cdot & \cdot \\ -0\cdot654 & -0\cdot509 & 1 \\ 0\cdot311 & 1 & \cdot \end{bmatrix} \begin{bmatrix} -1\cdot529 & 0\cdot585 & -1\cdot200 \\ \cdot & \cdot & -0\cdot086 \\ \cdot & -0\cdot753 & 1\cdot373 \end{bmatrix}$$

$$\qquad AL_2 \qquad\qquad\qquad L_3 \qquad\qquad\qquad R_3$$

$$\begin{bmatrix} -1\cdot346 & 0\cdot511 & -1\cdot200 \\ 1\cdot000 & 0 & 0 \\ -0\cdot654 & -0\cdot509 & 1\cdot000 \end{bmatrix} \rightarrow \begin{bmatrix} 1 & \cdot & \cdot \\ -0\cdot743 & -0\cdot502 & 1 \\ 0\cdot486 & 1 & \cdot \end{bmatrix} \begin{bmatrix} -1\cdot346 & 0\cdot511 & -1\cdot200 \\ \cdot & \cdot & -0\cdot097 \\ \cdot & -0\cdot757 & 1\cdot583 \end{bmatrix}$$

$$\qquad AL_3 \qquad\qquad\qquad L_4 \qquad\qquad\qquad R_4$$

$$\begin{bmatrix} -1\cdot257 & 0\cdot502 & -1\cdot200 \\ 1\cdot000 & 0 & 0 \\ -0\cdot743 & -0\cdot502 & 1\cdot000 \end{bmatrix} \rightarrow \begin{bmatrix} 1 & \cdot & \cdot \\ -0\cdot796 & -0\cdot500 & 1 \\ 0\cdot591 & 1 & \cdot \end{bmatrix} \begin{bmatrix} -1\cdot257 & 0\cdot502 & -1\cdot200 \\ \cdot & \cdot & -0\cdot101 \\ \cdot & -0\cdot799 & 1\cdot709 \end{bmatrix}$$

Orthogonalization process

$$\qquad AI \qquad\qquad\qquad Q_1 \qquad\qquad\qquad R_1$$

$$\begin{bmatrix} -2\cdot1 & -1\cdot2 & -0\cdot1 \\ 1 & 0 & 0 \\ 0 & 1 & 0 \end{bmatrix} = \begin{bmatrix} -0\cdot903 & -0\cdot196 & -0\cdot395 \\ 0\cdot430 & -0\cdot414 & -0\cdot789 \\ 0 & -0\cdot889 & -0\cdot474 \end{bmatrix} \begin{bmatrix} 2\cdot326 & 1\cdot084 & 0\cdot090 \\ \cdot & 1\cdot125 & 0\cdot020 \\ \cdot & \cdot & 0\cdot038 \end{bmatrix}$$

$$\qquad AQ_2 \qquad\qquad\qquad Q_2 \qquad\qquad\qquad R_2$$

$$\begin{bmatrix} 1\cdot380 & 0\cdot820 & 1\cdot824 \\ -0\cdot903 & -0\cdot196 & -0\cdot395 \\ 0\cdot430 & -0\cdot414 & -0\cdot789 \end{bmatrix} = \begin{bmatrix} 0\cdot810 & 0\cdot425 & 0\cdot402 \\ -0\cdot530 & 0\cdot235 & 0\cdot816 \\ 0\cdot252 & -0\cdot875 & 0\cdot425 \end{bmatrix} \begin{bmatrix} 1\cdot704 & 0\cdot664 & 1\cdot488 \\ \cdot & 0\cdot664 & 1\cdot373 \\ \cdot & 0 & 0\cdot087 \end{bmatrix}$$

$$\qquad AQ_2 \qquad\qquad\qquad Q_3 \qquad\qquad\qquad R_3$$

$$\begin{bmatrix} -1\cdot090 & -1\cdot087 & -1\cdot866 \\ 0\cdot810 & 0\cdot425 & 0\cdot402 \\ -0\cdot530 & 0\cdot235 & 0\cdot816 \end{bmatrix} = \begin{bmatrix} -0\cdot748 & -0\cdot523 & -0\cdot439 \\ 0\cdot556 & -0\cdot159 & -0\cdot806 \\ -0\cdot364 & 0\cdot837 & -0\cdot388 \end{bmatrix} \begin{bmatrix} 1\cdot458 & 0\cdot964 & 1\cdot322 \\ \cdot & 0\cdot700 & 1\cdot595 \\ \cdot & \cdot & 0\cdot098 \end{bmatrix}$$

$$\qquad AQ_3$$

$$\begin{bmatrix} 0\cdot940 & 1\cdot205 & 1\cdot928 \\ -0\cdot748 & -0\cdot523 & -0\cdot439 \\ 0\cdot556 & -0\cdot159 & -0\cdot806 \end{bmatrix}$$

In the application of treppen-iteration pivoting has been used in the triangularizations. The pivotal positions remain constant in each step. This will invariably be true when A has real eigenvalues except possibly for the first few stages. It will be observed that the second column of the L_s is tending rapidly to a limit. This limit is the vector having a zero first component and lying in the subspace corresponding to the quadratic divisor. The first column is tending to the unique eigenvector corresponding to $\lambda = -1$, but of course convergence is slow (§ 16). Since A is of order three the last column is fixed in any case as soon as the pivotal positions become constant, but $r_{23}^{(s)}$ tends rapidly to λ_3. The fact that $r_{23}^{(s)}$ converges and the second column of L_s soon tends to a true limit, while the first column does not, gives a strong indication that we have a non-linear elementary divisor. (If we had two equal linear elementary divisors both columns would tend to a limit, while if we had complex conjugate eigenvalues neither would tend to a limit.) It can be verified that columns 1 and 2 of L_s form an invariant subspace by showing that the first two columns of AL_3 and the first two columns of L_3 are almost linearly dependent. In fact by the method of least squares we have

$$
\begin{bmatrix} -1 \cdot 257 & 0 \cdot 502 \\ 1 \cdot 000 & 0 \\ -0 \cdot 743 & -0 \cdot 502 \end{bmatrix} - \begin{bmatrix} 1 & 0 \\ -0 \cdot 743 & -0 \cdot 502 \\ 0 \cdot 486 & 1 \cdot 000 \end{bmatrix} \begin{bmatrix} -1 \cdot 257 & 0 \cdot 502 \\ -0 \cdot 132 & -0 \cdot 743 \end{bmatrix}
$$

$$
= 10^{-3} \begin{bmatrix} 0 & 0 \\ -0 \cdot 215 & 0 \\ -0 \cdot 098 & -2 \cdot 972 \end{bmatrix}. \quad (41.1)
$$

The eigenvalues are found from the 2×2 matrix in equation (41.1). The characteristic equation is

$$
\lambda^2 + 2 \cdot 000\lambda + 1 \cdot 000215 = 0, \quad (41.2)
$$

but the determination of the eigenvalues is of course an ill-conditioned problem. We recall that the same was true of the example of § 13.

Turning now to the orthogonalization technique we find that the third column of the Q_s tends to a limit, the vector orthogonal to the subspace corresponding to the quadratic divisor. This vector should be $[6^{-\frac{1}{2}}, 2.6^{-\frac{1}{2}}, 6^{-\frac{1}{2}}]$, and it is not as accurate in our example as might be expected. This is because we used the Schmidt orthogonalization instead of orthogonalization by elementary Hermitians. In this example an accurate third column may easily be obtained by computing the (unique) vector orthogonal to the first two columns. Neither of the first two columns tends rapidly to a limit. The first column is tending slowly to the unique eigenvector corresponding to $\lambda = -1$. In fact, apart from the normalizing factor, the first column of Q_s is the same as the first column of L_s. The second column is tending equally slowly to the vector in the 2-space which is orthogonal to the eigenvector. The element $r_{33}^{(s)}$ tends rapidly to $|\lambda_3|$. The sign may be inferred from the fact that the third column of Q_s is changing sign with each iteration.

We do not have such a clear indication of the presence of a non-linear divisor as in the case of treppen-iteration, but the convergence of the third vector shows that the other two vectors are already spanning a 2-space. The corresponding eigenvalues are determined from the first two columns of Q_3 and AQ_3 which are almost linearly dependent. From the method of least squares we have

$$\begin{bmatrix} 0\cdot940 & 1\cdot205 \\ -0\cdot748 & -0\cdot523 \\ 0\cdot556 & -0\cdot159 \end{bmatrix} - \begin{bmatrix} -0\cdot748 & -0\cdot523 \\ 0\cdot556 & -0\cdot159 \\ -0\cdot364 & 0\cdot837 \end{bmatrix} \begin{bmatrix} -1\cdot320 & -1\cdot134 \\ 0\cdot090 & -0\cdot683 \end{bmatrix}$$

$$= 10^{-3} \begin{bmatrix} -0\cdot290 & -0\cdot441 \\ 0\cdot230 & -1\cdot093 \\ 0\cdot190 & -0\cdot105 \end{bmatrix}, \quad (41.3)$$

and the characteristic equation of the 2×2 matrix is

$$\lambda^2 + 2\cdot003\lambda + 1\cdot003620. \quad (41.4)$$

Again we have an ill-conditioned determination of the eigenvalues.

When we have a complex conjugate pair of eigenvalues instead of a non-linear divisor, treppen-iteration has the unsatisfactory feature that the pivotal positions will change throughout the course of the iteration. A vector corresponding to a real eigenvalue of smaller modulus than the complex pair will therefore not tend to a limit since its zero elements will not be in a fixed position, while if we do not use a pivotal strategy we may have severe numerical instability from time to time. There is no corresponding weakness with the orthogonalization technique.

The method of bi-iteration is rather unsatisfactory in the case of a non-linear divisor. The first columns of X_s and Y_s tend slowly to

x_1 and y_1 respectively and these are orthogonal (Chapter 1, § 7); hence from the definitions of the r_{ki} and u_{ki} of § 40 the bi-orthogonalization of later columns becomes increasingly numerically unstable.

Richardson's purification process

42. Richardson (1950) has described a process which can be used to find the eigenvectors when good approximations to the eigenvalues are known. The result is based on the following observation. If the eigenvalues were known exactly and if

$$u = \sum \alpha_i x_i \qquad (42.1)$$

is an arbitrary vector, then

$$\prod_{i \neq k} (A - \lambda_i I) u = \alpha_k \prod_{i \neq k} (\lambda_k - \lambda_i) x_k, \qquad (42.2)$$

all other components being eliminated. Hence we may obtain each of the eigenvectors in turn. If we have accurate approximations λ_i' rather than the exact values then pre-multiplication by $(A - \lambda_i' I)$ severely attenuates the component of x_i.

Although this appears to be quite a powerful tool its straightforward use in the manner we have described is very disappointing with some distributions. Consider the determination of x_{20} for a matrix having the eigenvalues $\lambda_r = 1/r$ ($r = 1, ..., 20$). Let us assume that $\lambda_1' = 1 - 10^{-10}$, $\lambda_2' = 1/r$ ($r = 2, ..., 20$) and that computation may be performed exactly. If we pre-multiply an arbitrary vector u by $\prod_{i \neq 20} (A - \lambda_i' I)$ all components are eliminated except those of x_1 and x_{20} which become

$$\alpha_1 10^{-10} (1 - \tfrac{1}{2})(1 - \tfrac{1}{3}) ... (1 - \tfrac{1}{19}) x_1 = \alpha_1 (10^{-10}/19) x_1, \qquad (42.3)$$

and

$$\alpha_{20} (\tfrac{1}{20} - 1 + 10^{-10})(\tfrac{1}{20} - \tfrac{1}{2})(\tfrac{1}{20} - \tfrac{1}{3}) ... (\tfrac{1}{20} - \tfrac{1}{19}) x_{20} \doteqdot -\alpha_{20} (20)^{-19} x_{20}, \qquad (42.4)$$

respectively. The x_1 component is actually *increased* in magnitude relative to that of x_{20} by a factor which is greater than 10^4. If we were to perform three initial iterations with $(A - \lambda_1' I)$ this would decrease the component of x_1 by a factor of 10^{-30} if exact arithmetic were performed, but rounding errors prevent the reduction of any component much below $2^{-t} x_i$. When the eigenvalues are known accurately one can often determine integers r such that $\prod_{i \neq k} (A - \lambda_i' I)^{r_i} u$

gives an accurate x_k, but in general we have found the useful range of application of Richardson's process to be a little disappointing.

Matrix squaring

43. Instead of forming the sequence $A^s u_0$ for an arbitrary u_0 we may construct the sequence of matrix powers directly. This has the advantage that when A^s is computed we may obtain A^{2s} in one matrix multiplication. Hence we may build up the sequence A, A^2, $A^4, \ldots, A^{2^s}$ and we have

$$A^{2^s} = \sum \lambda_i^{2^s} x_i y_i^T = \lambda_1^{2^s} \left\{ x_i y_1^T + \sum_2^n (\lambda_i/\lambda_1)^{2^s} x_i y_i^T \right\}. \qquad (43.1)$$

When the $|\lambda_i|$ are distinct the matrix powers are dominated by the term $\lambda_1^{2^s} x_1 y_1^T$ so that all rows become parallel to y_1 and all columns to x_1. The speed of convergence is determined by the rate at which $(\lambda_2/\lambda_1)^{2^s}$ tends to zero; hence a comparatively small number of iterations is required even when $|\lambda_1|$ and $|\lambda_2|$ are fairly poorly separated.

Generally the successive matrices A^{2^s} will either increase or decrease rapidly in size and even in floating-point arithmetic overspill or underspill will occur. This can easily be avoided by scaling each matrix by a power of two, chosen so that the element of maximum modulus has index zero. This introduces no additional rounding errors. The numbers of iterations required to eliminate the component of $x_2 y_2^T$ to working accuracy by matrix squaring and the simple power method are determined respectively by

$$|\lambda_2/\lambda_1|^{2^s} < 2^{-t} \quad \text{and} \quad |\lambda_2/\lambda_1|^s < 2^{-t}. \qquad (43.2)$$

If A does not contain any appreciable number of zero elements there is about n times as much work in one step of matrix squaring as in the simpler process. Hence matrix squaring is the more efficient if

$$t \log 2/\log |\lambda_1/\lambda_2| > n \log_2\{t \log 2/\log |\lambda_1/\lambda_2|\}. \qquad (43.3)$$

For a given separation matrix squaring will always be the less efficient for sufficiently large n. Alternatively, keeping n fixed, matrix squaring is the more efficient if the separation of λ_2 and λ_1 is sufficiently small. When A is symmetric all matrix powers are symmetric; we therefore gain from symmetry in matrix squaring but not in the simple iterative process.

As an example, if $|\lambda_2/\lambda_1| = 1 - 2^{-10}$ and $t = 40$, 15 steps of matrix squaring or about 28000 steps of the simpler process are required to eliminate the component of x_2 to working accuracy. Hence matrix squaring is the more efficient for matrices of orders up to about 2000, while for matrices of order 100 matrix squaring is almost 20 times as efficient.

However, such a comparison is a little misleading for two reasons.

(i) It is unlikely that we would carry out 28000 steps of the simple process without applying some method of accelerating convergence and it is not so convenient to apply analogous techniques in the matrix squaring process.

(ii) Most matrices of orders greater than 50 encountered in practice are quite sparse. Sparseness is usually destroyed by matrix squaring and hence our simple estimate of the amount of work involved in the two processes is unrealistic.

Numerical stability

44. As far as the dominant eigenvalue and eigenvector is concerned, matrix squaring is a very accurate method, particularly if inner-products can be accumulated. It is interesting to consider what happens if there is more than one eigenvector corresponding to the eigenvalue or eigenvalues of maximum modulus. If we write

$$A = [a_1 \mid a_2 \mid \ldots \mid a_n] \quad \text{then} \quad A^{2^s} = A^{2^s-1}[a_1 \mid a_2 \mid \ldots \mid a_n]. \quad (44.1)$$

Now each of the vectors a_k is expressible in terms of the x_i. Hence if, for example, we have a real dominant eigenvalue with r independent eigenvectors, all r of these will be represented in the columns of A^2. If, on the other hand, A has λ_1 and $-\lambda_1$ as its dominant eigenvalues, each column of A^{2^s} will ultimately consist of a vector in the subspace spanned by the two corresponding eigenvectors. In any case the eigenvalues of A are best obtained by using the simple power method on the independent column vectors of A^{2^s}.

Since the separation of the eigenvalues steadily improves as we form higher powers of A it is at first sight attractive to combine the matrix squaring method with the simple power method and deflation. In this combination we would compute $A^{2^r} = B_r$ for some comparatively modest value of r and then perform simple iteration using B_r. The weakness of this is that the eigenvalues of smaller modulus are comparatively poorly represented in B_r. In fact we have

$$A^{2^s} = \lambda_1^{2^s}\left\{x_1 y_1^T + \sum_2^n (\lambda_i/\lambda_1)^{2^s} x_i \, y_i^T + O(2^{-t})\right\}, \quad (44.2)$$

where $O(2^{-t})$ is used to denote the effect of rounding errors. If $|x_k| < \frac{1}{2}|\lambda_1|$ the term in $x_k y_k^T$ is already smaller than the rounding errors when $s = 6$, and *no* significant figures of x_k and λ_k can be obtained from B_6. However, a combination of r steps of matrix squaring followed by simple iteration will often be the most economical method for finding the dominant eigenvalue. In the case when several of the other eigenvalues are close to the dominant eigenvalue these may be found quite accurately from B_r using simple iteration and deflation. Thus if, for example, the eigenvalues of a matrix are given in decreasing order by

$$1\cdot00, \quad 0\cdot99, \quad 0\cdot98, \quad 0\cdot50, \ldots, 0\cdot00,$$

then in six steps of matrix squaring all eigenvalues except the first three are virtually eliminated. These become 1, $(0\cdot99)^{64}$, $(0\cdot98)^{64}$, respectively. The three corresponding eigenvectors are fully represented in B_r and can be obtained accurately by simple iteration and deflation; since they are now comparatively well separated few iterations are needed.

Use of Chebyshev polynomials

45. Suppose we know that all the eigenvalues λ_2 to λ_n lie in some interval (a, b) and that λ_1 lies outside this interval. For an effective iterative scheme we wish to determine a function $f(A)$ of the matrix A, such that $|f(\lambda_1)|$ is as large as possible relative to the maximum value assumed by $|f(\lambda)|$ in the interval (a, b). If we restrict ourselves to polynomial functions then the solution to this problem is provided by the Chebyshev polynomials.

Applying the transformation

$$x = [2\lambda - (a+b)]/(b-a), \tag{45.1}$$

the interval (a, b) becomes $(-1, 1)$, and the polynomial $f_n(\lambda)$ of degree n satisfying our requirement is given by

$$f_n(\lambda) = T_n(x) = \cos(n\cos^{-1}x). \tag{45.2}$$

Hence if we define B by the relation

$$B = [2A - (a+b)I]/(b-a), \tag{45.3}$$

the most effective polynomial of degree s is $T_s(B)$. Now the Chebyshev polynomials satisfy the recurrence relation

$$T_{s+1}(B) = 2BT_s(B) - T_{s-1}(B), \tag{45.4}$$

and hence starting with an arbitrary vector v_0 we may form the vectors $v_s = T_s(B)v_0$ for $s = 1, 2,\ldots$ from the relations

$$T_{s+1}(B)v_0 = 2BT_s(B)v_0 - T_{s-1}(B)v_0, \qquad (45.5)$$

$$v_{s+1} = 2Bv_s - v_{s-1}. \qquad (45.6)$$

Only one multiplication is required for each vector, no more than in the simple iterative procedure. Obviously the eigenvector corresponding to λ_n may be found in a similar manner.

There is an extensive literature on the use of orthogonal polynomials for the computation of eigenvectors. Such methods have been widely used for computing the smaller eigenvalues of large sparse matrices derived from finite-difference approximations to partial differential equations. In general, success depends rather critically on the details of the practical procedure. The reader is referred to papers by Stiefel (1958) and Engeli *et al.* (1959).

General assessment of methods based on direct iteration

46. We are now in a position to assess the methods we have discussed which are based on iteration with A itself or with $(A - pI)$. The outstanding weakness of all such methods is the limitation on the rate of convergence. Their main advantages are their simplicity and the fact that they enable one to take full advantage of sparseness in the matrix provided one does not perform explicit deflations.

I have found their most useful field of application the determination of a small number of the algebraically largest or smallest eigenvalues and the corresponding eigenvectors of large sparse matrices. Indeed for *very* large sparse matrices they are usually the only methods that can be used, since methods involving the reduction of the matrix to some condensed form usually destroy sparseness and take a prohibitively long time. If the matrix is sparse and the non-zero elements are defined in some systematic manner it is often possible to avoid storage of the matrix altogether by constructing the elements as required during each iteration. If more than one eigenvector is required the deflation must not be performed explicitly but rather as in § 30.

The methods in which several vectors (or even a complete set of n vectors) are used are of great theoretical interest because of their relationship with the LR and QR algorithms. In practice there do not appear to be many situations in which their use can be strongly recommended, though when it is known (or suspected) that there are non-linear divisors they are useful for computing the relevant invariant subspaces with reasonable accuracy.

It has been suggested that they should be used for refining eigensystems determined by techniques which rely on similarity transformations of A and which therefore introduce rounding errors. However, as we have pointed out, the more stable of the methods using similarity transformations are remarkably good and the superior stability of methods which work directly with A is not as marked as might be expected. I consider the method which we shall discuss in § 54 to be much superior for the refinement of complete eigensystems than treppen-iteration and related techniques. Nevertheless it is my experience that minor modifications of an algorithm sometimes improve its effectiveness to quite a remarkable extent and this could well prove to be true of treppen-iteration, the orthogonalization technique and bi-iteration.

Inverse iteration

47. We have already remarked that the LR and QR algorithms depend effectively on the power method. It might therefore seem strange that whereas a very high rate of convergence is attainable with these algorithms the main weakness of the methods based directly on the power method is their slow rate of convergence.

When converging to λ_n using a shift of origin p, the speed of convergence of the LR and QR techniques is determined by the speed at which $[(\lambda_n-p)/(\lambda_{n-1}-p)]^s$ tends to zero. In this process p is chosen as an approximation to λ_n. We can obtain a similar high rate of convergence with the power method if we iterate with $(A-pI)^{-1}$ rather than $(A-pI)$. Apart from rounding errors the process is defined by

$$v_{s+1} = (A-pI)^{-1}u_s, \qquad u_{s+1} = v_{s+1}/\max(v_{s+1}). \qquad (47.1)$$

If we write

$$u_0 = \sum \alpha_i x_i \qquad (47.2)$$

then apart from a normalizing factor we have

$$u_s = \sum \alpha_i(\lambda_i-p)^{-s}x_i. \qquad (47.3)$$

Hence if $|\lambda_k-p| \ll |\lambda_i-p|$ $(i \neq k)$ the components of x_i $(i \neq k)$ in u_s decrease rapidly as s increases. We have already described this process in connexion with the computation of eigenvectors of symmetric tridiagonal matrices, but because of its great practical value we shall now analyse it further in this more general context.

Error analysis of inverse iteration

48. The argument of the previous section shows that, in general, the nearer p is to the eigenvalue λ_k the faster u_s tends to x_k. However, as p tends to λ_k the matrix $(A-pI)$ becomes more and more ill-conditioned until when $p = \lambda_k$ it is singular. It is important to understand the effect of this on the iteration process.

In practice the matrix $(A-pI)$ is triangularly decomposed into the product LU using partial pivoting and a record is kept of the elements of L and U and of the pivotal positions (Chapter 4, § 39). This means that apart from rounding errors we have

$$P(A-pI) = LU \tag{48.1}$$

where P is some permutation matrix. With the stored information we may solve the equations $(A-pI)x = b$ effectively by solving the two triangular sets of equations

$$Ly = Pb \quad \text{and} \quad Ux = y. \tag{48.2}$$

Apart from giving numerical stability the interchanges play little part. The effect of rounding errors is that we have

$$P(A-pI+F) = LU, \tag{48.3}$$

and the computed x satisfies

$$P(A-pI+F+G_b)x = Pb \quad \text{or} \quad (A-pI+F+G_b)x = b, \tag{48.4}$$

where F and G_b are matrices of small elements depending on the rounding procedure (Chapter 4, § 63). Obviously the matrix F is independent of b but G_b is not.

Hence having obtained the triangular decomposition we can compute the sequences u_s and v_s defined by

$$(A-pI+F+G_{u_s})v_{s+1} = u_s, \qquad u_{s+1} = v_{s+1}/\|v_{s+1}\|_2, \tag{48.5}$$

where we have normalized u_{s+1} according to the 2-norm for convenience in the analysis. If the matrices G_{u_s} were null then we would be iterating exactly with $(A-pI+F)^{-1}$, and hence u_s would tend to the eigenvector of $(A+F)$ corresponding to the eigenvalue nearest to p. The accuracy of the computed vector would be high if $\|F\|/\|A\|$ were small unless the *eigenvector* were ill-conditioned. Now as we have pointed out, the computed solutions of triangular sets of equations are usually exceptionally accurate (Chapter 4, §§ 61, 62), so that in practice v_{s+1} is usually remarkably close to the solution of

$$(A-pI+F)x = u_s$$

at each stage and the iterated vectors do indeed converge eventually to the eigenvectors of $(A+F)$.

49. However, if we ignore this rather special feature of triangular matrices we can still show that inverse iteration gives good results even when p is very close to λ_k, or actually equal to it, unless x_k is ill-conditioned. We write

$$p = \lambda_k + \epsilon_1, \qquad F + G_{u_s} = H, \qquad \|H\|_2 = \epsilon_2, \qquad (49.1)$$

and assume without essential loss of generality that $\|A\|_2 = 1$. Hence we may write

$$(A - pI + H)v_{s+1} = u_s. \qquad (49.2)$$

We now show that in order to establish our result it is sufficient to prove that v_{s+1} is large. From (49.2) we have

$$(A - \lambda_k I - \epsilon_1 I + H)v_{s+1}/\|v_{s+1}\|_2 = u_s/\|v_{s+1}\|_2, \qquad (49.3)$$

$$(A - \lambda_k I)u_{s+1} = (\epsilon_1 I - H)u_{s+1} + u_s/\|v_{s+1}\|_2 = \eta, \qquad (49.4)$$

where

$$\|\eta\|_2 < (|\epsilon_1| + \epsilon_2) + 1/\|v_{s+1}\|_2. \qquad (49.5)$$

This shows that if $\|v_{s+1}\|_2$ is large and ϵ_1 and ϵ_2 are small, u_{s+1} and λ_k give a small residual and hence u_{s+1} is a good eigenvector except in so far as x_k is ill-conditioned. In fact, writing

$$u_s = \sum \alpha_i x_i, \qquad v_{s+1} = \sum \beta_i x_i, \qquad (49.6)$$

and assuming $\|x_i\|_2 = \|y_i\|_2 = 1$, we obtain from equation (49.2), on pre-multiplication by y_k^T on both sides,

$$\epsilon_1 \beta_k + y_k^T H v_{s+1}/s_k = \alpha_k. \qquad (49.7)$$

Hence

$$|\alpha_k| < |\epsilon_1| \, |\beta_k| + \epsilon_2 \, \|v_{s+1}\|_2/|s_k|$$

$$< |\epsilon_1| \, \|v_{s+1}\|_2/|s_k| + \epsilon_2 \, \|v_{s+1}\|_2/|s_k|, \qquad (49.8)$$

giving

$$\|v_{s+1}\|_2 > |\alpha_k s_k|/(|\epsilon_1| + \epsilon_2). \qquad (49.9)$$

Substitution in (49.5) gives

$$\|\eta\| < (|\epsilon_1| + \epsilon_2)(1 + 1/|\alpha_k s_k|). \qquad (49.10)$$

Hence provided neither α_k nor s_k is small the residual vector η is small.

General comments on the analysis

50. Inverse iteration has the following potential weaknesses.

(i) Since L and U are obtained using partial pivoting there is a *slight* danger that elements of U may be considerably larger than those of A, in which case ϵ_2 would not be small. As we remarked in Chapter 4, § 57, this is a comparatively remote danger. It could be reduced still further by using complete pivoting or removed altogether by performing an orthogonal triangularization of $(A-pI)$ using elementary Hermitians. I take the view that this danger is negligible. (See also §§ 54, 56.)

(ii) Our analysis did not actually show that the component $\beta_k x_k$ is large but only that $\|v_{s+1}\|$ is large if u_s is not deficient in x_k. It should be appreciated that we *cannot* show that β_k is large unless we make some further assumptions about the distribution of eigenvalues. If there are several eigenvalues very close to p the 'largeness' of $\|v_{s+1}\|$ may well result from large components of any of the corresponding vectors. This weakness is inevitable; if there are other eigenvalues very close to p this means that x_k is ill-conditioned and we must expect to suffer to some extent from this. What we can prove is that v_{s+1} cannot contain a large component of any x_i for which $(\lambda_i - p)$ is not small. For equating the components of x_i on both sides of (49.2) we have

$$(\lambda_i - p)\beta_i + y_i^T H v_{s+1}/s_i = \alpha_i, \tag{50.1}$$

$$|\beta_i| < \{\epsilon_2 \|v_{s+1}\|_2/|s_i| + |\alpha_i|\}/|\lambda_i - p|. \tag{50.2}$$

If λ_k is the only eigenvalue near to p then a large v_{s+1} *must* imply that the x_k component is large.

(iii) If the initial u_0 has a very small component of all x_i corresponding to those λ_i which are close to p then rounding errors may prevent these components from being enriched and we may never reach a u_s in which they are substantial. This does indeed appear to be possible but experience suggests that it is extremely unlikely. If p is very close to λ_k then in my experience initial vectors which are extraordinarily deficient in the x_k component usually have a substantial component after one iteration.

Further refinement of eigenvectors

51. I have found inverse iteration to be by far the most powerful and accurate of methods I have used for computing eigenvectors. By way of illustration of its power we describe how it may be used to give results which are correct to working accuracy. It is clear from our analysis that working with a fixed triangular decomposition of $(A-pI)$ we cannot expect to do better than to obtain an accurate eigenvector of the matrix $(A+F)$ of equation (48.3). Now we know that the matrix F corresponding to triangular decomposition is usually remarkably small, but if there are other eigenvalues moderately close to λ_k the computed eigenvector of $(A+F)$ will inevitably contain components of the corresponding eigenvectors which are far from negligible. As an example, if $\|A\| = O(1)$, $\lambda_1 = 1$, $\lambda_2 = 1-2^{-m}$ and all other eigenvalues are well separated from these two, the computed eigenvector u will be essentially of the form

$$x_1 + \epsilon x_2 + \sum_3^n \epsilon_i x_i$$

where ϵ may be of the order of magnitude $2^m \|F\|/|s_2|$ and the ϵ_i are of order $\|F\|/|s_i|$ (Chapter 2, § 10).

It is interesting to discuss whether we can obtain a more accurate eigenvector than u while still using the same triangular decomposition and the same precision of computation. Consider now the *exact* solution of the equation

$$(A-pI)x = u. \tag{51.1}$$

Clearly we have

$$x = \frac{x_1}{p-1} + \frac{\epsilon x_2}{p-1-2^{-m}} + \sum_3^n \frac{\epsilon_i x_i}{p-\lambda_i} \tag{51.2}$$

and hence if p is much closer to 1 than to $1-2^{-m}$, the normalized x will have a much smaller component of x_2 than had u. Now in Chapter 4, §§ 68, 69, we discussed the problem of solving a set of equations accurately and showed that (51.1) could be solved correct to working accuracy, provided inner-products can be accumulated and $(A-pI)$ is not too ill-conditioned.

Accordingly we propose a scheme consisting of two essentially different stages. In the first stage we compute the sequence u_s defined by

$$(A-pI)v_{s+1} = u_s, \qquad u_{s+1} = v_{s+1}/\max(v_{s+1}). \tag{51.3}$$

We continue with this iteration until u_s shows no further improvement. Renaming the last iterate u, we now solve accurately the equation $(A-pI)x = u$ using the same triangular decomposition. This too is an iterative process and has much in common with the iteration of the first stage. It has, however, an entirely different objective.

52. An interesting consideration is how to choose p so as to achieve maximum overall efficiency. The nearer p is taken to λ_1 the more rapidly does the first stage reach the u_s of limiting accuracy. In fact if we take $p = \lambda_1 + O(2^{-t})$ then the limiting accuracy will almost invariably be attained in two iterations. The ill-condition of $(A-pI)$ is irrelevant at this stage. Turning now to the second stage we find that if we can indeed solve $(A-pI)x = u$ accurately there is again an advantage in having p as close as possible to λ_1 since this ensures that the relative size of the components of x_2 in x is as small as possible. On the other hand the closer p is taken to λ_1 the more ill-conditioned $(A-pI)$ becomes and the slower the convergence of x to the correct solution of $(A-pI)x = u_1$ until ultimately $(A-pI)$ becomes too ill-conditioned for the correct solution to be obtained at all without working to higher precision.

In order to avoid the complication caused by having ill-conditioned eigenvalues let us take A to be symmetric so that all eigenvalues are perfectly conditioned. We assume that $\|A\| = 1$ for convenience and by way of illustration we take $\lambda_1 = 1$, $\lambda_2 = 1-2^{-10}$, $|\lambda_i| < 0.5$ $(i > 2)$, $t = 48$. The size of the error matrix F will depend on the rounding procedure; we shall assume that $\|F\|_2 = n^{\frac{1}{2}}2^{-48}$. If we denote the eigenvalues and eigenvectors of $(A+F)$ by λ_i' and x_i' then we have

$$|\lambda_i'-\lambda_i| < n^{\frac{1}{2}}2^{-48}, \qquad x_1' = x_1 + \sum_{2}^{n} \epsilon_i x_i, \qquad (52.1)$$

where
$$|\epsilon_2| < n^{\frac{1}{2}}2^{-48}/2^{-10} = n^{\frac{1}{2}}2^{-38}, \qquad |\epsilon_i| < n^{\frac{1}{2}}2^{-48}/0.5 = n^{\frac{1}{2}}2^{-47}$$
$$(i > 2). \quad (52.2)$$

Provided p is appreciably closer to λ_1 than λ_2 the first stage will give convergence to x_1'; only the *speed* of convergence is affected, not the ultimate accuracy. The component of x_2' will diminish relative to that of x_1' by a factor $(\lambda_1'-p)/(\lambda_2'-p)$ per iteration and the other components will diminish even faster. We now assess the effectiveness of the combined process for two different values of p.

(i) $p = 1-2^{-48}$. In the first stage the component of x_2' diminishes relative to that of x_1' by a factor which is smaller than

$$(n^{\frac{1}{2}}-1)2^{-48}/[2^{-10}-(n^{\frac{1}{2}}-1)2^{-48}].$$

Unless n is very large, 2 iterations will almost certainly be adequate. In the second stage $(A-pI)$ is so ill-conditioned that we cannot solve $(A-pI)x = u$ at all.

(ii) $p = 1-2^{-30}$. In the first stage x_2' diminishes relative to x_1' by a factor which is smaller than

$$(2^{-30}+n^{\frac{1}{2}}2^{-48})/(2^{-10}-2^{-30}-n^{\frac{1}{2}}2^{-48}).$$

Hence we gain about 20 binary digits per iteration and three iterations are almost certainly adequate. In the second stage the relative error in the successive solution of $(A - pI)x = u$ diminishes by a factor of approximately $n^{\frac{1}{2}}2^{30}2^{-48} = n^{\frac{1}{2}}2^{-18}$ per step. With $n = 64$ for example we certainly obtain the correctly rounded solution of these equations in four iterations. The exact solution is given by

$$x = \frac{x_1}{\lambda_1 - p} + \sum_2^n \frac{\epsilon_i x_i}{\lambda_i - p}, \qquad (52.3)$$

and hence after normalization

$$x = x_1 + \sum_2^n \eta_i x_i, \qquad |\eta_2| < n^{\frac{1}{2}}2^{-38}2^{-20}, \qquad |\eta_i| < n^{\frac{1}{2}}2^{-47}2^{-30} \quad (i > 2).$$
$$(52.4)$$

This shows that the computed x will almost certainly be a correctly rounded version of x_1 and the ill-condition of the eigenvector and the errors in the triangular decomposition have been completely overcome.

If λ_1 and λ_2 are even closer than in this example the vector x obtained in this way may not be x_1 to working accuracy. In this case the second stage of the process can then be repeated to obtain the accurate solution y of $(A - pI)y = x$, and so on. If $|\lambda_1 - \lambda_2| < 2n^{\frac{1}{2}}2^{-t}$ the eigenvector x_1 cannot be isolated without working to higher precision.

On DEUCE a number of programmes have been developed based on this technique. The most sophisticated was designed to find double-precision eigenvectors of complex matrices starting with approximate eigenvalues. Only one single-precision triangular decomposition was used in computing each eigenvector! It is possible to determine double-precision vectors using the technique described in Chapter 4, § 69, because, as pointed out in section (i) of that paragraph, we may use a double-precision right-hand side in the equations $(A - pI)x = b$ without performing any double-precision arithmetic. The DEUCE programme was something of a *tour-de-force* and on a computer with a large high-speed store it would probably have been a little more economical to use true double-precision computation. However, the programme gives considerable insight into the nature of inverse iteration. It also provides extremely reliable indications of the accuracy of the computed vectors. It is interesting that this reliable estimate may be obtained for λ_i and x_i without having anything more than crude estimates of the other eigenvalues.

Non-linear elementary divisors

53. When A has a non-linear elementary divisor this is also true of $(A-pI)^{-1}$ for all values of p, and we saw in § 16 that convergence to the dominant eigenvector is 'slow' when this is associated with a non-linear divisor. At first sight it would appear that little can be gained from using inverse iteration but this is not true.

Consider the simple Jordan submatrix

$$A = \begin{bmatrix} a & 1 \\ 0 & a \end{bmatrix}. \tag{53.1}$$

If we take $p = a - \epsilon$, we have

$$(A-pI)^{-1} = \begin{bmatrix} \epsilon^{-1} & -\epsilon^{-2} \\ 0 & \epsilon^{-1} \end{bmatrix}, \tag{53.2}$$

and since we are not interested in a constant multiplier, inverse iteration is essentially equivalent to direct iteration with the matrix

$$\begin{bmatrix} \epsilon & -1 \\ 0 & \epsilon \end{bmatrix}. \tag{53.3}$$

When ϵ is small, starting with an arbitrary vector we immediately obtain a good approximation to the unique eigenvector in one iteration though it is true that subsequent iterations give no substantial improvement. (This corresponds in the analysis of § 16 to the case when a is small.)

Inverse iteration with Hessenberg matrices

54. We have discussed inverse iteration in connexion with a full matrix A but in practice it is more convenient to find the eigenvectors of a condensed similarity transform B of the original matrix and to derive the eigenvectors of A from them. Since there are very stable methods for reducing a matrix to upper Hessenberg form H we consider this form first.

The triangular decomposition of a Hessenberg matrix using partial pivoting has been discussed in Chapter 4, § 33. It is particularly simple since, speaking in terms appropriate to Gaussian elimination, there is only one non-zero multiplier at each stage of the reduction. There are $\frac{1}{2}n^2$ multiplications in all compared with $\frac{1}{3}n^3$ for a full matrix. Since L contains only $(n-1)$ sub-diagonal elements the solution of $Ly = Pb$ (see equation (48.2)) involves only n multiplications.

There are therefore approximately $\frac{1}{2}n^2$ multiplications in each iteration compared with n^2 for a full matrix.

As in Chapter 5, § 54, we do not choose the initial vector u directly but assume it is such that

$$Le = Pu, \qquad (54.1)$$

where e is the vector of unit elements. Hence v_1 is obtained as the solution of $Uv_1 = e$. If p is a very good approximation to an eigenvalue then two iterations are usually adequate. If p were an exact eigenvalue and the decomposition were performed exactly then one of the diagonal elements of U (the last, if no sub-diagonal elements of H are zero) would be zero. In practice, however, none of the computed diagonal elements need be at all small. If a diagonal element of U is zero it may be replaced by $2^{-t}\|H\|_\infty$, but it is more economical to use the technique described in § 55.

We mention here that the vector x obtained by Hyman's method (Chapter 7, § 11) is an exact eigenvector when the quantity z is an exact eigenvalue and the computation is performed exactly. However, although Hyman's method determines the eigenvalues in a remarkably stable manner, this corresponding determination of the eigenvector is often extremely unstable and Hyman's method should not be used for this purpose. In fact when H happens to be tri-diagonal the vector given by Hyman's method is precisely that given by the method of Chapter 5, § 48, and we have shown why this is unstable. By an exactly similar analysis we may show that in general Hyman's method gives a poor approximation to x_r when the normalized left-hand eigenvector y_r has a small first component. This is true even when Hyman's method is performed exactly using a value of λ_r which is correct to working accuracy.

Degenerate cases

55. When one of the sub-diagonal elements $h_{i+1,i}$ is zero we may write H in the form

$$H = \begin{bmatrix} H_1 & B \\ O & H_2 \end{bmatrix}, \qquad (55.1)$$

where H_1 and H_2 are of Hessenberg form. The eigenvalues of H are then those of H_1 and H_2. The eigenvector corresponding to an eigenvalue belonging to H_1 is given by the corresponding eigenvector of H_1 augmented by zeros in the last $(n-i)$ positions, and hence less work is involved in its computation. The eigenvector corresponding to an eigenvalue of H_2 will not in general be specialized and the usual method of inverse iteration is the most economical method of determining it.

If H has an eigenvalue of multiplicity r with r independent eigenvectors it must have at least $(r-1)$ zero sub-diagonal elements (cf. Chapter 7, § 41). Nevertheless H may have eigenvalues which are equal to working accuracy without any of these elements being at all small. Independent eigenvectors of H may be obtained in this case by performing inverse iteration with slightly different values of p, the considerations being the same as those in Chapter 5, § 59.

When H has a non-linear elementary divisor $(\lambda - \lambda_1)^r$ the vectors obtained using different values of p close to λ_1 will themselves be quite close to one another. Inverse iteration will not display the characteristic 'instability' associated with a repeated linear divisor which enables us to obtain independent eigenvectors. Commonly when H has a quadratic divisor the computed eigenvalues are not equal but differ by a quantity of the order of magnitude of $2^{-t/2} \|H\|$ (cf. Chapter 2, § 23). The eigenvectors x_1 and x_2 determined by inverse iteration using these two different eigenvalues differ by quantities of order $2^{-t/2}$, and $(x_1 - x_2)$ gives a principal vector of grade two, but naturally it has at best only $\frac{1}{2}t$ correct figures (cf. Chapter 3, § 63).

Inverse iteration with band matrices

56. Inverse iteration is even more economical in the case of tri-diagonal matrices. We have already discussed this in Chapter 5 in connexion with symmetric matrices but since no advantage was taken of the symmetry the comments apply immediately to the unsymmetric case except that it is now possible for some eigenvalues to be complex even when the matrix is real.

As we saw in Chapter 6 none of the usual methods for reducing a general matrix to tri-diagonal form is unconditionally guaranteed to be numerically stable. For this reason, when the tri-diagonal matrix has been derived from a general matrix via an intermediate Hessenberg matrix, we recommend that the eigenvectors be found from the Hessenberg matrix rather than the tri-diagonal matrix. Moreover when an eigenvector of the Hessenberg matrix has been computed it should be used in direct iteration to improve the eigenvalue, since this may be comparatively inaccurate if the reduction from Hessenberg form to tri-diagonal form has been moderately unstable. We emphasize that if we find the eigenvector of the tri-diagonal matrix we have in any case to compute from it the eigen-vector of the Hessenberg matrix in order to obtain the eigenvector of the original matrix. It is therefore not so inefficient to find the eigenvector of the Hessenberg matrix as it might appear.

In eigenvalue problems associated with partial differential equations band matrices of width greater than three are involved. Usually such matrices are symmetric and positive definite but since we must work with $(A - pI)$ positive definiteness is inevitably lost unless p is an approximation to the smallest eigenvalue. Interchanges must therefore be used in the triangular decomposition so that one obtains the decomposition of $P(A - pI)$, where P is a permutation matrix. If we write

$$P(A - pI) = LU, \tag{56.1}$$

then it may readily be verified that if A is a band of width $(2m+1)$ the matrix L has only m non-zero sub-diagonal elements in each column (except of course the last m columns), while U has up to $(2m+1)$ non-zero elements in each row. It is tempting but unwise to omit the interchanges, particularly if a number of intermediate eigenvalues are required. We notice that these remarks are true whether or not A is symmetric.

This method has been used on ACE to find several eigenvectors of symmetric matrices of order 1500 and band width 31, starting from moderately good eigenvalues. The use of the Rayleigh quotient enabled eigenvalues of very high accuracy to be obtained with very little additional labour.

Complex conjugate eigenvectors

57. The calculation of a complex eigenvector of a real matrix is particularly interesting from the point of view of the error analysis. Suppose $(\xi_1 + i\eta_1)$ is a complex eigenvalue, x_1 is the corresponding complex eigenvector and we have an approximation $\lambda = \xi + i\eta$ to $(\xi_1 + i\eta_1)$. Inverse iteration may be carried out in a number of different ways.

(i) We may compute the sequence of complex vectors z_s and w_s given by

$$(A - \xi - i\eta)w_{s+1} = z_s, \qquad z_{s+1} = w_{s+1}/\max(w_{s+1}), \tag{57.1}$$

where the whole computation is performed using complex numbers where necessary. In general w_s tends to a complex eigenvector.

(ii) We may work entirely with real arithmetic as follows. We write $w_{s+1} = u_{s+1}+iv_{s+1}$, $z_s = p_s+iq_s$ and equate real and imaginary parts in the first of equations (57.1). This gives

$$(A-\xi I)u_{s+1}+\eta v_{s+1} = p_s, \tag{57.2}$$

$$-\eta u_{s+1}+(A-\xi I)v_{s+1} = q_s, \tag{57.3}$$

which we may write in the form

$$C(\lambda)\begin{bmatrix} u_{s+1} \\ \hline v_{s+1} \end{bmatrix} = \begin{bmatrix} A-\xi I & \eta I \\ \hline -\eta I & A-\xi I \end{bmatrix}\begin{bmatrix} u_{s+1} \\ \hline v_{s+1} \end{bmatrix} = \begin{bmatrix} p_s \\ \hline q_s \end{bmatrix}. \tag{57.4}$$

This is a set of real equations in $2n$ variables.

(iii) From equations (57.2) and (57.3) we may derive the equations

$$[(A-\xi I)^2+\eta^2 I]u_{s+1} = (A-\xi I)p_s-\eta q_s, \tag{57.5}$$

$$[(A-\xi I)^2+\eta^2 I]v_{s+1} = \eta p_s+(A-\xi I)q_s. \tag{57.6}$$

The matrix of both these sets of equations is the same and it is real and of order n. Hence only one triangular decomposition is required and that involves a real matrix of order n.

(iv) The vectors u_{s+1} and v_{s+1} may be determined either by solving (57.5) and using (57.2) or by solving (57.6) and using (57.3).

If $(\xi+i\eta)$ is not an exact eigenvalue none of the matrices of the equations occurring in these four methods is singular. Each of the solutions is unique and all are identical. In so far as they differ in practice this is entirely the result of rounding errors. The remarkable thing is that methods (i), (ii) and (iv) give convergence to the complex eigenvector but (iii) does not!

58. Before giving the error analysis we determine the relevant eigenvectors of the matrices of the various equations.

The matrix $(A-\lambda I)$ regarded as a matrix in the complex field has eigenvalues $(\lambda_1-\lambda)$ and $(\bar{\lambda}_1-\lambda)$ corresponding to the eigenvectors x_1 and $\bar{x}_1$. As λ tends to λ_1 the first eigenvalue tends to zero but the second does not. Hence $(A-\lambda_1 I)$ is of rank $(n-1)$ in general.

The matrix $B(\lambda) = (A-\xi I)^2+\eta^2 I$ is a polynomial in A and hence also has the eigenvectors x_1 and $\bar{x}_1$ corresponding to eigenvalues $(\lambda_1-\xi)^2+\eta^2$ and $(\bar{\lambda}_1-\xi)^2+\eta^2$. As λ tends to λ_1 both these eigenvalues tend to zero and hence, in general, $B(\lambda_1)$ is of rank $(n-2)$ since it has the two independent eigenvectors x_1 and $\bar{x}_1$.

Turning now to the matrix $C(\lambda)$ of (57.4) we verify immediately that

$$\begin{bmatrix} A-\xi I & \eta I \\ \hline -\eta I & A-\xi I \end{bmatrix}\begin{bmatrix} x_1 \\ \hline \pm ix_1 \end{bmatrix} = [(\lambda_1-\xi)\pm i\eta]\begin{bmatrix} x_1 \\ \hline \pm ix_1 \end{bmatrix}, \quad (58.1)$$

and

$$\begin{bmatrix} A-\xi I & \eta I \\ \hline -\eta I & A-\xi I \end{bmatrix}\begin{bmatrix} \bar{x}_1 \\ \hline \pm i\bar{x}_1 \end{bmatrix} = [(\bar{\lambda}_1-\xi)\pm i\eta]\begin{bmatrix} \bar{x}_1 \\ \hline \pm i\bar{x}_1 \end{bmatrix}. \quad (58.2)$$

Hence $C(\lambda)$ has the four eigenvalues $(\lambda_1-\xi)\pm i\eta$ and $(\bar{\lambda}_1-\xi)\pm i\eta$ and four corresponding independent eigenvectors. Now we have

$$\left.\begin{aligned} (\lambda_1-\xi)-i\eta &\to 0 \\ (\bar{\lambda}_1-\xi)+i\eta &\to 0 \end{aligned}\right\} \quad \text{as } \xi+i\eta \to \lambda_1, \quad (58.3)$$

showing that $C(\lambda_1)$ has two zero eigenvalues corresponding to the two independent eigenvectors $[x_1^T \mid -ix_1^T]$ and $[\bar{x}_1^T \mid i\bar{x}_1^T]$, and hence that $C(\lambda_1)$ is of rank $(2n-2)$.

Error analysis

59. These simple results enable us to see in quite a simple manner why it is that method (iii) fails in practice. When we perform inverse iteration with *any* matrix having two eigenvalues very close to zero the components of all vectors except those corresponding to the two small eigenvalues are rapidly attenuated. The iterates therefore lie accurately in the subspace determined by the two relevant eigenvectors. The particular vector obtained in this space depends intimately on the precise rounding errors and can even vary from iteration to iteration.

Method (i) presents no special points of interest. The matrix has only one small eigenvalue and we have rapid convergence to x_1. All the arithmetic is complex but there are only $\frac{1}{3}n^3$ complex multiplications in the triangularization and then n^2 complex multiplications in each iteration.

In method (ii) the vector $[u_s^T \mid v_s^T]$ obtained at each stage is necessarily real. It will converge to a vector in the subspace spanned by $[x_1^T \mid -ix_1^T]$ and $[\bar{x}_1^T \mid i\bar{x}_1^T]$ and hence must be of the form

$$\alpha[x_1^T \mid -ix_1^T]+\bar{\alpha}[\bar{x}_1^T \mid i\bar{x}_1^T].$$

It therefore gives as the complex eigenvector

$$(\alpha x_1+\bar{\alpha}\bar{x}_1)+i(-i\alpha x_1+i\bar{\alpha}\bar{x}_1) = 2\alpha x_1, \quad (59.1)$$

which is perfectly satisfactory.

In method (iii) the vectors u_s and v_s will obviously be real and will rapidly tend to vectors spanned by x_1 and $\bar{x}_1$. Hence when components of other vectors have been eliminated (which should be almost immediately when λ is very close to λ_1) we have

$$u_s = \alpha x_1 + \bar{\alpha}\bar{x}_1, \qquad v_s = \beta x_1 + \bar{\beta}\bar{x}_1, \tag{59.2}$$

but we have no reason to expect that α and β will be related in any way. The complex eigenvector given by this method is

$$u_s + iv_s = (\alpha + i\beta)x_1 + (\bar{\alpha} + i\bar{\beta})\bar{x}_1 \tag{59.3}$$

and this is not a multiple of x_1. Although u_s and v_s are real vectors spanning the same subspace as x_1 and $\bar{x}_1$ the vector $(u_s + iv_s)$ is not a multiple of x_1 and to obtain the latter further computation is required.

In method (iv) we really do obtain the correct real vector v_{s+1} to accompany u_{s+1} by making use of equation (57.2). In the limiting case when λ is equal to λ_1 to working accuracy we find that the solution u_{s+1} to equation (57.5) is larger than the p_s and q_s from which it is derived by a factor of the order $1/\|F\|$, where F is the error matrix arising from the decomposition of $(A - \xi I)^2 + \eta^2 I$. Hence p_s is negligible in equation (57.2) and the computed v_{s+1} essentially satisfies

$$\begin{aligned}
\eta v_{s+1} &= -(A - \xi I)u_s \\
&= -(A - \xi I)(\alpha x_1 + \bar{\alpha}\bar{x}_1) \\
&= \alpha(\xi - \lambda_1)x_1 + \bar{\alpha}(\xi - \bar{\lambda}_1)\bar{x}_1,
\end{aligned} \tag{59.4}$$

and (59.2) then provides

$$\eta(u_{s+1} + iv_{s+1}) = \alpha\eta x_1 + \bar{\alpha}\eta\bar{x}_1 + \alpha\eta x_1 - \bar{\alpha}\eta\bar{x}_1 = 2\alpha\eta x_1, \tag{59.5}$$

showing that $(u_{s+1} + iv_{s+1})$ is indeed a true multiple of x_1.

The fact that (iii) will give an incorrectly related u_{s+1} and v_{s+1} is fairly obvious in the case of a 2×2 matrix. In fact it is easy to see that even if $(p_s + iq_s)$ is a correct eigenvector, $(u_{s+1} + iv_{s+1})$ will be arbitrarily poor. This follows because in the 2×2 case, $B(\lambda_1) = 0$. Hence if we compute $B(\lambda)$ for a value of λ which is correct to working accuracy this matrix will consist only of rounding errors. It may therefore be expressed in the form $2^{-t}P$ where P is an arbitrary matrix of the same order of magnitude as A^2. Now if $(p_s + iq_s)$ is an eigenvector the right-hand sides of (57.5) and (57.6) are $-2\eta_1 q_s$ and $2\eta_1 p_s$ respectively. Apart from a constant factor we have

$$u_{s+1} = -P^{-1}q_s, \qquad v_{s+1} = P^{-1}p_s, \tag{59.6}$$

and since the matrix P is arbitrary we cannot expect u_{s+1} and v_{s+1} to be correctly related.

We have discussed the extreme case when λ is correct to working accuracy. When $\|A\| = 1$ and $|\lambda - \lambda_1| \doteq 2^{-k}$, then even in the most favourable circumstances method (iii) will give an eigenvector of the form $(x_1 + \epsilon \bar{x}_1)$ where ϵ is of the order of 2^{k-t}. As k approaches t the component of $\bar{x}_1$ becomes of the same order of magnitude as that of x_1.

Numerical example

60. In Table 3 we show the application of methods (ii), (iii) and (iv) of § 57 to the calculation of a complex eigenvector of a real matrix of order three. Inverse iteration has been performed using $\lambda = 0 \cdot 82197 + 0 \cdot 71237i$, which is correct to working accuracy. We first give the triangular decomposition with interchanges of the 6×6 real matrix $C(\lambda)$ of (57.4); the triangular matrices are denoted by L_1 and U_1. The fact that $C(\lambda)$ has two eigenvalues very close to zero is revealed in the most obvious way, the last two diagonal elements of U_1 being zero. In the first iteration we may start with the back-substitution and take e as the right-hand side. Clearly we may take the last two components of the solution to be arbitrarily large and in any ratio. The components of the right-hand side are virtually ignored. We have computed two solutions, taking the last two components to be 0 and 10^6 in one case and 10^6 and 10^6 in the other. Although the two real solutions are entirely different, the corresponding normalized complex eigenvectors are both equal to the true eigenvector to working accuracy.

Next we give the triangular decomposition of the 3×3 real matrix $B(\lambda)$; the triangles are denoted by L_2 and U_2. Again the fact that it has two eigenvalues very close to zero is revealed by the emergence of two almost zero diagonal elements in U_2. We have performed one step taking $(e + ei)$ as the initial complex vector. In method (iii) the right-hand sides $(A - \xi I)e - \eta e$ and $(A - \xi I)e + \eta e$ are labelled b_1 and b_2 respectively. The vectors obtained by method (iv) are both correct to within two units in the last figure retained. That obtained using method (iii) is completely inaccurate; it is a combination of x_1 and $\bar{x}_1$.

The generalized eigenvalue problem

61. The method of inverse iteration extends immediately to the generalized eigenvalue problem (Chapter 1, § 30). It will suffice to consider the most common case, the determination of λ and x such that

$$(A\lambda^2 + B\lambda + C)x = 0. \tag{61.1}$$

TABLE 3

$$A = \begin{bmatrix} 0.59797 & 0.29210 & 0.15280 \\ -0.95537 & 0.75756 & -1.23999 \\ 0.50608 & 0.13706 & 0.40117 \end{bmatrix} \qquad \lambda = 0.82197 + 0.71237i$$

L_1

$$\begin{bmatrix} 1.00000 & & & & & \\ 0.00000 & 1.00000 & & & & \\ -0.52972 & -0.14450 & 1.00000 & & & \\ 0.00000 & 0.00000 & 0.66104 & 1.00000 & & \\ 0.23446 & -0.43124 & -0.41157 & 0.40774 & 1.00000 & \\ 0.74565 & -0.06742 & -0.85798 & -0.68110 & 0.00000 & 1.00000 \end{bmatrix}$$

U_1

$$\begin{bmatrix} -0.95537 & -0.06441 & -1.23999 & 0.00000 & 0.71237 & 0.00000 \\ & -0.71237 & 0.00000 & -0.95537 & -0.06441 & -1.23999 \\ & & -1.07765 & -0.13805 & 0.36805 & 0.53319 \\ & & & 0.59734 & -0.10624 & -0.77326 \\ & & & & 0.00000 & 0.00001 \\ & & & & & 0.00000 \end{bmatrix}$$

(1st solution) $\times 10^{-6}$	(2nd solution) $\times 10^{-6}$	Normalized 1st and 2nd solutions
−0.19254	0.16158	0.05538 − 0.37233i
−3.47674	−3.80567	1.00000 + 0.00000i
0.32894	0.64769	−0.09461 − 0.28763i
1.29451	1.47236	
0.00000	1.00000	
1.00000	1.00000	

$(A - \xi I)^2 + \eta^2 I$ b_1 b_2

$$\begin{bmatrix} 0.35591 & -0.06330 & -0.46073 \\ -0.35200 & 0.06260 & 0.45568 \\ -0.45726 & 0.08132 & 0.59192 \end{bmatrix} \qquad \begin{matrix} -0.49147 \\ -2.97214 \\ -0.49003 \end{matrix} \qquad \begin{matrix} 0.93327 \\ -1.54740 \\ 0.93471 \end{matrix}$$

L_2 U_2

$$\begin{bmatrix} 1.00000 & & \\ -1.28476 & 1.00000 & \\ -0.98901 & 0.00000 & 1.00000 \end{bmatrix} \begin{bmatrix} 0.35591 & -0.06330 & -0.46073 \\ & -0.00001 & -0.00001 \\ & & 0.00001 \end{bmatrix}$$

Solution 1 from equations (57.5) *and* (57.2)	*Normalized solution 1*
−366219 − 228762i	0.05538 − 0.37234i
457966 − 1051690i	1.00000 + 0.00000i
−345821 − 32222i	−0.09461 − 0.28763i

Solution 2 from equations (57.6) *and* (57.3)	*Normalized solution 2*
−41426 − 107670i	0.05540 − 0.37232i
266728 − 150935i	1.00000 + 0.00000i
−68649 − 62439i	−0.09461 − 0.28763i

Solution 3 from equations (57.5) *and* (57.6)	
−366219 − 107670i	Normalized solution 3 is
457966 − 150935i	not an approximate
−345821 − 62439i	eigenvector.

Since this case is equivalent to the standard eigenvalue problem

$$\begin{bmatrix} O & I \\ \hline -A^{-1}C & -A^{-1}B \end{bmatrix} \begin{bmatrix} x \\ \hline y \end{bmatrix} = \lambda \begin{bmatrix} x \\ \hline y \end{bmatrix} \qquad (61.2)$$

the relevant iterative procedure is given by

$$\begin{bmatrix} -\lambda I & I \\ \hline -A^{-1}C & -A^{-1}B-\lambda I \end{bmatrix} \begin{bmatrix} x_{s+1} \\ \hline y_{s+1} \end{bmatrix} = \begin{bmatrix} x_s \\ \hline y_s \end{bmatrix}, \qquad (61.3)$$

where λ is an approximate eigenvalue. (We have omitted the normalization for convenience.) From equation (61.3) we have

$$\left. \begin{aligned} -\lambda x_{s+1} + y_{s+1} &= x_s \\ -A^{-1}Cx_{s+1} - A^{-1}By_{s+1} - \lambda y_{s+1} &= y_s \end{aligned} \right\} \qquad (61.4)$$

and hence

$$-C(y_{s+1}-x_s) - \lambda By_{s+1} - \lambda^2 Ay_{s+1} = \lambda Ay_s, \qquad (61.5)$$

$$(\lambda^2 A + \lambda B + C)y_{s+1} = Cx_s - \lambda Ay_s. \qquad (61.6)$$

The vectors y_{s+1} and x_{s+1} may then be computed from equations (61.6) and (61.4) respectively. Only one triangular decomposition is required, that of $(\lambda^2 A + \lambda B + C)$.

This process is remarkably effective when A, B and C are functions of a parameter V and we wish to trace the history of an eigenvalue $\lambda_1(V)$ and its corresponding eigenvector $x_1(V)$ as V is varied. The values of $\lambda_1(V)$ and $x_1(V)$ for previous values of V may then be used to provide good initial approximations to their current values.

Variation of approximate eigenvalues

62. So far we have discussed inverse iteration in terms of using a fixed approximation to each eigenvalue so that only one triangular decomposition is used in the determination of each eigenvector. This is an extremely efficient procedure when the approximation to an eigenvalue is very accurate. The speed of convergence is then such that the extra triangular decompositions involved in taking advantage of the best approximation that is currently available cannot be justified.

There is an alternative approach in which inverse iteration is used to locate the eigenvalue *ab initio*. When A is symmetric a convenient iterative process is defined (cf. Chapter 3, § 54) by the equations

$$\left. \begin{aligned} (A - \mu_s I)v_{s+1} &= u_s, \qquad \mu_{s+1} = v_{s+1}^T A v_{s+1} / v_{s+1}^T v_{s+1} \\ u_{s+1} &= v_{s+1}/(v_{s+1}^T v_{s+1})^{\frac{1}{2}} \end{aligned} \right\}. \qquad (62.1)$$

At each stage the new approximation μ_{s+1} to the eigenvalue is given by the Rayleigh quotient corresponding to the current vector. It is easy to show that the process is cubically convergent in the neighbourhood of each eigenvector. Suppose we have

$$u_s = x_k + \sum{}' \epsilon_i x_i, \tag{62.2}$$

where $\sum{}'$ denotes that the suffix k is omitted in the summation. Then

$$\mu_s = (\lambda_k + \sum{}'\lambda_i \epsilon_i^2)/(1 + \sum{}'\epsilon_i^2), \tag{62.3}$$

and hence

$$v_{s+1} = x_k/(\lambda_k - \mu_s) + \sum{}'\epsilon_i x_i/(\lambda_i - \mu_s). \tag{62.4}$$

Apart from a normalizing factor we have

$$
\begin{aligned}
u_{s+1} &= x_k + (\lambda_k - \mu_s) \sum{}'\epsilon_i x_i/(\lambda_i - \mu_s) \\
&= x_k + \left\{ \frac{\sum{}'(\lambda_i - \lambda_k)\epsilon_i^2}{1 + \sum{}'\epsilon_i^2} \right\} \sum{}'\epsilon_i x_i/(\lambda_i - \mu_s), \tag{62.5}
\end{aligned}
$$

and all coefficients other than that of x_k are cubic in the ϵ_i.

There is an analogous process for unsymmetric matrices in which approximate left-hand and right-hand eigenvectors are found at each stage. The process is defined by the equations

$$
\left.
\begin{aligned}
(A - \mu_s I)v_{s+1} = u_s, \qquad (A - \mu_s I)^T q_{s+1} = p_s, \\
\mu_{s+1} = q_{s+1}^T A v_{s+1}/q_{s+1}^T v_{s+1} \\
u_{s+1} = v_{s+1}/\max(v_{s+1}), \qquad p_{s+1} = q_{s+1}/\max(q_{s+1})
\end{aligned}
\right\}. \tag{62.6}
$$

The triangular decomposition of $(A - \mu_s I)$ enables us to determine both v_{s+1} and q_{s+1}. The current value of μ_{s+1} is now the generalized Rayleigh quotient corresponding to q_{s+1} and v_{s+1} (Chapter 3, § 58).

The process is again ultimately cubically convergent. Suppose we have

$$u_s = x_k + \sum{}'\epsilon_i x_i, \qquad p_s = y_k + \sum{}'\eta_i y_i. \tag{62.7}$$

Then

$$\mu_s = (\lambda_k y_k^T x_k + \sum{}'\lambda_i \epsilon_i \eta_i y_i^T x_i)/(y_k^T x_k + \sum{}'\epsilon_i \eta_i y_i^T x_i), \tag{62.8}$$

and hence apart from normalizing factors

$$u_{s+1} = x_k + (\lambda_k - \mu_s)\sum{}'\epsilon_i x_i/(\lambda_i - \mu_s), \tag{62.9}$$

$$p_{s+1} = y_k + (\lambda_k - \mu_s)\sum{}'\eta_i y_i/(\lambda_i - \mu_s). \tag{62.10}$$

All coefficients other than those of x_k and y_k are cubic in the ϵ_i and η_i. If the condition number $y_k^T x_k$ of λ_k is small the onset of cubic convergence is delayed.

The limiting rate of convergence is obviously very satisfactory, though quite often by the time the domain of cubic convergence has been reached the precision of computation will be limiting the possible gain in accuracy. If accurate eigenvalues are required they may be obtained by careful computation of the final Rayleigh quotient or generalized Rayleigh quotient. Usually the exact Rayleigh quotients at this stage will be correct to more than single-precision. The quotients can be found accurately without using double-precision if we compute the quantities

$$(A-\mu_s I)u_{s+1}, \qquad \mu_{s+1} = -\mu_s + u_{s+1}^T(A-\mu_s I)u_{s+1}/u_{s+1}^T u_{s+1}, \qquad (62.11)$$

or in the unsymmetric case

$$(A-\mu_s I)u_{s+1}, \qquad \mu_{s+1} = \mu_s + p_{s+1}^T(A-\mu_s I)u_{s+1}/p_{s+1}^T u_{s+1}. \qquad (62.12)$$

The vector $(A-\mu_s I)u_{s+1}$ should be computed using accumulation of inner-products. Considerable cancellation will normally occur and the single-precision version of $(A-\mu_s I)u_{s+1}$ may then be used in the determination of u_{s+1}.

Ostrowski (1958, 1959) has given a very detailed analysis of these techniques in a series of papers to which the reader is referred for further information.

The determination of both the left-hand and right-hand eigenvectors is time consuming. If a fairly good approximation μ_s to an eigenvalue is known then convergence of the simple inverse iteration process is in any case more than adequate. Direct iteration, inverse iteration and deflation can be used in combination in the following way. Direct iteration is used until a reasonably good value of μ_s is obtained and incidentally quite a good approximate eigenvector. Direct iteration is then replaced by inverse iteration with $(A-\mu_s I)$, the μ_s being kept fixed, starting with the final vector obtained from direct iteration. Finally, when the eigenvector has been determined, deflation is carried out. This is a particularly effective technique when A is a complex matrix. For an excellent account of its use see Osborne (1958).

Refinement of eigensystems

63. As our final topic we consider the refinement of a computed eigensystem. In § 51 we have already shown that we can normally determine the eigenvector corresponding to the eigenvalue nearest to an assigned value p to any desired accuracy by inverse iteration. Although in practice this method gives convincing evidence of the accuracy of the computed eigenvector, it is not possible to use it to obtain rigorous bounds.

In Chapter 3, §§ 59–64, we discussed the problem of the refinement of a complete eigensystem from a theoretical standpoint. We now describe in detail a practical procedure which makes use of the ideas described there. On ACE the procedure has been programmed in fixed-point arithmetic and we shall describe it in this form; the modifications appropriate to floating-point computation are quite straightforward.

Suppose an approximate eigensystem has been determined by some means. We denote the computed eigenvalues and eigenvectors by λ_i' and x_i' and the matrix of eigenvectors by X', and define the residual matrix F by the equation

$$AX' = X' \operatorname{diag}(\lambda_i') + F, \qquad (X')^{-1}AX' = \operatorname{diag}(\lambda_i') + (X')^{-1}F. \quad (63.1)$$

We assume that $\|F\| \ll \|A\|$, otherwise the λ_i' and x_i' can scarcely be regarded as an approximate eigensystem. On ACE each column f_i of F is computed *exactly* by accumulating inner-products; the elements of f_i are therefore double-precision numbers, though from our assumption about F most of the digits in the first word of each component are zeros. Each vector f_i is then expressed in the form $2^{-k_i}g_i$ where k_i is chosen so that the maximum element of $|g_i|$ is between $\frac{1}{2}$ and 1; in other words, each f_i is converted to a normalized double-precision block-floated vector.

So far no rounding errors have been incurred. We next wish to compute $(X')^{-1}F$ as accurately as is convenient. On ACE this is done using the iterative method described in Chapter 4, § 69. The matrix X' is first triangularly decomposed using interchanges and then each of the equations $X'h_i = g_i$ is solved by iteration. Notice that the method described in Chapter 4 uses single-precision computation with accumulation of inner-products but, as we pointed out, double-precision right-hand sides can be used. Provided $n2^{-t}\|(X)^{-1}\| < \frac{1}{2}$ the process converges and gives solutions h_i of the equations $X'h_i = g_i$ which are correct to working accuracy. If we write $p_i = 2^{-k_i}h_i$ we have

$$(X')^{-1}AX' = \operatorname{diag}(\lambda_i') + P + Q = B \quad \text{(say)}, \quad (63.2)$$

where each column p_i of P is known and is in the form of a single-precision block-floated vector; the matrix Q is not known explicitly but we have

$$\|q_i\|_\infty < 2^{-t-1}\|p_i\|_\infty \quad (63.3)$$

from our assumption that the iterative procedure gave the correctly rounded solutions. The eigenvalues of A are *exactly* those of B.

64. We can now use Gerschgorin's theorem (Chapter 2, § 13) to locate the eigenvalues of B. Since B is scaled by columns it is convenient on ACE to apply the usual Gerschgorin result to B^T rather than B itself, but we have used the customary discs here so that direct reference can be made to earlier work. A straightforward application of Gerschgorin's theorem shows that the eigenvalues of A lie in circular discs with centres $\lambda_i' + p_{ii} + q_{ii}$ and radii $\sum_{j \neq i} |p_{ij} + q_{ij}|$, but in general this gives quite a weak result. On ACE improved localization of the eigenvalues is obtained as follows.

Suppose the disc associated with λ_1' is isolated. Let us multiply row 1 of B by 2^{-k} and column 1 by 2^k, where k is the largest integer such that the first Gerschgorin disc is still isolated. From our assumption, $k \geqslant 0$. With this value of k Gerschgorin's theorem states that there is one eigenvalue in the disc with centre $\lambda_1' + p_{11} + q_{11}$ and radius $2^{-k} \sum_{j \neq 1} |p_{1j} + q_{1j}|$ and therefore a fortiori in the disc with centre $\lambda_1' + p_{11}$ and radius $|q_{11}| + 2^{-k} \sum_{j \neq 1} (|p_{1j}| + |q_{1j}|)$. We can replace the $|q_{1j}|$ by their upper bounds since they are in any case smaller than the general level of the $|p_{1j}|$ by a factor 2^{-t}.

The following observations make it possible to choose the appropriate k in a very simple manner. The radius of the first Gerschgorin disc is approximately $2^{-k} \sum_{j \neq 1} |p_{1j}|$ while the dominant parts of the radii of the other discs are approximately $2^k |p_{i1}|$ if k is appreciably larger than unity. Hence on ACE k is chosen in the first instance to be the largest integer such that

$$2^k |p_{i1}| < |\lambda_1' - \lambda_i'| \quad (i = 2, \ldots, n), \tag{64.1}$$

a very simple requirement. A check is then made that the first disc is indeed isolated; this is almost invariably true, but if not, k is decreased until the first disc is isolated. A similar process can be used to localize each eigenvalue in turn provided the corresponding Gerschgorin disc of B is isolated.

Numerical example

65. Before describing the analogous process for the eigenvectors we illustrate the eigenvalue technique by a simple numerical example. In Table 4 the approximate eigensystem consisting of the λ_i' and x_i' gives exactly the displayed residual vectors f_i; these have been expressed in block-floating *decimal* form. The matrix X' has been triangularly decomposed using interchanges to give the matrices L and U and the row permutation has been recorded. The matrix P is the correctly rounded version of $(X')^{-1}F$; as often happens, the same scale factor is appropriate for each of the columns of P. The elements of the unknown matrix Q are bounded by $\frac{1}{2}10^{-10}$. For the isolation of each eigenvalue factors of the form 10^{-k} have been used. The

TABLE 4

A

$$\begin{bmatrix} -0\cdot05811 & -0\cdot11695 & 0\cdot51004 & -0\cdot31330 \\ -0\cdot04291 & 0\cdot56850 & -0\cdot07041 & -0\cdot68747 \\ -0\cdot01652 & 0\cdot38953 & 0\cdot01203 & -0\cdot52927 \\ 0\cdot06140 & 0\cdot32179 & -0\cdot22094 & -0\cdot42448 \end{bmatrix}$$

x_1'	x_2'	x_3'	x_4'
0·38463	0·59641	1·00000	1·00000
0·91066	1·00000	0·74534	0·34879
0·67339	0·81956	0·46333	0·52268
1·00000	0·23761	0·61755	0·10392

λ'	$10^5 f_1$	$10^5 f_2$	$10^5 f_3$	$10^5 f_4$
−0·25660	−0·06427	0·32258	0·09052	−0·14193
0·32185	−0·22972	0·30806	−0·17442	0·13411
−0·10244	−0·19421	0·13829	−0·10132	0·05223
0·13513	−0·12232	0·05163	−0·09736	0·15437

L

$$\begin{bmatrix} 1\cdot00000 & & & \\ 0\cdot91066 & 1\cdot00000 & & \\ 0\cdot38463 & 0\cdot64447 & 1\cdot00000 & \\ 0\cdot67339 & 0\cdot84168 & -0\cdot16525 & 1\cdot00000 \end{bmatrix}$$

Permutation:— rows 4, 2, 1, 3 become 1, 2, 3, 4.

U

$$\begin{bmatrix} 1\cdot00000 & 0\cdot23761 & 0\cdot61755 & 0\cdot10392 \\ & 0\cdot78362 & 0\cdot18296 & 0\cdot25415 \\ & & 0\cdot64456 & 0\cdot79624 \\ & & & 0\cdot37037 \end{bmatrix}$$

$10^5 P = $ *rounded version of* $(X')^{-1}F$

$$\begin{bmatrix} -0\cdot14241 & -0\cdot31415 & -0\cdot11093 & 0\cdot18613 \\ -0\cdot17215 & 0\cdot29352 & -0\cdot18228 & 0\cdot07080 \\ 0\cdot09990 & 0\cdot52207 & 0\cdot06180 & -0\cdot04284 \\ -0\cdot00672 & -0\cdot25371 & 0\cdot18010 & -0\cdot21291 \end{bmatrix}$$

Elements $|q_{ij}|$ bounded by $\frac{1}{2}10^{-10}$.

Root	Permissible multiplication factor	Eigenvalue	Error bound
1	10^{-5}	−0·25660 14241	$10^{-10}(0\cdot61121)+2\times10^{-15}$
2	10^{-4}	0·32185 29352	$10^{-9}(0\cdot42523)+2\times10^{-14}$
3	10^{-5}	−0·10243 93820	$10^{-10}(0\cdot66481)+2\times10^{-15}$
4	10^{-5}	0·13512 78709	$10^{-10}(0\cdot44053)+2\times10^{-15}$

1st eigenvector of $\Lambda' + (X)^{-1}F$	1st eigenvector of A
1·00000	0·38462 67268 $\pm12\times10^{-10}$
$10^{-5}(0\cdot29760)\pm4\times10^{-10}$	0·91066 11902 $\pm15\times10^{-10}$
$10^{-5}(-0\cdot64801)\pm7\times10^{-10}$	0·67339 17329 $\pm12\times10^{-10}$
$10^{-5}(0\cdot01716)\pm3\times10^{-10}$	1·00000 00000

conditions corresponding to (64.1) do indeed give the optimum value of 10^{-k} correctly in each case and the corresponding bounds for the improved eigenvalues are displayed. The whole computation was performed in 5-digit decimal arithmetic with accumulation of inner-products wherever relevant, and yet it is possible to give error bounds which are generally of the order of magnitude 10^{-10}.

Refinement of the eigenvectors

66. If the eigenvectors of B are u_i, then the eigenvectors of A are $X'u_i$. Hence we need only find the eigenvectors of B to obtain those of A. The computation of these eigenvectors and the derivation of bounds for the errors is only slightly more difficult, but since the notation is rather complicated we shall merely illustrate the technique by means of a numerical example. Suppose we have

$$B = \begin{bmatrix} 0.93256 & & \\ & 0.61321 & \\ & & 0.11763 \end{bmatrix} +$$

$$+10^{-5}\begin{bmatrix} 0.31257 & 0.41321 & 0.21653 \\ 0.41357 & 0.21631 & -0.40257 \\ 0.31765 & 0.41263 & 0.71654 \end{bmatrix} +Q, \quad |q_{ik}| < \tfrac{1}{2}10^{-10}. \quad (66.1)$$

By the eigenvalue technique we may show that the first eigenvalue μ satisfies
$$|\mu - 0.93256\ 31257| < 10^{-9}. \qquad (66.2)$$

Let u be the corresponding eigenvector normalized so that its element of maximum modulus is unity. This component must be the first since if we assume that it is the second (or third), the second (or third) equation of $Bu = \mu u$ immediately leads to a contradiction.

If we write $u^T = (1, u_2, u_3)$ the second and third equations give

$$10^{-5}(0.41375) + q_{21} + (0.61321\ 21631 + q_{22})u_2 +$$
$$+[-10^{-5}(0.40257) + q_{23}]u_3 = \mu u_2, \quad (66.3)$$

$$10^{-5}(0.31765) + q_{31} + [10^{-5}(0.41263) + q_{32}]u_2 +$$
$$+(0.11763\ 71654 + q_{33})u_3 = \mu u_3. \quad (66.4)$$

Since $|u_2|, |u_3| < 1$ these equations together with (66.2) give

$$|u_2| < \frac{10^{-5}(0.41375) + \tfrac{1}{2}10^{-10} + 10^{-5}(0.40257) + \tfrac{1}{2}10^{-10}}{\lambda_1' - \lambda_2' - \tfrac{1}{2}10^{-10} - 10^{-9}}$$
$$< 3 \times 10^{-5}, \qquad (66.5)$$

using quite crude estimates. Similarly they give
$$|u_3| < 2 \times 10^{-5}. \tag{66.6}$$
Using these crude bounds we may now return to equations (66.3) and (66.4) respectively and obtain much closer approximations to u_2 and u_3. It will be simpler if we rewrite (66.3) in the form
$$(10^{-5}a \pm \tfrac{1}{2}10^{-10}) + (\lambda' \pm \tfrac{1}{2}10^{-10})u_2 + (10^{-5}c \pm \tfrac{1}{2}10^{-10})u_3 = (\lambda'_1 \pm 10^{-9})u_2, \tag{66.7}$$
where a and c are of order unity and the elements $\pm 10^{-k}$ indicate the uncertainty in each coefficient. From (66.7) we have
$$u_2 = \frac{(10^{-5}a \pm \tfrac{1}{2}10^{-10}) + (10^{-5}c \pm \tfrac{1}{2}10^{-10})u_3}{\lambda'_1 - \lambda'_2 \pm 10 \cdot 5 \times 10^{-10}}, \tag{66.8}$$
and we already know that u_3 lies between $\pm 2 \times 10^{-5}$. Taking the most unfavourable combination of signs we have
$$u_2 < \frac{10^{-5}a + \tfrac{1}{2}10^{-10} + 2 \times 10^{-5}(-10^{-5}c + \tfrac{1}{2}10^{-10})}{\lambda'_1 - \lambda'_2 - 10 \cdot 5 \times 10^{-10}}$$
$$< \frac{10^{-5}(0 \cdot 41377)}{0 \cdot 31935} < 10^{-5}(1 \cdot 2957), \tag{66.9}$$
$$u_2 > \frac{10^{-5}a - \tfrac{1}{2}10^{-10} - 2 \times 10^{-5}(-10^{-5}c + \tfrac{1}{2}10^{-10})}{\lambda'_1 - \lambda'_2 + 10 \cdot 5 \times 10^{-10}}$$
$$> \frac{10^{-5}(0 \cdot 41373)}{0 \cdot 31936} > 10^{-5}(1 \cdot 2954). \tag{66.10}$$
Similarly
$$10^{-5}(0 \cdot 38982) > u_3 > 10^{-5}(0 \cdot 38976). \tag{66.11}$$

Hence we have error bounds which are less than 2×10^{-9} for each component of the first eigenvector.

The calculation of the eigenvectors on a desk computer is not particularly lengthy, but it is of such an unpalatable nature that we have computed only the first eigenvector. We give the first eigenvectors of both B and A with error bounds for each. The bounds for the errors in the components of the vectors are nowhere greater than $1\frac{1}{2}$ units in the ninth decimal and these may be expected to be substantial overestimates. In fact the maximum error in any component is 2 in the tenth figure. It is difficult to avoid having overestimates of this nature without an unjustifiably detailed error analysis. On a computer such as ACE which has a word of 48 binary digits the sacrifice of a few binary digits in 96 is comparatively unimportant. The final bounds are usually far more precise than we are likely to require.

Complex conjugate eigenvalues

67. The analysis of the previous section may be modified in the obvious way to deal with complex matrices; the whole computation is in the complex domain but no essentially new feature is involved. When we have a real matrix with one or more complex conjugate eigenvalues we naturally wish to work in the real domain as much as possible. The appropriate technique is adequately illustrated by considering a matrix of order four with one complex conjugate pair of eigenvalues. We have then

$$AX' = X' \begin{bmatrix} \lambda' & \mu' & & \\ -\mu' & \lambda' & & \\ & & \lambda'_3 & \\ & & & \lambda'_4 \end{bmatrix} + F, \tag{67.1}$$

where the first two columns of X' are the real and imaginary parts of the computed complex eigenvector. As before we may compute a matrix P so that

$$(X')^{-1}AX' = \begin{bmatrix} \lambda' & \mu' & & \\ -\mu' & \lambda' & & \\ & & \lambda'_3 & \\ & & & \lambda'_4 \end{bmatrix} + P + Q, \tag{67.2}$$

where P is known explicitly and we have bounds for the elements of Q. So far all elements are real. If we write

$$D = \begin{bmatrix} 1 & 1 & & \\ i & -i & & \\ \hline & & 1 & \\ & & & 1 \end{bmatrix}, \qquad D^{-1} = \tfrac{1}{2} \begin{bmatrix} 1 & -i & & \\ 1 & i & & \\ \hline & & 1 & \\ & & & 1 \end{bmatrix}, \tag{67.3}$$

then

$$D^{-1}(X')^{-1}AX'D = \begin{bmatrix} \lambda'+i\mu' & & & \\ & \lambda'-i\mu' & & \\ & & \lambda'_3 & \\ & & & \lambda'_4 \end{bmatrix} + D^{-1}PD + D^{-1}QD, \tag{67.4}$$

and

$$D^{-1}PD = \tfrac{1}{2}\left[\begin{array}{cc|cc} p_{11}+p_{22}+i(p_{12}-p_{21}) & p_{11}-p_{22}-i(p_{12}+p_{21}) & p_{13}-ip_{23} & p_{14}-ip_{24} \\ p_{11}-p_{22}+i(p_{12}+p_{21}) & p_{11}+p_{22}-i(p_{12}-p_{21}) & p_{13}+ip_{23} & p_{14}+ip_{24} \\ \hline p_{31}+ip_{32} & p_{31}-ip_{32} & 2p_{33} & 2p_{34} \\ p_{41}+ip_{42} & p_{41}-ip_{42} & 2p_{43} & 2p_{44} \end{array}\right].(67.5)$$

The centres of the two relevant Gerschgorin discs are

$$\lambda' + \tfrac{1}{2}(p_{11}+p_{22}) \pm i[\mu' + \tfrac{1}{2}(p_{12}-p_{21})]$$

and these are the improved eigenvalues. We can reduce the radius of the first disc as before by multiplying the first row by 2^{-k} and the first column by 2^{k}. In analogy with equation (64.1) we now choose k so that

$$\left.\begin{array}{l} \tfrac{1}{2}2^{k}\,|p_{11}-p_{22}+i(p_{12}+p_{21})| < 2\,|\mu'|, \\ \tfrac{1}{2}2^{k}\,|p_{j1}+ip_{j2}| < |\lambda'-\lambda_j+i\mu'| \quad (j > 2) \end{array}\right\}. \tag{67.6}$$

Coincident and pathologically close eigenvalues

68. The techniques we have described so far give small error bounds corresponding to any eigenvalue which is reasonably well separated from the others. As the separation between two or more eigenvalues becomes progressively poorer, so the error bounds deteriorate. These bounds may be improved in the following way.

Consider the matrix $(\Lambda' + \epsilon B)$ where Λ' is a diagonal matrix and Λ' and B have elements of order unity. Suppose the first two diagonal elements λ'_1 and λ'_2 of Λ' are separated by a quantity of order ϵ. We partition $(\Lambda' + \epsilon B)$ in the form

$$\left[\begin{array}{cc|c} \lambda'_1 & & \\ & \lambda'_2 & \\ \hline & & \Lambda'_{n-2} \end{array}\right] + \epsilon \left[\begin{array}{c|c} B_{22} & B_{2,n-2} \\ \hline B_{n-2,2} & B_{n-2,n-2} \end{array}\right]. \tag{68.1}$$

If we write

$$C_{22} = \left[\begin{array}{cc} b_{11} & b_{12} \\ b_{21} & b_{22}+(\lambda'_2-\lambda'_1)\epsilon^{-1} \end{array}\right], \tag{68.2}$$

(68.1) may be expressed in the form

$$\left[\begin{array}{cc|c} \lambda'_1 & & \\ & \lambda'_1 & \\ \hline & & \Lambda'_{n-2} \end{array}\right] + \epsilon \left[\begin{array}{c|c} C_{22} & B_{2,n-2} \\ \hline B_{n-2,2} & B_{n-2,n-2} \end{array}\right], \tag{68.3}$$

where C_{22} has elements of order unity. Suppose now we compute the eigensystem of C_{22}, so that we have

$$C_{22}X_{22} = X_{22}\begin{bmatrix} \mu_1 & \\ & \mu_2 \end{bmatrix} + \epsilon G_{22}, \qquad (68.4)$$

where G_{22} is of order unity. Then we have

$$X_{22}^{-1}C_{22}X_{22} = \begin{bmatrix} \mu_1 & \\ & \mu_2 \end{bmatrix} + \epsilon H_{22}, \quad \text{where } H_{22} = X_{22}^{-1}G. \quad (68.5)$$

Again we will expect the elements of H_{22} to be of order unity unless C_{22} has an ill-conditioned *eigenvalue* problem. Hence writing

$$Y = \begin{bmatrix} X_{22} & O \\ \hline O & I \end{bmatrix} \qquad (68.6)$$

we have

$$Y^{-1}(\Lambda' + \epsilon B)Y = \begin{bmatrix} \lambda_1' + \epsilon\mu_1 & & & \\ & \lambda_1' + \epsilon\mu_2 & & \\ & & & \Lambda_{n-2} \end{bmatrix} +$$

$$+ \epsilon \begin{bmatrix} \epsilon H_{22} & X_{22}^{-1}B_{2,n-2} \\ \hline B_{n-2,2}X_{22} & B_{n-2,n-2} \end{bmatrix}. \quad (68.7)$$

The second of the matrices has elements of order ϵ^2 in the top left-hand corner and elements of order ϵ elsewhere.

Consider now a specific example. Suppose the matrix (68.7) is of the form

$$\begin{bmatrix} 0.92563\ 41653 & & \\ & 0.92563\ 21756 & \\ & & 0.11325 \end{bmatrix} +$$

$$+ 10^{-5}\begin{bmatrix} 10^{-5}(0.23127) & 10^{-5}(0.31257) & 0.31632 \\ 10^{-5}(0.41365) & 10^{-5}(0.41257) & 0.21753 \\ \hline 0.28316 & 0.10175 & 0.20375 \end{bmatrix}. \quad (68.8)$$

If we multiply the first *two* rows by 10^{-5} and the first *two* columns by 10^5 the Gerschgorin discs are

Centre 0·92563 41653 23127   radius $10^{-10}(0·31257+0·31632)$

Centre 0·92563 21756 41257   radius $10^{-10}(0·41365+0·21753)$

Centre 0·11325 20375   radius 10^{-5} $(0·28316+0·10175)$.

The three discs are now disjoint and the first two eigenvalues are given with errors of order 10^{-10}. In this example we have omitted the error matrix corresponding to Q; in line with our earlier treatment this would have elements of the order 10^{-10} in magnitude. This merely gives contributions of order 10^{-10} to our error bounds.

In general when a group of r eigenvalues are pathologically close we will be able to resolve them by solving an independent eigenvalue problem of order r using the same precision of computation as in the main problem. If two or more eigenvalues are actually coincident the corresponding discs will continue to overlap, but the error bound will still improve in the same way.

Comments on the ACE programmes

69. Several programmes based on the techniques we have just described have been completed on ACE. These are of varying degrees of sophistication. The most sophisticated includes a section which gives a more refined resolution of pathologically close eigenvalues as described in § 68.

For most matrices, even those with eigenvalue separations which would normally be regarded as exceptionally severe, the error bounds are far smaller than anything which is likely to be required. For example on a matrix of order 15 having elements of order unity and with eigenvalue separations of magnitudes 10^{-8}, 10^{-7}, 10^{-7}, 10^{-6}, 10^{-5}, 10^{-5}, 10^{-5}, 10^{-5}, 10^{-2}, 10^{-2}, 10^{-2}, 10^{-2}, 10^{-1}, 1, all eigenvalues and eigenvectors were guaranteed to 17 decimals, with much better guarantees for most of them. Notice that the fundamental limitation on all the programmes is the requirement that it should be possible to compute $(X')^{-1}F$ correct to working accuracy. As the condition number of X' with respect to inversion approaches $2^t/n$ this becomes progressively more difficult, and when this value is attained the process collapses and higher precision computation is essential. On ACE this failure occurs only for matrices which have extremely ill-conditioned eigenvalue problems.

The programme may be used iteratively, in which case after computing the improved eigensystem it rounds all values to single-precision and applies the process again. When used in this way the extra computations needed to give the error bounds are omitted on all steps except the last and we may start the process with a comparatively poor eigensystem. This iterative technique has much in common with the methods of Magnier (1948), Jahn (1948) and Collar (1948).

In these methods the matrix P defined by

$$(X')^{-1}AX' = \text{diag}(\lambda_i') + P = B \qquad (69.1)$$

is computed without any special precautions to ensure accuracy. The improved eigenvalues of A are taken to be $(\lambda_i' + p_{ii})$ and the ith eigenvector of B (Chapter 2, § 6) to be

$$[p_{1i}/(\lambda_i' - \lambda_1'); \ p_{2i}/(\lambda_i' - \lambda_2'); \ldots; \ 1; \ldots; \ p_{ni}/(\lambda_i' - \lambda_n')], \qquad (69.2)$$

the improved eigenvectors of A being determined in the obvious way.

The methods of the previous section were developed independently (Wilkinson, 1961a) mainly in an attempt to determine rigorous bounds by computational means. It should be emphasized that when an approximate eigensystem has been determined by one of the more accurate methods we have discussed, the errors will usually be of the same order of magnitude as those which will result from a straightforward application of equations (69.1) and (69.2). If an improved eigensystem is to be obtained the careful attention to detail embodied in the ACE programmes is essential.

Additional notes

The use of iterative methods in which a number of vectors are used simultaneously has been described in papers by Bauer (1957 and 1958) and generally I have followed his treatment. The method of § 38 has been discussed more recently by Voyevodin (1962). A good deal of the material in this chapter is also discussed in Chapter 7 of Householder's book (1964), in many instances from a rather different point of view.

Treppen-iteration, the orthogonalization technique and bi-iteration may be performed using the inverse matrix. We have typically in the case of treppen-iteration

$$A X_{s+1} = L_s, \qquad X_{s+1} = L_{s+1} R_{s+1},$$

and it is obvious that A^{-1} need not be determined explicitly. We need only carry out a triangular decomposition of A, and this is particularly important if A is of Hessenberg form. We may incorporate shifts of origin, but a triangular decomposition of $(A - pI)$ is required for each different value of p. Again we have the weakness referred to in § 34 that p cannot be chosen freely after a value of λ_i has been determined.

Bibliography

AITKEN, A. C., 1937, The evaluation of the latent roots and latent vectors of a matrix. *Proc. roy. Soc. Edinb.* **57**, 269–304.

ARNOLDI, W. E., 1951, The principle of minimized iterations in the solution of the matrix eigenvalue problem. *Quart. appl. Math.* **9**, 17–29.

BAIRSTOW, L., 1914, The solution of algebraic equations with numerical coefficients. *Rep. Memor. Adv. Comm. Aero., Lond.*, No. 154, 51–64.

BAUER, F. L., 1957, Das Verfahren der Treppeniteration und verwandte Verfahren zur Lösung algebraischer Eigenwertprobleme. *Z. angew. Math. Phys.* **8**, 214–235.

BAUER, F. L., 1958, On modern matrix iteration processes of Bernoulli and Graeffe type. *J. Ass. comp. Mach.* **5**, 246–257.

BAUER, F. L., 1959, Sequential reduction to tridiagonal form. *J. Soc. industr. appl. Math.* **7**, 107–113.

BAUER, F. L., 1963, Optimally scaled matrices. *Numerische Math.* **5**, 73–87.

BAUER, F. L. and FIKE, C. T., 1960, Norms and exclusion theorems. *Numerische Math.* **2**, 137–141.

BAUER, F. L. and HOUSEHOLDER, A. S., 1960, Moments and characteristic roots. *Numerische Math.* **2**, 42–53.

BELLMAN, R., 1960, *Introduction to Matrix Analysis*. McGraw-Hill, New York.

BIRKHOFF, G. and MACLANE, S., 1953, *A Survey of Modern Algebra* (revised edition). Macmillan, New York.

BODEWIG, E., 1959, *Matrix Calculus* (second edition). Interscience Publishers, New York; North-Holland Publishing Company, Amsterdam.

BROOKER, R. A. and SUMNER, F. H., 1956, The method of Lanczos for calculating the characteristic roots and vectors of a real symmetric matrix. *Proc. Instn elect. Engrs* B.103 Suppl. 114–119.

CAUSEY, R. L., 1958, Computing eigenvalues of non-Hermitian matrices by methods of Jacobi type. *J. Soc. industr. appl. Math.* **6**, 172–181.

CAUSEY, R. L. and GREGORY, R. T., 1961, On Lanczos' algorithm for tridiagonalizing matrices. *SIAM Rev.* **3**, 322–328.

CHARTRES, B. A., 1962, Adaptation of the Jacobi method for a computer with magnetic-tape backing store. *Computer J.* **5**, 51–60.

COLLAR, A. R., 1948, Some notes on Jahn's method for the improvement of approximate latent roots and vectors of a square matrix. *Quart. J. Mech.* **1**, 145–148.

COLLATZ, L., 1948, *Eigenwertprobleme und Ihre Numerische Behandlung*. Chelsea, New York.

COURANT, R. and HILBERT, D., 1953, *Methods of Mathematical Physics*, Vol. 1. Interscience Publishers, New York.

DANILEWSKI, A., 1937, O čislennom rešenii vekovogo uravnenija, *Mat. Sb.* **2**(44), 169–171.

DEAN, P., 1960, Vibrational spectra of diatomic chains. *Proc. roy. Soc.* A. **254**, 507–521.

DEKKER, T. J., 1962, *Series AP 200 Algorithms*. Mathematisch Centrum, Amsterdam.

DURAND, E., 1960, *Solutions Numériques des Équations Algébriques*. Tome 1. Masson, Paris.

DURAND, E., 1961, *Solutions Numériques des Équations Algébriques*. Tome 2. Masson, Paris.

EBERLEIN, P. J., 1962, A Jacobi-like method for the automatic computation of eigenvalues and eigenvectors of an arbitrary matrix. *J. Soc. industr. appl. Math.* **10**, 74–88.

ENGELI, M., GINSBURG, T., RUTISHAUSER, H. and STIEFEL, E., 1959, *Refined Iterative Methods for Computation of the Solution and the Eigenvalues of Self-Adjoint Boundary Value Problems*. Birkhäuser, Basel/Stuttgart.

FADDEEV, D. K. and FADDEEVA, V. N., 1963, *Vyčislitel'nye Metody Lineǐnoǐ Algebry*, Gos. Izdat. Fiz. Mat. Lit., Moskow.

FADDEEVA, V. N., 1959, *Computational Methods of Linear Algebra*. Dover, New York.

FELLER, W. and FORSYTHE, G. E., 1951, New matrix transformations for obtaining characteristic vectors. *Quart. appl. Math.* **8**, 325–331.

FORSYTHE, G. E., 1958, Singularity and near singularity in numerical analysis. *Amer. math. Mon.* **65**, 229–240.

FORSYTHE, G. E., 1960, Crout with pivoting. *Commun. Ass. comp. Mach.* **3**, 507–508.

FORSYTHE, G. E. and HENRICI, P., 1960, The cyclic Jacobi method for computing the principal values of a complex matrix. *Trans. Amer. math. Soc.* **94**, 1–23.

FORSYTHE, G. E. and STRAUS, E. G., 1955, On best conditioned matrices. *Proc. Amer. math. Soc.* **6**, 340–345.

FORSYTHE, G. E. and STRAUS, L. W., 1955, The Souriau-Frame characteristic equation algorithm on a digital computer. *J. Maths Phys.* **34**, 152–156.

FORSYTHE, G. E. and WASOW, W. R., 1960, *Finite-Difference Methods for Partial Differential Equations*. Wiley, New York, London.

FOX, L., 1954, Practical solution of linear equations and inversion of matrices. *Appl. Math. Ser. nat. Bur. Stand.* **39**, 1–54.

FOX, L., 1964, *An Introduction to Numerical Linear Algebra*. Oxford University Press, London.

FOX, L., HUSKEY, H. D. and WILKINSON, J. H., 1948, Notes on the solution of algebraic linear simultaneous equations. *Quart. J. Mech.* **1**, 149–173.

FRANCIS, J. G. F., 1961, 1962, The QR transformation, Parts I and II. *Computer J.* **4**, 265–271, 332–345.

FRANK, W. L., 1958, Computing eigenvalues of complex matrices by determinant evaluation and by methods of Danilewski and Wielandt. *J. Soc. industr. appl. Math.* **6**, 378–392.

FRAZER, R. A., DUNCAN, W. J. and COLLAR, A. R., 1947, *Elementary Matrices and Some Applications to Dynamics and Differential Equations*. Macmillan, New York.

GANTMACHER, F. R., 1959a, *The Theory of Matrices*, translated from the Russian with revisions by the author. Vols. 1 and 2. Chelsea, New York.

GANTMACHER, F. R., 1959b. *Applications of the Theory of Matrices*, translated from the Russian and revised by J. L. Brenner. Interscience, New York.

GASTINEL, N., 1960, *Matrices du second degré et normes générales en analyse numérique linéaire*. Thesis, University of Grenoble.

GERSCHGORIN, S., 1931, Über die Abgrenzung der Eigenwerte einer Matrix. *Izv. Akad. Nauk SSSR, Ser. fiz.-mat.* **6**, 749–754.

GIVENS, J. W., 1953, A method of computing eigenvalues and eigenvectors suggested by classical results on symmetric matrices. *Appl. Math. Ser. nat. Bur. Stand.* **29**, 117–122.

GIVENS, J. W., 1954, Numerical computation of the characteristic values of a real symmetric matrix. Oak Ridge National Laboratory, *ORNL*-1574.

GIVENS, J. W., 1957, The characteristic value-vector problem. *J. Ass. comp. Mach.* **4**, 298–307.

GIVENS, J. W., 1958, Computation of plane unitary rotations transforming a general matrix to triangular form. *J. Soc. industr. appl. Math.* **6**, 26–50.

GIVENS, J. W., 1959, The linear equations problem. Applied Mathematics and Statistics Laboratories, Stanford University, *Tech. Rep.* No. 3.

GOLDSTINE, H. H. and HORWITZ, L. P., 1959, A procedure for the diagonalization of normal matrices. *J. Ass. comp. Mach.* **6**, 176–195.

GOLDSTINE, H. H., MURRAY, F. J. and VON NEUMANN, J., 1959, The Jacobi method for real symmetric matrices. *J. Ass. comp. Mach.* **6**, 59–96.

GOLDSTINE, H. H. and VON NEUMANN, J., 1951, Numerical inverting of matrices of high order, II. *Proc. Amer. math. Soc.* **2**, 188–202.

GOURSAT, E., 1933, *Cours d'Analyse Mathématique, Tome II* (fifth edition). Gauthier Villars, Paris.

GREENSTADT, J., 1955, A method for finding roots of arbitrary matrices. *Math. Tab., Wash.* **9**, 47–52.

GREENSTADT, J., 1960, The determination of the characteristic roots of a matrix by the Jacobi method. *Mathematical Methods for Digital Computers* (edited by Ralston and Wilf), 84–91. Wiley, New York.

GREGORY, R. T., 1953, Computing eigenvalues and eigenvectors of a symmetric matrix on the ILLIAC. *Math. Tab., Wash.* **7**, 215–220.

GREGORY, R. T., 1958, Results using Lanczos' method for finding eigenvalues of arbitrary matrices. *J. Soc. industr. appl. Math.* **6**, 182–188.

HANDSCOMB, D. C., 1962, Computation of the latent roots of a Hessenberg matrix by Bairstow's method. *Computer J.* **5**, 139–141.

HENRICI, P., 1958a, On the speed of convergence of cyclic and quasicyclic Jacobi methods for computing eigenvalues of Hermitian matrices. *J. Soc. industr. appl. Math.* **6**, 144–162.

HENRICI, P., 1958b, The quotient-difference algorithm. *Appl. Math. Ser. nat. Bur. Stand.* **49**, 23–46.

HENRICI, P., 1962, Bounds for iterates, inverses, spectral variation and fields of values of non-normal matrices. *Numerische Math.* **4**, 24–40.

HESSENBERG, K., 1941, Auflösung linearer Eigenwertaufgaben mit Hilfe der Hamilton-Cayleyschen Gleichung, Dissertation, Technische Hochschule, Darmstadt.

HOFFMAN, A. J. and WIELANDT, H. W., 1953, The variation of the spectrum of a normal matrix. *Duke Math. J.* **20**, 37–39.

HOTELLING, H., 1933, Analysis of a complex of statistical variables into principal components. *J. educ. Psychol.* **24**, 417–441, 498–520.

HOUSEHOLDER, A. S., 1958a, The approximate solution of matrix problems. *J. Ass. comp. Mach.* **5**, 205–243.

HOUSEHOLDER, A. S., 1958b, Unitary triangularization of a nonsymmetric matrix. *J. Ass. comp. Mach.* **5**, 339–342.

HOUSEHOLDER, A. S., 1958c, Generated error in rotational tridiagonalization. *J. Ass. comp. Mach.* **5**, 335–338.

HOUSEHOLDER, A. S., 1961, On deflating matrices. *J. Soc. industr. appl. Math.* **9**, 89–93.

HOUSEHOLDER, A. S., 1964, *The Theory of Matrices in Numerical Analysis*. Blaisdell, New York.

HOUSEHOLDER, A. S. and BAUER, F. L., 1959, On certain methods for expanding the characteristic polynomial. *Numerische Math.* **1**, 29–37.

HYMAN, M. A., 1957, Eigenvalues and eigenvectors of general matrices. *Twelfth National meeting A.C.M., Houston, Texas.*

JACOBI. C. G. J., 1846, Über ein leichtes Verfahren die in der Theorie der Säculärstörungen vorkommenden Gleichungen numerisch aufzulösen. *Crelle's J.* **30**, 51–94.

JAHN, H. A., 1948, Improvement of an approximate set of latent roots and modal columns of a matrix by methods akin to those of classical perturbation theory. *Quart. J. Mech.* **1**, 131–144.

JOHANSEN, D. E., 1961, A modified Givens method for the eigenvalue evaluation of large matrices. *J. Ass. comp. Mach.* **8**, 331–335.

KATO, T., 1949, On the upper and lower bounds of eigenvalues. *J. phys. Soc. Japan* **4**, 334–339.

KINCAID, W. M., 1947, Numerical methods for finding characteristic roots and vectors of matrices. *Quart. appl. Math.* **5**, 320–345.

KRYLOV, A. N., 1931, O čislennom rešenii uravnenija, kotorym v techničeskih voprasah opredeljajutsja častoty malyh kolebaniĭ material'nyh sistem, *Izv. Akad. Nauk SSSR. Ser. fiz.-mat.* **4**, 491–539.

KUBLANOVSKAYA, V. N., 1961, On some algorithms for the solution of the complete eigenvalue problem. *Zh. vych. mat.* **1**, 555–570.

LANCZOS, C., 1950, An iteration method for the solution of the eigenvalue problem of linear differential and integral operators. *J. Res. nat. Bur. Stand.* **45**, 255–282.

LOTKIN, M., 1956, Characteristic values of arbitrary matrices. *Quart. appl. Math.* **14**, 267–275.

MAGNIER, A., 1948, Sur le calcul numérique des matrices. *C.R. Acad. Sci., Paris* **226**, 464–465.

MAEHLY, H. J., 1954, Zur iterativen Auflösung algebraischer Gleichungen. *Z. angew. Math, Phys.* **5**, 260–263.

MARCUS, M., 1960, Basic theorems in matrix theory. *Appl. Math. Ser. nat. Bur. Stand.* **57.**

McKEEMAN, W. M., 1962, Crout with equilibration and iteration. *Commun. Ass. comp. Mach.* **5**, 553–555.

MULLER, D. E., 1956, A method for solving algebraic equations using an automatic computer. *Math. Tab., Wash.* **10**, 208–215.

VON NEUMANN, J. and GOLDSTINE, H. H., 1947, Numerical inverting of matrices of high order. *Bull. Amer. math. Soc.* **53**, 1021–1099.

NEWMAN, M., 1962, Matrix computations. *A Survey of Numerical Analysis* (edited by Todd), 222–254. McGraw-Hill, New York.

OLVER, F. W. J., 1952, The evaluation of zeros of high-degree polynomials. *Phil. Trans. roy. Soc.* A, **244**, 385–415.

ORTEGA, J. M., 1963, An error analysis of Householder's method for the symmetric eigenvalue problem. *Numerische Math.* **5**, 211–225.

ORTEGA, J. M. and KAISER, H. F., 1963. The LL^T and QR methods for symmetric tridiagonal matrices. *Computer J.* **6**, 99–101.

OSBORNE, E. E., 1958, On acceleration and matrix deflation processes used with the power method. *J. Soc. industr. appl. Math.* **6**, 279–287.

OSBORNE, E. E., 1960, On pre-conditioning of matrices. *J. Ass. comp. Mach.* **7**, 338–345.

OSTROWSKI, A. M., 1957, Über die Stetigkeit von charakteristischen Wurzeln in Abhängigkeit von den Matrizenelementen. *Jber. Deutsch. Math. Verein.* **60**, Abt. 1, 40–42.

OSTROWSKI, A. M., 1958–59, On the convergence of the Rayleigh quotient iteration for the computation of characteristic roots and vectors. I, II, III, IV, V, VI, *Arch. rat. Mech. Anal.* **1**, 233–241; **2**, 423–428; **3**, 325–340; **3**, 341–347; **3**, 472–481; **4**, 153–165.

OSTROWSKI, A. M., 1960, *Solution of Equations and Systems of Equations.* Academic Press, New York and London.

PARLETT, B., 1964, Laguerre's method applied to the matrix eigenvalue problem. *Maths Computation* **18**, 464–485.

POPE, D. A. and TOMPKINS, C., 1957, Maximizing functions of rotations—experiments concerning speed of diagonalization of symmetric matrices using Jacobi's method. *J. Ass. comp. Mach.* **4**, 459–466.

RICHARDSON, L. F., 1950, A purification method for computing the latent columns of numerical matrices and some integrals of differential equations. *Phil. Trans. roy. Soc.* A, **242**, 439–491.

ROLLETT, J. S. and WILKINSON, J. H., 1961, An efficient scheme for the co-diagonalization of a symmetric matrix by Givens' method in a computer with a two-level store. *Computer J.* **4**, 177–180.

ROSSER, J. B., LANCZOS, C., HESTENES, M. R. and KARUSH, W. 1951, The separation of close eigenvalues of a real symmetric matrix. *J. Res. nat. Bur. Stand.* **47**, 291–297.

RUTISHAUSER, H., 1953, Beiträge zur Kenntnis des Biorthogonalisierungs-Algorithmus von Lanczos, *Z. angew. Math. Phys.* **4**, 35–56.

RUTISHAUSER, H., 1955, Bestimmung der Eigenwerte und Eigenvektoren einer Matrix mit Hilfe des Quotienten-Differenzen-Algorithmus. *Z. angew. Math. Phys.* **6**, 387–401.

RUTISHAUSER, H., 1957, *Der Quotienten-Differenzen-Algorithmus.* Birkhäuser, Basel/Stuttgart.

RUTISHAUSER, H., 1958, Solution of eigenvalue problems with the *LR*-transformation. *Appl. Math. Ser. nat. Bur. Stand.* **49**, 47–81.

RUTISHAUSER, H., 1959, Deflation bei Bandmatrizen. *Z. angew. Math. Phys.* **10**, 314–319.

RUTISHAUSER, H., 1960, Über eine kubisch konvergente Variante der *LR*-Transformation. *Z. angew. Math. Mech.* **40**, 49–54.

RUTISHAUSER, H., 1963, On Jacobi rotation patterns. *Proceedings A.M.S. Symposium in Applied Mathematics.* **15**, 219–239.

RUTISHAUSER, H. and SCHWARZ, H. R., 1963 Handbook Series Linear Algebra. The *LR* transformation method for symmetric matrices. *Numerische Math.* **5**, 273–289.

SCHÖNHAGE, A., 1961, Zur Konvergenz des Jacobi-Verfahrens. *Numerische Math.* **3**, 374–380.

STIEFEL, E. L., 1958, Kernel polynomials in linear algebra and their numerical applications. *Appl. Math. Ser. nat. Bur. Stand.* **49**, 1–22.

STRACHEY, C. and FRANCIS, J. G. F., 1961, The reduction of a matrix to co-diagonal form by eliminations. *Computer J.* **4**, 168–176.

TARNOVE, I., 1958, Determination of eigenvalues of matrices having polynomial elements. *J. Soc. industr. appl. Math.* **6**, 163–171.

TAUSSKY, O., 1949, A recurring theorem on determinants. *Amer. math. Mon.* **56**, 672–676.

TAUSSKY, O., 1954, Contributions to the solution of systems of linear equations and the determination of eigenvalues. *Appl. Math. Ser. nat. Bur. Stand.* **39**.

TAUSSKY, O. and MARCUS, M., 1962, Eigenvalues of finite matrices. *A Survey of Numerical Analysis* (edited by Todd), 279–313. McGraw-Hill, New York.

TEMPLE, G., 1952, The accuracy of Rayleigh's method of calculating the natural frequencies of vibrating systems. *Proc. roy. Soc.* A. **211**, 204–224.

TODD, J., 1949, The condition of certain matrices. I. *Quart. J. Mech.* **2**, 469–472.

TODD, J., 1950, The condition of a certain matrix. *Proc. Camb. phil. Soc.* **46**, 116–118.

TODD, J., 1954a, The condition of certain matrices. *Arch. Math.* **5**, 249–257.

TODD, J., 1954b, The condition of the finite segments of the Hilbert matrix. *Appl. Ser. Math. nat. Bur. Stand.* **39**, 109–116.

TODD, J., 1961, Computational problems concerning the Hilbert matrix. *J. Res. nat. Bur. Stand.* **65**, 19–22.

TRAUB, J. F., 1964, *Iterative Methods for the Solution of Equations.* Prentice-Hall, New Jersey.

TURING, A. M., 1948, Rounding-off errors in matrix processes. *Quart. J. Mech.* **1**, 287–308.

TURNBULL, H. W., 1950, *The Theory of Determinants, Matrices, and Invariants.* Blackie, London and Glasgow.

TURNBULL, H. W. and AITKEN, A. C., 1932, *An Introduction to the Theory of Canonical Matrices.* Blackie, London and Glasgow.

VARGA, R. S., 1962, *Matrix Iterative Analysis.* Prentice-Hall, New Jersey.

LE VERRIER, U. J. J., 1840, Sur les variations séculaires des éléments elliptiques des sept planètes principales. *Liouville J. Maths* **5**, 220–254.

VOYEVODIN, V. V., 1962, A method for the solution of the complete eigenvalue problem. *Zh. vych. Mat.* **2**, 15–24.

WHITE, P. A., 1958, The computation of eigenvalues and eigenvectors of a matrix. *J. Soc. industr. appl. Math.* **6**, 393–437.

WIELANDT, H., 1944a, Bestimmung höherer Eigenwerte durch gebrochene Iteration. *Ber, B44/J/37 der Aerodynamischen Versuchsanstalt Göttingen.*

WIELANDT, H., 1944b, Das Iterationsverfahren bei nicht selbstadjungierten linearen Eigenwertaufgaben. *Math. Z.* **50**, 93–143.

WILKINSON, J. H., 1954, The calculation of the latent roots and vectors of matrices on the pilot model of the ACE. *Proc. Camb. phil. Soc.* **50**, 536–566.

WILKINSON, J. H., 1958a, The calculation of the eigenvectors of codiagonal matrices. *Computer J.* **1**, 90–96.

WILKINSON, J. H., 1958b, The calculation of eigenvectors by the method of Lanczos. *Computer J.* **1**, 148–152.

WILKINSON, J. H., 1959a, The evaluation of the zeros of ill-conditioned polynomials, Parts I and II. *Numerische Math.* **1**, 150–180.

WILKINSON, J. H., 1959b, Stability of the reduction of a matrix to almost triangular and triangular forms by elementary similarity transformations. *J. Ass. comp. Mach.* **6**, 336–359.

WILKINSON, J. H., 1960a, Householder's method for the solution of the algebraic eigenproblem. *Computer J.* **3**, 23–27.

WILKINSON, J. H., 1960b, Error analysis of floating-point computation. *Numerische Math.* **2**, 319–340.

WILKINSON, J. H., 1960c, Rounding errors in algebraic processes. *Information Processing*, 44–53. UNESCO, Paris; Butterworths, London.

WILKINSON, J. H., 1961a, Rigorous error bounds for computed eigensystems. *Computer J.* **4**, 230–241.

WILKINSON, J. H., 1961b, Error analysis of direct methods of matrix inversion. *J. Ass. comp. Mach.* **8**, 281–330.

WILKINSON, J. H., 1962a, Instability of the elimination method of reducing a matrix to tri-diagonal form. *Computer J.* **5**, 61–70.

WILKINSON, J. H., 1962b, Error analysis of eigenvalue techniques based on orthogonal transformations. *J. Soc. industr. appl. Math.* **10**, 162–195.

WILKINSON, J. H., 1962c, Note on the quadratic convergence of the cyclic Jacobi process. *Numerische Math.* **4**, 296–300.

WILKINSON, J. H., 1963a, Plane-rotations in floating-point arithmetic. *Proceedings A.M.S. Symposium in Applied Mathematics* **15**, 185–198.

WILKINSON, J. H., 1963b, *Rounding Errors in Algebraic Processes, Notes on Applied Science No.* 32, Her Majesty's Stationery Office, London; Prentice-Hall, New Jersey.

ZURMÜHL, R., 1961, *Matrizen und ihre technischen Anwendungen, Dritte, neubearbeitete Auflage.* Springer-Verlag, Berlin; Göttingen; Heidelberg.

Index

GANTMACHER, F. R., on the theory of
matrices, 61, 496; Theorem of Corresponding Matrices, 18.
GARWICK, J., on termination of iterative processes, 463.
GASTINEL, N., on error analysis, 264.
Gaussian elimination, 200, 484; breakdown of 204–205, 224; by columns,
370–371; error analysis of, 209–212,
214–215; on high-speed computer, 220–
221; relation with triangular decomposition, 221–223.
Gerschgorin's theorems, 71–72, 109; application to perturbation theory, 72–81;
638–646.
GINSBURG, T., (See ENGELI, M.)
GIVENS, J. W., on error analysis, 343,
reduction to tri-diagonal form, 282,
Sturm sequences, 300, 307, 344, triangularization, 240–241, Wielandt-Hoffman theorem, 105.
Givens' method, symmetric, 282–290,
error analysis of, 286–290; unsymmetric,
345–347, error analysis of, 350–351;
relationship with Householder's method,
351–353.
GOLDSTINE, H. H., and HORWITZ,
L. P., on the Jacobi method for normal
matrices, 486.
and MURRAY, F. J., and VON NEU-
MANN, J., on the Jacobi method, 343.
(See also VON NEUMANN, J.)
GOURSAT, E., on algebraic functions, 64.
Grade of vector, 37.
GREENSTADT, J., on Jacobi-type methods, 568.
GREGORY, R. T., on the Jacobi method,
343, Lanczos' method, 412.
(See also CAUSEY, R.)

HANDSCOMB, D. C., on Bairstow's
method for Hessenberg matrices, 452,
476.
HENRICI, P., on departure from normality, 168, 568, serial Jacobi method, 269,
270.
Hermitian, matrix, 24–26, elementary, 48–
50; eigenvalue problem for, 265–344.
transposed matrix, 25.
(See also Elementary Hermitian matrices.)
HESSENBERG, K., on reduction to
Hessenberg form, 379–382.
Hessenberg matrix, 218, determinants of,
426–431, direct reduction to, 357–365,
invariance with respect to LR algorithm, 502–504, QR algorithm, 524;
reduction to, 345–395; further reduction

of, to tri-diagonal form, 396–403, Frobenius form, 405–411.
method, 379–382, generalized, 377–395;
applied to Hessenberg matrix, 492–
505; relation to Lanczos' method,
402–403.
HESTENES, M. R., (See ROSSER, J. B.)
HILBERT, D., (See COURANT, R.)
Hilbert matrix, 233–234.
HOFFMAN, A. J., and WIELANDT,
H. W., on the Wielandt-Hoffman
theorem, 104.
HOTELLING, H., on matrix deflation,
585, 607.
HOUSEHOLDER, A. S., on the theory
of matrices in numerical analysis, 61,
109, 569, 647, unitary triangularization,
233, use of elementary Hermitian matrices, 343.
and BAUER, F. L., on Householder's
method, 343.
Householder's method, symmetric, 290–
299, error analysis of, 297, unsymmetric,
347–353, error analysis of, 350; relationship with Givens' method, 351–353.
HYMAN, M. A., on Hyman's method,
426–431, 469, 627.

Inner-product, 4, 25; accumulation of,
in fixed-point, 111–112, floating-point,
116–118.
Interchanges, 162, 205–207, 399–401, 498–
501.
Interpolation methods, 435–440, error
analysis of, 455–459.
Invariant factors, 22–24.
subspaces, 185, deflation using, 587,
599–602.
Inverse iteration, for complex conjugate
eigenvectors, 629–633, for eigenvectors
of band matrices, 628–629, of general
matrices, 619–633, of generalized eigenvalue problem, 633–635, of Hessenberg
matrices, 626–627, of symmetric tri-
diagonal matrices, 321–330.
Inversion of matrices, 252–254.
Iterative method for the solution of the
eigenvalue problem, 570–576.
refinement of, solution of linear equations, 255–263, eigensystem, 637–646.

JACOBI, C. G. J., on the Jacobi method,
266.
Jacobi's method, 270–282, 486, 568, classical, 266–269, serial, 269, threshold, 277–
278; convergence of, 270–273; error
analysis of, 279–281; for normal matrices, 486.